FUNCTIONS AND CHANGE
A Modeling Approach to College Algebra

FIFTH EDITION

Bruce Crauder | Benny Evans | Alan Noell

OKLAHOMA STATE UNIVERSITY

BROOKS/COLE
CENGAGE Learning

Australia • Brazil • Japan • Korea • Mexico • Singapore • Spain • United Kingdom • United States

BROOKS/COLE
CENGAGE Learning·

Functions and Change: A Modeling Approach to College Algebra, **Fifth Edition**
Bruce Crauder, Benny Evans, Alan Noell

Publisher: Liz Covello

Acquisitions Editor: Gary Whalen

Developmental Editor: Stacy Green

Assistant Editor: Cynthia Ashton

Editorial Assistant: Samantha Lugtu

Media Editor: Lynh Pham

Senior Market Development Manager: Danae April

Senior Brand Manager: Gordon Lee

Content Project Manager: Jennifer Risden

Art Director: Vernon Boes

Manufacturing Planner: Becky Cross

Rights Acquisitions Specialist: Tom McDonough

Production and Composition: MPS Limited

Photo Researcher: Bill Smith Group

Text Researcher: Pablo D'Stair

Copy Editor: Martha Williams

Illustrator: Precision Graphics; MPS Limited

Text and Cover Designer: Lisa Henry

Cover Image: Len DeLessio/Getty Images

For product information and technology assistance, contact us at **Cengage Learning Customer & Sales Support, 1-800-354-9706**

For permission to use material from this text or product, submit all requests online at **www.cengage.com/permissions**
Further permissions questions can be e-mailed to **permissionrequest@cengage.com**

Library of Congress Control Number: 2011942328

ISBN-13: 978-1-133-36555-6
ISBN-10: 1-133-36555-8

Brooks/Cole
20 Davis Drive
Belmont, CA 94002-3098
USA

Cengage Learning is a leading provider of customized learning solutions with office locations around the globe, including Singapore, the United Kingdom, Australia, Mexico, Brazil, and Japan. Locate your local office at **www.cengage.com/global**

Cengage Learning products are represented in Canada by Nelson Education, Ltd.

To learn more about Brooks/Cole, visit **www.cengage.com/brookscole**
Purchase any of our products at your local college store or at our preferred online store **www.cengagebrain.com**

Printed in Canada
1 2 3 4 5 6 7 16 15 14 13 12

CONTENTS

6 RATES OF CHANGE   395

APPENDIX A: ADDITIONAL TOPICS   A1

APPENDIX B: USING THE GRAPHING CALCULATOR   B1

Far too many students enter college today with an abiding distaste for mathematics. They may doubt their own ability to succeed, and they often consider mathematics to be irrelevant to their own experience. These ideas may have been reinforced by their experiences in high school. The dual goals of this text are to show students the importance of mathematics in everything from business to science to politics and to show them that they can not only succeed but also excel at sophisticated mathematics. As an added benefit, students see that mastery of basic mathematical thinking can be a powerful tool for success in many other areas.

The theme of rates of change, understood informally, pervades the text. This emphasis reflects both that the idea is fundamentally important to mathematics and that rates of change are the everyday language used to present many real applications of mathematics. The first encounter in the text is with average rates of change and their relation to data, as seen in the popular media as well as scientific presentations. Linear functions are defined not using a formula, but as those functions having a constant rate of change. Similarly, exponential functions are defined as those functions that show a constant percentage change.

The graphing calculator is an integral part of this course. Many beginning students never appreciate the beautiful and important ideas in mathematics because they get bogged down in what seem like meaningless algebraic manipulations. The graphing calculator helps avoid this distraction. For example, nonlinear equations are routinely solved by using the calculator to find where graphs cross. The same tool allows students to explore maxima, minima, points of inflection, and even limits at infinity. The authors have often described this course as "the calculus you can do with minimal algebraic manipulations."

The exposition here is driven by real applications of mathematics to many disparate areas. Functions are always presented in real-world contexts so that all variables involved have meanings that are clear to students. We almost never use abstract symbols that do not relate directly to real events because symbols devoid of context may seem pointless to beginning students. Explanations are presented in an intuitive fashion using everyday language. They are designed for students to read, not for professional mathematicians who have no need of our instruction.

CHANGES IN THE FIFTH EDITION

The changes in the fifth edition reflect input we have received from both students and educators across the country. Readers will find this edition streamlined, with extraneous material deleted, thus producing a better focus on the topics and presentations that work so well in the classroom. The previous *Algebraic Look* sections of the book have been trimmed and either grouped together at the end of chapters or placed in Appendix A. Instructors may certainly decide which of these materials they wish to use.

We have added the new feature *Test Your Understanding*, which follows examples in the text and challenges students to solve similar problems. Answers are provided to reinforce correct work.

Instructors and students alike have pointed out that earlier editions show a heavy focus on science applications. In response, we have added a significant number of non-science applications to each exercise set. This allows for an appeal to an even wider audience.

New exercises have been added to each exercise set. In addition many of the exercises have new data to reflect updated information, such as the 2010 U.S. Census. We moved the skill exercises to the end of the major exercises to emphasis the importance of the major exercises and the many applications they contain.

We have added two new sections to the fifth edition. The first new section is in Chapter 2 and is concerned with solving inequalities. It follows the section on solving equations by the method of crossing graphs. In this way the fundamental ideas are reinforced and expanded upon. The second new section is in Chapter 4. Many students find exponential functions to be a challenge. We have broken the introductory section on exponential functions into two pieces, thus slowing the pace and allowing students more time to master the material. We have moved the notion of constant percentage change into its own section because of its importance in applications of exponential functions.

We have revised the treatment of exponential regression and power regression to eliminate the use of the logarithm. We have also combined the treatment of quadratics, higher-degree polynomials, and rational functions into one section.

Material from the previous *Technology Guide* has been updated and placed in Appendices B and C.

The new exciting design, with additional color and photographs, nicely breaks up the text, adding interest and appeal. We think you'll be pleased with this new design and the way it brings more clarity to the text.

We want to thank both students and instructors who have corresponded with us. Your comments and suggestions allowed us to make the book better.

FLOW OF THE TEXT

The prologue prepares students for what is to come. It allows them to adapt to the intensive way in which calculators are used throughout the course.

Chapter 1 shows students the common ways in which functions are seen in the real world. Functions may be given by formulas, tables, graphs, or with a verbal description. Advantages and disadvantages of each are discussed.

Chapter 2 shows how to use the graphing calculator to generate a table and a graph from a formula. This tool is used to solve equations, inequalities, and optimization problems.

Chapter 3 focuses on linear functions and includes analysis of data via linear regression.

Chapter 4 studies exponential and logarithmic functions. Exponential regression is used to analyze data.

Chapter 5 looks at additional functions, including logistic and power functions as well as polynomial and rational functions.

Chapter 6, which treats rates of change in general, is the culmination of the book. Many of the ideas discussed throughout the text are tied together in this chapter.

FEATURES OF THE TEXT

Exercises

The collection of exercises is perhaps the greatest strength of the text. They come in two flavors.

Major Exercises   These are real applications of mathematics. They are often drawn from professional publications, news articles, or texts from other disciplines.

These exercises tend to be substantial, and many can be used effectively for group work. A typical example drawn from scientific literature is an exercise on Alexander's formula for the stride length of dinosaurs. Note in the exercise the source reference. A key feature is that this is not a "made-up" application. Rather, it is an example of how scientists actually use mathematics to study dinosaurs.

---

14. Alexander's Formula   One interesting problem in the study of dinosaurs is to determine from their tracks how fast they ran. The scientist R. McNeill Alexander developed a formula giving the velocity of any running animal in terms of its stride length and the height of its hip above the ground.[10]

The stride length of a dinosaur can be measured from successive prints of the same foot, and the hip height (roughly the leg length) can be estimated on the basis of the size of a footprint, so Alexander's formula gives a way of estimating from dinosaur tracks how fast the dinosaur was running. See Figure 2.45.

FIGURE 2.45

If the velocity v is measured in meters per second, and the stride length s and hip height h are measured in meters, then Alexander's formula is
$$v = 0.78s^{1.67}h^{-1.17}.$$
(For comparison, a length of 1 meter is 39.37 inches, and a velocity of 1 meter per second is about 2.2 miles per hour.)

a. First we study animals with varying stride lengths but all with a hip height of 2 meters (so $h = 2$).

  i. Find a formula for the velocity v as a function of the stride length s.

  ii. Make a graph of v versus s. Include stride lengths from 2 to 10 meters.

  iii. What happens to the velocity as the stride length increases? Explain your answer in practical terms.

  iv. Some dinosaur tracks show a stride length of 3 meters, and a scientist estimates that the hip height of the dinosaur was 2 meters. How fast was the dinosaur running?

[10]See his article "Estimates of speeds of dinosaurs," *Nature* **261** (1976), 129–130. See also his book *Animal Mechanics*, 2nd ed. (Oxford: Blackwell, 1983).

Many of the exercises involve data drawn from news articles or public records. A problem on the gross national product is typical. Emphasis is placed on the practical meaning of the slope, or rate of change, of a linear function.

---

12. Gross National Product   The United States gross national product, in trillions of dollars, is given in the table below.

Date	Gross national product
2005	12.7
2006	13.5
2007	14.2
2008	14.6

a. Find the equation of the regression line, and explain the meaning of its slope. (Round regression line parameters to two decimal places.)

b. Plot the data points and the regression line.

c. When would you predict that a gross national product of 15.3 trillion dollars would be reached? The actual gross national product in 2009 was 14.4 trillion dollars. What does that say about your prediction?

Many of the newer non-science exercises are just for fun—see the Crayola exercise below. Other subjects include determination of reading level, sailing, and movies.

---

4. Crayola Colors The table below shows the number C of Crayola colors available t years after 1900.

t = years since 1900	3	49	58	72	90	98	103
C = number of colors	8	48	64	72	80	120	120

a. Find the equation of the regression line for C as a function of t.

b. How many Crayola colors does the regression line indicate for 1993? (Round your answer to the nearest whole number. Note that the actual number is 96.)

c. Plot the data points and the regression line.

Skill Building Exercises   These exercises are designed to make sure students have mastered the basic skills of the section. Some students will be able to proceed directly to the major exercises, while others should be diverted to the skill building exercises before proceeding to the major exercises. A typical example is Skill Building Exercise S-7 from Section 2.2 on making graphs.

S-7. Finding a Window   Find an appropriate window setup that will show a good graph of $(x^4 + 1)/(x^2 + 1)$ with a horizontal span of 0 to 300.

Strong Examples   Each section includes examples that illustrate the key ideas presented in the section and prepare students for the exercises that follow.

EXAMPLE 2.12   **GROWTH OF FOREST STANDS**

In forestry management it is important to know the *growth* and the *yield* of a forest stand.[33] The growth G is the amount by which the volume of wood will increase in a unit of time, and the yield Y is the total volume of wood. A forest manager has determined that in a certain stand of age A, the growth $G = G(A)$ is given by the formula

$$G = 32A^{-2}e^{10 - 32A^{-1}}$$

and the yield $Y = Y(A)$ is given by the formula

$$Y = e^{10 - 32A^{-1}}.$$

Here G is measured in cubic feet per acre per year, Y in cubic feet per acre, and A in years.

Part 1 Draw a graph of growth as a function of age that includes ages up to 60 years.

Part 2 At what age is growth maximized?[34]

Part 3 Draw a graph of yield as a function of age. What is the physical meaning of the point on this graph that corresponds to your answer to part 2?

Part 4 To meet market demand, loggers are considering harvesting a relatively young stand of trees. This area was initially clear-cut[35] and left barren. The forest was replanted, with plans to get a new harvest within the next 14 years. At what time in this 14-year period will growth be maximized?

All examples are accompanied by complete solutions.

> **Solution to Part 1** The first step is to enter the growth function and record the appropriate correspondences:
>
> $Y_1 = G$, growth, in cubic feet per acre per year, on vertical axis
>
> $X = A$, age, in years, on horizontal axis.
>
> We need to look at a table of values to help us set the window size. We made the table in Figure 2.94 using a starting value of 0 and a table increment of 10. We were told specifically that our graph should include ages up to 60, so we made the graph in Figure 2.95 using a window with a horizontal span from $A = 0$ to $A = 60$ and a vertical span from $G = 0$ to $G = 500$.
>
> **Solution to Part 2** Growth is maximized at the peak of the graph in Figure 2.95. We have used the calculator to locate this point in Figure 2.96. We see from the prompt at the bottom of the screen that a maximum growth of 372.62 cubic feet per acre per year occurs when the stand is 16 years old.

Test Your Understanding

Many of the examples are followed by a challenge to the student to solve a similar problem. Such is the case for the growth of forest stands.

> **TEST YOUR UNDERSTANDING** | **FOR EXAMPLE 2.12**
>
> Replace the growth formula in the example by $G = 40 A^{-2} e^{11 - 40 A^{-1}}$. At what age is growth maximized? What is the maximum growth rate? ∎

Students can check their answers at the end of each section.

> **ANSWERS FOR TEST YOUR UNDERSTANDING**
>
> **2.12** 20 years; 810.31 cubic feet per acre per year

Calculator Screens

The graphing calculator is an integral part of this text. Although specific keystrokes are not shown, generic calculator instructions such as those shown in the example above are always included. Typical calculator screen displays are also shown so that students can check their work.

We should note that this text would work just as well with a spreadsheet program like Excel® in place of a graphing calculator.

A Further Look

At the end of each chapter there is additional material that may be used by instructors who wish to delve a bit deeper. This material may also be used for strong students who want to find out more. An example from the end of Chapter 1 involves areas associated with functions.

> If $f(x)$ is a positive function between $x = a$ and $x = b$, then its graph is above the horizontal axis and determines a region bounded by the graph, the vertical line $x = a$, the vertical line $x = b$, and the horizontal axis. This region is shown in Figure 1.56. The area is commonly called the *area under a curve*.
>
> The problem of calculating the area shown in Figure 1.56 is an important topic in calculus. We will consider only restricted versions of the problem here. The following reminder about areas will be useful.

Chapter Summary

Each chapter is completed by a chapter summary that refreshes and codifies the key ideas presented in the chapter.

Chapter Review Exercises

At the end of each chapter summary is a collection of chapter review exercises. These give students the opportunity to test their understanding of the material covered in the chapter.

ACKNOWLEDGEMENTS

In writing this book, we have also relied on the help of other mathematicians as well as specialists from agriculture, biology, business, chemistry, ecology, economics, engineering, physics, political science, and zoology. We offer our thanks to Bruce Ackerson, Brian Adam, Robert Darcy, Joel Haack, Stanley Fox, Adrienne Hyle, Smith Holt, Jerry Johnson, Lionel Raff, Scott Turner, and Gary Young. Any errors and inaccuracies in applications are due to the authors' misrepresentation of correct information provided by our able consultants. We are grateful to the National Science Foundation for its foresight and support of initial development and to Oklahoma State University for its support. We very much appreciate Charles Hartford's continuing patience and good humor through some trying times.

The most important participants in the development of this work are the students at Oklahoma State University, particularly those in the fall of 1995 and spring of 1996, who suffered through very early versions of this text, and whose input has shaped the current version. This book is written for entering mathematics students, and further student reaction will direct the evolution of the text into a better product. Students and teachers at Oklahoma State University have had fun and learned with this material. We hope the same happens for others.

Bruce Crauder | Benny Evans | Alan Noell

Reviewers

We authors are truly indebted to the many reviewers' kind and constructive comments over the years. Reviewers of the fifth edition include:

Stephen J. Nicoloff, *Paradise Valley Community College*

Christopher Stuart, *New Mexico State University*

Miles Harris, *Northampton Community College*

Mitsue Nakamura, *University of Houston Downtown*

Cassie Firth, *Northern Oklahoma College*

Meghan McIntyre, *Wake Technical Community College*

Lisa Hodge, *Wake Technical Community College*

SUPPLEMENTS

Instructor Supplements

Complete Solutions Manual—ISBN-13: 978-1-133-95497-2   This manual contains the complete worked-out solutions for all the exercises in the text in an easy-to-use online format. Written by the authors, the Complete Solutions Manual uses exactly the same style and format as the worked examples of the text.

Instructor's Guide—ISBN-13: 978-1-285-05232-8   The Instructor's Guide contains valuable teaching tools for both new and experienced instructors. Written by the authors, the Instructor's Guide includes teaching tips for each section of the text as well as sample syllabi that reflect the authors' experience using the text.

PowerLecture with ExamView—ISBN-13: 978-1-133-36556-3   This CD-ROM provides the instructor with dynamic media tools for teaching. Create, deliver, and customize tests (both print and online) in minutes with ExamView® Computerized Testing Featuring Algorithmic Equations. Easily build solution sets for homework or exams using Solution Builder's online solutions manual. Microsoft® PowerPoint® lecture slides and figures from the book are also included on this CD-ROM.

Test Bank—ISBN-13: 978-1-133-95499-6   The Test Bank includes test forms for each chapter of the text in an easy-to-edit electronic format.

Solution Builder—www.cengage.com/solutionbuilder   This online instructor database offers complete worked solutions to all exercises in the text, allowing you to create customized, secure solution printouts (in PDF format) matched exactly to the problems you assign in class.

Enhanced WebAssign—ISBN-13: 978-0-538-73810-1   Exclusively from Cengage Learning, Enhanced WebAssign® offers an extensive online program for precalculus to encourage the practice that's so critical for concept mastery. The meticulously crafted pedagogy and exercises in this text become even more effective in Enhanced WebAssign, supplemented by multimedia tutorial support and immediate feedback as students complete their assignments. Algorithmic problems allow you to assign unique versions to each student. The Practice Another Version feature (activated at your discretion) allows students to attempt the questions with new sets of values until they feel confident enough to work the original problem. Students benefit from a new Premium eBook with highlighting and search features; Personal Study Plans (based on diagnostic quizzing) that identify chapter topics they still need to master; and links to video solutions, interactive tutorials, and even live online help.

Student Supplements

Student Solutions Manual—ISBN-13: 978-1-133-36558-7   This manual includes worked-out solutions to every odd-numbered exercise in the text. Written by the authors, the Student Solutions Manual contains complete and carefully written solutions to odd-numbered exercises in the same style and format as the worked examples of the text. In particular the solutions give additional instruction to allow students to find success doing similar even-numbered exercises.

Enhanced WebAssign-ISBN-13: 978-0-538-73810-1   Exclusively from Cengage Learning, Enhanced WebAssign® offers an extensive online program for precalculus to encourage the practice that's so critical for concept mastery. You'll receive multimedia tutorial support as you complete your assignments. You'll also benefit from a new Premium eBook with highlighting and search features; Personal Study Plans (based on diagnostic quizzing) that identify chapter topics you still need to master; and links to video solutions, interactive tutorials, and even live online help.

CengageBrain.com   Visit www.cengagebrain.com to access additional course materials and companion resources. At the CengageBrain.com home page, search for the ISBN of your title (from the back cover of your book) using the search box at the top of the page. This will take you to the product page where free companion resources can be found.

TO THE STUDENT

There are two important messages that the authors would like to convey to you. First, in spite of what your previous experiences with mathematics may have been, you are perfectly capable of understanding and doing sophisticated mathematics. This text aims to show you what you *can* do—not what you can't. It has already helped thousands of students succeed at mathematics, and it can aid in your achievement as well.

Second, you do not have to rely entirely on someone else to tell you how to deal with the confusing wealth of analytical data that is so much a part of the modern world. This course will equip you with tools for data analysis that are commonly used by professionals. These tools will help you as a student to successfully apply mathematics to your chosen field. They will also help you as a citizen to sort truth from fiction in the media blitz that assails you daily.

Mathematics can help you understand and deal with the world around you. We hope you enjoy the experience that is in store for you. We always like to hear from you. Your feedback is essential to make future editions of this book better.

PROLOGUE: CALCULATOR ARITHMETIC

P

Len DeLessio/Getty Images

Astrofoto/Peter Arnold/Getty Images

To describe concisely the vast distances in space, astronomers use scientific notation. See, for example, Exercise 17 on page 11.

GRAPHING CALCULATORS are powerful tools for mathematical analysis, and this power has profound effects on how modern mathematics and its applications are done. Many mathematical applications that traditionally required sophisticated mathematical development can now be successfully analyzed at an elementary level. Indeed, modern calculating power enables entering students to attack problems that in the past would have been considered too complicated. The first step is to become proficient with arithmetic on the calculator. In this chapter we discuss key mathematical ideas associated with calculator arithmetic. Appendix B is intended to provide additional help for those who are new to the operation of the calculator and those who need a brief refresher on arithmetic operations.

Student resources are available on the website **www.cengagebrain.com**

Typing Mathematical Expressions

When we write expressions such as $\frac{71}{7} + 3^2 \times 5$ using pen and paper, the paper serves as a two-dimensional display, and we can express fractions by putting one number on top of another and exponents by using a superscript. When we enter such expressions on a computer, calculator, or typewriter, however, we must write them on a single line, using special symbols and (often) additional parentheses. The *caret* symbol $\wedge$ is commonly used to denote an exponent, so in *typewriter notation* $\frac{71}{7} + 3^2 \times 5$ comes out as

$$71 \div 7 + 3 \wedge 2 \times 5.$$

In Figure P.1, we have entered this expression, and the resulting answer 55.14285714 is shown in Figure P.2. You should use your calculator to verify that we did it correctly.

FIGURE P.1 Entering $\frac{71}{7} + 3^2 \times 5$

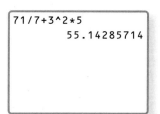

FIGURE P.2 The value of $\frac{71}{7} + 3^2 \times 5$

Rounding

When a calculation yields a long answer such as the 55.14285714 shown in Figure P.2, we will commonly shorten it to a more manageable size by *rounding*. Rounding means that we keep a few of the digits after the decimal point, possibly changing the last one, and discard the rest. There is no set rule for how many digits after the decimal point you should keep; in practice, it depends on how much accuracy you need in your answer, as well as on the accuracy of the data you input. As a general rule, in this text we will round to two decimal places. Thus for

$$\frac{71}{7} + 3^2 \times 5 = 55.14285714$$

we would report the answer as 55.14.

In order to make the abbreviated answer more accurate, it is standard practice to increase the last decimal entry by 1 if the following entry is 5 or greater. Verify with your calculator that

$$\frac{58.7}{6.3} = 9.317460317.$$

0, 1, 2, 3, 4 always round down.

5, 6, 7, 8, 9 always round up.

In this answer the next digit after 1 is 7, which indicates that we should round up, so we would report the answer rounded to two decimal places as 9.32. Note that in reporting 55.14 as the rounded answer above, we followed this same rule. The next digit after 4 in 55.14285714 is 2, which does not indicate that we should round up.

```
12/7
          1.714285714
```

FIGURE P.3 An answer that will be reported as 1.71

```
17/9
          1.888888889
```

FIGURE P.4 An answer that will be reported as 1.89

Parentheses tell us that a group of numbers all go together.

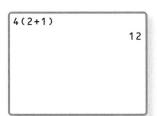

```
4(2+1)
                12
```

FIGURE P.5 A correct calculation of 4(2 + 1) when parentheses are properly entered

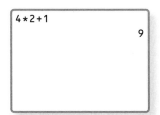

```
4*2+1
                 9
```

FIGURE P.6 An incorrect calculation of 4(2 + 1) caused by omitting parentheses

To provide additional emphasis for this idea, Figure P.3 shows a calculation where rounding does not change the last reported digit, and Figure P.4 shows a calculation where rounding requires that the last reported digit be changed.

> **KEY IDEA P.1 ROUNDING**
>
> When reporting complicated answers, we will adopt the convention of rounding to two places beyond the decimal point. The last digit is increased by 1 if the next following digit is 5 or greater.

Although we generally round to two decimal places, there will be times when it is appropriate to use fewer or more decimal places. These circumstances will be explicitly noted in the text or will be clear from the context of the calculation.

 ## Parentheses and Grouping

When parentheses appear in a calculation, the operations inside are to be done first. Thus $4(2 + 1)$ means that we should first add $2 + 1$ and then multiply the result by 4, getting an answer of 12. This is correctly entered and calculated in Figure P.5. *Where parentheses appear, their use is essential.* If we had entered the expression as $4 \times 2 + 1$, leaving out the parentheses, the calculator would have interpreted it to mean first to multiply 4 times 2 and then to add 1 to the result, giving an incorrect answer of 9. This incorrect entry is shown in Figure P.6.

Sometimes parentheses do not appear, but we must supply them. For example, $\frac{17}{5+3}$ means $17 \div (5 + 3)$. The parentheses are there to show that the whole expression $5 + 3$ goes in the denominator. To do this on the calculator, we must supply these parentheses. Figure P.7 shows the result. If the parentheses are not used, and $\frac{17}{5+3}$ is entered as $17 \div 5 + 3$, the calculator will interpret it to mean that only the 5 goes in the denominator of the fraction. This error is shown in Figure P.8. Similarly, $\frac{8+9}{7+2}$ means $(8 + 9) \div (7 + 2)$; the parentheses around $8 + 9$ indicate that the entire expression goes in the numerator, and the parentheses around $7 + 2$ indicate that the entire expression goes in the denominator. Enter the expression on your calculator and check that the answer rounded to two places is 1.89. The same problem can occur with exponents. For example, $3^{2.7 \times 1.8}$ in typewriter notation is $3 \wedge (2.7 \times 1.8)$. Check to see that the answer rounded to two places is 208.36.

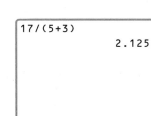

```
17/(5+3)
                2.125
```

FIGURE P.7 Proper use of parentheses in the calculation of $\frac{17}{5+3}$

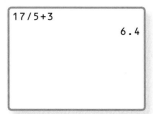

```
17/5+3
                  6.4
```

FIGURE P.8 An incorrect calculation of $\frac{17}{5+3}$ caused by omitting parentheses

In general, we advise that if you have trouble entering an expression into your calculator, or if you get an answer that you know is incorrect, go back and re-enter the expression after first writing it out in typewriter notation, and be careful to supply all needed parentheses.

Minus Signs

The minus sign used in arithmetic calculations actually has two different meanings. If you have $9 in your wallet and spend $3, then you will have $9 - 3 = 6$ dollars left. Here the minus sign means that we are to perform the operation of subtracting 3 from 9. Suppose in another setting that you receive news from the bank that your checking account is overdrawn by $30, so your balance is -30 dollars. Here the minus sign is used to indicate that the number we are dealing with is negative; it does not signify an operation between two numbers. In everyday usage, the distinction is rarely emphasized and may go unnoticed. But most calculators actually have different keys for the two operations, and they cannot be used interchangeably. Thus differentiating between the two becomes crucial to using the calculator correctly.

Once the problem is recognized, it is usually easy to spot when the minus sign denotes subtraction (when two numbers are involved) and when it indicates a change in sign (when only one number is involved). The following examples should help clarify the situation:

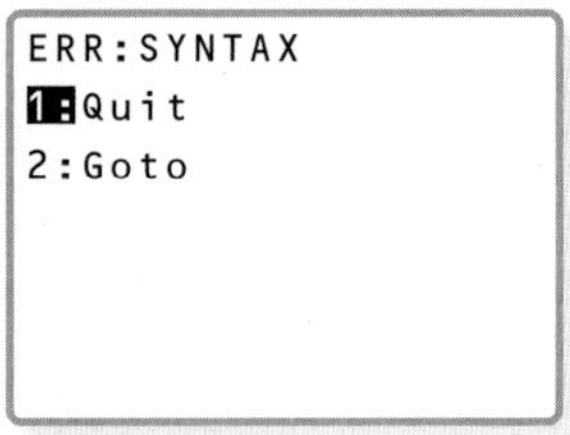

$$-8 - 4 \quad \text{means} \quad \textit{negative } 8 \textit{ subtract } 4$$

$$\frac{3 - 7}{-2 \times 3} \quad \text{means} \quad \frac{3 \textit{ subtract } 7}{\textit{negative } 2 \times 3}$$

$$2^{-3} \quad \text{means} \quad 2^{\textit{negative } 3}.$$

The calculation of 2^{-3} is shown in Figure P.9. If we try to use the calculator's subtraction key, the calculator will not understand the input and will produce an error message such as the one in Figure P.10.

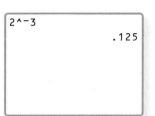

FIGURE P.9 Calculation of 2^{-3} using the negative key

FIGURE P.10 Syntax error when subtraction operation is used in 2^{-3}

EXAMPLE P.1   **SOME SIMPLE CALCULATIONS**

Make the following calculations, and report the answer rounded to two digits beyond the decimal point.

Part 1 $\dfrac{\sqrt{11.4 - 3.5}}{26.5}$

Part 2 $\dfrac{7 \times 3^{-2} + 1}{3 - 2^{-3}}$

Solution to Part 1 To make sure everything we want is included under the square root symbol, we need to use parentheses. In typewriter notation, this looks like

$$\sqrt{(11.4 - 3.5)} \div 26.5.$$

We have calculated this in Figure P.11. Since the third digit beyond the decimal point, 6, is 5 or larger, we report the answer as 0.11.

Solution to Part 2 We need to take care to use parentheses to ensure that the numerator and denominator are right, and we must use the correct keys for negative signs and subtraction.

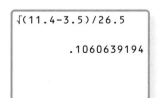

FIGURE P.11 Solution to part 1

```
(7*3^-2+1)/(3-2^
-3)
            .6183574879
```

FIGURE P.12 Solution to part 2

In expanded typewriter notation,

$$\frac{7 \times 3^{-2} + 1}{3 - 2^{-3}} = (7 \times 3 \wedge \textit{negative } 2 + 1) \div (3 \textit{ subtract } 2 \wedge \textit{negative } 3).$$

The result 0.6183574879 is shown in Figure P.12. We round this to 0.62.

TEST YOUR UNDERSTANDING | **FOR EXAMPLE P.1**

Calculate the following expression and round to two places:

$$\frac{2^{-3} + 4}{-0.6 \times 0.4}.$$

Special Numbers π and e

Two numbers, π and e, occur so often in mathematics and its applications that they deserve special mention. The number π is familiar from the formulas for the circumference and area of a circle:

$$\text{Area of a circle of radius } r = \pi r^2$$

$$\text{Circumference of a circle of radius } r = 2\pi r.$$

The approximate value of π is 3.14159, but its exact value cannot be expressed by a simple decimal, and that is why it is normally written using a special symbol. Most calculators allow you to enter the symbol π directly, as shown in Figure P.13. When we ask the calculator for a numerical answer, we get the decimal approximation of π shown in Figure P.13.

The number e may not be as familiar as π, but it is just as important. Like π, it cannot be expressed exactly as a decimal, but its approximate value is 2.71828. In Figure P.14 we have entered e, and the calculator has responded with the decimal approximation shown. Often expressions that involve the number e include exponents, and most calculators have features to make entering such expressions easy. For example, when we enter $e^{1.02}$ we obtain 2.77 after rounding.

$\pi = 3.14159 \ldots .$

$e = 2.71828 \ldots .$

```
π
            3.141592654
```

FIGURE P.13 A decimal approximation of π

```
e
            2.718281828
```

FIGURE P.14 A decimal approximation of e

Chain Calculations

Some calculations are most naturally done in stages. Many calculators have a special key that accesses the result of the last calculation, allowing you to enter your work in pieces. To show how this works, let's look at

$$(\sqrt{13} - \sqrt{2})^3 + \frac{17}{2 + \pi}.$$

We will make the calculation in pieces. First we calculate $(\sqrt{13} - \sqrt{2})^3$. Enter this to get the answer in Figure P.15. To finish the calculation, we need to add this answer to $17/(2 + \pi)$:

$$(\sqrt{13} - \sqrt{2})^3 + \frac{17}{2 + \pi} = \text{First answer} + \frac{17}{2 + \pi}.$$

In Figure P.16 we have used the answer from Figure P.15 to complete the calculation.

```
(√(13)-√(2))^3
          10.52271818
```

FIGURE P.15 The first step in a chain calculation

```
(√(13)-√(2))^3
          10.52271818
Ans+17/(2+π)
          13.82908668
```

FIGURE P.16 Completing a chain calculation

```
7/9
          .7777777778
(3^Ans+2^Ans)/(5
^Ans-4^Ans)
          7.295922612
```

FIGURE P.17 Accessing previous results to get an accurate answer

Accessing the results of one calculation for use in another can be particularly helpful when the same thing appears several times in an expression. For example, let's calculate

$$\frac{3^{7/9} + 2^{7/9}}{5^{7/9} - 4^{7/9}}.$$

Since $7/9$ occurs several times, we have calculated it first in Figure P.17. Then we have used the results to complete the calculation. We would report the final answer rounded to two decimal places as 7.30.

There is an additional advantage to accessing directly the answers of previous calculations. It might seem reasonable to calculate $7/9$ first, round it to two decimal places, and then use that to complete the calculation. Thus we would be calculating

$$\frac{3^{0.78} + 2^{0.78}}{5^{0.78} - 4^{0.78}}.$$

```
7/9
          .7777777778
(3^.78+2^.78)/(5
^.78-4^.78)
          7.265871182
```

FIGURE P.18 Inaccurate answer caused by early rounding

This is done in Figure P.18, which shows the danger in this practice. We got an answer, rounded to two decimal places, of 7.27—somewhat different from the more accurate answer, 7.30, that we got earlier. In many cases, errors caused by early rounding can be much more severe than is shown by this example. In general, if you are making a calculation in several steps, you should not round until you get the final answer. An important exception to this general rule occurs in applications where the result of an intermediate step must be rounded because of the context. For example, in a financial computation dollar amounts would be rounded to two decimal places.

EXAMPLE P.2 **COMPOUND INTEREST AND APR**

There are a number of ways in which lending institutions report and charge interest.

Part 1 Paying *simple interest* on a loan means that you wait until the end of the loan before calculating or paying any interest. If you borrow $5000 from a bank that charges 7% simple interest, then after t years you will owe

$$5000 \times (1 + 0.07t) \text{ dollars.}$$

Under these conditions, how much money will you owe after 10 years?

Part 2 Banks more commonly *compound the interest*. That is, at certain time periods the interest you have incurred is calculated and added to your debt. From that time on, you incur interest not only on your principal (the original debt) but on the added interest as well. Suppose the interest is compounded yearly, but you make no payments and there are no finance charges. Then, again with a principal of $5000 and 7% interest, after t years you will owe

$$5000 \times 1.07^t \text{ dollars.}$$

Under these conditions, how much will you owe after 10 years?

Part 3 For many transactions such as automobile loans and home mortgages, interest is compounded monthly rather than yearly. In this case, the amount owed is calculated each month using the *monthly interest rate*. If r (as a decimal) is the monthly interest rate, then after m months, the amount owed is

$$5000 \times (1 + r)^m \text{ dollars,}$$

assuming the principal is $5000.

The value of r is usually not apparent from the loan agreement. But lending institutions are required by the *Truth in Lending Act* to report the *annual percentage rate,* or APR, in a prominent place on all loan agreements. The same statute requires that the value of r be calculated using the formula

$$r = \frac{\text{APR}}{12}.$$

If the annual percentage rate is 7%, what is the amount owed after 10 years?[1]

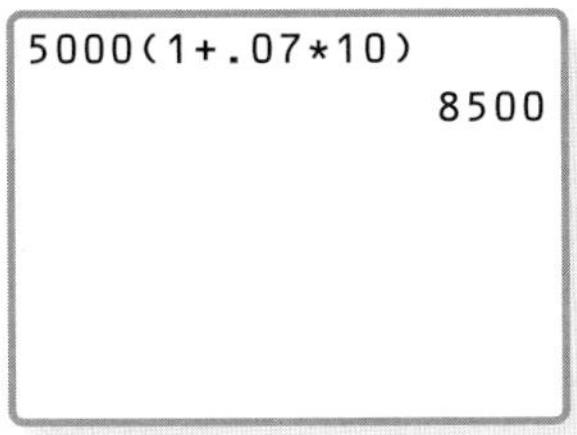

Solution to Part 1 To find the amount owed after 10 years, we use $t = 10$ to get

$$5000 \times (1 + 0.07 \times 10).$$

Entering this on the calculator as we have done in Figure P.19 reveals that the amount owed in 10 years will be $8500.

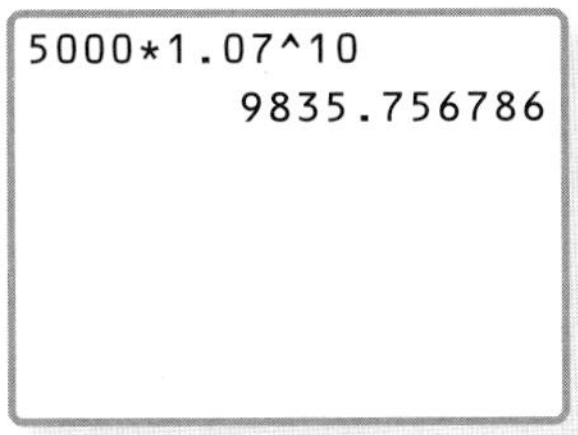

Solution to Part 2 This time we use

$$5000 \times 1.07^{10}.$$

From Figure P.20 we see that, rounded to the nearest cent, the amount owed will be $9835.76. Comparison with part 1 shows the effect of compounding interest. We should note that at higher interest rates the effect is more dramatic.

Solution to Part 3 The first step is to use the formula

$$r = \frac{\text{APR}}{12} = \frac{0.07}{12}$$

to get the value of r as we have done in Figure P.21. Ten years is 120 months, and this is the value we use for m. Using this value for m and incorporating the value of r that we just calculated, we entered $5000 \times (1 + r)^{120}$ in Figure P.22, and we conclude that the amount owed will be $10,048.31. Comparing this with the answer from part 2, we see that the difference between yearly and monthly compounding is significant. It is important that you know how interest on your loan is calculated, and this may not be

---

The APR is a measure of the annual interest rate charged on consumer loans.

```
5000(1+.07*10)
              8500
```

FIGURE P.19 Balance after 10 years using simple interest

```
5000*1.07^10
       9835.756786
```

FIGURE P.20 Balance after 10 years using yearly compounding

Monthly rate $= \dfrac{\text{APR}}{12}$

---

[1]Many consider the relationship between the monthly interest rate and the APR mandated by the Truth in Lending Act to be misleading. If, for example, you borrow $100 at an APR of 10%, then if no payments are made, you may expect to owe $110 at the end of 1 year. If interest is compounded monthly, however, you will in fact owe somewhat more. For more information, see the discussion in Section 5.6 of *Fundamentals of Corporate Finance* by S. Ross, R. Westerfield, and B. Jordan (Chicago: Richard D. Irwin, 1995). See also Exercise 16.

```
.07/12
              .0058333333
```

FIGURE P.21 Getting the monthly interest rate from the APR

```
.07/12
              .0058333333
5000(1+Ans)^120
              10048.30688
```

FIGURE P.22 Balance after 10 years using monthly compounding

easy to find out from the paperwork you get from a lending institution. The APR will be reported, but the compounding periods may not be shown at all.

TEST YOUR UNDERSTANDING | **FOR EXAMPLE P.2**

If the annual percentage rate is 8% and interest is compounded monthly, what is the amount owed on a principal of $5000 after 15 years? ■

Scientific Notation

It is cumbersome to write down all the digits of some very large or very small numbers. A prime example of such a large number is *Avogadro's number*, which is the number of atoms in 12 grams of carbon 12. Its value is about

$$602,000,000,000,000,000,000,000.$$

An example of a small number that is awkward to write is the mass in kilograms of an electron:

$$0.000\,000\,000\,000\,000\,000\,000\,000\,000\,000\,000\,911 \text{ kilogram.}$$

```
2^50
              1.125899907E15
```

FIGURE P.23 Scientific notation for a large number

Scientists and mathematicians usually express such numbers in a more compact form using *scientific notation*. In this notation, numbers are written in a form with one nonzero digit to the left of the decimal point times a power of 10. Examples of numbers written in scientific notation are 2.7×10^4 and 2.7×10^{-4}. The power of 10 tells how the decimal point should be moved in order to write the number out in longhand. The 4 in 2.7×10^4 means that we should move the decimal point four places to the right. Thus

$$2.7 \times 10^4 = 27,000$$

since we move the decimal point four places to the right. When the exponent on 10 is negative, the decimal point should be moved to the left. Thus

$$2.7 \times 10^{-4} = 0.00027$$

```
7/3^20
              2.007580394E-9
```

FIGURE P.24 Scientific notation for a small number

E*n* means to move the decimal point *n* places to the right.

E-*n* means to move the decimal point *n* places to the left.

since we move the decimal point four places to the left. With this notation, Avogadro's number comes out as 6.02×10^{23}, and the mass of an electron as 9.11×10^{-31} kilogram.

Many times calculators display numbers like this but use a different notation for the power of 10. For example, Avogadro's number 6.02×10^{23} is displayed as 6.02E23, and the mass in kilograms of an electron 9.11×10^{-31} is shown as 9.11E-31. In Figure P.23 we have calculated 2^{50}. The answer reported by the calculator written in longhand is 1,125,899,907,000,000. In presenting the answer in scientific notation, it would in many settings be appropriate to round to two decimal places as 1.13×10^{15}. In Figure P.24 we have calculated $7/3^{20}$. The answer reported there equals 0.000 000 002 007 580 394. If we write it in scientific notation and round to two decimal places, we get 2.01×10^{-9}.

ANSWERS FOR TEST YOUR UNDERSTANDING

P.1 -17.19

P.2 $16,534.61

P EXERCISES

Reminder   Round all answers to two decimal places unless otherwise indicated.

1. **Valentine's Day**   According to the National Retail Federation, on Valentine's Day 2010 American men spent an average of $129.95 on their sweethearts, and women spent only $72.28 on their heroes. What percentage of the average male expenditure was the average female expenditure?

2. **Cat Owners**   According to the Humane Society, in 2009 33% of U.S. households owned at least one cat, and 56% of households who did own cats owned at least two. The U.S. Census Bureau tells us that there were 116 million households in 2009. How many households owned at least two cats? Report your answer in millions rounded to two places.

3. **A Billion Dollars**   A one-dollar bill is 0.0043 inch thick. If you had a billion one-dollar bills and made a stack of them, how high in miles would the stack be? Remember that there are 12 inches in a foot and 5280 feet in a mile.

4. **National Debt**   In early 2010 the U.S. population was about 308 million. The national debt was well over $12 trillion. In millions of dollars the debt was $12,367,728. How much did each American owe in early 2010? Report your answer in thousands of dollars rounded to the nearest whole number.

5. **10% Discount and 10% Tax**   Suppose you want to buy a great pair of designer jeans that were originally priced at $75, but are now on sale for 10% off. When you buy the jeans, you need to pay sales tax of 10% on the sales price. How much will you have to pay for the jeans?

6. **A Good Investment**   You have just received word that your original investment of $850 has increased in value by 13%. What is the value of your investment today?

7. **A Bad Investment**   You have just received word that your original investment of $720 has decreased in value by 7%. What is the value of your investment today?

8. **An Uncertain Investment**   Suppose you invested $1300 in the stock market 2 years ago. During the first year, the value of the stock increased by 12%. During the second year, the value of the stock decreased by 12%. How much money is your investment worth at the end of the two-year period? Did you earn money or lose money? (*Note:* The answer to the first question is *not* $1300.)

9. **Pay Raise**   You receive a raise in your hourly pay from $9.25 per hour to $9.50 per hour. What percent increase in pay does this represent?

10. **Heart Disease**   In a certain county, the number of deaths due to heart disease decreased from 235 in one year to 221 in the next year. What percent decrease in deaths due to heart disease does this represent?

11. **Trade Discount**   Often retailers sell merchandise at a suggested retail price determined by the manufacturer. The *trade discount* is the percentage discount given to the retailer by the manufacturer. The resulting price is the retailer's net cost and so is called the *cost price*. For example, if the suggested retail price is $100.00 and the trade discount is 45%, then the cost price is $100.00 - 45\% \times 100.00 = 55.00$ dollars.

    a. If an item has a suggested retail price of $9.99 and the trade discount is 40%, what is the retailer's cost price?

    b. If an item has a cost price of $37.00 and a suggested retail price of $65.00, what trade discount was used?

12. **Series Discount**   *This is a continuation of Exercise 11.* Sometimes manufacturers give more than one discount instead of a single trade discount—for example, in trading with large-volume retailers.

(continued)

Such a *series discount* is quoted as a sequence of discounts, taken one after another. Suppose a manufacturer normally gives a trade discount of 45%, but it has too much of the item in inventory and so wants to sell more. In this case, the manufacturer may give all retailers another discount of 15% and may perhaps extend yet *another* discount of 10% to a specific retailer it wants to land as a client. In this example, the series discount would be 45%, 15%, 10%, calculated one after another, like this: For an item with a suggested retail price of $100.00, applying the first discount gives $100.00 − 45\% \times 100.00 = 55.00$ dollars. The second discount of 15% is applied to the $55.00 as follows: $55.00 − 15\% \times 55.00 = 46.75$ dollars. Now the third discount gives the final cost price of $46.75 − 10\% \times 46.75 = 42.08$ dollars.

a. Suppose an item has a suggested retail price of $80.00 and the manufacturer is giving a series discount of 25% and 10%. What is the resulting cost price?

b. Suppose an item has a suggested retail price of $100.00 and the manufacturer is giving a series discount of 35%, 10%, 5%. What is the resulting cost price?

c. What single trade discount would give the same cost price as a series discount of 35%, 10%, 5%? (*Note:* The answer is *not* 50%.)

d. Explain why we could have calculated the same answer as in part b by multiplying

$$100.00 \times 0.65 \times 0.90 \times 0.95.$$

In this case, what do the 0.65, 0.90, and 0.95 represent?

13. **Present Value**   *Present value* is the amount of money that must be invested now at a given rate of interest to produce a given future value. For a 1-year investment, the present value can be calculated using

$$\text{Present value} = \frac{\text{Future value}}{1 + r},$$

where r is the yearly interest rate expressed as a decimal. (Thus, if the yearly interest rate is 8%, then $1 + r = 1.08$.) If an investment yielding a yearly interest rate of 12% is available, what is the present value of an investment that will be worth $5000 at the end of 1 year? That is, how much must be invested today at 12% in order for the investment to have a value of $5000 at the end of a year?

14. **Future Value**   Business and finance texts refer to the value of an investment at a future time as its *future value*. If an investment of P dollars is compounded yearly at an interest rate of r as a decimal, then the value of the investment after t years is given by

$$\text{Future value} = P \times (1 + r)^t.$$

In this formula, $(1 + r)^t$ is known as the *future value interest factor*, so the formula above can also be written

$$\text{Future value} = P \times \text{Future value interest factor}.$$

Financial officers normally calculate this (or look it up in a table) first.

a. What future value interest factor will make an investment double? Triple?

b. Say you have an investment that is compounded yearly at a rate of 9%. Find the future value interest factor for a 7-year investment.

c. Use the results from part b to calculate the 7-year future value if your initial investment is $5000.

15. **The Rule of 72**   *This is a continuation of Exercise 14*. Financial advisors sometimes use a rule of thumb known as the *Rule of 72* to get a rough estimate of the time it takes for an investment to double in value. For an investment that is compounded yearly at an interest rate of $r\%$, this rule says it will take about $72/r$ years for the investment to double. In this calculation, r is the integer interest rate rather than a decimal. Thus, if the interest rate is 8%, we would use $72/8$ rather than $72/0.08$.

For the remainder of this exercise, we will consider an investment that is compounded yearly at an interest rate of 13%.

a. According to the Rule of 72, how long will it take the investment to double in value?

Parts b and c of this exercise will check to see how accurate this estimate is for this particular case.

b. Using the answer you got from part a of this exercise, calculate the future value interest factor (as defined in Exercise 14). Is it exactly the same as your answer to the first question in part a of Exercise 14?

c. If your initial investment was $5000, use your answer from part b to calculate the future value. Did your investment exactly double?

16. **The Truth in Lending Act**   Many lending agencies compound interest more often than yearly, and as we noted in Example P.2, they are required to report the annual percentage rate, or APR, in a prominent place on the loan agreement. Furthermore, they are

required to calculate the APR in a specific way. If r is the monthly interest rate, then the APR is calculated using

$$\text{APR} = 12 \times r.$$

a. Suppose a credit card company charges a monthly interest rate of 1.9%. What APR must the company report?

b. The phrase *annual percentage rate* leads some to believe that if you borrow $6000 from a credit card company which quotes an APR of 22.8%, and if no payments are made, then at the end of 1 year interest would be calculated as 22.8% simple interest on $6000. How much would you owe at the end of a year if interest is calculated in this way?

c. If interest is compounded monthly (which is common), then the actual amount you would owe in the situation of part b is given by

$$6000 \times 1.019^{12}.$$

What is the actual amount you would owe at the end of a year?

17. **The Size of the Earth** The radius of the Earth is approximately 4000 miles.

a. How far is it around the equator? (*Hint:* You are looking for the circumference of a circle.)

b. What is the volume of the Earth? (*Note:* The volume of a sphere of radius r is given by $\frac{4}{3}\pi r^3$.)

c. What is the surface area of the Earth? (*Note:* The surface area of a sphere of radius r is given by $4\pi r^2$.)

18. **When the Radius Increases**

a. A rope is wrapped tightly around a wheel with a radius of 2 feet. If the radius of the wheel is increased by 1 foot to a radius of 3 feet, by how much must the rope be lengthened to fit around the wheel?

b. Consider a rope wrapped around the Earth's equator. We noted in Exercise 17 that the radius of the Earth is about 4000 miles. That is 21,120,000 feet. Suppose now that the rope is to be suspended exactly 1 foot above the equator. By how much must the rope be lengthened to accomplish this?

19. **The Length of Earth's Orbit** The Earth is approximately 93 million miles from the sun. For this exercise we will assume that the Earth's orbit is a circle.[2]

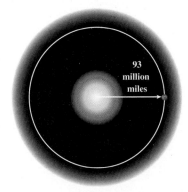

93 million miles

a. How far does the Earth travel in a year?

b. What is the velocity in miles per year of the Earth in its orbit? (*Hint:* Recall that Velocity $= \frac{\text{Distance}}{\text{Time}}$.)

c. How many hours are there in a year? (*Note:* Assume a year is 365 days.)

d. What is the velocity in miles per hour of the Earth in its orbit?

20. **A Population of Bacteria** Some populations, such as bacteria, can be expected under the right conditions to show *exponential growth*. If 2000 bacteria of a certain type are incubated under ideal conditions, then after t hours we expect to find 2000×1.07^t bacteria present. How many bacteria would we expect to find after 8 hours? How many after 2 days?

21. **Newton's Second Law of Motion** Newton's second law of motion states that the force F on an object is the product of its mass m with its acceleration a:

$$F = ma.$$

If mass is measured in kilograms and acceleration in meters per second per second, then the force is

(continued)

---

[2]The orbit of the Earth is in fact an ellipse, but for many practical applications the assumption that it is a circle yields reasonably accurate results.

given in *newtons*. Another way to measure force is in pounds; in fact, 1 newton is 0.225 pound (which is about a quarter of a pound). In the case of an object near the surface of the Earth, the force due to gravity is its weight. Near the surface of the Earth, acceleration due to gravity is 9.8 meters per second per second. What is the weight in newtons of a man with a mass of 75 kilograms? What is his weight in pounds?

22. **Weight on the Moon**   *This is a continuation of Exercise 21.* Acceleration due to gravity near the surface of the Earth's moon is only 1.67 meters per second per second. Thus an object has a different weight on the Earth than it would on the moon. What is the weight of the 75-kilogram man from Exercise 21 if he is standing on the moon? Give your answer first in newtons and then in pounds.

23. **Frequency of Musical Notes**   Counting sharps and flats, there are 12 notes in an octave on a standard piano. If one knows the frequency of a note, then one can find the frequency of the next higher note by multiplying by the 12th root of 2:

    Frequency of next higher note
           = Frequency of given note $\times\ 2^{1/12}$.

    The frequency of middle C is 261.63 cycles per second. What is the frequency of the next higher note (which is C#) on a piano? What is the frequency of the D note just above middle C? (The D note is two notes higher than middle C.)

24. **Lean Body Weight in Males**   A person's *lean body weight L* is the amount he or she would weigh if all body fat were magically to disappear. One text[3] gives the "equation that practitioners can use most feasibly in the field to predict lean body weight in young adult males." The equation is

    $$L = 98.42 + 1.08W - 4.14A.$$

Here L is lean body weight in pounds, W is weight in pounds, and A is abdominal circumference in inches. Find the approximate lean body weight of a young adult male who weighs 188 pounds and has an abdominal circumference of 35 inches. What is the weight of his body fat? What is his body fat percent?

25. **Lean Body Weight in Females**   *This is a continuation of Exercise 24.* The text cited in Exercise 24 gives a more complex method of calculating lean body weight for young adult females:

    $$L = 19.81 + 0.73W + 21.2R - 0.88A \\ - 1.39H + 2.43F.$$

Here L is lean body weight in pounds, W is weight in pounds, R is wrist diameter in inches, A is abdominal circumference in inches, H is hip circumference in inches, and F is forearm circumference in inches. According to this formula, what is the approximate lean body weight of a young adult female who weighs 132 pounds and has wrist diameter of 2 inches, abdominal circumference of 27 inches, hip circumference of 37 inches, and forearm circumference of 7 inches? What is the weight of her body fat? What is her body fat percent?

26. **Manning's Equation**   Hydrologists sometimes use *Manning's equation* to calculate the velocity v, in feet per second, of water flowing through a pipe. The velocity depends on the *hydraulic radius R* in feet, which is one-quarter of the diameter of the pipe when the pipe is flowing full; the *slope S* of the pipe, which gives the vertical drop in feet for each horizontal foot; and the *roughness coefficient n*,

---

[3]D. Kirkendall, J. Gruber, and R. Johnson, *Measurement and Evaluation for Physical Educators*, 2nd ed. (Champaign, IL: Human Kinetics Publishers, 1987).

which depends on the material of which the pipe is made. The relationship is given by

$$v = \frac{1.486}{n} R^{2/3} S^{1/2}.$$

For a certain brass pipe, the roughness coefficient has been measured to be $n = 0.012$. The pipe has a diameter of 3 feet and a slope of 0.2 foot per foot. (That is, the pipe drops 0.2 foot for each horizontal foot.) If the pipe is flowing full, find the hydraulic radius of the pipe, and find the velocity of the water flowing through the pipe.

27. **Relativistic Length**   A rocket ship traveling near the speed of light appears to a stationary observer to shorten with speed. A rocket ship with a length of 200 meters will appear to a stationary observer to have a length of

$$200\sqrt{1 - r^2} \text{ meters,}$$

where r is the ratio of the velocity of the ship to the speed of light. What is the apparent length of the rocket ship if it is traveling at a speed that is 99% of the speed of light?

28. **Equity in a Home**   When you purchase a home by securing a mortgage, the total paid toward the principal is your *equity* in the home. (Technically, the lending agency calculates your equity by subtracting the amount you still owe on your mortgage from the *current value* of your home, which may be higher or lower than your principal.) Assume that your mortgage is for $350,000 at a monthly rate of 0.007 as a decimal and that the term of the mortgage is 30 years. Then your equity after k monthly payments is

$$350,000 \times \frac{1.007^k - 1}{1.007^{360} - 1} \text{ dollars.}$$

Calculate the equity in your home after 10 years.

29. **The Advantage Cash Card**   At the Jot Travis Student Union on the campus of the University of Nevada, Reno, you can save on food purchases by using the *Advantage Cash* card. You deposit money into an Advantage Cash account and are issued a credit card that you use to purchase food. The card has several advantages:

- If you open your Advantage Cash account for $200 or more, a 5% bonus is added to your account balance.

- When you use your Advantage Cash card, you receive 5% off the retail price of any food purchase.

- When you buy food with cash you must pay a sales tax of 7.375%. With the Advantage Cash card, you pay no sales tax.

   a. An item retails for $1.00. What do you pay if you use your Advantage Cash card?

   b. An item retails for $1.00. What do you pay if you use cash? Round your answer to five decimal places for use in part d.

   c. What retail value of food will you be able to purchase if you open an Advantage Cash account for $300? (*Suggestion:* Don't forget your 5% bonus, and use the results of part a.)

   d. What retail value of food would you be able to purchase with $300 if you spend it as cash at the food court? (*Suggestion:* Use the results of part b.)

   e. Calculate the percentage increase from your answer for part d to your answer for part c. Explain in practical terms the meaning of this percentage.

P SKILL BUILDING EXERCISES

S-1. **Basic Calculations** $\dfrac{2.6 \times 5.9}{6.3}$

S-2. **Basic Calculations** $3^{3.2} - 2^{2.3}$

S-3. **Basic Calculations** $\dfrac{e}{\sqrt{\pi}}$

S-4. **Basic Calculations** $\dfrac{7.6^{1.7}}{9.2}$

S-5. **Parentheses and Grouping** $\dfrac{7.3 - 6.8}{2.5 + 1.8}$

S-6. **Parentheses and Grouping** $3^{2.4 \times 1.8 - 2}$

S-7. **Parentheses and Grouping** $\dfrac{\sqrt{6 + e} + 1}{3}$

S-8. **Parentheses and Grouping** $\dfrac{\pi - e}{\pi + e}$

S-9. **Subtraction Versus Sign** $\dfrac{-3}{4 - 9}$

S-10. **Subtraction Versus Sign** $-2 - 4^{-3}$

S-11. **Subtraction Versus Sign** $-\sqrt{8.6 - 3.9}$

S-12. **Subtraction Versus Sign** $\dfrac{-\sqrt{10} + 5^{-0.3}}{17 - 6.6}$

(continued)

S-13. Chain Calculations   The following are intended to provide practice with chain calculations.

a. $\dfrac{3}{7.2 + 5.9} + \dfrac{7}{6.4 \times 2.8}$

b. $\left(1 + \dfrac{1}{36}\right)^{\left(1 - \frac{1}{36}\right)}$

S-14. Evaluate Expression   Evaluate $e^{-3} - \pi^2$.

S-15. Evaluate Expression   Evaluate $\dfrac{5.2}{7.3 + 0.2^{4.5}}$.

Arithmetic   In Exercises S-16 through S-20, perform the calculation and report the answer rounded to two decimal places. For some of the calculations, you may wish to use the chain calculation facility of your calculator to help avoid errors.

S-16. $(4.3 + 8.6)(8.4 - 3.5)$

S-17. $\dfrac{2^{3.2} - 1}{\sqrt{3} + 4}$

S-18. $\sqrt{2^{-3} + e}$

S-19. $(2^{-3} + \sqrt{7} + \pi)\left(e^2 + \dfrac{7.6}{6.7}\right)$

S-20. $\dfrac{17 \times 3.6}{13 + \frac{12}{3.2}}$

Evaluating Formulas   In Exercises S-21 through S-30 you are given a formula that you are asked to evaluate with given values for some of the variables. Report your answers rounded to 2 decimal places, except for Exercise S-29, where you should round to four decimal places.

S-21. Evaluate the formula $\dfrac{A - B}{A + B}$ using $A = 4.7$ and $B = 2.3$.

S-22. Evaluate the formula $\dfrac{p(1 + r)}{\sqrt{r}}$ using $p = 144$ and $r = 0.13$.

S-23. Evaluate the formula $\sqrt{x^2 + y^2}$ using $x = 1.7$ and $y = 3.2$.

S-24. Evaluate the formula $p^{1 + 1/q}$ using $p = 4$ and $q = 0.3$.

S-25. Evaluate the formula $(1 - \sqrt{A})(1 + \sqrt{B})$ using $A = 3$ and $B = 5$.

S-26. Evaluate the formula $\left(1 + \dfrac{1}{x}\right)^2$ using $x = 20$.

S-27. Evaluate the formula $\sqrt{b^2 - 4ac}$ using $b = 7$, $a = 2$, and $c = 0.07$.

S-28. Evaluate the formula $\dfrac{1}{1 + \frac{1}{x}}$ using $x = 0.7$.

S-29. Evaluate the formula $(x + y)^{-x}$ using $x = 3$ and $y = 4$.

S-30. Evaluate the formula $\dfrac{A}{\sqrt{A} + \sqrt{B}}$ using $A = 5$ and $B = 6$.

S-31. Lending Money   For a certain loan, the interest I due at the end of a loan period is given by $I = Prt$, where P is the principal borrowed, r is the yearly interest rate as a decimal, and t is the number of years since the money was borrowed. What interest is accrued if 3 years ago we borrowed $5000 at an interest rate of 5%?

S-32. Monthly Payment   For a certain installment loan, the monthly payment M is given by

$$M = \frac{Pr(1 + r)^t}{(1 + r)^t - 1},$$

where P is the original amount borrowed, r is the monthly interest rate as a decimal, and t is the number of months required to pay off the loan. What is the monthly payment if the monthly interest rate as a decimal is 0.05, the amount borrowed was $12,000, and the loan is paid off in 36 months?

S-33. Temperature   If the Celsius temperature is C, then the Fahrenheit temperature F is given by $F = \frac{9}{5}C + 32$. What is the Fahrenheit temperature when the Celsius temperature is 32 degrees?

S-34. A Skydiver   When a skydiver jumps from an airplane, his downward velocity, in feet per second, before he opens his parachute is given by $v = 176(1 - 0.834^t)$, where t is the number of seconds that have elapsed since he jumped from the airplane. What is the velocity after 5 seconds?

S-35. Future Value   In certain savings scenarios the value F of an investment after t years, the future value, is given by $F = P(1 + r)^t$. Here r is the yearly interest rate as a decimal, P is the amount of the original investment, and t is the term of the investment. If we invest $1000 at an interest rate of 0.06 per year as a decimal, and if the term of the investment is 5 years, what is the future value?

S-36. A Population of Deer   The number N of deer in a certain population t years after observation began is given by $N = \dfrac{12.36}{0.03 + 0.55^t}$. What is the deer population after 10 years? Round your answer to the nearest whole number.

S-37. Carbon 14   The amount C, in grams, of carbon 14 remaining in a certain sample after t years is given by $C = 5 \times 0.5^{t/5730}$. How much remains after 5000 years?

S-38. Getting Three Sixes   If we roll n fair dice, then the probability of getting exactly 3 sixes (not more and not less) is given by

$$P = \frac{n(n-1)(n-2)}{750}\left(\frac{5}{6}\right)^n.$$

What is the probability of getting exactly 3 sixes if we roll 7 fair die?

CHAPTER P | SUMMARY

Modern graphing calculators are well designed for ease of use, but care must be taken when entering expressions. The most common errors occur when parentheses are omitted or misused. Also, rounding and scientific notation are significant concepts when you use a calculator. The special numbers e and π are important.

P.1   TYPING EXPRESSIONS AND PARENTHESES

When entering an expression in the calculator, you must enter it not as one would write it on paper but, rather, in *typewriter notation*. If you have trouble getting an expression into the calculator properly, first write it out on paper in typewriter notation and then enter it into the calculator. Parentheses are essential when you need to tell the calculator that a certain operation is to be applied to a group of numbers.

P.2   ROUNDING

In order to do accurate calculations, the calculator uses decimals with many digits. Often only a few digits after the decimal point are needed for the final answer. We limit the number of decimal places by *rounding*. There is no set number of digits used in rounding; that depends on the accuracy of the data entered and on the accuracy needed for the answer. In general, however, answers are reported in this text rounded to two decimal places.

   Rounding Convention for This Text: Unless otherwise specified, answers should be rounded to two decimal places. If the third digit beyond the decimal point is less than 5, discard all digits beyond the second. If the third digit is 5 or larger, increase the second digit by 1 before discarding additional digits.

P.3   SPECIAL NUMBERS

There are two special numbers, π and e, that occur so often in mathematics and its applications that their use cannot be avoided. Modern calculators allow for their direct entry. The number π is the familiar ratio of the circumference of a circle to its diameter. The number e is perhaps less familiar but is just as important, and it often arises in certain exponential contexts. Neither of these numbers can be expressed exactly as a finite decimal, but their approximate values are given below:

$$\pi \approx 3.14159$$

$$e \approx 2.71828.$$

P.4 SCIENTIFIC NOTATION

Some numbers use so many digits that it is more convenient to express them in *scientific notation*. This simply means to write the number using only a few digits and multiply by a power of 10 that tells how the decimal point should be adjusted. The adjustment required depends on the sign on the power of 10. Scientific notation can be entered in the calculator using 10 to a power, but when the calculator reports an answer in scientific notation, a special notation is common.

Entry	Calculator Display	Meaning
number $\times\ 10^{+k}$	*number* E + k	Move decimal point k places right.
number $\times\ 10^{-k}$	*number* E − k	Move decimal point k places left.

CHAPTER P   REVIEW EXERCISES

1. **Parentheses and Grouping**   Evaluate $\dfrac{5.7 + 8.3}{5.2 - 9.4}$.

2. **Evaluate Expression**   Evaluate $\dfrac{8.4}{3.5 + e^{-6.2}}$.

3. **Evaluate Expression**   Evaluate $\left(7 + \dfrac{1}{e}\right)^{\left(\frac{5}{2+\pi}\right)}$.

4. **Gas Mileage**   For a truck that gets gas mileage of 15 miles per gallon, the number g of gallons required to travel m miles is

$$g = \frac{m}{15}.$$

How many gallons are required to travel 27 miles? How many gallons are required to travel 250 miles?

5. **Kepler's Third Law**   According to Kepler's third law of planetary motion, the mean distance D, in millions of miles, from a planet in our solar system to the sun is related to the time P, in years, it takes for the planet to complete a revolution around the sun, and the relationship is

$$D = 93P^{2/3}.$$

It takes the planet Pluto 249 years to complete a revolution around the sun. What is the mean distance from Pluto to the sun? What is the mean distance from Earth to the sun? Give your answers to the nearest million miles.

6. **Traffic Signal**   Traffic engineers study how long the yellow light for a traffic signal should be. For one intersection, the number of seconds n required for a yellow light is related to the average approach speed v, in feet per second, by

$$n = 1 + \frac{v}{30} + \frac{100}{v}.$$

If the approach speed is 80 feet per second (about 55 miles per hour), how long should the yellow light be?

FUNCTIONS

The heights of winning pole vaults in the Olympic Games can be modeled by functions. See Exercise 6 on page 22.

A FUNDAMENTAL IDEA in mathematics and its applications is that of a *function*, which tells how one thing depends on others. One example of a function is the interest incurred on a loan after a certain number of years. In this case there is a formula[1] that allows you to calculate precisely how much you owe, and the formula makes explicit how the debt depends on time. Another example is the value of the Dow Jones Industrial Average at the close of each business day. In this case the value depends on the date, but there is no known formula. This idea of a function is the cornerstone to understanding and using mathematics.

In applications of mathematics, functions are often representations of real phenomena or events. Thus we say that they are *models*. Obtaining a function or functions to act as a model is commonly the key to understanding physical, natural, and social science phenomena. In this chapter we look at functions given by formulas, by tables, by graphs, and by words. Analyzing a function from each of these perspectives will be essential as we progress.

---

[1]See Example P.2 of the Prologue.

Student resources are available on the website **www.cengagebrain.com**

17

1.1 FUNCTIONS GIVEN BY FORMULAS

We look first at functions given by formulas, since this provides a natural context for explaining how a function works.

Functions of One Variable

If your job pays $9.00 per hour, then the money M, in dollars, that you make depends on the number of hours h that you work, and the relationship is given by a simple formula:

$$\text{Money} = 9 \times \text{Hours worked}, \quad \text{or} \quad M = 9h \text{ dollars}.$$

The formula $M = 9h$ shows how the money M that you earn depends on the number of hours h that you work, and we say that M *is a function of* h. In this context we are thinking of h as a *variable* whose value we may not know until the end of the week. Once the value of h is known, the formula $M = 9h$ can be used to calculate the value of M. To emphasize that M is a function of h, it is common to write $M = M(h)$ and to write the formula as $M(h) = 9h$.

Functions given in this way are very easy to use. For example, if you work 30 hours, then in *functional notation*, $M(30)$ is the money you earn. To calculate that, you need only replace h in the formula by 30:

$$M(30) = 9 \times 30 = 270 \text{ dollars}.$$

It is important to remember that h is measured in hours and M is measured in dollars. You will not be very happy if your boss makes a mistake and pays you $9 \times 10 = 90$ cents for 10 hours worked. You may be happier if she pays you $9 \times 30 = 270$ dollars for 30 minutes of work, but both calculations are incorrect. The formula is not useful unless you state in words the units you are using. A proper presentation of the formula for this function would be $M = 9h$, where h is measured in hours and M is measured in dollars. *The words that give the units are as important as the formula.*

We should also note that you can use different letters for variables if you want. Whatever letters you use, it is critical that you explain in words what the letters mean. We could, for example, use the letter t instead of h to represent the number of hours worked. If we did that, we would emphasize the functional relationship with $M = M(t)$ and present the formula as $M = 9t$, where t is the number of hours worked, and M is the money earned in dollars.

> The parentheses in functional notation indicate the dependence of the function on the variable. They do not represent multiplication. For example $M(30)$ is not the same as $M \times 30$.

> **KEY IDEA 1.1**   **NOTATION FOR FUNCTIONS**
>
> To represent a function in an economical way, choose letters to stand for the function and variables. Be sure to explain what the letters represent and to include units.

Functions of Several Variables

Sometimes functions depend on more than one variable. Your grocery bill G may depend on the number a of apples you buy, the number s of sodas you buy, and the number

p of frozen pizzas you put in your basket. If apples cost 80 cents each, sodas cost \$1.25 each, and pizzas cost \$4.25 each, then we can express $G = G(a, s, p)$ as

Grocery bill $=$ Total cost of apples $+$ Total cost of sodas $+$ Total cost of pizzas

$$G = 0.8a + 1.25s + 4.25p,$$

where G is measured in dollars. The notation $G = G(a, s, p)$ is simply a way of emphasizing that G is a function of the variables a, s, and p—that is, that the value of G depends on a, s, and p. We could also give a correct formula for the function as $G = 80a + 125s + 425p$, where G is measured this time in cents. Either expression is correct as long as we explicitly say what units we are using.

EXAMPLE 1.1   A GROCERY BILL

Suppose your grocery bill is given by the function $G = G(a, s, p)$ above (with G measured in dollars). Recall that you are purchasing apples at 80 cents each, sodas at \$1.25 each, and pizzas at \$4.25 each.

Part 1 Use functional notation to show the cost of buying 4 apples, 2 sodas, and 3 pizzas, and then calculate that cost.

Part 2 Explain the meaning of $G(2, 6, 1)$.

Part 3 Calculate the value of $G(2, 6, 1)$.

$G(4, 2, 3)$ is the grocery bill expressed in functional notation.

\$18.45 is the value of $G(4, 2, 3)$.

Solution to Part 1 Since we are buying 4 apples, we use $a = 4$. Similarly, we are buying 2 sodas and 3 pizzas, so $s = 2$ and $p = 3$. Thus in functional notation our grocery bill is $G(4, 2, 3)$. To calculate this we use the formula $G = 0.8a + 1.25s + 4.25p$, replacing a by 4, s by 2, and p by 3:

$$G(4, 2, 3) = 0.8 \times 4 + 1.25 \times 2 + 4.25 \times 3$$

$$= 18.45 \,\text{dollars}.$$

Thus the cost is \$18.45.

Solution to Part 2 The expression $G(2, 6, 1)$ is the value of G when $a = 2$, $s = 6$, and $p = 1$. It is your grocery bill when you buy 2 apples, 6 sodas, and 1 frozen pizza.

Solution to Part 3 We calculate $G(2, 6, 1)$ just as we did in part 1, but this time we use $a = 2$, $s = 6$, and $p = 1$:

$$G(2, 6, 1) = 0.8 \times 2 + 1.25 \times 6 + 4.25 \times 1$$

$$= 13.35 \,\text{dollars}.$$

Thus the cost is \$13.35.

TEST YOUR UNDERSTANDING | FOR EXAMPLE 1.1

Explain the meaning of $G(4, 3, 2)$ and calculate its value. ■

Even when the formula for a function is complicated, the idea of how you use it remains the same. Let's look, for example, at $f = f(x)$, where f is determined as a function of x by the formula

$$f = \frac{x^2 + 1}{\sqrt{x}}.$$

The value of f when x is 3 is expressed in functional notation as $f(3)$. To calculate $f(3)$, we simply replace x in the formula by 3:

$$f(3) = \frac{3^2 + 1}{\sqrt{3}}.$$

You should check to see that the calculator gives an answer of 5.773502692, which we round to 5.77. Do not allow formulas such as this one to intimidate you. With the aid of the calculator, it is easy to deal with them.

EXAMPLE 1.2   BORROWING MONEY

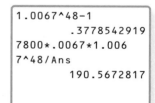

When you borrow money to buy a home or a car, you pay off the loan in monthly payments, but interest is always accruing on the outstanding balance. This makes the determination of your monthly payment on a loan more complicated than you might expect. If you borrow P dollars at a monthly interest rate[2] of r (as a decimal) and wish to pay off the note in t months, then your monthly payment $M = M(P, r, t)$ in dollars can be calculated using

$$M = \frac{Pr(1 + r)^t}{(1 + r)^t - 1}.$$

Part 1 Explain the meaning of $M(7800, 0.0067, 48)$ and calculate its value.

Part 2 Suppose you borrow \$5000 to buy a car and wish to pay off the loan over 3 years. Take the prevailing monthly interest rate to be 0.58%. (That is an annual percentage rate, APR, of $12 \times 0.58 = 6.96\%$.) Use functional notation to show your monthly payment, and then calculate its value.

Solution to Part 1 The expression $M(7800, 0.0067, 48)$ gives your monthly payment on a \$7800 loan that you pay off in 48 months (4 years) at a monthly interest rate of 0.67%. (That is an APR of $12 \times 0.67 = 8.04\%$.) To get its value, we use the formula above, putting 7800 in place of P, 0.0067 in place of r, and 48 in place of t:

$$M(7800, 0.0067, 48) = \frac{7800 \times 0.0067 \times 1.0067^{48}}{1.0067^{48} - 1}.$$

This can be entered all at once on the calculator, but to avoid typing errors, we do the calculation in pieces. The calculation of the denominator $1.0067^{48} - 1$ is shown in Figure 1.1. To complete the calculation we need to get

$$\frac{7800 \times 0.0067 \times 1.0067^{48}}{\text{Answer from first calculation}}.$$

We round the answer shown in Figure 1.2 to get the monthly payment of \$190.57.

$M(7800, 0.0067, 48)$ is the monthly payment expressed in functional notation. \$190.57 is its value.

```
1.0067^48-1
        .3778542919
```

FIGURE 1.1 The first step in calculating a loan payment

```
1.0067^48-1
            .3778542919
7800*.0067*1.006
7^48/Ans
            190.5672817
```

FIGURE 1.2 Completing the calculation

---

[2]Here we are assuming monthly payment and interest compounding. If you use the annual percentage rate (APR) reported on your loan agreement, then you have $r = \text{APR}/12$. See also Exercise 16 at the end of the Prologue.

Solution to Part 2 We borrow $5000, so we use $P = 5000$. The monthly interest rate is 0.58%, so we use $r = 0.0058$, and we pay off the loan in 3 years, or 36 months, so $t = 36$. In functional notation, the monthly payment is $M(5000, 0.0058, 36)$. To calculate it we use

$$M(5000, 0.0058, 36) = \frac{5000 \times 0.0058 \times 1.0058^{36}}{1.0058^{36} - 1}.$$

Once again we make the calculation in two stages. First we get $1.0058^{36} - 1$ as shown in Figure 1.3. As before, we use this answer to complete the calculation as follows:

$$\frac{5000 \times 0.0058 \times 1.0058^{36}}{\text{Answer from the first calculation}}.$$

The result in Figure 1.4 shows that we will have to make a monthly payment of $154.29.

```
1.0058^36-1
          .2314555099
```

FIGURE 1.3 The first step in calculating the payment on a $5000 loan

```
1.0058^36-1
          .2314555099
5000*.0058*1.005
8^36/Ans
          154.2940576
```

FIGURE 1.4 Completing the calculation

TEST YOUR UNDERSTANDING | **FOR EXAMPLE 1.2**

What is your monthly payment if you borrow $5000 at a monthly rate of 0.61% and pay it off in 5 years? ■

ANSWERS FOR TEST YOUR UNDERSTANDING

1.1 $G(4, 3, 2)$ is our grocery bill if we buy 4 apples, 3 sodas, and 2 pizzas. Its value is $15.45.
1.2 $99.76

1.1 EXERCISES

Reminder Round all answers to two decimal places unless otherwise indicated.

Note Some of the formulas below use the special number e, which was presented in the Prologue.

1. **Speed from Skid Marks** When a car makes an emergency stop on dry pavement, it leaves skid marks on the pavement. The speed S, in miles per hour, of the car when the brakes were applied is

related to the length L, in feet, of the skid mark. The relationship is

$$S(L) = 5.05\sqrt{L}.$$

   a. Use functional notation to express the speed at which the skid mark will be 60 feet. Then calculate that speed.

   b. Explain in practical terms the meaning of $S(100)$.

2. **Harris-Benedict Formula** Your *basal metabolic rate* is the amount of energy (in calories) your body needs to function at rest. The *Harris-Benedict formula* is used to estimate the basal metabolic rate. There is one formula for adult males and another for adult females. In these formulas w is your body

(continued)

weight in pounds, h is your height in inches, a is your age in years, $M = M(w, h, a)$ is the basal metabolic rate for adult males, and $F = F(w, h, a)$ is the basal metabolic rate for adult females:

$$M = 66 + 6.3w + 12.7h - 6.8a$$
$$F = 655 + 4.3w + 4.7h - 4.7a.$$

Use functional notation to express your own basal metabolic rate, and then calculate its value.

3. Adult Weight from Puppy Weight There is a formula that estimates how much your puppy will weigh when it reaches adulthood. The method we present applies to medium-sized breeds. First find your puppy's weight w, in pounds, at an age of a weeks, where a is 16 weeks or less. Then the predicted adult weight $W = W(a, w)$, in pounds, is given by the formula

$$W = 52\frac{w}{a}.$$

PAUL ATKINSON/Shutterstock.com

a. Use functional notation to express the adult weight of a puppy that weighs 6 pounds at 14 weeks.

b. Calculate the predicted adult weight for the puppy from part a.

4. Gross Profit Margin The *gross profit margin* is a measurement of a company's manufacturing and distribution efficiency during the production process. If G is the gross profit and T is the total revenue, both in dollars, then the gross profit margin $M = M(G, T)$ is given by the formula

$$M = \frac{G}{T}.$$

a. Use functional notation to express the gross profit margin for a company that has a gross profit of $335,000 and a total revenue of $540,000.

b. Calculate the gross profit margin in part a. The gross profit margin is often expressed as a percent. Give your answer as both a decimal and a percent.

c. If the gross profit stays the same but total revenue increases, would the gross profit margin increase or decrease?

5. Tax Owed The income tax T owed in a certain state is a function of the taxable income I, both measured in dollars. The formula is

$$T = 0.11I - 500.$$

a. Express using functional notation the tax owed on a taxable income of $13,000, and then calculate that value.

b. If your taxable income increases from $13,000 to $14,000, by how much does your tax increase?

c. If your taxable income increases from $14,000 to $15,000, by how much does your tax increase?

6. Pole Vault The height of the winning pole vault in the early years of the modern Olympic Games can be modeled as a function of time by the formula

$$H = 0.05t + 3.3.$$

Here t is the number of years since 1900, and H is the winning height in meters. (One meter is 39.37 inches.)

Jim Parkin/Shutterstock.com

a. Calculate $H(4)$ and explain in practical terms what your answer means.

b. By how much did the height of the winning pole vault increase from 1900 to 1904? From 1904 to 1908?

7. Flying Ball A ball is tossed upward from a tall building, and its upward velocity V, in feet per second, is a function of the time t, in seconds, since the ball was thrown. The formula is

$$V = 40 - 32t$$

if we ignore air resistance. The function V is positive when the ball is rising and negative when the ball is falling.

a. Express using functional notation the velocity 1 second after the ball is thrown, and then calculate that value. Is the ball rising or falling then?

b. Find the velocity 2 seconds after the ball is thrown. Is the ball rising or falling then?

c. What is happening 1.25 seconds after the ball is thrown?

d. By how much does the velocity change from 1 to 2 seconds after the ball is thrown? From 2 to 3 seconds? From 3 to 4 seconds? Compare the answers to these three questions and explain in practical terms.

8. Flushing Chlorine   City water, which is slightly chlorinated, is being used to flush a tank of heavily chlorinated water. The concentration $C = C(t)$ of chlorine in the tank t hours after flushing begins is given by

$$C = 0.1 + 2.78e^{-0.37t} \text{ milligrams per gallon.}$$

a. What is the initial concentration of chlorine in the tank?

b. Express the concentration of chlorine in the tank after 3 hours using functional notation, and then calculate its value.

9. A Population of Deer   When a breeding group of animals is introduced into a restricted area such as a wildlife reserve, the population can be expected to grow rapidly at first but to level out when the population grows to near the maximum that the environment can support. Such growth is known as *logistic population growth*, and ecologists sometimes use a formula to describe it. The number N of deer present at time t (measured in years since the herd was introduced) on a certain wildlife reserve has been determined by ecologists to be given by the function

$$N = \frac{12.36}{0.03 + 0.55^t}.$$

Sandro V. Maduell/Shutterstock.com

a. How many deer were initially on the reserve?

b. Calculate $N(10)$ and explain the meaning of the number you have calculated.

c. Express the number of deer present after 15 years using functional notation, and then calculate it.

d. How much increase in the deer population do you expect from the 10th to the 15th year?

10. A Car That Gets 32 Miles per Gallon   The cost C of operating a certain car that gets 32 miles per gallon is a function of the price g, in dollars per gallon, of gasoline and the distance d, in miles, that you drive. The formula for $C = C(g, d)$ is $C = gd/32$ dollars.

a. Use functional notation to express the cost of operation if gasoline costs 98 cents per gallon and you drive 230 miles. Calculate the cost.

b. Calculate $C(3.53, 172)$ and explain the meaning of the number you have calculated.

11. Radioactive Substances change form over time. For example, carbon 14, which is important for radiocarbon dating, changes through radiation into nitrogen. If we start with 5 grams of carbon 14, then the amount $C = C(t)$ of carbon 14 remaining after t years is given by

$$C = 5 \times 0.5^{t/5730}.$$

a. Express the amount of carbon 14 left after 800 years in functional notation, and then calculate its value.

b. How long will it take before half of the carbon 14 is gone? Explain how you got your answer. (*Hint:* You might use trial and error to solve this, or you might solve it by looking carefully at the exponent.)

12. A Roast is taken from the refrigerator (where it had been for several days) and placed immediately in a preheated oven to cook. The temperature $R = R(t)$ of the roast t minutes after being placed in the oven is given by

$$R = 325 - 280e^{-0.005t} \text{ degrees Fahrenheit.}$$

a. What is the temperature of the refrigerator?

b. Express the temperature of the roast 30 minutes after being put in the oven in functional notation, and then calculate its value.

c. By how much did the temperature of the roast increase during the first 10 minutes of cooking?

d. By how much did the temperature of the roast increase from the first hour to 10 minutes after the first hour of cooking?

13. What If Interest Is Compounded More Often Than Monthly?   Some lending institutions compound interest daily or even continuously. (The term *continuous compounding* is used when interest is

(continued)

being added as often as possible—that is, at each instant in time.) The point of this exercise is to show that, for most consumer loans, the answer you get with monthly compounding is very close to the right answer, even if the lending institution compounds more often. In part 1 of Example 1.2, we showed that if you borrow $7800 from an institution that compounds monthly at a monthly interest rate of 0.67% (for an APR of 8.04%), then in order to pay off the note in 48 months, you have to make a monthly payment of $190.57.

a. Would you expect your monthly payment to be higher or lower if interest were compounded daily rather than monthly? Explain why.

b. Which would you expect to result in a larger monthly payment, daily compounding or continuous compounding? Explain your reasoning.

c. When interest is compounded continuously, you can calculate your monthly payment $M = M(P, r, t)$, in dollars, for a loan of P dollars to be paid off over t months using

$$M = \frac{P(e^r - 1)}{1 - e^{-rt}},$$

where $r = \text{APR}/12$ if the APR is written in decimal form. Use this formula to calculate the monthly payment on a loan of $7800 to be paid off over 48 months with an APR of 8.04%. How does this answer compare with the result in Example 1.2?

14. **Present Value**   The amount of money originally put into an investment is known as the *present value* P of the investment. For example, if you buy a $50 U.S. Savings Bond that matures in 10 years, the present value of the investment is the amount of money you have to pay for the bond today. The value of the investment at some future time is known as the *future value F*. Thus, if you buy the savings bond mentioned above, its future value is $50.

If the investment pays an interest rate of r (as a decimal) compounded yearly, and if we know the future value F for t years in the future, then the present value $P = P(F, r, t)$, the amount we have to pay today, can be calculated using

$$P = F \times \frac{1}{(1 + r)^t}$$

if we measure F and P in dollars. The term $1/(1 + r)^t$ is known as the *present value factor,* or the *discount rate,* so the formula above can also be written as

$$P = F \times \text{discount rate}.$$

a. Explain in your own words what information the function $P(F, r, t)$ gives you.

For the remainder of this problem, we will deal with an interest rate of 9% compounded yearly and a time t of 18 years in the future.

b. Calculate the discount rate.

c. Suppose you wish to put money into an account that will provide $100,000 to help your child attend college 18 years from now. How much money would you have to put into savings today in order to attain that goal?

15. **How Much Can I Borrow?**   The function in Example 1.2 can be rearranged to show the amount of money $P = P(M, r, t)$, in dollars, that you can afford to borrow at a monthly interest rate of r (as a decimal) if you are able to make t monthly payments of M dollars:

$$P = M \times \frac{1}{r} \times \left(1 - \frac{1}{(1 + r)^t}\right).$$

Suppose you can afford to pay $350 per month for 4 years.

a. How much money can you afford to borrow for the purchase of a car if the prevailing monthly interest rate is 0.75%? (That is 9% APR.) Express the answer in functional notation, and then calculate it.

b. Suppose your car dealer can arrange a special monthly interest rate of 0.25% (or 3% APR). How much can you afford to borrow now?

c. Even at 3% APR you find yourself looking at a car you can't afford, and you consider extending the period during which you are willing to make payments to 5 years. How much can you afford to borrow under these conditions?

16. **Financing a New Car**   You are buying a new car, and you plan to finance your purchase with a loan you will repay over 48 months. The car dealer offers two options: either dealer financing with a low APR, or a $2000 rebate on the purchase price. If you use dealer financing, you will borrow $14,000 at an APR of 3.9%. If you take the rebate, you will reduce the amount you borrow to $12,000, but you will have to go to the local bank for a loan at an APR of 8.85%. Should you take the dealer financing or the rebate? How much will you save over the life of the loan by taking the option you chose? To answer the first question, you may need the formula

$$M = \frac{Pr(1 + r)^{48}}{(1 + r)^{48} - 1}.$$

Here M is your monthly payment, in dollars, if you borrow P dollars with a term of 48 months at a monthly interest rate of r (as a decimal), and $r = \text{APR}/12$.

17. **Brightness of Stars**   The *apparent magnitude m* of a star is a measure of its apparent brightness as the star is viewed from Earth. Larger magnitudes correspond to dimmer stars, and magnitudes can be negative, indicating a very bright star. For example, the brightest star in the night sky is Sirius, which has an apparent magnitude of −1.45. Stars with apparent magnitude greater than about 6 are not visible to the naked eye. The magnitude scale is not *linear* in that a star that is double the magnitude of another does not appear to be twice as dim. Rather, the relation goes as follows: If one star has an apparent magnitude of m_1 and another has an apparent magnitude of m_2, then the first star is t times as bright as the second, where t is given by

$$t = 2.512^{m_2 - m_1}.$$

The North Star, Polaris, has an apparent magnitude of 2.04. How much brighter than Polaris does Sirius appear?

18. **Stellar Distances**   *This is a continuation of Exercise 17.* The *absolute magnitude M* of a star is a measure of its true brightness and is not dependent on its distance from Earth or on any other factors that might affect its apparent brightness. If both the absolute and the apparent magnitude of a star are known, the distance from Earth can be calculated as follows:

$$d(m, M) = 3.26 \times 10^{(m - M + 5)/5}.$$

Here m is the apparent magnitude, M is the absolute magnitude, and d is the distance measured in *light-years*.[3]

a. Explain in words the meaning of $d(2.2, 0.7)$.

b. In Exercise 17 we noted that Sirius has an apparent magnitude of −1.45. The absolute magnitude of Sirius is +1.45. How far away is Sirius from Earth?

19. **Parallax**   If we view a star now, and then view it again 6 months later, our position will have changed by the diameter of the Earth's orbit around the sun. For *nearby* stars (within 100 light-years or so), the change in viewing location is sufficient to make the star appear to be in a slightly different location in the sky. Half of the angle from one location to the next is known as the *parallax angle* (see Figure 1.5). Parallax can be used to measure the distance to the star. An approximate relationship is given by

$$d = \frac{3.26}{p},$$

where d is the distance in light-years, and p is the parallax measured in seconds of arc.[4] Alpha Centauri is the star nearest to the sun, and it has a parallax angle of 0.751 second. How far is Alpha Centauri from the sun?

Note: Parallax is used not only to measure stellar distances. Our binocular vision actually provides the brain with a parallax angle that it uses to estimate distances to objects we see.

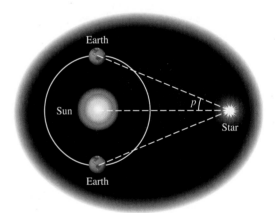

FIGURE 1.5

20. **Sound Pressure and Decibels**   Sound exerts a pressure P on the human ear.[5] This pressure increases as the loudness of the sound increases. If the loudness D is measured in *decibels* and the pressure P in dynes[6] per square centimeter, then the relationship is given by

$$P = 0.0002 \times 1.122^D.$$

a. Ordinary conversation has a loudness of about 65 decibels. What is the pressure exerted on the human ear by ordinary conversation?

---

[3]One light-year is the distance light travels in 1 year, or about 5,879,000,000,000 miles.
[4]One degree is divided into 60 minutes of arc, and 1 minute of arc is divided into 60 seconds of arc. One second of arc is very small indeed. If you hold a sheet of paper edgewise at arm's length, the thickness of the paper subtends an arc of about 30 seconds.
[5]In fact, the human ear detects variation in pressure on the ear drum. The amplitude of this variation determines the intensity of the sound, and the perceived loudness of the sound is related to this intensity.
[6]The dyne is a very small unit of force. It takes 444,800 dynes to make a pound. A small insect egg has a weight of about 1 dyne.

(continued)

b. A decibel level of 120 causes pain to the ear and can result in damage. What is the corresponding pressure level on the ear?

21. Mitscherlich's Equation   An important agricultural problem is to determine how a quantity of nutrient, such as nitrogen, affects the growth of plants. We consider the situation wherein sufficient quantities of all but one nutrient are present. One *baule*[7] of a nutrient is the amount needed to produce 50% of maximum possible yield. In 1909 E. A. Mitscherlich proposed the following relation, which is known as Mitscherlich's equation:[8]

$$Y = 1 - 0.5^b.$$

Here b is the number of baules of nutrient applied, and Y is the percentage (as a decimal) of maximum yield produced.

a. Verify that the formula predicts that 50% of maximum yield will be produced if 1 baule of nutrient is applied.

b. Use functional notation to express the percentage of maximum yield produced by 3 baules of nutrient, and calculate that value.

c. The exact value of a baule depends on the nutrient in question. For nitrogen, 1 baule is 223 pounds per acre. What percentage of maximum yield will be produced if 500 pounds of nitrogen per acre is present?

22. Yield Response to Several Growth Factors   *This is a continuation of Exercise 21.* If more than one nutrient is considered, the formula for percentage of maximum yield is a bit more complex. For three nutrients, the formula is

$$Y(b, c, d) = (1 - 0.5^b)(1 - 0.5^c)(1 - 0.5^d).$$

Here Y is the percentage of maximum yield as a decimal, b is the number of baules of the first nutrient, c is the number of baules of the second nutrient, and d is the number of baules of the third nutrient.

a. Express using functional notation the percentage of maximum yield produced from 1 baule of the first nutrient, 2 baules of the second nutrient, and 3 baules of the third nutrient, and then calculate that value.

b. One baule of nitrogen is 223 pounds per acre, 1 baule of phosphorus is 45 pounds per acre, and 1 baule of potassium is 76 pounds per acre. What percentage of maximum yield will be obtained from 200 pounds of nitrogen per acre, 100 pounds of phosphorus per acre, and 150 pounds of potassium per acre?

23. Thermal Conductivity   The heat flow Q due to conduction across a rectangular sheet of insulating material can be calculated using

$$Q = \frac{k(t_1 - t_2)}{d}.$$

Here Q is measured in watts per square meter, d is the thickness of the insulating material in meters, t_1 is the temperature in degrees Celsius of the warm side of the insulating material, t_2 is the temperature of the cold side, and k is the *coefficient of thermal conductivity*, which is measured experimentally for different insulating materials. For glass, the coefficient of thermal conductivity is 0.85. Suppose the temperature inside a home is 24 degrees Celsius (about 75 degrees Fahrenheit) and the temperature outside is 5 degrees Celsius (about 41 degrees Fahrenheit; see Figure 1.6).

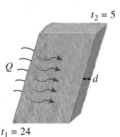

FIGURE 1.6

a. What is the heat flow through a glass window that is 0.007 meter (about one-quarter of an inch) thick?

b. The total heat loss due to conduction is the product of the heat flow with the area of the insulating material. In the situation of part a, what is the total heat loss due to conduction if the window has an area of 2.5 square meters?

24. Reynolds Number   The *Reynolds number* is very important in such fields as fluid flow and aerodynamics. In the case of a fluid flowing through a pipe, the Reynolds number R is given by

$$R = \frac{vdD}{\mu}.$$

---

[7]After the German mathematician who proposed the unit.
[8]A few years later the relation was noted independently by W. J. Spillman, and the relation is sometimes referred to as Spillman's equation. It should also be noted that the validity of Mitscherlich's equation is a source of controversy among modern agricultural scientists.

Here v is the velocity of the fluid in meters per second, d is the diameter of the pipe in meters, D is the density of the fluid in kilograms per cubic meter, and μ is the viscosity of the fluid measured in newton-seconds per square meter. Generally, when the Reynolds number is above 2000, the flow becomes turbulent, and rapid mixing occurs.[9] When the Reynolds number is less than 2000, the flow is streamline. Consider a fluid flowing through a pipe of diameter 0.05 meter at a velocity of 0.2 meter per second.

a. If the fluid in the pipe is toluene, its viscosity is 0.00059 newton-seconds per square meter, and its density is 867 kilograms per cubic meter. Is the flow turbulent or streamline?

b. If the toluene is replaced by glycerol, then the viscosity is 1.49 newton-seconds per square meter, and the density is 1216.3 kilograms per cubic meter. Is the glycerol flow turbulent or streamline?

25. Fault Rupture Length Earthquakes can result in various forms of damage, but many result in fault ruptures, or cracks in the Earth's surface. In a 1958 study[10] of earthquakes in California and Nevada, D. Tocher found the following relationship between the fault rupture length L, in kilometers, and the magnitude M (on the Richter scale) of an earthquake:

$$L = 0.0000017 \times 10.47^M.$$

What fault rupture length would be expected from an earthquake measuring 6.5 on the Richter scale?

26. Tubeworm An article in *Nature* reports on a study of the growth rate and life span of a marine *tubeworm*.[11] These tubeworms live near hydrocarbon seeps on the ocean floor and grow very slowly

littlesam/Shutterstock.com

indeed. Collecting data for creatures at a depth of 550 meters is extremely difficult. But for tubeworms living on the Louisiana continental slope, scientists developed a model for the time T (measured in years) required for a tubeworm to reach a length of L meters. From this model the scientists concluded that this tubeworm is the longest-lived noncolonial marine invertebrate known. The model is

$$T = 14e^{1.4L} - 20.$$

A tubeworm can grow to a length of 2 meters. How old is such a creature? (Round your answer to the nearest year.)

27. Equity in a Home When you purchase a home by securing a mortgage, the total paid toward the principal is your *equity* in the home. (Technically, the lending agency calculates your equity by subtracting the amount you still owe on your mortgage from the *current value* of your home, which may be higher or lower than your principal.) If your mortgage is for P dollars, and if the term of the mortgage is t months, then your equity E, in dollars, after k monthly payments is given by

$$E = P \times \frac{(1 + r)^k - 1}{(1 + r)^t - 1}.$$

Here r is the monthly interest rate as a decimal, with $r = \text{APR}/12$.

Suppose you have a home mortgage of $400,000 for 30 years at an APR of 6%.

a. What is the monthly rate as a decimal? Round your answer to three decimal places.

b. Express, using functional notation, your equity after 20 years of payments, and then calculate that value.

c. Find a formula that gives your equity after y years of payments.

28. Adjustable Rate Mortgage—Approximating Payments An *adjustable rate mortgage,* or ARM, is a mortgage whose interest rate varies over the life of the loan. The interest rate is often tied in some fashion to the *prime rate,* which may go up or down. One advantage of an ARM is that it usually has an initial rate that is lower than that of a fixed rate mortgage. In the summer of 2007 defaults on home mortgages led to a crisis in the U.S. economy. At least part of the blame was placed on ARMs.

---

[9]Reynolds numbers are dimensionless. That is, they have no units, such as grams or meters, associated with them.
[10]D. Tocher, "Earthquake energy and ground breakage," *Bull. Seism. Soc. Am.* **48** (1958), 48.
[11]D. Bergquist, F. Williams, and C. Fisher, "Longevity record for deep-sea invertebrate," *Nature* **403** (2000), 499–500.

(continued)

This exercise illustrates the difficulties that many homeowners faced during this period. We make use of the following formula for the monthly payment:

$$M = \frac{Pr(1 + r)^t}{(1 + r)^t - 1}.$$

Here M is the monthly payment, in dollars, P is the amount borrowed, in dollars, t is the term of the loan, in months, and r is the monthly interest rate as a decimal, with $r = \text{APR}/12$. In this exercise, round r to five decimal places.

Suppose you purchased a home in 2005, securing a mortgage of \$325,000 with a 30-year ARM.

a. In 2005 interest rates were at historical lows. Suppose that at the time of the loan the rate for your ARM was 4.5% APR. Calculate your monthly payment.

b. Suppose you earn \$6000 per month. What percentage of your income is going toward your house payment?

c. Suppose that after 24 payments your ARM rate adjusted to 7% APR. We will assume that after 24 months your loan balance is still \$325,000. (This is not as unreasonable an assumption as it may appear. The correct calculation is shown in Exercise 29.) What is your monthly payment now? *Be careful:* The term of the loan is now 28 years, not 30 years.

d. Using the assumptions of part c, what percentage of your income is going to your house payment now?

29. **Adjustable Rate Mortgage—Exact Payments** *This is a continuation of Exercise 28.* In Exercise 28 we only approximated the increased payment when the rate for an ARM increases: We assumed that over the first 24 months you accrue 0 equity in your home. (Your equity in a home is the total you have paid toward the principal.) We address that point here. We make use of the equity formula

$$E = P \times \frac{(1 + r)^k - 1}{(1 + r)^t - 1}.$$

Here E is your equity, in dollars, after k monthly payments. The quantities P, t, and r are defined as in Exercise 28. Round r to five decimal places.

We assume as in Exercise 28 that you borrowed \$325,000 at an initial APR of 4.5% with a term of 30 years.

a. What equity have you accrued after 24 months?

b. When your rate adjusts to 7% after 24 months, the new amount borrowed is \$325,000 less your

equity. The term is now 28 years. What is your new monthly payment?

30. **Research Project** Look in a textbook for another class to find a function interesting to you that is given by a formula. Identify all the variables used in the formula, explaining the meaning of each variable. Explain how this formula is used.

1.1 SKILL BUILDING EXERCISES

Evaluating Formulas In Exercises S-1 through S-24, evaluate the given function as required.

S-1. $f(x) = \dfrac{\sqrt{x + 1}}{x^2 + 1}$ at $x = 2$

S-2. $f(x) = \left(3 + x^{1.2}\right)^{x + 3.8}$ at $x = 4.3$

S-3. $g(x, y) = \dfrac{x^3 + y^3}{x^2 + y^2}$ at $x = 4.1$, $y = 2.6$

S-4. Calculate $f(1.3)$ if $f(t) = 87.1 - e^{4t}$.

S-5. Calculate $f(6.1)$ if $f(s) = \dfrac{s^2 + 1}{s^2 - 1}$.

S-6. Calculate $f(2, 5, 7)$ if
$$f(r, s, t) = \sqrt{r + \sqrt{s + \sqrt{t}}}.$$

S-7. Calculate $h(3, 2.2, 9.7)$ if $h(x, y, z) = x^y/z$.

S-8. Calculate $H(3, 4, 0.7)$ if
$$H(p, q, r) = \dfrac{2 + 2^{-p}}{q + r^2}.$$

S-9. Calculate $f(3, 4)$ if $f(x, y) = \dfrac{1.2^x + 1.3^y}{\sqrt{x + y}}$.

S-10. Calculate $g(2, 3, 4)$ if $g(s, t, u) = (1 + s/t)^u$.

S-11. Calculate $W(2.2, 3.3, 4.4)$ if
$$W(a, b, c) = \dfrac{a^b - b^a}{c^a - a^c}.$$

S-12. Calculate $f(3)$ if $f(x) = 3x + \dfrac{1}{x}$.

S-13. Calculate $f(3)$ if $f(x) = 3^{-x} - \dfrac{x^2}{x + 1}$.

S-14. Calculate $f(3)$ if $f(x) = \sqrt{2x + 5}$.

S-15. Calculate $C(0)$ and $C(10)$ if
$$C(t) = 0.1 + 2.78e^{-0.37t}.$$

S-16. Calculate $N(0)$ and $N(10)$ if
$$N(t) = \frac{12.36}{0.03 + 0.55^t}.$$

S-17. Calculate $C(52.3, 13.5)$ if $C(g, d) = \dfrac{gd}{32}$.

S-18. Calculate $C(0)$ and $C(3000)$ if
$$C(t) = 5 \times 0.5^{t/5730}.$$

S-19. Calculate $R(0)$ and $R(30)$ if
$$R(t) = 325 - 280e^{-0.005t}.$$

S-20. Calculate $M(5000, 0.04, 36)$ if
$$M(P, r, t) = \frac{P(e^r - 1)}{1 - e^{-rt}}.$$

S-21. Calculate $P(500, 0.06, 30)$ if
$$P(F, r, t) = F/(1 + r)^t.$$

S-22. Calculate $P(300, 0.005, 40)$ if
$$P(M, r, t) = \frac{M}{r}\left(1 - \frac{1}{(1 + r)^t}\right).$$

S-23. Calculate $d(3.1, 0.5)$ if
$$d(m, M) = 3.26 \times 10^{(m - M + 5)/5}.$$

S-24. Calculate $A(5.1, 3.2)$ if $A(u, v) = \dfrac{\sqrt{u} + \sqrt{v}}{\sqrt{u} - \sqrt{v}}$.

What Formulas Mean   In Exercises S-25 through S-31 you are asked to relate functional notation to practical explanations of what certain functions mean.

S-25. **Profit**   If $p(t)$ is the profit I expect my business to earn t years after opening, use functional notation to express expected profit after 2 years and 6 months.

S-26. **Grocery Bill**   If $c(p, s, h)$ is the cost of buying p bags of potato chips, s sodas, and h hot dogs, use functional notation to express the cost of buying 2 bags of potato chips, 3 sodas, and 5 hot dogs.

S-27. **Speed of Fish**   If $s(L)$ is the top speed of a fish that is L inches long, what does $s(13)$ mean?

S-28. **Time to Pay Off a Loan**   The function $T(P, r, m)$ gives the time required to pay off an installment loan of P dollars if the APR is r as a decimal and we make monthly payments of m dollars. Use functional notation to indicate the time required to pay off a \$12,000 automobile loan at an APR of 5% if the monthly payment is \$450.

S-29. **Doubling Time**   The function $D(P, r)$ gives the time required for an investment of P dollars at an APR of r as a decimal to double in value. Explain in practical terms the meaning of $D(5000, 0.06)$.

S-30. **Monthly Payment**   The function $M(P, r, t)$ gives the monthly payment for a loan of P dollars at an APR of r as a decimal if the loan is to be paid off in t months. You borrow \$23,000 at an APR of 5% to buy a car. Use functional notation to express your monthly payment if you are to pay off the loan in 4 years.

S-31. **A Bird Population**   The function $N(t)$ gives the number of birds in a certain population t years after observation began. Explain in practical terms the meaning of $N(7)$.

1.2 FUNCTIONS GIVEN BY TABLES

Long before the idea of a function was formalized, it was used in the form of tables of values. Some of the earliest surviving samples of mathematics are from Babylon and Egypt and date from 2000 to 1000 B.C. They contain a variety of tabulated functions such as tables of squares of numbers. Functions given in this way nearly always leave gaps and are incomplete. In this respect they may appear less useful than functions given by formulas. On the other hand, tables often clearly show trends that are not easily discerned from formulas, and in many cases tables of values are much easier to obtain than a formula.

 | ## Reading Tables of Values

The population N of the United States depends on the date d. That is, $N = N(d)$ is a function of d. Table 1.1 was taken from information supplied by the U.S. Census Bureau.

It shows the population in millions each decade from 1970 through 2010. This is a common way to express functions when data are gathered by *sampling*, in this case by census takers.

In order to get the population in 1980, we look at the column corresponding to $d = 1980$ and read $N = 226.54$ million people. In functional notation this is $N(1980) = 226.54$ million people. Similarly, we read from the table that $N(1970) = 203.30$, which indicates that the U.S. population in 1970 was 203.30 million people.

TABLE 1.1 Population of the United States

d = Year	1970	1980	1990	2000	2010
N = Population in millions	203.30	226.54	248.71	281.42	308.75

Filling Gaps by Averaging

Functions given by tables of values have their limitations in that they nearly always leave gaps. Sometimes it is appropriate to fill these gaps by *averaging*. For example, Table 1.1 does not give the population in 1975. In the absence of further information, a reasonable guess for the value of $N(1975)$ would be the average of the populations in 1970 and 1980:

$$\frac{N(1970) + N(1980)}{2} = \frac{203.30 + 226.54}{2} = 214.92 \text{ million people.}$$

> The average of two function values gives an estimate of the function value halfway between.

Thus the population in 1975 was approximately 214.92 million people. Population records show the actual population of the United States in 1975 as 215.97 million people, so it seems that the idea of averaging to estimate the value of $N(1975)$ worked pretty well in this case.

We emphasize here that, in the absence of further data, we have no way of determining the exact value of $N(1975)$, and so you should not assume that the answer we gave is the only acceptable one. If, for example, you had reason to believe that the population of the United States grew faster in the earlier part of the decade of the 1970s than it did in the latter part, then it would be reasonable for you to give a larger value for $N(1975)$. Such an answer, supported by an appropriate argument, would have as much validity as the one given here.

Robert F. Balazik/Shutterstock.com

Average Rates of Change

A key tool in the analysis of functions is the idea of an *average rate of change*. To illustrate this idea, let's get the best estimate we can for $N(1972)$, the U.S. population in 1972.

Since 1972 is not halfway between 1970 and 1980, it does not make sense here to average the two as we did above to estimate the population in 1975, but a simple extension of this idea will help. From 1970 to 1980 the population increased from 203.30 million to 226.54 million people. That is an increase of $226.54 - 203.30 = 23.24$ million people in 10 years. Thus, on average, during the decade of the 1970s the population was increasing by $23.24/10 = 2.324$ million people per year. This is the *average yearly rate of change in N* during the 1970s, and there is a natural way to use it to estimate the population in 1972. In 1970 the population was 203.30 million people, and during this period the population was growing by about 2.324 million people per year. Thus in 1972 the population $N(1972)$ was approximately

$$\text{Population in 1972} = \text{Population in 1970} + \text{Two years of growth}$$

$$= 203.30 + 2 \times 2.324 = 207.95 \, \text{million.}$$

The average rate of change estimates the growth rate from one value to the next.

> **KEY IDEA 1.2 EVALUATING FUNCTIONS GIVEN BY TABLES**
>
> You can evaluate functions given by tables by locating the appropriate entry in the table. It is sometimes appropriate to fill in gaps in the table by using averages or average rates of change.

EXAMPLE 1.3 WOMEN EMPLOYED OUTSIDE THE HOME

Table 1.2 shows the number W of women in the United States employed outside the home as a function of the date d. It was taken from the *Statistical Abstract of the United States*.

TABLE 1.2 Number of Women Employed Outside the Home in the United States

d = Year	1970	1980	1990	2000	2009
W = Number in millions	31.5	45.5	56.8	66.3	72.0

Part 1 Explain the meaning of $W(1970)$ and give its value.

Part 2 Explain the meaning of $W(1975)$ and estimate its value.

Part 3 Express the number of women employed outside the home in 1972 in functional notation, and use the average yearly rate of change from 1970 to 1980 to estimate its value.

Part 4 According to the *1950 Statistical Abstract of the United States*, in 1943 there were 18.7 million women employed outside the home, and in 1946 the number was 16.8 million. On the basis of this information, find the average yearly rate of change in W from 1943 to 1946 and the average decrease per year over this time interval. Use the results to estimate $W(1945)$.

Solution to Part 1 The expression $W(1970)$ represents the number of women in the United States employed outside the home in 1970. Consulting the second column of the table, we find that $W(1970) = 31.5$ million women.

Solution to Part 2 The expression $W(1975)$ represents the number of women in the United States employed outside the home in 1975. This value is not given in the table, so we must estimate it. Since 1975 is halfway between 1970 and 1980, it is reasonable to estimate $W(1975)$ as the average of $W(1970)$ and $W(1980)$:

$$\frac{W(1970) + W(1980)}{2} = \frac{31.5 + 45.5}{2} = 38.5 \text{ million.}$$

Thus we estimate that $W(1975)$ is about 38.5 million women.

Solution to Part 3 In functional notation, the number of women employed outside the home in 1972 is $W(1972)$. To estimate this function value, we first calculate the average rate of change during the decade of the 1970s. From 1970 to 1980, the number of women employed outside the home increased from 31.5 million to 45.5 million, an increase of 14.0 million over the 10-year period. Thus in the decade of the 1970s the number increased on average by $14.0/10 = 1.40$ million per year. In the 2-year period from 1970 to 1972, the increase was about $2 \times 1.40 = 2.80$ million. We estimate that

$$
\begin{aligned}
W(1972) &= W(1970) + 2 \text{ years' growth} \\
&= W(1970) + 2.80 \\
&= 31.5 + 2.80 \\
&= 34.3 \text{ million.}
\end{aligned}
$$

Thus we estimate that the number of women employed outside the home in 1972 was about 34.3 million.

Solution to Part 4 At the outset we observe that apparently the function is decreasing over this period, so we expect the rate of change to be negative. From 1943 to 1946 the change was $16.8 - 18.7 = -1.9$. Thus over this 3-year period the number changed on average by $-1.9/3$ million per year, or about -0.63 million per year. This is the average yearly rate of change from 1943 to 1946. The average decrease per year is obtained by dividing the amount of decrease, which is $18.7 - 16.8$, by the time interval, which is 3 years. The result is about 0.63 million per year. Note that the average decrease per year is just the magnitude of the average yearly rate of change.

To estimate $W(1945)$ we proceed as in part 3, using the average yearly rate of change. We expect that in the 2-year period from 1943 to 1945 the change was about $2 \times -0.63 = -1.26$ million. Thus we estimate that

$$
\begin{aligned}
W(1945) &= W(1943) - 1.26 \\
&= 18.7 - 1.26 \\
&= 17.44 \text{ million.}
\end{aligned}
$$

Our estimate for $W(1945)$ is 17.4 million women.

TEST YOUR UNDERSTANDING │ **FOR EXAMPLE 1.3**

Use the notation of the example. Explain the meaning of $W(1973)$, and use the average rate of change to estimate its value. ■

There are features of this problem that are worth emphasizing. First, mathematics involves common-sense ideas that are used on a regular basis. Second, many times applications of mathematics do not involve a set way to solve the problem, and often there is not a simple "right answer." What is important is clear thinking leading to helpful information. Third, in applying mathematics we should be aware of the context. For example, in part 4 the time interval began in 1943, when the United States was heavily involved in

World War II and thus many women were working outside the home. By 1946 (at the end of this interval) the war was over, and this is reflected in the decrease in the function W. Because the war continued well into 1945, we should be skeptical that our estimate of 17.4 million women for that year is reliable. In fact, the number was near its peak then, and the actual value of $W(1945)$ is 19.0 million (slightly higher than the value in 1943).

Spotting Trends

In some situations, tables of values show clear trends or *limiting values* for functions. A good example of this is provided by the famous[12] yeast experiment that Tor Carlson performed in the early 1900s. In this experiment, a population of yeast cells growing in a confined area was carefully monitored as it grew. Each day, the amount of yeast[13] was found and recorded. We let t be the number of hours since the experiment began and $N = N(t)$ the amount of yeast present. Carlson presented data through only 18 hours of growth. In Table 1.3 we have partially presented Carlson's data and added data that modern studies indicate he might have recorded if the experiment had continued. Thus, for example, $N(5) = 119$ is the amount of yeast present 5 hours after the experiment began.

TABLE 1.3 Amount of Yeast

Time t	0	5	10	15	20	25	30
Amount of yeast N	10	119	513	651	662	664	665

We want to estimate the value of $N(35)$—that is, the amount of yeast present after 35 hours. Before we look at the data, let's think how our everyday experience tells us that a similar, but more familiar, situation would progress. The population of mold on a slice of bread[14] will show behavior much like that of yeast. On bread, the mold spreads rapidly at first, but eventually the bread is covered and no room is left for further population growth. We would expect the amount of mold to stabilize at a *limiting value*. The same sort of growth seems likely for yeast growing in a confined area. In terms of the function N, that means we expect to see its values increase at first but eventually stabilize, and this is borne out by the data in Table 1.3, where evidently the yeast population is leveling out at about 665. Thus it is reasonable to expect that $N(35)$ is about 665.

The limiting value of a function shows its long-term behavior.

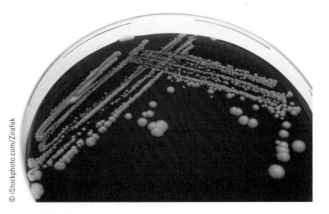

© iStockphoto.com/Zirafek

---

[12]This experiment is described by R. Pearl in "The growth of populations," *Quart. Rev. Biol* **2** (1927), 532–548. It served to illustrate the now widely used *logistic population growth model,* which we will examine in more detail in subsequent chapters.
[13]The actual unit of measure is unclear from the original reference. It is simply reported as *amount of yeast.*
[14]See also Exercise 27 at the end of this section.

We note here that the significance of 665 is not just that it is the last entry in the table. Rather, this entry is indicative of the trend shown by the last few entries of the table: 662, 664, and 665. We also emphasize that we did not propose a value for $N(35)$ by looking at the data divorced from their meaning. Before we looked at the data, the physical situation led us to believe that N would level out at *some value*, and our everyday experience with moldy bread tells us that once mold begins to grow, this level will be reached in a relatively short time. The data simply enabled us to make a good guess at the limiting value we expected to see.

KEY IDEA 1.3   **LIMITING VALUES**

Information about physical situations can sometimes show that limiting values are to be expected for functions that model those physical situations. Under such conditions, the limiting value may be estimated from a trend established by the data.

The next example shows how using rates of change can confirm the expectation that there is a limiting value.

EXAMPLE 1.4   **A SKYDIVER**

During the period of a skydiver's free fall, he is pulled downward by gravity, but his velocity is retarded by air resistance, which physicists believe increases as velocity does. Table 1.4 shows the downward velocity $v = v(t)$, in feet per second, of an average-sized man t seconds after jumping from an airplane.

TABLE 1.4 Velocity in Feet per Second t Seconds into Free Fall

Time t	0	10	20	30	40	50	60
Velocity v	0	147	171	175	175.8	176	176

Part 1 From the table, describe in words how the velocity of the skydiver changes with time.

Part 2 Use functional notation to give the velocity of the skydiver 15 seconds into the fall. Estimate its value.

Germanskydiver/Shutterstock.com

Part 3 Explain how the physical situation leads you to believe that the function v will approach a limiting value.

Part 4 Make a table showing the average rate of change per second in velocity v over each of the 10-second intervals in Table 1.4. Explain how your table confirms the conclusion of part 3.

Part 5 Estimate the *terminal velocity* of the skydiver—that is, the greatest velocity that the skydiver can attain.

Solution to Part 1 The velocity increases rapidly during the first part of the fall, but the rate of increase slows as the fall progresses.

Solution to Part 2 In functional notation, the velocity 15 seconds into the fall is $v(15)$. It is not given in Table 1.4, so we must estimate its value. One reasonable estimate is the average of the velocities at 10 and 20 seconds into the fall:

$$\frac{v(10) + v(20)}{2} = \frac{147 + 171}{2} = 159 \text{ feet per second.}$$

With the information we have, this is a reasonable estimate for $v(15)$, but a little closer look at the data might lead us to adjust this. Note that velocity increases very rapidly (by $147 - 0 = 147$ feet per second) during the first 10 seconds of the fall but increases by only $171 - 147 = 24$ feet per second during the next 10 seconds. It appears that the rate of increase in velocity is slowing dramatically. Thus it is reasonable to expect that the velocity increased more from 10 to 15 seconds than it did from 15 to 20 seconds. We might be led to make an upward revision in our estimate of $v(15)$. It turns out that the actual velocity at $t = 15$ is 164.44 feet per second, but with the information we have, we have no way of discovering that. What is important here is to obtain an estimate that is reasonable and is supported by an appropriate argument.

Solution to Part 3 The force of gravity causes the skydiver's velocity to increase, but this increase is retarded by air resistance, which increases along with velocity. When downward velocity reaches a certain level, we would expect the force of retardation to match the downward pull of gravity. From that point on, velocity will not change.

Solution to Part 4 Over the interval from $t = 0$ to $t = 10$, the change in v is $147 - 0 = 147$ feet per second, so the average rate of change per second in v over this 10-second interval is $\frac{147}{10} = 14.7$ feet per second per second. From $t = 10$ to $t = 20$, the change in v is $171 - 147 = 24$ feet per second, so the average rate of change per second in v over this interval is $\frac{24}{10} = 2.4$ feet per second per second. If we repeat this computation for each of the intervals, we get the values shown in Table 1.5. Here the average rate of change is measured in feet per second per second.

TABLE 1.5 Average Rate of Change in Velocity v over Each 10-Second Interval

Time interval	0 to 10	10 to 20	20 to 30	30 to 40	40 to 50	50 to 60
Average rate of change in v	14.7	2.4	0.4	0.08	0.02	0

Note that Table 1.5 is consistent with our observations in parts 1 and 2. The table shows that the average rate of change in velocity decreases to zero as time goes on. This confirms our conclusion in part 3 that after a time, velocity will not change.

Solution to Part 5 From parts 3 and 4, we have good reason to believe that the data should show a limiting value for velocity. From $t = 30$ to $t = 60$, the velocity seems to be inching up toward 176 feet per second, and it appears to level out there. This is confirmed by the fact that the average rate of change in velocity decreases to zero. The limiting value of 176 for v is the terminal velocity of the skydiver. You may be interested to know that this is about 120 miles per hour.

TEST YOUR UNDERSTANDING | **FOR EXAMPLE 1.4**

Use the table of values you made in part 4 of the example to find the limiting value of the average rate of change in velocity.   ∎

ANSWERS FOR TEST YOUR UNDERSTANDING

1.3 The expression $W(1973)$ represents the number of women in the United States employed outside the home in 1973. Its value is approximately 35.7 million.

1.4 0

1.2 EXERCISES

Reminder   Round all answers to two decimal places unless otherwise indicated.

1. **Box Office Hits**   The table below shows the highest grossing movies of the given year. The amount is the domestic box office gross in millions of dollars.

Year	Movie	Amount (millions)
2002	*Spider-Man*	$403.71
2003	*Lord of the Rings: ROTK*	377.03
2004	*Shrek 2*	441.23
2005	*Star Wars Ep. III*	380.27
2006	*Pirates of the Caribbean: Dead Man's Chest*	423.32
2007	*Spider-Man 3*	336.53
2008	*The Dark Knight*	533.35
2009	*Avatar*	760.51
2010	*Toy Story 3*	415.00

Let $M = M(y)$ denote the highest grossing movie in year y, and let $B = B(y)$ denote the gross for that movie.

a. Give the values of $M(2005)$ and $B(2005)$.

b. Use functional notation to indicate the amount for the movie with the highest gross in 2003.

2. **Mobile Phone Sales**   In 2000 mobile handset sales totaled $414.99 million. In 2005 the total was $778.75 million. Let $M = M(t)$ denote total mobile handset sales in year t. What was the average rate of change per year in $M(t)$ from 2000 to 2005? Be sure to include proper units with your answer.

3. **Choosing a Bat**   A chart from Dick's Sporting Goods gives the recommended bat length B in inches for a man weighing between 161 and 170 pounds as a function of his height h in inches. The table is partially reproduced below.

h = Height	B = Bat length
45–48	30
49–52	31
53–56	31
57–60	32
61–64	32
65–68	33
69–72	33
73+	33

a. Explain in practical terms the meaning of $B(55)$ and give its value.

b. Use functional notation to express the recommended bat length for a man weighing between 161 and 170 pounds if his height is 63 inches.

4. **Freight on Class I Railroads**   According to the Association of American Railroads, Class I freight railroads are the line-haul freight railroads with

2006 operating revenue in excess of $346.8 million.[15] Let $F = F(t)$ denote the freight revenue in billions of dollars of Class I railroads in year t. In 2005 Class I railroads had a freight revenue of $44.5 billion. In 2007 the revenue was $52.9 billion. Calculate the average rate of change per year in F from 2005 to 2007 and explain in practical terms its meaning.

5. **The American Food Dollar** The following table shows the percentage $P = P(d)$ of the American food dollar that was spent on eating away from home (at restaurants, for example) as a function of the date d.

d = Year	P = Percent spent away from home
1969	25%
1989	30%
2009	34%

   a. Find $P(1989)$ and explain what it means.

   b. What does $P(1999)$ mean? Estimate its value.

   c. What is the average rate of change per year in percentage of the food dollar spent away from home for the period from 1989 to 2009?

   d. What does $P(2004)$ mean? Estimate its value. (*Hint:* Your calculation in part c should be useful.)

   e. Predict the value of $P(2014)$ and explain how you made your estimate.

6. **Gross Domestic Product** The following table[16] shows the U.S. gross domestic product (GDP) G, in trillions of dollars, as a function of the year t.

t = Year	2004	2008	2010
G = GDP (trillions of dollars)	11.87	14.37	14.66

   a. Explain in practical terms what $G(2004)$ means, and find its value.

   b. Use functional notation to express the gross domestic product in 2006, and estimate that value.

   c. What is the average yearly rate of change in G from 2008 to 2010?

   d. Use your answer to part c to predict the gross domestic product in the year 2018.

7. **Internet Access** The following table gives the number $I = I(t)$, in millions, of adult Americans with Internet access in year t.

t = Year	2000	2003	2009
I = Millions of Americans	113	166	196

   a. Find $I(2000)$ and explain what it means.

   b. Find the average rate of change per year during the period from 2003 to 2009.

   c. Estimate the value of $I(2007)$. Explain how you got your answer.

8. **A Cold Front** At 4 P.M. on a winter day, an arctic air mass moved from Kansas into Oklahoma, causing temperatures to plummet. The temperature $T = T(h)$ in degrees Fahrenheit h hours after 4 P.M. in Stillwater, Oklahoma, on that day is recorded in the following table.

h = Hours since 4 P.M.	T = Temperature
0	62
1	59
2	38
3	26
4	22

---

[15]In 2006 the Class I freight railroads in the United States were BNSF Railway, CSX Transportation, Grand Trunk Corporation, Kansas City Southern Railway, Norfolk Southern Combined Railroad Subsidiaries, Soo Line Railroad, and Union Pacific Railroad.
[16]From the *Bureau of Economic Analysis*.

(continued)

a. Use functional notation to express the temperature in Stillwater at 5:30 P.M., and then estimate its value.

b. What was the average rate of change per minute in temperature between 5 P.M. and 6 P.M.? What was the average decrease per minute over that time interval?

c. Estimate the temperature at 5:12 P.M.

d. At about what time did the temperature reach the freezing point? Explain your reasoning.

9. **A Troublesome Snowball** One winter afternoon, unbeknownst to his mom, a child brings a snowball into the house, lays it on the floor, and then goes to watch TV. Let $W = W(t)$ be the volume of dirty water that has soaked into the carpet t minutes after the snowball was deposited on the floor. Explain in practical terms what the limiting value of W represents, and tell what has happened physically when this limiting value is reached.

10. **Falling with a Parachute** If an average-sized man jumps from an airplane with a properly opening parachute, his downward velocity $v = v(t)$, in feet per second, t seconds into the fall is given by the following table.

t = Seconds into the fall	v = Velocity
0	0
1	16
2	19.2
3	19.84
4	19.97

a. Explain why you expect v to have a limiting value and what this limiting value represents physically.

b. Estimate the terminal velocity of the parachutist.

11. **Carbon-14** Carbon-14 is a radioactive substance that decays over time. One of its important uses is in dating relatively recent archaeological events. In the following table, time t is measured in thousands of years, and $C = C(t)$ is the amount, in grams, of carbon-14 remaining.

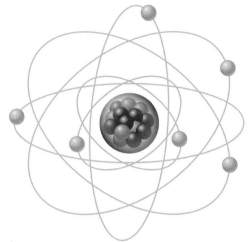

Carbon-14
unstable (radioactive)

t = Thousands of years	C = Grams remaining
0	5
5	2.73
10	1.49
15	0.81
20	0.44

a. What is the average yearly rate of change of carbon-14 during the first 5000 years?

b. How many grams of carbon-14 would you expect to find remaining after 1236 years?

c. What would you expect to be the limiting value of C?

12. **Newton's Law of Cooling** says that a hot object cools rapidly when the difference between its temperature and that of the surrounding air is large, but it cools more slowly when the object nears room temperature. Suppose a piece of aluminum is removed from an oven and left to cool. The following table gives the temperature $A = A(t)$, in degrees Fahrenheit, of the aluminum t minutes after it is removed from the oven.

a. Explain the meaning of $A(75)$ and estimate its value.

b. Find the average decrease per minute of temperature during the first half-hour of cooling.

t = Minutes	A = Temperature
0	302
30	152
60	100
90	81
120	75
150	73
180	72
210	72

n = Compounding periods	E = EAR
1	12%
2	12.36%
12	12.683%
365	12.747%
8760	12.750%
525,600	12.750%

c. Find the average decrease per minute of temperature during the first half of the second hour of cooling.

d. Explain how parts b and c support Newton's law of cooling.

e. Use functional notation to express the temperature of the aluminum after 1 hour and 13 minutes. Estimate the temperature at that time. (*Note:* Your work in part c should be helpful.)

f. What is the temperature of the oven? Express your answer using functional notation, and give its value.

g. Explain why you would expect the function A to have a limiting value.

h. What is room temperature? Explain your reasoning.

13. Effective Percentage Rate for Various Compounding Periods   We have seen that in spite of its name, the annual percentage rate (APR) does not generally tell directly how much interest accrues on a loan in a year. That value, known as the *effective annual rate*,[17] or EAR, depends on how often the interest is compounded. Consider a loan with an annual percentage rate of 12%. The following table gives the EAR, $E = E(n)$, if interest is compounded n times each year. For example, there are 8760 hours in a year, so that column corresponds to compounding each hour.

a. State in everyday language the type of compounding that each row represents.

b. Explain in practical terms what $E(12)$ means and give its value.

c. Use the table to calculate the interest accrued in 1 year on an $8000 loan if the APR is 12% and interest is compounded daily.

d. Estimate the EAR if compounding is done continuously—that is, if interest is added at each moment in time. Explain your reasoning.

14. New Construction   The following table[18] shows the value B, in billions of dollars, of new construction put in place in the United States during year t.

t = Year	B = Value (billions of dollars)
2000	831.1
2003	891.5
2006	1167.6
2009	935.6

a. Make a table showing, for each of the 3-year periods, the average yearly rate of change in B.

b. Explain in practical terms what $B(2008)$ means, and estimate its value.

c. Over what period was the growth in value of new construction the greatest?

d. According to the table, in what year was the value of new construction the greatest?

---

[17]The EAR is the same as the *annual percentage yield,* or APY. The APY is usually used in advertising investment returns. The APR, which is smaller, is used in advertising loan rates, although the EAR would be a more accurate representation.
[18]From the *Statistical Abstract of the United States.*

(*continued*)

15. Growth in Height   The following table gives, for a certain man, his height $H = H(t)$ in inches at age t in years.

t = Age (years)	H = Height (inches)
0	21.5
5	42.5
10	55.0
15	67.0
20	73.5
25	74.0

a. Use functional notation to express the height of the man at age 13, and then estimate its value.

b. Now we study the man's growth rate.

  i. Make a table showing, for each of the 5-year periods, the average yearly growth rate—that is, the average yearly rate of change in H.

  ii. During which 5-year period did the man grow the most in height?

  iii. Describe the general trend in the man's growth rate.

c. What limiting value would you estimate for the height of this man? Explain your reasoning in physical terms.

16. Growth in Weight   The following table gives, for a certain man, his weight $W = W(t)$ in pounds at age t in years.

t = Age (years)	W = Weight (pounds)
4	36
8	54
12	81
16	128
20	156
24	163

a. Make a table showing, for each of the 4-year periods, the average yearly rate of change in W.

b. Describe in general terms how the man's gain in weight varied over time. During which 4-year period did the man gain the most in weight?

c. Estimate how much the man weighed at age 30.

d. Use the average rate of change to estimate how much he weighed at birth. Is your answer reasonable?

17. Tax Owed   The following table shows the income tax T owed in a certain state as a function of the taxable income I, both measured in dollars.

I = Taxable income	T = Tax owed
16,000	870
16,200	888
16,400	906
16,600	924

a. Make a table showing, for each of the intervals in the tax table above, the average rate of change in T.

b. Describe the general trend in the average rate of change. What does this mean in practical terms?

c. Would you expect T to have a limiting value? Be sure to explain your reasoning.

18. Sales Income   The following table shows the net monthly income N for a real estate agency as a function of the monthly real estate sales s, both measured in dollars.

s = Sales	N = Net income
450,000	4000
500,000	5500
550,000	7000
600,000	8500

a. Make a table showing, for each of the intervals in the table above, the average rate of change in N. What pattern do you see?

b. Use the average rate of change to estimate the net monthly income for monthly real estate sales of $520,000. In light of your answer to part a, how confident are you that your estimate is an accurate representation of the actual income?

c. Would you expect N to have a limiting value? Be sure to explain your reasoning.

19. Yellowfin Tuna   Data were collected comparing the weight W, in pounds, of a yellowfin tuna to its length L, in centimeters.[19] These data are presented in the table below.

Brian J. Skerry/Getty Images

L = Length	W = Weight
70	14.3
80	21.5
90	30.8
100	42.5
110	56.8
120	74.1
130	94.7
140	119
160	179
180	256

a. What is the average rate of change, in weight per centimeter of length, in going from a length of 100 centimeters to a length of 110 centimeters?

b. What is the average rate of change, in weight per centimeter, in going from 160 to 180 centimeters?

c. Judging from the data in the table, does an extra centimeter of length make more difference in weight for a small tuna or for a large tuna?

d. Use the average rate of change to estimate the weight of a yellowfin tuna that is 167 centimeters long.

e. What is the average rate of change, in length per pound of weight, in going from a weight of 179 pounds to a weight of 256 pounds?

f. What would you expect to be the length of a yellowfin tuna weighing 225 pounds?

20. Arterial Blood Flow   Medical evidence shows that a small change in the radius of an artery can indicate a large change in blood flow. For example, if one artery has a radius only 5% larger than another, the blood flow rate is 1.22 times as large. Further information is given in the table below.

Increase in radius	Times greater blood flow rate
5%	1.22
10%	1.46
15%	1.75
20%	2.07

a. Use the average rate of change to estimate how many times greater the blood flow rate is in an artery that has a radius 12% larger than another.

b. Explain why if the radius is increased by 12% and then we increase the radius of the new artery by 12% again, the total increase in the radius is 25.44%.

c. Use parts a and b to answer the following question: How many times greater is the blood flow rate in an artery that is 25.44% larger in radius than another?

d. Answer the question in part c using the average rate of change.

21. Widget Production   The following table shows, for a certain manufacturing plant, the number W of

---

[19]See Michael R. Cullen, *Mathematics for the Biosciences* (Fairfax, VA: TechBooks, 1983).

(*continued*)

widgets, in thousands, produced in a day as a function of n, the number of full-time workers.

n = Number of workers	W = Thousands of widgets produced
10	25.0
20	37.5
30	43.8
40	46.9
50	48.4

a. Make a table showing, for each of the 10-worker intervals, the average rate of change in W per worker.

b. Describe the general trend in the average rate of change. Explain in practical terms what this means.

c. Use the average rate of change to estimate how many widgets will be produced if there are 55 full-time workers.

d. Use your answer to part b to determine whether your estimate in part c is likely to be too high or too low.

22. **Timber Stumpage Prices**   The following table shows timber stumpage prices for pine pulpwood in two regions of the American South.[20] Prices are in dollars per ton and were recorded at the start of the indicated year.

Year	Southeast	Mid-Atlantic
2002	6.50	4.00
2005	7.50	6.70
2007	7.00	10.00

a. For the Mid-Atlantic, what is the average rate of change per year in price from 2002 to 2005?

b. Use your answer to part a to estimate the price in the Mid-Atlantic region at the start of 2004. (The actual price was $6.40 per ton.)

c. For the Southeast, what is the average rate of change per year in price from 2005 to 2007?

d. Use your answer to part c to estimate the price in the Southeast at the start of 2008.

e. For each region find the percentage increase in price from 2002 to 2007.

f. On the basis of your answer to part e, in the absence of other factors, would an investor in timber be better advised to choose the Southeast or the Mid-Atlantic?

23. **The Margaria-Kalamen Test**   The Margaria-Kalamen test is used by physical educators as a measure of leg strength. An individual runs up a staircase, and the elapsed time from the third to the ninth step is recorded. The power score is calculated using a formula involving the individual's weight, the height of the stairs, and the running time. The following table[21] shows the power scores required of men for an excellent rating for selected ages.

Mark Stout Photography/Shutterstock.com

Age	Power score for excellent rating
18	224
25	210
35	168
45	125
55	98

a. What is the average rate of change per year in excellence level from 25 years to 35 years old?

b. What power score would merit an excellent rating for a 27-year-old man?

c. During which 10-year period from age 25 to 55 would you expect to see the greatest decrease in leg power?

---

[20]The table is adapted from an article by Marshall Thomas in the *F&W Forestry Report* (Fall 2007).
[21]Adapted from D. K. Mathews and E. L. Fox, *The Physiological Basis of Physical Education and Athletics*, 2nd ed. (Philadelphia: Saunders, 1976).

24. **ACRS and MACRS** In 1981 Congress introduced the Accelerated Cost Recovery System (ACRS) tax depreciation. ACRS allowed for uniform recovery periods at fixed percentage rates each year. Part of the intent of the law was to encourage businesses to invest in new equipment by allowing faster depreciation. For example, a delivery truck is designated by ACRS as a 3-year property, and the depreciation percentages allowed each year are given in the table below.

Year	ACRS depreciation
1	25%
2	38%
3	37%

Tax depreciation allows expenses in one year for durable goods, such as a truck, to be allocated and deducted from taxable income over several years. The depreciation percentage refers to percentage of the original expense. For example, if an item costs $10,000, then in the first year 25%, or $2500, is deducted; in the second year 38%, or $3800, is deducted; and in the third year 37%, or $3700, is deducted.

In the Tax Reform Act of 1986, the Modified Accelerated Cost Recovery System (MACRS) was mandated. This law designated light trucks, such as delivery trucks, as 5-year property and (curiously) required depreciation over 6 years, as shown in the following table.

Year	MACRS depreciation
1	20%
2	32%
3	19.2%
4	11.52%
5	11.52%
6	5.76%

Suppose the delivery truck in question costs $14,500.

a. Under either ACRS or MACRS, what is the total depreciation allowed over the life of the truck?

b. Under ACRS, what dollar tax deduction for depreciation on the truck is allowed in each of the first 3 years?

c. Under MACRS, what dollar tax deduction for depreciation on the truck is allowed in each of the first 3 years?

d. Suppose your company has a 28% *marginal tax rate*. That is, a 1 dollar tax deduction results in a reduction of 28 cents in actual taxes paid. Under ACRS, what is the total tax savings due to depreciation of the truck during the second year?

25. **Home Equity** When you purchase a home by securing a mortgage, the total paid toward the principal is your *equity* in the home. The accompanying table shows the equity E, in dollars, accrued after t years of payments on a mortgage of $170,000 at an APR of 6% and a term of 30 years.

t = Years of payments	E = Equity in dollars
0	0
5	11,808
10	27,734
15	49,217
20	78,194
25	117,279
30	170,000

a. Explain the meaning of $E(10)$ and give its value.

b. Make a new table showing the average yearly rate of change in equity over each 5-year period.

c. Judging on the basis of your answer to part b, does your equity accrue more rapidly early or late in the life of a mortgage?

d. Use the average rate of change to estimate the equity accrued after 17 years.

e. Would it make sense to use the average rate of change to estimate any function values beyond the limits of this table?

26. **Defense Spending** Data about recent federal defense spending are given in the accompanying *Statistical Abstract of the United States* table. Here t denotes the time, in years, since 1990 and D denotes federal defense spending, in billions of dollars.

a. Calculate the average yearly rate of change in defense spending from 1990 to 1995.

b. Use your answer from part a to estimate $D(3)$, and explain what it means.

(continued)

$t = $ Years since 1990	$D = $ Spending (billions of dollars)
0	328.4
5	310.0
10	341.5
15	565.5
20	843.8

c. Calculate the average yearly rate of change in defense spending from 2005 to 2010.

d. Use your answer from part c to estimate the value of $D(22)$.

27. A Home Experiment In our discussion of Carlson's experiment with yeast, we indicated that there should be similarities between the growth of yeast and the growth of mold on a slice of bread. In this exercise, you will verify that. Begin with a slice of bread that has a few moldy spots on it. Put it in a plastic bag and leave it in a warm place such as your kitchen counter. Estimate the percentage of the bread surface that is covered by mold two or three times each day until the bread is entirely covered with mold. (This may take several days to a week.) Record your data and provide a written report describing the growth of the mold.

28. Research Project Look on the web for data of interest to you, and analyze the average rate of change of the data. Try to relate your findings to economic or social trends during the given period.

1.2 SKILL BUILDING EXERCISES

For these exercises, round all *estimates* to one decimal place.

A Tabulated Function The following table gives values for a function $N = N(t)$.

t	$N = N(t)$
10	17.6
20	23.8
30	44.6
40	51.3
50	53.2
60	53.7
70	53.9

Exercises S-1 through S-20 refer to this function.

S-1. Function Values Find the value of $N(10)$.

S-2. Function Values Find the value of $N(20)$.

S-3. Function Values Find the value of $N(30)$.

S-4. Function Values Find the value of $N(40)$.

S-5. Function Values Find the value of $N(50)$.

S-6. Function Values Find the value of $N(60)$.

S-7. Function Values Find the value of $N(70)$.

S-8. Averaging Use averaging to estimate the value of $N(15)$.

S-9. Averaging Use averaging to estimate the value of $N(25)$.

S-10. Averaging Use averaging to estimate the value of $N(35)$.

S-11. Averaging Use averaging to estimate the value of $N(45)$.

S-12. Averaging Use averaging to estimate the value of $N(55)$.

S-13. Averaging Use averaging to estimate the value of $N(65)$.

S-14. Average Rate of Change Calculate the average rate of change from $t = 10$ to $t = 20$. Use your answer to estimate the value of $N(13)$.

S-15. Average Rate of Change Calculate the average rate of change from $t = 20$ to $t = 30$. Use your answer to estimate the value of $N(27)$.

S-16. Average Rate of Change Calculate the average rate of change from $t = 30$ to $t = 40$. Use your answer to estimate the value of $N(36)$.

S-17. Average Rate of Change Calculate the average rate of change from $t = 40$ to $t = 50$. Use your answer to estimate the value of $N(42)$.

S-18. Average Rate of Change Calculate the average rate of change from $t = 50$ to $t = 60$. Use your answer to estimate the value of $N(58)$.

S-19. Average Rate of Change Calculate the average rate of change from $t = 60$ to $t = 70$. Use your answer to estimate the value of $N(64)$.

S-20. Limiting Values Assuming the function N describes a physical situation for which a limiting value is expected, estimate the limiting value of N.

Another Table   The following is a partial table of values for $f = f(x)$.

x	$f = f(x)$
0	5.7
5	4.3
10	1.1
15	−3.6
20	−7.9

Use this table to complete Exercises S-21 through S-30.

S-21. Function Values   What is the value of $f(0)$?

S-22. Function Values   What is the value of $f(5)$?

S-23. Function Values   What is the value of $f(10)$?

S-24. Function Values   What is the value of $f(15)$?

S-25. Function Values   What is the value of $f(20)$?

S-26. Average Rate of Change   Calculate the average rate of change from $x = 0$ to $x = 5$. Use your answer to estimate the value of $f(3)$.

S-27. Average Rate of Change   Calculate the average rate of change from $x = 5$ to $x = 10$. Use your answer to estimate the value of $f(7)$.

S-28. Average Rate of Change   Calculate the average rate of change from $x = 10$ to $x = 15$. Use your answer to estimate the value of $f(13)$.

S-29. Average Rate of Change   Calculate the average rate of change from $x = 15$ to $x = 20$. Use your answer to estimate the value of $f(19)$.

S-30. Average Rate of Change   Use the results from Exercise S-29 to estimate the value of $f(25)$.

S-31. When Limiting Values Occur   Suppose $S(t)$ represents the average speed, in miles per hour, for a 100-mile trip that requires t hours. Explain why we expect S to have a limiting value.

S-32. Does a Limiting Value Occur?   A rocket ship is flying away from Earth at a constant velocity, and it continues on its course indefinitely. Let $D(t)$ denote its distance from Earth after t years of travel. Do you expect that D has a limiting value?

1.3 FUNCTIONS GIVEN BY GRAPHS

The idea of picturing a function as a graph is usually credited to the 17th-century mathematicians Fermat and Descartes. But some give credit to Nicole Oresme, who preceded Fermat and Descartes by 300 years. In his study of velocity, he hit upon the idea of representing the velocity of an object using two dimensions. He drew a *base line* to represent time and then added *perpendiculars* from this line whose lengths represented the velocity at each instant. The velocity, he said, "cannot be known any better, more clearly, or more easily than by such mental images and relations to figures."[22] Oresme was expressing his belief, which modern mathematicians share, that the graph provides deeper insight into a function and makes mathematics easier. It is one more illustration of the old cliché "A picture is worth a thousand words." One of the best features of a graph is that it provides an overall view of a function and thus makes it easy to deduce important properties.

Reading Graphs

Many factors affect fuel economy, but a website maintained by the U.S. government warns that "gas mileage usually decreases rapidly at speeds above 60 miles per hour. . . ." That

---

[22]From Marshall Clagett, trans. and ed., *Nicole Oresme and the Medieval Geometry of Qualities and Motions* (Madison: University of Wisconsin Press, 1968).

site includes the graph in Figure 1.7, which shows fuel economy $M = M(s)$, in miles per gallon (mpg), for a typical American car as a function of the speed s, in miles per hour (mph). It is customary to describe this as a graph of M *versus* s or as a graph of M *against s*. Either of these expressions indicates that the horizontal axis corresponds to s and the vertical axis corresponds to M. To find the fuel economy at 30 mph, we locate 30 on the horizontal axis, move along the vertical line shown in Figure 1.8 up to the graph and then move horizontally to locate the corresponding fuel economy on the vertical axis. In this case we see that it is about 28 mpg, so in functional notation $M(30) = 28$.

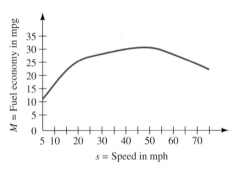

FIGURE 1.7 Fuel economy versus speed

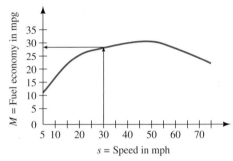

FIGURE 1.8 Finding M at a speed of 30 mph from the graph

As we can see in this example, graphical presentations sometimes allow only rough approximations of function values, but interesting features of a function are often easy to spot when it is presented graphically. For example, in Figure 1.9 we have marked the *maximum*, or highest point on the graph, as well as regions where the graph is *increasing* and where it is *decreasing*. This shows clearly that fuel economy increases as the speed increases from 5 to 50 mph, where fuel economy reaches a maximum of about 30 mpg, and that fuel economy decreases rapidly at higher speeds. This is typical of the shape of a graph near a maximum value; the graph increases up to the maximum value and decreases after that. Similarly, the lowest point, or *minimum*, typically occurs where the graph changes from decreasing to increasing. Sometimes, however, maximum or minimum values occur at the ends of a graph. For example, we see in Figure 1.7 that, for speeds between 5 and 75 mph, the minimum fuel economy of about 11 mpg occurs at 5 mph, at the left-hand end of the graph.

> Graphs can show maximum and minimum values, along with intervals of increase or decrease.

A great deal more information is available from the graph. Suppose, for example, that we want to know what speeds give a fuel economy of 25 mpg. In functional notation, we want to solve the equation $M(s) = 25$ for s. To do so, we locate 25 on the vertical axis and move horizontally until we get to the graph. We see from Figure 1.10 that this line crosses the graph twice, which indicates that two different speeds will give a fuel economy of 25 mpg. From a point where this line crosses the graph we drop down to the horizontal axis as illustrated in Figure 1.10, and we see that we get a fuel economy of 25 mpg at about 20 mph and at about 65 mph. In functional notation this is $M(20) = 25$ and $M(65) = 25$. We might also ask for the average decrease in fuel economy per mph from 50 to 75 mph. From the graph we read $M(50) = 30$ and $M(75) = 22$. That is a decrease of 8 mpg over a span of 25 mph, so the average decrease is $8/25 = 0.32$ mpg for each mph.

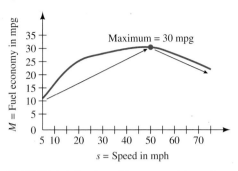

FIGURE 1.9 Region of increase, region of decrease, and the maximum value

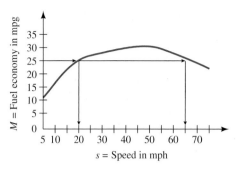

FIGURE 1.10 Finding when fuel economy is 25 mpg

KEY IDEA 1.4   **EVALUATING FUNCTIONS GIVEN BY GRAPHS**

To evaluate a function given by a graph, locate the point of interest on the horizontal axis, move vertically to the graph, and then move horizontally to the vertical axis. The function value is the location on the vertical axis.

EXAMPLE 1.5   **CURRENCY EXCHANGE**

Figure 1.11 shows the value $E = E(d)$, in U.S. dollars, of the euro as a function of the date d. We have added grid lines to the graph to aid in reading function values from the graph.

FIGURE 1.11 The value of the euro in U.S. dollars

Part 1 Explain the meaning of $E(2003)$ and estimate its value.

Part 2 From 2000 through 2011, what was the largest value the euro attained? When did that happen?

Part 3 What was the average yearly increase in the value of the euro from 2006 to 2009?

Part 4 During which one-year period was the graph increasing most rapidly?

Part 5 As an American investor, would you have made money if you bought euros in 2002 and sold them in 2008?

Solution to Part 1 The expression $E(2003)$ is the value in U.S. dollars of the euro in 2003. That is just over $1.00.

Solution to Part 2 The euro reached its largest value of about $1.50 in 2008.

Solution to Part 3 From 2006 to 2009, the euro increased in value from $E(2006) = \$1.20$ to $E(2009) = \$1.40$. That is an increase of $0.20 over the 3-year period. Thus the value of the euro increased by $0.20/3 = 0.07$ dollar per year.

Solution to Part 4 The value of the euro is increasing most rapidly where the graph is the steepest going upward. That appears to be from 2003 to 2004.

Solution to Part 5 The investor would have bought euros for about $0.90 each and sold them for about $1.50 each. That is a tidy profit margin.

TEST YOUR UNDERSTANDING | **FOR EXAMPLE 1.5**

From 2000 through 2011, what was the lowest value the euro attained? When did that happen? ∎

FIGURE 1.12 A graph that is concave up and increasing

Concavity and Rates of Change

Important features of a graph include places where it is increasing or decreasing, maxima and minima, and places where the graph may cross the horizontal axis. But a feature called *concavity* is often as important as any of these. A graph may be *concave up*, meaning that it has the shape of a wire whose ends are bent upward, or it may be *concave down*, meaning that it has the shape of a wire whose ends are bent downward. Figures 1.12 and 1.13 show graphs that are concave up, and Figures 1.14 and 1.15 show graphs that are concave down.

FIGURE 1.13 A graph that is concave up and decreasing

Note that the concavity of a graph does not determine whether it is increasing or decreasing, but it does give additional information about the rate of increase or decrease. The graphs in Figures 1.12 and 1.14 are both increasing, but they increase in different ways. As we move to the right in Figure 1.12, the graph gets steeper. That is, the graph is increasing at a faster rate. As we move to the right in Figure 1.14, the graph becomes less steep, showing that the rate of increase is slowing. This is typical. Increasing graphs that are concave up increase at an increasing rate, whereas increasing graphs that are concave down increase at a decreasing rate. A similar analysis for Figures 1.13 and 1.15 shows that decreasing graphs that are concave up decrease at a decreasing rate, whereas decreasing graphs that are concave down decrease at an increasing rate. Concavity, if it persists, can have important long-term effects. For example, if prices were increasing, it would be a good deal more troubling in the long term if the graph of prices against time were concave up than if it were concave down.

FIGURE 1.14 A graph that is concave down and increasing

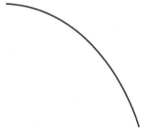

FIGURE 1.15 A graph that is concave down and decreasing

A steeper graph indicates more rapid increase or decrease.

EXAMPLE 1.6 **CONCAVITY AND CURRENCY EXCHANGE**

Consider once more the value of the euro as given by the graph in Figure 1.11.

Part 1 From 2000 to late 2003, is the graph concave up or concave down?

Part 2 Explain in practical terms what the concavity means about the value of the euro during this period.

Solution to Part 1 The graph is "bent upward" during this period, so it is concave up.

Solution to Part 2 We divide our analysis here into the places where the graph is increasing and decreasing. From 2000 to 2002, the graph is decreasing and concave up. The graph is leveling off as we move toward the right. This means that the value of the euro is decreasing, but at a decreasing rate.

From 2002 to late 2003, the graph is increasing and concave up. The graph gets steeper as we move toward the right. This means that the value of the euro is increasing at an increasing rate from 2002 to late 2003.

TEST YOUR UNDERSTANDING | **FOR EXAMPLE 1.6**

Locate a region where the graph is increasing and concave down. Explain what is happening to the value of the euro during the period you select. ■

Inflection Points

Inflection points are points where concavity changes.

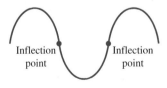

FIGURE 1.16 A graph with changing concavity

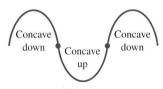

FIGURE 1.17 Inflection points

We refer to points on a graph where concavity changes, up to down or down to up, as *inflection points*. In Figure 1.16 we have shown a graph where concavity changes. These changes in concavity determine the two inflection points marked in Figure 1.17. The first inflection point in Figure 1.17 is where the graph changes from concave down to concave up, and the second is where the graph changes from concave up to concave down. Note that the first inflection point in Figure 1.17 is also the point where the graph is decreasing most rapidly. Similarly, the second inflection point is where the graph is increasing most rapidly. Points of maximum increase or decrease are often found at inflection points, and in many applications this is where we should look. The graph in Figure 1.11 has an inflection point in late 2003, where concavity changes from up to down.

EXAMPLE 1.7 **DRIVING DOWN A STRAIGHT ROAD**

You leave home at noon and drive your car down a straight road for a visit to a friend's house. Your distance D, in miles, from home as a function of time t, in hours since noon, is shown in the graph in Figure 1.18.

Part 1 During what time period are you at your friend's house?

Part 2 Over what time period is the graph decreasing? Explain in practical terms what this portion of the graph represents.

Part 3 At what value of t is the graph rising most steeply? Explain in terms of your speedometer what your answer means.

Part 4 At what time do you get back home?

When you are not moving, the graph of distance is level.

Solution to Part 1 During the time you are at the friend's house, your distance from home is unchanging, or constant, so the graph of D versus t is level over that period. Since the graph is level between about $t = 2$ and $t = 4\frac{1}{2}$ hours after noon, you are at the friend's

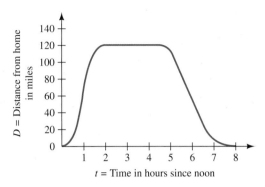

FIGURE 1.18 Distance from home versus time

house between about 2:00 P.M. and 4:30 P.M. Note that the constant value of D over this period is about 120, so the friend lives about 120 miles from your home.

Solution to Part 2 The graph is decreasing from around $t = 4\frac{1}{2}$ to around $t = 7\frac{3}{4}$, or roughly from 4:30 P.M. to 7:45 P.M. Since D is your distance from home, this portion of the graph represents your trip back home.

Solution to Part 3 The steepest rise in the graph occurs at the first inflection point, where the graph changes from concave up to concave down. Examination of the graph shows that this point occurs at about $t = 1$. Before $t = 1$, the graph is concave up, so the distance D is increasing at an *increasing* rate; for a time after $t = 1$, the graph is concave down, and the distance is increasing at a *decreasing* rate. This means that the greatest rate of increase in distance occurs at $t = 1$. Now on the way to your friend's house, the speedometer reading tells you how fast you are going away from home; in other words, it gives the rate of increase in distance. Therefore, you interpret the time when the graph is rising most steeply as the time on your way to the friend's house when the speedometer reading is the greatest. In short, on the first part of your trip, you reach your maximum speed at about 1:00 P.M.

Solution to Part 4 When you get back home, the distance D reaches 0, so the graph touches the horizontal axis. This occurs at around $t = 7\frac{3}{4}$ hours after noon, or 7:45 P.M.

> Inflection points often indicate points of most rapid increase or decrease.

TEST YOUR UNDERSTANDING   │   **FOR EXAMPLE 1.7**

At what time is the graph decreasing most rapidly? Explain in terms of your speedometer what your answer means.   ■

■ │ Drawing a Graph

A graph is a pictorial representation of a function, and it is generally not difficult to make such pictures. Here is an example. According to the *Information Please Almanac*, in 1900 about 9.3 Americans per thousand were married each year. This number reached an all-time high of about 12.2 per thousand in 1945 (in the aftermath of World War II). This number then decreased and reached a low of 8.5 per thousand in 1960. This information describes a function $M = M(d)$ that gives the number of marriages per thousand Americans in year d, and we want to draw its graph.

To make a graph of the function M means that we graph M against d. That is, the horizontal axis will correspond to the date, and the vertical axis will correspond to the number of marriages per thousand. These axes are so labeled in Figure 1.19. We note that the verbal description of the function gave us three data points. For example, in functional notation the first point is $M(1900) = 9.3$.

The next step is to mark these points on the graph as shown in Figure 1.19. To complete the graph, we join these points with a curve incorporating the additional information we have. In particular, the verbal description of *M* tells us that the graph should decrease after 1945. Our completed graph is shown in Figure 1.20.

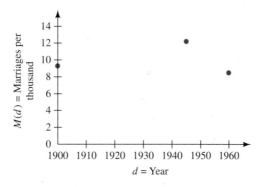

FIGURE 1.19 Locating given points for *M* versus *d*

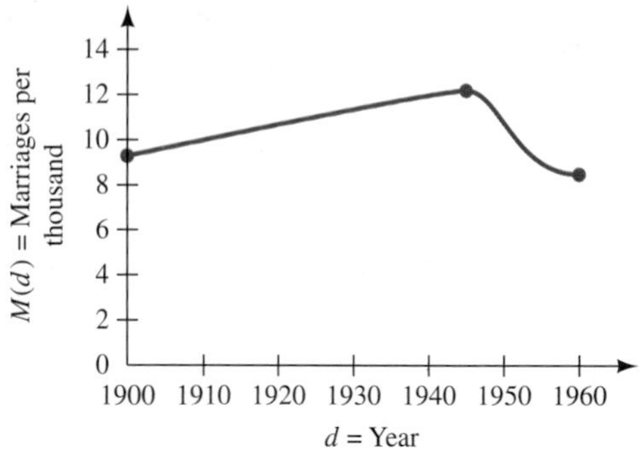

FIGURE 1.20 Marriages per thousand Americans

It is important to note that there was a certain amount of guesswork done to produce the graph in Figure 1.20. The points we located in Figure 1.19 are not guesses, but the way we chose to join these points to complete the graph in Figure 1.20 does involve a good deal of guessing. For example, we were told that the marriage rate per thousand decreased from 1945 to 1960, but we were not told how it decreased. The true graph might, for example, be concave down in this region (rather than up, as we drew it), or it might be a jagged line (rather than the smooth curve we have drawn). When graphs are used to represent functions given by sampling data, some guessing is inevitably involved in connecting known data points. You should be aware that artistic flair can dramatically affect the appearance of such graphs. For example, if we had wanted to emphasize the high marriage rate at the end of World War II, we might have drawn the graph as in Figure 1.21. If we had wanted to de-emphasize it, we might have made the graph as in Figure 1.22. Note that both of these graphs agree with the verbal description we were given, but they convey quite different information to the eye. Further emphasis or de-emphasis of particular features of a function can be attained by adjusting the scale on the horizontal or vertical axis.

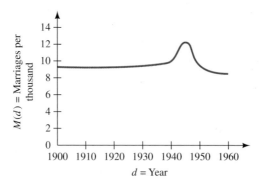

FIGURE 1.21 A representation of the marriage rate that emphasizes the peak in 1945

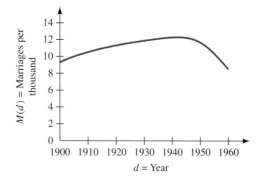

FIGURE 1.22 A representation of the marriage rate that de-emphasizes the peak in 1945

Some graphs reflect the author's interpretation.

How do we know which of the pictures, if any, among Figures 1.20, 1.21, and 1.22 provide a true representation of the function *M*? The answer is that graphs give a more accurate representation of a function when more data points are used. To find the true nature of the American marriage rate in the first half of the 20th century, you should consult the almanac referred to earlier, where a good deal more information is available. See also Exercise 28 at the end of this section. As an informed citizen, you should be especially leery of free-hand graphs that do not provide references for data sources or for which the artist may be an advocate of a particular point of view.

EXAMPLE 1.8 **EDUCATION IN THE UNITED STATES**

According to United States census figures, in 1940 about 74% of Americans 25 years old and over had completed less than 12 years of school. This figure decreased to about 31% in 1980. From that point on, the rate of decrease was smaller, and the figure was 13% in 2009. Let $E = E(d)$ be the percentage of Americans 25 years old and over who have completed less than 12 years of school in year d.

Part 1 Three exact data points were given. Express them in functional notation.

Part 2 Draw a graph of E against d. Be sure to label your axes and to make the graph as accurate as the data allow.

Solution to Part 1 We are told that 74% of the target population had less than 12 years of school in 1940. Thus $E(1940) = 74$. Similarly, $E(1980) = 31$, and $E(2009) = 13$.

Solution to Part 2 To graph E against d means that the date d goes on the horizontal axis and the percentage E goes on the vertical axis. In Figure 1.23 we labeled both years and percentage points in 10-year increments and noted what the numbers on the axes represent. Next we located the three given function values in Figure 1.23. To complete the graph as shown in Figure 1.24, we follow the verbal description and draw a decreasing curve through these points. Note that the fact that the rate of decrease was less pronounced after 1980 is reflected by the fact that the graph does not go down so steeply from that point on. Remember, though, that there are a number of related graphs, all of which could be correct.

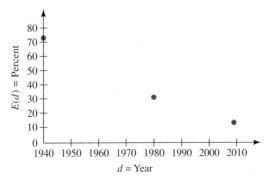

FIGURE 1.23 Locating three data points

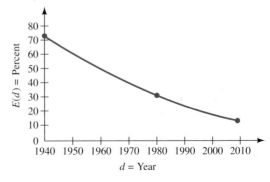

FIGURE 1.24 Percentage of Americans 25 years old and over with less than 12 years of school

TEST YOUR UNDERSTANDING | **FOR EXAMPLE 1.8**

The given information includes the fact that the rate of decrease was smaller from 1980 to 2009 than from 1940 to 1980. Draw a new graph that shows a faster rate of decrease from 1980 to 2009 than from 1940 to 1980. (Use a lower value for 2009 than the actual value.) ■

ANSWERS FOR TEST YOUR UNDERSTANDING

1.5 The euro was at its lowest value of about $0.90 in 2002.

1.6 The graph is increasing and concave down from late 2003 to late 2004, from early 2007 to 2008, and again from mid 2009 to 2010. In each of these periods the value of the euro is increasing, but at a decreasing rate.

1.7 The graph appears to be decreasing most rapidly at approximately $t = 6$. This is the place on your way home where your speedometer reading is the greatest. In short, on the second part of the trip you reach your maximum speed at about 6:00 P.M.

1.8 Figure 1.25 shows a graph with the desired rates of decrease.

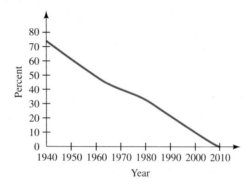

FIGURE 1.25 A faster rate of decrease from 1980 on

1.3 EXERCISES

Reminder Round all answers to two decimal places unless otherwise indicated.

1. **Sketching a Graph with Given Concavity:**

   a. Sketch a graph that is always decreasing but starts out concave down and then changes to concave up. There should be a point of inflection in your picture. Mark and label it.

   b. Sketch a graph that is always decreasing but starts out concave up and then changes to concave down. There should be a point of inflection in your picture. Mark and label it.

2. **An Investment** In 2010 an investor put money into a fund. The graph in Figure 1.26 shows the value $v = v(d)$ of the investment, in dollars, as a function of the date d.

   a. Express the original investment using functional notation and give its value.

   b. Is the graph concave up or concave down? Explain what this means about the growth in value of the account.

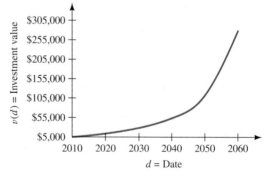

FIGURE 1.26 An investment

   c. When will the value of the investment reach $55,000?

   d. What is the average yearly increase from 2050 to 2060?

   e. Which is larger, the average yearly increase from 2050 to 2060 or the average yearly increase from 2010 to 2020? Explain your reasoning.

(continued)

3. **Skirt Length** One author[23] analyzed skirt lengths in the United Kingdom (UK) and Germany. To get a standardized measure of skirt length she calculated the ratio

$$\frac{\text{Length from shoulder to skirt hem}}{\text{Length from shoulder to ankle}}.$$

The article includes the following graph of the ratio for skirt length in terms of the year. In the graph the solid line is the ratio in the UK, and the dashed line is the ratio in Germany.

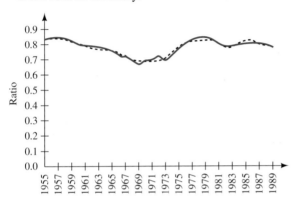

FIGURE 1.27 Ratio for skirt length

a. Does a larger ratio indicate a longer or shorter skirt?

b. When were skirt lengths shortest in the UK?

c. In this period did skirt lengths in either Germany or the UK ever reach to the ankle?

4. **Number of Weddings** The graph in Figure 1.28 is taken from the website of *The Wedding Report*.[24] It shows the number $W = W(y)$ of weddings (in millions) in year y.

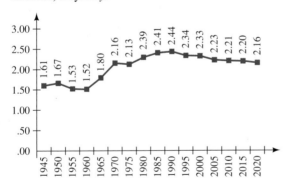

FIGURE 1.28 Number of weddings

a. In what year did W reach a maximum?

b. What is the minimum value of W?

c. Over what 5-year period was the number of weddings increasing at the fastest rate?

5. **Unemployment** The graph in Figure 1.29 shows the unemployment percentage $U = U(d)$ in the United States in year d.

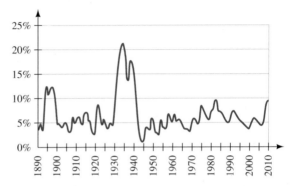

FIGURE 1.29 Unemployment as a function of the date

a. Explain the meaning of $U(1990)$ and give its value.

b. The Great Depression was a 10-year period in which unemployment was at historical highs. Judging from the graph, when was the Great Depression?

c. The high unemployment rate in 2010 is attributed to a severe recession. What was the most recent date before 2010 when unemployment was as high as in 2010?

6. **Cancer Mortality Rates** The graph in Figure 1.30 shows mortality rates per 100,000 *person-years* by year, race, and gender. A person-year is the number of people times the number of years. For example, if we look at a group of 50 people over a 10-year period, that is 500 person-years.

a. From 1970 through 1994, which group showed the largest rate of increase in cancer mortality?

b. When was cancer mortality at a maximum for white males?

[23]From L. Curran, "An analysis of cycles in skirt lengths and widths in the UK and Germany, 1954–1990," *Clothing and Textiles Research Journal* **17** (1999), 65–72. Copyright © 1999 Sage Publication. Used with permission. The online version can be found at http://ctr.sagepub.com/content/17/2/65.full.pdf

[24]Copyright © 2012 The Wedding Report, Inc., All rights reserved. http://www.theweddingreport.com/?brand=google&gclid=CP--yeem_Z8CFQOL5wod116nnA

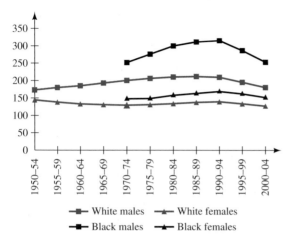

White males — White females
Black males — Black females

FIGURE 1.30 Cancer mortality rates

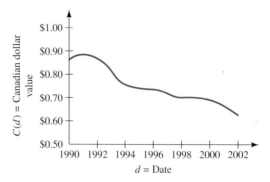

FIGURE 1.31 The value in American dollars of the Canadian dollar

7. **A Stock Market Investment** A stock market investment of $10,000 was made in 1970. During the decade of the 1970s, the stock lost half its value. Beginning in 1980, the value increased until it reached $35,000 in 1990. After that its value has remained stable. Let $v = v(d)$ denote the value of the stock, in dollars, as a function of the date d.

   a. What are the values of $v(1970)$, $v(1980)$, $v(1990)$, and $v(2010)$?

   b. Make a graph of v against d. Label the axes appropriately.

   c. Estimate the time when your graph indicates that the value of the stock was most rapidly increasing.

8. **The Value of the Canadian Dollar** The value $C = C(d)$, in American dollars, of the Canadian dollar is given by the graph in Figure 1.31.

a. Describe how the value of the Canadian dollar fluctuated from 1990 to 2002. Give specific function values in your description where they are appropriate.

b. When was the Canadian dollar worth 80 American cents?

c. What was the average yearly decrease in the value of the Canadian dollar from 1992 to 1994?

d. For an American investor trading in Canadian dollars, what would have been ideal buy and sell dates? Explain your answer.

9. **River Flow** The graph in Figure 1.32 shows the mean flow F for the Arkansas River, in cubic feet of water per second, as a function of the time t, in months, since the start of the year. The flow is measured near the river's headwaters in the Rocky Mountains.

a. Use functional notation to express the flow at the end of July, and then estimate that value.

b. When is the flow at its greatest?

c. At what time is the flow increasing the fastest?

(continued)

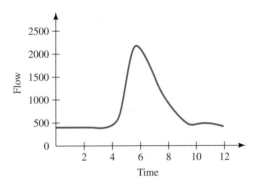

FIGURE 1.32 Flow for the Arkansas River

d. Estimate the average rate of change per month in the flow during the first 2 months of the year.

e. In light of the source of the Arkansas River, interpret your answers to parts b, c, and d.

10. **Logistic Population Growth** The graph in Figure 1.33 shows the population $N = N(d)$, in thousands of animals, of a particular species on a protected reserve as a function of the date d. Ecologists refer to growth of the type shown here as *logistic population growth*.

a. Explain in general terms how the population N changes with time.

b. When did the population reach 80 thousand?

c. Your answer in part b is the solution of an equation involving $N(d)$. Which equation?

d. During which period is the graph concave up? Explain what this means about population growth during this period.

e. During which period is the graph concave down? Explain what this means about population growth during this period.

f. When does a point of inflection occur on the graph? Explain how this point may be interpreted in terms of the growth rate.

g. What is the *environmental carrying capacity* of this reserve for this species of animal? That is, what is the maximum number of individuals that the environment can support?

h. Make a new graph of $N(d)$ under the following new scenario. The population grew as in Figure 1.33 until 1990, when a natural disaster caused half of the population to die. After that the population resumed logistic growth.

11. **Cutting Trees** In forestry management it is important to know the *net stumpage value* of a stand (that is, a group) of trees. This is the commercial value of the trees minus the costs of felling, hauling, etc. The graph in Figure 1.34 shows the net stumpage value V, in dollars per acre, of a Douglas fir stand in the Pacific Northwest as a function of the age t, in years, of the stand.[25]

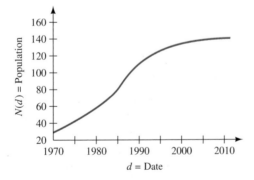

FIGURE 1.33 A population, in thousands, showing logistic population growth

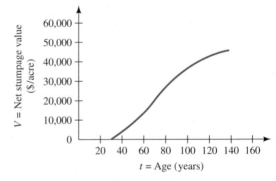

FIGURE 1.34 Net stumpage value of a Douglas fir stand

---

[25]The normal-yield table used here (with site index 140) is from a study by R. E. McArdle et al., as presented by Thomas E. Avery and Harold E. Burkhart, *Forest Measurements*, 4th ed. (New York: McGraw-Hill, 1994).

a. Estimate the net stumpage value of a Douglas fir stand that is 60 years old.

b. Estimate the age of a Douglas fir stand whose net stumpage value is $40,000 per acre.

c. At what age does the commercial value of the stand equal the costs of felling, hauling, etc.?

d. At what age is the net stumpage value increasing the fastest?

e. This graph shows V only up to age $t = 160$ years, but the Douglas fir lives for hundreds of years. Draw a graph to represent what you expect for V over the life span of the tree. Explain your reasoning.

12. Wind Chill   The graph in Figure 1.35 shows the temperature $T = T(v)$ adjusted for wind chill as a function of the velocity v of the wind when the thermometer reads 30 degrees Fahrenheit. The adjusted temperature T shows the temperature that has an equivalent cooling power when there is no wind.

a. At what wind speed is the temperature adjusted for wind chill equal to 0?

b. Your answer in part a is the solution of an equation involving $T(v)$. Which equation?

c. At what value of v would a small increase in v have the greatest effect on $T(v)$? In other words, at what wind speed could you expect a small increase in wind speed to cause the greatest change in wind chill? Explain your reasoning.

d. Suppose the wind speed is 45 miles per hour. Judging from the shape of the graph, how significant would you expect the effect on $T(v)$ to be if the wind speed increased?

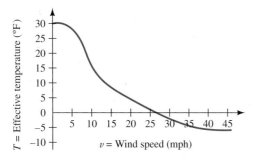

FIGURE 1.35 Temperature adjusted for wind chill when the thermometer reads 30 degrees Fahrenheit

13. Tornadoes in Oklahoma   The graph in Figure 1.36 shows the number $T = T(d)$ of tornadoes reported by the Oklahoma Climatological Survey.

a. When were the most tornadoes reported? How many were reported in that year?

b. When were the fewest tornadoes reported? How many were reported in that year?

c. What was the average yearly rate of decrease in tornadic activity from 2001 to 2002?

d. What was the average yearly rate of increase in tornadic activity from 2002 to 2004?

e. What was the average yearly *rate of change* (that is, the average yearly rate of total increase or total decrease) in tornadic activity from 2001 to 2004?

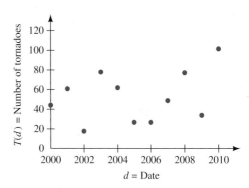

FIGURE 1.36 Tornadoes in Oklahoma

14. Inflation   During a period of high inflation, a political leader was up for re-election. Inflation had

(continued)

been increasing during his administration, but he announced that the *rate of increase* of inflation was decreasing. Draw a graph of inflation versus time that illustrates this situation. Would this announcement convince you that economic conditions were improving?

15. **Driving a Car** You are driving a car. The graph in Figure 1.37 shows your distance D, in miles, from home as a function of the time t, in minutes, since 1:00 P.M. Make up a driving story that matches the graph. Be sure to explain how your story incorporates the times when the graph is increasing, when it is decreasing, and when it is constant.

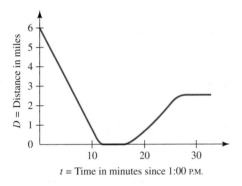

FIGURE 1.37 Distance from home versus time

16. **Walking to School** You walk from home due east to school to get a book, and then you walk back west to visit a friend. Figure 1.38 shows your distance D, in yards, east of home as a function of the time t, in minutes, since you left home.

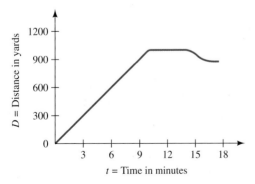

FIGURE 1.38 Distance east of home versus time

a. How far away is school?

b. At what time do you reach school?

c. At what time(s) are you a distance of 900 yards from home?

d. Compute the average rate of change in D over the intervals from $t = 0$ to $t = 3$, from $t = 3$ to $t = 6$, and from $t = 6$ to $t = 9$.

e. What does your answer to part d tell you about how fast you are walking to school? How is this related to the shape of the graph over the interval from $t = 0$ to $t = 9$?

f. At what time are you walking back west the fastest?

17. **Photosynthesis** During photosynthesis, plants both absorb and emit carbon dioxide. The net exchange (output or input) depends on many factors. Two of these factors are light and temperature. Figure 1.39 on the following page shows[26] the net carbon dioxide exchange rates for 5-week-old rice plants at 40 degrees, 60 degrees, and 80 degrees Fahrenheit. The vertical axis shows the net exchange, in milligrams, of carbon dioxide per hour per plant. Values above the horizontal axis indicate that the plant is taking in more carbon dioxide than it releases. Values below the horizontal axis indicate that the plant is releasing more carbon dioxide than it takes in. The horizontal axis shows amount of light, measured in thousands of foot-candles.[27] The following questions pertain to the 5-week-old rice plants at the indicated temperatures.

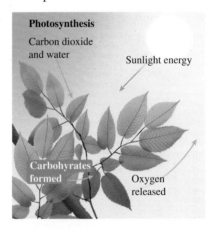

a. At a temperature of 80 degrees, about how many foot-candles of light will cause a plant to emit the same amount of carbon dioxide that it absorbs?

[26]Adapted from D. P. Ormrod, "Photosynthesis rates of young rice plants as affected by light intensity and temperature," *Agron. J.* **53** (1961), 93.
[27]The foot-candle is a measure of luminance. Modern physicists use the *candela*, which is defined in terms of black-body radiation.

b. At about how many foot-candles will plants at 80 degrees and 40 degrees show the same net exchange of carbon dioxide?

c. At 0 foot-candles—that is, in the dark—which temperature (80 degrees, 60 degrees, or 40 degrees) will result in the largest emission of carbon dioxide?

d. At which temperature (80 degrees, 60 degrees, or 40 degrees) is the net exchange of carbon dioxide least sensitive to light? Explain your reasoning.

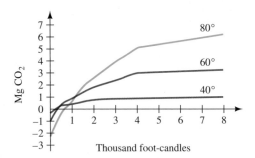

FIGURE 1.39 Net carbon dioxide exchange in young rice plants

18. **Protein Content of Wheat Grain**   Protein content of wheat grain is affected by soil moisture and the amount of available nitrogen (among other things). Figure 1.40 shows[28] the percent of protein content of wheat grain versus pounds of nitrogen per acre applied in three separate situations. In each case soil moisture refers to moisture at the soil depth of 2 inches to 12 inches.

- **Situation 1:** Irrigation was used when soil moisture dropped to 49%.
- **Situation 2:** Irrigation was used when soil moisture dropped to 34%.
- **Situation 3:** Irrigation was used when soil moisture dropped to 1%.

a. If irrigation begins when soil moisture reaches 49%, what application of nitrogen will result in the lowest percentage of protein in wheat grain?

b. If irrigation begins when soil moisture reaches 34%, what application of nitrogen will result in the same protein content of wheat grain as beginning irrigation when soil moisture reaches 1%?

c. If you irrigate when soil moisture reaches 34%, how much nitrogen should you apply to achieve a 13% protein content in wheat grain?

d. Does Figure 1.40 indicate that, for nitrogen levels at 45 pounds per acre or higher, increased protein content in wheat grain is associated with higher or lower soil moisture?

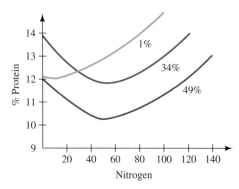

FIGURE 1.40 Protein content versus availability of nitrogen

19. **Carbon Dioxide Concentrations**   The amount of carbon dioxide in the air is influenced by several factors, including respiration of plants. Figure 1.41 on the following page shows[29] the concentration of carbon dioxide in parts per million (PPM) in a closed (that is, unventilated) greenhouse on a cold, clear day during a 24-hour period beginning at 6 A.M.

a. At what time is carbon dioxide in the greenhouse at its lowest concentration?

b. When is carbon dioxide in the greenhouse at its highest concentration?

---

[28]Adapted from R. Fernandez and R. J. Laird, "Yield and protein content of wheat in central Mexico as affected by available soil moisture and nitrogen fertilization," *Agron. J.* **51** (1959), 33.
[29]Adapted from S. H. Wittwer and W. Robb, "Carbon dioxide enrichment of greenhouse atmospheres for vegetable crop production," *Econ. Botany* **18** (1964), 34.

(continued)

c. Photosynthesis results in a carbon dioxide exchange between plants and the surrounding air. During what times do the plants in the greenhouse show a net absorption of carbon dioxide?

d. What is happening in terms of carbon dioxide exchange from 6 A.M. to 9 A.M.?

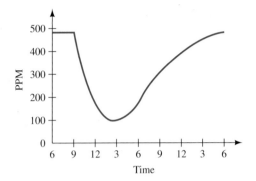

FIGURE 1.41 Carbon dioxide concentrations in a greenhouse

20. Hydrographs When a rainfall brings more water than the soil can absorb, runoff occurs, and hydrologists refer to the event as a *rainfall excess*. The easiest way to envision runoff is to think of a watershed that drains into the mouth of a single stream. The runoff is the number of cubic feet per minute (cfpm) being dumped into the mouth of the stream. An important way of depicting runoff is the *hydrograph*, which is simply the graph of total discharge, in cubic feet per minute, versus time. A typical runoff hydrograph is shown in Figure 1.42. The horizontal axis is hours since rainfall excess began. A hydrograph displays a number of important features.

a. *Time to peak* is the elapsed time from the start of rainfall excess to peak runoff. What is the time to peak shown by the hydrograph in Figure 1.42?

b. *Time of concentration* is the elapsed time from the end of rainfall excess to the inflection point after peak runoff. The end of rainfall excess is not readily apparent from a hydrograph, but it occurs before the peak. If the end of rainfall excess occurred 5 hours after the start of rainfall excess, estimate the time of concentration from Figure 1.42.

c. *Recession time* is the time from peak runoff to the end of runoff. Estimate the recession time for the hydrograph in Figure 1.42.

d. *Time base* is the time from beginning to end of surface runoff. What is the time base for the hydrograph in Figure 1.42?

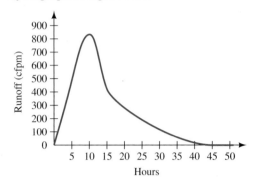

FIGURE 1.42 A runoff hydrograph

21. Profit from Fertilizer Fertilizer can increase crop production, but the cost of the fertilizer must be taken into account. The curved graph in Figure 1.43 shows[30] the added bushels per acre of corn that resulted from fertilizer application versus pounds per acre of nitrogen applied. The straight line shows the cost (in bushels of corn, not dollars) versus pounds per acre of nitrogen applied.

a. How many pounds per acre of nitrogen should be applied to produce maximum crop yield?

b. The profit is the crop yield minus the fertilizer cost. For each level of fertilizer application, how can the profit be read off the graph?

c. How many pounds per acre of nitrogen should be applied to achieve maximum profit?

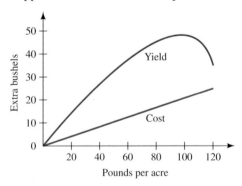

FIGURE 1.43 Effect of nitrogen on corn production.

22. H-R Diagrams The *luminosity* of a star is the total amount of energy the star radiates (visible light as well as X-rays and all other wavelengths) in 1 second. In practice, astronomers compare

[30]Adapted from an article by Allstetter in the *National Fertilizer Review*, January–February–March, 1953.

the luminosity of a star with that of the sun and speak of *relative luminosity*. Thus a star of relative luminosity 5 is five times as luminous as the sun. One of the most important graphical representations in astronomy is the *Hertzsprung-Russell diagram*, or *H-R diagram*, which plots relative luminosity[31] versus surface temperature in thousands of *kelvins* (degrees on the Kelvin scale). Figure 1.44 shows an H-R diagram. Note that the temperature scale *decreases* as we read from left to right.

a. What is the surface temperature of a main sequence star that is 10,000 times as luminous as the sun?

b. What is the relative luminosity of the sun?

c. About 90% of all stars, including the sun, lie on or near the main sequence. Approximate the surface temperature of the sun.

d. The H-R diagram in Figure 1.44 shows that white dwarfs lie well below the main sequence. Are white dwarfs more or less luminous than main sequence stars of the same surface temperature?

e. If one star is three times as luminous as another, yet they have the same surface temperature, then the brighter star must have three times the surface area of the dimmer star. How would the surface area of a supergiant star with the same surface temperature as the sun compare with the surface area of the sun?

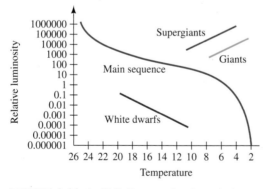

FIGURE 1.44 An H-R diagram showing relative luminosity versus surface temperature

23. Home Equity   When you purchase a home by securing a mortgage, the total paid toward the principal is your *equity* in the home. The graph in Figure 1.45 shows for a certain 30-year mortgage the equity, in dollars, accrued after a given number of monthly payments.

a. Is the graph concave up or concave down?

b. The graph is increasing. What does the concavity you identified in part a say about how the equity grows?

c. Use the graph to estimate the amount of the mortgage.

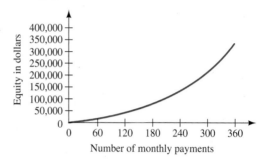

FIGURE 1.45 Home equity growth

24. Relativistic Time   A correctly functioning hourglass on board a rocket ship would appear to a stationary observer to take longer than an hour to empty. The graph in Figure 1.46 shows the time T, in hours, that it appears to a stationary observer to take an hourglass to empty if it is on a rocket ship moving at a given percentage r, as a decimal, of the speed of light.

a. Use the graph to estimate the value of $T(0.6)$ and explain in practical terms what it means.

b. Is the graph concave up or concave down?

c. Does a small change in velocity cause a greater change in apparent time at slower speeds or at faster speeds?

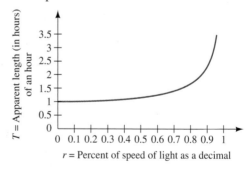

FIGURE 1.46 Relativistic length of an hour

Exercises 25 through 27 are concerned with *survivorship curves*, which provide a visual presentation of the age structure of a population. Suppose we start with 1000 individuals in the population, all born at the same time. Let l_x denote the number of individuals still alive at age x. A survivorship curve is a graph of l_x as a

---

[31]Strictly speaking, an H-R diagram plots absolute magnitude versus surface temperature.

(continued)

function of x. It is common to look at these graphs on a semilogarithmic scale, which means that the vertical axis represents the power of 10 that gives l_x, denoted $\log l_x$. (We will study logarithms in Chapter 4.)

25. Grain Beetles   This exercise illustrates how the survivorship curves of two strains of grain beetles depend on environmental conditions. (Here x is measured in weeks.)

  a. In Figure 1.47 two survivorship curves for the small strain of beetles grown in wheat are given. For one the ambient temperature was 29.1 degrees Celsius (about 84 degrees Fahrenheit), and for the other the ambient temperature was 32.3 degrees Celsius (about 90 degrees Fahrenheit). Use the curves to decide at which of these two temperatures you would store your wheat. Explain your answer.

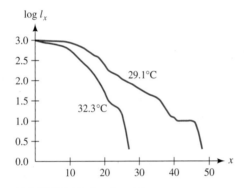

FIGURE 1.47 Survivorship curves at varying temperatures

  b. In Figure 1.48 two survivorship curves for the large strain of beetles grown at 29.1 degrees Celsius are given. For one the food was wheat, and for the other the food was maize. Use the curves to decide which of these two grains the large strain would be more likely to infest. Explain your answer.

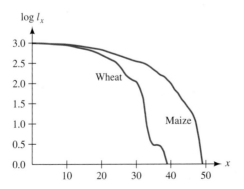

FIGURE 1.48 Survivorship curves in varying grains

26. Life Insurance Rates   Consider the survivorship curves given in Figure 1.49 for males and females in the human population of the United States.

  a. On the basis of these curves, explain why life insurance rates are higher for men than for women.

  b. On the basis of these curves, explain why life insurance rates increase so rapidly at high ages.

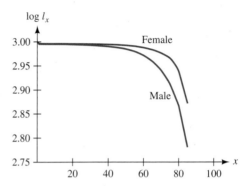

FIGURE 1.49 Survivorship curves for the U.S. population

27. Span of Life   Many scientists believe that, for humans at least, there is a natural span of life. That is, there is a limit beyond which human life cannot extend, regardless of advances in medical science. Currently the maximum life span in developed countries is around 120 years, and evidence suggests that it has been that way since Roman times. However, with improvements in medical care, the *average* life span has increased significantly during the past century. Draw a sequence of survivorship curves that illustrates increasing average life span with fixed maximum life span.

28. Research Project   In the discussion about Figures 1.21 and 1.22, we pointed out that the right way to get an accurate graph of marriage rates was to include more data points. In the *Information Please Almanac*, the marriage rate per thousand Americans is given for several years. This information will also be available in other reference sources in your library. Pursue an appropriate reference and, by producing your own graph, settle the question of which (if any) of the graphs we presented is an accurate representation of American marriage rates.

1.3 SKILL BUILDING EXERCISES

A Function Given by a Graph   The following is the graph of a function $f = f(x)$. Exercises S-1 through S-14 refer to this graph.

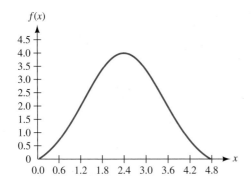

S-1. What is the value of $f(1.8)$?

S-2. What is the value of $f(2.4)$?

S-3. What is the value of $f(4.2)$?

S-4. What is the smallest value of x for which $f(x) = 1.5$?

S-5. What is the largest value of x for which $f(x) = 1.5$?

S-6. What is the smallest value of x for which $f(x) = 3$?

S-7. Where does the graph reach a maximum, and what is that maximum value?

S-8. Where is the graph increasing?

S-9. Where is the graph decreasing?

S-10. What is the concavity of the graph between $x = 1.8$ and $x = 2.4$?

S-11. What is the concavity of the graph between $x = 0$ and $x = 1.8$?

S-12. What is the concavity of the graph between $x = 2.4$ and $x = 3.0$?

S-13. What is the concavity of the graph between $x = 3.0$ and $x = 4.8$?

S-14. Where on the graph are there points of inflection?

A Function Given by a Graph   The following is the graph of a function $f = f(x)$. Exercises S-15 through S-23 refer to this graph.

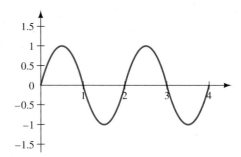

S-15. At what values of x does the graph reach a maximum value?

S-16. What is the maximum value of $f(x)$?

S-17. At what values of x does the graph reach a minimum value?

S-18. What is the minimum value of $f(x)$?

S-19. Find the x value of the inflection points of $f(x)$.

S-20. What is the concavity of the graph between $x = 1$ and $x = 2$?

S-21. What is the concavity of the graph between $x = 2$ and $x = 3$?

S-22. Where is the graph decreasing and concave up?

S-23. Where is the graph increasing and concave up?

S-24. Concavity Again   A certain graph is increasing, but at a decreasing rate. Discuss its concavity.

S-25. Special Points   What is the mathematical term for points where concavity changes?

Sketching Graphs   In Exercises S-26 through S-32, sketch a graph with the given properties. We note that for each exercise there are many correct solutions.

S-26. Sketch a graph that has a maximum value of 3 when $x = 2$ and a zero at $x = 4$.

S-27. Sketch a graph that increases from $x = 0$ to $x = 3$ and decreases after that.

S-28. Sketch a graph with a maximum at $x = 1$ and a minimum at $x = 3$.

S-29. Sketch a graph that attains maximum values at $x = 1$ and $x = 3$ and a minimum value at $x = 2$.

S-30. Sketch a graph that is always increasing but is concave down from $x = 0$ to $x = 2$ and concave up after that.

S-31. Sketch a graph that has a minimum at $x = 2$ and points of inflection at $x = 1$ and $x = 3$.

S-32. Sketch a graph that is always increasing and concave down.

1.4 FUNCTIONS GIVEN BY WORDS

Many times mathematical functions are described verbally. In order for us to work effectively with such functions, it is essential that the verbal description be clearly understood. Sometimes this is enough, but more often it is necessary to supplement the verbal description with a formula, graph, or table of values.

 ## Comparing Formulas and Words

For example, there are initially 2000 bacteria in a petri dish. The bacteria reproduce by cell division, and each hour the number of bacteria doubles. This is a verbal description of a function $N = N(t)$, where N is the number of bacteria present at time t. It is common in situations like this to begin at time $t = 0$. Thus $N(0)$ is the number of bacteria we started with, 2000. One hour later the population doubles, so $N(1) = 4000$ bacteria. In one more hour the population doubles again, giving $N(2) = 8000$ bacteria. Continuing, we see that $N(3) = 16{,}000$ bacteria.

Note that although we can calculate $N(4)$ or $N(5)$, it is not so easy to figure out $N(4.5)$, the number of bacteria we have after 4 hours and 30 minutes. To make such a calculation, we need another description of the function, a formula. This situation is an example of *exponential growth*, which we will study in more detail in Chapter 4; there we will learn how to find formulas to match verbal descriptions of this kind. For now, we only want to make a comparison to indicate that the verbal description matches the formula $N = 2000 \times 2^t$ without showing where the formula came from. That is, we want to show that the formula gives the same answers as the verbal description. You can verify the following calculations by hand or by using your calculator:

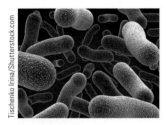

$$N(0) = 2000 \times 2^0 = 2000 \times 1 = 2000$$

$$N(1) = 2000 \times 2^1 = 2000 \times 2 = 4000$$

$$N(2) = 2000 \times 2^2 = 2000 \times 4 = 8000$$

$$N(3) = 2000 \times 2^3 = 2000 \times 8 = 16{,}000.$$

Checking several values can lend evidence that a formula is correct.

Note that these are the same values we got using the function's verbal description. You may wish to do further calculations to satisfy yourself that the formula agrees with the verbal description. This formula enables us to calculate the number of bacteria present after 4 hours and 30 minutes. Since that is 4.5 hours after the experiment began, we use $t = 4.5$:

$$N(4.5) = 2000 \times 2^{4.5} = 45{,}255 \text{ bacteria.}$$

We have rounded here to the nearest whole number since we don't expect to see fractional parts of bacteria.

We should point out that comparisons via the few calculations we made cannot establish for a certainty that the formula and the verbal description match. In fact, such a certain conclusion could not be drawn no matter how many individual calculations we made. But such calculations can establish patterns and provide partial evidence of agreement, and that is what we are after here.

EXAMPLE 1.9   **PURIFYING WATER**

Water that is initially contaminated with a concentration of 9 milligrams of pollutant per liter of water is subjected to a cleaning process. The cleaning process is able to reduce

the pollutant concentration by 25% each hour. Let $C = C(t)$ denote the concentration, in milligrams per liter, of pollutant in the water t hours after the purification process begins.

Part 1 What is the concentration of pollutant in the water after 3 hours?

Part 2 Compare the results predicted by the verbal description at the end of each of the first 3 hours with those given by the formula $C = 9 \times 0.75^t$.

Part 3 The formula given in part 2 is in fact the correct one. Use it to find the concentration of pollutant after $4\frac{1}{4}$ hours of cleaning.

Solution to Part 1 There is initially 9 milligrams per liter of pollutant in the water. One hour later, this has been reduced by 25%. That is a reduction of $0.25 \times 9 = 2.25$ milligrams per liter, leaving $9 - 2.25 = 6.75$ milligrams per liter in the water. During the second hour, the level is reduced by 25% again, leaving $6.75 - 0.25 \times 6.75 = 5.0625$ milligrams per liter. After one more hour there will be $5.0625 - 0.25 \times 5.0625 = 3.80$ milligrams per liter (rounded to two decimal places).

Solution to Part 2 We calculated the appropriate concentrations in part 1. Use your calculator to verify the following and check part 1 for agreement:

$$9 \times 0.75^1 = 6.75$$

$$9 \times 0.75^2 = 5.0625$$

$$9 \times 0.75^3 = 3.80 \text{ (rounded to two decimal places).}$$

Solution to Part 3 To get the concentration after $4\frac{1}{4}$ hours, we put 4.25 into the formula 9×0.75^t:

$$9 \times 0.75^{4.25} = 2.65 \text{ milligrams per liter.}$$

TEST YOUR UNDERSTANDING | **FOR EXAMPLE 1.9**

Suppose we start with 9 milligrams of pollutant per liter of water, as in the example. Assume that the cleaning process decreases the concentration of pollutant by 30% each hour. What is the pollutant concentration after each of the first 3 hours? ■

EXAMPLE 1.10 **A RISKY STOCK MARKET INVESTMENT**

An entrepreneur invested $2000 in a risky stock. Unfortunately, over the next year the value of the stock decreased by 7% each month. Let $V = V(t)$ denote the value in dollars of the investment t months after the stock purchase was made.

Part 1 Make a table of values showing the initial value of the stock and its value at the end of each of the first 5 months.

Part 2 Make a graph of investment value versus time.

Part 3 Verify that the formula $V = 2000 \times 0.93^t$ gives the same values you found in your table in part 1.

Part 4 The formula given in part 3 is in fact a valid description of the function V. What is the value of the investment after $9\frac{1}{2}$ months?

A graph can show what happens in the long run.

Part 5 If the trend for this stock continues—that is, if it continues to decrease in value by 7% each month—what will happen to the value in the long run?

Solution to Part 1 We use the verbal description to find the needed function values. The original investment was 2000, so $V(0) = 2000$. During the first month, the investment lost 7% of its value. That is, its value decreased by $0.07 \times 2000 = 140$ dollars. Thus $V(1) = 2000 - 140 = 1860$ dollars. During the second month, the value again decreased by 7%—that is, by $0.07 \times 1860 = 130.20$ dollars. Thus $V(2) = 1860 - 130.20 = 1729.80$ dollars.

Continuing the calculation in this manner yields

$$V(3) = \$1608.71$$

$$V(4) = \$1496.10$$

$$V(5) = \$1391.38.$$

Arranging these in a table of values, we get

t = Months	0	1	2	3	4	5
$V(t)$ = Investment value	$2000	$1860	$1729.80	$1608.71	$1496.10	$1391.38

Solution to Part 2 Since we are graphing investment value versus time, we use time for the horizontal axis and investment value for the vertical axis. The first step is to locate the data points from the table as in Figure 1.50. Next we join the dots as in Figure 1.51.

Solution to Part 3 We use $t = 0, 1, 2, 3, 4, 5$ to make the calculations

$$V(0) = 2000 \times 0.93^0 = 2000 \times 1 = 2000$$

$$V(1) = 2000 \times 0.93^1 = 2000 \times 0.93 = 1860$$

$$V(2) = 2000 \times 0.93^2 = 1729.80$$

$$V(3) = 2000 \times 0.93^3 = 1608.71$$

$$V(4) = 2000 \times 0.93^4 = 1496.10$$

$$V(5) = 2000 \times 0.93^5 = 1391.38.$$

We note that these calculations agree with the table of values we calculated in part 1 using the verbal description of the function. This is evidence that $V = 2000 \times 0.93^t$ gives a correct formula for this function.

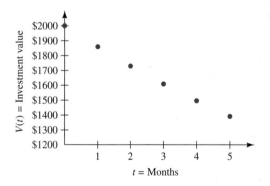

FIGURE 1.50 Plotting points

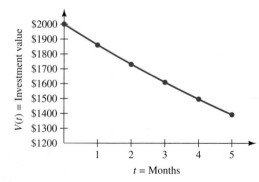

FIGURE 1.51 Completing the graph

Solution to Part 4 We use the formula from part 3 with $t = 9.5$:

$$V(9.5) = 2000 \times 0.93^{9.5} = 1003.73 \text{ dollars.}$$

Solution to Part 5 The value of the investment decreases by 7% each month. By calculating $V(t)$ for some large values of t using the formula from part 3, we can see whether a trend emerges. Here we use $t = 10$, $t = 50$, and $t = 100$ months:

$$V(10) = 2000 \times 0.93^{10} = 967.96 \, \text{dollars}$$

$$V(50) = 2000 \times 0.93^{50} = 53.11 \, \text{dollars}$$

$$V(100) = 2000 \times 0.93^{100} = 1.41 \, \text{dollars}.$$

In the long run, the investment will have virtually no value at all.

> **TEST YOUR UNDERSTANDING** | **FOR EXAMPLE 1.10**
>
> An initial investment of $2000 increases in value by 7% each month. Make a table of values that shows the value of the investment at the end of each of the first 5 months. ■

Getting Formulas from Words

Many times verbal descriptions can be directly translated into formulas for functions. Suppose, for example, that an engineering firm invests $78,000 in the design and development of a more efficient computer hard disk drive. For each disk drive sold, the engineering firm makes a profit of $98. We want to get a formula that shows the company's net profit as a function of the number of disk drives sold, taking into account the initial investment. This is not a difficult problem, and you may find that you can do it without further help, but we want to show a general method that may be useful for more complicated descriptions.

Step 1 *Identify the function and the things on which it depends, and write the relationships you know in a formula using words.* The function we are interested in is net profit. That is the profit on disk drives sold minus money spent on initial development. The profit on drives sold depends in turn on the number of sales:

$$\text{Net profit} = \text{Profit from sales} - \text{Initial investment} \tag{1.1}$$

$$= \text{Profit per item} \times \text{Number sold} - \text{Initial investment.} \tag{1.2}$$

Step 2 *Select and record letter names for the function and for each of the variables involved, and state their units.* We can use any letters we want as long as we identify them and state clearly what they mean. To say that we want to express net profit as a function of the number of disk drives sold means that net profit is the function and the number of disk drives is the variable. We will use N to denote our function, the net profit in dollars, and d to represent the variable, the number of disk drives sold:

$$N = \text{Net profit in dollars}$$

$$d = \text{Number of drives sold.}$$

Step 3 *Replace the words in step 1 by the letters identified in step 2 and appropriate information from the verbal description.* Here we simply replace the words *Net profit* in Equation (1.2) by N, *Profit per item* by 98, *Number sold* by d, and *Initial investment* by 78,000:

Net profit	=	Profit per item	×	Number sold	−	Initial investment
N		98		d		78,000

Thus we get the formula $N = 98d - 78{,}000$, where d is the number of disk drives sold and N is the net profit in dollars. We can use this formula to provide information about the engineering firm's disk drive project. For example, in functional notation, $N(1200)$ represents the net profit in dollars if 1200 disk drives are sold. We can calculate its value by replacing d with 1200: $N(1200) = 98 \times 1200 - 78{,}000 = 39{,}600$ dollars.

It is important to state the meaning of letters that you use as variables.

> **KEY IDEA 1.5   GOING FROM WORDS TO FORMULAS**
>
> To get a formula from a verbal description, first identify the function and the variables, and then write the relationships you know in a formula using words. Next choose letters to stand for the function and variables, and be sure to state their units. Then replace the words in the first step by the letters chosen and other quantities given in the verbal description.

EXAMPLE 1.11   CUTTING A DIAMOND

Arsgera/Shutterstock.com

The total investment a jeweler has in a gem-quality diamond is the price paid for the rough stone plus the amount paid to work the stone. Suppose the gem cutter earns $40 per hour.

Part 1 Choose variable and function names, and give a formula for a function that shows the total investment in terms of the cost of the stone and the number of hours required to work the stone.

Part 2 Use functional notation to express the jeweler's investment in a stone that costs $320 and requires 5 hours and 15 minutes of labor by the gem cutter.

Part 3 Use the formula you made in part 1 to calculate the value from part 2.

Solution to Part 1 To say that we want the total investment as a function of the cost of the stone and the hours of labor means that *total investment* is our function and that *cost of stone* and *hours of labor* are our variables. The total investment is the cost of the stone plus the cost of labor, which in turn depends on the number of hours needed to work the stone. We write out the relationship using a formula with words. We will accomplish this in two steps:

$$\text{Investment} = \text{Cost of stone} + \text{Cost of labor} \tag{1.3}$$

$$= \text{Cost of stone} + \text{Hourly wage} \times \text{Hours of labor.} \tag{1.4}$$

The next step is to choose letters to represent the function and the variables. We let c be the cost in dollars of the rough stone and h the number of hours of labor required. (It is fine to use any letters you like as long as you say in words what they represent.) We let $I = I(c, h)$ be the total investment in dollars in the diamond. To get the formula we want, we use Equation (1.4), replacing *Investment* by I, *Cost of stone* by c, *Hourly wage* by 40, and *Hours of labor* by h:

Investment		Cost of stone		Hourly wage		Hours of labor
I	$=$	c	$+$	40	$\times$	h

The result is $I = c + 40h$, where I is the total investment in dollars, c is the cost in dollars of the rough stone, and h is the number of hours required to work the stone.

Note that, in our final presentation of the formula, we were careful to identify the meaning of each letter, noting the appropriate units. Such descriptions are as important as the formula. Also, in this case you may find it easy enough to bypass much of what we presented here and go directly to the final answer. You are very much encouraged, however, to put in the intermediate step using words, just as it appears here. You will encounter many *word problems*, and this practice will make things much easier for you as the course progresses.

Solution to Part 2 The cost of the rough stone is $320, so we use $c = 320$. To get the value of h, we need to remember that h is measured in hours. Because 5 hours and 15 minutes is $5\frac{1}{4}$ hours, $h = 5.25$ hours, and that is what we will put in for h. Thus in functional notation, the total investment is $I(320, 5.25)$.

Solution to Part 3 In the formula we put 320 in place of c and 5.25 in place of h:

$$I(320, 5.25) = 320 + 40 \times 5.25 = 530 \text{ dollars.}$$

TEST YOUR UNDERSTANDING | **FOR EXAMPLE 1.11**

An automobile costs A dollars. You spend $45 per month on insurance and $70 per month on gasoline. Give a formula for the total cost C, in dollars, of buying the car and operating it for m months. Use your formula to find the cost of buying a $12,000 car and operating it for 36 months. ■

 Proportion

It is common in many applications of mathematics to use the term *proportional* to indicate that one thing is a multiple of another. For example, the total salary of an hourly wage earner can be calculated using

$$\text{Salary} = \text{Hourly wage} \times \text{Hours worked.}$$

Another way of expressing this is to say that salary is *proportional to*[32] the number of hours worked. If we use S to denote the salary, w for the hourly wage, and h for the hours worked, then the formula is

$$S = wh.$$

Many texts use the special symbol $\propto$ to denote a proportionality relation. Thus $S \propto h$ is shorthand notation for the phrase "S is proportional to h." In this context, the hourly wage w would be termed the *constant of proportionality*. It is what we need to multiply by in order to change the proportionality relation into an equation. The statement that S is proportional to h usually carries with it the implicit understanding that h is the variable but that the constant of proportionality w does not change. Thus we would be discussing a worker with a fixed hourly wage but whose working hours might differ from pay period to pay period.

Since the salary S is a multiple of the hourly wage w, it is also correct to say that S is proportional to w or, in symbols, that $S \propto w$. Looking at it this way, we are thinking of w as the variable and of h (the number of hours worked) as the constant of proportionality. Implicit in this description is that h does not change but w does. We might, for example, be discussing the long-term salary of someone who always works 40 hours each week but is anticipating raises.

When a function is given verbally, it can sometimes be a challenge to find a formula that represents it. But when such verbal descriptions are in terms of a proportion, it is always an easy task to get the formula. This can be done even without understanding the

---

[32]Sometimes the phrase *directly proportional to* is used.

meaning of the letters involved. For example, if you are told that a quantity denoted by the letter T is proportional to another quantity denoted by z, then you know that $T \propto z$ and that there is a constant of proportionality k that makes the equation $T = kz$ true. Determining the value of k and making sense of the equation do, of course, depend on an understanding of what the letters mean.

EXAMPLE 1.12   TOTAL COST

You are paying for a group of people to visit the zoo. Your total cost is proportional to the number of people in the group. Let c denote your total cost in dollars and n the number of people in the group. Assume that the price of admission is $7.50 per person.

Part 1 What is your total cost if there are 12 people in the group?

Part 2 Write an equation that shows the proportionality relation. What is the constant of proportionality? Explain in practical terms what it represents.

Part 3 Use the equation you found in part 2 to calculate your total cost if there are 15 people in the group.

Solution to Part 1 Since the price of admission is $7.50 per person and there are 12 people in the group, the total cost is $7.50 \times 12 = 90$ dollars. The purpose of this part of the example is to illustrate that to get the total cost, we multiply the admission price of $7.50 by the number of people in the group.

Proportional means *is a multiple of.*

Solution to Part 2 The total cost c is proportional to the number n of people in the group. That is, $c \propto n$. We noted in part 1 that to get the total cost we multiply $7.50 by the number of people in the group. Written as an equation, this is $c = 7.50n$. Thus the constant of proportionality is 7.50 dollars per person. In practical terms it is the admission price per person. Note that the units fit together here: The total cost in dollars is the product of the admission price in dollars per person with the number of people. It is a good idea to check the units in proportionality relations like this.

Solution to Part 3 We use the formula we got in part 2, putting in 15 for n:

$$c = 7.50n = 7.50 \times 15 = 112.50 \, \text{dollars}.$$

The total cost is $112.50.

TEST YOUR UNDERSTANDING | FOR EXAMPLE 1.12

Assume that the number of pages of a book is proportional to the number of words. If each page contains 200 words, express the number p of pages in terms of the number w of words.   ■

ANSWERS FOR TEST YOUR UNDERSTANDING

1.9 After 1 hour: 6.30 milligrams per liter
   After 2 hours: 4.41 milligrams per liter
   After 3 hours: 3.09 milligrams per liter

1.10

Months	1	2	3	4	5
Investment value	$2140.00	$2289.80	$2450.09	$2621.60	$2805.11

(*Note:* The appropriate formula will give slightly different values for months 4 and 5 than those in the table.)

1.11 $C = A + 45m + 70m = A + 115m$. The cost is $12,000 + 115 \times 36 = \$16,140$.

1.12 $p = w/200$.

1.4 EXERCISES

Reminder Round all answers to two decimal places unless otherwise indicated.

1. **Fat in Fast Food** You are shopping for dinner for your family in a fast-food restaurant but are concerned about fat. An informative menu gives you the following information.

Item	Hamburger	Chicken sandwich	Fries	Onion rings
Grams of fat	24	13	30	25

   a. How much total fat is in an order of 3 hamburgers, 1 chicken sandwich, 2 orders of fries, and 2 orders of onion rings?

   b. Use a formula to express the total grams of fat F in an order of h hamburgers, c chicken sandwiches, f orders of fries, and o orders of onion rings.

2. **Thanksgiving Dinner** It takes 10 minutes to preheat your oven to 325 degrees. For an oven preheated to 325 degrees, the recommended cooking time for a turkey is about 15 minutes per pound.[33]

   a. Use a formula to express the total time T, in minutes, needed to preheat the oven and then bake a turkey weighing p pounds.

   b. Use your formula from part a to find the approximate time required to prepare a turkey weighing 18 pounds.

3. **United States Population Growth** In 1960 the population of the United States was about 180 million. Since that time the population has increased by approximately 1.2% each year. This is a verbal description of the function $N = N(t)$, where N is the population, in millions, and t is the number of years since 1960.

   a. Express in functional notation the population of the United States in 1963. Calculate its value.

   b. Use the verbal description of N to make a table of values that shows U.S. population in millions from 1960 through 1965.

   c. Make a graph of U.S. population versus time. Be sure to label your graph appropriately.

   d. Verify that the formula 180×1.012^t million people, where t is the number of years since 1960, gives the same values as those you found in the table in part b. (*Note:* Because t is the

number of years since 1960, you would use $t = 2$ to get the population in 1962.)

   e. Assuming that the population has been growing at the same percentage rate since 1960, what value does the formula above give for the population in 2000? (*Note:* The actual population in 2000 was about 281 million.)

4. **Education and Income** According to the U.S. Census Bureau, in 2008 the median (middle of the range) annual income of a high school graduate with no further education was about $40,000 per year. If we assume that it takes 4 years to earn a bachelor's degree and 2 additional years to earn a master's degree, the median annual income increases by 15% for each year spent toward successful completion of a degree. This describes a function $I = I(y)$, where I is the median income in 2008 for a person with y years of college education.

   a. Express in functional notation the median income in 2008 of an individual with an associate degree, which requires 2 years of college. Calculate that value.

   b. Make a table of values that shows the median income in 2008 for individuals completing 0 through 4 years of college.

   c. Make a graph of median income versus years of college completed. Be sure to provide appropriate labels.

   d. Verify that the formula $I = 40 \times 1.15^y$ thousand dollars, where y is the number of years of college completed, gives the same values as those you found in the table you made.

   e. Using the formula given in part d, find the median income in 2008 of an individual who has a master's degree.

   f. Assuming it takes 3 years beyond the master's degree to complete a Ph.D., and assuming the formula in part d applies, what was the median income of a Ph.D. in 2008?

   g. In fact, the same rate of increase does *not* apply for the years spent on a Ph.D. The actual median income for a Ph.D. in 2008 was $100,000. Does that mean that the increase in median income for years spent earning a Ph.D. is higher or lower than 15% per year?

---

[33]Cooking times are approximate. A food thermometer must be used to insure the turkey is properly cooked.

(continued)

5. **Altitude** A helicopter takes off from the roof of a building that is 200 feet above the ground. The altitude of the helicopter increases by 150 feet each minute.

   a. Use a formula to express the altitude of a helicopter as a function of time. Be sure to explain the meaning of the letters you choose and the units.

   b. Express using functional notation the altitude of the helicopter 90 seconds after takeoff, and then calculate that value.

   c. Make a graph of altitude versus time covering the first 3 minutes of the flight. Explain how the description of the function is reflected in the shape of the graph.

6. **Swimming Records** The world record time for a certain swimming event was 63.2 seconds in 1950. Each year thereafter, the world record time decreased by 0.4 second.

   a. Use a formula to express the world record time as a function of the time since 1950. Be sure to explain the meaning of the letters you choose and the units.

   b. Express using functional notation the world record time in the year 1955, and then calculate that value.

   c. Would you expect the formula to be valid indefinitely? Be sure to explain your answer.

7. **A Rental** A rental car agency charges $49.00 per day and 25 cents per mile.

   a. Calculate the rental charge if you rent a car for 2 days and drive 100 miles.

   b. Use a formula to express the cost of renting a car as a function of the number of days you keep it and the number of miles you drive. Identify the function and each variable you use, and state the units.

   c. It is about 250 miles from Dallas to Austin. Use functional notation to express the cost to rent a car in Dallas, drive it to Austin, and return it in Dallas 1 week later. Use the formula from part b to calculate the cost.

8. **Preparing a Letter** You pay your secretary $9.25 per hour. A stamped envelope costs 50 cents, and paper costs 3 cents per page.

   a. How much does it cost to prepare and mail a 3-page letter if your secretary spends 2 hours on typing and corrections?

   b. Use a formula to express the cost of preparing and mailing a letter as a function of the number of pages in the letter and the time it takes your secretary to type it. Identify the function and each of the variables you use, and state the units.

   c. Use the function you made in part b to find the cost of preparing and mailing a 2-page letter that it takes your secretary 25 minutes to type. (*Note:* 25 minutes is 25/60 hour.)

9. **Preparing a Letter, Continued** *This is a continuation of Exercise 8.* You pay your secretary $9.25 per hour. A stamped envelope costs 50 cents, and regular stationery costs 3 cents per page, but fancy letterhead stationery costs 16 cents per page. Assume that a letter requires fancy letterhead stationery for the first page but that regular paper will suffice for the rest of the letter.

   a. How much does the stationery alone cost for a 3-page letter?

   b. How much does it cost to prepare and mail a 3-page letter if your secretary spends 2 hours on typing and corrections?

   c. Use a formula to express the cost of the stationery alone for a letter as a function of the number of pages in the letter. Identify the function and each of the variables you use, and state the units.

   d. Use a formula to express the cost of preparing and mailing a letter as a function of the number of pages in the letter and the time it takes your secretary to type it. Identify the function and each of the variables you use, and state the units.

   e. Use the function you made in part d to find the cost of preparing and mailing a 2-page letter that it takes your secretary 25 minutes to type.

10. **A Car That Gets *m* Miles per Gallon** The cost of operating a car depends on the gas mileage m that your car gets, the cost g per gallon of gasoline, and the distance d that you drive.

a. How much does it cost to drive 100 miles if your car gets 25 miles per gallon and gasoline costs 349 cents per gallon?

Monkey Business Images/Shutterstock.com

b. Find a formula that gives the cost *C* as a function of *m*, *g*, and *d*. Be sure to state the units of each variable.

c. Use functional notation to show the cost of driving a car that gets 28 miles per gallon a distance of 138 miles if gasoline costs $3.69 per gallon. Use the formula from part b to calculate the cost.

11. Stock Turnover Rate   In a retail store the stock turnover rate of an item is the number of times that the average inventory of the item needs to be replaced as a result of sales in a given time period. It is an important measure of sales demand and merchandising efficiency. Suppose a retail clothing store maintains an average inventory of 50 shirts of a particular brand.

a. Suppose that the clothing store sells 350 shirts of that brand each year. How many orders of 50 shirts will be needed to replace the items sold?

b. What is the annual stock turnover rate for that brand of shirt if the store sells 350 shirts each year?

c. What would be the annual stock turnover rate if 500 shirts were sold?

d. Write a formula expressing the annual stock turnover rate as a function of the number of shirts sold. Identify the function and the variable, and state the units.

12. Stock Turnover Rate, Continued   *This is a continuation of Exercise 11.* As we saw earlier, the stock turnover rate of an item is the number of times that the average inventory of the item needs to be replaced as a result of sales in a given time period. Suppose that a hardware store sells 80 shovels each year.

a. Suppose that the hardware store maintains an average inventory of 5 shovels. What is the annual stock turnover rate for the shovels? How is this related to the yearly number of orders to the wholesaler needed to restock inventory?

b. What would be the annual stock turnover rate if the store maintained an average inventory of 20 shovels?

c. Write a formula expressing the annual stock turnover rate as a function of the average inventory of shovels. Identify the function and the variable, and state the units.

13. Total Cost   The *total cost C* for a manufacturer during a given time period is a function of the number *N* of items produced during that period. To determine a formula for the total cost, we need to know the manufacturer's *fixed costs* (covering things such as plant maintenance and insurance), as well as the cost for each unit produced, which is called the *variable cost*. To find the total cost, we multiply the variable cost by the number of items produced during that period and then add the fixed costs.

Suppose that a manufacturer of widgets has fixed costs of $9000 per month and that the variable cost is $15 per widget (so it costs $15 to produce 1 widget).

a. Use a formula to express the total cost *C* of this manufacturer in a month as a function of the number of widgets produced in a month. Be sure to state the units you use.

b. Express using functional notation the total cost if there are 250 widgets produced in a month, and then calculate that value.

14. Total Revenue and Profit   *This is a continuation of Exercise 13.* The *total revenue R* for a manufacturer during a given time period is a function of the number *N* of items produced during that period. To determine a formula for the total revenue, we need to know the selling price per unit of the item. To find the total revenue, we multiply this selling price by the number of items produced.

The *profit P* for a manufacturer is the total revenue minus the total cost. If this number is positive, then the manufacturer *turns a profit*, whereas if this number is negative, then the manufacturer *has a loss*. If the profit is zero, then the manufacturer is at a *break-even point*.

Suppose the manufacturer of widgets in Exercise 13 sells the widgets for $25 each.

a. Use a formula to express this manufacturer's total revenue *R* in a month as a function of the number of widgets produced in a month. Be sure to state the units you use.

(continued)

b. Use a formula to express the profit P of this manufacturer as a function of the number of widgets produced in a month. Be sure to state the units you use.

c. Express using functional notation the profit of this manufacturer if there are 250 widgets produced in a month, and then calculate that value.

d. At the production level of 250 widgets per month, does the manufacturer turn a profit or have a loss? What about at the production level of 1000 widgets per month?

15. More on Revenue *This is a continuation of Exercises 13 and 14.* In general, the highest price p per unit of an item at which a manufacturer can sell N items is not constant but is rather a function of N. The total revenue R is still the product of p and N, but the formula for R is more complicated when p depends on N.

Suppose the manufacturer of widgets in Exercises 13 and 14 no longer sells widgets for $25 each. Rather, the manufacturer has developed the following table showing the highest price p, in dollars, of a widget at which N widgets can be sold.

a. Verify that the formula $p = 50 - 0.01N$, where p is the price in dollars, gives the same values as those in the table.

N = Number of widgets sold	p = Price
100	49
200	48
300	47
400	46
500	45

b. Use the formula from part a and the fact that R is the product of p and N to find a formula expressing the total revenue R as a function of N for this widget manufacturer.

c. Express using functional notation the total revenue of this manufacturer if there are 450 widgets produced in a month, and then calculate that value.

16. More on Profit *This is a continuation of Exercises 13, 14, and 15.* In this exercise we use the formula for the total cost of the widget manufacturer found

in Exercise 13 and the formula for the total revenue found in Exercise 15.

a. Use a formula to express the profit P of this manufacturer as a function of N.

b. Consider the three production levels: $N = 200$, $N = 700$, and $N = 1200$. For each of these, determine whether the manufacturer has a loss, turns a profit, or is at a break-even point.

17. Renting Motel Rooms You own a motel with 30 rooms and have a pricing structure that encourages rentals of rooms in groups. One room rents for $85.00, two for $83.00 each, and in general the group rate per room is found by taking $2 off the base of $85 for each extra room rented.

a. How much money do you charge per room if a group rents 3 rooms? What is the total amount of money you take in?

b. Use a formula to give the rate you charge for each room if you rent n rooms to an organization.

c. Find a formula for a function $R = R(n)$ that gives the total revenue from renting n rooms to a convention host.

d. Use functional notation to show the total revenue from renting a block of 9 rooms to a group. Calculate that value.

18. A Cattle Pen A rancher wants to use a fence as an enclosure for a rectangular cattle pen with area 400 square feet.

a. Suppose he decides to make one side of the pen 40 feet long. Draw a picture and label the length of each side of the pen. What is the length of each side? What is the total amount of fence needed?

b. What would be the total amount of fence needed if the pen were a square?

c. Can two rectangles with the same area have different perimeters?

d. Find a formula for a function $F = F(l)$ that gives total amount of fence, in feet, required in terms of the length l, in feet, of one of its sides. (*Hint:* First draw a picture of the pen and label one side l. Next figure out the lengths of the other sides in terms of l.)

19. Catering a Dinner You are having a dinner catered. You pay a rental fee of $150 for the dining hall, and you pay the caterer $10 for each person who attends the dinner.

a. Suppose you just want to break even.

   i How much should you charge *per ticket* if you expect 50 people to attend?

   ii. Use a formula to express the amount you should charge per ticket as a function of the number of people attending. Be sure to explain the meaning of the letters you choose and the units.

   iii. You expect 65 people to attend the dinner. Use your answer to part ii to express in functional notation the amount you should charge per ticket, and then calculate that amount.

b. Suppose now that you want to make a profit of $100 from the dinner. Use a formula to express the amount you should charge per ticket as a function of the number of people attending. Again, be sure to explain the meaning of the letters you choose and the units.

20. A Car The distance d, in miles, that a car travels on a 3-hour trip is proportional to its speed s (which we assume remains the same throughout the trip), in miles per hour.

a. What is the constant of proportionality in this case?

b. Write a formula that expresses d as a function of s.

21. Production Rate The total number t of items that a manufacturing company can produce is directly proportional to the number n of employees.

a. Choose a letter to denote the constant of proportionality, and write an equation that shows the proportionality relation.

b. What in practical terms does the constant of proportionality represent in this case?

22. Density The total weight of a rock depends on its size and is proportional to its *density*. In this context, density is the weight per cubic inch. Let w denote the weight of the rock in pounds, s the size of the rock in cubic inches, and d the density of the rock in pounds per cubic inch.

a. What is the total weight of a 3-cubic-inch rock that weighs 2 pounds per cubic inch?

b. Write an equation that shows the proportionality relation. What is the constant of proportionality?

c. Use the equation you found in part b to find the total weight of a 14-cubic-inch rock with density 0.3 pound per cubic inch.

23. Head and Pressure Determining the water pressure at a given location employs the concept of the *head*, which is the vertical distance, in feet, from the surface of a source body of water to the location. The pressure exerted by water is proportional to the head. If we measure head in feet and pressure in pounds per square inch, then the constant of proportionality is the weight of a column of water that is 1 foot high and 1 inch square at the base. That much water weighs 0.434 pound. (See Figure 1.52.)

      12 ft
      12 ft
      12 ft
      12 ft

FIGURE 1.52

a. Write an equation that expresses the proportionality relationship between pressure p and head h.

b. For a pumper truck pumping water to a fire, the *back pressure* is the additional pressure on the pump caused by the height of the nozzle. Consider a pumper at street level pumping water through a hose to firefighters on the top of the eighth floor of a building. If each floor is 12 feet high, what is the head of water at the mouth of the nozzle? What is the back pressure on the pumper? (Another way of thinking of back pressure is as the minimum pressure the pumper must produce in order to make water flow out the end of the nozzle.)

c. Head (and therefore back pressure) depends only on the height of the nozzle above the pumper. It is affected neither by the volume of the water nor by horizontal distance. A pumper in a remote location is pumping water to firefighters on the far slope of a hill. At its peak, the hill is 185 feet higher than the pumper. The hose goes over the hill and then down the hill to a point 40 feet below the peak. Find the head and the back pressure on the pumper.

(continued)

24. Head and Aquifers *This is a continuation of Exercise 23.* In underground water supplies such as aquifers, the water normally permeates some other medium such as sand or gravel. The head for such water is determined by first drilling a well down to the water source. When the well reaches the aquifer, pressure causes the water to rise in the well. The head is the height to which the water rises. In this setting, we get the pressure using

$$\text{Pressure} = \text{Density} \times 9.8 \times \text{Head}.$$

Here density is in kilograms per cubic meter, head is in meters, and pressure is in newtons per square meter. (One newton is about a quarter of a pound.) A sandy layer of soil has been contaminated with a dangerous fluid at a density of 1050 kilograms per cubic meter. Below the sand there is a rock layer that contains water at a density of 990 kilograms per cubic meter. This aquifer feeds a city water supply. Test wells show that the head in the sand is 4.3 meters, whereas the head in the rock is 4.4 meters. A liquid will flow from higher pressure to lower pressure. Is there a danger that the city water supply will be polluted by the material in the sand layer?

25. Darcy's Law The French hydrologist Henri Darcy discovered that the velocity *V* of underground water is proportional to *S*, the magnitude of the slope of the water table (see Figure 1.53). The constant of proportionality in this case is the *permeability* of the medium through which the water is flowing. This proportionality relationship is known as *Darcy's law*, and it is important in modern hydrology.

Sandstone

FIGURE 1.53

a. Using *K* as the constant of proportionality, express Darcy's law as an equation.

b. Sandstone has a permeability of about 0.041 meter per day. If an underground aquifer is seeping through sandstone, and if *S* = 0.03 for the water table, what is the velocity of the water flow? Be sure to use appropriate units for velocity and keep all digits.

c. Sand has a permeability of about 41 meters per day. If the aquifer from part b were flowing through sand, what would be its velocity?

26. Hubble's Constant Astronomers believe that the universe is expanding and that stellar objects are moving away from us at a radial velocity *V* proportional to the distance *D* from Earth to the object.

a. Write *V* as a function of *D* using *H* as the constant of proportionality.

b. The equation in part a was first discovered by Edwin Hubble in 1929 and is known as *Hubble's law*. The constant of proportionality *H* is known as *Hubble's constant*. The currently accepted value of Hubble's constant is 70 kilometers per second per megaparsec. (One megaparsec is about 3.086×10^{19} kilometers.) With these units for *H*, the distance *D* is measured in megaparsecs, and the velocity *V* is measured in kilometers per second. The galaxy G2237 + 305 is about 122.7 megaparsecs from Earth. How fast is G2237 + 305 receding from Earth?

c. One important feature of Hubble's constant is that scientists use it to estimate the age of the universe. The approximate relation is

$$y = \frac{10^{12}}{H},$$

where *y* is time in years. Hubble's constant is extremely difficult to measure, and Edwin Hubble's best estimate in 1929 was about 530 kilometers per second per megaparsec. What is the approximate age of the universe when this value of *H* is used?

d. The calculation in part c would give scientists some concern since Earth is thought to be about 4.6 billion years old. What estimate of the age of the universe does the more modern value of 70 kilometers per second per megaparsec give?

27. Loan Origination Fee Lending institutions often charge fees for mortgages. One type of fee consists of *closing costs.* Closing costs often comprise a fixed fee for a confusing list of necessities associated with the loan. Additionally, some mortgages require a *loan origination fee* or *points,* which is a fixed percentage of the mortgage. At a certain institution the closing costs are $2500 and the points are 2% of the mortgage amount.

a. What are the fees for securing a mortgage of $322,000?

b. Use a formula to express the loan fees F, in dollars, associated with a mortgage of M dollars.

28. The 3x + 1 Problem Here is a mathematical function $f(n)$ that applies only to whole numbers n. If a number is even, divide it by 2. If it is odd, triple it and add 1. For example, 16 is even, so we divide by 2: $f(16) = 16/2 = 8$. On the other hand, 15 is odd, so we triple it and add 1: $f(15) = 3 \times 15 + 1 = 46$.

a. Apply the function f repeatedly beginning with $n = 1$. That is, calculate $f(1)$, f(the answer from the first part), f(the answer from the second part), and so on. What pattern do you see?

b. Apply the function f repeatedly beginning with $n = 5$. How many steps does it take to get to 1?

c. Apply the function f repeatedly beginning with $n = 7$. How many steps does it take to get to 1?

d. Try several other numbers of your own choosing. Does the process always take you back to 1? (*Note:* We can't be sure what your answer will be here. Every number that anyone has tried so far leads eventually back to 1, and it is conjectured that this happens no matter what number you start with. This is known to mathematicians as the $3x + 1$ *conjecture,* and it is, as of the writing of this book, an unsolved problem. If you can find a starting number that does not lead back to 1, or if you can somehow show that the path *always* leads back to 1, you will have solved a problem that has eluded mathematicians for a number of years. Good hunting!)

29. Research Project For this project you are to find and describe a function that is commonly used.

Find a patient person whose job is interesting to you. Ask that person what types of calculations he or she makes. These calculations could range from how many bricks to order for building a wall to lifetime wages lost for a wrongful-injury settlement to how much insulin to inject. Be creative and persistent—don't settle for "I look it up in a table." Write a description, in words, of the function and how it is calculated. Then write a formula for the function, carefully identifying variables and units.

1.4 SKILL BUILDING EXERCISES

S-1. A Description You have $5000 in a cookie jar. Each month you spend half of the balance. How much do you have after 4 months?

S-2. Light It is 93,000,000 miles from the Earth to the sun. Light travels 186,000 miles per second. How long does it take light to travel from the sun to the Earth?

S-3. A Description The initial value of a function $f = f(x)$ is 5. That is, $f(0) = 5$. Each time x is increased by 1, the value of f triples. What is the value of $f(4)$? Verify that the formula $f(x) = 5 \times 3^x$ gives the same answer.

Getting a Formula In Exercises S-4 through S-13, a verbal description of a function is given. Use a formula to express the function.

S-4. You sell lemonade for 25 cents per glass. You invested $2.00 in the ingredients. Write a formula that gives the profit $P = P(n)$ as a function of the number n of glasses you sell.

S-5. Each time a certain balding gentleman showers, he loses 67 strands of hair down the shower drain. Write a formula that gives the total number $N = N(s)$ of hairs lost by this man after s showers.

S-6. You pay $500 to rent a special area in a restaurant. In addition, you pay $10 for each guest. Write a formula that gives your total cost C, in dollars, as a function of the number n of dinner guests.

S-7. You currently have $500 in a piggy bank. You add $37 to the bank each month. Find a formula that gives the balance B, in dollars, in the piggy bank after t months.

S-8. An object is removed from a hot oven and left to cool. After t minutes, the difference between

(continued)

the temperature of the object and room temperature, 75 degrees, is 325×0.07^t degrees. Find a formula for the temperature T, in degrees, of the object t minutes after it is removed from the oven.

S-9. You paid $249 for your iPod. You download songs to your iPod for 99 cents each. Find a formula that gives the cost C, in dollars, of purchasing the iPod and downloading s songs.

S-10. Your new 60-inch plasma TV cost $2700, and you bought a new DVD player for $622. The accompanying surround-sound system cost $850. You paid $322 for a nice stand to accommodate the entire business. Since getting the nice home video system, you subscribed to a cable TV package, which includes many high-definition channels and costs $122 per month. Write a formula that gives the cost C, in dollars, of purchasing your home video system and paying for m months of cable service.

S-11. For office supplies you buy s boxes of staples, r reams of paper, and p boxes of ink pens. Staples cost $1.46 per box, paper costs $3.50 per ream, and ink pens cost $2.40 per box. Write a formula that expresses the cost C, in dollars, of all the office supplies as a function of s, r, and p.

S-12. With no advertising your newspaper has a circulation of 8000 papers. For each dollar you spend on advertising, your newspaper circulation increases by 5 papers. You earn a profit (taking into account advertising expense) of 7 cents for each paper sold, but you must pay $400 to have the papers delivered. Find a formula that gives the total profit P, in dollars, that you earn if you spend a dollars on advertising.

S-13. You pay $300 an ounce for high-quality silver for making jewelry. You pay an average of $21 per ounce for turquoise. You sell your completed jewelry for $500 per ounce. Find the net profit P, in dollars, you make using s ounces of silver and t ounces of turquoise.

S-14. Proportionality For a certain function $f = f(x)$, we know that f is proportional to x and that the constant of proportionality is 8. Find a formula for f.

S-15. Constant of Proportionality If $g(t) = 16t$, then g is proportional to t. What is the constant of proportionality?

S-16. Proportionality and Initial Value The function y is proportional to x. What is the value of y when x is 0?

Is It Proportional? In Exercises S-17 through S-25 determine whether or not the relationship described is a proportionality relationship.

S-17. Pizza Is the weight of a cheese pizza of a fixed diameter proportional to its thickness?

S-18. More Pizza Is the weight of a pizza proportional to its diameter?

S-19. A Tire Is the circumference of a (properly inflated) tire proportional to its radius?

S-20. Snowfall If there is no snow on the ground and then it begins to snow at a constant rate, is the amount of snowfall proportional to the time since it started snowing?

S-21. A Square Box Is the volume of a cube proportional to the length of a side?

S-22. A Man's Height Is a man's height proportional to his age?

S-23. Wages A man makes $16 per hour. Is his monthly salary proportional to the number of hours he works?

S-24. Sales Tax Is the amount of sales tax you owe on a given purchase proportional to the purchase price?

S-25. Sodas Is the cost of buying sodas (at a fixed price per soda) proportional to the number of sodas bought?

CHAPTER 1 SUMMARY

The idea of a *function* is as old as mathematics itself, and it is central to mathematics and applications. It is certainly a key topic in this text, where it occurs in one form or another throughout. A function is nothing more than a clear description of how one thing depends on another (or on several other things), and functions are presented in various ways. The most common ways of presenting functions are the section topics of this chapter.

1.1 FUNCTIONS GIVEN BY FORMULAS

This is perhaps the way in which most people think of a function, and it is an important one. An example of a function given by a formula is

$$M = 9h,$$

where h is the number of hours an employee may work, and M is the money earned, in dollars. The formula simply says that one can calculate the money earned by multiplying the number of hours worked by 9. In other words, the formula describes the pay of an employee who earns $9.00 an hour. Sometimes formulas for functions are quite complicated. The calculator makes functions given by formulas easy to deal with in spite of their apparent complexity.

In applications of mathematics, functions are often representations of real phenomena or events. Thus we say that they are *models*. Obtaining a function or functions to act as a model is commonly the key to understanding physical, natural, and social science phenomena. This applies to business and many other areas as well.

1.2 FUNCTIONS GIVEN BY TABLES

One of the most common ways in which functions are encountered in daily life is in terms of *tables of values*. Such tables can be found everywhere, presenting census data, payment schedules, college enrollment statistics, lists of species occupying a certain region, and a myriad of other familiar types of data. An example is the U.S. census data that appears in the following table.

d = Year	1970	1980	1990	2000	2010
N = Population in millions	203.30	226.54	248.71	281.42	308.75

This table gives U.S. population $N = N(d)$ as a function of the date. Tables are almost always incomplete; that is, some information is left out of the table. Here, for example, the population in 1994 is not reported. A common way of estimating function values that are not given in a table is by using the *average rate of change*. For example, from 1990 to 2000, the U.S. population grew from 248.71 million to 281.42 million. That is an increase of 32.71 million over a 10-year period. Thus from 1990 to 2000, the U.S. population grew at a rate of approximately

$$\frac{32.71}{10} = 3.271 \text{ million per year.}$$

This is the average yearly rate of change during the 1990s. It is reasonable to estimate the population in 1994 by

$$N(1994) = 248.71 + 4 \times 3.271 = 261.794 \text{ million,}$$

or about 261.79 million.

Sometimes, a function has a *limiting value* that may be estimated from the tabular form of a function. For example, the following table shows the amount of yeast present in an enclosed area t hours after observations began.

Time t	0	5	10	15	20	25	30
Amount of yeast N	10	119	513	651	662	664	665

Since an enclosed area is being observed, we expect that there is a limit to the amount of yeast that will ever be present. Looking at the table, it is reasonable to expect that this limiting value is about 665.

1.3 FUNCTIONS GIVEN BY GRAPHS

Another way in which functions are commonly presented is with graphs. Generally, it is more difficult to get exact function values from a graph than from a formula or table, but a graph has the advantage of clearly showing certain overall features of a function. It clearly shows, for example, when a function is increasing or decreasing, when it reaches maxima or minima, any concavity of the graph, and (often) limiting values of the function. These function properties are displayed by the graph according to the following table.

Function property	**Graph display**
Increasing	Rising graph
Decreasing	Falling graph
Maximum	Graph reaches a peak
Minimum	Graph reaches a valley
Concave up	Graph is bent upward (holds water)
Concave down	Graph is bent downward (spills water)
Has a limiting value	Graph levels out on the right-hand side

The concavity of a graph gives important information about the rate of change of the function. Increasing graphs that are concave up represent functions that increase at an increasing rate, whereas increasing graphs that are concave down represent functions that increase at a decreasing rate. Similarly, decreasing graphs that are concave up represent functions that decrease at a decreasing rate, whereas decreasing graphs that are concave down represent functions that decrease at an increasing rate.

1.4 FUNCTIONS GIVEN BY WORDS

Very often a function is presented with a verbal description, and the key to understanding it may well be to translate this verbal description into a formula, table, or graph. For example, suppose that a company invests $78,000 in the design and development of a more efficient computer hard drive and that, for each drive sold, the firm makes a profit of $98. This can be thought of as a verbal description of the net profit $N = N(d)$ as a function of the number d of drives sold. Furthermore, it is not difficult to translate this verbal description into a formula:

$$\text{Net profit} = \text{Profit from sales} - \text{Initial investment}$$

$$N = 98d - 78{,}000 \text{ dollars.}$$

One way in which verbal descriptions of functions are commonly given is in terms of *proportion*. This simply indicates that one thing is a multiple of another. For example, money earned by a wage employee is proportional to the number of hours worked. In terms of a formula, this proportionality statement means

$$\text{Money earned} = \text{Hourly wage} \times \text{Hours worked.}$$

In this context, the hourly wage is known as the *proportionality constant*.

CHAPTER 1 REVIEW EXERCISES

Reminder Round all answers to two decimal places unless otherwise indicated.

1. **Evaluating Formulas** If
$$M(P, r, t) = \frac{Pr(1 + r)^t}{(1 + r)^t - 1},$$
calculate $M(9500, 0.01, 24)$.

2. **U.S. Population** The population of the United States from 1790 to 1860 is given by the formula $N = 3.93 \times 1.03^t$, where t is years since 1790 and N is the population, in millions.

   a. What was the population in 1790?

   b. Express the population in 1810 using functional notation.

   c. Calculate the value of the population in 1810.

3. **Averages and Average Rate of Change** The following is a partial table of values for $f = f(x)$.

x	0	2	4	6
$f = f(x)$	32.3	36.0	40.1	43.7

   a. Estimate the value of $f(5)$ by averaging.

   b. Find the average rate of change for f between $x = 4$ and $x = 6$.

4. **High School Graduates** The following table shows the number, in millions, graduating from high school in the United States, in the given year.

t = Year	N = Number graduating
1985	2.83
1987	2.65
1989	2.47
1991	2.29

a. Explain in practical terms what $N(1989)$ means, and find its value.

b. Use functional notation to express the number of graduates in 1988, and estimate its value.

c. Find the average rate of change per year during the period 1989 to 1991.

d. Estimate the value of $N(1994)$.

5. **Increasing, Decreasing, and Concavity** The graph in Figure 1.54 shows a population $N = N(d)$, where d is the date.

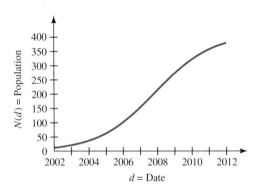

FIGURE 1.54

a. Is the function increasing, decreasing, or neither for d from 2002 to 2012?

b. When is the function concave down, and when is it concave up?

c. At what date is there an inflection point?

6. **Logistic Population Growth** The graph in Figure 1.54 shows a population as a function of the date d.

a. Explain in general terms how the population N changes with time.

b. When does the population reach 300?

c. When is the population increasing most rapidly?

d. What is the name of the point of most rapid population increase?

7. **Getting a Formula** You currently have $780 in your bank account, which pays no interest. You withdraw $39 each week. Find a formula for the balance B, in dollars, in the account after t weeks.

8. **Cell Phone Charges** One cell phone plan charges a flat monthly rate of $39.95 with extra charges of $0.10 per text message after the first 100 text messages.

(continued)

a. Choose letters to represent the variables.

b. Write a formula to express the cell phone charges as a function of the number of text messages (assume that the number is at least 100).

c. Use functional notation to show the cost of the cell phone if you have 450 text messages. Use the formula from part b to calculate the cost.

d. Write a formula to express the cell phone charges, this time assuming that the number of text messages is less than 100.

9. Cell Phone Charges Again One cell phone plan charges a flat monthly rate of $34.95 with extra charges of $0.35 per minute for each minute after the first 4000 minutes and $0.10 per text message after the first 100 text messages.

a. Choose letters to represent the variables.

b. Write a formula to express the cell phone charges as a function of the number of minutes used (assume that the number is at least 4000) and the number of text messages (assume that the number is at least 100).

c. What are your cell phone charges if you use 6000 minutes and 450 text messages?

d. Write a formula to express the cell phone charges, this time assuming that the minutes are at least 4000, but the number of text messages is less than 100.

e. What are your cell phone charges if you use 4200 minutes and 88 text messages?

10. Practicing Calculations For each of the following functions $C = C(t)$, find the value of $C(0)$, reporting the answer rounded to two decimal places. Use your calculator where it is appropriate.

a. $C = 0.2 + 2.77e^{-0.37t}$

b. $C = \dfrac{12.36}{0.03 + 0.55^t}$

c. $C = \dfrac{t - 1}{\sqrt{t + 1}}$

d. $C = 5 \times 0.5^{t/5730}$

11. Amortization If you borrow P dollars at a monthly interest rate of r (as a decimal) and wish to pay off the loan in t months, then the monthly payment $M = M(P, r, t)$ can be calculated using

$$M = \frac{Pr(1 + r)^t}{(1 + r)^t - 1},$$

in dollars.

a. Calculate $M(5500, 0.01, 24)$ and explain in practical terms what your answer means.

b. Express the monthly payment for a loan of $8000 at a monthly rate of 0.6% paid over 36 months in functional notation, and then calculate its value.

12. Using Average Rate of Change The following is a partial table of values for $f = f(x)$.

x	0	3	6
$f = f(x)$	50	55	61

a. Find the average rate of change for f between $x = 0$ and $x = 3$.

b. Find the average rate of change for f between $x = 3$ and $x = 6$.

c. Use your answer to part b to estimate the value of $f(4)$.

13. Timber Values Under Scribner Scale The following table compares dollar values per standard cord (128 cubic feet) to values per thousand board-feet (MBF) under the Scribner scale for trees 12 inches in diameter.[34]

Value per cord	Value per MBF Scribner
$20	$68.00
$24	$81.60
$28	$95.20
$36	$122.40

a. Make a table showing the average rate of change for each interval of values.

---

[34]From the *Service Forester's Handbook*, USDA Forest Service, July 1986.

b. Would you expect the value per MBF Scribner to have a limiting value?

c. If you are selling timber from trees 12 inches in diameter, which is the better price: $25 per cord or $71 per MBF Scribner? What if you are buying the timber?

14. Concavity A certain graph is decreasing, but at an increasing rate. Discuss its concavity. How would the concavity change if it were decreasing, but at a decreasing rate?

15. Longleaf Pines The graph in Figure 1.55 gives the height H, in feet, of second-growth longleaf pines for various ages a, in years.[35]

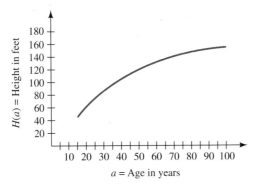

FIGURE 1.55

a. Describe how the height of the trees changes with age. Why is this reasonable?

b. What is the tree height for a 60-year-old tree?

c. Is there a limiting value to the height of these trees? Why?

d. Describe the concavity of the graph and explain what it means in practical terms.

16. Getting a Formula You pay $56 to rent a hotel room, but if you rent more than one room, you get a discount of $2 per room for each additional room rented.

a. If you rent 3 rooms, how much will each one cost?

b. If you rent 3 rooms, how much will you pay for all three combined?

c. Say you rent n rooms. Find a formula showing how much each one costs as a function of n.

d. Say you rent n rooms. Find a formula showing how much you will pay for all n rooms combined as a function of n.

17. A Wedding Reception You rent a wedding venue for a cost of $3200. The cost includes a catered lunch for 50 guests. For each additional guest, though, the catered lunch costs $31.

a. What is the cost of the venue and lunch if you invite 100 guests?

b. Find a formula showing the cost of the venue and lunch as a function of n, the number of guests. Assume that n is at least 50.

c. The amount you have budgeted for the venue and catered lunch is $5500. How many guests can you invite?

18. Limiting Values

a. Do all tables show limiting values?

b. Explain how you can identify a limiting value from a table.

c. Do all graphs show limiting values?

d. Explain how you can identify a limiting value from a graph.

---

[35]Reproduced from USDA Forestry Service Misc. Pub. No. 50 for longleaf pine stands with a site index of 120.

A FURTHER LOOK

Average Rates of Change with Formulas

Average rates of change can be calculated not only for functions given by tables but also for functions given by formulas. In the latter case, we can calculate the rate of change over any interval in the domain that we choose. In general, the average rate of change in f from $x = a$ to $x = b$ is given by

$$\text{Average rate of change} = \frac{\text{Change in function value}}{\text{Change in } x \text{ value}} = \frac{f(b) - f(a)}{b - a}.$$

For example, if we take $f(x) = x^2$, we can calculate the average rate of change from $x = 3$ to $x = 5$ as follows:

$$\text{Average rate of change} = \frac{f(5) - f(3)}{5 - 3} = \frac{25 - 9}{2} = 8.$$

Average Rate of Change in Practical Settings

In practical settings, the average rate of change often has an important physical meaning. If, for example, $S(t)$ gives distance traveled as a function of time t, then the average rate of change of S is

$$\frac{\text{Change in } S}{\text{Change in } t} = \frac{\text{Change in distance}}{\text{Elapsed time}} = \text{Average velocity}.$$

EXAMPLE 1.13 A FALLING ROCK

A rock dropped near the surface of Earth travels $D(t) = 16t^2$ feet during the first t seconds of the fall.

Part 1 What is the average velocity of the rock from time $t = 2$ to time $t = 3$?

Part 2 What is the average velocity of the rock over the time interval from $t = 2$ to $t = 2 + h$?

Solution to Part 1 The average velocity we want is the average rate of change in D from $t = 2$ to $t = 3$:

$$\text{Average velocity} = \frac{D(3) - D(2)}{3 - 2} = \frac{16 \times 3^2 - 16 \times 2^2}{1} = 80 \text{ feet per second}.$$

Solution to Part 2 The average velocity is the average rate of change in D from $t = 2$ to $t = 2 + h$:

$$\text{Average velocity} = \frac{D(2 + h) - D(2)}{(2 + h) - 2}$$

$$= \frac{16(2 + h)^2 - 64}{h}.$$

To proceed we recall the squaring identity $(a + b)^2 = a^2 + 2ab + b^2$. We apply this identity with $a = 2$ and $b = h$: $(2 + h)^2 = 4 + 4h + h^2$. Thus

$$\text{Average velocity} = \frac{16(4 + 4h + h^2) - 64}{h}$$

$$= \frac{64 + 64h + 16h^2 - 64}{h} = \frac{64h + 16h^2}{h}$$

$$= \frac{h(64 + 16h)}{h} = \frac{\cancel{h}(64 + 16h)}{\cancel{h}}$$

$$= 64 + 16h.$$

TEST YOUR UNDERSTANDING | **FOR EXAMPLE 1.13**

What is the average velocity of the rock from $t = 4$ to $t = 6$? ∎

ANSWERS FOR TEST YOUR UNDERSTANDING

1.13 160 feet per second

EXERCISES

Reminder Round all answers to two decimal places unless otherwise indicated.

1. **Calculating Rates of Change** Find the average rate of change for $f(x) = 1/x$ from $x = 2$ to $x = 4$.

2. **Calculating Rates of Change** Find the average rate of change for $f(x) = x^2 + 1$ from $x = 1$ to $x = 2$.

3. **Calculating Rates of Change** Find the average rate of change for $f(x) = \sqrt{x}$ from $x = 4$ to $x = 9$.

4. **Average Rates of Change with Variables** Calculate the average rate of change for $f(x) = 2x + 1$ from $x = 3$ to $x = 3 + h$.

5. **Average Rates of Change with Variables** Calculate the average rate of change for $f(x) = x^2$ from $x = 0$ to $x = h$.

6. **Difference Quotients** Calculate the average rate of change for $f(x) = 3x + 1$ from x to $x + h$.

7. **Difference Quotients** Calculate the average rate of change for $f(x) = x^2 + x$ from x to $x + h$.

8. **Linear Functions** A function of the form $f(x) = mx + b$ is known as a *linear function*. Show that for a linear function, the average rate of change from $x = p$ to $x = q$ does not depend on either p or q.

9. **The Effect of Adding a Constant** How does the average rate of change from $x = a$ to $x = b$ for a function $f(x)$ compare with the average rate of change over the same interval for the function $g(x) = f(x) + c$? (Here c is a constant.)

10. **A Fish** A certain fish grows so that after t years of life its length is given by $L(t) = 10 - 1/2^t$ inches. Calculate the average rate of growth over the first year of life.

11. **Radioactive Decay** For a certain radioactive substance, the amount remaining after t minutes is given by $A(t) = 20/2^t$ grams.

    a. Calculate the average rate of change from $t = 0$ to $t = 2$.

    b. Your answer from part a should be negative. What is the physical meaning of this fact?

12. **A Derivative** For $f(x) = x^2$, the average rate of change from x to $x + h$ is $2x + h$. The *derivative* or *instantaneous rate of change* is obtained by finding what the average rate of change is close to when h is close to 0. What is the derivative of x^2?

13. **A Derivative** For $f(x) = x^3$, the average rate of change from x to $x + h$ is $3x^2 + 3xh + h^2$. The *derivative* or *instantaneous rate of change* is obtained by finding what the average rate of change is close to when h is close to 0. What is the derivative of x^3?

A FURTHER LOOK

Areas Associated with Graphs

If $f(x)$ is a positive function between $x = a$ and $x = b$, then its graph is above the horizontal axis and determines a region bounded by the graph, the vertical line $x = a$, the vertical line $x = b$, and the horizontal axis. This region is shown in Figure 1.56. The area is commonly called the *area under a curve*.

The problem of calculating the area shown in Figure 1.56 is an important topic in calculus. We will consider only restricted versions of the problem here. The following reminder about areas will be useful.

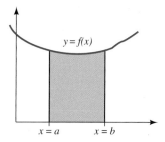

FIGURE 1.56 The area under a curve from $x = a$ to $x = b$

REMINDER

The area of a rectangle is given by

$$\text{Area} = \text{Base} \times \text{Height}.$$

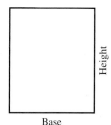

The area of a right triangle is given by

$$\text{Area} = \frac{1}{2}\,\text{Base} \times \text{Height}.$$

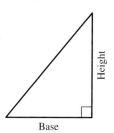

EXAMPLE 1.14 **FINDING AREAS**

In this example we find areas associated with graphs.

Part 1 Figure 1.57 shows the graph of $f(x) = 3x$. Find the shaded area.

Part 2 Figure 1.58 shows the graph of $f(x) = x^2$. Find the shaded area.

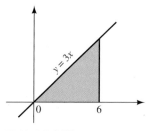

FIGURE 1.57 The graph of $f(x) = 3x$ for part 1

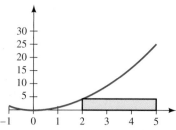

FIGURE 1.58 The graph of $f(x) = x^2$ for part 2

Solution to Part 1 The shaded region in Figure 1.57 is a right triangle. The base is the distance from $x = 0$ to $x = 6$, or 6 units. The height of the triangle is the height of the graph of $f(x) = 3x$ at $x = 6$. That is, the height is $3 \times 6 = 18$ units. Applying the formula for the area of a triangle, we find

$$\text{Area of triangle} = \frac{1}{2} \text{ Base} \times \text{Height}$$

$$= \frac{1}{2} \times 6 \times 18$$

$$= 54 \text{ square units.}$$

Solution to Part 2 The shaded region in Figure 1.58 is a rectangle. Its base is the distance from $x = 2$ to $x = 5$, or 3 units. The height of the rectangle is the height of the graph of $f(x) = x^2$ at $x = 2$. That is $2^2 = 4$ units. Therefore,

$$\text{Area of rectangle} = \text{Base} \times \text{Height}$$

$$= 3 \times 4$$

$$= 12 \text{ square units.}$$

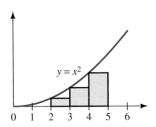

FIGURE 1.59 Figure for TEST YOUR UNDERSTANDING 1.14

TEST YOUR UNDERSTANDING | **FOR EXAMPLE 1.14**

Find the area of the shaded region associated with the graph of $f(x) = 2x$ in Figure 1.59. This is a *trapezoid. Suggestion*: You can think of the area of the shaded region as the difference between the areas of two triangles. ■

Often we may calculate areas using several pieces, as is shown in the following example.

EXAMPLE 1.15 **UPPER AND LOWER SUMS**

Find the shaded area under the curve $y = x^2$ in Figure 1.60. This is a special case of an important concept from calculus known as a *lower sum*.

Solution The area shown in Figure 1.60 is made up of three rectangles. Each of the three rectangles has base 1. The heights of the rectangles are determined by the graph.

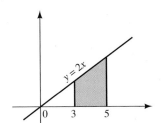

FIGURE 1.60 Figure for Example 1.15

Left-Hand Rectangle Height $= 2^2 = 4$, so Area $= 1 \times 4 = 4$.

Middle Rectangle Height $= 3^2 = 9$, so Area $= 1 \times 9 = 9$.

Right-Hand Rectangle Height $= 4^2 = 16$, so Area $= 1 \times 16 = 16$.

We find the total area by adding the areas of the pieces:

$$\text{Total Area} = 4 + 9 + 16 = 29.$$

TEST YOUR UNDERSTANDING | **FOR EXAMPLE 1.15**

Find the shaded area under the curve $y = x^2$ in Figure 1.61. This is a special case of an important concept from calculus known as an *upper sum*. ■

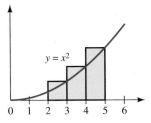

FIGURE 1.61 Figure for TEST YOUR UNDERSTANDING 1.15

ANSWERS FOR TEST YOUR UNDERSTANDING

1.14 16

1.15 50

EXERCISES

Reminder   Round all answers to two decimal places unless otherwise indicated.

1. **Rectangle**   Find the area of the shaded region associated with the curve $y = x^2$ in Figure 1.62.

2. **Rectangle**   Find the area of the shaded region associated with the curve $y = x^2$ in Figure 1.63.

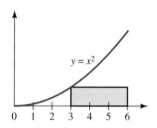

FIGURE 1.62 Picture for Exercise 1

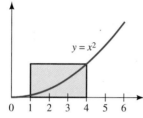

FIGURE 1.63 Picture for Exercise 2

3. **Triangle**   Find the area of the shaded region associated with the curve $y = x - 2$ in Figure 1.64.

4. **Triangle**   Find the area of the shaded region associated with the curve $y = 10 - 2x$ in Figure 1.65.

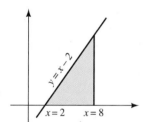

FIGURE 1.64 Picture for Exercise 3

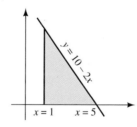

FIGURE 1.65 Picture for Exercise 4

5. **Lower Sum**   Find the area of the shaded region associated with the curve $y = x^3$ in Figure 1.66.

6. **Upper Sum**   Find the area of the shaded region associated with the curve $y = x^3$ in Figure 1.67.

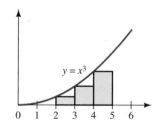

FIGURE 1.66 Picture for Exercise 5

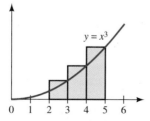

FIGURE 1.67 Picture for Exercise 6

7. **Trapezoid**   Find the area of the shaded region associated with the curve $y = 4x$ in Figure 1.68.

8. **Trapezoid**   Find the area of the shaded region associated with the curve $y = 18 - 3x$ in Figure 1.69.

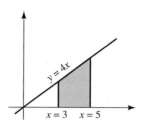

FIGURE 1.68 Picture for Exercise 7

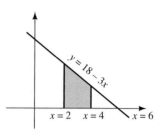

FIGURE 1.69 Picture for Exercise 8

9. **Describing an Area**   Find the area of the region bounded by the horizontal axis, the graph of $y = 6x$, and the vertical line $x = 4$. *Suggestion*: Draw and label a picture.

10. **Describing an Area**   Find the area of the region bounded by the horizontal axis, the graph of $y = 2x$, and the vertical lines $x = 4$ and $x = 8$. *Suggestion*: Draw and label a picture.

11. **A Lower Sum**   Find the area of the shaded region in Figure 1.70. The graph in the figure is that of $f(x) = 40 - x^2$.

12. **An Upper Sum**   Find the area of the shaded region in Figure 1.71. The graph in the figure is that of $f(x) = 40 - x^2$.

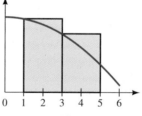

FIGURE 1.70 Figure for Exercise 11

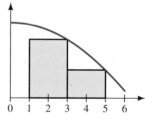

FIGURE 1.71 Figure for Exercise 12

GRAPHICAL AND TABULAR ANALYSIS

2

Len DeLessio/Getty Images

Peter Hulla/Shutterstock.com

Dinosaur tracks give information regarding the size and speed of the animals that made them. See Exercise 14 on page 115.

EACH TYPE of function presentation that we studied in Chapter 1 has advantages and disadvantages, and one of the most useful methods of analysis is to look at functions in more than one way. Tables and graphs often show information that is difficult to obtain directly from formulas. The graphing calculator makes it easy to go from a formula to a table or graph and hence becomes a tool for solving significant problems.

Student resources are available on the website **www.cengagebrain.com**

89

2.1 TABLES AND TRENDS

The advantage of functions given by formulas is that they allow for the calculation of any function value. Tables of values always leave gaps, but they may be more helpful than formulas for seeing trends, predicting future values, or discerning other interesting information about the function.

 ## Getting Tables from Formulas

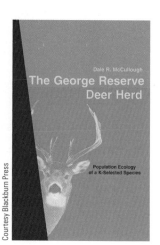

Courtesy Blackburn Press

Let's look at an example to illustrate this. Proper management of wildlife depends on the ability of ecologists to monitor and predict population growth. In many situations, it is reasonable to expect animal populations to exhibit *logistic growth*. A special formula that is studied extensively by ecologists describes this type of growth.

The circumstances surrounding the *George Reserve* in Michigan have made it particularly easy for ecologists to monitor accurately the growth of the deer population on the reserve and to develop a logistic growth formula for the number $N = N(t)$ of deer expected to be present after t years:[1]

$$N = \frac{6.21}{0.035 + 0.45^t} \text{ deer.}$$

When a breeding group of animals is introduced into a limited area, one expects that it will over time grow to the largest size that the environment can support. Wildlife managers refer to this as the *environmental carrying capacity*. Let's find the carrying capacity for deer of the George Reserve. That is, we want to know the deer population after a long period of time. This is not an easy question to answer by looking at the formula, but if we use the formula to make a table of values, then the trend will become apparent. In the accompanying table, we have calculated $N(0)$, $N(5)$, $N(10)$, . . . , $N(30)$. (These values represent the initial population, the population after 5 years, the population after 10 years, etc.)

Scanning down the right-hand column of the table, we see the growth of the deer population with time. From the first row of the table, we see that there were initially 6 deer on the reserve. During the first 10 years, the population increases rapidly, but the rate of increase slows down dramatically after that. It appears that after about 20 years, the population levels out at approximately 177 deer. (We have rounded to the nearest whole number since we don't expect to see parts of deer on the reserve.) This is the carrying capacity of the reserve.

Year t	$N(t)$ Population in Year t
0	6
5	116.18
10	175.72
15	177.4
20	177.43
25	177.43
30	177.43

Formulas can be used to make tables of values.

> **KEY IDEA 2.1** **GETTING TABLES FROM FORMULAS**
>
> Many times, needed information is difficult to obtain directly from a formula. Supplementing the formula with a table of values can provide deeper insight into what is happening. This includes studying long-term behavior of the function.

Tables of values can be made by calculating each wanted function value one at a time, but many calculators have a built-in feature that generates such tables automatically. Making tables of values using a calculator is a skill that will be needed often in what follows, so you are strongly encouraged to consult Appendix B for instructions on how to do this. To become familiar with the procedure, you should work through the practice problems that are presented there.

---

[1]Dale R. McCullough, *The George Reserve Deer Herd* (Ann Arbor: University of Michigan Press, 1979).

When you make a table of values with a calculator, there are three key bits of information that you must input. You must tell your calculator which function you want to use; you must decide on a place to begin the table, the *table starting value;* and you must decide on the periods, the *table increment value*, when you want to see additional data. In making the preceding table for deer population, we used the function $6.21/(0.035 + 0.45^t)$ with a table starting value of $t = 0$, and we viewed the data in 5-year periods. That is, we used a table increment value of 5. In what follows, we will refer to these latter two items as the *table setup.* Thus, if we wanted to instruct you to make the table above in exactly the same way we did, we would say "Enter the function $6.21/(0.035 + 0.45^t)$, and for table setup use a starting value of $t = 0$ and a table increment value of 5." The function entry screen on a graphing calculator will typically appear as in Figure 2.1 and the completed function entry as in Figure 2.2, but displays will vary from calculator to calculator. The table setup will typically appear as in Figure 2.3, and the completed table of values will typically appear as in Figure 2.4.

When you make a table of values, you must supply the *starting value* and the *increment value.*

FIGURE 2.1 A typical function entry screen

FIGURE 2.2 The properly entered function

FIGURE 2.3 A typical TABLE SETUP menu

FIGURE 2.4 A table of values for the deer population

We are working with a function whose name is N and with a variable t, but as we see in Figures 2.2 and 2.4, the calculator has chosen its own name, Y_1, for the function N and X for the variable t. Your calculator may use other letters to represent the function and the variable, but whatever letters your calculator uses, it is important to keep track of the proper associations. It is good practice to write down the correspondence, and we will always do that in the examples we present:

$$Y_1 = N, \text{population} \qquad (2.1)$$

$$X = t, \text{time in years.} \qquad (2.2)$$

It is always important to say what your variables represent.

Now if we want to use the table in Figure 2.4 to find the value of N when t is 20, we note from Equation (2.2) that we should look in the X column to find 20 and from Equation (2.1) that the corresponding Y_1 value 177.43 is the function value for N. That is, $N(20) = 177.43$.

EXAMPLE 2.1 **A SKYDIVER**

A falling object is pulled downward by gravity, but its fall is retarded by air resistance, which under appropriate conditions is directly proportional to velocity. When a skydiver jumps from an airplane, her downward velocity $v = v(t)$ before she opens her parachute is given by

$$v = 176(1 - 0.834^t) \text{ feet per second,}$$

where t is the number of seconds that have elapsed since she jumped from the airplane.

Part 1 Express the velocity of the skydiver 2 seconds into the fall using functional notation, and calculate its value.

Part 2 Describe how the velocity of the skydiver changes with time. Include in your description the average rate of increase in velocity during the first 5 seconds and the average rate of increase in velocity during the next 5 seconds.

Part 3 What is the *terminal velocity*? That is, what is the maximum speed the skydiver can attain?

Part 4 How long does it take the skydiver to reach 99% of terminal velocity?

Solution to Part 1 In functional notation, the downward velocity 2 seconds into the fall is $v(2)$. To make the calculation, we put 2 in for t:

$$v(2) = 176(1 - 0.834^2) = 53.58 \text{ feet per second.}$$

```
Plot1 Plot2 Plot3
\Y1=176(1-0.834^
X)
\Y2=
\Y3=
\Y4=
\Y5=
\Y6=
```

FIGURE 2.5 Entering the function for velocity

```
TABLE SETUP
 TblStart=0
 ΔTbl=5
Indpnt: Auto Ask
Depend: Auto Ask
```

FIGURE 2.6 Setting up the table

Solution to Part 2 We want to see how the velocity increases with time. This is difficult to see from the formula, but a table of values showing velocity in 5-second intervals will give us the information we need. Thus we want to enter the function, and we want a table setup with a starting value of 0 and an increment of 5. The correctly entered function is shown in Figure 2.5, and the correctly configured table setup menu is shown in Figure 2.6.

Once again, the calculator is using its own choices of letters, so we record the appropriate correspondences:

$$Y_1 = v\text{, velocity, in feet per second}$$

$$X = t\text{, time in seconds.}$$

X	Y_1
0	0
5	104.99
10	147.35
15	164.44
20	171.34
25	174.12
30	175.24
X=0	

FIGURE 2.7 A table of values for velocity

When we view the table shown in Figure 2.7, we can read down the right-hand column to see that the velocity increases rapidly to begin with but seems to be leveling off near 30 seconds into the fall. It appears that the downward pull of gravity makes the skydiver accelerate rapidly at first, but air resistance seems to have a greater effect at high velocities. This is a consequence of the fact that air resistance is in this case directly proportional to velocity.

The table shows that the velocity increased from 0 to 104.99 feet per second during the first 5 seconds of the fall. Thus, during this period, velocity increased at an average rate of $104.99/5 = 21$ feet per second per second. During the next 5 seconds of the fall, the velocity increased by $147.35 - 104.99 = 42.36$ feet per second. That gives an average increase in velocity of $42.36/5 = 8.47$ feet per second per second. As we should have expected, the rate of increase in velocity is much less during the second 5-second period than in the first 5 seconds. What would this say about the concavity of the graph of velocity against time?

X	Y_1
35	175.69
40	175.88
45	175.95
50	175.98
55	175.99
60	176
65	176
X=65	

FIGURE 2.8 Extending the table

Solution to Part 3 To get the terminal velocity, we want to know what the velocity of the skydiver would be if she continued in free fall for a long time without opening her parachute. That is, we want to look at the table for large values of t. Your graphing calculator may have a feature that lets you extend the table without going back to the table setup menu. We have shown the table for $t = 35$ seconds to $t = 65$ seconds in Figure 2.8. This table shows clearly that velocity levels out at 176 feet per second. (Look further down the table for more evidence of this.) This is the terminal velocity, where the downward pull of gravity matches air resistance.

Limiting values are often indicated when a table of values levels out.

Solution to Part 4 Now 99% of terminal velocity is $0.99 \times 176 = 174.24$ feet per second. Consulting the table in Figure 2.7, we see that this velocity is reached about 25 seconds into the fall. You may wish to improve the accuracy of this answer by changing the table setup so that the increment is 0.5.

We should also note that questions about terminal velocity in particular—and about limiting values of functions in general—can lead to unexpected difficulties that require more advanced mathematical analysis than is appropriate here. See Exercise 28 at the end of this section.

TEST YOUR UNDERSTANDING | **FOR EXAMPLE 2.1**

If a coffee filter is dropped, its velocity after t seconds is given by $v(t) = 4(1 - 0.0003^t)$ feet per second. What is the terminal velocity, and how long does it take the filter to reach 99% of terminal velocity? Use a table increment of 0.1 and give your answer to the nearest tenth of a second. ■

Tables of values can show maximum and minimum values.

Serp/Shutterstock.com

```
Plot1 Plot2 Plot3
\Y₁◻X(X-1)(X-2)/
750*(5/6)^X
\Y₂=
\Y₃=
\Y₄=
\Y₅=
\Y₆=
```

FIGURE 2.9 Entering a probability function

```
TABLE SETUP
 TblStart=1
 ΔTbl=1
Indpnt: Auto Ask
Depend: Auto Ask
```

FIGURE 2.10 Setting up the table

X	Y₁
1	0
2	0
3	.00463
4	.01543
5	.03215
6	.05358
7	.07814
X=3	

FIGURE 2.11 Probability of 3 sixes when 1 through 7 dice are used

It is worth emphasizing the significance of what we have done here. The table of values enabled us to make an in-depth analysis of what happens in the free-fall period of a skydiver's fall. None of the conclusions we drew in parts 2 and 3 are apparent from the formula, but they are easily discernible from the table of values. The technology we are using gives us the power to attack and resolve real problems.

Optimizing with Tables of Values

We can also use tables of values to find maximum and minimum values for functions. To illustrate this, let's suppose we roll several dice hoping to get exactly 3 sixes, not more or less. The probability that this will occur depends on how many dice we roll. If we use only 4 dice, then it seems unlikely that we will get as many as 3 sixes. If we roll 100 dice, then we will expect to get more than 3 sixes. Elementary probability theory can be used to show that if we roll N dice, then the probability $p = p(N)$ of getting exactly 3 sixes is given by the formula

$$p = \frac{N(N - 1)(N - 2)}{750} \times \left(\frac{5}{6}\right)^N.$$

Thus if we want to know the probability of getting exactly 3 sixes when we roll 10 dice, we put $N = 10$ into the formula for p:

$$p(10) = \frac{10 \times 9 \times 8}{750} \times \left(\frac{5}{6}\right)^{10} = 0.15505.$$

If we round this to two decimal places, we get $p = 0.16$. This means that if you roll 10 dice, you can expect to get exactly 3 sixes about 16 times in 100 rolls.

How many dice should we roll so that we have the best possible chance of getting exactly 3 sixes? To answer this, we want to make a table of values showing the probability of getting exactly 3 sixes for various values of N. First we enter the function as shown in Figure 2.9. We record the letter correspondences:

$$Y_1 = p, \text{ probability of exactly 3 sixes}$$

$$X = N, \text{ number of dice.}$$

Next we set up the table using a starting value of 1 and an increment of 1. The correctly configured table setup menu is shown in Figure 2.10. When we view the table of values, we get the display in Figure 2.11. We see that if we use 1 or 2 dice, the probability is 0, indicating (as we expected) that it is impossible to get 3 sixes by rolling fewer than 3 dice. If we roll 3 dice, we see that the probability of getting exactly 3 sixes is 0.00463, or 0.0046 rounded to four decimal places. This means that we would expect to get 3 sixes only 46 times out of 10,000 rolls. As the number of dice increases from 3 up

X	Y$_1$
15	.23626
16	.24231
17	.2452
18	.2452
19	.24264
20	.23789
21	.23128

X=21

FIGURE 2.12 Probability of 3 sixes when 15 through 21 dice are used

AVAVA/Shutterstock.com

We find a formula from the verbal description. Then we obtain a table from the formula.

X	Y$_1$
7	203
8	224
9	243
10	260
11	275
12	288
13	299

X=7

FIGURE 2.13 Renting to groups that number 7 to 13

Information given in the problem can restrict the values we need to consider.

X	Y$_1$
14	308
15	315
16	320
17	323
18	324
19	323
20	320

X=20

FIGURE 2.14 Renting to groups that number 14 to 20

through 7, the probability of getting exactly 3 sixes gets larger. In Figure 2.12 we have extended the table so that we can see what happens when we roll between 15 and 21 dice. This table shows us that the probability of 3 sixes increases as the number of dice increases up to 17 but decreases for more than 18 dice. The decrease occurs because if we roll too many dice, we would expect to get more than 3 sixes. Thus the probability is at its largest, 0.2452, if we use either 17 or 18 dice. You should look both forward and backward in the table to ensure that there are not larger values outside the ranges shown in Figures 2.11 and 2.12.

EXAMPLE 2.2 **RENTING CANOES**

A small business has 20 canoes that it rents for float trips down the Illinois River. The pricing structure offers a discount for group rentals. One canoe rents for $35, two rent for $34 each, three rent for $33 each, and in general the group rate per canoe is found by taking $1 off the base of $35 for each extra canoe rented.

Part 1 How much money is taken in if 4 canoes are rented to a group?

Part 2 Write a formula that gives the price charged for each canoe if n canoes are rented to a group.

Part 3 Find a formula $R = R(n)$ that shows how much money is taken in from renting n canoes to a group.

Part 4 How large a rental to a single group will bring the most income?

Solution to Part 1 One canoe rents for $35, two rent for $34 each, and three rent for $33 each. We expect that four rent for $32 each. If four are rented, then $4 \times 32 = 128$ dollars will be collected.

Solution to Part 2 We want a formula expressing the amount charged per canoe in terms of the number of canoes rented to a group. If the business rents 1 canoe, the charge is $35 = 36 - 1$ dollars per canoe. If the group rental is 2 canoes, the charge is $34 = 36 - 2$ dollars per canoe; for 3, the charge is $33 = 36 - 3$ dollars per canoe. If n canoes are rented to a group, the charge is $36 - n$ dollars per canoe.

Solution to Part 3 We know that n canoes are rented at the rate we found in part 2. Thus if the business rents n canoes, then it charges $36 - n$ dollars for each canoe. That means $n(36 - n)$ dollars are taken in:

$$R(n) = n(36 - n).$$

Solution to Part 4 We want a table that shows how much money is taken in from renting canoes in groups. Thus we need a table of values for R showing money taken in for $n = 1$ through $n = 20$. We enter the function and record the correspondence for function and variable names:

$$\text{Y}_1 = R, \text{ revenue, in dollars}$$

$$\text{X} = n, \text{ canoes rented.}$$

Next we set up the table using a table starting value of 1 and an increment value of 1. The tables for 7 through 13 and 14 through 20 canoe rentals are shown in Figure 2.13 and Figure 2.14. Since the business has only 20 canoes, we need not look further down the table. It shows that the most money, $324, is taken in when 18 canoes are rented to a group. You should view the entire table from $n = 1$ through $n = 20$ to make sure that nothing has been overlooked.

TEST YOUR UNDERSTANDING | **FOR EXAMPLE 2.2**

Suppose one canoe rents for $40, and $2 is taken off the price for each additional canoe rented by a group. What size group gives the most income? Assume that there are 20 canoes available. ∎

ANSWERS FOR TEST YOUR UNDERSTANDING

2.1 Terminal velocity is about 4 feet per second; 99% of terminal velocity is reached in about 0.6 second.

2.2 A group of 10 or 11 gives a maximum income of $220.

2.1 EXERCISES

Reminder   Round all answers to two decimal places unless otherwise indicated.

1. **Economic Efficiency**   *Marginal cost* is the additional cost imposed by the production of one additional item. *Marginal benefit* is the additional benefit of producing one additional item. For a certain company, the marginal cost C, in dollars, if you are currently producing n items, is given by $C(n) = 5 + n$, and the marginal benefit B, in dollars, is given by $B(n) = 25 \times 0.9^n$. *Economic efficiency* occurs at the production level where marginal cost and marginal benefit are the same. What production level n gives economic efficiency for this company? (Round the function values to one decimal place.)

2. **Flesch Reading Ease**   The *Flesch Reading Ease Test* is a test that ranks the reading level of passages.[2] Let W be the total number of words in the passage, S the number of sentences, and L the number of syllables. Then the Flesch test calculates the score F of the passage using the formula

$$F = 206.835 - 1.015\frac{W}{S} - 84.6\frac{L}{W}.$$

   a. Calculate F for the paragraph above (the one that describes the Flesch Reading Ease Test). Omit the formula itself. Round your answer to the nearest whole number. (We count 3 sentences, 51 words, and 71 syllables.)

   b. For the rest of this exercise we consider passages that have 10 sentences and 350 syllables. Find a formula expressing F as a function of W for such passages.

   c. A passage is thought to be easily understandable by 13- to 15-year-old students if the Flesch score, rounded to one decimal place, is 60 or higher. For the passages considered in part b, how many words should be in the passage to make it easily understood by 13- to 15-year-old students? *Suggestion*: First use a table increment of 20 to get an estimate of the answer. Then change to a table increment of 1.

   d. What is the maximum number of words in the passages considered in part b? (Can there be more words than syllables?)

   e. Bear in mind your answer from part d, and determine the maximum possible Flesch score for the passages considered in part b. Round your answer to the nearest whole number.

3. **Spache Readability Formula**   The *Spache Readability Formula*[3] is designed to categorize reading materials for grades 1 through 3. For a passage of text, the variables are S, the number of sentences, W, the number of words, and U, the number of *unfamiliar words* (words not on a specific list). The formula (as revised) for the grade level is

   $$\text{Grade level} = 0.121\frac{W}{S} + 0.082U + 0.659.$$

   The calculated grade level is rounded down to its whole number part. (For example, 2.6 is rounded down to 2.) In this exercise we consider passages having 200 words and using 10 unfamiliar words.

   a. As the number of sentences increases, does grade level increase or decrease?

   b. What is the maximum possible number of sentences? (Can there be more sentences than words?)

   c. What is the minimum number of sentences that would put this passage at the second-grade level?

---

[2]R. Flesch, "A new readability yardstick," *Journal of Applied Psychology* **32** (1948), 221–233.
[3]G. Spache, "A new readability formula for primary-grade reading materials," *The Elementary School Journal* **53**(7) (1953), 410–413.

(continued)

4. **Weight Lifting** *Brzycki's formula* is used by weight lifters to determine the weight M that is equivalent in a single repetition to n repetitions of a weight w. The formula is

$$M = \frac{w}{1.0278 - 0.0278n}.$$

It applies for up to $n = 10$ repetitions.

In this exercise we consider a weight lifter who bench-presses several repetitions of $w = 240$ pounds.

a. What single bench press weight M corresponds to 5 repetitions of this weight?

b. What number of repetitions of this weight corresponds to a single bench press of 278.75 pounds?

5. **Harvard Step Test** The Harvard Step Test was developed[4] in 1943 as a physical fitness test, and modifications of it remain in use today. The candidate steps up and down on a bench 20 inches high 30 times per minute for 5 minutes. The pulse is counted three separate times for 30 seconds each: at 1 minute,

© Fitness & Wellness, Inc.

2 minutes, and 3 minutes after the exercise is completed. If P is the sum of the three pulse counts, then the *physical efficiency index E* is calculated using

$$E = \frac{15,000}{P}.$$

The following table shows how to interpret the results of the test.

Efficiency Index	Interpretation
Below 55	Poor condition
55 to 64	Low average
65 to 79	High average
80 to 89	Good
90 & above	Excellent

a. Does the physical efficiency index increase or decrease with increasing values of P? Explain in practical terms what this means.

b. Express using functional notation the physical efficiency index of someone whose total pulse count is 200, and then calculate that value.

c. What is the physical condition of someone whose total pulse count is 200?

d. What pulse counts will result in an excellent rating?

6. **Public High School Enrollment** One model for the number of students enrolled in U.S. public high schools as a function of time since 1986 is

$$N = 0.05t^2 - 0.42t + 12.33.$$

Here N is the enrollment in millions of students, t is the time in years since 1986, and the model is relevant from 1986 to 1996.

a. Use functional notation to express the number of students enrolled in U.S. public high schools in the year 1989, and then calculate that value.

b. Explain in practical terms what $N(8)$ means and calculate that value.

c. In what year was the enrollment the smallest?

7. **Later Public High School Enrollment** Here is a model for the number of students enrolled in U.S. public high schools as a function of time since 2000:

$$N = -0.033t^2 + 0.46t + 13.37.$$

In this formula N is the enrollment in millions of students, t is the time in years since 2000, and the model is applicable from 2000 to 2010.

a. Calculate $N(10)$ and explain in practical terms what it means.

b. In what year was the enrollment the largest? What was the largest enrollment?

c. Find the average yearly rate of change in enrollment from 2004 to 2010. Is the result misleading, considering your answer to part b?

8. **Species-Area Relation** The number of species of a given taxonomic group within a given habitat (often an island) is a function of the area of the habitat. For islands in the West Indies, the formula

$$S(A) = 3A^{0.3}$$

---

[4]Adapted from L. Brouha, "The Step Test," *Research Quarterly* **14**(1) (1943), 31–35.

approximates the number S of species of amphibians and reptiles on an island in terms of the island area A in square miles. This is an example of a *species-area relation*.

a. Make a table giving the value of S for islands ranging in area from 4000 to 40,000 square miles.

b. Explain in practical terms what $S(4000)$ means and calculate that value.

c. Use functional notation to express the number of species on an island whose area is 8000 square miles, and then calculate that value.

d. Would you expect a graph of S to be concave up or concave down?

9. Competition Two friends enjoy competing with each other to see who has the best time in running a mile. Initially (before they ever raced each other), the first friend runs a mile in 7 minutes, and for each race that they run, his time decreases by 13 seconds. Initially, the second friend runs a mile in 7 minutes and 20 seconds, and for each race that they run, his time decreases by 16 seconds. Which will be the first race in which the second friend beats the first?

10. Profit The profit P, in thousands of dollars, that a manufacturer makes is a function of the number N of items produced in a year, and the formula is

$$P = -0.2N^2 + 3.6N - 9.$$

a. Express using functional notation the profit at a production level of 5 items per year, and then calculate that value.

b. Determine the two break-even points for this manufacturer—that is, the two production levels at which the profit is zero.

c. Determine the maximum profit if the manufacturer can produce at most 20 items in a year.

11. Counting when Order Matters The *factorial function* occurs often in probability and statistics. For a non-negative integer n, the factorial is denoted $n!$ (which is read "n factorial") and is defined as follows: First, $0!$ is defined to be 1. Next, if n is 1 or larger, then $n!$ means $n(n-1)(n-2)\cdots 3 \times 2 \times 1$. Thus $3! = 3 \times 2 \times 1 = 6$. Consult Appendix B to see how to enter the factorial operation on the calculator.

In some counting situations, order makes a difference. For example, if we arrange people into a line (first to last), then each different ordering is considered a different arrangement. The number of ways in which you can arrange n individuals in a line is $n!$.

a. In how many ways can you arrange 5 people in a line?

b. How many people will result in more than 1000 possible arrangements for a line?

c. Suppose you remember that your four-digit bank card PIN number uses 7, 5, 3, and 1, but you can't remember in which order they come. How many guesses would you need to ensure that you got the right PIN number?

d. There are 52 cards in an ordinary deck of playing cards. How many possible shufflings are there of a deck of cards?

12. Counting when Order Does Not Matter *This is a continuation of Exercise 11.* In many situations, the number of possibilities is not affected by order. For example, if a group of 4 people is selected from a group of 20 to go on a trip, then the order of selection does not matter. In general, the number C of ways to select a group of k things from a group of n things is given by

$$C = \frac{n!}{k!(n-k)!}$$

if k is not greater than n.

a. How many different groups of 4 people could be selected from a group of 20 to go on a trip?

b. How many groups of 16 could be selected from a group of 20?

c. Your answers in parts a and b should have been the same. Explain why this is true.

d. What group size chosen from among 20 people will result in the largest number of possibilities? How many possibilities are there for this group size?

13. APR and EAR Recall that the APR (the *annual percentage rate*) is the percentage rate on a loan that the Truth in Lending Act requires lending institutions to report on loan agreements. It does not tell directly what the interest rate really is. If you borrow money for 1 year and make no payments, then in order to calculate how much you owe at the end of the year, you must use another interest rate, the EAR (the *effective annual rate*), which is not normally reported on loan agreements. The calculation is made by adding the interest indicated by the EAR to the amount borrowed.

(continued)

The relationship between the APR and the EAR depends on how often interest is compounded. If you borrow money at an annual percentage rate APR (as a decimal), and if interest is compounded n times per year, then the effective annual rate EAR (as a decimal) is given by

$$\text{EAR} = \left(1 + \frac{\text{APR}}{n}\right)^n - 1.$$

For the remainder of this problem, we will assume an APR of 10%. Thus in the formula above, we would use 0.1 in place of APR.

a. Would you expect a larger or a smaller EAR if interest is compounded more often? Explain your reasoning.

b. Make a table that shows how the EAR depends on the number of compounding periods. Use your table to report the EAR if interest is compounded once each year, monthly, and daily. (*Note:* The formula will give the EAR as a decimal. You should report your answer as a percent with three decimal places.)

c. If you borrow $5000 and make no payments for 1 year, how much will you owe at the end of a year if interest is compounded monthly? If interest is compounded daily?

d. If interest is compounded as often as possible—that is, continuously—then the relationship between APR and EAR is given by

$$\text{EAR} = e^{\text{APR}} - 1.$$

Again using an APR of 10%, compare the EAR when the interest is compounded monthly with the EAR when the interest is compounded continuously.

14. An Amortization Table Suppose you borrow P dollars at a monthly interest rate of r (as a decimal) and wish to pay off the loan in t months. Then your monthly payment can be calculated using

$$M = \frac{Pr(1 + r)^t}{(1 + r)^t - 1} \text{ dollars.}$$

Remember that for monthly compounding, you get the monthly rate by dividing the APR by 12. Suppose you borrow $3500 at 9% APR (meaning that

you use $r = 0.09/12$ in the preceding formula) and pay it back in 2 years.

a. What is your monthly payment?

b. Let's look ahead to the time when the loan is paid off.

  i. What is the total amount you paid to the bank?

  ii. How much of that was interest?

c. The amount B that you still owe the bank after making k monthly payments can be calculated using the variables r, P, and t. The relationship is given by

$$B = P \times \left(\frac{(1 + r)^t - (1 + r)^k}{(1 + r)^t - 1}\right) \text{dollars.}$$

  i. How much do you still owe the bank after 1 year of payments?

  ii. An *amortization table* is a table that shows how much you still owe the bank after each payment. Make an amortization table for this loan.

15. An Amortization Table for Continuous Compounding *This is a continuation of Exercise 14.* Suppose you have borrowed P dollars from a lending institution that compounds interest as often as possible—that is, continuously. If the loan is to be paid off in t months, then you would calculate your monthly payment in dollars using

$$M = \frac{P(e^r - 1)}{1 - e^{-rt}},$$

where $r = \text{APR}/12$ if the APR is written in decimal form. Under these circumstances, the balance B in dollars that you owe the bank after k monthly payments is given by

$$B = \frac{P(e^{rt} - e^{rk})}{e^{rt} - 1}.$$

Suppose you borrow $3500 at an APR of 9% and pay off the note in 2 years.[5]

a. Calculate your monthly payment, and compare your answer with the answer you obtained in Exercise 14.

b. Make an amortization table and compare it with the answer you got in Exercise 14.

---

[5]It is worth pointing out that the formulas we see in this exercise are simpler than the corresponding ones in Exercise 14. Continuous compounding may appear at first sight to be more complicated than monthly compounding, but it is in fact easier to handle. And as you will see when you complete this exercise, in many applications it does not give significantly different answers.

16. **Renting Motel Rooms**   You own a motel with 30 rooms and have a pricing structure that encourages rentals of rooms in groups. One room rents for $85, two rent for $83 each, and in general the group rate per room is found by taking $2 off the base of $85 for each extra room rented.

    a. How much money do you take in if a family rents two rooms?

    b. Use a formula to give the rate you charge for each room if you rent n rooms to an organization.

    c. Find a formula for a function $R = R(n)$ that gives the revenue from renting n rooms to a convention host.

    d. What is the most money you can make from rental to a single group? How many rooms do you rent?

17. **Inventory**   For retailers who buy from a distributor or manufacturer and sell to the public, a major concern is the cost of maintaining unsold inventory. You must have appropriate stock to do business, but if you order too much at a time, your profits may be eaten up by storage costs. One of the simplest tools for analysis of inventory costs is the *basic order quantity model*. It gives the yearly inventory expense $E = E(c, N, Q, f)$ when the following inventory and restocking cost factors are taken into account:

Marcin Balcerzak/Shutterstock.com

- The *carrying cost* c, which is the cost in dollars per year of keeping a single unsold item in your warehouse.

- The number N of this item that you expect to sell in 1 year.

- The number Q of items you order at a time.

- The fixed costs f in dollars of processing a restocking order to the manufacturer. (*Note:* This is not the cost of the order; the price of an item does not play a role here. Rather, f is the cost you would incur with any order of any size. It might include the cost of processing the paperwork,

fixed costs you pay the manufacturer for each order, shipping charges that do not depend on the size of the order, the cost of counting your inventory, or the cost of cleaning and rearranging your warehouse in preparation for delivery.)

The relationship is given by

$$E = \left(\frac{Q}{2}\right)c + \left(\frac{N}{Q}\right)f \text{ dollars per year.}$$

A new-car dealer expects to sell 36 of a particular model car in the next year. It costs $850 per year to keep an unsold car on the lot. Fixed costs associated with preparing, processing, and receiving a single order from Detroit total $230 per order.

a. Using the information provided, express the yearly inventory expense $E = E(Q)$ as a function of Q, the number of automobiles included in a single order.

b. What is the yearly inventory expense if 3 cars at a time are ordered?

c. How many cars at a time should be ordered to make yearly inventory expenses a minimum?

d. Using the value of Q you found in part c, determine how many orders to Detroit will be placed this year.

e. What is the average rate of increase in yearly inventory expense from the number you found in part c to an order of 2 cars more?

18. **A Population of Foxes**   A breeding group of foxes is introduced into a protected area and exhibits logistic population growth. After t years the number of foxes is given by

$$N(t) = \frac{37.5}{0.25 + 0.76^t} \text{ foxes.}$$

Menno Schaefer/Shutterstock.com

a. How many foxes were introduced into the protected area?

b. Calculate $N(5)$ and explain the meaning of the number you have calculated.

(continued)

c. Explain how the population varies with time. Include in your explanation the average rate of increase over the first 10-year period and the average rate of increase over the second 10-year period.

d. Find the carrying capacity for foxes in the protected area.

e. As we saw in the discussion of terminal velocity for a skydiver, the question of when the carrying capacity is reached may lead to an involved discussion. We ask the question differently. When is 99% of carrying capacity reached?

19. Falling with a Parachute If an average-sized man jumps from an airplane with an open parachute, his downward velocity t seconds into the fall is $v(t) = 20(1 - 0.2^t)$ feet per second.

Marcel Jancovic/Shutterstock.com

a. Use functional notation to express the velocity 2 seconds into the fall, and then calculate it.

b. Explain how the velocity increases with time. Include in your explanation the average rate of change from the beginning of the fall to the end of the first second and the average rate of change from the fifth second to the sixth second of the fall.

c. Find the terminal velocity.

d. Compare the time it takes to reach 99% of terminal velocity here with the time it took to reach 99% of terminal velocity in Example 2.1. On the basis of the information we have, which would you expect to reach 99% of terminal velocity first, a feather or a cannonball?

20. Rolling 4 Sixes If you roll N dice, then the probability $p = p(N)$ that you will get exactly 4 sixes is given by

$$p = \frac{N(N - 1)(N - 2)(N - 3)}{24} \times \left(\frac{1}{6}\right)^4 \left(\frac{5}{6}\right)^{N-4}.$$

a. What is the probability, rounded to three decimal places, of getting exactly 4 sixes if 10 dice are rolled? How many times out of 1000 rolls would you expect this to happen?

b. How many dice should be rolled so that the probability of getting exactly 4 sixes is the greatest?

21. Profit *The background for this exercise can be found in Exercises 13 and 14 in Section 1.4.* A manufacturer of widgets has fixed costs of $150 per month, and the variable cost is $50 per widget (so it costs $50 to produce 1 widget). Let N be the number of widgets produced in a month.

a. Find a formula for the manufacturer's total cost C as a function of N.

b. The manufacturer sells the widgets for $65 each. Find a formula for the total revenue R as a function of N.

c. Use your answers to parts a and b to find a formula for the profit P of this manufacturer as a function of N.

d. Use your formula from part c to determine the break-even point for this manufacturer.

22. Profit with Varying Price *The background for this exercise can be found in Exercises 13, 14, 15, and 16 in Section 1.4.* A manufacturer of widgets has fixed costs of $1200 per month, and the variable cost is $40 per widget (so it costs $40 to produce 1 widget). Let N be the number of widgets produced in a month.

a. Find a formula for the manufacturer's total cost C as a function of N.

b. The highest price p, in dollars, of a widget at which N widgets can be sold is given by the formula $p = 53 - 0.01N$. Using this, find a formula for the total revenue R as a function of N.

c. Use your answers to parts a and b to find a formula for the profit P of this manufacturer as a function of N.

d. Use your formula from part c to determine the two break-even points for this manufacturer. Assume here that the manufacturer produces the widgets in blocks of 50, so a table setup showing N in multiples of 50 is appropriate.

e. Use your formula from part c to determine the production level at which profit is maximized if the manufacturer can produce at most 1500 widgets in a month. As in part d, assume that the manufacturer produces the widgets in blocks of 50.

23. A Precocious Child and Her Blocks   A child has 64 blocks that are 1-inch cubes. She wants to arrange the blocks into a solid rectangle h blocks long and w blocks wide. There is a relationship between h and w that is determined by the restriction that all 64 blocks must go into the rectangle. A rectangle h blocks long and w blocks wide uses a total of $h \times w$ blocks. Thus $hw = 64$. Applying some elementary algebra, we get the relationship we need:

$$w = \frac{64}{h}. \qquad (2.3)$$

a. Use a formula to express the perimeter P in terms of h and w.

b. Using Equation (2.3), find a formula that expresses the perimeter P in terms of the height only.

c. How should the child arrange the blocks if she wants the perimeter to be the smallest possible?

d. Do parts b and c again, this time assuming that the child has 60 blocks rather than 64 blocks. In this situation the relationship between h and w is $w = 60/h$. (*Note:* Be careful when you do part c. The child will not cut the blocks into pieces!)

24. Renting Paddleboats   An enterprise rents out paddleboats for all-day use on a lake. The owner knows that he can rent out all 27 of his paddleboats if he charges \$1 for each rental. He also knows that he can rent out only 26 if he charges \$2 for each rental and that, in general, there will be 1 less paddleboat rental for each extra dollar he charges per rental.

a. What would the owner's total revenue be if he charged \$3 for each paddleboat rental?

b. Use a formula to express the number of rentals as a function of the amount charged for each rental.

c. Use a formula to express the total revenue as a function of the amount charged for each rental.

d. How much should the owner charge to get the largest total revenue?

25. Growth in Length of Haddock   In 1933, Riatt found that the length L of haddock in centimeters as a function of the age t in years is given approximately by the formula

$$L = 53 - 42.82 \times 0.82^t.$$

a. Calculate $L(4)$ and explain what it means.

b. Compare the average yearly rate of growth in length from age 5 to 10 years with the average yearly rate of growth from age 15 to 20 years. Explain in practical terms what this tells you about the way haddock grow.

c. What is the longest haddock you would expect to find anywhere?

26. Discharge from a Fire Hose   The discharge of a fire hose depends on the diameter of the nozzle. Nozzle diameters are normally in multiples of $\frac{1}{8}$ inch. Sometimes it is important to replace several hoses with a single hose of equivalent discharge capacity. Hoses with nozzle diameters $d_1, d_2, \ldots, d_n$ have the same discharge capacity as a single hose with nozzle diameter D, where

$$D = \sqrt{d_1^2 + d_2^2 + \cdots + d_n^2}.$$

a. A nozzle of what diameter has the same discharge capacity as three combined nozzles of diameters $1\frac{1}{8}$ inches, $1\frac{5}{8}$ inches, and $1\frac{7}{8}$ inches? You should report your answer as an available nozzle size, that is, in multiples of $\frac{1}{8}$.

b. We have two 1-inch nozzles and wish to use a third so that the combined discharge capacity of the three nozzles is the same as the discharge capacity of a $2\frac{1}{4}$-inch nozzle. What should be the diameter of the third nozzle?

(continued)

c. If we wish to use n hoses each with nozzle size d in order to have the combined discharge capacity of a single hose with nozzle size D, then we must use

$$n = \left(\frac{D}{d}\right)^2 \text{ nozzles.}$$

How many half-inch nozzles are needed to attain the discharge capacity of a 2-inch nozzle?

d. We want to replace a nozzle of diameter $2\frac{1}{4}$ inches with 4 hoses each of the same nozzle diameter. What nozzle diameter for the 4 hoses will produce the same discharge capacity as the single hose?

27. California Earthquakes We most often hear of the power of earthquakes given in terms of the *Richter scale*, but this tells only the power of the earthquake at its epicenter. Of more immediate importance is how an earthquake affects the location where we are. Seismologists measure this in terms of ground movement, and for technical reasons they find the acceleration of ground movement most useful. For the purpose of this problem, a *major earthquake* is one that produces a ground acceleration of at least 5% of g, where g is the acceleration[6] due to gravity near the surface of the Earth. In California, the probability $p(n)$ of one's home being affected by exactly n major earthquakes over a 10-year period is given approximately[7] by

$$p(n) = 0.379 \times \frac{0.9695^n}{n!}.$$

See Exercise 11 for an explanation of $n!$.

a. What is the probability of a California home being affected by exactly 3 major earthquakes over a 10-year period?

b. What is the limiting value of $p(n)$? Explain in practical terms what this means.

c. What is the probability of a California home being affected by no major earthquakes over a 10-year period?

d. What is the probability of a California home being affected by *at least* one major earthquake over a 10-year period?

Hint: It is a certainty that an event either will or will not occur, and the probability assigned to a certainty is 1. Expressed in a formula, this is

Probability of an event occurring
+ Probability of an event not occurring $= 1$.

This, in conjunction with part c, may be helpful for part d.

28. Terminal Velocity Revisited In one of the early "Functions and Change" pilot courses at Oklahoma State University, the instructor asked the class to determine when in Example 2.1 terminal velocity would be reached. Three students gave the following three answers:

Student 1: 58 seconds into the fall.

Student 2: 147 seconds into the fall.

Student 3: Never.

Each student's answer was accompanied by what the instructor judged to be an appropriate supporting argument, and each student received full credit for the problem. What supporting arguments might the students have used to convince the instructor that these three different answers could all be deserving of full credit? (*Hint:* Consider the formula given in Example 2.1. For student 1, look at a table of values where the entries are rounded to two decimal places. For student 2, look at a table of values made by using all the digits beyond the decimal point that the calculator can handle. In this case that was nine. For student 3, consider what value 0.834^t must have to make $176(1 - 0.834^t)$ equal to 176.)

29. Research Project Find a function given by a formula in one of your textbooks for another class or some other handy source. If the formula involves more than one variable, assign reasonable values for all the variables except one so that your formula involves only one variable. Now make a table of values, using an appropriate starting value and increment value so that the table shows some interesting aspect of the function, such as a trend or significant values. Carefully describe the function, formula, variables (including units), and table, and explain how the table is useful.

---

[6]The value of g is 9.8 meters per second per second, or 32 feet per second per second.
[7]The probability here is calculated using earthquake frequency data and intensity distributions given on pages 80 and 81 of *Earthquake Engineering*, edited by Robert L. Wiegel (Englewood Cliffs, NJ: Prentice-Hall, 1970). The calculation also assumes that earthquakes are equally likely at any location in California. It is thus too low for highly seismic regions and too high for some less seismic regions.

2.1 SKILL BUILDING EXERCISES

Making Tables and Comparing Functions   In Exercises S-1 through S-10, make the table of values.

S-1. Make a table for $f(x) = x^2 - 1$ showing function values for $x = 4, 6, 8, \ldots$.

S-2. Make a table that shows the values of f from Exercise S-1 together with those of $g = 2^x$. (Use the same table setup values as in Exercise S-1.)

S-3. Make a table for $f(x) = 16 - x^3$ showing function values for $x = 3, 7, 11, \ldots$.

S-4. Make a table that shows the values of f from Exercise S-3 together with those of $g = 23 - 2^x$. (Use the same table setup values as in Exercise S-3.)

S-5. Make a table of values for $f(x) = 2^x + x^2$. Use $x = 0.1, 0.2, 0.3, \ldots$.

S-6. Make a table of values for $f(x) = x + 1/x$. Use $x = 10, 20, 30, \ldots$.

S-7. Make a table of values for $f(x) = \sqrt{x} - x/30$. Use $x = 5, 10, 15, \ldots$.

S-8. Make a table of values for $f(x) = x/(1 + \sqrt{x})$. Use $x = 100, 200, 300, \ldots$.

S-9. Make a table of values for $f(x) = (1 + 2^x)/(1 + 3^x)$. Use $x = 1, 2, 3, \ldots$.

S-10. Make a table of values for $f(x) = x^2/2^x$. Use $x = 3, 6, 9, \ldots$.

Finding Limiting Values   In Exercises S-11 through S-20, use a table of values to estimate the limiting value.

S-11. It is a fact that the function $(4x^2 - 1)/(7x^2 + 1)$ has a limiting value. Use a table of values to estimate the limiting value. (*Suggestion:* We suggest starting the table at 0 and using a table increment of 20.)

S-12. It is a fact that the function $(2 + 3^{-x})/(5 - 3^{-x})$ has a limiting value. Use a table of values to estimate the limiting value. (*Suggestion:* We suggest starting the table at 0 and using a table increment of 2.)

S-13. It is a fact that the function $\left(1 + \dfrac{1}{x}\right)^x$ has a limiting value. Use a table of values to estimate the limiting value.

S-14. It is a fact that the function $(1 + x)^{1/x}$ has a limiting value. Use a table of values to estimate the limiting value.

S-15. It is a fact that the function $x(2^{1/x} - 1)$ has a limiting value. Use a table of values to estimate the limiting value.

S-16. It is a fact that the function $x(3^{1/x} - 1)$ has a limiting value. Use a table of values to estimate the limiting value.

S-17. It is a fact that the function $\sqrt{9x^2 + x} - 3x$ has a limiting value. Use a table of values to estimate the limiting value.

S-18. It is a fact that the function $\sqrt{x + 1} - \sqrt{x}$ has a limiting value. Use a table of values to estimate the limiting value.

S-19. It is a fact that the function $\sqrt{x^2 + x + 1} - x$ has a limiting value. Use a table of values to estimate the limiting value.

S-20. It is a fact that the function $\left(\dfrac{x + 3}{x}\right)^x$ has a limiting value. Use a table of values to estimate the limiting value.

Finding Maxima and Minima   In Exercises S-21 through S-28, find the maximum or minimum. In each case we suppose the given function describes a physical situation that only makes sense for whole numbers on a given range (e.g., family expense as a function of the number of children).

S-21. Suppose the function $f(x) = x^2 - 8x + 21$ describes a physical situation that only makes sense for whole numbers between 0 and 20. For what value of x does f reach a minimum, and what is that minimum value? (*Suggestion:* We suggest a table starting at 0 with a table increment of 1.)

S-22. Suppose the function $f(x) = 50 - 9x + x^4/30$ describes a physical situation that only makes sense for whole numbers between 0 and 10. For what value of x does f reach a minimum, and what is the minimum value? (*Suggestion:* We suggest beginning with a table starting at 0 with a table increment of 1 and then panning further down the table.)

S-23. Suppose the function $f(x) = 9x^2 - 2^x + 1$ describes a physical situation that only makes sense for whole numbers between 0 and 10. For what value of x does f reach a maximum, and what is the maximum value? (*Suggestion:* We suggest beginning with a table starting at 0 with a table increment of 1 and then panning further down the table.)

(continued)

S-24. Suppose the function $f(x) = 3 \times 2^x - 2.15^x$ describes a physical situation that only makes sense for whole numbers between 0 and 15. For what value of x does f reach a maximum, and what is that maximum value? (*Suggestion:* We suggest a table starting at 0 with a table increment of 1.)

S-25. Suppose $f(x) = x^2/2^x$ describes a physical situation that only makes sense for whole numbers between 0 and 10. For what value of x does f reach a maximum, and what is that maximum value?

S-26. Suppose $f(x) = 9x/(1 + x^2)$ describes a physical situation that only makes sense for whole numbers

between 0 and 10. For what value of x does f reach a maximum, and what is that maximum value?

S-27. Suppose $f(x) = 2^x - x^3 + 300$ describes a physical situation that only makes sense for whole numbers between 0 and 10. For what value of x does f reach a maximum, and what is that maximum value?

S-28. Suppose $f(x) = \sqrt{x^2 + 10} - x$ describes a physical situation that only makes sense for whole numbers between 0 and 10. For what value of x does f reach a maximum, and what is that maximum value?

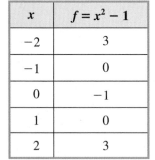

2.2 GRAPHS

A graph is a picture of a function, and just as tables do, graphs can show features of a function that are difficult to see by looking at formulas. For many applications, the graph is more useful than a table of values. The graphing calculator can generate the graph of a function as easily as it makes tables.

Hand-Drawn Graphs from Formulas

x	$f = x^2 - 1$
-2	3
-1	0
0	-1
1	0
2	3

If we have a function given by a formula, there is a standard procedure for generating the graph. We will make a hand-drawn picture of the graph of $f = f(x)$, where $f = x^2 - 1$. The first step is to make a table of values.

This table tells us that the points $(-2, 3), (-1, 0), (0, -1), (1, 0), (2, 3)$ lie on the graph. We plot the individual points as shown in Figure 2.15. We complete the graph by joining the dots with a smooth curve as shown in Figure 2.16.

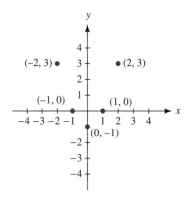

FIGURE 2.15 Plotting points

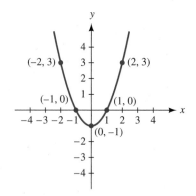

FIGURE 2.16 Completing the graph: $x^2 - 1$

Graphing with the Calculator

Calculators make graphs one point at a time just as you do. They just do the arithmetic very fast.

For more complicated formulas this process is tedious and subject to inaccuracy. But the graphing calculator can make graphs easily. It does exactly what we did above, but it doesn't mind doing all that arithmetic and can accurately plot many more points than we used. For calculator-specific instructions on how to make graphs from formulas, see Appendix B.

There are two steps involved in using a calculator to get a graph from a formula. We will look at them in the context of making the graph of $f = x^2 - 1$.

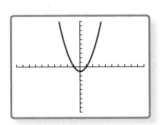

FIGURE 2.17 Entering $f = x^2 - 1$

Step 1, Entering the Function First we have to tell the calculator which function we want to use. We use the function entry screen to do this exactly as we did in Section 2.1 when we made tables of values. In Figure 2.17 we have cleared old formulas from the function entry screen and entered the new one. As expected, the calculator has chosen its own names for functions and variables, and it is important to record the associations:

$$\mathsf{Y}_1 = f, \text{ corresponding to the vertical axis}$$

$$\mathsf{X} = x, \text{ corresponding to the horizontal axis.}$$

FIGURE 2.18 The standard view of the graph

Step 2, Selecting the Viewing Window Like any picture, the graph looks different from different points of view. If you are sure that your viewing window is set up to show the graph as you wish to see it, you can skip this step and go directly to the graph. If this produces an unsatisfactory view, then it will be necessary to make adjustments in the viewing window. Figure 2.18 illustrates what we will refer to as the *standard view*, which shows the graph in a window extending from −10 to 10 in the horizontal direction and from −10 to 10 in the vertical direction. This standard display is satisfactory for some graphs, and we will use it occasionally in what follows. Below we will discuss what to do when the standard view is unsatisfactory.

Tracing the Graph

We can find function values from a graph by *tracing*.

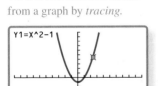

FIGURE 2.19 Tracing the graph

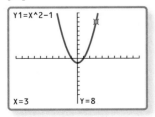

FIGURE 2.20 Getting $f(3)$

Once we have a graph on the screen, there are several ways to adjust the view or to get information from it. We will continue working with $f = x^2 - 1$, and, to follow the discussion, it is important for you to have a graph on your screen that matches the one in Figure 2.18. If you are having difficulty getting that picture, see Appendix B.

Most calculators allow you to *trace* a graph that appears on the screen. This means to put on the screen a movable cursor that follows the graph and is controlled by left and right arrow keys. In Figure 2.19 we have used this feature to add a cursor to the screen and move it along the graph. As the cursor moves, its location is recorded on the screen. In Figure 2.19 the X=2.3404255 prompt at the bottom of the screen shows where we are relative to the horizontal axis, and the Y=4.4775917 prompt shows where we are relative to the vertical axis. This tells us that the cursor is located at the point (2.3404255, 4.4775917). Since this point lies on the graph, it also tells us that $f(2.3404255) = 4.4775917$, or, rounded to two decimal places, $f(2.34) = 4.48$.

The trace feature cannot directly show all function values. For example, on our calculator we were unable, using the arrow keys, to make the cursor land exactly on X=3. Most calculators allow alternative ways to locate the cursor at X=3. We have used this feature in Figure 2.20, and we read from the prompt at the bottom of the screen that $f(3) = 8$ as expected. It is important that you be able to move the graphing cursor to any point on the graph, and you are encouraged to consult Appendix B for information on how to do this.

Choosing a Viewing Window in Practical Settings

The key to making a usable graph on the calculator is to make a proper choice of a viewing window. Finding a viewing window that shows a good picture can sometimes be a bit frustrating, but often appropriate settings can be determined from practical considerations.

Consider, for example, a leather craftsman who has produced 25 belts that he intends to sell at an upcoming art fair for $22.75 each. He has invested a total of $300 in leather, buckles, and other accessories for the belts. We want to look at his net profit $p = p(n)$ in dollars as a function of the number n of belts that he sells:

$$\text{Net profit} = \text{Profit from sales} - \text{Investment}$$
$$= \text{Price per item} \times \text{Number sold} - \text{Investment}$$
$$p = 22.75n - 300.$$

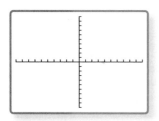

FIGURE 2.21 Entering a function for net profit on the sale of belts

Let's make a graph of p versus n—that is, a graph that shows net profit as a function of the number of belts sold. First we enter the function as shown in Figure 2.21. The proper variable associations are

$$Y_1 = p, \text{net profit, in dollars, on the vertical axis}$$
$$X = n, \text{number sold on the horizontal axis.}$$

If we look at the standard view of the function as shown in Figure 2.22, we see no graph at all! We need to choose a different viewing window that will show the graph. To do that, we note first of all that there are 25 belts available for sale, so we are interested in the function only for values of n between 0 and 25. As a consequence, the horizontal axis should extend from 0 to 25 rather than from the standard setting of -10 to 10. But it is not immediately apparent how to choose a vertical span. A good way to handle this is to look at a table of values, which we already know how to make. As shown in Figure 2.23, we choose a starting value of 0 and a table increment of 5. When we look at the table in Figure 2.24, we see that the possible values for net profit range from a low of $-\$300$ for no sales to $\$268.75$ for the sale of all 25 belts. This tells us how to choose the vertical span of the viewing window; we want it to go from -300 to 268.75. Allowing a little extra margin, we show in Figure 2.25 a window setup where the horizontal span goes from 0 to 25 and the vertical span goes from -325 to 300. Now when we graph we get a very good picture of the function, as shown in Figure 2.26. If we now trace the graph, we can get usable information from it. For example, in Figure 2.27 we have put the cursor at $X=10$, and we see that if 10 belts are sold, the net profit is -72.50 dollars. That is, the craftsman will lose $72.50. In Figure 2.28 we have put the cursor at $X=20$, and we see that if 20 belts are sold, there will be a net profit of $155. What is the practical significance of the place where the graph crosses the horizontal axis?

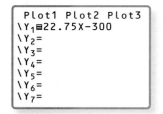

FIGURE 2.22 The standard view of the graph is unsatisfactory.

A table of values can aid in the selection of a viewing window.

TABLE SETUP
TblStart=0
ΔTbl=5
Indpnt: **Auto** Ask
Depend: **Auto** Ask

FIGURE 2.23 Setting up the table

X	Y_1
0	-300
5	-186.3
10	-72.5
15	41.25
20	155
25	268.75
30	382.5
X=0	

FIGURE 2.24 Table for belt sales

WINDOW
Xmin=0
Xmax=25
Xscl=1
Ymin=⁻325
Ymax=300
Yscl=1
Xres=1

FIGURE 2.25 Setting up the viewing window

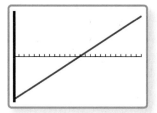

FIGURE 2.26 Graph of profit versus sales

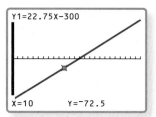

FIGURE 2.27 $72.50 lost if only 10 belts are sold

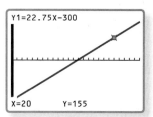

FIGURE 2.28 A net profit of $155 if 20 belts are sold

> **KEY IDEA 2.2** **SETTING THE GRAPHING WINDOW**
>
> In practical settings, to choose the horizontal span for the graphing window, think about what values of the variable are reasonable. Then look at a table of values for the function to determine a suitable vertical span.

The example of the leather craftsman illustrates that, in many situations, the horizontal span can be determined from practical considerations. We noted in the example that the reasonable values of the variable n were between 0 and 25. Traditionally this has been called the *domain* of the function, so the horizontal span of the graphing window is a way of visualizing the domain. The corresponding values taken by the function form what has traditionally been called the *range* of the function. We found this in the example by using a table for the function p to get the vertical span of -325 to 300. Again, the vertical span of the graphing window gives a helpful way of visualizing the range of a function.

EXAMPLE 2.3 **LABOR PRODUCTIVITY**

For a manufacturing company with w workers working h hours per day and producing a total of I items, *labor productivity* $P = P(w, h, I)$ is measured by the function

$$P = \frac{I}{wh} \text{ items per worker hour.}$$

Part 1 Suppose a company produces 750 items per day and each worker works 8 hours per day. The company budget allows for no more than 11 workers.

    a. Find a formula for labor productivity P as a function of the number of workers w, and make its graph.

    b. Use the graph to find labor productivity if there are 7 workers.

    c. What happens to labor productivity if the number of workers is increased? What happens if the number is decreased?

Part 2 Suppose your company has 5 workers, each of whom works 8 hours each day. Company resources and product demand dictate that between 500 and 1000 items per day must be produced. Use a graph to show how labor productivity changes when I increases.

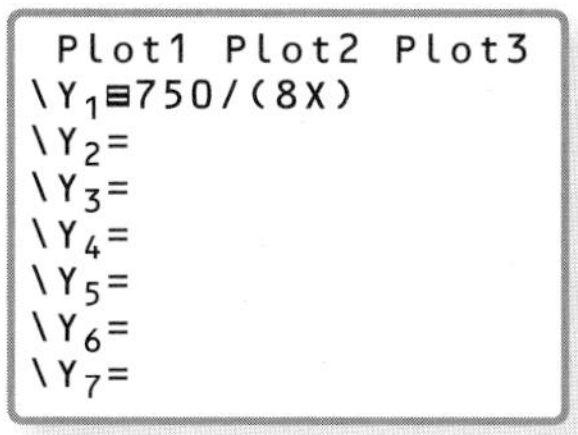

Solution to Part 1a Under the given conditions we have $I = 750$ and $h = 8$. Thus labor productivity is given by the formula $P = 750/8w$. To make the graph, we first enter the function as shown in Figure 2.29 and record the variable correspondences:

$$Y_1 = P, \text{ productivity on the vertical axis}$$

$$X = w, \text{ number of workers on the horizontal axis.}$$

```
Plot1 Plot2 Plot3
\Y₁≣750/(8X)
\Y₂=
\Y₃=
\Y₄=
\Y₅=
\Y₆=
\Y₇=
```

FIGURE 2.29 Entering the labor productivity function

It is important to understand the relationship between the table of values and the viewing window.

Next we need to choose a window for graphing. Since the company budget allows for at most 11 workers, we will set the horizontal span of the window from 1 to 11. To find the vertical span, we made the table of values shown in Figure 2.30 with a starting value of 1 and an increment value of 2. Consulting the table and adding a little margin at the top and bottom, we use a window setup with a horizontal span from 1 to 11 and a vertical span from 0 to 100. This is shown in Figure 2.31. These settings give the graph in Figure 2.32.

X	Y₁
1	93.75
3	31.25
5	18.75
7	13.393
9	10.417
11	8.5227
13	7.2115
X=1	

FIGURE 2.30 A table of values for labor productivity

```
WINDOW
 Xmin=1
 Xmax=11
 Xscl=1
 Ymin=0
 Ymax=100
 Yscl=1
 Xres=1
```

FIGURE 2.31 Setting up the window

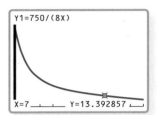

FIGURE 2.32 Graph of labor productivity versus number of workers

Solution to Part 1b We want to get $P(7)$ from the graph. To do this, we locate the cursor at X=7. We read from the Y= prompt at the bottom of the screen in Figure 2.32 that using 7 workers gives a labor productivity of about 13.39 items per worker hour.

Solution to Part 1c The graph in Figure 2.32 is decreasing. This shows that as the number of workers increases, labor productivity decreases. This is not a surprise, since we are increasing the number of worker hours but holding the number of items produced constant at 750 items. Thus each worker is producing less. We see from Figure 2.32 that if we decrease the number of workers, labor productivity increases. Once again, this is not surprising, since we are decreasing worker hours but holding the number of items produced constant. Thus each worker is producing more.

Graphs show clearly where functions increase or decrease—much more clearly than do formulas.

Solution to Part 2 In this scenario, we are holding the number of workers constant at 5 and the number of hours constant at 8. We are looking at labor productivity as a function of the number of items produced:

$$P = \frac{I}{5 \times 8} = \frac{I}{40} \text{ items per worker hour.}$$

We enter this function and record the variable correspondences:

$$Y_1 = P, \text{ productivity on the vertical axis}$$

$$X = I, \text{ number of items produced on the horizontal axis.}$$

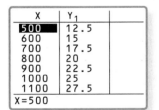

X	Y₁
500	12.5
600	15
700	17.5
800	20
900	22.5
1000	25
1100	27.5
X=500	

FIGURE 2.33 A table of values for productivity as a function of items produced

Since the company must produce between 500 and 1000 items per day, we will set the horizontal span of the graphing window from 500 to 1000. To get the vertical span for the window, we made the table of values in Figure 2.33 with a starting value of 500 and a table increment of 100. From this table, we choose a window setup with a horizontal span of 500 to 1000 and a vertical span of 0 to 30. This gives the graph in Figure 2.34. It is interesting to note that if productivity is viewed as a function of the number of items produced, then its graph is a straight line, but if productivity is viewed as a function of the number of worker hours, then the graph is curved.

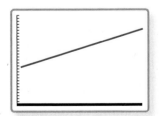

FIGURE 2.34 A graph of productivity versus items produced

TEST YOUR UNDERSTANDING | **FOR EXAMPLE 2.3**

Suppose a company has 15 workers and always produces 1000 items per day. Union agreements limit working time for each employee to no more than 10 hours per day. Make a graph of labor productivity versus hours worked per employee per day. In this setting, when the number of hours worked increases, does productivity increase or decrease? ◼

Getting Limiting Values from Graphs

We have seen that tables of values can be helpful in determining limiting values. Graphs can do that as well, and in addition they provide an informative picture of how limiting values may be approached. The height $h = h(t)$ of some plants as a function of time t closely follows a *logistic formula*. Interestingly enough, this is the same type of formula that is often used to study population growth. For a certain variety of sunflower[8] growing under ideal conditions, and starting at a time when the plant is already a few centimeters tall, its height may be given by the function

$$h = \frac{13}{0.93^t + 0.05},$$

where h is measured in centimeters, and t is measured in days. We want to make a graph of h versus t and see what information we can gain from it. In particular, we would like to be able to describe how the sunflower grows and figure out its maximum height.

First we enter the function as shown in Figure 2.35 and record the variable correspondences:

$Y_1 = h$, height, in centimeters, on the vertical axis

$X = t$, time, in days, on the horizontal axis.

Next we need to determine a viewing window. In contrast with what happened in Example 2.3, we are not provided with explicit information on the values of t to help us set up the window. But our everyday experience with annual plants can give us the information we need. To get an idea of what is happening over a possible 4-month growing period, we show in Figure 2.36 a table of values listing the heights of the sunflower from 60 to 120 days at 10-day intervals. We see that in 120 days the sunflower will be just over 259 centimeters tall, and if you pan farther down the table, you will see that almost no further growth occurs. Adding a bit of extra room, as usual, we have, in Figure 2.37, plotted the graph in a window where the horizontal span is from 0 to 120 and the vertical span is from 0 to 300. Also in this figure, we have traced and moved the cursor toward the right-hand end of the graph, where the height is just over 259 centimeters. From the graph and from the table of values, it appears that the maximum height of the sunflower is about 260 centimeters.

```
Plot1 Plot2 Plot3
\Y₁◻13/(0.93^X+0
.05)
\Y₂=
\Y₃=
\Y₄=
\Y₅=
\Y₆=
```

FIGURE 2.35 Entering a logistic growth function

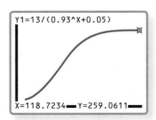

X	Y₁	
60	206.83	
70	231.23	
80	245.23	
90	252.64	
100	256.38	
110	258.24	
120	259.14	
X=60		

FIGURE 2.36 A table of values for four months of growth

```
Y1=13/(0.93^X+0.05)




X=118.7234  Y=259.0611
```

FIGURE 2.37 Sunflower growth

---

[8]This example is adapted from the presentation on pages 42 and 43 of *Differential Equations* by D. Lomen and D. Lovelock (New York: John Wiley, 1999).

Graphs can show limiting values.

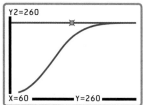

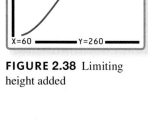

FIGURE 2.38 Limiting height added

Limiting values can be shown to good advantage if the corresponding horizontal line is added to the graph. We added $Y_2 = 260$ to the function list, and in Figure 2.38 both graphs are shown. Note the inflection point in the graph of h, where the graph changes from concave up to concave down. This represents the point of most rapid growth. This picture shows a growth model for sunflowers that agrees with our everyday experience. Growth is slow when the plant is very young, but once a healthy plant is established, it grows rapidly. When maturity is reached, growth slows and little if any additional height is attained.

EXAMPLE 2.4 BLOOD CHOLESTEROL

The amount $C = C(t)$ of cholesterol, in milligrams per deciliter, in the blood of a certain man on an unhealthy diet is given by

$$C = 235 - 105e^{-0.3t},$$

where t is time measured in months.

Part 1 Make a graph that shows the blood cholesterol level as a function of time if the unhealthy diet is continued.

Part 2 The doctor has issued a warning that this man may experience severe health problems if cholesterol levels in excess of 200 milligrams per deciliter of blood are reached. Is there a danger of exceeding this level?

Part 3 If the unhealthy diet is continued indefinitely, what eventual cholesterol level will be reached?

Part 4 Is the graph of C concave up or concave down? Explain in practical terms what your answer means.

Solution to Part 1 We first enter the function and record the variable correspondences:

$Y_1 = C$, blood cholesterol, in milligrams per deciliter, on vertical axis

$X = t$, time, in months, on horizontal axis.

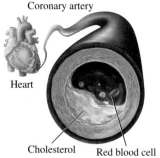

X	Y₁
0	130
1	157.21
2	177.37
3	192.31
4	203.37
5	211.57
6	217.64

X=0

FIGURE 2.39 A table for blood cholesterol in 1-month intervals

We have very little information beyond the formula itself to help us set up the graphing window. Thus we experiment a little with tables of values. First we make a table of values starting at $t = 0$ with a table increment of 1. This is shown in Figure 2.39, where we see that, after 6 months, cholesterol levels are still increasing. If you pan farther down the table, you will see the levels continue to increase. If you are patient and pan down far enough, you can see what eventually happens. A quicker way is to return to the table setup and increase the table increment value. In Figure 2.40 we have changed the starting value to 6 and viewed the table in 6-month intervals. We see that blood cholesterol has leveled out at 235 milligrams per deciliter by 36 months. As usual, we allow some margin and set up the graphing window in Figure 2.41 with a horizontal span from 0 to 36 and a vertical span from 100 to 250. The graph with these settings is shown in Figure 2.42.

Note how **Ymin** and **Ymax** are used to set the viewing window.

X	Y₁
6	217.64
12	232.13
18	234.53
24	234.92
30	234.99
36	235
42	235

X=6

FIGURE 2.40 A table for blood cholesterol in 6-month intervals

```
WINDOW
 Xmin=0
 Xmax=36
 Xscl=1
 Ymin=100
 Ymax=250
 Yscl=1
 Xres=1
```

FIGURE 2.41 Setting up the graphing window

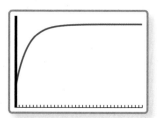

FIGURE 2.42 Blood cholesterol versus time

We can use the trace option to estimate the solutions of equations.

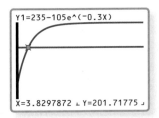

FIGURE 2.43 When blood cholesterol reaches 200 milligrams per deciliter

Concavity shows the *rate of change* in the increase or decrease.

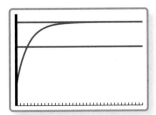

FIGURE 2.44 Limiting value for blood cholesterol

Solution to Part 2 We have already seen from our table of values that a blood cholesterol level of 200 will certainly be exceeded. But we can use the graph to show this in a striking way. In Figure 2.43 we have added the horizontal line corresponding to a blood level of 200 milligrams of cholesterol per deciliter. Note that the point where these graphs cross gives the time when the danger level will be exceeded. In Figure 2.43 we have used the trace option and moved the graphing cursor as close as we could to the crossing point, and we see from the prompt at the bottom of the screen that this individual may incur health risks in about $3\frac{1}{2}$ months.

We will return to this problem in Section 2.4, where we will show how to locate accurately crossing points such as this one.

Solution to Part 3 The table of values shows that the limiting value for blood cholesterol in this case is about 235 milligrams per deciliter. In Figure 2.44 we have added the horizontal line corresponding to this limiting value.

Solution to Part 4 The graph is concave down, so the function is increasing at a decreasing rate. Cholesterol levels increase rapidly at first, but the rate of increase slows near the limiting value.

TEST YOUR UNDERSTANDING │ **FOR EXAMPLE 2.4**

Changes in lifestyle result in a new formula for cholesterol: $C = 190 - 55e^{-0.4t}$, with units as before. Make a graph of this new function. Will cholesterol ever reach the danger level of 200 milligrams per deciliter? ∎

ANSWERS FOR TEST YOUR UNDERSTANDING

2.3 Horizontal, 0 to 10; vertical, 0 to 60

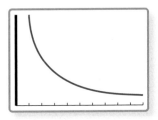

As hours increase, productivity decreases.

2.4 Horizontal, 0 to 20; vertical, 0 to 200

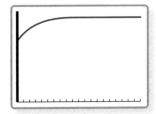

The danger level is never reached.

2.2 EXERCISES

Reminder Round all answers to two decimal places unless otherwise indicated.

In each of the following exercises, you are asked to produce a graph that you should turn in as part of the solution. Ideally you would transfer the graph via a computer link to a printer. If such technology is not available to you, you should provide hand-drawn copies of calculator-generated graphs. Be sure to label your graphs, to include identifying names for the horizontal and vertical axes, and to indicate the graphing window you use. Note also that the graphing windows you choose may not be the same as those we used to make the odd-answer key. Thus you should not expect your graphs to match ours exactly.

1. **Adult Weight from Puppy Weight** There is a formula that estimates how much your puppy will weigh when it reaches adulthood. The method we present applies to medium-sized breeds. First find your puppy's weight w at an age of a weeks, where a is 16 weeks or less. Then the predicted adult weight $W = W(a, w)$, in pounds, is given by the formula

$$W = 52\frac{w}{a}.$$

(continued)

In this exercise we consider puppies that weigh $w = 5$ pounds at age a.

a. Write a formula for W as a function of the age a.

b. Make a graph of W for ages up to 16 weeks.

c. Does a weight of 5 pounds at an early age indicate a larger or smaller adult weight than a weight of 5 pounds at a later age?

d. Is the graph concave up or concave down? Explain the meaning of the concavity in practical terms.

2. Mosteller Formula for Body Surface Area   Body surface area is an important piece of medical information because it is a factor in temperature regulation as well as some drug level determinations. The Mosteller formula[9] gives one way of estimating body surface area B in square meters. The formula uses the weight w in kilograms and the height h in centimeters. The relation is

$$B = \frac{\sqrt{hw}}{60}.$$

In this exercise we consider people who are 190 centimeters tall.

a. Write a formula for B as a function of w.

b. Make a graph of B versus w. Include weights up to 100 kilograms.

c. Is the graph you made in part b concave up or concave down?

d. In this part we consider the effect of weight gain on men who are 190 centimeters tall. Would the weight gain have a greater effect on surface area for a lighter man or for a heavier man?

3. Weekly Cost   The weekly cost of running a small firm is a function of the number of employees. Every week there is a fixed cost of $2500, and each employee costs the firm $350. For example, if there are 10 employees, then the weekly cost is $2500 + 350 \times 10 = 6000$ dollars.

a. What is the weekly cost if there are 3 employees?

b. Find a formula for the weekly cost as a function of the number of employees. (You need to choose variable and function names. Be sure to state the units.)

c. Make a graph of the weekly cost as a function of the number of employees. Include values of the variable up to 10 employees.

d. For what number of employees will the weekly cost be $4250?

4. Average Speed   A commuter regularly drives 70 miles from home to work, and the amount of time required for the trip varies widely as a result of road and traffic conditions. The average speed for such a trip is a function of the time required. For example, if the trip takes 2 hours, then the average speed is $70/2 = 35$ miles per hour.

a. What is the average speed if the trip takes an hour and a half?

b. Find a formula for the average speed as a function of the time required for the trip. (You need to choose variable and function names. Be sure to state the units.)

c. Make a graph of the average speed as a function of the time required. Include trips from 1 hour to 3 hours in length.

d. Is the graph concave up or concave down? Explain in practical terms what this means.

5. Resale Value   The resale value V, in dollars, of a certain car is a function of the number of years t since the year 2012. In the year 2012 the resale value is $18,000, and each year thereafter the resale value decreases by $1700.

a. What is the resale value in the year 2013?

b. Find a formula for V as a function of t.

c. Make a graph of V versus t covering the first 4 years since the year 2012.

d. Use functional notation to express the resale value in the year 2015, and then calculate that value.

6. Profit   The yearly profit P for a widget producer is a function of the number n of widgets sold. The formula is

$$P = -180 + 100n - 4n^2.$$

Here P is measured in thousands of dollars, n is measured in thousands of widgets, and the formula is valid up to a level of 20 thousand widgets sold.

a. Make a graph of P versus n.

b. Calculate $P(0)$ and explain in practical terms what your answer means.

c. What profit will the producer make if 15 thousand widgets are sold?

d. The break-even point is the sales level at which the profit is 0. Approximate the break-even point for this widget producer.

e. What is the largest profit possible?

---

[9]R. D. Mosteller, "Simplified calculation of body-surface area," *N. Engl. J. Med.* **317**(17) (Oct. 22, 1987), 1098 (letter).

7. **Baking a Potato** A potato is placed in a preheated oven to bake. Its temperature $P = P(t)$ is given by

$$P = 400 - 325e^{-t/50},$$

where P is measured in degrees Fahrenheit and t is the time in minutes since the potato was placed in the oven.

   a. Make a graph of P versus t. (*Suggestion:* In choosing your graphing window, it is reasonable to look at the potato over no more than a 2-hour period. After that, it will surely be burned to a crisp. You may wish to look at a table of values to select a vertical span.)

   b. What was the initial temperature of the potato?

   c. Did the potato's temperature rise more during the first 30 minutes or the second 30 minutes of baking? What was the average rate of change per minute during the first 30 minutes? What was the average rate of change per minute during the second 30 minutes?

   d. Is this graph concave up or concave down? Explain what that tells you about how the potato heats up, and relate this to part c.

   e. The potato will be done when it reaches a temperature of 270 degrees. Approximate the time when the potato will be done.

   f. What is the temperature of the oven? Explain how you got your answer. (*Hint:* If the potato were left in the oven for a long time, its temperature would match that of the oven.)

8. **Functional Response** The amount C of food consumed in a day by a sheep is a function of the amount V of vegetation available, and a model is

$$C = \frac{3V}{50 + V}.$$

Here C is measured in pounds and V in pounds per acre. This relationship is called the *functional response*.

Dragan Trifunovic/Photos.com

   a. Make a graph of C versus V. Include vegetation levels up to 1000 pounds per acre.

   b. Calculate $C(300)$ and explain in practical terms what your answer means.

   c. Is the graph concave up or concave down? Explain in practical terms what this means.

   d. From the graph it should be apparent that there is a limit to the amount of food consumed as more and more vegetation is available. Find this limiting value of C.

9. **Population Growth** The growth G of a population over a week is a function of the population size n at the beginning of the week. If both n and G are measured in thousands of animals, the formula is

$$G = -0.25n^2 + 5n.$$

   a. Make a graph of G versus n. Include values of n up to 25 thousand animals.

   b. Use functional notation to express the growth over a week if the population at the beginning is 4 thousand animals, and then calculate that value.

   c. Calculate $G(22)$ and explain in practical terms what your answer means.

   d. For what values of n is the function G increasing? Determine whether the graph is concave up or concave down for these values, and explain in practical terms what this means.

10. **Ohm's Law** says that when electric current is flowing across a resistor, the current i, measured in amperes, can be calculated from the voltage v, measured in volts, and the resistance R, measured in ohms. The relationship is given by

$$i = \frac{v}{R} \text{ amperes.}$$

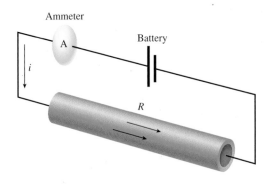

   a. A resistor in a radio circuit is rated at 4000 ohms.

      i. Find a formula for the current as a function of the voltage.

(continued)

ii. Plot the graph of i versus v. Include values of the voltage up to 12 volts.

iii. What happens to the current when voltage increases?

b. The lights on your car operate on a 12-volt battery.

i. Find a formula for the current in your car lights as a function of the resistance.

ii. Plot the graph of i versus R. We suggest a horizontal span here of 1 to 25.

iii. What happens to the current when resistance increases?

11. The Economic Order Quantity Model tells a company how many items at a time to order so that inventory costs will be minimized. The number $Q = Q(N, c, h)$ of items that should be included in a single order depends on the demand N per year for the product, the fixed cost c in dollars associated with placing a single order (not the price of the item), and the carrying cost h in dollars. (This is the cost of keeping an unsold item in stock.) The relationship is given by

$$Q = \sqrt{\frac{2Nc}{h}}.$$

a. Assume that the demand for a certain item is 400 units per year and that the carrying cost is \$24 per unit per year. That is, $N = 400$ and $h = 24$.

i. Find a formula for Q as a function of the fixed ordering cost c, and plot its graph. For this particular item, we do not expect the fixed ordering costs ever to exceed \$25.

ii. Use the graph to find the number of items to order at a time if the fixed ordering cost is \$6 per order.

iii. How should increasing fixed ordering cost affect the number of items you order at a time?

b. Assume that the demand for a certain item is 400 units per year and that the fixed ordering cost is \$14 per order.

i. Find a formula for Q as a function of the carrying cost h and make its graph. We do not expect the carrying cost for this particular item ever to exceed \$25.

ii. Use the graph to find the optimal order size if the carrying cost is \$15 per unit per year.

iii. How should an increase in carrying cost affect the optimal order size?

iv. What is the average rate of change per dollar in optimal order size if the carrying cost increases from \$15 to \$18?

v. Is this graph concave up or concave down? Explain what that tells you about how optimal order size depends on carrying costs.

12. Monthly Payment for a Home If you borrow \$120,000 at an APR of 6% in order to buy a home, and if the lending institution compounds interest continuously, then your monthly payment $M = M(Y)$, in dollars, depends on the number of years Y you take to pay off the loan. The relationship is given by

$$M = \frac{120{,}000(e^{0.005} - 1)}{1 - e^{-0.06Y}}.$$

a. Make a graph of M versus Y. In choosing a graphing window, you should note that a home mortgage rarely extends beyond 30 years.

b. Express in functional notation your monthly payment if you pay off the loan in 20 years, and then use the graph to find that value.

c. Use the graph to find your monthly payment if you pay off the loan in 30 years.

d. From part b to part c of this exercise, you increased the debt period by 50%. Did this decrease your monthly payment by 50%?

e. Is the graph concave up or concave down? Explain your answer in practical terms.

f. Calculate the average decrease per year in your monthly payment from a loan period of 25 to a loan period of 30 years.

13. An Annuity Suppose you are able to find an investment that pays a monthly interest rate of r as a decimal. You want to invest P dollars that will help support your child. If you want your child to be able to withdraw M dollars per month for t months, then the amount you must invest is given by

$$P = M \times \frac{1}{r} \times \left(1 - \frac{1}{(1 + r)^t}\right) \text{dollars}.$$

A fund such as this is known as an *annuity*. For the remainder of this problem, we suppose that you have found an investment with a monthly interest rate of 0.01 and that you want your child to be able to withdraw \$200 from the account each month.

a. Find a formula for your initial investment P as a function of t, the number of monthly withdrawals you want to provide, and make a graph of P versus t. Be sure your graph shows up through 40 years (480 months).

b. Use the graph to find out how much you need to invest so that your child can withdraw $200 per month for 4 years.

c. How much would you have to invest if you wanted your child to be able to withdraw $200 per month for 10 years?

d. A *perpetuity* is an annuity that allows for withdrawals for an indefinite period. How much money would you need to invest so that your descendants could withdraw $200 per month from the account forever? Be sure to explain how you got your answer.

14. Alexander's Formula   One interesting problem in the study of dinosaurs is to determine from their tracks how fast they ran. The scientist R. McNeill Alexander developed a formula giving the velocity of any running animal in terms of its stride length and the height of its hip above the ground.[10]

The stride length of a dinosaur can be measured from successive prints of the same foot, and the hip height (roughly the leg length) can be estimated on the basis of the size of a footprint, so Alexander's formula gives a way of estimating from dinosaur tracks how fast the dinosaur was running. See Figure 2.45.

If the velocity v is measured in meters per second, and the stride length s and hip height h are measured in meters, then Alexander's formula is

$$v = 0.78s^{1.67}h^{-1.17}.$$

FIGURE 2.45

(For comparison, a length of 1 meter is 39.37 inches, and a velocity of 1 meter per second is about 2.2 miles per hour.)

a. First we study animals with varying stride lengths but all with a hip height of 2 meters (so $h = 2$).

i. Find a formula for the velocity v as a function of the stride length s.

ii. Make a graph of v versus s. Include stride lengths from 2 to 10 meters.

iii. What happens to the velocity as the stride length increases? Explain your answer in practical terms.

iv. Some dinosaur tracks show a stride length of 3 meters, and a scientist estimates that the hip height of the dinosaur was 2 meters. How fast was the dinosaur running?

b. Now we study animals with varying hip heights but all with a stride length of 3 meters (so $s = 3$).

i. Find a formula for the velocity v as a function of the hip height h.

ii. Make a graph of v versus h. Include hip heights from 0.5 to 3 meters.

iii. What happens to the velocity as the hip height increases? Explain your answer in practical terms.

15. Artificial Gravity   To compensate for weightlessness in a space station, artificial gravity can be produced by rotating the station.[11] The required number N of rotations per minute is a function of two variables: the distance r to the center of rotation, and a, the desired acceleration (or magnitude of artificial gravity). See Figure 2.46. The formula is

$$N = \frac{30}{\pi} \times \sqrt{\frac{a}{r}}.$$

We measure r in meters and a in meters per second per second.

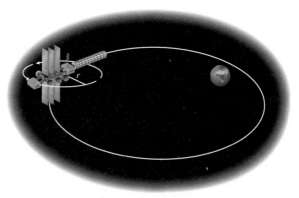

FIGURE 2.46

---

[10]See his article "Estimates of speeds of dinosaurs," *Nature* **261** (1976), 129–130. See also his book *Animal Mechanics*, 2nd ed. (Oxford: Blackwell, 1983).
[11]This exercise is based on B. Kastner's *Space Mathematics*, published by NASA in 1985.

(continued)

a. First we assume that we want to simulate the gravity of Earth, so $a = 9.8$ meters per second per second.

    i. Find a formula for the required number N of rotations per minute as a function of the distance r to the center of rotation.

    ii. Make a graph of N versus r. Include distances from 10 to 200 meters.

    iii. What happens to the required number of rotations per minute as the distance increases? Explain your answer in practical terms.

    iv. What number of rotations per minute is necessary to produce Earth gravity if the distance to the center is 150 meters?

b. Now we assume that the distance to the center is 150 meters (so $r = 150$).

    i. Find a formula for the required number N of rotations per minute as a function of the desired acceleration a.

    ii. Make a graph of N versus a. Include values of a from 2.45 (one-quarter of Earth gravity) to 9.8 meters per second per second.

    iii. What happens to the required number of rotations per minute as the desired acceleration increases?

16. **Plant Growth** The amount of growth of plants in an ungrazed pasture is a function of the amount of plant biomass already present and the amount of rainfall.[12] For a pasture in the arid zone of Australia, the formula

$$Y = -55.12 - 0.01535N - 0.00056N^2 + 3.946R$$

gives an approximation of the growth. Here R is the amount of rainfall, in millimeters, over a 3-month period; N is the plant biomass, in kilograms per hectare, at the beginning of that period; and Y is the growth, in kilograms per hectare, of the biomass over that period. (For comparison, 100 millimeters is about 3.9 inches, and 100 kilograms per hectare is about 89 pounds per acre.)

    For this exercise, assume that the amount of plant biomass initially present is 400 kilograms per hectare, so $N = 400$.

a. Find a formula for the growth Y as a function of the amount R of rainfall.

b. Make a graph of Y versus R. Include values of R from 40 to 160 millimeters.

c. What happens to Y as R increases? Explain your answer in practical terms.

d. How much growth will there be over a 3-month period if initially there are 400 kilograms per hectare of plant biomass and the amount of rainfall is 100 millimeters?

17. **More on Plant Growth** *This is a continuation of Exercise 16.* Now we consider the amount of growth Y as a function of the amount of plant biomass N already present.

a. First we assume that the rainfall is 100 millimeters, so $R = 100$.

    i. Find a formula for the growth Y as a function of the amount N of plant biomass already present.

    ii. Make a graph of Y versus N. Include biomass levels N from 0 to 800 kilograms per hectare.

    iii. What happens to the amount of growth Y as the amount N of plant biomass already present increases? Explain your answer in practical terms.

b. Next we assume that the rainfall is 80 millimeters, so $R = 80$.

    i. With this lower rainfall level, find a formula for the growth Y as a function of the amount N of plant biomass already present.

    ii. Add the graph of Y versus N for the lower rainfall to the graph you found in part a.

    iii. According to the graph you just found, what is the effect of lowered rainfall on plant growth? Is your answer consistent with that in part c of Exercise 16?

18. **Viewing Earth** Astronauts looking at Earth from a spacecraft can see only a portion of the surface.[13] See Figure 2.47. The fraction F of the surface of Earth that is visible at a height h, in kilometers, above the surface is given by the formula

$$F = \frac{0.5h}{R + h}.$$

Here R is the radius of Earth, about 6380 kilometers. (For comparison, 1 kilometer is about 0.62 mile, and the moon is about 380,000 kilometers from Earth.)

---

[12]This exercise and the next are based on the work of G. Robertson, "Plant dynamics," in G. Caughley, N. Shepherd, and J. Short, eds., *Kangaroos* (Cambridge, England: Cambridge University Press, 1987).
[13]This exercise is based on B. Kastner's *Space Mathematics*, published by NASA in 1985.

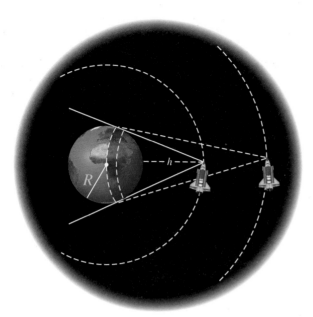

FIGURE 2.47

a. Make a graph of F versus h covering heights up to 100,000 kilometers.

b. A value of F equal to 0.25 means that 25%, or one-quarter, of Earth's surface is visible. At what height is this fraction visible?

c. During one flight of a space shuttle, astronauts performed an extravehicular activity at a height of 280 kilometers. What fraction of the surface of Earth is visible at that height?

d. Is the graph of F concave up or concave down? Explain your answer in practical terms.

e. Determine the limiting value for F as the height h gets larger. Explain your answer in practical terms.

19. **Magazine Circulation** The circulation C of a certain magazine as a function of time t is given by the formula

$$C = \frac{5.2}{0.1 + 0.3^t}.$$

Here C is measured in thousands, and t is measured in years since the beginning of 2006, when the magazine was started.

a. Make a graph of C versus t covering the first 6 years of the magazine's existence.

b. Express using functional notation the circulation of the magazine 18 months after it was started, and then find that value.

c. Over what time interval is the graph of C concave up? Explain your answer in practical terms.

d. At what time was the circulation increasing the fastest?

e. Determine the limiting value for C. Explain your answer in practical terms.

20. **Growth** The length L, in inches, of a certain flatfish is given by the formula

$$L = 15 - 19 \times 0.6^t,$$

and its weight W, in pounds, is given by the formula

$$W = (1 - 1.3 \times 0.6^t)^3.$$

Here t is the age of the fish, in years, and both formulas are valid from the age of 1 year.

a. Make a graph of the length of the fish against its age, covering ages 1 to 8.

b. To what limiting length does the fish grow? At what age does it reach 90% of this length?

c. Make a graph of the weight of the fish against its age, covering ages 1 to 8.

d. To what limiting weight does the fish grow? At what age does it reach 90% of this weight?

e. One of the graphs you made in parts a and c should have an inflection point, whereas the other is always concave down. Identify which is which, and explain in practical terms what this means. Include in your explanation the approximate location of the inflection point.

21. **Buffalo** Waterton Lakes National Park of Canada, where the Great Plains dramatically meet the Rocky Mountains in Alberta, has a migratory buffalo (bison) herd that spends falls and winters in the park. The herd is currently managed and so kept small; however, if it were unmanaged and allowed to grow, then the number N of buffalo in the herd could be estimated by the logistic formula

$$N = \frac{315}{1 + 14e^{-0.23t}}.$$

Here t is the number of years since the beginning of 2002, the first year the herd is unmanaged.

a. Make a graph of N versus t covering the next 30 years of the herd's existence (corresponding to dates up to 2032).

b. How many buffalo are in the herd at the beginning of 2002?

c. When will the number of buffalo first exceed 300?

d. How many buffalo will there eventually be in the herd?

(continued)

e. When is the graph of N, as a function of t, concave up? When is it concave down? What does this mean in terms of the growth of the buffalo herd?

22. Doppler Effect   Motion toward or away from us distorts the pitch of sound, and it also distorts the wavelength of light. This phenomenon is known as the *Doppler effect*. In the case of light, the distortion is measurable only for objects moving at extremely high velocities. Motion of objects toward us produces a *blue shift* in the spectrum, whereas motion of objects away from us produces a *red shift*. Quantitatively, the red shift S is the change in wavelength divided by the unshifted wavelength, and thus red shift is a pure number that has no units associated with it. Cosmologists believe that the universe is expanding and that many stellar objects are moving away from Earth at *radial velocities* sufficient to produce a measurable red shift. Particularly notable among these are quasars, which have a number of important properties (some of which remain poorly understood).[14] Quasars are moving rapidly away from us and thus produce a large red shift. The radial velocity V can be calculated from the red shift S using

$$V = c \times \left(\frac{(S+1)^2 - 1}{(S+1)^2 + 1} \right),$$

where c is the speed of light.

a. Most known quasars have a red shift greater than 1. What would be the radial velocity of a quasar showing a red shift of 2? Report your answer as a multiple of the speed of light.

b. Make a graph of the radial velocity (as a multiple of the speed of light) versus the red shift. Include values of the red shift from 0 to 5.

c. The quasar 3C 48 shows a red shift of 0.37. How fast is 3C 48 moving away from us?

d. Approximate the red shift that would indicate a radial velocity of half the speed of light.

e. What is the maximum theoretical radial velocity that a quasar could achieve?

23. Research Project   Find a function given by a formula in one of your textbooks for another class or some other handy source. If the formula involves more than one variable, assign reasonable values for all the variables except one so that your formula involves only one variable. Now make a graph, using an appropriate horizontal and vertical span so that the graph shows some interesting aspect of the function, such as a trend, significant values, or concavity. Carefully describe the function, formula, variables (including units), and graph, and explain how the graph is useful.

2.2 SKILL BUILDING EXERCISES

S-1. Graphs and Function Values   Get the standard view of the graph of $f = 2 - x^2$. Use the graph to get the value of $f(3)$.

S-2. Graphs and Function Values   Get the standard view of the graph of $f = x^3/30 + 1$. Use the graph to get the value of $f(3)$.

S-3. Graphs and Function Values   Get the standard view of the graph of $f = (x^2 + 2^x)/(x + 10)$. Use the graph to get the value of $f(3)$.

S-4. Zooming In   Get the standard view of $x - x^4/75$. Then zoom in once near the peak of the graph.

S-5. Finding a Window   Find an appropriate window setup that will show a good graph of $x^3/500$ with a horizontal span of -3 to 3.

S-6. Finding a Window   Find an appropriate window setup that will show a good graph of $2^x - x^2$ with a horizontal span of 0 to 5.

S-7. Finding a Window   Find an appropriate window setup that will show a good graph of $(x^4 + 1)/(x^2 + 1)$ with a horizontal span of 0 to 300.

S-8. Finding a Window   Find an appropriate window setup that will show a good graph of $\sqrt{x^2 + 10} - \sqrt{x^2 + 5}$ with a horizontal span of 0 to 10.

---

[14]The universe is thought to be 15 to 20 billion years old. Quasars have been located as far away as 15 billion light-years, so their light reaching Earth shows the universe when it was relatively young.

S-9. Finding a Window Find an appropriate window setup that will show a good graph of $1/(x^2 + 1)$ with a horizontal span of -2 to 2.

S-10. Two Graphs Show the graphs of $f = x + 1$ and $g = 3 - x$ together on the same screen. (Use the standard viewing window.)

Finding Windows and Making Graphs In Exercises S-11 through S-26, find an appropriate window setup and produce the graph.

S-11. Find an appropriate window and make a good graph of $3x/(50 + x)$ with a horizontal span of 0 to 100.

S-12. Find an appropriate window and make a good graph of $400 - 325e^{-x/50}$ with a horizontal span of 0 to 120.

S-13. Find an appropriate window and make a good graph of $-180 + 100x - 4x^2$ with a horizontal span of 0 to 20.

S-14. Find an appropriate window and make a good graph of $-0.25x^2 + 5x$ with a horizontal span of 0 to 25.

S-15. Find an appropriate window and make a good graph of $120{,}000(e^{0.005} - 1)/(1 - e^{-0.06x})$ with a horizontal span of 5 to 30.

S-16. Find an appropriate window and make a good graph of $20{,}000(1 - 1/1.01^t)$ with a horizontal span of 0 to 480.

S-17. Find an appropriate window and make a good graph of $(30/\pi)\sqrt{9.8/x}$ with a horizontal span of 10 to 200.

S-18. Find an appropriate window and make a good graph of $0.5x/(6380 + x)$ with a horizontal span of 0 to 100,000.

S-19. Find an appropriate window and make a good graph of $x/(1 + x^2)$ with a horizontal span of 0 to 20.

S-20. Find an appropriate window and make a good graph of $(1 - x)/(1 + x)$ with a horizontal span of 0 to 20.

S-21. Find an appropriate window and make a good graph of $x/(1 + 1.1^x)$ with a horizontal span of 0 to 20.

S-22. Find an appropriate window and make a good graph of $\sqrt{x} + \sqrt{20 - x}$ with a horizontal span of 0 to 20.

S-23. Find an appropriate window and make a good graph of $x/(\sqrt{x} + \sqrt{20 - x})^2$ with a horizontal span of 0 to 20.

S-24. Find an appropriate window and make a good graph of $\sqrt{x}\sqrt{20 - x}$ with a horizontal span of 0 to 20.

S-25. Find an appropriate window and make a good graph of $(x - 10)/(x + 10)$ with a horizontal span of 0 to 20.

S-26. Find an appropriate window and make a good graph of $(1 + 1/x)^x$ with a horizontal span of 1 to 20.

2.3 SOLVING LINEAR EQUATIONS

Many times we will need to find when two functions are equal or when a function value is zero. To do this we will need to solve an equation. For our purposes, equations come in two types, *linear equations* and *nonlinear equations*. Linear equations are the simplest kind of equation; they are equations that do not involve powers, square roots, or other complications of the variable for which we want to solve.

 | **The Basic Operations**

As you may recall from your elementary algebra course, linear equations can always be solved using two basic operations: subtraction (or addition) and division (or multiplication). The following is presented as a reminder.

Subtracting from Both Sides of an Equation You may subtract (or add) the same thing from (or to) both sides of an equation. This can be thought of as moving a term from one side of the equation to the other, provided that you change its sign, positive to negative or negative to positive.

Divide (or Multiply) Both Sides of an Equation by Any Nonzero Number The same thing may be accomplished if you divide (or multiply) each term of an equation by any nonzero number; however, the results may sometimes appear to be different.

We will show how to use these operations to solve $5x + 7 = 18 - 2x$. The steps in solving a linear equation are always the same.

Step 1 Move all terms that include the variable to one side of the equation and all terms that do not involve the variable to the other side of the equation, changing the sign of each moved term. This is accomplished by adding (or subtracting) the same thing to (or from) both sides of an equation.

In our case we need to move $-2x$ to the left side of the equation and 7 to the right side. We can do this by adding $2x$ to both sides of the equation and subtracting 7 from both sides. We get

$$5x + 7 \quad = \quad 18 - 2x$$

<small>Add 2x and subtract 7   Add 2x and subtract 7</small>

$$5x + 2x = 18 - 7.$$

Step 2 Combine terms. Combining $5x + 2x$ into a single term is an easy task. For example, 5 cars + 2 cars is 7 cars, and it works the same with x's: $5x + 2x = 7x$. We get

$$5x + 2x \quad = \quad 18 - 7$$

<small>Combine to 7x   Combine to 11</small>

$$7x = 11.$$

Step 3 Divide by the coefficient of x. In our case, we want to divide both sides of the equation by 7:

$$7x \quad = \quad 11$$

<small>Divide by 7   Divide by 7</small>

$$x = \frac{11}{7} = 1.57,$$

rounding to two decimal places.

A word of caution is in order concerning division in equations. Sometimes we may have more than two terms in our equation when we need to divide, and the results will *appear* to be different, depending on whether we divide both sides of the equation or divide each term of the equation. For example, suppose that in step 3 above we had $7x = 11 + 8y$, and our goal was to solve for x. We could divide both sides of the equation by 7 to get $x = (11 + 8y)/7$. Alternatively, we could divide each term by 7, but in that case we must remember to divide both 11 and $8y$. The result is $x = 11/7 + 8y/7$, or (rounding to two decimal places) $x = 1.57 + 1.14y$. Even though the two answers, $x = (11 + 8y)/7$ and $x = 1.57 + 1.14y$, may appear to be different, they are the same (allowing for round-off error).

EXAMPLE 2.5 RENTAL CARS

A car rental company charges \$38.00 in insurance and other fees plus a flat rate of \$12.00 per day.

Part 1 Use a formula to express the cost of renting a car as a function of the number of days that you keep it.

Part 2 Your expense account allows you $200 to spend on a rental car. How many days can you keep the car without exceeding your expense account?

Solution to Part 1 We need to choose variable and function names. We let d be the number of days that we keep the car and C the cost (in dollars) of renting the car.

The cost is the initial fee of $38 plus $12 per day:

$$\text{Cost} = \text{Initial fee} + 12 \times \text{Number of days}, \quad \text{or} \quad C = 38 + 12d.$$

It is always important to record the names of variables you use.

Solution to Part 2 We want to know how many rental days will make the cost be $200. That is, we want to solve the following equation for d:

$$C = 200, \quad \text{or} \quad 38 + 12d = 200.$$

We solve it using the steps described above.

$$12d = 200 - 38 \qquad \text{Subtract 38 from both sides.}$$

$$12d = 162$$

$$d = \frac{162}{12} = 13.5. \quad \text{Divide by 12.}$$

Thus you can rent the car for $13\frac{1}{2}$ days without exceeding your expense account. Since the rental company will probably charge you a full day's price for half a day's rental, you may be able to rent the car for only 13 days.

TEST YOUR UNDERSTANDING | **FOR EXAMPLE 2.5**

In the situation of the example, how long could you keep the car if your expense account allowed $242? ∎

◼ Reversing the Roles of Variables

Sometimes when we are given formulas expressing one thing in terms of another, we can gain additional information by changing the roles of the variables. In formal mathematical terms, we are finding the *inverse* of a function.

For example, the relationship between the temperature in degrees Fahrenheit F and the temperature in degrees Celsius C is given by the formula

$$F = 1.8C + 32. \tag{2.4}$$

This formula tells us that a room where the temperature is 30 degrees Celsius has a temperature of $F = 1.8 \times 30 + 32 = 86$ degrees Fahrenheit.

But Equation (2.4) does not immediately tell us the temperature in degrees Celsius of a room that is 75 degrees Fahrenheit. To find this, we need to rearrange the formula so that it shows C in terms of F. That is, we need to solve Equation (2.4) for C. Since the equation is linear, we know how to do that.

$$F = 1.8C + 32$$

$$F - 32 = 1.8C \qquad \text{Subtract 32 from both sides.}$$

$$\frac{F - 32}{1.8} = C. \qquad \text{Divide by 1.8.}$$

We can use this formula to find out the temperature in degrees Celsius of a room where the temperature is 75 degrees Fahrenheit:

$$C = \frac{75 - 32}{1.8} = 23.89 \text{ degrees Celsius.}$$

In the derivation above, we divided both sides of the equation by 1.8. Alternatively, we could have divided each term by 1.8, getting

$$C = \frac{F}{1.8} - \frac{32}{1.8}.$$

The two expressions for C in terms of F look different, but they are in fact the same, and either is acceptable.

EXAMPLE 2.6 A MOVING CAR

If a car moves at a constant velocity, then the distance traveled d can be expressed as a function of velocity v and the time traveled t. If we measure d in miles, v in miles per hour, and t in hours, then the relationship is

$$\text{Distance} = \text{Velocity} \times \text{Time}$$

$$d = vt.$$

Part 1 What distance have you traveled if you drive 55 miles per hour for $2\frac{1}{2}$ hours?

Part 2 Use a formula to express v as a function of d and t. Use your function to find your velocity if it takes you 3 hours to drive 172 miles.

Part 3 Use a formula to express time as a function of velocity and distance traveled. Use your function to find the time it takes to travel 230 miles at a velocity of 40 miles per hour.

Solution to Part 1 Simply plug in the given numbers for v and t:

$$d = 55 \times 2.5 = 137.5 \text{ miles.}$$

Solution to Part 2 You need to solve the equation $d = vt$ for v. Divide both sides by t to obtain

$$v = \frac{d}{t} \text{ miles per hour.}$$

Note that this is the familiar formula that says velocity is distance divided by time:

$$\text{Velocity} = \frac{\text{Distance}}{\text{Time}}.$$

If you take 3 hours to drive 172 miles, then your velocity is $v = 172/3 = 57.33$ miles per hour.

Solution to Part 3 This time you want to solve the equation $d = vt$ for t. Divide both sides by v and obtain

$$t = \frac{d}{v}.$$

This says

$$\text{Time} = \frac{\text{Distance}}{\text{Velocity}}.$$

Driving 230 miles at 40 miles per hour, it takes $230/40 = 5.75$ hours to complete the trip.

TEST YOUR UNDERSTANDING │ **FOR EXAMPLE 2.6**

If you travel 300 miles on the first day and then drive v miles per hour for t hours on the second day, then the total distance traveled over the two-day period is given by $d = 300 + vt$ miles. Use a formula to express v as a function of d and t for this two-day event. ■

KEY IDEA 2.3 **REVERSING THE ROLES OF VARIABLES**

Sometimes a function depends on a variable in a linear way. To reverse the roles, solve for that variable in the linear formula.

EXAMPLE 2.7 **DOWNLOADING MUSIC**

Company Alpha charges a monthly fee of $5.00 plus 99 cents per download for songs. Company Beta charges a monthly fee of $7.00 plus 79 cents per download for songs.

Part 1 Use a formula to express the monthly cost of downloading music from Company Alpha as a function of the number of songs you download in a month.

Part 2 Use a formula to express the monthly cost of downloading music from Company Beta as a function of the number of songs you download in a month.

Part 3 For what number of downloads will the monthly cost for Company Alpha and Company Beta be the same?

Part 4 On the same screen, plot the graphs of the cost of downloading music from Company Alpha and the cost of downloading music from Company Beta.

Part 5 How do you decide which company to use as your music source?

Solution to Part 1 First choose variable and function names: Let s be the number of songs downloaded in a month and A the cost, in dollars, of downloading music from Company Alpha. You pay a monthly fee of $5.00 plus 99 cents for each song you buy:

$$\text{Cost for Alpha} = \text{Monthly fee} + 0.99 \times \text{Songs downloaded}$$

$$A = 5 + 0.99s.$$

Solution to Part 2 Let B be the cost, in dollars, of downloading music from Company Beta. You pay a monthly fee of $7.00 plus 79 cents for each song you buy:

$$\text{Cost for Beta} = \text{Monthly fee} + 0.79 \times \text{Songs downloaded}$$

$$B = 7 + 0.79s.$$

The number of songs that gives the same cost is the solution of an equation.

Solution to Part 3 We want to find when the cost for the two companies is the same:

$$\text{Cost for Alpha} = \text{Cost for Beta}$$

$$A = B$$

$$5 + 0.99s = 7 + 0.79s \qquad \text{Subtract 5 and } 0.79s \text{ from both sides.}$$

$$0.99s - 0.79s = 7 - 5$$

$$0.2s = 2$$

$$s = \frac{2}{0.2} = 10. \qquad \text{Divide by 0.2.}$$

```
Plot1  Plot2  Plot3
\Y₁⊟5+.99X
\Y₂⊟7+.79X
\Y₃=
\Y₄=
\Y₅=
\Y₆=
\Y₇=
```

FIGURE 2.48 Entering both functions

Solution to Part 4 We need to enter both functions in the calculator. This entry is shown in Figure 2.48, and we list the appropriate correspondences:

$$Y_1 = A, \text{ cost for Alpha, in dollars, on vertical axis}$$

$$Y_2 = B, \text{ cost for Beta, in dollars, on vertical axis}$$

$$X = s, \text{ songs downloaded on horizontal axis.}$$

Before we make the graphs, let's think about how to set up the window. We surely want the picture to show where the cost is the same, so we want to include 10 in the horizontal span. We set the horizontal span to be from 0 to 15 songs. To choose the vertical span, we look at the table of values shown in Figure 2.49. We used a starting value of 0 and a table increment of 3. Note that the Y_1 column corresponds to Company Alpha and the Y_2 column corresponds to Company Beta. Allowing a little extra room, we choose a vertical span of 0 to 20.

The properly configured window is shown in Figure 2.50. When we graph, both plots appear as in Figure 2.51. In this figure, the thicker line represents Company Alpha, and the thinner line represents Company Beta.

X	Y₁	Y₂
0	5	7
3	7.97	9.37
6	10.94	11.74
9	13.91	14.11
12	16.88	16.48
15	19.85	18.85
18	22.82	21.22
X=0		

FIGURE 2.49 A table of values for music downloads

```
WINDOW
 Xmin=0
 Xmax=15
 Xscl=1
 Ymin=0
 Ymax=20
 Yscl=1
 Xres=1
```

FIGURE 2.50 Configuring the window

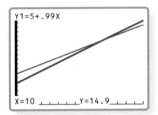

FIGURE 2.51 Thick graph for Company Alpha, thin graph for Company Beta

Solution to Part 5 We want to spend only as much money as is necessary. Figure 2.51 shows that the graph of the cost of Alpha is initially on the bottom, which indicates that it costs less. But after the graphs cross at $s = 10$ songs, the graph of the cost of Beta is on the bottom, which indicates that it costs less. We found the exact value of this crossing point in part 3. Thus, if we are going to download fewer than 10 songs per month, we should use Company Alpha. If we intend to download more music, we should use Company Beta.

TEST YOUR UNDERSTANDING │ **FOR EXAMPLE 2.7**

Company Gamma charges a monthly fee of $11 and charges 59 cents per download. How many songs do you need to download to make the cost for Gamma the same as for Company Alpha? ■

ANSWERS FOR TEST YOUR UNDERSTANDING

2.5 17 days

2.6 $v = \dfrac{d - 300}{t}$

2.7 15 songs

2.3 EXERCISES

Reminder Round all answers to two decimal places unless otherwise indicated.

1. **Sweater Sales** Initially you have an inventory of 2000 sweaters on hand. You sell 400 sweaters each week.

   a. Use s for the total number of sweaters remaining in inventory and w for the number of weeks, and find a formula that gives s as a function of w.

   b. If you do not replenish your inventory, and if sales continue at this rate, how long will it be before your inventory is depleted?

2. **Trans Fat** You are ordering fast food for your family. The burgers you are ordering have a total of 10 grams of trans fat.[15] Fries have 8 grams of trans fat per order.

   a. Find a formula that gives the total grams T of trans fat in your order if you add f orders of fries to your hamburger order.

   b. How many orders of fries can you get if you want to limit your family's trans fat intake in this meal to 42 grams?

3. **Gross Domestic Product** The *gross domestic product* is used by economists as a measure of the nation's economic position. The gross domestic product P is calculated as the sum of personal consumption expenditures C, gross private domestic investment I, government consumption expenditures and gross investment G, and net exports E of goods and services (exports minus imports). All of these are measured in billions of dollars.

   a. Find a formula that gives the gross domestic product P in terms of C, I, G, and E.

   b. In 2009, personal consumption expenditures was 10,089 billion dollars, gross private domestic investment was 1629 billion dollars, government consumption expenditures and gross investment was 2931 billion dollars, and net exports of goods and services was -392 billion dollars.

   i. Which was larger in 2009, U.S. imports or exports?

   ii. Calculate the gross domestic product for 2009.

   c. Solve the equation in part a for E.

   d. Suppose that in another year the values for C, I, and G remain the same as the 2009 figures, but the gross domestic product is 14,876 billion dollars. What is the net export of goods and services for this year?

4. **Juice Sales** The number J, in thousands, of cans of frozen orange juice sold weekly is a function of the price P, in dollars, of a can. In a certain grocery store, the formula is

   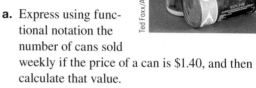

   $$J = 11 - 2.5P.$$

   a. Express using functional notation the number of cans sold weekly if the price of a can is $1.40, and then calculate that value.

   b. At what price will there be 7.25 thousand cans sold weekly?

   c. Solve for P in the formula above to obtain a formula expressing P as a function of J.

   d. At what price will there be 7.75 thousand cans sold weekly?

5. **Resale Value** The resale value V, in thousands of dollars, of a boat is a function of the number of years t since the start of 2011, and the formula is

   $$V = 12.5 - 1.1t.$$

   a. Calculate $V(3)$ and explain in practical terms what your answer means.

   b. In what year will the resale value be 7 thousand dollars?

   c. Solve for t in the formula above to obtain a formula expressing t as a function of V.

   d. In what year will the resale value be 4.8 thousand dollars?

6. **Aerobic Power** Aerobic power can be thought of as the maximum oxygen consumption attainable per kilogram of body mass. There are a number of ways in which physical educators estimate this. One method uses the *Queens College Step Test*.[16] In this test, males step up and down a 16-inch bleacher step 24 times per minute for

---

[15]Trans fat is the common name for unsaturated fat that is partially hydrogenated.
[16]Adapted from F. I. Katch and W. D. McArdle, *Nutrition, Weight Control, and Exercise* (Philadelphia: Lea & Febiger, 1983).

(continued)

3 minutes, and females do 22 steps per minute for 3 minutes. Five seconds after the exercise is complete, a 15-second pulse count P is taken. Maximum oxygen consumption M, in milliliters per kilogram, for males is approximated by

iStockphoto.com/Johnnyhetfield

$$M = 111.3 - 1.68P.$$

For females the recommended formula is

$$F = 65.81 - 0.74P.$$

a. Calculate the maximum oxygen consumption for both a male and a female who show a 15-second pulse count of 40.

b. What 15-second pulse count for a male will indicate a maximum oxygen consumption of 35.7 milliliters per kilogram?

c. What 15-second pulse count will indicate the same maximum oxygen consumption for a male as for a female?

d. What maximum oxygen consumption is associated with your answer in part c?

7. **Gas Mileage**   The distance d, in miles, that you can travel without stopping for gas depends on the number of gallons g of gasoline in your tank and the gas mileage m, in miles per gallon, that your car gets. The relationship is

$$d = gm. \qquad (2.5)$$

a. How far can you drive if you have 12 gallons of gas in your tank and your car gets 24 miles per gallon?

b. Solve Equation (2.5) for m.

   i. Explain in everyday terms what this new equation means.

   ii. Use this equation to determine the gas mileage of your car if you can drive 335 miles on a full 13-gallon tank of gas.

c. A Detroit engineer wants to be sure that the car she is designing can go 425 miles on a full tank of gas, and she must design a gas tank to ensure that. She does not yet know what gas mileage this new-model car will get, and so she decides

to make a graph of the size of the gas tank as a function of gas mileage.

   i. Solve Equation (2.5) for g using 425 for the distance.

   ii. Make the graph that the engineer made. Is it a straight line?

8. **Isocost Equation**   We are to buy quantities of two items: n_1 units of item 1 and n_2 units of item 2.

a. If item 1 costs \$3.50 per unit and item 2 costs \$2.80 per unit, find a formula that gives the total cost C, in dollars, of the purchase.

b. An *isocost equation* shows the relationship between the number of units of each of two items to be purchased when the total purchase price is predetermined. If the total purchase price is predetermined to be $C = 162.40$ dollars, find the isocost equation for items 1 and 2 from part a.

c. Solve for n_1 in the isocost equation you found in part b.

d. Using the equation from part c, determine how many units of item 1 are to be purchased if 18 units of item 2 are purchased.

9. **Supply and Demand**   The quantity S of barley, in billions of bushels, that barley suppliers in a certain country are willing to produce in a year and offer for sale at a price P, in dollars per bushel, is determined by the relation

$$P = 1.9S - 0.7.$$

The quantity D of barley, in billions of bushels, that barley consumers are willing to purchase in a year at price P is determined by the relation

$$P = 2.8 - 0.6D.$$

The *equilibrium price* is the price at which the quantity supplied is the same as the quantity demanded. Find the equilibrium price for barley.

10. **Stock Turnover at Retail**   In retail sales an important marker of retail activity is the *stock turnover at retail*. This figure is calculated for a specific period of time as the total net sales divided by the retail value of the average stock during that time, where both are measured in dollars.[17] As a formula this is written

$$\text{Stock turnover} = \frac{\text{Net sales}}{\text{Average stock at retail}}.$$

[17]See A. P. Kneider, *Mathematics of Merchandising*, 4th ed. (Englewood Cliffs, NJ: Regents/Prentice Hall, 1994).

This formula expresses stock turnover as a function of net sales and average stock at retail.

a. Suppose that your store had net sales of $682,000 in men's shoes over the past six months and that the retail value of the average stock of men's shoes was $163,000. What was the stock turnover at retail for that time period?

b. Suppose that in a certain month your store's net sales of women's dresses were $83,000 and that the usual stock turnover at retail is 0.8 per month. What do you estimate to be your store's average stock at retail?

c. Solve the equation for average stock at retail— that is, write a formula giving average stock at retail as a function of stock turnover and net sales.

d. Suppose that in a certain time period your store had an average stock of socks with a retail value of $45,000 and a stock turnover at retail of 1.6. What were the store's net sales of socks during that time period?

e. Solve the equation for net sales—that is, write a formula giving net sales as a function of stock turnover and average stock at retail.

11. **Fire Engine Pressure**   For firefighting, an important concept is that of *engine pressure*, which is the pressure that the pumper requires to overcome friction loss from the hose and to maintain the proper pressure at the nozzle of the hose. One formula[18] for engine pressure is

$$EP = NP(1.1 + K \times L),$$

where EP is the engine pressure in pounds per square inch (psi), NP is the nozzle pressure in psi, K is the "K" factor (which depends on which tip is used on the nozzle), and L is the length of the hose in feet divided by 50. This formula expresses EP as a function of NP, K, and L.

a. Find the engine pressure if the nozzle pressure is 80 psi, the "K" factor is 0.51, and the length of the hose is 80 feet.

b. Find the nozzle pressure if the "K" factor is 0.73, the length of the hose is 45 feet, and the engine pressure is 150 psi.

c. Solve the equation for NP—that is, write a formula expressing NP as a function of EP, K, and L.

d. Find the "K" factor for a nozzle tip if the length of the hose is 190 feet, the engine pressure is 160 psi, and the nozzle pressure is 90 psi.

e. Solve the equation for K—that is, write a formula expressing K as a function of NP, EP, and L.

12. **Sales Strategy**   A small business is considering hiring a new sales representative (sales rep) to market its product in a nearby city. Two pay scales are under consideration.

> **Pay Scale 1** Pay the sales rep a base yearly salary of $10,000 plus a commission of 8% of total yearly sales.

> **Pay Scale 2** Pay the sales rep a base yearly salary of $13,000 plus a commission of 6% of total yearly sales.

a. For each of the pay scales above, use a formula to express the total yearly earnings for the sales rep as a function of total yearly sales. Be sure to identify clearly what the letters that you use mean.

b. What amount of total yearly sales would result in the same total yearly earnings for the sales rep no matter which of the two pay scales is used?

c. On the same screen, plot the graphs of the functions you made in part a. Copy the picture onto the paper you turn in. Be sure to label the horizontal and vertical axes, and be sure your picture includes the number you found in part b.

d. If you were a sales rep negotiating for the new position, under what conditions would you prefer pay scale 1? Under what conditions would you prefer pay scale 2?

13. **Net Profit**   Suppose you pay R dollars per month to rent space for the production of dolls. You pay c dollars in material and labor to make each doll, which you then sell for d dollars.

a. If you produce n dolls per month, use a formula to express your net profit p per month as a function of R, c, d, and n. (*Suggestion:* First make a formula using the words *rent*, *cost of a doll*, *selling price*, and *number of dolls*. Then replace the words by appropriate letters.)

b. What is your net profit per month if the rent is $1280 per month, it costs $2 to make each doll, which you sell for $6.85, and you produce 826 dolls per month?

c. Solve the equation you got in part a for d.

d. Your accountant tells you that you need to make a net profit of $4000 per month. Your rent is

---

[18]See Eugene Mahoney, *Fire Department Hydraulics* (Boston: Allyn and Bacon, 1980).

(*continued*)

$1200 per month, it costs $2 to make each doll, and your production line can make only 700 of them in a month. Under these conditions, what price do you need to charge for each doll?

14. **Ohm's Law** says that when electric current is flowing across a resistor, then the voltage v, measured in volts, is the product of the current i, measured in amperes, and the resistance R, measured in ohms. That is, $v = iR$.

   a. What is the voltage if the current is 20 amperes and the resistance is 15 ohms?

   b. Find a formula expressing resistance as a function of current and voltage. Use your function to find the resistance if the current is 15 amperes and the voltage is 12 volts.

   c. Find a formula expressing current as a function of voltage and resistance. Use your function to find the current if the voltage is 6 volts and the resistance is 8 ohms.

15. **Temperature Conversions**   In everyday experience, the measures of temperature most often used are Fahrenheit F and Celsius C. Recall that the relationship between them is given by

$$F = 1.8C + 32.$$

Physicists and chemists often use the Kelvin temperature scale.[19] You can get *kelvins K* from degrees Celsius by using

$$K = C + 273.15.$$

   a. Explain in practical terms what $K(30)$ means, and then calculate that value.

   b. Find a formula expressing the temperature C in degrees Celsius as a function of the temperature K in kelvins.

   c. Find a formula expressing the temperature F in degrees Fahrenheit as a function of the temperature K in kelvins.

   d. What is the temperature in degrees Fahrenheit of an object that is 310 kelvins?

16. **The Ideal Gas Law**   A *mole* of a chemical compound is a fixed number,[20] like a dozen, of molecules (or atoms in the case of an element) of that compound. A mole of water, for example, is about 18 grams, or just over a half an ounce in your kitchen. Chemists often use the mole as the measure of the amount of a chemical compound.

A mole of carbon dioxide has a fixed mass, but the volume V that it occupies depends on pressure p and temperature T; greater pressure tends to compress the gas into a smaller volume, whereas increasing temperature tends to make the gas expand into a larger volume. If we measure the pressure in atmospheres (1 atm is the pressure exerted by the atmosphere at sea level), the temperature in kelvins, and the volume in liters, then the relationship is given by the *ideal gas law*:

$$pV = 0.082T.$$

   a. Solve the ideal gas law for the volume V.

   b. What is the volume of 1 mole of carbon dioxide under 3 atm of pressure at a temperature of 300 kelvins?

   c. Solve the ideal gas law for pressure.

   d. What is the pressure on 1 mole of carbon dioxide if it occupies a volume of 0.4 liter at a temperature of 350 kelvins?

   e. Solve the ideal gas law for temperature.

   f. At what temperature will 1 mole of carbon dioxide occupy a volume of 2 liters under a pressure of 0.3 atm?

17. **Running Ants**   A scientist observed that the speed S at which certain ants ran was a function of T, the ambient temperature.[21] He discovered the formula

$$S = 0.2T - 2.7,$$

where S is measured in centimeters per second and T is in degrees Celsius.

   a. Using functional notation, express the speed of the ants when the ambient temperature is 30 degrees Celsius, and calculate that speed using the formula above.

   b. Solve for T in the formula above to obtain a formula expressing the ambient temperature T as a function of the speed S at which the ants run.

   c. If the ants are running at a speed of 3 centimeters per second, what is the ambient temperature?

---

[19]With this temperature scale, physicists traditionally do not use the term *degrees*. Rather, if the temperature is 100 on the Kelvin scale, they say that it is "100 kelvins." The temperature scale is named after Lord Kelvin, and its name is capitalized, but the unit of temperature, the kelvin, is not.

[20]You may be interested to know that this number is 6.02217×10^{23}. It may appear to be a strange number, but this is a natural unit of measure for chemists to use.

[21]See the study by H. Shapley, "Note on the thermokinetics of Dolichoderine ants," *Proc. Nat. Acad. Sci.* **10** (1924), 436–439.

18. **Tax Owed**   According to the Oklahoma Income Tax Table for 2010, the income tax owed by an Oklahoma resident who is married and filing jointly with a taxable income above $100,000 is $5070 plus 5.5% of the taxable income over $100,000.

    a. Use a formula to express the tax owed as a function of the taxable income, assuming that the taxable income is above $100,000. Be sure to explain the meaning of the letters you choose and the units.

    b. What is the tax on a taxable income of $108,650? Round your answer to the nearest dollar.

    c. For what taxable income would the tax be $5686?

19. **Profit**   *The background for this exercise can be found in Exercises 13 and 14 in Section 1.4.* A manufacturer of widgets has fixed costs of $200 per month, and the variable cost is $55 per thousand widgets (so it costs $55 to produce a thousand widgets). Let N be the number, in thousands, of widgets produced in a month.

    a. Find a formula for the manufacturer's total cost C as a function of N.

    b. The manufacturer sells the widgets for $58 per thousand widgets. Find a formula for the total revenue R as a function of N.

    c. Use your answers to parts a and b to find a formula for the profit P of this manufacturer as a function of N.

    d. Use your formula from part c to determine the break-even point for this manufacturer.

20. **Growth in Weight and Height**   Between the ages of 7 and 11 years, the weight w, in pounds, of a certain girl is given by the formula

$$w = 8t.$$

    Here t represents her age in years.

    a. Use a formula to express the age t of the girl as a function of her weight w.

    b. At what age does she attain a weight of 68 pounds?

    c. The height h, in inches, of this girl during the same period is given by the formula

$$h = 1.8t + 40.$$

    i. Use your answer to part b to determine how tall she is when she weighs 68 pounds.

    ii. Use a formula to express the height h of the girl as a function of her weight w.

    iii. Answer the question in part i again, this time using your answer to part ii.

21. **Plant Growth**   The amount of growth of plants in an ungrazed pasture is a function of the amount of plant biomass already present and the amount of rainfall.[22] For a pasture in the arid zone of Australia, the formula

$$Y = -55.12 - 0.01535N$$
$$-0.00056N^2 + 3.946R \qquad (2.6)$$

gives an approximation of the growth. Here R is the amount of rainfall (in millimeters) over a 3-month period, N is the plant biomass (in kilograms per hectare) at the beginning of that period, and Y is the growth (in kilograms per hectare) of the biomass over that period. (For comparison, 100 millimeters is about 3.9 inches, and 100 kilograms per hectare is about 89 pounds per acre.)

    a. Solve Equation (2.6) for R.

    b. Ecologists are interested in the relationship between the amount of rainfall and the initial plant biomass if there is to be no plant growth over the period. Put $Y = 0$ in the equation you found in part a to get a formula for R in terms of N that describes this relationship.

    c. Use the formula you found in part b to make a graph of R versus N (again with $Y = 0$). Include values of N from 0 to 800 kilograms per hectare. This graph is called the *isocline for zero growth*. It shows the amount of rainfall needed over the 3-month period just to maintain a given initial plant biomass.

    d. With regard to the isocline for zero growth that you found in part c, what happens to R as N increases? Explain your answer in practical terms.

    e. How much rainfall is needed just to maintain the initial plant biomass if that biomass is 400 kilograms per hectare?

    f. A point below the zero isocline graph corresponds to having less rainfall than is needed to sustain the given initial plant biomass, and in this situation the plants will die back. A point above the zero isocline graph corresponds to having more rainfall than is needed to sustain

---

[22]This exercise is based on the work of G. Robertson, "Plant dynamics," in G. Caughley, N. Shepherd, and J. Short, eds. *Kangaroos* (Cambridge, England: Cambridge University Press, 1987).

(continued)

the given initial plant biomass, and in this situation the plants will grow. If the initial plant biomass is 500 kilograms per hectare and there are 40 millimeters of rain, what will happen to the plant biomass over the period?

22. **Collision**   In reconstructing an automobile accident, investigators study the *total momentum*, both before and after the accident, of the vehicles involved. The total momentum of two vehicles moving in the same direction is found by multiplying the weight of each vehicle by its speed and then adding the results.[23] For example, if one vehicle weighs 3000 pounds and is traveling at 35 miles per hour, and another weighs 2500 pounds and is traveling at 45 miles per hour in the same direction, then the total momentum is $3000 \times 35 + 2500 \times 45 = 217{,}500$.

In this exercise we study a collision in which a vehicle weighing 3000 pounds ran into the rear of a vehicle weighing 2000 pounds.

a. After the collision, the larger vehicle was traveling at 30 miles per hour, and the smaller vehicle was traveling at 45 miles per hour. Find the total momentum of the vehicles after the collision.

b. The smaller vehicle was traveling at 30 miles per hour before the collision, but the speed V, in miles per hour, of the larger vehicle before the collision is unknown. Find a formula expressing the total momentum of the vehicles before the collision as a function of V.

c. The *principle of conservation of momentum* states that the total momentum before the collision equals the total momentum after the collision. Using this principle with parts a and b,

determine at what speed the larger vehicle was traveling before the collision.

23. **Competition Between Populations**   In this exercise we consider the question of competition between two populations that vie for resources but do not prey on each other. Let m be the size of the first population and n be the size of the second (both measured in thousands of animals), and assume that the populations coexist eventually. Here is an example of one common model for the interaction:

Per capita growth rate for $m = 5(1 - m - n)$,

Per capita growth rate for $n = 6(1 - 0.7m - 1.2n)$.

a. An *isocline* is formed by the points at which the per capita growth rate for m is zero. These are the solutions of the equation $5(1 - m - n) = 0$. Find a formula for n in terms of m that describes this isocline.

b. The points at which the per capita growth rate for n is zero form another isocline. Find a formula for n in terms of m that describes this isocline.

c. At an *equilibrium point* the per capita growth rates for m and for n are both zero. If the populations reach such a point, they will remain there indefinitely. Use your answers to parts a and b to find the equilibrium point.

24. **Research Project**   Find a function from a textbook in auditing, marketing, or merchandising given by a formula that involves several variables. For each of the variables, try to reverse the role of variable and function by appropriately rewriting the formula. For example, if the function f is given by a formula involving three variables x, y, and z, then you are to try to write three new formulas: one for x involving f, y, and z; one for y involving f, x, and z; and one for z involving f, x, and y. For some formulas this task is too difficult; in that case find another formula to use. Carefully describe the original formula and its variables (including units), and explain how these new formulas could be used.

2.3 SKILL BUILDING EXERCISES

Solving Linear Equations   In Exercises S-1 through S-10, solve the given linear equation.

S-1. Solve $3x + 7 = x + 21$ for x.

S-2. Solve $2x + 4 = 5x - 12$ for x.

---

[23]See J. C. Collins, *Accident Reconstruction* (Springfield, IL: Charles C Thomas Publisher, 1979). In physics the mass is used instead of the weight, but in this exercise we ignore the distinction.

S-3. Solve $3 - 5x = 23 + 4x$ for x.

S-4. Solve $13x + 4 = 5x + 33$ for x.

S-5. Solve $12x + 4 = 55x + 42$ for x.

S-6. Solve $6 - x = 16 + x$ for x.

S-7. Solve for x: $cx + d = 12$.

S-8. Solve for p: $cp = r$.

S-9. Solve for k: $2k + m = 5k + n$.

S-10. Solve for t: $tx^2 = t + 1$.

Reversing Roles of Variables   In Exercises S-11 through S-24, reverse the roles of the variables by solving for the indicated variable.

S-11. Solve for C: $F = (9/5)C + 32$

S-12. Solve for q: $p = 3q - 9$.

S-13. Solve for S: $T = 5S + 15$.

S-14. Solve for u: $w = uv + t$.

S-15. Solve for v: $w = uv + t$.

S-16. Solve for t: $w = uv + t$.

S-17. Solve for b: $a = cb + db - e$.

S-18. Solve for c: $a = cb + db - e$.

S-19. Solve for b: $a = bc + c$.

S-20. Solve for b: $a = bc + b$.

S-21. Solve for b: $a = bc + ba$.

S-22. Solve for b: $a = bc + ab$.

S-23. Solve for b: $a = bc + bd$.

S-24. Solve for b: $a = bcde$.

2.4 SOLVING NONLINEAR EQUATIONS

Nonlinear equations—those that involve powers, square roots, or other complicating factors—are in general much more difficult to solve than linear equations. In fact, most nonlinear equations cannot be solved by simple hand calculation. There are some notable exceptions. For instance, you may recall from elementary algebra that the *quadratic formula* can be used to solve *second-degree* equations. We will nonetheless treat all nonlinear equations the same and solve them using the graphing calculator.

Let's look, for example, at an aluminum bar that must be heated before it can be properly worked. The bar is placed in a preheated oven, and its temperature $A = A(t)$, t minutes later, is given by

$$A = 800 - 730e^{-0.06t} \text{ degrees Fahrenheit.}$$

The aluminum bar will be ready for bending and shaping when it reaches a temperature of 600 degrees. How long should the bar remain in the oven?

We want to know when the temperature A is 600 degrees. That is, we want to solve the equation

$$800 - 730e^{-0.06t} = 600$$

for t. This equation is not a linear equation like the ones we solved in Section 2.3. Rather, it is *nonlinear*, and it cannot be solved via the methods of Section 2.3. We will show two methods for solving equations of this type using the graphing calculator.

 The Crossing-Graphs Method

We illustrate the crossing-graphs method using the equation

$$800 - 730e^{-0.06t} = 600.$$

For this method we will make two graphs, the graph of the left-hand side of the equation, $800 - 730e^{-0.06t}$, and the graph of the right-hand side of the equation, the function

```
Plot1 Plot2 Plot3
\Y₁≡800-730e^(⁻0
.06X)
\Y₂≡600
\Y₃=
\Y₄=
\Y₅=
\Y₆=
```

FIGURE 2.52 Entering left-hand and right-hand sides of the equation for the crossing-graphs method

---

The crossing-graphs method gives more accurate answers than a rough table of values.

The crossing-graphs method can be used to solve any equation you are likely to encounter.

with a constant value of 600. Then we will find out where they are the same—that is, where the graphs cross.

The first step is to enter both functions as shown in Figure 2.52. We record the appropriate correspondences:

$Y_1 = A$, temperature, in degrees Fahrenheit, on vertical axis

$Y_2 = 600$, target temperature, in degrees Fahrenheit

$X = t$, minutes on horizontal axis.

To get a rough idea of when the temperature will reach 600 degrees, we look at the table of values for $800 - 730e^{-0.06t}$ in Figure 2.53, where we used a starting value of 0 and an increment of 5 minutes. We see that the temperature will reach 600 degrees between 20 and 25 minutes after the bar is placed in the oven.

On the basis of the table in Figure 2.53, we use a window setup with a horizontal span from 0 to 30 minutes and a vertical span from 0 to 700 degrees. The graphs are shown in Figure 2.54. We want the point where the graphs cross, or intersect. We can get reasonably close to the answer if we trace one graph and move the cursor as close as we can to the crossing point, but a calculator can locate the crossing point quite accurately, and only a few keystrokes are required. The result is shown in Figure 2.55, and we see from the prompt at the bottom of the screen that the temperature will be 600 degrees at the time 21.58 minutes after the bar is placed in the oven. Consult Appendix B to find out exactly what keystrokes are required to execute the crossing-graphs method on your calculator.

X	Y₁	Y₂
0	70	600
5	259.2	600
10	399.37	600
15	503.2	600
20	580.13	600
25	637.11	600
30	679.33	600
X=0		

FIGURE 2.53 A table of values for temperature

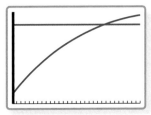

FIGURE 2.54 Graphs of temperature and the target temperature

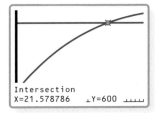

```
Intersection
X=21.578786    ↓Y=600
```

FIGURE 2.55 When the bar reaches 600 degrees

The Single-Graph Method

A slightly different method for solving nonlinear equations, the *single-graph method*, finds where a function equals 0. To illustrate it, we consider the account for a health care plan a firm offers its employees. When the firm instituted the plan, it pledged that for 5 years there would be no increase in the insurance premium, but since then health expenses have risen dramatically. The plan manager estimates that the plan's account balance B, in millions of dollars, as a function of the number of years t since the plan was instituted, is given by the formula

$$B = 102 + 12t - 100e^{0.1t}.$$

When will the plan's account run out of money? Will this happen before the 5-year period is over? To answer these questions, we need to know when the balance B is 0, so we must solve the equation

$$102 + 12t - 100e^{0.1t} = 0$$

for t. This is not a linear equation, and we will use the calculator to solve it.

An alternative to the single-graph method is to use crossing-graphs with one function being $y = 0$.

We want to see where $102 + 12t - 100e^{0.1t}$ is zero, and this is where the graph crosses the horizontal axis. Such points are known as *zeros* or *roots*. We enter the function as shown in Figure 2.56. As in the crossing-graphs method, we first look at a table of values. We are interested in the period between 0 and 5 years after the plan was instituted, and from the table in Figure 2.57, we see that the account will run out of money before the end of the 5-year period.

```
Plot1 Plot2 Plot3
\Y₁◼102+12X-100e
^(0.1X)
\Y₂=
\Y₃=
\Y₄=
\Y₅=
\Y₆=
```

X	Y₁
0	2
1	3.4829
2	3.8597
3	3.0141
4	.81753
5	-2.872
6	-8.212

X=0

FIGURE 2.56 Entering the function for the single-graph method

FIGURE 2.57 Getting an estimate with a table of values

Now we set up the graphing window. We want to see the graph near the place where it crosses the horizontal axis, so, on the basis of the table, we use a window setup with a horizontal span from 0 to 5 and a vertical span from -5 to 6. The graph is shown in Figure 2.58. As in the crossing-graphs method, we can get a pretty good estimate of where the graph crosses the horizontal axis by tracing. But once again, your calculator can find the point quickly and accurately, as shown in Figure 2.59. Consult Appendix B for specific instructions regarding your calculator. We see from the prompt at the bottom of the screen in Figure 2.59 that the account will run out of money 4.26 years, or about 4 years and 3 months, after the plan is instituted.

The graphical skills developed in Section 2.2 are crucial here and in the remainder of the text.

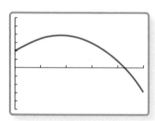

FIGURE 2.58 Picture for the single-graph method

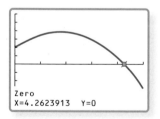

Zero
X=4.2623913   Y=0

FIGURE 2.59 Finding the root

> **KEY IDEA 2.4** **SOLVING EQUATIONS USING GRAPHS**
>
> The crossing-graphs method applies when we want to find where a function takes a particular value or where two functions are equal. The single-graph method applies directly when we want to find where a single function is equal to zero. Since more general equations can be put in this form, the single-graph method can be applied in those situations as well. The crossing-graphs method is widely applicable and is often easier to use.

We should note that whenever the calculator executes a routine to solve an equation, it is seldom able to find exactly the value it is looking for. Rather, it provides an approximation that is normally so accurate that we need not be concerned. But the presentation of approximate rather than exact answers is not a failing of the calculator. It is, rather, a mathematical necessity reflected in how the calculator functions.

EXAMPLE 2.8 **THE GEORGE RESERVE AGAIN**

For this example we look again at the logistic population growth formula introduced in Section 2.1 for deer on the George Reserve. The number N of deer expected to be present on the reserve after t years has been determined by ecologists to be

$$N = \frac{6.21}{0.035 + 0.45^t} \text{ deer.}$$

```
Plot1 Plot2 Plot3
\Y₁⬛6.21/(0.035+
0.45^X)
\Y₂=
\Y₃=
\Y₄=
\Y₅=
\Y₆=
```

FIGURE 2.60 Entering a logistic function for population growth

Part 1 Plot the graph of N versus t and explain how the deer population increases with time.

Part 2 For planning purposes, the wildlife manager for the reserve needs to know when to expect there to be 85 deer on the reserve. The answer to that is the solution of an equation. Which equation?

Part 3 Solve the equation you found in part 2.

Solution to Part 1 We enter the population function as shown in Figure 2.60 and record the proper variable associations:

$$Y_1 = N, \text{ population on vertical axis}$$

$$X = t, \text{ years on horizontal axis.}$$

To determine a good window size, we look at the table of values in Figure 2.61. This table leads us to choose the window setup in Figure 2.62. Now when we graph, we see in Figure 2.63 the classic S-shaped curve that is characteristic of logistic population growth. This graph shows that the deer population increases rapidly for the first few years, but as population size nears the *carrying capacity* of the reserve, the rate of increase slows, and the population levels off at around 177 deer.

X	Y₁
0	6
5	116.18
10	175.72
15	177.4
20	177.43
25	177.43
30	177.43

X=0

FIGURE 2.61 Searching for a good window size

```
WINDOW
Xmin=0
Xmax=15
Xscl=1
Ymin=0
Ymax=200
Yscl=1
Xres=1
```

FIGURE 2.62 Selecting the window size

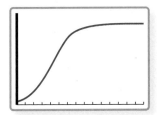

FIGURE 2.63 The graph of deer population versus time

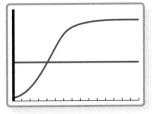

FIGURE 2.64 Adding the line at height 85

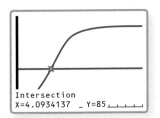

Intersection
X=4.0934137 _ Y=85

FIGURE 2.65 Solution using the crossing-graphs method

Solution to Part 2 We want to know when the deer population will reach 85. We first write the equation in words and then replace the words by the appropriate letters:

$$\text{Deer population} = 85$$

$$N = 85$$

$$\frac{6.21}{0.035 + 0.45^t} = 85.$$

The time when the deer population will reach 85 is the value of t that makes this equation true. That is, we need to solve this equation for t.

Solution to Part 3 To solve this equation we will use the crossing-graphs method. Since we have already entered the left-hand side of the equation, we need only enter the right-hand side of the equation, 85. Now we graph to see the picture in Figure 2.64. We use the calculator to find the crossing point shown in Figure 2.65. We see that 85 deer can be expected to be present in about 4.09 years, or about 4 years and 1 month.

We should note that this equation is difficult to solve by hand calculation. That is why we do it with the calculator. We should also note that acquiring the skills necessary to solve such an equation is a significant step forward in your mathematical development.

Find when there will be 100 deer on the reserve.   ■

When we use graphs to solve nonlinear equations in practical settings, it is important for the context to guide us in our choice of the graphing window. The next example illustrates this.

EXAMPLE 2.9 **A FLOATING BALL**

According to *Archimedes' law*, the weight of water that is displaced by a floating ball is equal to the weight of the ball. A certain wooden ball of diameter 4 feet weighs 436 pounds (see Figure 2.66). If the ball is allowed to float in pure water, Archimedes' law can be used to show that d feet of its diameter will be below the surface of the water, where d is a solution of the equation

$$62.4\pi d^2\left(2 - \frac{d}{3}\right) = 436.$$

(The left-hand side of this equation represents the weight of the displaced water, and the right-hand side is the weight of the ball.) How much of this ball's diameter is below the surface of the water?

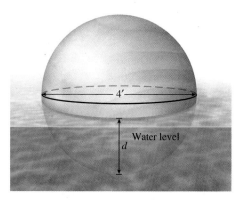

FIGURE 2.66

Solution To use the crossing-graphs method, we graph the function from the left-hand side of the equation, namely

$$62.4\pi d^2\left(2 - \frac{d}{3}\right),$$

and we graph the constant function 436 from the right-hand side of the equation. The first step is to enter the functions into the calculator and record the appropriate correspondences:

Y_1 = Weight of displaced water, in pounds, on vertical axis

Y_2 = Weight of ball, in pounds

X = d, diameter below surface, in feet, on horizontal axis.

It is important to remember that the diameter of the ball is 4 feet. Thus we are interested in viewing the graphs only on the span from $d = 0$ to $d = 4$. (See Exercise 7 at the end of this section.) To get the proper vertical span of the viewing screen, we consult the table in Figure 2.67. On the basis of this, we set the viewing window as shown in

X	Y₁
0	⁻436
1	⁻109.3
2	609.52
3	1328.3
4	1655
5	1197.6
6	⁻436

X = 4

FIGURE 2.67 A table for setting vertical screen span

The diameter of the ball gives us the horizontal range we need. We find the vertical span from a table of values.

Figure 2.68. Now when we graph, we get the picture shown in Figure 2.69. We use the calculator to find where the graphs cross as shown in Figure 2.70. We read the answer $d = 1.18$ feet from the X= display in Figure 2.70.

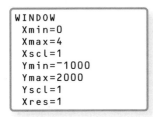

```
WINDOW
 Xmin=0
 Xmax=4
 Xscl=1
 Ymin=⁻1000
 Ymax=2000
 Yscl=1
 Xres=1
```

FIGURE 2.68 The properly configured WINDOW menu

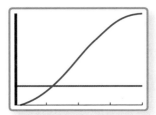

FIGURE 2.69 Preparing to solve using the crossing-graphs method

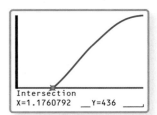

```
Intersection
X=1.1760792 __Y=436 ____
```

FIGURE 2.70 The part of a floating ball that is below the water line

TEST YOUR UNDERSTANDING | **FOR EXAMPLE 2.9**

What would be the value of d if the ball weighed 500 pounds? ■

ANSWERS FOR TEST YOUR UNDERSTANDING

2.8 At 4.52 years

2.9 $d = 1.27$ feet

2.4 EXERCISES

Reminder Round all answers to two decimal places unless otherwise indicated.

1. **Engine Displacement** The *displacement* for an automobile engine is the total volume displaced by the cylinders in one cycle of the engine. It is calculated in terms of the *stroke S* and the *bore B*. The stroke is the total distance traveled by the piston from the top of the cylinder to the bottom of the cylinder. The bore is the diameter of the cylinder. If bore and stroke are given in inches, then the displacement D is in cubic inches. The displacement is given by the formula

$$D = n\pi S(B/2)^2,$$

where n is the number of cylinders.

a. Subaru's 2.5L H4 model 2000 engine has 4 cylinders with a bore of 3.92 inches and a stroke of 3.11 inches. What is the displacement for this engine?

b. A 6-cylinder engine with a stroke of 3.01 inches is to be built so that the total displacement is 175 cubic inches. What is the proper bore for this engine?

2. **Altman's z-score** Altman's z-score is a financial tool for predicting insolvency in a business. For a certain company, the z-score formula is

$$z = 0.84 + 0.6 \times \frac{\text{Stock price} \times \text{Outstanding shares}}{\text{Total liability}}.$$

In this exercise we assume that there are 45,000 outstanding shares at a price of $4.80 each.

a. Use L for total liability, in dollars, and give a formula for the z-score for this company.

b. Would a larger total liability indicate a better or worse outlook for the company?

c. A z-score of 1.81 or lower indicates a high probability of insolvency. What is the smallest value of L that will give a z-score that indicates a high probability of insolvency?

3. **Admiralty Coefficient** The *Admiralty Coefficient* A is a nautical value that relates a ship's (top) speed S, horsepower H, and displacement D. The relationship is

$$A = \frac{D^{2/3}S^3}{H}.$$

Here speed is measured in knots and displacement is measured in tons. For *Victory Ships* produced in World War II, the Admiralty Coefficient was about 427. Their engines produced 6000 horsepower, and they had a displacement of 14,800 tons. What was the speed of a typical Victory Ship?

4. **Propeller Size** An ideal diameter d, in feet, of a ship's propeller is given by the formula

$$d = 50\left(\frac{h}{r^3}\right)^{1/5}.$$

Here h is the horsepower of the engine driving the propeller, and r is the (maximum) revolutions per minute of the propeller.

   a. The *Victory Ships* produced in World War II had an engine that produced 6000 horsepower and rotated the propeller at 100 revolutions per minute. According to the formula, what should have been the diameter of their propellers? (The actual diameter was 18.2 feet.)

   b. If a ship's engine has a horsepower of 8500 turning a propeller with a diameter of 20 feet, what should be the (maximum) revolutions per minute of the propeller?

5. **A Population of Foxes** A breeding group of foxes is introduced into a protected area, and the population growth follows a logistic pattern. After t years the population of foxes is given by

$$N = \frac{37.5}{0.25 + 0.76^t} \text{ foxes.}$$

   a. How many foxes were introduced into the protected area?

   b. Make a graph of N versus t and explain in words how the population of foxes increases with time.

   c. When will the fox population reach 100 individuals?

6. **Profit** The monthly profit P for a widget producer is a function of the number n of widgets sold. The formula is

$$P = -15 + 10n - 0.2n^2.$$

Here P is measured in thousands of dollars, n is measured in thousands of widgets, and the formula is valid up to a level of 15 thousand widgets sold.

   a. Make a graph of P versus n.

   b. Calculate $P(1)$ and explain in practical terms what your answer means.

   c. Is the graph concave up or concave down? Explain in practical terms what this means.

   d. The break-even point is the sales level at which the profit is 0. Find the break-even point for this widget producer.

7. **Revisiting the Floating Ball** In Example 2.9 we solved the equation

$$62.4\pi d^2\left(2 - \frac{d}{3}\right) = 436$$

on the interval from $d = 0$ to $d = 4$. If you graph the function $62.4\pi d^2(2 - d/3)$ and the constant function 436 on the span from $d = -2$ to $d = 7$, you will see that this equation has two other solutions. Find these solutions. Is there a physical interpretation of these solutions that makes sense?

8. **The Skydiver Again** When a skydiver jumps from an airplane, her downward velocity, in feet per second, before she opens her parachute is given by $v = 176(1 - 0.834^t)$, where t is the number of seconds that have elapsed since she jumped from the airplane. We found earlier that the terminal velocity for the skydiver is 176 feet per second. How long does it take to reach 90% of terminal velocity?

9. **Falling with a Parachute** If an average-sized man with a parachute jumps from an airplane, he will fall $12.5(0.2^t - 1) + 20t$ feet in t seconds. How long will it take him to fall 140 feet?

10. **A Cup of Coffee** The temperature C of a fresh cup of coffee t minutes after it is poured is given by

$$C = 125e^{-0.03t} + 75 \text{ degrees Fahrenheit.}$$

Tiplyashin Anatoly/Shutterstock.com

   a. Make a graph of C versus t.

   b. The coffee is cool enough to drink when its temperature is 150 degrees. When will the coffee be cool enough to drink?

   c. What is the temperature of the coffee in the pot? (*Note:* We are assuming that the coffee pot is being kept hot and is the same temperature as the cup of coffee when it was poured.)

   d. What is the temperature in the room where you are drinking the coffee? (*Hint:* If the coffee is left to cool a long time, it will reach room temperature.)

(continued)

11. **Reaction Rates**   In a chemical reaction, the reaction rate R is a function of the concentration x of the product of the reaction. For a certain second-order reaction between two substances, we have the formula

$$R = 0.01x^2 - x + 22.$$

Here x is measured in moles per cubic meter and R is measured in moles per cubic meter per second.

   a. Make a graph of R versus x. Include concentrations up to 100 moles per cubic meter.

   b. Use functional notation to express the reaction rate when the concentration is 15 moles per cubic meter, and then calculate that value.

   c. The reaction is said to be in *equilibrium* when the reaction rate is 0. At what two concentrations is the reaction in equilibrium?

12. **Population Growth**   The growth G of a population of lower organisms over a day is a function of the population size n at the beginning of the day. If both n and G are measured in thousands of organisms, the formula is

$$G = -0.03n^2 + n.$$

   a. Make a graph of G versus n. Include values of n up to 40 thousand organisms.

   b. Calculate $G(35)$ and explain in practical terms what your answer means.

   c. For what two population levels will the population grow by 5 thousand over a day?

   d. If there is no population to start with, of course there will be no growth. At what other population level will there be no growth?

13. **Van der Waals Equation**   In Exercise 16 at the end of Section 2.3, we discussed the *ideal gas law*, which shows the relationship among volume V, pressure p, and temperature T for a fixed amount (1 mole) of a gas. But chemists believe that in many situations, the *van der Waals equation* gives more accurate results. If we measure temperature T in kelvins, volume V in liters, and pressure p in atmospheres (1 atm is the pressure exerted by the atmosphere at sea level), then the relationship for carbon dioxide is given by

$$p = \frac{0.082T}{V - 0.043} - \frac{3.592}{V^2} \text{ atm.}$$

What volume does this equation predict for 1 mole of carbon dioxide at 500 kelvins and 100 atm? (*Suggestion:* Consider volumes ranging from 0.1 to 1 liter.)

14. **Radioactive Decay**   The *half-life* of a radioactive substance is the time H that it takes for half of the substance to change form through radioactive decay. This number does not depend on the amount with which you start. For example, carbon-14 is known to have a half-life of $H = 5770$ years. Thus if you begin with 1 gram of carbon-14, then 5770 years later you will have $\frac{1}{2}$ gram of carbon-14. And if you begin with 30 grams of carbon-14, then after 5770 years there will be 15 grams left. In general, radioactive substances decay according to the formula

$$A = A_0 \times 0.5^{t/H},$$

where H is the half-life, t is the elapsed time, A_0 is the amount you start with (the amount when $t = 0$), and A is the amount left at time t.

   a. Uranium-228 has a half-life H of 9.3 minutes. Thus the decay function for this isotope of uranium is

$$A = A_0 \times 0.5^{t/9.3},$$

   where t is measured in minutes. Suppose we start with 8 grams of uranium-228.

   i. How much uranium-228 is left after 2 minutes?

   ii. How long will you have to wait until there are only 3 grams left?

   b. Uranium-235 is the isotope of uranium that can be used to make nuclear bombs. It has a half-life of 713 million years. Suppose we start with 5 grams of uranium-235.

   i. How much uranium-235 is left after 200 million years?

   ii. How long will you have to wait until there are only 3 grams left?

15. **Radiocarbon Dating**   *This is a continuation of Exercise 14.* We rewrite the decay formula as

$$\frac{A}{A_0} = 0.5^{t/H}.$$

This formula shows the fraction of the original amount that is present as a function of time.

   Carbon-14 has a half-life of 5.77 thousand years. It is thought that in recent geological time (the last few million years or so), the amount of ^{14}C in the atmosphere has remained constant. As a consequence, all organisms that take in air (trees, people, and so on) maintain the same level of ^{14}C so long as they are alive. When a living organism dies, it no longer takes in ^{14}C, and the amount present at death begins to decay according to the formula above. This phenomenon can be used to date some archaeological objects.

Suppose that the amount of carbon-14 in charcoal from an ancient campfire is 1/3 of the amount in a modern, living tree. In terms of the formula above, this means $A/A_0 = 1/3$. When did the tree that was used to make the campfire die? Be sure to explain how you got your answer.

16. **Monthly Payment on a Loan** If you borrow $5000 at an APR of r (as a decimal) from a lending institution that compounds interest continuously, and if you wish to pay off the note in 3 years, then your monthly payment M, in dollars, can be calculated using

$$M = \frac{5000(e^{r/12} - 1)}{1 - e^{-3r}}.$$

Your budget will allow a payment of $150 per month, and you are shopping for an interest rate that will give a payment of this size. What interest rate do you need to find?

17. **Grazing Kangaroos**
The amount of vegetation eaten in a day by a grazing animal is a function of the amount V of food available (measured as biomass, in units such as pounds per acre).[24] This relationship is called the *functional response*. If there is little vegetation available, the daily intake will be small, since the animal will have difficulty finding and eating the food. As the food biomass increases, so does the daily intake. Clearly, though, there is a limit to the amount the animal will eat, regardless of the amount of food available. This maximum amount eaten is the *satiation level*.

Rusty Dodson/Shutterstock.com

a. For the western grey kangaroo of Australia, the functional response is

$$G = 2.5 - 4.8e^{-0.004V},$$

where $G = G(V)$ is the daily intake (measured in pounds) and V is the vegetation biomass (measured in pounds per acre).

i. Draw a graph of G against V. Include vegetation biomass levels up to 2000 pounds per acre.

ii. Is the graph you found in part i concave up or concave down? Explain in practical terms what your answer means about how this kangaroo feeds.

iii. There is a *minimal* vegetation biomass level below which the western grey kangaroo will eat nothing. (Another way of expressing this is to say that the animal cannot reduce the food biomass below this level.) Find this minimal level.

iv. Find the satiation level for the western grey kangaroo.

b. For the red kangaroo of Australia, the functional response is

$$R = 1.9 - 1.9e^{-0.033V},$$

where R is the daily intake (measured in pounds) and V is the vegetation biomass (measured in pounds per acre).

i. Add the graph of R against V to the graph of G you drew in part a.

ii. A simple measure of the grazing efficiency of an animal involves the minimal vegetation biomass level described above: The lower the minimal level for an animal, the more efficient it is at grazing. Which is more efficient at grazing, the western grey kangaroo or the red kangaroo?

18. **Grazing Rabbits and Sheep** *This is a continuation of Exercise 17.* In addition to the kangaroos, the major grazing mammals of Australia include merino sheep and rabbits. For sheep the functional response is

$$S = 2.8 - 2.8e^{-0.01V},$$

and for rabbits it is

$$H = 0.2 - 0.2e^{-0.008V}.$$

Here S and H are the daily intake (measured in pounds), and V is the vegetation biomass (measured in pounds per acre).

a. Find the satiation level for sheep and that for rabbits.

---

[24]This exercise and the next are based on the work of J. Short, "Factors affecting food intake of rangelands herbivores," in G. Caughley, N. Shepherd, and J. Short, eds., *Kangaroos* (Cambridge, England: Cambridge University Press, 1987).

(continued)

b. One concern in the management of range-lands is whether the various species of grazing animals are forced to compete for food. It is thought that competition will not be a problem if the vegetation biomass level provides at least 90% of the satiation level for each species. What biomass level guarantees that competition between sheep and rabbits will not be a problem?

19. Growth Rate The per capita growth rate r (on an annual basis) of a population of grazing animals is a function of V, the amount of vegetation available. A positive value of r means that the population is growing, whereas a negative value of r means that the population is declining. For the red kangaroo of Australia, the relationship has been given[25] as

$$r = 0.4 - 2e^{-0.008V}.$$

Here V is the vegetation biomass, measured in pounds per acre.

a. Draw a graph of r versus V. Include vegetation biomass levels up to 1000 pounds per acre.

b. The population size will be stable if the per capita growth rate is zero. At what vegetation level will the population size be stable?

20. Hosting a Convention You are hosting a convention for a charitable organization. You pay a rental fee of $25,000 for the convention center, plus you pay the caterer $12 for each person who attends the convention. Suppose you just want to break even.

a. Use a formula to express the amount you should charge per ticket as a function of the number of people attending. Be sure to explain the meaning of the letters you choose and the units.

b. Make a graph of the function that gives the amount you should charge per ticket. Include attendance sizes up to 1000.

c. How many must attend if you are to break even with a ticket price of $50?

21. Breaking Even *The background for this exercise can be found in Exercises 13, 14, 15, and 16 in Section 1.4.* A manufacturer of widgets has fixed costs of $700 per month, and the variable cost is $65 per thousand widgets (so it costs $65 to produce 1 thousand widgets). Let N be the number, in thousands, of widgets produced in a month.

a. Find a formula for the manufacturer's total cost C as a function of N.

b. The highest price p, in dollars per thousand widgets, at which N can be sold is given by the formula $p = 75 - 0.02N$. Using this, find a formula for the total revenue R as a function of N.

c. Use your answers to parts a and b to find a formula for the profit P of this manufacturer as a function of N.

d. Use your formula from part c to determine the two break-even points for this manufacturer. Assume that the manufacturer can produce at most 500 thousand widgets in a month.

22. Tubeworms In a study of a marine tubeworm, scientists developed a model for the time T (measured in years) required for the tubeworm to reach a length of L meters.[26] On the basis of their model, they estimate that 170 to 250 years are required for the organism to reach a length of 2 meters, making this tubeworm the longest-lived noncolonial marine invertebrate known. The model is

$$T = \frac{20e^{bL} - 28}{b}.$$

Here b is a constant that requires estimation. What is the value of b if the lower estimate of 170 years to reach a length of 2 meters is correct? What is the value of b if the upper estimate of 250 years to reach a length of 2 meters is correct?

23. Water Flea F. E. Smith has reported on population growth of the water flea.[27] In one experiment, he found that the time t, in days, required to reach a population of N is given by the relation

$$e^{0.44t} = \frac{N}{N_0}\left(\frac{228 - N_0}{228 - N}\right)^{4.46}.$$

Here N_0 is the initial population size. If the initial population size is 50, how long is required for the population to grow to 125?

---

[25]This formula is based on the work of P. Bayliss and G. Caughley, in G. Caughley, *Ibid.*
[26]D. Bergquist, F. Williams, and C. Fisher, "Longevity record for deep-sea invertebrate," *Nature* **403** (2000), 499–500.
[27]See "Population dynamics in *Daphnia magna* and a new model for population growth," *Ecology* **44** (1963), 651–663. The population size is measured per unit volume.

24. Holling's Functional Response Curve The total number P of prey taken by a predator depends on the availability of prey. C. S. Holling proposed a function of the form $P = cn/(1 + dn)$ to model the number of prey taken in certain situations.[28] Here n is the density of prey available, and c and d are constants that depend on the organisms involved as well as on other environmental features. Holling took data gathered earlier by T. Burnett on the number of sawfly cocoons found by a small wasp parasite at given host density. In one such experiment conducted, Holling found the relationship

$$P = \frac{21.96n}{1 + 2.41n},$$

where P is the number of cocoons parasitized and n is the density of cocoons available (measured as number per square inch).

a. Draw a graph of P versus n. Include values of n up to 2 cocoons per square inch.

b. What density of cocoons will ensure that the wasp will find and parasitize 6 of them?

c. There is a limit to the number of cocoons that the wasp is able to parasitize no matter how readily available the prey may be. What is this upper limit?

25. Radius of a Shock Wave An explosion produces a spherical shock wave whose radius R expands rapidly. The rate of expansion depends on the energy E of the explosion and the elapsed time t since the explosion. For many explosions, the relation is approximated closely by

$$R = 4.16E^{0.2}t^{0.4}.$$

Here R is the radius in centimeters, E is the energy in ergs, and t is the elapsed time in seconds. The relation is valid only for very brief periods of time, perhaps a second or so in duration.

a. An explosion of 50 pounds of TNT produces an energy of about 10^{15} ergs. See Figure 2.71. How long is required for the shock wave to reach a point 40 meters (4000 centimeters) away?

b. A nuclear explosion releases much more energy than conventional explosions. A small nuclear device of yield 1 kiloton releases approximately 9×10^{20} ergs. How long would it take for the shock wave from such an explosion to reach a point 40 meters away?

c. The shock wave from a certain explosion reaches a point 50 meters away in 1.2 seconds. How much energy was released by the explosion? The values of E in parts a and b may help you set an appropriate window.

Note: In 1947 the government released film of the first nuclear explosion in 1945, but the *yield* of the explosion remained classified. Sir Geoffrey Taylor used the film to determine the rate of expansion of the shock wave and so was able to publish a scientific paper[29] concluding correctly that the yield was in the 20-kiloton range.

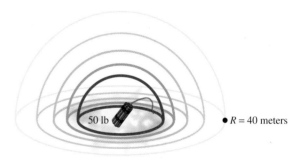

FIGURE 2.71

26. Friction Loss in Fire Hoses When water flows inside a hose, the contact of the water with the wall of the hose causes a drop in pressure from the pumper to the nozzle. This drop is known as *friction loss*. Although it has come under criticism for lack of accuracy, the most commonly used method for calculating friction loss for flows under

---

[28]"Some characteristics of simple types of predation and parasitism," *Can. Ent.* **91** (1959), 385–398.
[29]"The formation of a blast wave by a very intense explosion, II: The atomic explosion of 1945," *Proc. Roy. Soc. A* **201** (1950), 175–186.

(continued)

100 gallons per minute uses what is called the *underwriter's formula*:

$$F = \left(2\left(\frac{Q}{100}\right)^2 + \frac{Q}{200}\right)\left(\frac{L}{100}\right)\left(\frac{2.5}{D}\right)^5.$$

Here F is the friction loss in pounds per square inch, Q is the flow rate in gallons per minute, L is the length of the hose in feet, and D is the diameter of the hose in inches.

a. In a 500-foot hose of diameter 1.5 inches, the friction loss is 96 pounds per square inch. What is the flow rate?

b. In a 500-foot hose, the friction loss is 80 pounds per square inch when water flows at 65 gallons per minute. What is the diameter of the hose? Round your answer to the nearest $\frac{1}{8}$ inch.

27. Home Equity When you purchase a home by securing a mortgage, the total paid toward the principal is your *equity* in the home. If your mortgage is for P dollars, and if the term of the mortgage is t months, then your equity E, in dollars, after k monthly payments is given by

$$E = P \times \frac{(1 + r)^k - 1}{(1 + r)^t - 1}.$$

Here r is the monthly interest rate as a decimal, with $r = APR/12$.

Suppose you have a mortgage of $425,000 at an APR of 9% and a term of 30 years. How long does it take for your equity to reach half of the amount of the original mortgage? (Round r to four decimal places.)

28. Research Project Find an equation involving one variable from a textbook in some other subject (not mathematics) or other handy source. Solve that equation using the techniques you've learned in this section. Compare your solution with the solution found in your source. Write a careful description of the original equation, of its use, and of your solution, and add a brief description of the solution found in your source.

2.4 SKILL BUILDING EXERCISES

The Crossing-Graphs Method In Exercises S-1 through S-12, use the crossing-graphs method to solve the given equation.

S-1. $\dfrac{20}{1 + 2^x} = x$

S-2. $x^2 - x^3 + 3 = x^5/20$

S-3. $3^x + x = 2^x + 1$

S-4. $\sqrt{x^2 + 1} = x^3 + \sqrt{x^4 + 2}$

S-5. $\dfrac{5}{x^2 + x + 1} = 1$. There are two solutions. Find them both.

S-6. $x + 2^x = 5$

S-7. $x + \sqrt{x + 1} = \sqrt{x + 2}$

S-8. $\dfrac{1}{x} + \dfrac{x}{x + 2} = \dfrac{1}{x + 1}$

S-9. $2^x + 3^x = 4^x$

S-10. $x^4 = 2^x$. There are three solutions. Find all three.

S-11. $4 + x = (1 + x)(1 + x^2)$

S-12. $\dfrac{1 + 2^x}{1 + 3^x} = x^2$. There are two solutions. Find them both.

The Single-Graph Method In Exercises S-13 through S-23, use the single-graph method to solve the given equation.

S-13. $\dfrac{20}{1 + 2^x} - x = 0$.

S-14. $\dfrac{5}{x^2 + x + 1} - 1 = 0$. There are two solutions. Find them both.

S-15. $3^x - 8 = 0$.

S-16. $\dfrac{-x^4}{x^2 + 1} + 1 = 0$. There are two solutions. Find them both.

S-17. $x^3 - x - 5 = 0$

S-18. $x^3 - 2^x = 0$

S-19. $\sqrt{x^2 + 1} = 2x$

S-20. $e^{-x} - x = 0$

S-21. $\dfrac{x^4}{x^2 + 1} - 0.2 = 0$. There are two solutions. Find them both.

S-22. $e^{3x} - 5e^{2x} = 0$

S-23. $x^4 + x^5 = x^6$. There are three solutions. Find all three.

S-24. $37.5/(0.25 + 0.76^x) = 100$

2.5 INEQUALITIES

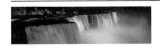

As we have seen, problems in science, business, industry, and the world around us often are modeled by equations. The graphing calculator (or computer) can solve the equations we encounter quickly and accurately. But it often happens that the required model is an *inequality* rather than an equality. The same technology that allows us to solve equations can also solve inequalities.

As an example, a certain drug is injected into the bloodstream. The concentration of the drug initially increases rapidly but then declines over time. Suppose the concentration after t hours is given by $C = 70t \times 0.7^t$ milligrams per deciliter. The drug is known to be effective as long as there are 50 milligrams per deciliter or more in the bloodstream. We wish to find the time period when the drug is effective. That is, we want to know when the concentration is 50 milligrams per deciliter or more. Expressed in terms of formulas, this condition is an inequality:

$$70t \times 0.7^t \geq 50.$$

```
Plot1 Plot2 Plot3
\Y1=70X*0.7^X
\Y2=50
\Y3=
\Y4=
\Y5=
\Y6=
\Y7=
```

FIGURE 2.72 Two functions entered

We make a graph in order to solve this inequality. The first step is to enter the two functions involved, as shown in Figure 2.72. We record the appropriate correspondences:

$Y_1 = C$, concentration, in milligrams per deciliter, on the vertical axis

$Y_2 = 50$, target concentration, in milligrams per deciliter

$X = t$, time in hours on the horizontal axis.

X	Y_1	Y_2
0	0	50
2	68.6	50
4	67.228	50
6	49.413	50
8	32.283	50
10	19.773	50
12	11.627	50
X=0		

FIGURE 2.73 A table of values for concentration

We consult the table of values in Figure 2.73 to help select a viewing window for the graph. The table uses a starting value of 0 and an increment of 2 hours.

Note once again how a table helps us choose a viewing window.

On the basis of this table we choose a horizontal span of 0 to 10 and a vertical span of 0 to 80 as shown in Figure 2.74. With this window we get the graph in Figure 2.75.

The crossing-graphs method can help solve inequalities as well as equalities.

The inequality is satisfied (and hence the drug effective) when the graph of C is at or above the line $C = 50$. The time period is clearly pictured in Figure 2.75. To get the correct time period we need to find the two crossing points. We find these using the crossing-graphs method presented in the previous section. From Figure 2.75 we see that the first crossing point is at 1.03 hours, and Figure 2.76 shows that the second crossing point at 5.94 hours. Hence the drug is effective in the time period from 1.03 hours to 5.94 hours. If wanted to express the solution as an inequality, we would write $1.03 \leq t \leq 5.94$.

Inequalities are solved by finding when one graph is above another.

```
WINDOW
 Xmin=0
 Xmax=10
 Xscl=1
 Ymin=0
 Ymax=80
 Yscl=1
 Xres=1
```

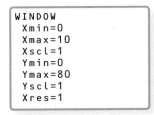

FIGURE 2.74 Setting the viewing window

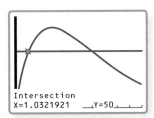

FIGURE 2.75 The graph and first point of intersection

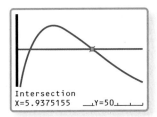

FIGURE 2.76 The second point of intersection

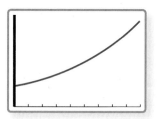

X	Y₁
4.0000	494.39
5.0000	583.38
6.0000	688.39
7.0000	812.30
8.0000	958.51
9.0000	1131.0
10.000	1334.6

X=10

FIGURE 2.77 A table of values for gold prices

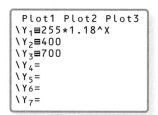

FIGURE 2.78 A graph of gold prices

EXAMPLE 2.10 **GOLD PRICES**

Beginning in 2001 the price of gold was modeled approximately by $G(t) = 255 \times 1.18^t$ dollars per ounce, where t is the time in years since the beginning of 2001.

Part 1 What price of gold does this model give for 2006?

Part 2 Make a graph of gold prices for 2001 through 2010.

Part 3 A certain broker felt that gold was a good buy as long as its price was between $400 and $700 per ounce. This time period is the solution of an inequality involving G. State the inequality.

Part 4 Solve the inequality from part 3 to find the time period the broker thought was good for investment in gold.

Part 5 This same broker counseled his customers to sell their gold when the price was at or above $900 per ounce. When in the period from 2001 to 2010 was this broker advising customers to sell?

Solution to Part 1 Now 2006 is 5 years after 2001, so we use $t = 5$ in the formula for G:

$$G(5) = 255 \times 1.18^5$$
$$= 583.38.$$

Thus the model gives $583.38 per ounce as the price of gold in 2006. It is interesting to note that the actual value was $603.46 per ounce.

Solution to Part 2 We record the following variable associations:

$$Y_1 = G, \text{ gold price, in dollars per ounce, on vertical axis}$$

$$X = t, \text{ time in years on horizontal axis.}$$

The year 2010 corresponds to $t = 9$. The table of values in Figure 2.77 shows the value of gold increasing up to just over $1131 dollars per ounce by 2010. So we choose a horizontal span of 0 to 9 and a vertical span of 0 to 1200. The graph is in Figure 2.78.

Solution to Part 3 We want to know when gold prices are between $400 and $700 per ounce. That is when $G(t)$ is at least 400 but no more than 700:

$$400 \leq G(t) \leq 700.$$

Solution to Part 4 To solve the inequality we add the target values of 400 and 700 as shown in Figure 2.79:

$$Y_2 = 400$$
$$Y_3 = 700.$$

The new graph is shown in Figure 2.80. Gold is at its target price between the two horizontal lines. We find the crossing points using the crossing-graphs method. The lower crossing point $t = 2.72$ is shown in Figure 2.80. The upper crossing point $t = 6.10$ is shown in Figure 2.81. This corresponds to a buying period from late 2003 to early 2007.

Solution to Part 5 Now we want to know when the price of gold is $900 or more. That is when $G(t) \geq 900$. Using $Y_3 = 900$, we enter this new target value. This new line is shown in Figure 2.82. The broker is counseling sale of gold when the graph is at or above the top line in the figure. That is, we sell on or after $t = 7.62$, which is late 2008.

```
Plot1 Plot2 Plot3
\Y₁⊟255*1.18^X
\Y₂⊟400
\Y₃⊟700
\Y₄=
\Y₅=
\Y₆=
\Y₇=
```

FIGURE 2.79 Adding target values

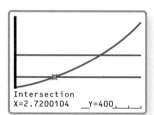

Intersection
X=2.7200104   Y=400

FIGURE 2.80 The lower crossing value

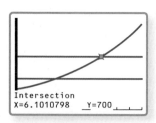

Intersection
X=6.1010798   Y=700

FIGURE 2.81 The upper crossing value

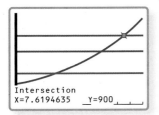

FIGURE 2.82 A time to sell gold

TEST YOUR UNDERSTANDING | **FOR EXAMPLE 2.10**

During what time period would the broker advise buying if he thought $500 to $800 was an appropriate buy range? ■

Sometimes the inequalities we need to solve involve more than one function, as is shown in the next example.

EXAMPLE 2.11 **PRODUCING FINGERLINGS**

A fish farm makes its money by selling fingerlings. Hence, income depends on the growth rate of the fish population. For a certain farm the income, in dollars per month, is given by

$$I = 5000(0.3n - 0.02n^2),$$

where n is the fish population in hundreds of fish. But the expense of maintaining the farm also depends on the population. It is given by

$$E = 1000 + 250n \text{ dollars per month.}$$

Part 1 Use a table of values to find the population range over which income I is positive.

Part 2 Make a graph of income over the range you found in part 1.

Part 3 The fish farm is profitable when income exceeds expenses. Add the graph of expenses to the picture and explain what the picture shows in terms of profitability.

Part 4 What range of fish population results in a profitable fish farm?

Solution to Part 1 We enter the income function using the following assignments.

$$Y_1 = I, \text{ income in dollars}$$

$$X = n, \text{ population of fish, in hundreds.}$$

Consulting the tables of values shown in Figure 2.83 and Figure 2.84, we see that income is positive when the population is between 0 and 15 hundred fish.

X	Y_1
-2.000	-3400
-1.000	-1600
0.0000	0.0000
1.0000	1400.0
2.0000	2600.0
3.0000	3600.0
4.0000	4400.0
X=-2	

FIGURE 2.83 A table of values for income near $n = 0$

X	Y_1
11.000	4400.0
12.000	3600.0
13.000	2600.0
14.000	1400.0
15.000	0.0000
16.000	-1600
17.000	-3400
X=17	

FIGURE 2.84 A table of values for income near $n = 15$

Solution to Part 2 We use the horizontal span of 0 to 15 hundred fish found in part 1. The table of values we used in part 1 helps us choose a vertical span of 0 to 6000 dollars per month. The graph is shown in Figure 2.85.

Solution to Part 3 We add the expenses E to the picture using the correspondence

$$Y_2 = E, \text{ expense on the vertical axis}$$

$$X = n, \text{ population on the horizontal axis.}$$

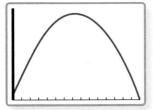

FIGURE 2.85 Graph of income

The graph of *I* is above the graph of *E* where the inequality *I* > *E* is true.

The two graphs are shown together in Figure 2.86. The farm is profitable when the graph of income is above the graph of expenses. That is where the inequality *I* > *E* is true.

Solution to Part 4 To find where income is greater than expenses, we need to find the two crossing points. We find these using the crossing-graphs method as shown in Figure 2.87 and Figure 2.88. Thus the farm is profitable when $0.86 < n < 11.64$. That is when the fish population is between 0.86 and 11.64 hundred fish.

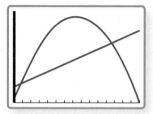

FIGURE 2.86 Graph of expenses added

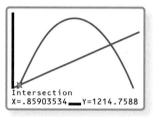

FIGURE 2.87 The left-hand crossing point

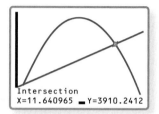

FIGURE 2.88 The right-hand crossing point

TEST YOUR UNDERSTANDING | **FOR EXAMPLE 2.11**

What population range would provide profitability if insurance costs rise by $200 per month? The new expense function is $1200 + 250n$. ■

KEY IDEA 2.5 **SOLVING INEQUALITIES**

To solve the inequality $f(x) \geq g(x)$ we make the graph of the two functions on the same screen. The inequality is true when the graph of *f* is at or above the graph of *g*. Typically we will need to find crossing points using the crossing-graphs method.

ANSWERS FOR TEST YOUR UNDERSTANDING

2.10 From $t = 4.07$ to $t = 6.91$, so early 2005 to late 2007.

2.11 The population should be between 1.05 and 11.45 hundred fish.

2.5 EXERCISES

Reminder Round all answers to two decimal places unless otherwise indicated.

1. **Estimating Wave Height** Sailors use the following function to estimate wave height *h*, in feet, from wind speed *w*, in miles per hour:

$$h = 0.02w^2.$$

   a. Make a graph of wave height versus wind speed. Include wind speeds of up to 25 miles per hour.

   b. A small boat can sail safely provided wave heights are no more than 4 feet. What range of wind speed will give safe sailing for this boat?

2. **Speed from Skid Marks** When a car makes an emergency stop on dry pavement, it leaves skid marks on the pavement. The length *L*, in feet, of the skid mark is related to the speed *S*, in miles per hour, when the brakes were applied. The relationship is

$$S(L) = 5.06\sqrt{L}.$$

   a. Make a graph of *S* versus *L*. Include lengths of up to 200 feet.

   b. A man involved in an emergency stop claimed his speed was between 30 and 40 miles per hour. What skid mark length would support this claim?

3. **Adult Weight from Puppy Weight**   There is a formula that estimates how much your puppy will weigh when it reaches adulthood. The method we present applies to medium-sized breeds. First find your puppy's weight w at an age of a weeks, where a is 16 weeks or less. Then the predicted adult weight W, in pounds, is given by the formula

$$W(a, w) = 52\frac{w}{a}.$$

In this exercise we consider medium-sized dogs whose puppy weight is $w = 12$ pounds.

   a. Make a graph of adult weight W versus puppy age a. Include ages from 2 to 16 weeks.

   b. You are hoping for an adult dog weighing between 50 and 75 pounds. What age range corresponds to this adult weight?

4. **Death of the Dinosaurs**   An object of mass m kilograms moving at a velocity of v meters per second has a kinetic energy k given by the formula

$$k(m, v) = \frac{1}{2}mv^2 \text{ joules.}$$

The *Chicxulub* object, which many scientists believe resulted in the extinction of the dinosaurs (along with many other species), struck Earth near the Yucatán region of Mexico about 65 million years ago.[30] It had a mass of about 8×10^{15} kilograms.

   The largest nuclear weapon ever exploded released an energy of about 4×10^{15} joules. Scientists estimate that the Chicxulub object released energy equivalent to at least $10^8 = 100,000,000$ of these super bombs. That is, its kinetic energy must have been at least 4×10^{23} joules.

   Use the information given here to find the range of possible velocities of the Chicxulub object. In making your graph we suggest a horizontal span from 0 to 20,000 meters per second and a vertical span from 0 to 10^{24} joules.

5. **Drug Concentration**   When a drug is administered orally, it takes some time before the blood concentration reaches its maximum level. After that time, concentration levels decrease. When 500 milligrams of procainamide is administered orally, one model

for a particular patient gives blood concentration C, in milligrams per liter, after t hours as

$$C = 2.65(e^{-0.2t} - e^{-2t}).$$

During what time period is the drug concentration level at least 1.5 milligrams per liter?

6. **Flesch-Kincaid Formula**   The *Flesch-Kincaid Grade Level Formula*[31] gives the grade reading level K of a passage of text. For this formula, W is the number of words in the passage, S is the number of sentences, and L is the number of syllables. The formula is

$$K = 0.39\frac{W}{S} + 11.8\frac{L}{W} - 15.59.$$

Note that many word-processing programs calculate the reading level of a document using this formula. In this exercise we consider passages with 20 sentences and 600 syllables.

   a. Write a formula for K for the passages described above.

   b. Make a graph of K versus W. Include passages from 300 to 600 words.

   c. Over the range considered in part b, does having more words result in a higher or lower reading level?

   d. What number of words gives a reading level of 10 to 12? Round your answers to the nearest whole numbers.

7. **Mosteller Formula for Body Surface Area**   Body surface area is an important piece of medical information because it is a factor in temperature regulation as well as some drug level determinations. The Mosteller formula[32] gives one way of estimating body surface area B, in square meters. The formula uses the weight w, in kilograms, and the height h, in centimeters. The relation is

$$B = \frac{\sqrt{hw}}{60}.$$

Consider a man who weighs 80 kilograms.

   a. Write a formula for B as a function of h.

   b. What height range will result in a body surface area range from 1.9 to 2 square meters?

8. **Cohort Biomass**   A *cohort* of fish is a group of fish born at the same time. The *biomass* of a cohort is

---

[30]The object is thought to have been a comet or small asteroid.
[31] J. P. Kincaid, R. P. Fishburne, Jr., R. L. Rogers, and B. S. Chissom (1975), *Derivation of New Readability Formulas (Automated Readability Index, Fog Count and Flesch Reading Ease Formula) for Navy Enlisted Personnel*, Research Branch Report 8–75 (Memphis-Millington, TN: Naval Technical Training Command, U. S. Naval Air Station).
[32]R. D. Mosteller, "Simplified calculation of body-surface area," *N. Engl. J. Med.* **317**(17) (Oct. 22, 1987), 1098 (letter).

(continued)

the total weight of the cohort. For a certain type of North Sea flatfish the total biomass B, in pounds, of a cohort that is t years old is given by

$$B = 31{,}600e^{-0.1t}(1 - 0.93e^{-0.095t})^3.$$

a. Make a graph of B versus t. Include ages through 30 years.

b. It is safe to allow limited fishing as long as the biomass is at least 2000 pounds. What age range is safe for fishing?

c. It is decided to protect fish over age 15 for breeding purposes. Add this restriction to the one in part b, and determine the allowable age for fishing.

9. A Train Each night at midnight a train leaves the station. The distance from a house to the train after t hours is given by

$$D = \sqrt{(30t)^2 + (5 - 90t)^2} \text{ miles.}$$

It is too noisy to sleep when the train is within 2 miles of the house. When is the train too near the house each night? Round your answers to the nearest minute. *Note:* Be sure to convert your decimal answers to minutes.

10. A Support Bar A metal bar is to act as a furniture support. It is heated with a torch and then laid on a workbench for bending. The temperature B, in degrees Fahrenheit, of the bar t minutes after heat is applied is given by

$$B = 70 + 600(0.96^t - 0.34^t).$$

a. Make a graph of the temperature of the bar over the first 20 minutes.

b. The bar is hot enough to bend as long as its temperature is at least 400 degrees. During what time period is the bar at the proper temperature for bending?

c. What is the limiting value of B?

11. Two Savings Accounts There are two savings accounts. The balances, in dollars, of the two accounts after t years are given by the formulas

$$B_1 = 500 \times 1.03^t$$
$$B_2 = 350 \times 1.05^t.$$

a. Make a graph of both balances on the same screen. Include times up to 25 years.

b. Find the time when the two account balances will be the same.

c. Explain how the two account balances compare over time.

12. North Sea Cod The rate of growth G in the weight of a fish is a function of the weight w of the fish. For North Sea cod, the relationship is given by

$$G = 2.1w^{2/3} - 0.6w.$$

Here w is measured in pounds and G in pounds per year. The maximum size for a North Sea cod is about 40 pounds.

a. Make a graph of G versus w.

b. What weight range gives a growth rate of at least 3 pounds per year?

13. Fireworks If an object is shot upward from the ground with an initial velocity of v_0 feet per second, then its height after t seconds (until it hits the ground) is given by

$$H = v_0 t - 16t^2 \text{ feet.}$$

A firework device is fired upward at a rate of 50 feet per second. You can see the device from your apartment window when it is 30 feet or higher. During what time span can you view the device?

14. Logistic Growth with a Threshold For a certain population the growth rate G, measured in thousands per year, depends on the population n, measured in thousands. The relation is

$$G = -0.76n\left(1 - \frac{n}{8}\right)\left(1 - \frac{n}{50}\right).$$

This equation determines what is known as *logistic growth* with a *threshold* of 8 thousand and a *carrying capacity* of 50 thousand.

a. Make a graph of G versus n. Include population sizes up to $n = 60$ thousand.

b. Find all places where the growth rate is 0. What is happening to the population at those points?

c. Over what population range is the growth rate positive? What is happening to the population over that range?

d. Over what ranges is the growth rate negative? What is happening to the population over those ranges?

15. More on Logistic Growth with a Threshold *This is a continuation of Exercise 14.*

a. Over what population ranges is the growth rate at least 20 thousand per year?

b. Over what population ranges is the growth rate -1 thousand per year or less?

 2.5 SKILL BUILDING EXERCISES

Solve the following inequalities.

S-1. $x^3 + x \geq 4$

S-2. $x + \dfrac{1}{1 + x^2} \geq 5$

S-3. $xe^x \geq 8$

S-4. $\pi^x \geq x^2 + 3$

S-5. $x \geq 1 + \sqrt{x}$

S-6. $\sqrt{x} + \sqrt{x + 1} \geq x + 1$ *Be careful.* The graphs cross twice.

S-7. $5 - x^2 \geq 2 + x^4$ *Be careful.* The graphs cross twice.

S-8. $x^3 \geq x$ *Be careful.* The graphs cross three times.

S-9. $x^2 + 3x \leq x^3$ *Be careful.* The graphs cross three times.

S-10. $5e^x - e^{2x} \geq x^2$ *Be careful.* The graphs cross twice.

S-11. $x^2 + 4 \geq x$ *Be careful.* The graphs do not cross.

S-12. $x^2 \geq 3^x + 3^{-x}$ *Be careful.* The graphs do not cross.

S-13. $6 \leq x^3 + x \leq 20$

S-14. $4 \leq 2^x + 3^x \leq 5$

S-15. $2^x \leq x + 4 \leq 3^x$

2.6 OPTIMIZATION

Formulas don't readily show maximum values, but graphs do.

One key feature of a graph is its *zeros*, the places where it crosses or touches the horizontal axis. We learned how to find these in Section 2.4. Other important features of a graph include its *maxima* and *minima*. The need to locate these will arise, for example, when you want to find the number of items you can produce that will yield a maximum profit or when you wish to lay a pipeline in such a way that the cost is a minimum. Graphs make it easy to find maxima and minima. Thus if we have a function given by a formula, we can *optimize* it (find the optimal values) by first getting the graph.

▌ Optimizing at Peaks and Valley

In many instances optimal values for a function are located at the peaks or valleys of its graph. For example, if a cannon is placed at the origin (X=0, Y=0) and elevated at an angle of 45 degrees (see Figure 2.89), then when it is fired, the cannonball will follow the graph of

$$y = x - g\left(\frac{x}{m}\right)^2.$$

FIGURE 2.89

X	Y₁
0	0
500	372
1000	488
1500	348
2000	⁻48
2500	⁻700
3000	⁻1608

X=0

FIGURE 2.90 A table of values for a cannonball flight

```
WINDOW
 Xmin=0
 Xmax=2500
 Xscl=1
 Ymin=-100
 Ymax=600
 Yscl=1
 Xres=1
```

FIGURE 2.91 Window settings from the table

Maxima often occur where graphs reach a peak.

(This simple model ignores air resistance.) Here x represents the number of feet downrange, y is the height in feet, and m is the *muzzle velocity,* in feet per second, of the cannonball. The constant g is the acceleration due to gravity near the surface of the Earth, about 32 feet per second per second. Thus, if the cannon is fired with a muzzle velocity of 250 feet per second, it will follow the graph of

$$y = x - 32\left(\frac{x}{250}\right)^2.$$

Let's use the graph to analyze the flight of the cannonball. The first step is to enter the function and record the appropriate correspondences:

$$Y_1 = y, \text{height, in feet, on vertical axis}$$

$$X = x, \text{distance downrange, in feet, on horizontal axis.}$$

We need to look at a table of values to help us choose a window. Since we expect the cannonball to travel several hundreds of feet, we made the table in Figure 2.90 with a starting value of 0 and a table increment of 500. Allowing a little extra room, we set the window as shown in Figure 2.91.

Now graphing produces the cannonball path shown in Figure 2.92. In this picture the horizontal axis shows feet downrange, and the vertical axis shows height above the ground. Thus the horizontal axis is ground level. In Figure 2.92 we have located the place where the graph crosses the horizontal axis—that is, where the cannonball strikes the ground—and we see that this will happen about 1953 feet downrange. (That is just over a third of a mile.)

Let's find the maximum height of the cannonball, which is at the top of the graph in Figure 2.92. We can get a reasonable estimate for the maximum if we trace the graph and move the cursor as close to the peak as possible. But most graphing calculators have the ability to find maximum values such as this in a fashion similar to the way they find zeros or where graphs cross. You should consult Appendix B for the exact keystrokes needed to accomplish this. In Figure 2.93 we have used this feature to locate the peak of the graph, and we see from the prompt at the bottom of the screen that the cannonball reaches a maximum height of 488.28 feet at 976.56 feet downrange.

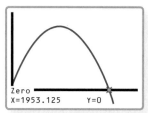

FIGURE 2.92 Where the cannonball lands

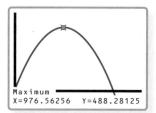

FIGURE 2.93 The peak of the cannonball's flight

EXAMPLE 2.12 **GROWTH OF FOREST STANDS**

In forestry management it is important to know the *growth* and the *yield* of a forest stand.[33] The growth G is the amount by which the volume of wood will increase in a unit of time, and the yield Y is the total volume of wood. A forest manager has determined that in a certain stand of age A, the growth $G = G(A)$ is given by the formula

$$G = 32A^{-2}e^{10 - 32A^{-1}}$$

[33] See Thomas E. Avery and Harold E. Burkhart, *Forest Measurements*, 4th ed. (New York: McGraw-Hill, 1994).

and the yield $Y = Y(A)$ is given by the formula

$$Y = e^{10 - 32A^{-1}}.$$

Here G is measured in cubic feet per acre per year, Y in cubic feet per acre, and A in years.

Part 1 Draw a graph of growth as a function of age that includes ages up to 60 years.

Part 2 At what age is growth maximized?[34]

Part 3 Draw a graph of yield as a function of age. What is the physical meaning of the point on this graph that corresponds to your answer to part 2?

Part 4 To meet market demand, loggers are considering harvesting a relatively young stand of trees. This area was initially clear-cut[35] and left barren. The forest was replanted, with plans to get a new harvest within the next 14 years. At what time in this 14-year period will growth be maximized?

Solution to Part 1 The first step is to enter the growth function and record the appropriate correspondences:

$\mathsf{Y}_1 = G$, growth, in cubic feet per acre per year, on vertical axis

$\mathsf{X} = A$, age, in years, on horizontal axis.

We need to look at a table of values to help us set the window size. We made the table in Figure 2.94 using a starting value of 0 and a table increment of 10. We were told specifically that our graph should include ages up to 60, so we made the graph in Figure 2.95 using a window with a horizontal span from $A = 0$ to $A = 60$ and a vertical span from $G = 0$ to $G = 500$.

Solution to Part 2 Growth is maximized at the peak of the graph in Figure 2.95. We have used the calculator to locate this point in Figure 2.96. We see from the prompt at the bottom of the screen that a maximum growth of 372.62 cubic feet per acre per year occurs when the stand is 16 years old.

Solution to Part 3 Now we want to enter the yield function

$$Y = e^{10 - 32A^{-1}} \text{ cubic feet per acre.}$$

In Figure 2.97 we see in the Y_2 column values for the yield. It would be good to show graphs of both functions on the same screen, but the relative sizes of function values for

X	Y$_1$	
0	ERROR	
10	287.31	
20	355.77	
30	269.53	
40	197.94	
50	148.66	
60	114.86	
X=0		

FIGURE 2.94 A table of values for growth

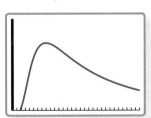

FIGURE 2.95 A graph of forest growth versus age

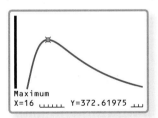

Maximum
X=16   Y=372.61975

FIGURE 2.96 The age of maximum growth

X	Y$_1$	Y$_2$
0	ERROR	ERROR
10	287.31	897.85
20	355.77	4447.1
30	269.53	7580.5
40	197.94	9897.1
50	148.66	11614
60	114.86	12922
X=0		

FIGURE 2.97 A table of values for yield

---

[34]In practice, forest managers are interested in the maximum *mean annual growth* (see Exercise 16 at the end of this section). The age of maximum growth can be used to determine the rotation age that will give the largest total wood production over a series of harvests.
[35]Clear-cutting is the practice of harvesting by cutting *all* trees in an area. The result shows as large swaths of land that are almost barren.

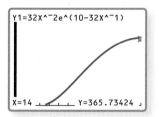

FIGURE 2.98 Yield at maximum growth

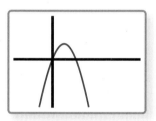

FIGURE 2.99 Maximum growth for a young stand of trees

growth and yield make this impractical. (Try it to see what happens.) Thus we turn off the growth function and graph the yield in a window with horizontal span of 0 to 60 and vertical span of 0 to 13,000. The result is shown in Figure 2.98.

In Figure 2.98 we have put the cursor at the age $A = 16$, corresponding to the maximum growth that we found in part 2. We see that if we harvest at the time of maximum growth, we will get a yield of $Y = 2980.96$ cubic feet per acre. Examination of the graph of the yield function shows that it is the steepest at age $A = 16$, which is where the inflection point occurs. The physical meaning of this is that maximum growth corresponds to the fastest increase in yield.

Solution to Part 4 For this part of the problem, we want to look at growth again. Thus we go to the function entry screen and turn Y off and G on. We reset the window so that the horizontal span is from 0 to 14 and the vertical span is from 0 to 500. The resulting graph is in Figure 2.99. We see that the graph of growth will be increasing over this span, and so the maximum growth will occur at the right-hand endpoint. That is, we get a maximum growth of $G = 365.73$ cubic feet per acre per year if we wait to the end of the 14-year period to harvest. Do you think it is appropriate to proceed with harvesting at this time?

TEST YOUR UNDERSTANDING | **FOR EXAMPLE 2.12**

Replace the growth formula in the example by $G = 40\,A^{-2}e^{11-40A^{-1}}$. At what age is growth maximized? What is the maximum growth rate?   ■

Optimizing at Endpoints

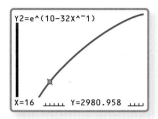

FIGURE 2.100 The formula represents the cannonball path only on a restricted range.

Let's look back at the graph of the flight of the cannonball in Figure 2.92. If we adjust the window, we see the view in Figure 2.100. This shows that the graph of $y = x - 32(x/250)^2$ extends far below the horizontal axis and to the left of the vertical axis. This indicates, among other things, that the cannonball would travel underground, which is nonsense. In fact, the function $y = x - 32(x/250)^2$ represents the path of the cannonball only during the period when it is flying through the air. This is from the place where the cannonball was fired ($x = 0$) to the location $x = 1953.125$ feet downrange where it strikes the ground. Beyond these limits, the graph in Figure 2.100 is not a representation of the physical situation. A more accurate picture would show the graph starting at $x = 0$ and ending where the cannonball lands at $x = 1953.125$. That is, an accurate graphical model would have endpoints, and if we wanted to find the *minimum* height of the cannonball, we would not find it at a peak or valley of the graph. Rather, we would find it at ground level: at the endpoints of the graph.

A similar situation occurs in part 4 of Example 2.12, where we are finding the maximum growth for a clear-cut forest. Because of market demand, loggers want to harvest the stand within 14 years from planting, and under these conditions, growth is maximized at the end of the 14-year period. This is at the right endpoint of the graph in Figure 2.99.

> **KEY IDEA 2.6**   **OPTIMIZING AT ENDPOINTS**
>
> Often the maximum or minimum value of a function occurs at the endpoints of the graph being considered, instead of occurring at peaks or valleys.

For the cannonball, we don't need a graph to tell us that the minimum height will be at ground level, but there are many physical situations where the optimal value is not so obvious but occurs at endpoints of the graph. Let's look, for example, at a simple geometry problem. Suppose we have 100 yards of fence from which we wish to construct two pens. We will use part of the fence to make a square pen and the rest to make a circular pen, as illustrated in Figure 2.101.

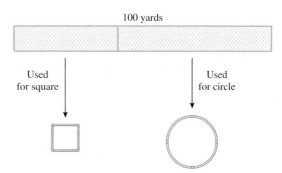

FIGURE 2.101 Cutting fence to make two pens

If we use s yards of the 100-yard stretch of fence for the square, then the total area $A = A(s)$ enclosed by the two pens turns out to be

$$A = \left(\frac{s}{4}\right)^2 + \frac{(100 - s)^2}{4\pi} \text{ square yards.}$$

We want to investigate how much fence to use for the square, s yards, and how much to use for the circle, $100 - s$ yards, to optimize the total area A. The first step is to enter the function and list the variable correspondences:

$$\mathsf{Y_1} = A, \text{ total area, in square yards, on vertical axis}$$

$$\mathsf{X} = s, \text{ yards used for square on horizontal axis.}$$

Minima often occur where graphs reach the bottom of a valley.

Bearing in mind that we have only 100 yards of fence, so that s is between 0 and 100, we made the table of values in Figure 2.102 with a starting value of 0 and an increment of 20. This led us to choose a window with a horizontal span from 0 to 100 and a vertical span from 0 to 1000, which gives the graph shown in Figure 2.103. In Figure 2.103 we have located the minimum value for area. We see from the prompt at the bottom of the screen that we enclose the minimum amount of area, $A = 350$ square yards, if we use about $s = 56$ yards of fence to make the square. That leaves $100 - 56 = 44$ yards of fence to make the circle.

X	Y₁
0	795.77
20	534.3
40	386.48
60	352.32
80	431.83
100	625
120	931.83

X=0

FIGURE 2.102 A table of values for area

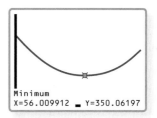

Minimum
X=56.009912 ▪ Y=350.06197

FIGURE 2.103 The minimum area enclosed by a square and a circle

Now, let's find how we should use the fence to enclose the maximum area. In Figure 2.103, we found the minimum at a *valley* of the graph, but there is no *peak* that corresponds to the maximum. The critical factor here is that we have only 100 yards of fence to work with, so *s* is between 0 and 100. The graph shows clearly that the maximum occurs at an endpoint.

We have traced the graph to locate the cursor at the right-hand endpoint in Figure 2.104 and at the left-hand endpoint in Figure 2.105. We see in Figure 2.105 that we get the maximum area of *A* = 795.77 square yards if we use *s* = 0 yards for the square. That is, all the fence goes to make the circle.

Maxima and minima sometimes occur at endpoints.

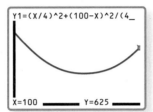

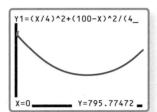

FIGURE 2.104 The right-hand endpoint: the area when all the fence is used for the square

FIGURE 2.105 The maximum area at the left-hand endpoint, when all the fence is used for the circle

ANSWERS FOR TEST YOUR UNDERSTANDING

2.12 20 years; 810.31 cubic feet per acre per year

2.6 EXERCISES

Reminder   Round all answers to two decimal places unless otherwise indicated.

1. **Builder's Old Measurement**   The *Builder's Old Measurement* was instituted by law in England in 1773 as the way to estimate the total tonnage *T* of a wooden ship from its beam width *W* and length *L*, both measured in feet. The formula is

$$T = \frac{(L - 0.6W)W^2}{188}.$$

In this exercise we consider wooden ships of length 150 feet.

   a. Make a graph of *T* versus *W* including beam widths up to 250 feet.

   b. What is the maximum tonnage for a ship of this length?

   c. Comment on the dimensions of the ship from part b.

   d. What is the maximum tonnage of a ship whose width is no more than half its length? That is, we assume that the beam width is at most 75 feet.

2. **Drug Concentration**   When a drug is administered orally it takes some time before the blood

concentration reaches its maximum level. After that time, concentration levels decrease. When 500 milligrams of procainamide is administered orally, one model for a particular patient gives blood concentration *C*, in milligrams per liter, after *t* hours as

$$C = 2.65(e^{-0.2t} - e^{-2t}).$$

What is the maximum blood-level concentration, and when does that level occur?

3. **The Cannon at a Different Angle**   Suppose a cannon is placed at the origin and elevated at an angle of 60 degrees. If the cannonball is fired with a muzzle velocity of 0.15 mile per second, it will follow the graph of $y = x\sqrt{3} - 160x^2/297$, where distances are measured in miles.

   a. Make a graph that shows the path of the cannonball.

   b. How far downrange does the cannonball travel? Explain how you got your answer.

   c. What is the maximum height of the cannonball, and how far downrange does that height occur?

4. **Profit** The weekly profit P for a widget producer is a function of the number n of widgets sold. The formula is

$$P = -2 + 2.9n - 0.3n^2.$$

Here P is measured in thousands of dollars, n is measured in thousands of widgets, and the formula is valid up to a level of 7 thousand widgets sold.

   a. Make a graph of P versus n.

   b. Calculate $P(0)$ and explain in practical terms what your answer means.

   c. At what sales level is the profit as large as possible?

5. **Marine Fishery** One class of models for population growth rates in marine fisheries assumes that the harvest from fishing is proportional to the population size. For one such model, we have

$$G = 0.3n\left(1 - \frac{n}{2}\right) - 0.1n.$$

Here G is the growth rate of the population, in millions of tons of fish per year, and n is the population size, in millions of tons of fish.

   a. Make a graph of G versus n. Include values of n up to 1.5 million tons.

   b. Use functional notation to express the growth rate if the population size is 0.24 million tons, and then calculate that value.

   c. Calculate $G(1.42)$ and explain in practical terms what your answer means.

   d. At what population size is the growth rate the largest?

6. **Enclosing a Field** You have 16 miles of fence that you will use to enclose a rectangular field.

   a. Draw a picture to show that you can arrange the 16 miles of fence into a rectangle of width 3 miles and length 5 miles. What is the area of this rectangle?

   b. Draw a picture to show that you can arrange the 16 miles of fence into a rectangle of width 2 miles and length 6 miles. What is the area of this rectangle?

   c. The first two parts of this exercise are designed to show you that you can get different areas for rectangles of the same perimeter, 16 miles. In general, if you arrange the 16 miles of fence into a rectangle of width w miles, then it will enclose an area of $A = w(8 - w)$ square miles.

   i. Make a graph of area enclosed as a function of w, and explain what the graph is showing.

   ii. What width w should you use to enclose the most area?

   iii. What is the length of the maximum-area rectangle that you have made, and what kind of figure do you have?

7. **Forming a Pen** We want to form a rectangular pen of area 100 square feet. One side of the pen is to be formed by an existing building and the other three sides by a fence (see Figure 2.106). Let W be the length, in feet, of the sides of the rectangle perpendicular to the building, and let L be the length, in feet, of the other side.

FIGURE 2.106

   a. Find a formula for the total amount of fence needed in terms of W and L.

   b. Express, as an equation involving W and L, the requirement that the total area formed be 100 square feet.

   c. Solve the equation you found in part b for L.

   d. Use your answers to parts a and c to find a formula for F, the total amount, in feet, of fence needed, as a function of W alone.

   e. Make a graph of F versus W.

   f. Determine the dimensions of the rectangle that requires a minimum amount of fence.

8. **Sales Growth** In this exercise we develop a model for the growth rate G, in thousands of dollars per year, in sales of a product as a function of the sales level s, in thousands of dollars.[36] The model

---

[36]The model we develop for sales growth can be applied in other settings where there is a limit to growth. Examples include the spread of a new technology and population growth under environmental constraints.

(continued)

assumes that there is a limit to the total amount of sales that can be attained. In this situation we use the term *unattained sales* for the difference between this limit and the current sales level. For example, if we expect sales to grow to 3 thousand dollars in the long run, then $3 - s$ gives the unattained sales. The model states that the growth rate G is proportional to the product of the sales level s and the unattained sales. Assume that the constant of proportionality is 0.3 and that the sales grow to 2 thousand dollars in the long run.

a. Find a formula for unattained sales.

b. Write an equation that shows the proportionality relation for G.

c. On the basis of the equation from part b, make a graph of G as a function of s.

d. At what sales level is the growth rate as large as possible?

e. What is the largest possible growth rate?

9. **Maximum Sales Growth** *This is a continuation of Exercise 8.* In this exercise we determine how the sales level that gives the maximum growth rate is related to the limit on sales. Assume, as above, that the constant of proportionality is 0.3, but now suppose that sales grow to a level of 4 thousand dollars in the limit.

a. Write an equation that shows the proportionality relation for G.

b. On the basis of the equation from part a, make a graph of G as a function of s.

c. At what sales level is the growth rate as large as possible?

d. Replace the limit of 4 thousand dollars with another number, and find at what sales level the growth rate is as large as possible. What is the relationship between the limit and the sales level that gives the largest growth rate? Does this relationship change if the proportionality constant is changed?

e. Use your answers in part d to explain how to determine the limit if we are given sales data showing the sales up to a point where the growth rate begins to decrease.

10. **An Aluminum Can** The cost of making a can is determined by how much aluminum A, in square inches, is needed to make it. This in turn depends on the radius r and the height h of the can, both measured in inches. You will need some basic facts about cans. See Figure 2.107.

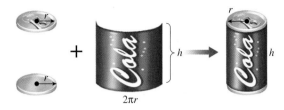

FIGURE 2.107

The surface of a can may be modeled as consisting of three parts: two circles of radius r and the surface of a cylinder of radius r and height h. The area of each circle is πr^2, and the area of the surface of the cylinder is $2\pi rh$. The volume of the can is the volume of a cylinder of radius r and height h. That is $V = \pi r^2 h$.

In what follows, we assume that the can must hold 15 cubic inches, and we will look at various cans holding the same volume.

a. Explain why the height of any can that holds a volume of 15 cubic inches is given by

$$h = \frac{15}{\pi r^2}.$$

b. Make a graph of the height h as a function of r, and explain what the graph is showing.

c. Is there a value of r that gives the least height h? Explain.

d. If A is the amount of aluminum needed to make the can, explain why

$$A = 2\pi r^2 + 2\pi rh.$$

e. Using the formula for h from part a, explain why we may also write A as

$$A = 2\pi r^2 + \frac{30}{r}.$$

11. **An Aluminum Can, Continued** *This is a continuation of Exercise 10.* The cost of making a can is determined by how much aluminum A, in square inches, is needed to make it. As we saw in Exercise 10, we can express both the height h and the amount of aluminum A in terms of the radius r:

$$h = \frac{15}{\pi r^2}$$

$$A = 2\pi r^2 + \frac{30}{r}.$$

a. What is the height, and how much aluminum is needed to make the can, if the radius is 1 inch? (This is a tall, thin can.)

b. What is the height, and how much aluminum is needed to make the can, if the radius is 5 inches? (This is a short, fat can.)

c. The first two parts of this problem are designed to illustrate that for an aluminum can, different surface areas can enclose the same volume of 15 cubic inches.

    i. Make a graph of A versus r and explain what the graph is showing.

    ii. What radius should you use to make the can using the least amount of aluminum?

    iii. What is the height of the can that uses the least amount of aluminum?

12. **Cost for a Can** *This is a continuation of Exercises 10 and 11.* Suppose now that we use different materials in making different parts of the can. The material for the side of the can costs $0.10 per square inch, and the material for both the top and bottom costs $0.05 per square inch.

a. Use a formula to express the cost C, in dollars, of the material for the can as a function of the radius r.

b. What radius should you use to make the least expensive can?

13. **Profit** *The background for this exercise can be found in Exercises 13, 14, 15, and 16 in Section 1.4.* A manufacturer of widgets has fixed costs of $600 per month, and the variable cost is $60 per thousand widgets (so it costs $60 to produce a thousand widgets). Let N be the number, in thousands, of widgets produced in a month.

a. Find a formula for the manufacturer's total cost C as a function of N.

b. The highest price p, in dollars per thousand widgets, at which N can be sold is given by the formula $p = 70 - 0.03N$. Using this, find a formula for the total revenue R as a function of N.

c. Use your answers to parts a and b to find a formula for the profit P of this manufacturer as a function of N.

d. Use your formula from part c to determine the production level at which profit is maximized if the manufacturer can produce at most 300 thousand widgets in a month.

14. **Laying Phone Cable** City A lies on the north bank of a river that is 1 mile wide. You need to run a phone cable from City A to City B, which lies on the opposite bank 5 miles down the river. You will lay L miles of the cable along the north shore of the river, and from the end of that stretch of cable you will lay W miles of cable running under water directly toward City B. (See Figure 2.108.)

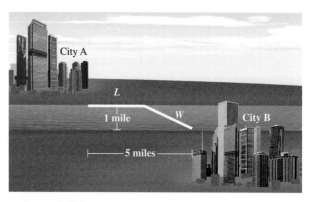

FIGURE 2.108 Laying phone cable

You will need the following fact about right triangles: A right triangle has two legs, which meet at the right angle, and the hypotenuse, which is the longest side. An ancient and beautiful formula, the *Pythagorean theorem*, relates the lengths of the three sides:

Length of hypotenuse

$$= \sqrt{\text{Length of one leg}^2 + \text{Length of other leg}^2}.$$

a. Find an appropriate right triangle that shows that $W = \sqrt{1 + (5 - L)^2}$.

b. Find a formula for the length of the total phone cable P from City A to City B as a function of L.

c. Make a graph of the total phone cable length P as a function of L, and explain what the graph is showing.

d. What value of L gives the least length for the total phone cable? Draw a picture showing the least-length total phone cable.

15. **Laying Phone Cable, Continued** *This is a continuation of Exercise 14.* Suppose it costs $300 per mile to run cable on land but $500 per mile to lay it under water.

a. Let C be the total cost of the project, in dollars. Explain why $C = 300L + 500W$.

b. Write a formula for C as a function of L. (*Hint:* Use part a of Exercise 14.)

c. Find the amount of cable that runs under water and the total cost of the project if you choose L to be 1 mile long. Draw and properly label a picture that shows this cable plan.

(continued)

d. Find the amount of cable that runs under water and the total cost of the project if you choose L to be 3 miles long. Draw and properly label a picture that shows this cable plan.

e. Make a graph of the cost C as a function of L, and explain what the graph is showing.

f. What value of L gives the least cost for the project?

g. Find the value of W that corresponds to your answer in part f, and draw a picture showing the least-cost project.

h. Potential legal disputes about easements have caused the cost of laying cable on land to increase to $700 per mile. With this change, find a new formula for the cost C. How does this affect your plans for the least-cost project?

16. Mean Annual Growth *This is a continuation of Example 2.12.* Forest managers study the *mean annual growth* $M = M(A)$, defined as the yield at age A divided by A.

a. Use the formula for Y in Example 2.12 to find a formula for M as a function of A.

b. Add the graph of M as a function of A to the graph of G.

c. The *rotation age* may be determined as the age of maximum mean annual growth. Determine the rotation age using the graph of M.

d. From your two graphs in part b, can you suggest a different way to find the rotation age?

17. Growth of Fish Biomass An important model for commercial fisheries is that of Beverton and Holt.[37] It begins with the study of a single *cohort* of fish— that is, all the fish in the study are born at the same time. For a cohort of the North Sea plaice (a type of flatfish), the number $N = N(t)$ of fish in the population is given by
$$N = 1000e^{-0.1t},$$
and the weight $w = w(t)$ of each fish is given by
$$w = 6.32(1 - 0.93e^{-0.095t})^3.$$
Here w is measured in pounds and t in years. The variable t measures the so-called recruitment age, which we refer to simply as the age.

The *biomass* $B = B(t)$ of the fish cohort is defined to be the total weight of the cohort, so it is obtained by multiplying the population size by the weight of a fish.

a. If a plaice weighs 3 pounds, how old is it?

b. Use the formulas for N and w given above to find a formula for $B = B(t)$, and then make a graph of B against t. (Include ages through 20 years.)

c. At what age is the biomass the largest?

d. In practice, fish below a certain size can't be caught, so the biomass function becomes relevant only at a certain age.

   i. Suppose we want to harvest the plaice population at the largest biomass possible, but a plaice has to weigh 3 pounds before we can catch it. At what age should we harvest?

   ii. Work part i under the assumption that we can catch plaice weighing at least 2 pounds.

18. Spawner-Recruit Model In fish management it is important to know the relationship between the abundance of the *spawners* (also called the parent stock) and the abundance of the *recruits*—that is, those hatchlings surviving to maturity.[38] According to the *Ricker model*, the number of recruits R as a function of the number of spawners P has the form
$$R = APe^{-BP}$$
for some positive constants A and B. This model describes well a phenomenon observed in some fisheries: A large spawning group can actually lead to a small group of recruits.[39]

In a study of the sockeye salmon, it was determined that $A = 4$ and $B = 0.7$. Here we measure P and R in thousands of salmon.

a. Make a graph of R versus P for the sockeye salmon. (Assume there are at most 3000 spawners.)

b. Find the maximum number of salmon recruits possible.

c. If the number of recruits R is greater than the number of spawners P, then the difference $R - P$ of the recruits can be removed by fishing, and next season there will once again be P spawners surviving to renew the cycle. What value of P gives the maximum value of $R - P$, the number of fish available for removal by fishing?

---

[37]See R. J. H. Beverton and S. J. Holt, *On the Dynamics of Exploited Fish Populations*, Fishery Investigations, Series 2, Volume 19 (London: Ministry of Agriculture, Fisheries and Food, 1957).
[38]See W. E. Ricker, "Stock and recruitment," *J. Fish. Res. Board Can.* **11** (1954), 559–623.
[39]Biological mechanisms that contribute to this phenomenon are suspected to include competition and cannibalism of the young.

19. **Rate of Growth** The rate of growth G in the weight of a fish is a function of the weight w of the fish. For the North Sea cod, the relationship is given by

$$G = 2.1w^{2/3} - 0.6w.$$

Here w is measured in pounds and G in pounds per year. The maximum size for a North Sea cod is about 40 pounds.

   a. Make a graph of G versus w.
   b. Find the greatest rate of growth among all cod weighing at least 5 pounds.
   c. Find the greatest rate of growth among all cod weighing at least 25 pounds.

20. **Health Plan** The manager of an employee health plan for a firm has studied the balance B, in millions of dollars, in the plan account as a function of t, the number of years since the plan was instituted. He has determined that the account balance is given by the formula $B = 60 + 7t - 50e^{0.1t}$.

   a. Make a graph of B versus t over the first 7 years of the plan.
   b. At what time is the account balance at its maximum?
   c. What is the smallest value of the account balance over the first 7 years of the plan?

21. **Size of High Schools** The farm population has declined dramatically in the years since World War II, and with that decline, rural school districts have been faced with consolidating in order to be economically efficient. One researcher studied data from the early 1960s on expenditures for high schools ranging from 150 to 2400 in enrollment.[40] He considered the cost per pupil as a function of the number of pupils enrolled in the high school, and he found the approximate formula

$$C = 743 - 0.402n + 0.00012n^2,$$

where n is the number of pupils enrolled and C is the cost, in dollars, per pupil.

   a. Make a graph of C versus n.
   b. What enrollment size gives a minimum per-pupil cost?
   c. If a high school had an enrollment of 1200, how much in per-pupil cost would be saved by increasing enrollment to the optimal size found in part b?

22. **Radioactive Decay** Radium-223 is a radioactive substance that is itself a product of the radioactive decay of thorium-227. For one experiment, the amount A of radium-223 present, as a function of the time t since the experiment began, is given by the formula $A = 3(e^{-0.038t} - e^{-0.059t})$, where A is measured in grams and t in days.

   a. Make a graph of A versus t covering the first 60 days of the experiment.
   b. What was the largest amount of radium-223 present over the first 60 days of the experiment?
   c. What was the largest amount of radium-223 present over the first 10 days of the experiment?
   d. What was the smallest amount of radium-223 present over the first 60 days of the experiment?

23. **Water Flea** F. E. Smith has studied population growth for the water flea.[41] Let N denote the population size. In one experiment, Smith found that G, the rate of growth per day in the population, can be modeled by

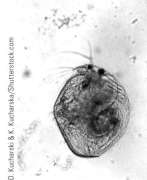

$$G = \frac{0.44N(228 - N)}{228 + 3.46N}.$$

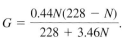

   a. Draw a graph of G versus N. Include values of N up to 350.
   b. At what population level does the greatest rate of growth occur?
   c. There are two values of N where G is zero. Find these values of N and explain what is occurring at these population levels.
   d. What is the rate of population growth if the population size is 300? Explain what is happening to the population at this level.

24. **An Epidemic** One model for the spread of epidemics gives the number of newly infected individuals t days after the outbreak of the epidemic as

$$\text{New cases} = \frac{\beta n(n+1)^2 e^{(n+1)\beta t}}{(n + e^{(n+1)\beta t})^2}.$$

Here n is the total number of people we expect to be infected over the course of the epidemic, and β

---

[40]This exercise is based on the study by J. Riew, "Economies of scale in high school operation," *Review of Economics and Statistics* **48** (August 1966), 280–287. In our presentation of his results, we have suppressed variables such as teacher salaries.
[41]See "Population dynamics in *Daphnia magna* and a new model for population growth," *Ecology* **44** (1963), 651–663. The population size is measured per unit volume.

(continued)

depends on the nature of the infection as well as on other environmental factors. For a certain epidemic, the number of new cases is

$$\text{New cases} = \frac{75{,}150e^{0.3t}}{(500 + e^{0.3t})^2}.$$

a. Make a graph of the number of new cases versus days since the outbreak. Include times up to 30 days.

b. What is the greatest number of new cases we expect to see in 1 day, and when does that occur?

c. The local medical facilities can handle no more than 25 new cases per day. During what time period will it be necessary to recruit help from outside sources?

25. Research Project   Locate a calculus textbook and look in the section entitled optimization, max/min problems, or something similar. Find a worked-out example in the text where a maximum or minimum for a function is found. Solve that same problem using the optimization procedure you've learned in this section. Compare your solution with the solution provided in the calculus text. Write a careful description of the function, its use, and your solution to the problem, and add a very brief description of the solution found in your source.

2.6 SKILL BUILDING EXERCISES

Finding Maxima and Minima   In Exercises S-1 through S-19, find all maxima and minima as instructed. You should find both horizontal and vertical coordinates.

S-1. Find the maximum value of $5x + 4 - x^2$ on the horizontal span of 0 to 5.

S-2. Find the minimum value of $2^x - x^2 + 5$ on the horizontal span of 0 to 5.

S-3. Find the minimum value of $x + \dfrac{x + 5}{x^2 + 1}$ on the horizontal span of 0 to 5.

S-4. Find the maximum value of $5x^2 - e^x$ on the horizontal span of 0 to 5.

S-5. Find the maximum value of $x^{1/x}$ on the horizontal span of 0 to 10.

S-6. Find the minimum value of $4e^{2x} + 3x^2$ on the horizontal span of -2 to 1.

S-7. Find all maxima and minima of $f(x) = x^3 - 6x + 1$ with a horizontal span of -2 to 2 and a vertical span of -10 to 10.

S-8. Find all maxima and minima of $f(x) = \dfrac{8x}{1 + x^2}$. Use a horizontal span of -5 to 5.

S-9. Find all maxima and minima of $f(x) = x^3 - 2^x$ on the horizontal span of 0 to 10.

S-10. Find the minimum of $5\sqrt{x^2 + 1} - x$ on the horizontal span of -5 to 5.

S-11. Find the minimum of $e^x - 2^x - x$ on the horizontal span of -2 to 2.

S-12. Find all maxima and minima of $f(x) = x^3 - 10x + 1$ on the horizontal span of -3 to 3.

S-13. Find the maximum of $\dfrac{50}{1 + 1.2^{-x}} - x$ on the horizontal span of 0 to 20.

S-14. Find the minimum of $\dfrac{1}{x(5 - x)}$ on the horizontal span of 1 to 4.

S-15. Find the maximum of $1 - \dfrac{1}{\sqrt{x}(5 - \sqrt{x})}$ on the horizontal span of 1 to 10.

S-16. Find the maximum of $\sqrt{x} + \sqrt{x + 1} - \dfrac{x}{5}$ on the horizontal span of 0 to 50.

S-17. Find the minimum and maximum of $(x - 3)^2 e^{-1/x^2}$ on the horizontal span of 1 to 4.

S-18. Find the minimum of $x - 3\sqrt{x} + 6$ on the horizontal span of 0 to 5.

S-19. Find the maximum of $x^3 e^{-x}$ on the horizontal span of 0 to 5.

S-20. Endpoint Maximum   Find the maximum value of $x^3 + x$ on the horizontal span of 0 to 5.

S-21. Endpoint Minimum   Find the minimum value of $x^3 + x$ on the horizontal span of 0 to 5.

S-22. Endpoint Minimum   Find the minimum value of $200 - x^3$ on the horizontal span of 0 to 5.

S-23. Endpoint Maximum   Find the maximum value of $200 - x^3$ on the horizontal span of 0 to 5.

S-24. Maximum and Minimum   Find the maximum and minimum values of $x^3 - 9x^2 + 6$ on the horizontal span of 0 to 10.

CHAPTER 2 | SUMMARY

Each type of function presentation has its advantages and disadvantages, and one of the most effective methods of mathematical analysis is to take a function presented in one way and express it in another form. It is particularly tedious to get a graph or table of values from a function given by a complicated formula, but doing so can also be very illuminating. The graphing calculator takes the tedium out of this procedure by producing graphs and tables on demand.

2.1 TABLES AND TRENDS

N	P
15	0.23626
16	0.24231
17	0.2452
18	0.2452
19	0.24264
20	0.23789
21	0.23128

It is often difficult to determine limiting values or maximum and minimum values from a formula, but as we saw in Chapter 1, this is easily done from a table of values. The calculator produces such tables on demand, and in this respect it is an important problem-solving tool. For example, the probability of getting exactly 3 sixes when you roll N dice is given by the formula

$$P = \frac{N(N-1)(N-2)}{750}\left(\frac{5}{6}\right)^N.$$

It is by no means clear from this formula how many dice should be rolled to make the probability of achieving exactly 3 sixes the greatest. A calculator can be used to produce quickly the accompanying partial table of values.

The table shows clearly that the maximum probability is achieved by rolling 17 or 18 dice.

2.2 GRAPHS

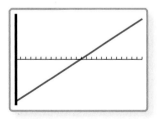

FIGURE 2.109 A table of values for net profit on belt sales

As with tables, the graphing calculator can quickly produce graphs from formulas, and effective analysis can proceed. Graphs viewed from different perspectives can differ considerably in appearance. Effective graphical analysis depends on being able to view the graph in an appropriate window. This is done through an intelligent choice of *horizontal span* and *vertical span*. When the function is a model for some physical phenomenon, choosing the horizontal span, or *domain*, may be a matter of common sense. The vertical span, or *range*, can then be chosen by looking at a table of values.

As an example, suppose a leather craftsman has produced 25 belts that he intends to sell at an upcoming art fair for $22.75 each. He has invested a total of $300 in materials. The net profit p, in dollars, from the sale of n belts is given by

$$p = 22.75n - 300.$$

In order to graph this function, we need to choose a horizontal and a vertical span. Since he has only 25 belts, he will sell somewhere between 0 and 25 belts. This is the horizontal span, or domain, for our function. To get an appropriate vertical span, we look at the table of values shown in Figure 2.109. This shows that we should choose a vertical span of about -325 to 300. The resulting graph, which is now available for further analysis, is shown in Figure 2.110.

FIGURE 2.110 A graph of net profit on belt sales

2.3 SOLVING LINEAR EQUATIONS

Many times the analysis required to understand a phenomenon involves solving an equation. For many equations, this can be a difficult process, but for *linear equations*, solutions are easy. Linear equations are those that do not involve powers, roots, or other complications of the variable; they are the simplest of all equations. Any linear equation can be solved using two basic rules.

Moving Terms Across the Equals Sign You may add (or subtract) the same thing to (or from) both sides of the equation. This can be thought of as moving a term from one side of the equation to the other, provided that you change its sign.

Dividing Each Term of an Equation You may divide (or multiply) both sides of an equation by any nonzero number.

As an example, we show how to solve $5x + 7 = 18 - 2x$ using these two rules.

$$5x + 7 = 18 - 2x$$
$$5x = 18 - 7 - 2x \qquad \text{Move 7 across and change its sign.}$$
$$5x + 2x = 11 \qquad\qquad \text{Move } 2x \text{ across and change its sign.}$$
$$7x = 11$$
$$x = \frac{11}{7} = 1.57 \qquad \text{Divide each term by 7.}$$

2.4 SOLVING NONLINEAR EQUATIONS

Equations that are not linear may be difficult or impossible to solve exactly by hand, but the graphing calculator can provide approximate (yet highly accurate) solutions to virtually any equation quickly and easily.

The most useful method for solving equations with a calculator is referred to in this text as the *crossing-graphs method*. We will illustrate it with an example. Suppose the temperature, after t minutes, of an aluminum bar being heated in an oven is given by

$$A = 800 - 730e^{-0.06t} \text{ degrees Fahrenheit.}$$

We want to know when the temperature will reach 600 degrees. In terms of an equation, that means we need to solve

$$800 - 730e^{-0.06t} = 600$$

for t. This is done by entering the functions $800 - 730e^{-0.06t}$ and 600 into the calculator. We want to know when these functions are the same—that is, where the graphs cross. We see in Figure 2.111 that this occurs 21.58 minutes after the bar is put in the oven.

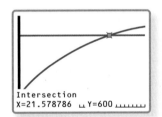

Intersection
X=21.578786 ⎵⎵Y=600 ⎵⎵⎵⎵⎵

FIGURE 2.111 Finding when the temperature is 600 degrees

2.5 SOLVING INEQUALITIES

Inequalities may be solved in much the same fashion as equalities. Consider once again the aluminum bar being heated in an oven whose temperature after t minutes in the oven is given by

$$A = 800 - 730e^{-0.06t} \text{ degrees Fahrenheit.}$$

The metal is suitable for working when the temperature is between 500 and 600 degrees Fahrenheit. The time when the temperature is in this range is the solution for t of the inequality

$$500 \leq A \leq 600.$$

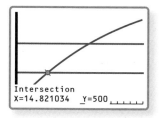

FIGURE 2.112 Solving an inequality

To solve this inequality we first graph the function A and include temperatures from 500 to 600 degrees. We add the lines $A = 500$ and $A = 600$. The inequality is true when the graph is between these two lines. Referring to Figure 2.111 and Figure 2.112, we see that the solution of the inequality is

$$14.82 \le t \le 21.58.$$

Hence the metal is suitable for working from 14.82 to 21.58 minutes after it has been put in the oven.

2.6 OPTIMIZATION

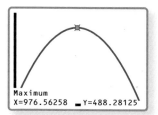

FIGURE 2.113 The maximum height of a cannonball

A strategy similar to that used to solve nonlinear equations can be used to optimize functions given by formulas. We use the calculator to make the graph and then look for peaks or valleys. To illustrate the method, we look for the maximum height of a cannonball following the graph of

$$x - 32\left(\frac{x}{250}\right)^2,$$

where x is the distance downrange, measured in feet. We graph the function as shown in Figure 2.113. That figure shows that the cannonball reaches its maximum height of 488.28 feet when it is 976.56 feet downrange.

CHAPTER 2 REVIEW EXERCISES

Reminder   Round all answers to two decimal places unless otherwise indicated.

1. **Finding a Minimum**   Suppose the function $f = x^3 - \frac{9}{2}x^2 + 6x + 1$ describes a physical situation that makes sense only for whole numbers between 1 and 5. For what value of x does f reach a minimum, and what is the minimum value?

2. **A Population of Foxes**   A breeding group of foxes is introduced into a protected area and exhibits logistic population growth. After t years the number of foxes is given by

$$N(t) = \frac{37.5}{0.25 + 0.76^t} \text{ foxes.}$$

   a. Make a graph of N versus t covering the first 20 years.

   b. Calculate $N(9)$ and explain in practical terms what your answer means.

   c. For what values of t is the graph concave up? For what values is it concave down?

   d. What is the carrying capacity (the limiting value of N)?

3. **Linear Equations**   Solve $11 - x = 3 + x$ for x.

4. **Water Jug**   If a completely full 5-gallon water jug is drained through a spigot, the depth is given by the formula

$$D = 0.265t^2 - 4.055t + 15.5,$$

   where t is the time, in minutes, that the spigot is open and D is the depth, in inches, above the spigot.

   a. Make a graph of D versus t. Include values of t up to 7.5 minutes.

   b. Calculate $D(3)$ and explain in practical terms what your answer means.

   c. When will the water jug be completely drained?

   d. Does the water drain faster near the beginning or when the water is mostly drained?

5. **Maxima and Minima**   Find the maximum and minimum values of $x^2 + 100/x$ on the horizontal span of 1 to 5. Be sure to include endpoint maxima or minima.

6. **George Reserve Population**   The number of deer on the George Reserve t years after introduction is given by

$$N = \frac{6.21}{0.035 + 0.45^t} \text{ deer.}$$

(*continued*)

a. How many deer were introduced into the deer reserve?

b. Calculate $N(4)$ and explain the meaning of the number you have calculated.

c. Find the carrying capacity for deer in the reserve.

d. Explain how the population varies with time. Include in your explanation the average rate of increase over each 2-year period for the first 8 years.

7. **Making a Graph**   For the function

$$N = \frac{315}{1 + 14e^{-0.23t}},$$

find an appropriate window to make a good graph of N versus t for t from 0 to 30.

8. **Forming a Pen**   We want to form a free-standing rectangular pen. Let W be the width, in feet, and L the length, in feet. Let F be the total amount of fence needed. Assume the total area of the pen will be 144 square feet.

a. Write a formula for F in terms of W and L.

b. Express, as an equation involving W and L, the requirement that the total area of the pen be 144 square feet.

c. Solve the equation you found in part b for W.

d. Use your answers to parts a and c to find a formula for F in terms of L alone.

e. Make a graph of F versus L for values of L up to 24 feet.

f. Determine the dimensions of the pen that require the minimum amount of fence. Explain how your graph from part e illustrates or confirms your answer.

9. **The Crossing-Graphs Method**   Solve using the crossing-graphs method: $6 + 69 \times 0.96^t = 32$.

10. **Gliding Pigeons**   A gliding bird's movement is measured in terms of the rate at which the bird's altitude decreases, the *sinking speed s*, and the airspeed u. For a gliding pigeon, these are related by

$$s = \frac{u^3}{2500} + \frac{25}{u}.$$

Here s and u are measured in meters per second.

a. Make a graph of the sinking speed s as a function of the airspeed u for values of u up to 20 meters per second.

b. At what airspeed does the smallest sinking speed occur?

c. If the airspeed is very slow, does the pigeon sink quickly or slowly?

11. **Finding a Maximum**   Suppose the function $f = 12x - x^2 - 25$ describes a physical situation that makes sense only for whole numbers between 3 and 9 feet. For what value of x does f reach a maximum, and what is the maximum value?

12. **Thrown Ball**   If we stand on top of a 150-foot-tall building and throw a ball up into the air with an initial velocity of 22 feet per second, then the height of the ball above the ground is given by

$$H = -16t^2 + 22t + 150.$$

Here t is time in seconds, and H is measured in feet.

a. Make a table of H versus t for $t = 0, 0.25, 0.50, \ldots, 3.75, 4$.

b. Make a graph of H versus t for t from 0 to 4.

c. Find the value of t that maximizes H.

d. Find the maximum value of H and explain the meaning of the number in practical terms.

e. At what times t is the height of the ball 155 feet? (*Hint:* There are two answers.)

13. **Linear Equations**   Solve for W: $L = 98.42 + 1.08W - 4.14A$.

14. **Growth of North Sea Sole**   The length of North Sea sole, a species of fish, can be determined using the von Bertalanffy model by the formula

$$L = 14.8 - 19.106 \times 0.631^t.$$

Here t is age, in years, and L is length, in inches.

a. Make a graph of L versus t. Include values of t from 1 to 10 years.

b. Use functional notation to express the length of a 4-year-old sole, and then calculate that value.

c. At what age will a sole be 10 inches long?

d. What is the limiting length for a North Sea sole?

15. **Minimum**   Find the minimum value of $x^2 + 20/(x + 1)$ on the horizontal span of 0 to 10.

16. **Profit**   The profit P, in millions of dollars, that a manufacturer makes is a function of the number N, in millions, of items produced in a year, and the formula is

$$P = 10N - N^2 - 6.34.$$

A negative quantity for P represents a negative profit—that is, a loss—and the formula is valid up to a level of 10 million items produced.

a. Express using functional notation the profit at a production level of 7 million items.

b. What is the loss at a production level of $N = 0$ million items?

c. Determine the two break-even points for this manufacturer—that is, the two production levels at which the profit is zero.

d. Determine the production level that gives maximum profit, and determine the amount of the maximum profit.

17. Making a Graph For the function
$$F = \frac{0.5h}{6380 + h},$$
find an appropriate window to make a good graph of F versus h for h from 0 to 100,000.

18. Temperature Conversions The three principal measures of temperature are Fahrenheit F, Celsius C, and Kelvin K. These are related by the formulas below:
$$F = 1.8C + 32$$
$$K = C + 273.15.$$

a. Solve each of the equations for C.

b. Find a formula expressing F in terms of K.

c. Find a formula expressing K in terms of F.

d. Express the temperature 0 degrees Celsius in terms of degrees Fahrenheit and in terms of kelvins (degrees on the Kelvin scale).

e. Express the temperature 72 degrees Fahrenheit in terms of degrees Celsius and in terms of kelvins.

19. Lidocaine Lidocaine is a drug used to treat irregular heartbeats. After an injection of a 100-mg dose of lidocaine, the amount of the drug in a patient's bloodstream is
$$33.67(e^{-0.0075731t} - e^{-0.12043t}) \text{ mg}$$
at time t minutes after the injection.

a. Make a graph of the amount of drug in the bloodstream for t up to 4 hours (240 minutes).

b. When does the drug reach its maximum level in the bloodstream?

c. For a person of typical size, the drug is effective as long as the amount in the bloodstream is at least 7.5 mg. For how long is the drug at or above that level? (*Hint*: The drug is at that level twice.)

d. For a person of typical size, the lethal level is when the amount in the bloodstream exceeds 30 mg. Is this dose lethal for such a person?

e. For a small person, the lethal level is when the amount in the bloodstream exceeds 15 mg. Is this dose lethal for such a person? If so, after how many minutes will it be lethal?

20. The Single-Graph Method Use the single-graph method to solve $0.3x - 0.2x^3 = 0$. (*Note*: There are three solutions. Find all three.)

21. Inequality Solve the inequality $x \geq 5e^{-x}$.

A Further Look

Limits

We have used tables to spot trends or limiting values. More formally, we are looking at a mathematical concept known as *limits*. In Example 2.1 of Section 2.1, we looked at the velocity of a skydiver t seconds into a fall. The velocity is given by

$$v = 176(1 - 0.834^t).$$

In part 3 of that example, we considered the notion of *terminal velocity*. That is, we asked what would be the velocity of the skydiver after a long period of falling. Formally, this limiting value is denoted by $\lim_{t \to \infty} v$, and it represents the value that v gets close to when t increases toward infinity. In that example, we estimated the limiting velocity to be 176 feet per second. We will see that, in this case, this is more than an estimate and is in fact the exact limiting value.

KEY ALGEBRAIC IDEA

The precise definition of a limit is better left to a more advanced mathematics course, but informally $\lim_{x \to \infty} f(x) = b$ means that when x is very large, $f(x)$ is "close to" b.

Basic Limits

Some limits are difficult to calculate, but many are not. In a number of important cases, the exact limiting value can be calculated directly from a formula. There are two cases that occur often and allow us to calculate many limiting values directly.

You will be asked in the algebraic exercises to supply numerical evidence for the following two basic limits:

Basic exponential limit: If $a < 1$ is a positive number, then

$$\lim_{t \to \infty} a^t = 0.$$

Basic power limit: If n is a positive number, then

$$\lim_{t \to \infty} \frac{1}{t^n} = 0.$$

Let's see how to use the basic exponential limit to calculate the terminal velocity of the skydiver: $\lim_{t \to \infty} 176(1 - 0.834^t)$. Because $0.834 < 1$, the basic exponential limit tells us that 0.834^t is near 0 when t is very large. Thus

$$\lim_{t \to \infty} 176(1 - 0.834^t) = 176(1 - 0) = 176.$$

EXAMPLE 2.13 **USING THE BASIC LIMITS**

Part 1 The number of deer on the George Reserve after t years is given by

$$N = \frac{6.21}{0.035 + 0.45^t}.$$

Explain in practical terms the meaning of $\lim_{t \to \infty} N$, and use the basic exponential limit to calculate its value.

Part 2 Use the basic power limit to calculate $\lim_{t \to \infty}\left(5 + \frac{4}{t^3}\right)$.

Solution to Part 1 In practical terms, the limiting value is the number of deer eventually on the reserve; this limiting value is known as the *carrying capacity*. Because 0.45 is less than 1, the basic exponential limit tells us that $\lim_{t \to \infty} 0.45^t = 0$. Thus

$$\lim_{t \to \infty} \frac{6.21}{0.035 + 0.45^t} = \frac{6.21}{0.035 + 0} = 177.43.$$

This is the exact value of the limit, but in the practical application to the number of deer, we would probably round this to 177 deer.

Solution to Part 2 The basic power limit tells us that when t is large, $\dfrac{1}{t^3}$ is near 0. Thus

$$\lim_{t \to \infty}\left(5 + \frac{4}{t^3}\right) = \lim_{t \to \infty}\left(5 + 4\frac{1}{t^3}\right) = 5 + 4 \times 0 = 5.$$

Quotients of Polynomials

A *polynomial function* is one of the form $f(x) = a_0 + a_1 x + a_2 x^2 + \cdots + a_n x^n$, where n is a non-negative whole number and $a_0, \ldots, a_n$ are constants. If $a_n \neq 0$ then we say that $a_n x^n$ is the *leading term*. For example, $1 + 2x + 3x^2$ is a polynomial with leading term $3x^2$. On the other hand, $1 + x^{1/2}$ is not a polynomial because the exponent $1/2$ is not a whole number. We will study polynomials in much more detail in Chapter 5; for now, little beyond the definition is needed.

The following rule allows us to calculate the limiting value of a quotient of polynomials.

IMPORTANT FACT FOR QUOTIENTS OF POLYNOMIALS

If $g(x) = \dfrac{P(x)}{Q(x)}$ is a quotient of polynomials, then

$$\lim_{x \to \infty} \frac{P(x)}{Q(x)} = \lim_{x \to \infty} \frac{\text{Leading term of } P(x)}{\text{Leading term of } Q(x)},$$

provided either limit exists.

EXAMPLE 2.14 **CALCULATING LIMITS OF QUOTIENTS OF POLYNOMIALS**

Part 1 Calculate $\lim\limits_{x \to \infty} \dfrac{3x^2 + 1}{2x^2 + x + 3}$.

Part 2 Calculate $\lim\limits_{x \to \infty} \dfrac{4x^3 - 1}{3x^5 + 4}$.

Part 3 Calculate $\lim\limits_{x \to \infty} \dfrac{5x^3 + x + 3}{2x + 5}$ if the limit exists.

Solution to Part 1 We examine the quotient of the leading terms:

$$\lim_{x \to \infty} \frac{3x^2 + 1}{2x^2 + x + 3} = \lim_{x \to \infty} \frac{3x^2}{2x^2} = \lim_{x \to \infty} \frac{3}{2} = \frac{3}{2}.$$

Solution to Part 2 Now

$$\lim_{x \to \infty} \frac{4x^3 - 1}{3x^5 + 4} = \lim_{x \to \infty} \frac{4x^3}{3x^5} = \lim_{x \to \infty} \frac{4}{3x^2}.$$

Our basic power limit tells us that this limit is 0, and that is our final answer.

Solution to Part 3 If the limit exists,

$$\lim_{x \to \infty} \frac{5x^3 + x + 3}{2x + 5} = \lim_{x \to \infty} \frac{5x^3}{2x} = \lim_{x \to \infty} \frac{5x^2}{2}.$$

Our basic power limit does not tell us the answer here, but a little thought makes it clear. We want to know what happens to $5x^2/2$ when x is very large. The answer is clear: When x is large, $5x^2/2$ is a very large number. As x increases, $5x^2/2$ increases without bound. Thus the limit does not exist.

EXERCISES

Reminder Round all answers to two decimal places unless otherwise indicated.

1. **Verifying the Basic Exponential Limit** Choose several values of a that are less than 1, and for each choice make a table of values to give evidence that $\lim\limits_{t \to \infty} a^t = 0$.

2. **Verifying the Basic Power Limit** Choose several values of n larger than 0, and make a table of values to give evidence that

$$\lim_{t \to \infty} \frac{1}{t^n} = 0$$

   for each choice.

Calculating with the Basic Exponential Limit Use the basic exponential limit to calculate the limits in Exercises 3 through 5.

3. $\lim\limits_{t \to \infty} 15(1 - 3 \times 0.5^t)$

4. $\lim\limits_{t \to \infty} \sqrt{7 + 0.5^t}$

5. $\lim\limits_{x \to \infty} \dfrac{\sqrt{4 + 0.5^x}}{\sqrt{9 + 0.25^x}}$

Calculating with the Basic Power Limit Use the basic power limit to calculate the limits in Exercises 6 through 8.

6. $\lim\limits_{t \to \infty} \dfrac{2 + \dfrac{1}{t}}{3 + \dfrac{4}{t^2}}$

7. $\lim\limits_{t \to \infty} (5 + t^{-1/3})$

8. $\lim\limits_{x \to \infty} \dfrac{\sqrt{1 + 1/x}}{\sqrt{1 - 1/x}}$

Calculating with Quotients of Polynomials   Exercises 9 through 12 involve limits of quotients of polynomials. Either calculate the given limit or state that the limit does not exist.

9. $\lim\limits_{t \to \infty} \dfrac{6t^3 + 5}{2t^3 + 3t + 1}$

10. $\lim\limits_{x \to \infty} \dfrac{x^2 + 1}{x + 1}$

11. $\lim\limits_{x \to \infty} \dfrac{7x^2 + 9x + 1}{2x^4 + x + 2}$

12. $\lim\limits_{x \to \infty} \dfrac{5x^3 + x + 4}{10x^3 + 9}$

Applications   Exercises 13 through 15 give applications of limiting values.

13. **Carrying Capacity for the General Logistic Function** In general, a logistic function can be written in the form

$$\frac{k}{1 + pa^t},$$

where $0 < a < 1$. If this function describes population growth for a certain species in a certain area, what is the carrying capacity?

14. **Water in a Tank**   The volume V of water in a tank at time t is given by $V = a - bc^t$ cubic feet, where c is positive and $c < 1$. How much water will be in the tank after a long period of time?

15. **Weight of a Fish**   A certain fish grows so that after t years its weight W is given by $W = 25 - 12 \times 0.7^t$ pounds. What is the largest weight of such a fish?

A FURTHER LOOK

Shifting and Stretching

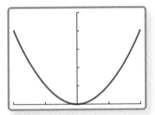

FIGURE 2.114 The graph of x^2

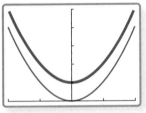

FIGURE 2.115 Shifting up 1 unit to get the graph of $x^2 + 1$

It is important to understand how certain parameters affect graphs. Perhaps the most important of these are *shifts* and *stretches*.

We look first at shifts. In Figure 2.114 we have used the calculator to get the familiar parabolic graph of $f(x) = x^2$. (We have used a horizontal span of -2 to 2 and a vertical span of 0 to 5.) We want to know how the graph is affected if we add 1 to the function. In Figure 2.115 we have included the graph of $g(x) = x^2 + 1$. We see that the new graph is obtained from the old one by moving it up 1 unit. This is not a surprise since we get the graph of $g(x)$ by adding 1 unit to each point on the graph of $f(x)$. That is, we move the graph of $f(x)$ up 1 unit to get the graph of $g(x)$. The same idea works in general if we add a constant to a function.

Vertical shifts: Let a be a positive constant. The graph of $g(x) = f(x) + a$ is obtained by shifting the graph of $f(x)$ up a units. The graph of $h(x) = f(x) - a$ is obtained by shifting the graph of $f(x)$ down a units.

We see another kind of shift if we look at the graph of $(x - 1)^2$. In Figure 2.116 we have included this graph with the graph of x^2. We see that the new graph is obtained from the old by shifting to the right 1 unit. This is as expected since the value of x^2 is the same as the value of $(x - 1)^2$ at the point 1 unit to the right. Similarly, we see in Figure 2.117 that we get the graph of $(x + 1)^2$ by moving the graph of x^2 to the left 1 unit.

As with vertical shifts, this works with complete generality.

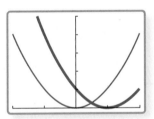

FIGURE 2.116 Shifting 1 unit to the right to get the graph of $(x - 1)^2$

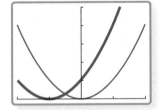

FIGURE 2.117 Shifting 1 unit to the left to get the graph of $(x + 1)^2$

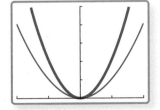

FIGURE 2.118 Stretching vertically to obtain the graph of $2x^2$

Horizontal shifts: Let a be a positive constant. The graph of $g(x) = f(x - a)$ is obtained by shifting the graph of $f(x)$ to the right a units. The graph of $h(x) = f(x + a)$ is obtained by shifting the graph of $f(x)$ to the left a units.

If instead of adding a constant we multiply by a constant, then we get a vertical rescaling or stretching of the graph. Let's look, for example, at the graph of $2x^2$, which is shown in Figure 2.118. We see that we get the new graph by stretching vertically by a factor of 2. That is because the value of $2x^2$ is twice that of x^2.

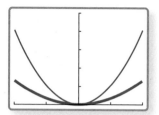

FIGURE 2.119 Stretching horizontally to obtain the graph of $(x/2)^2$

We get a different kind of stretch if we look at $(2x)^2$. Note that the value of $(2x)^2$ is the same as the value of x^2 twice as far down the horizontal axis. Thus we get the graph of $(2x)^2$ by compressing the graph of x^2 horizontally by a factor of 2. Similarly, we get the graph of $(x/2)^2$ by stretching the graph horizontally by a factor of 2. This stretch is shown in Figure 2.119.

Vertical stretches: Let $a > 1$ be a constant. We get the graph of $g(x) = af(x)$ by stretching the graph of $f(x)$ vertically by a factor of a. We get the graph of $h(x) = (1/a)f(x)$ by compressing the graph of $f(x)$ vertically by a factor of a.

Horizontal stretches: Let $a > 1$ be a constant. We get the graph of $g(x) = f(ax)$ by compressing horizontally by a factor of a. We get the graph of $h(x) = f(x/a)$ by stretching horizontally by a factor of a.

Many times, we need to use a combination of shifts and stretches to get a graph. Suppose, for example, that the graph of $f(x)$ is as shown in Figure 2.120. We want to sketch the graph of $2f(x + 1) - 1$. We do this in three steps. First we get the graph of $f(x + 1)$ by shifting the graph of $f(x)$ to the left 1 unit as shown in Figure 2.121.

Next we stretch the graph of $f(x + 1)$ vertically by a factor of 2 to get the graph of $2f(x + 1)$ shown in Figure 2.122. Finally, we move the graph of $2f(x + 1)$ down 1 unit to get the graph of $2f(x + 1) - 1$ shown in Figure 2.123.

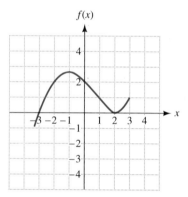

FIGURE 2.120 The graph of $f(x)$

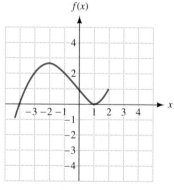

FIGURE 2.121 The graph of $f(x + 1)$

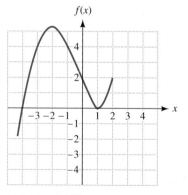

FIGURE 2.122 The graph of $2f(x + 1)$

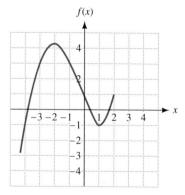

FIGURE 2.123 The graph of $2f(x + 1) - 1$

EXERCISES

Reminder   Round all answers to two decimal places unless otherwise indicated.

Shifting and Stretching   Exercises 1 through 13 refer to the graph in Figure 2.124. Sketch the graph of each of the following functions.

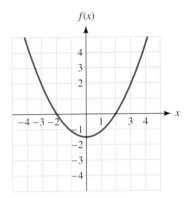

FIGURE 2.124 The graph of $f(x)$

1. **Shifting** $f(x + 2)$

2. **Shifting** $f(x - 2)$

3. **Shifting** $f(x + 2) + 2$

4. **Shifting** $f(x - 2) + 2$

5. **Stretching** $2f(x)$

6. **Stretching** $f(2x)$

7. **Stretching** $2f(x/2)$

8. **Stretching** $2f(2x)$

9. **Shifting and Stretching** $f(2x) + 1$

10. **Shifting and Stretching** $f(x/2) + 1$

11. **Shifting and Stretching** $2f(x/2) + 1$

12. **Shifting and Stretching** $2f(x - 1) + 1$

13. **Shifting and Stretching** $\dfrac{1}{2}f(x - 1) - 1$

Reflections   Exercises 14 and 15 refer to the function $g(x)$ whose graph is shown in Figure 2.125.

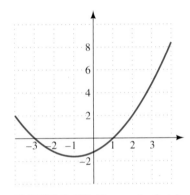

FIGURE 2.125 The graph of $g(x)$

14. Sketch the graph of $-g(x)$. If you do this properly, you will produce the reflection of the graph through the horizontal axis.

15. Sketch the graph of $-g(-x)$. Describe the result in terms of reflections of the graph of g.

A FURTHER LOOK

Optimizing with Parabolas

In the general case, finding maxima and minima exactly requires the use of calculus. But for quadratic functions the graph is a *parabola*, and there is an easy way to find the maximum or minimum.

In general, a quadratic expression has the form $ax^2 + bx + c$. In the case where a is positive, its graph is a parabola that opens upward and has a minimum but no maximum. An example is shown in Figure 2.126. When a is negative, the graph is a parabola that opens downward and has a maximum but no minimum. A typical example is shown in Figure 2.127. In general, we refer to the maximum or minimum point of a parabola as its *vertex*.

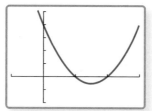

FIGURE 2.126 When the leading coefficient is positive, the parabola has a minimum.

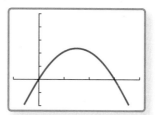

FIGURE 2.127 When the leading coefficient is negative, the parabola has a maximum.

We get the graph of $ax^2 + bx$ from the graph of $ax^2 + bx + c$ by shifting the graph up or down, depending on whether c is positive or negative. Thus the height of the vertex for $ax^2 + bx$ may be different from the height of the vertex for $ax^2 + bx + c$, but the x value, or horizontal location, is the same. Parabolas are symmetric about a vertical line through the vertex. Thus when a parabola crosses the horizontal axis twice, the horizontal location of its vertex is halfway between the crossing points. Since $ax^2 + bx = x(ax + b)$, this parabola crosses the horizontal axis at $x = 0$ and at $x = -b/a$. Its vertex occurs halfway between, at $x = -b/(2a)$. Since the vertex of $ax^2 + bx$ has the same horizontal location as the vertex of $ax^2 + bx + c$, we have located the vertex of a general quadratic.

Locating the vertex: The vertex of the graph of $ax^2 + bx + c$ is located at $x = -b/(2a)$.

Thus, for example, the vertex of $2x^2 - 4x + 7$ occurs at

$$x = -\frac{-4}{2 \times 2} = 1.$$

Since the leading coefficient 2 is positive, this is a minimum. The vertical location of the minimum is found by getting the function value at $x = 1$:

$$\text{Vertical location of minimum} = 2 \times 1^2 - 4 \times 1 + 7 = 5.$$

173

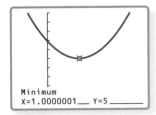

FIGURE 2.128 The graph of $2x^2 - 4x + 7$, showing the vertex at $(1, 5)$

Thus the vertex of the parabola occurs at the point $(1, 5)$. To verify our calculations, we have made the graph of $2x^2 - 4x + 7$ in Figure 2.128.

We can use what we know about the vertex of a parabola to solve certain optimization problems exactly.

EXAMPLE 2.14 **MAXIMIZING AREA**

One hundred yards of fence are to be used to make three sides of a rectangle. The fourth side of the rectangle is the bank of a straight river, as shown in Figure 2.129. The river bank should not be fenced. Let x denote the width of the rectangle, in yards, and assume that the length of the rectangle is along the river.

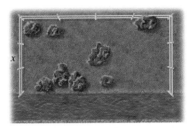

FIGURE 2.129 A rectangle with a river bank on one side

Part 1 Express the length of the rectangle in terms of x.

Part 2 Express the area of the rectangle in terms of x.

Part 3 What value of x will give the rectangle of maximum area?

Part 4 What is the largest area we can achieve?

Solution to Part 1 We use two sections of x yards of fence to make the left and right sides of the rectangle. That leaves $100 - 2x$ yards for the length of the rectangle.

Solution to Part 2 The area is the width x times the length $100 - 2x$:

$$\text{Area} = x(100 - 2x).$$

Solution to Part 3 To find the maximum of $x(100 - 2x)$, we first write it in standard form as $-2x^2 + 100x$. Thus the vertex occurs at

$$x = -\frac{100}{2 \times (-2)} = 25 \text{ yards}.$$

Since the leading coefficient is negative, this is a maximum.

Solution to Part 4 To find the largest area, we calculate the value of the area function $x(100 - 2x)$ for $x = 25$:

$$\text{Maximum area} = (25)(100 - 2 \times 25) = 1250 \text{ square yards}.$$

In Figure 2.130 we have plotted the graph of $x(100 - 2x)$ and located the maximum in order to verify our work.

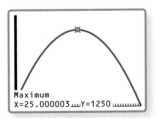

FIGURE 2.130 Locating the vertex of $x(100 - 2x)$

EXERCISES

Reminder Round all answers to two decimal places unless otherwise indicated.

Locating the Vertex of a Parabola In Exercises 1 through 4, find the vertex of the parabola that is the graph of the given quadratic function of x. (Give the horizontal and vertical coordinates.) Determine whether the vertex represents a maximum or a minimum for the function.

1. $x^2 + 6x - 4$

2. $3x^2 - 30x + 1$

3. $-2x^2 + 12x - 3$

4. $a^2x^2 + 4$ if $a \neq 0$

Applications Exercises 5 through 13 illustrate applications of optimization using parabolas.

5. **Finding the Least Area of a Triangle** One hundred yards of fence are to be used to construct a right triangular pen with a straight river serving as the hypotenuse of the triangle, as shown in Figure 2.131 on the following page. No fence is placed along the river. Let x denote the height, in yards, of the triangle.

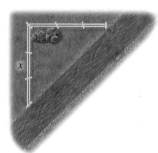

FIGURE 2.131 Making a triangle next to a river

   a. Express the base (the horizontal side in Figure 2.133) of the triangle in terms of x.
   b. Express the area of the triangle in terms of x.
   c. What value of x gives the maximum area?
   d. What is the maximum area that can be achieved?

6. **Distance from a Line** A typical point on a certain line in the plane has the form $(x, x + 1)$. The square of the distance from this point to the origin is given by
$$D = x^2 + (x + 1)^2.$$
   Find the x value that gives the point on the line nearest the origin.

7. **Size of High Schools** A study found that the cost C per pupil of operating a high school in a certain country depends on the number n of students enrolled. The cost is given by $C = 65{,}000 - 500n + n^2$ dollars per pupil. What enrollment produces the minimum cost per pupil, and what is the minimum cost?

8. **Bending Metal** A newly formed piece of metal is allowed to cool and then is reheated for bending. Its temperature after t minutes is given by $T = 4t^2 - 16t + 130$ degrees Fahrenheit. When is the temperature at a minimum, and what is the minimum temperature?

9. **Concentration of Medicine** The concentration C in milligrams per liter of a certain drug in the bloodstream t hours after ingestion is given by $C = 6 + 6t - t^2$. When does the maximum concentration occur, and what is the maximum concentration?

10. **A Rock** A rock is tossed upward. Its height after t seconds is given by $S = -16t^2 + 20t + 5$ feet. What is its maximum height, and when does it reach that height?

11. **A Fish Farm** On a certain fish farm, the growth rate G of the population depends on the population size n in tons. The relationship is $G = n(1 - n)/2$ tons per year. What population gives the maximum growth rate, and what is that growth rate?

12. **Net Profit** The net profit N from sales depends on the amount d of advertising dollars spent. The relationship is given by $N = 300 + 1000d - d^2$ dollars. What advertising expenditure results in the maximum profit, and what is that profit?

13. **A Window** A window consists of a rectangle of height h and width d with half circles of diameter d on top and bottom.
   a. Show that the perimeter of the opening is $\pi d + 2h$.
   b. Show that the area is $dh + \pi d^2/4$.
   c. If the perimeter is to be 10 feet, show that
   $$h = \frac{10 - \pi d}{2}.$$
   d. If the perimeter is to be 10 feet, what width d and height h will produce the maximum area?

STRAIGHT LINES AND LINEAR FUNCTIONS

3

Techniques for modeling linear data can be applied to the analysis of tuition increases at universities such as Columbia. See Exercise 4 on page 209.

STRAIGHT LINES in the form of city streets, directions, boundaries, and many other things are among the most obvious mathematical objects that we experience in daily life. Historically, mathematicians made extensive studies of the *geometry* of straight lines several centuries before these lines were associated with a *linear formula*. Today it is understood that it is the combination of pictures and formulas that make lines so useful and so easy to handle.

Len DeLessio/Getty Images

Robert Brenner/PhotoEdit

Student resources are available on the website
www.cengagebrain.com

3.1 THE GEOMETRY OF LINES

 | ## Characterizations of Straight Lines

One way in which straight lines are often characterized is that they are determined by two points. For example, to describe a straight ramp, it is only necessary to give the locations of the ends of the ramp. In Figure 3.1 we have depicted a ramp with one end atop a 4-foot-high retaining wall and the other on the ground 10 feet away. It is often convenient to represent such lines on coordinate axes. If we choose the horizontal axis to be ground level and let the vertical axis follow the retaining wall, we get the picture in Figure 3.2. The ends of the ramp in Figure 3.1 match the points in Figure 3.2 where the line crosses the horizontal and vertical axes. These crossing points are known as the *horizontal and vertical intercepts*. (Sometimes they are called the *x-intercept* and the *y-intercept*.) Thus in Figure 3.2, the horizontal intercept is 10 and the vertical intercept is 4.

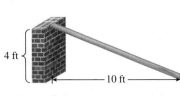

FIGURE 3.1 A ramp on a retaining wall

FIGURE 3.2 Representing the ramp line on coordinate axes

To illustrate these ideas further, in Figure 3.3 we have drawn a roof line that is 8 feet high at the outside wall and 8.5 feet high 1 foot toward the interior of the structure. If we want to represent the roof line on coordinate axes, it is natural to let the horizontal axis correspond to ground level and to let the vertical axis follow the outside wall. We have done this in Figure 3.4. We note that the vertical intercept of this line is 8, but in this case it is not immediately apparent what the horizontal intercept is or what its physical significance might be. We will return to this in Example 3.1.

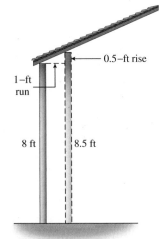

FIGURE 3.3 The roof of a building

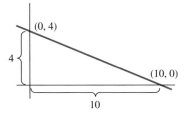

FIGURE 3.4 Representing the roof line

A line always rises or falls at the same rate

Another way of characterizing a straight line is to say that it rises or falls at the same rate everywhere on the line. (Curved graphs rise or fall at different rates at different places.) If we stand at the outside wall of the house depicted in Figure 3.3 and move 1 foot toward its interior, the roof rises from 8 feet to 8.5 feet. That is, in 1 horizontal foot, the roof rises by 0.5 vertical foot. Since the roof line is straight, we know that if we move 1 more foot toward the interior, the roof will rise by that same amount, 0.5 foot, and the roof at that point will be 9 feet high. Each 1-foot horizontal movement toward the interior results in a 0.5-foot vertical rise in the roof line. In this setting, horizontal change is often referred to as the *run* and vertical change as the *rise*, and we would say that the roof line rises 0.5 foot for each 1-foot run.

EXAMPLE 3.1 **FURTHER EXAMINATION OF THE ROOF LINE**

In Figure 3.5 we have extended the roof to its peak 14 horizontal feet toward the interior of the structure.

Part 1 How high is the roof line at its peak?

Part 2 Locate the horizontal intercept in Figure 3.6 and explain its physical significance.

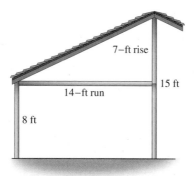

FIGURE 3.5 Extending the roof line to its peak

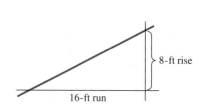

FIGURE 3.6 Finding the horizontal intercept

Solution to Part 1 We know that the roof rises 0.5 foot for each 1-foot run. To get to the peak of the roof, we need to make a 14-foot run. That will result in 14 half-foot rises. Thus the roof rises by $14 \times 0.5 = 7$ feet. Since the roof is 8 feet high at the outside wall, that makes it $8 + 7 = 15$ feet high at its peak. This is illustrated in Figure 3.5.

Solution to Part 2 In Figure 3.6, movement 1 foot to the right results in a 0.5-foot rise in the line. Thus a 1-foot movement to the left would give a fall of 0.5 foot. To get down to the horizontal axis, we need to drop a total of 8 units. Since we do that in 0.5-foot steps, we need to move 16 units to the left of the vertical axis to get to the place in Figure 3.6 where the line crosses the horizontal axis. Thus the horizontal intercept is −16 (negative, since it's to the left of zero). This is where an extended roof line would meet the ground. It might, for example, be important if we wanted to build an "A-frame" structure. The minus sign here means that an extended roof line would meet the ground 16 feet in the exterior direction from the wall.

TEST YOUR UNDERSTANDING | **FOR EXAMPLE 3.1**

Suppose that the peak of the roof occurs 20 (rather than 14) horizontal feet toward the interior of the structure. How high would the peak of the roof be in this setting?

◼ | The Slope of a Line

The slope of a line is the rate of change in height with respect to horizontal distance.

The number 0.5 associated with the roof line in Example 3.1 is the *rate of change* in height with respect to horizontal distance. For straight lines, this is universally termed the *slope* and is commonly denoted by the letter *m*. The slope is one of the most important and useful features of a line. It tells the vertical change along the line when there is a horizontal change of 1 unit. In other words, a line of slope m exhibits m units of rise for each unit of run. Another way to express this is to say that the rise is proportional to the run, and the constant of proportionality is called the slope. Horizontal change is sometimes denoted by Δx and vertical change by Δy. For a line of slope m, the following three equations say exactly the same thing using these different terms:

$$\text{Rise} = m \times \text{Run} \tag{3.1}$$

$$\text{Vertical change} = m \times \text{Horizontal change} \tag{3.2}$$

$$\Delta y = m \times \Delta x. \tag{3.3}$$

Figure 3.7 shows lines of slopes 0, 1, and 2. Note that the horizontal line has slope $m = 0$. This makes sense because a horizontal change results in no vertical change at all. The line with slope $m = 1$ rises 1 unit for each unit of run. Thus it slants upward at a 45-degree angle. The line of slope $m = 2$ is much steeper, showing a rise of 2 units for each unit of run. In general, larger positive slopes mean steeper lines that point upward from left to right.

If a line has negative slope, then a run of 1 unit will result in a *drop* rather than a rise. The line in Figure 3.8 with $m = -1$ drops 1 unit for each unit of run. Similarly, the line with $m = -2$ drops 2 units for each unit of run. In general, large negative slopes mean steep lines that point downward.

> Larger positive slopes indicate steeper lines. A slope of 0 indicates a horizontal line. Negative slopes that are larger in size indicate lines that fall more steeply.

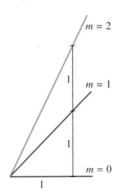

FIGURE 3.7 Some lines with positive slope

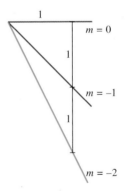

FIGURE 3.8 Some lines with negative slope

Getting Slope from Points

Equations (3.1), (3.2), and (3.3) show us how to use the slope of a line to determine vertical change. These are all linear equations, so we can use division to solve each of them for m. This will give us equations that tell how to find the slope m of a line. Once again, all three equations say the same thing using different notations:

$$\text{Rate of change} = \text{slope} = m = \frac{\text{Rise}}{\text{Run}} \tag{3.4}$$

$$\text{Rate of change} = \text{slope} = m = \frac{\text{Vertical change}}{\text{Horizontal change}} \tag{3.5}$$

$$\text{Rate of change} = \text{slope} = m = \frac{\Delta y}{\Delta x}. \tag{3.6}$$

> **KEY IDEA 3.1** **THE SLOPE OF A LINE**
>
> - The slope, or rate of change, m, of a line shows how steeply it is increasing or decreasing. It tells the vertical change along the line when there is a horizontal change of 1 unit.
>
>   If m is positive, the line is rising from left to right. Larger positive values of m mean steeper lines.
>
>   If $m = 0$, the line is horizontal.
>
>   If m is negative, the line is falling from left to right. Negative values of m that are larger in size correspond to lines that fall more steeply.
>
> - The slope m of a line can be calculated using
>
> $$m = \frac{\text{Vertical change}}{\text{Horizontal change}} = \frac{\text{Rise}}{\text{Run}}.$$
>
> - The slope m can be used to calculate vertical change:
>
> $$\text{Vertical change} = m \times \text{Horizontal change}$$
>
> $$\text{Rise} = m \times \text{Run}.$$

EXAMPLE 3.2 **A CIRCUS TENT**

The outside wall of the circus tent depicted in Figure 3.9 is 10 feet high. Five feet toward the center pole, the tent is 12 feet high. The center pole of the tent is 60 feet from the outside wall.

Part 1 What is the slope of the line that follows the roof of the circus tent?

Part 2 Use the slope to find the height of the tent at its center pole.

Part 3 A rope is attached to the roof of the tent at the outside wall and anchors the tent to a stake in the ground. The rope follows the line of the tent roof as shown in Figure 3.10. How far away from the tent wall is the anchoring stake?

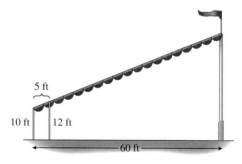

FIGURE 3.9 A circus tent

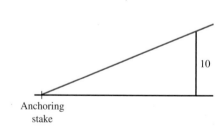

FIGURE 3.10 An anchor rope

When we know two points on a line we can find the slope by calculating Rise/Run.

Solution to Part 1 To calculate the slope we use Equation (3.4). At the outside wall, the circus tent is 10 feet high. If we run 5 feet toward the center pole, the height increases to 12 feet; that is a rise of 2 feet. Thus

$$m = \frac{\text{Rise}}{\text{Run}} = \frac{2}{5} = 0.4 \text{ foot per foot.}$$

If we know the slope we can find the rise by calculating Slope × Run.

Solution to Part 2 To find the height at the center pole, we first find the rise from the top of the outside wall to the center pole. We do this by using Equation (3.1) and the value of the slope we found in part 1. From the outside wall to the center pole is a run of 60 feet, so

$$\text{Rise} = m \times \text{Run} = 0.4 \times 60 = 24 \text{ feet}.$$

Thus from the top of the outside wall to the center pole, the height increases by 24 feet. To find the height at the center pole, we need to add the height of the outside wall. We find that the height of the center pole is $24 + 10 = 34$ feet.

Solution to Part 3 Since the slope of the line is 0.4, if we stand at the anchoring stake and move toward the tent, each horizontal foot that we move results in a 0.4-foot rise in the rope. We need the total rise to be 10 feet:

$$\text{Rise} = m \times \text{Run}$$

$$10 = 0.4 \times \text{Run}$$

$$\frac{10}{0.4} = \text{Run}$$

$$25 = \text{Run}.$$

Thus the anchoring stake is located 25 feet outside the wall of the tent. If the vertical axis corresponds to the outside wall of the tent, then the horizontal intercept of the roof line is −25.

TEST YOUR UNDERSTANDING | **FOR EXAMPLE 3.2**

Re-do the example under the assumption that the height of the tent 5 feet toward the center from the outside wall is 13 (not 12) feet. ∎

ANSWERS FOR TEST YOUR UNDERSTANDING

3.1 18 feet

3.2 The slope is 0.6 foot per foot. The center pole is 46 feet high. The anchoring stake is 16.67 feet outside the wall of the tent.

3.1 EXERCISES

Reminder Round all answers to two decimal places unless otherwise indicated.

1. **A Line with Given Intercepts**   On coordinate axes, draw a line with vertical intercept 4 and horizontal intercept 3. Do you expect its slope to be positive or negative? Calculate the slope.

2. **A Line with Given Vertical Intercept and Slope**   On coordinate axes, draw a line with vertical intercept 3 and slope 1. What is its horizontal intercept?

3. **Another Line with Given Vertical Intercept and Slope**   A line has vertical intercept 8 and slope −2. What is its horizontal intercept?

4. **A Line with Given Horizontal Intercept and Slope**   A line has horizontal intercept 6 and slope 3. What is its vertical intercept?

5. **Lines with the Same Slope**   On the same coordinate axes, draw two lines, each of slope 2. The first line has vertical intercept 1, and the second has vertical intercept 3. Do the lines cross? In general, what can you say about different lines with the same slope?

6. **Where Lines with Different Slopes Meet**   On the same coordinate axes, draw one line with vertical intercept 2 and slope 3 and another with vertical intercept 4 and slope 1. Do these lines cross? If so, do

they cross to the right or left of the vertical axis? In general, if one line has its vertical intercept below the vertical intercept of another, what conditions on the slope will ensure that the lines cross to the right of the vertical axis?

7. A Wrap Skirt   Figure 3.11 shows a simplified pattern for a wrap skirt that is 20 inches long. The bottom hem for this pattern has a length of 63 inches. Suppose you decide to alter the pattern to make a skirt that is 24 inches long. What should be the length of the bottom hem?

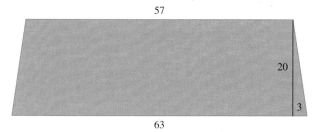

FIGURE 3.11 Wrap skirt pattern

8. View from the Top   Your office window is 35 feet high. Looking out your window, you find that the top of a statue lines up exactly with the bottom of a building that is 600 horizontal feet from your office. You know the statue is 125 feet from the building. How tall is the statue? (See Figure 3.12.)

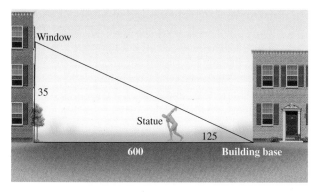

FIGURE 3.12 A view from your office window

9. A Topographical Map   In making a topographical map, it is not practical to measure directly heights of structures such as mountains. This exercise illustrates how some such measurements are taken. A surveyor whose eye is 6 feet above the ground views a mountain peak that is 2 horizontal miles distant. (See Figure 3.13.) Directly in his line of sight is the top of a surveying pole that is 10 horizontal feet distant and 8 feet high. How tall is the mountain peak? *Note*: One mile is 5280 feet.

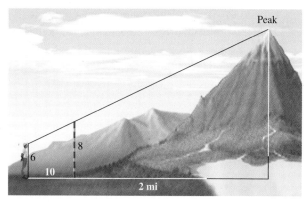

FIGURE 3.13 A mountain peak

10. An Ice Cream Cone   An ice cream cone is 4 inches deep and 2 inches across the top. (See Figure 3.14.) If we wanted to make a king-size cone that has the same shape but is 2.5 inches across the top, how deep would the cone be?

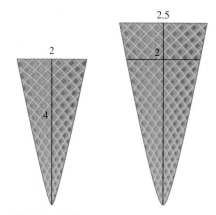

FIGURE 3.14 Ice cream cones

11. A Ramp to a Building   The base of a ramp sits on the ground (see Figure 3.15). Its slope is 0.4, and it

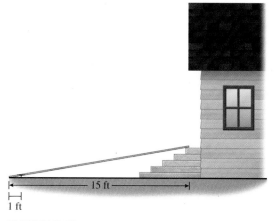

FIGURE 3.15

(continued)

extends to the top of the front steps of a building 15 horizontal feet away.

a. How high is the ramp 1 horizontal foot toward the building from the base of the ramp?

b. How high is the top of the steps relative to the ground?

12. A Wheelchair Service Ramp The *Americans with Disabilities Act* (ADA) requires, among other things, that wheelchair service ramps have a slope not exceeding 1/12.

a. Suppose the front steps of a building are 2 feet high. You want to make a ramp conforming to ADA standards that reaches from the ground to the top of the steps. How far away from the building is the base of the ramp?

b. Another way to give specifications on a ramp is to give allowable inches of rise per foot of run. In these terms, how many inches of rise does the ADA requirement allow in 1 foot of run?

13. A Cathedral Ceiling A cathedral ceiling shown in Figure 3.16 is 8 feet high at the west wall of a room. As you go from the west wall toward the east wall, the ceiling slants upward. Three feet from the west wall, the ceiling is 10.5 feet high.

a. What is the slope of the ceiling?

b. The width of the room (the distance from the west wall to the east wall) is 17 feet. How high is the ceiling at the east wall?

c. You want to install a light in the ceiling as far away from the west wall as possible. You intend to change the bulb, when required, by standing near the top of your small stepladder. If you stand on the highest safe step of your stepladder, you can reach 12 feet high. How far from the west wall should you install the light?

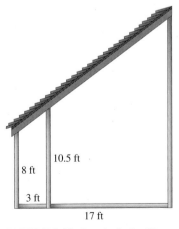

FIGURE 3.16 A cathedral ceiling

14. Roof Trusses Trusses as shown in Figure 3.17 are to be constructed to support the roof of a building. The truss is to have a 16-foot horizontal base (joist) that spans from one wall to the opposite wall. The vertical center strut is 4 feet long.

a. Vertical struts are located 3 horizontal feet inside each wall. How long are these vertical struts?

b. The rafter extends 1.5 horizontal feet outside the wall. If the top of the wall is 8 feet above the floor, how high above the floor is the outside tip of the rafter?

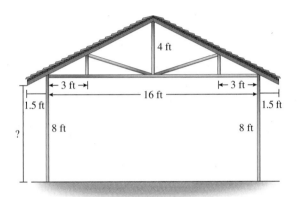

FIGURE 3.17 A roof truss

15. Cutting Plywood Siding Plywood siding is to be used to cover the exterior wall of a house. Plywood siding comes in sheets 4 feet wide and 8 feet high. On the exterior wall shown in Figure 3.18, three pieces of siding are shown cut to conform to the roof line. The piece on the far right is 1 foot high on the shorter side and 2 feet 6 inches high on the longer side (toward the peak of the roof). To make proper cuts on the next two sheets of plywood, we need to know the lengths *h* and *k* shown in Figure 3.18. Find these lengths.

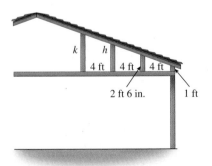

FIGURE 3.18 Plywood siding for a house

16. An Overflow Pipeline An overflow pipeline for a pond is to run in a straight line from the pond at maximum water level a distance of 96 horizontal feet to a drainage area that is 5 vertical feet below the maximum water level (see Figure 3.19). How much lower is the pipe at the end of each 12-foot horizontal stretch?

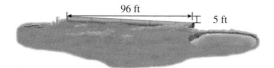

FIGURE 3.19

17. Looking over a Wall Twenty horizontal feet north of a 50-foot building is a 35-foot wall (see Figure 3.20). A man 6 feet tall wishes to view the top of the building from the north side of the wall. How far north of the wall must he stand in order to view the top of the building?

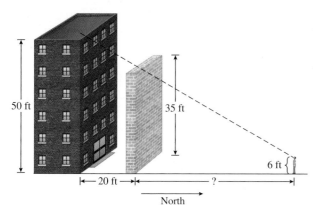

FIGURE 3.20

18. The Mississippi River For purposes of this exercise, we will think of the Mississippi River as a straight line beginning at its headwaters, Lake Itasca, Minnesota, at an elevation of 1475 feet above sea level, and sloping downward to the Gulf of Mexico 2340 miles to the south.

a. Think of the southern direction as pointing to the right along the horizontal axis. What is the slope of the line representing the Mississippi River? Be sure to indicate what units you are using.

b. Memphis, Tennessee, sits on the Mississippi River 1982 miles south of Lake Itasca. What is the elevation of the river as it passes Memphis?

c. How many miles south of Lake Itasca would you find the elevation of the Mississippi to be 200 feet?

19. A Road up a Mountain You are driving on a straight road that is inclined upward toward the peak of a mountain (see Figure 3.21). You pass a sign that reads "Elevation 4130 Feet." Three horizontal miles farther you pass a sign that reads "Elevation 4960 Feet."

a. Think of the direction in which you are driving as the positive direction. What is the slope of the road? (*Note:* Units are important here.)

b. What is the elevation of the road 5 horizontal miles from the first sign?

c. You know that the peak of the mountain is 10,300 feet above sea level. How far in horizontal distance is the first sign from the peak of the mountain?

FIGURE 3.21

20. An Underground Water Source An underground aquifer near Seiling, Oklahoma, sits on an impermeable layer of limestone. West of Seiling the limestone layer is thought to slope downward in a straight line. In order to map the limestone layer, hydrologists started at Seiling heading west and drilled *sample wells.* Two miles west of Seiling the limestone layer was found at a depth of 220 feet. Three miles west of Seiling the limestone layer was found at a depth of 270 feet.

a. What would you expect to be the depth of the limestone layer 5 miles west of Seiling?

b. You want to drill a well down to the limestone layer as far west from Seiling as you can. Your budget will allow you to drill a well that is 290 feet deep. How far west of Seiling can you go to drill the well?

c. Four miles west of Seiling someone drilled a well and found the limestone layer at 273 feet. Were the hydrologists right in saying that the limestone layer slopes downward in a straight line west of Seiling? Explain your reasoning.

(continued)

21. **Earth's Umbra**   Earth has a shadow in space, just as people do on a sunny day. The darkest part[1] of that shadow is a conical region in space known as the *umbra*. A representation of Earth's umbra is shown in Figure 3.22. Earth has a radius of about 3960 miles, and the umbra ends at a point about 860,000 miles from Earth. The moon is about 239,000 miles from Earth and has a radius of about 1100 miles. Consider a point on the opposite side of Earth from the sun and at a distance from Earth equal to the moon's distance from Earth. What is the radius of the umbra at that point? Can the moon fit inside Earth's umbra? What celestial event occurs when this happens?

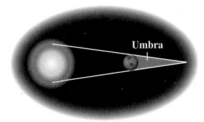

FIGURE 3.22 The Earth's umbra

22. **Earth's Penumbra**   *This is a continuation of Exercise 21.* Earth also has a partial shadow known as the *penumbra*. The penumbra is represented in Figure 3.23. The penumbra has a radius of about 10,000 miles at a point on the opposite side of Earth from the sun and at a distance from Earth equal to the moon's distance from Earth. The penumbra has a conical shape and, if extended toward the sun, would reach an apex shown in Figure 3.23. How far away from Earth is this apex?

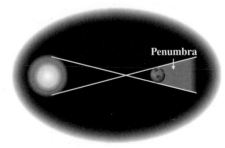

FIGURE 3.23 The Earth's penumbra

23. **The Umbra of the Moon**   *This is a continuation of Exercise 21.* A total eclipse of the sun occurs when we are in the umbra of the moon. The size of the moon's umbra on Earth's surface is so small[2] that we can consider that the umbra reaches its apex at Earth's center. The sun is 93,498,600 miles away from Earth. What is the radius of the sun? (The actual radius is 434,994 miles. Because of our simplified assumptions, you will get a slightly different, though relatively close, answer.)

24. **The Sonora Pass**   California State Highway 108 crosses the Sierra Nevada mountain range at Sonora Pass, the second highest highway pass[3] in the Sierra range. The route features striking natural beauty as well as some very steep highway grades. Just west of the pass one encounters a sign that says "28% grade next two miles." Of course, this means that the *maximum* grade over the next two miles is 28%; nonetheless, this is an extraordinarily steep state highway. Suppose that the highway is in fact a straight line with a slope of 0.28 and that the elevation at the beginning is 6000 feet. What would be the elevation after 2 horizontal miles? (One mile is 5280 feet.)

3.1 SKILL BUILDING EXERCISES

S-1. **Slope from Rise and Run**   One end of a ladder is on the ground. The top of the ladder rests at the top of an 8-foot wall. The wall is 2 horizontal feet from the base of the ladder. What is the slope of the line made by the ladder? (Assume that the positive direction points from the base of the ladder toward the wall.)

S-2. **Slope from Rise and Run**   One end of a ladder is on the ground. The top of the ladder rests at the top of a 15-foot wall. The wall is 3 horizontal feet from the base of the ladder. What is the slope of the line made by the ladder? (Assume that the positive direction points from the base of the ladder toward the wall.)

S-3. **Height from Slope and Horizontal Distance**   The base of a ladder is 3 horizontal feet from the wall where its top rests (see Figure 3.24). The slope of the line made by the

height

├── 3 ft ──┤

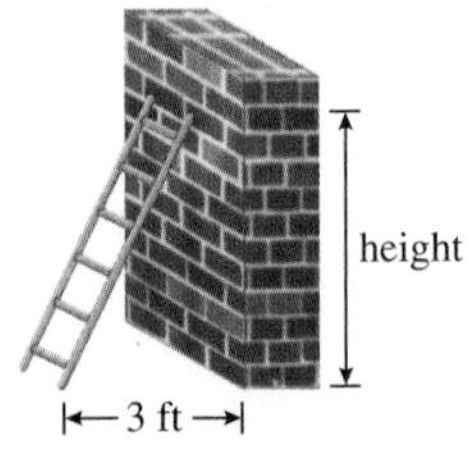

FIGURE 3.24

---

[1]In this part of the shadow the sun is not visible, and it would be totally dark but for reflected light from the moon and light bent by Earth's atmosphere. Blue light is scattered by Earth's atmosphere more easily than red light. This is why the moon in total eclipse has a coppery color.
[2]At its largest, it has a radius of about 84 miles. If the moon were just a little farther away from Earth, total solar eclipses would never occur.
[3]The Sonora Pass has an elevation of 9624 feet. The highest pass is a few miles to the south at Tioga pass, which is the entrance to Yosemite National Park.

ladder is 2.5. What is the vertical height of the top of the ladder? (Assume that the positive direction points from the base of the ladder toward the wall.)

S-4. Height from Slope and Horizontal Distance The base of a ladder is 4 horizontal feet from the wall where its top rests. The slope of the line made by the ladder is 1.7. What is the vertical height of the top of the ladder? (Assume that the positive direction points from the base of the ladder toward the wall.)

S-5. Horizontal Distance from Height and Slope A ladder leans against a wall so that its slope is 1.75. The top of the ladder is 9 vertical feet above the ground. What is the approximate horizontal distance from the base of the ladder to the wall? (Assume that the positive direction points from the base of the ladder toward the wall.)

S-6. Horizontal Distance from Height and Slope A ladder leans against a wall so that its slope is 2.1. The top of the ladder is 12 vertical feet above the ground. What is the horizontal distance from the base of the ladder to the wall? (Assume that the positive direction points from the base of the ladder toward the wall.)

S-7. Slope from Two Points Take west to be the positive direction. The height of a sloped roof above the place where you stand is 12 feet (see Figure 3.25). If you move 3 feet west, the height is 10 feet. What is the slope?

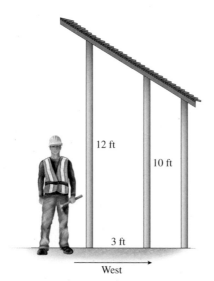

FIGURE 3.25

S-8. Continuation of Exercise S-7 If you move 5 additional feet west, what is the height of the roof above the place where you stand?

S-9. A Circus Tent You are at the center of a circus tent, where the height is 22 feet (see Figure 3.26). You are facing due west, which you take to be the positive direction. The slope of the tent line is -0.8. If you walk 7 feet west, how high is the tent?

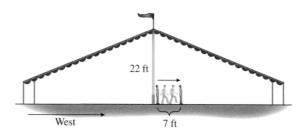

FIGURE 3.26

S-10. More on the Circus Tent Assume that the roof of the circus tent in Exercise S-9 extends in a straight line to the ground. How far from the center of the tent does the roof meet the ground?

S-11. Slope If a building is 100 feet tall and is viewed from a spot on the ground 70 feet away from the base of the building, what is the slope of a line from the spot on the ground to the top of the building?

Slopes of Lines in the Coordinate Plane For Exercises S-12 through S-23, use the fact that for points (a_1, b_1) and (a_2, b_2) in the coordinate plane, we can calculate the slope of the line through these points using

$$\text{Slope} = \frac{\Delta y}{\Delta x} = \frac{b_2 - b_1}{a_2 - a_1}.$$

S-12. Find the slope of the line through the points $(5, 2)$ and $(6, 9)$.

S-13. Find the slope of the line through the points $(2, 2)$ and $(4, 1)$.

S-14. Find the slope of the line through the points $(3.1, 4.2)$ and $(6.3, 5.5)$.

S-15. Find the slope of the line through the points $(-3.6, 2.5)$ and $(2.4, -1.6)$.

S-16. Find the point where the line through $(1, 4)$ with slope -2 crosses the horizontal axis. (*Suggestion:* The line must fall by 4 units to reach the horizontal axis.)

S-17. Find the point where the line through $(1.2, 3.1)$ with slope -0.8 crosses the horizontal axis.

(continued)

S-18. Find the point where the line through (6, 2) with slope 2 crosses the vertical axis.

S-19. Find the point where the line through (−4, 3) with slope 3 crosses the vertical axis.

S-20. Find the point where the line through (1.1, 3.6) with slope 2.3 crosses the vertical axis.

S-21. Find the point where the line through (3.2, −1.5) with slope −2.3 crosses the vertical axis.

S-22. Find the point with vertical coordinate 17 that lies on the line through (2, 2) with slope 3.

S-23. Find the point with horizontal coordinate 6.6 that lies on the line through (1, 5) with slope −1.1.

3.2 LINEAR FUNCTIONS

 ## Constant Rates of Change

The rate of change or slope of a linear function is always the same.

We now turn to some special types of functions, *linear functions*, that are intimately related to straight lines. A linear function is one whose *rate of change,* or *slope,* is always the same. Let's look at an example to show what we mean. Suppose the CEO of a company wants to have a dinner catered for employees. He finds that he must pay a dining hall rental fee of $275 and an additional $28 for each meal served. Then the cost $C = C(n)$, in dollars, is a function of the number n of meals served. If an unexpected guest arrives for dinner, then the CEO will have to pay an additional $28. This is the rate of change in C, and it is always the same no matter how many dinner guests are already seated. In more formal terms, when the variable n increases by 1, the function C changes by 28. This means that C is a linear function of n with slope or rate of change 28.

The rate of change, or slope, of a linear function can be used in ways that remind us of how slope is used for straight lines. Suppose, for example, that the caterer, having anticipated that 35 people would attend the dinner, sent the CEO a bill for $1255. But 48 people actually attended the dinner. What should the total price of the dinner be under these circumstances? Since 13 additional people attended the dinner, and the price per meal is $28, we can calculate the additional charge over $1255:

$$\text{Additional charge} = 28 \times \text{Additional people}$$

$$= 28 \times 13$$

$$= 364 \text{ dollars.}$$

Thus the total cost for 48 people is $1255 + 364 = 1619$ dollars.

In general, the slope or rate of change of a linear function is the amount the function changes when the variable increases by 1 unit. Furthermore, we can use the slope m just as we did above to calculate the change in function value that corresponds to a given change in the variable. We state the relationship in two equivalent ways:

$$\text{Change in function} = m \times \text{Change in variable}$$

$$\text{Additional function value} = m \times \text{Additional variable value.}$$

We can use division to restate these so that they show how to calculate m from changes in function value corresponding to changes in the variable. Once again, the following two equations say exactly the same thing using different words:

$$m = \frac{\text{Change in function}}{\text{Change in variable}}$$

$$m = \frac{\text{Additional function value}}{\text{Additional variable value}}.$$

> **KEY IDEA 3.2** **LINEAR FUNCTIONS**
>
> A function is linear if it has a constant slope or rate of change. This slope is then the amount the function changes when the variable increases by 1 unit. Suppose $y = y(x)$ is a linear function of x. Then
>
> - The slope m, or rate of change in y with respect to x, can be calculated from the change in y corresponding to a given change in x:
>
> $$m = \frac{\text{Change in function}}{\text{Change in variable}} = \frac{\text{Change in } y}{\text{Change in } x}.$$
>
> - The slope m can be used to calculate the change in y resulting from a given change in x:
>
> $$\text{Change in function} = m \times \text{Change in variable}.$$
>
> Or, using letters,
>
> $$\text{Change in } y = m \times \text{Change in } x.$$

EXAMPLE 3.3 **OKLAHOMA INCOME TAX**

The amount of income tax $T = T(I)$, in dollars, owed to the state of Oklahoma is a linear function of the taxable income I, in dollars, at least over a suitably restricted range of incomes. According to the Oklahoma Income Tax table for the year 2010, a single Oklahoma resident taxpayer with a taxable income of $15,000 owes $594 in Oklahoma income tax. In functional notation this is $T(15,000) = 594$. If the taxable income is $15,500, then the table shows a tax liability of $622.

Part 1 Calculate the rate of change in T with respect to I, and explain in practical terms what it means.

Part 2 How much does the taxpayer owe if the taxable income is $15,350?

Solution to Part 1 Since we are thinking of the tax T as a function of the variable I representing income, we calculate the slope as follows:

$$m = \frac{\text{Change in } T}{\text{Change in } I} = \frac{\text{Change in tax}}{\text{Change in income}} = \frac{622 - 594}{15{,}500 - 15{,}000} = \frac{28}{500} = 0.056.$$

In practical settings the rate of change of a linear function always has an important physical meaning.

This means that for each additional dollar earned, the taxpayer can expect to pay 5.6 cents more in Oklahoma income tax. In economics this is known as the *marginal tax rate*, because it is the rate at which new money that you earn is taxed. It is worth noting that the marginal tax rate normally does not appear directly in tax tables but must be calculated as we did here. It can provide crucial information for financial planning.

The rate of change is key to solving problems involving linear functions.

Solution to Part 2 An income of $15,350 is an additional $350 income over the $15,000 level. Now we can use the marginal tax rate m calculated in part 1 to get the additional tax:

$$\text{Additional tax} = m \times \text{Additional income} = 0.056 \times 350 = 19.60.$$

Thus we would owe the tax on $15,000—that is, $594—plus an additional tax of $19.60, for a total tax liability of $613.60, or about $614.

TEST YOUR UNDERSTANDING │ **FOR EXAMPLE 3.3**

The State of New York also has an income tax. The 2010 tax table for New York shows that a single New York resident with a taxable income of $15,000 owes $679 in New York income tax. If the taxable income is $15,500, then the table shows a tax liability of $709. Assume that the state tax is a linear function of taxable income, and find the tax owed if the taxable income is $15,350. ∎

Linear Functions and Straight Lines

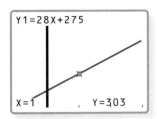

FIGURE 3.27 The graph of the linear function $C = 28n + 275$ is a straight line.

If we look more carefully at the catered dinner, we can write a formula for the total cost $C = C(n)$ when there are n dinner guests:

$$\text{Cost} = \text{Cost of food} + \text{Rent}$$

$$\text{Cost} = \text{Price per meal} \times \text{Number of guests} + \text{Rent}$$

$$C = 28n + 275.$$

If we graph this function as we did in Figure 3.27, we see the fundamental relationship between linear functions and straight lines: The graph of a linear function is a straight line. (We used a horizontal span of -1 to 3 and a vertical span of 250 to 400.)

But there is still more to find out. In Figure 3.27 we placed the cursor at X=0, and we see that the vertical intercept of the graph is 275. That is the rental fee for the dining hall, or, more to the point, it is the value of the linear function C when the variable n is zero. This value $C(0)$ is often referred to as the *initial value* of the function. It is true in general that the vertical intercept of the graph corresponds to the initial value of the linear function and is denoted b.

In Figure 3.28 we have put the cursor at X=1, and we see that a horizontal change of 1 unit causes the graph to rise from 275 to 303. That means the slope of the straight-line graph is $303 - 275 = 28$, and that is the same as the rate of change of the linear function. Once again, this is true in general: The slope of a linear function is the same as the slope of its graph.

The graph of a linear function is a straight line, and the rate of change of the function is the slope of the line.

Finally, we note that the horizontal intercept of the graph (not shown in Figure 3.27) is the solution of the equation $C(n) = 0$. The form of C and the relationships we have observed here between linear functions and straight-line graphs are typical.

```
Y1=28X+275




X=1          , Y=303 ,
```

FIGURE 3.28 The slope of the graph is the same as the rate of change of the function.

KEY IDEA 3.3 **THE RELATIONSHIPS BETWEEN LINEAR FUNCTIONS AND STRAIGHT LINES**

1. The formula for a linear function is

   $$y = \text{Slope} \times x + \text{Initial value}$$

   $$y = mx + b.$$

2. The graph of a linear function is a straight line.

3. The slope of a linear function is the same as the slope of its graph.

4. The vertical intercept of the graph corresponds to the initial value b of the linear function—that is, the value of the function when the variable is 0.

5. The horizontal intercept of the graph corresponds to the value of the variable when the linear function is zero. This is the solution of the equation

   $$\text{Linear function} = 0.$$

EXAMPLE 3.4   **SELLING JEWELRY AT AN ART FAIR**

Suppose you pay $192 to rent a booth for selling necklaces at an art fair. The necklaces sell for $32 each.

Part 1 Explain why the function that shows your net income (revenue from sales minus rental fee) as a function of the number of necklaces sold is a linear function.

Part 2 Write a formula for this function.

Part 3 Use functional notation to show your net income if you sell 25 necklaces, and then calculate that value.

Part 4 Make the graph of the net income function.

Part 5 Identify the vertical intercept and slope of the graph, and explain in practical terms what they mean.

Part 6 Find the horizontal intercept and explain its meaning in practical terms.

> We can show that a function is linear by observing that its rate of change is constant.

Solution to Part 1 We choose variable and function names: Let n be the number of necklaces sold, and let $P = P(n)$ be the net income in dollars. Each time n increases by 1—that is, when 1 additional necklace is sold—the value of P, the net income, increases by the same amount, $32. Thus the rate of change for P is always the same, and hence it is a linear function.

Solution to Part 2 You pay $192 to rent the booth and take in $32 for each necklace sold.

$$\text{Net income} = \text{Income from sales} - \text{Rent}$$

$$\text{Net income} = \text{Price} \times \text{Number sold} - \text{Rent}$$

$$P = 32n - 192 \text{ dollars.}$$

Solution to Part 3 If 25 necklaces are sold, then in functional notation the net income is $P(25)$. To calculate that, we put 25 into the formula in place of n.

$$P(25) = 32 \times 25 - 192 = 608 \text{ dollars.}$$

Solution to Part 4 First we enter the function and record appropriate correspondences:

$$Y_1 = P, \text{ net income, in dollars, on vertical axis}$$

$$X = n, \text{ necklaces sold on horizontal axis.}$$

X	Y₁
0	⁻192
5	⁻32
10	128
15	288
20	448
25	608
30	768
X=0	

FIGURE 3.29 A table for setting the window

To choose a window size, we made the table in Figure 3.29 using a starting value of 0 and an increment of 5. This led us to choose a horizontal span of $n = 0$ to $n = 30$ and a

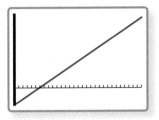

FIGURE 3.30 Net income for selling necklaces

We can find the equation of a line from its slope and initial value.

vertical span of $P = -200$ to $P = 800$. This makes the graph in Figure 3.30, and we note that, as expected, the graph of our linear function P is a straight line.

Solution to Part 5 We know in general that the vertical intercept of the graph of $y = mx + b$ is b and the slope is m. Thus for $P = 32n - 192$, the vertical intercept is -192, and the slope is 32. The vertical intercept -192 is the net income if no necklaces are sold. That is, you lost $192 because you had to pay rent for the booth but sold no necklaces. The slope of 32 dollars per necklace indicates that for each additional necklace sold, your net income increases by $32. Thus the slope represents the price of each necklace. We should note that economists refer to the slope here as the *marginal income*. It shows the additional income taken when 1 additional item is sold.

Solution to Part 6 The horizontal intercept occurs where the graph crosses the horizontal axis. That is where $P(n) = 0$. We will show how to find that by direct calculation.

$$P(n) = 0$$
$$32n - 192 = 0$$
$$32n = 192 \qquad \text{Add 192 to both sides}$$
$$n = \frac{192}{32} = 6 \quad \text{Divide by 32.}$$

Thus the horizontal intercept occurs at $n = 6$ necklaces. That is where $P(n) = 0$, and in this case it is the "break-even" point. You need to sell 6 necklaces to avoid losing money on this venture.

We found the horizontal intercept here by hand calculation, but we should note that it could also be found with the calculator by looking at the graph or a table.

TEST YOUR UNDERSTANDING | **FOR EXAMPLE 3.4**

You pay a broker a flat fee of $35 to buy stock, and you purchase a stock that costs $67 per share. Find a formula that gives the cost of buying n shares of stock. Make a graph of the linear function you find. Include the cost of purchasing up to 100 shares. ■

Linear Equations from Data

There are a number of ways in which information about a linear function may be given, but the following three situations are quite common. We will illustrate each case using variations on Example 3.4.

Getting a linear equation if you know the slope and initial value In Example 3.4, we were effectively told the slope, 32, of the linear function P and its initial value, -192. As we have already noted, this enables us immediately to write a formula:

$$P = \text{Slope} \times n + \text{Initial value}$$
$$P = 32n - 192.$$

This works in general. If we know the slope m of a linear function and its initial value b, we can immediately write down the formula as $y = mx + b$.

Getting a linear equation if you know the slope and one data point Suppose that we were given the information for Example 3.4 in a different way: We are told that the price for each necklace is \$32 and that if we sell $n = 8$ necklaces, we will have a net income of $P = 64$ dollars. Now we know the slope, 32, but we don't know the initial value. We can get it by solving an equation. The information tells us that the formula for P is

$$P = 32n + \text{Initial value}. \qquad (3.7)$$

We also know that when $n = 8$, $P = 64$. We put these values into Equation (3.7) and proceed to solve for the initial value:

$$64 = 32 \times 8 + \text{Initial value}$$

$$64 = 256 + \text{Initial value}$$

$$-192 = \text{Initial value}.$$

Now we have the slope and initial value, and we can write the formula for P as $P = 32n - 192$. This works in general. If we know the slope and one data point for a linear function, we can solve an equation as we did above to find the initial value and then write down the formula.

Getting a linear equation from two data points Suppose the information in Example 3.4 were given as two data points. For example, suppose that we make a net income $P = 64$ dollars when we sell $n = 8$ necklaces and a net income of $P = 160$ dollars when $n = 11$ necklaces are sold. In this case, we know neither the slope nor the initial value. Since the slope is so important, we get it first, using a familiar formula:

$$\text{Slope} = \frac{\text{Change in function}}{\text{Change in variable}} = \frac{160 - 64}{11 - 8} = 32 \text{ dollars per necklace}.$$

Now we know the slope, 32, and the data point $P = 64$ when $n = 8$. With this information, we can proceed exactly as we did above to get the initial value and then the formula for P.

We also know that $P = 160$ when $n = 11$. What would happen if we used this data point with the slope rather than $n = 8$ and $P = 64$? Would the equation we find be different? Let's see:

$$P = \text{Slope} \times n + \text{Initial value}$$

$$P = 32n + \text{Initial value}$$

$$160 = 32 \times 11 + \text{Initial value}$$

$$160 = 352 + \text{Initial value}$$

$$-192 = \text{Initial value}.$$

We got the same initial value as we did before, and hence we will get the same formula for P, namely $P = 32n - 192$.

In general, if you are given two data points and need to find the linear function that they determine, you proceed in two steps. First, use the formula

$$\text{Slope} = \frac{\text{Change in function}}{\text{Change in variable}}$$

to compute the slope. Next, use the slope you found and put *either* of the given data points into the equation

$$y = \text{Slope} \times x + \text{Initial value}$$

to solve for the initial value.

> **KEY IDEA 3.4**   **HOW TO GET THE EQUATION OF A LINEAR FUNCTION**
>
> **1. If you know the slope and the initial value,** use
>
> $$y = \text{Slope} \times x + \text{Initial value}.$$
>
> **2. If you know the slope and one data point,** put the given slope and data point into the equation
>
> $$y = \text{Slope} \times x + \text{Initial value}$$
>
> and solve it for the initial value. You can now get the formula as in part 1 above.
>
> **3. If you know two data points,** first use the formula
>
> $$\text{Slope} = \frac{\text{Change in function}}{\text{Change in variable}}$$
>
> to get the slope. Now use either of the data points and proceed as in part 2 above.

Two points can give us the slope of a linear function.

EXAMPLE 3.5   **CHANGING CELSIUS TO FAHRENHEIT**

The temperature $F = F(C)$ in Fahrenheit is a linear function of the temperature C in Celsius. A lab assistant placed a Fahrenheit thermometer beside a Celsius thermometer and observed the following: When the Celsius thermometer reads 30 degrees ($C = 30$), the Fahrenheit thermometer reads 86 degrees ($F = 86$). When the Celsius thermometer reads 40 degrees, the Fahrenheit thermometer reads 104 degrees.

Part 1 Use a formula to express F as a linear function of C.

Part 2 At sea level, water boils at 212 degrees Fahrenheit. What temperature in degrees Celsius makes water boil?

Part 3 Explain in practical terms what the slope means in this setting.

Solution to Part 1 Since we know two points, the first step is to calculate the slope. When C changes from 30 to 40, F changes from 86 to 104. Since we are thinking of F as the function and of C as the variable, we use

$$\text{Slope} = \frac{\text{Change in function}}{\text{Change in variable}} = \frac{\text{Change in } F}{\text{Change in } C} = \frac{104 - 86}{40 - 30} = 1.8.$$

Thus the slope is 1.8 degrees Fahrenheit per degree Celsius. Now we use the slope $m = 1.8$, and the data point $F = 86$ when $C = 30$, to make an equation that we solve for the initial value:

$$F = \text{Slope} \times C + \text{Initial value}$$

$$86 = 1.8 \times 30 + \text{Initial value}$$

$$86 = 54 + \text{Initial value}$$

$$32 = \text{Initial value}.$$

Now we know the slope, 1.8, and the initial value, 32, so $F = 1.8C + 32$. It is worth noting the significance of the initial value we found. The temperature 32 degrees Fahrenheit is the freezing point for water, and this corresponds to a temperature of 0 degrees Celsius. That is, on the Celsius scale, a temperature of 0 is the freezing point for water.

Solution to Part 2 We want to know the value of C when $F = 212$. Thus we put in 212 for F in the equation we found in part 1 and then solve for C:

$$F = 1.8C + 32$$
$$212 = 1.8C + 32$$
$$180 = 1.8C$$
$$\frac{180}{1.8} = C$$
$$100 = C.$$

Thus water boils at 100 degrees Celsius.

Solution to Part 3 As we found in part 1, the slope is 1.8. Remember that the slope tells how much F changes when C changes by 1. Thus a change of 1 degree Celsius results in a change of 1.8 degrees Fahrenheit.

TEST YOUR UNDERSTANDING | **FOR EXAMPLE 3.5**

A donation to the university is required for the privilege of purchasing premium season football passes. The cost is a linear function of the number of season passes. It costs $8000 to get season passes for myself and my spouse. To get passes for my entire family of 5 it costs $17,000. Use a formula to express the total cost C, in dollars, of n season passes.   ■

ANSWERS FOR TEST YOUR UNDERSTANDING

3.3 $700

3.4 If we denote the total cost by C, measured in dollars, the formula is $C = 67n + 35$.
Horizontal, 0 to 100; vertical, 0 to 7000

3.5 $C = 3000n + 2000$

3.2 EXERCISES

Reminder Round all answers to two decimal places unless otherwise indicated.

1. **Walking** If you take a brisk walk on a flat surface, you will burn about 258 calories per hour. You have just finished a hard workout that used 700 calories.

   a. Find a formula that gives the total calories burned if you finish your workout with a walk of h hours.

   b. How long do you need to walk at the end of your workout in order to burn a total of 1100 calories?

2. **Cricket Chirps** The temperature T in degrees Fahrenheit is (approximately) a linear function of the number C of cricket chirps per minute. Twenty cricket chirps per minute corresponds to a temperature of 42 degrees Fahrenheit. Each additional chirp per minute corresponds to an increase of 0.25 degree.

   a. Use a formula to express T as linear function of C.

   b. What temperature is indicated by 100 cricket chirps per minute?

(continued)

3. **Nail Growth** The rate of fingernail growth depends on many factors, but in adults nails grow at an average rate of 3 millimeters per month. If a nail is initially 12 millimeters long, find a formula that gives the length L, in millimeters, of the nail (if left unclipped) after t months.

4. **Hair Growth** When you are 18 years old you have a hair that is 14 centimeters long, and your hair grows about 12 centimeters each year. Let H(t) be the length, in centimeters, of that hair t years after age 18.

   a. Find a formula that gives H as a linear function of t.

   b. How long will it take for the hair to reach a length of 90 centimeters?

5. **Getting Celsius from Fahrenheit** Water freezes at 0 degrees Celsius, which is the same as 32 degrees Fahrenheit. Also water boils at 100 degrees Celsius, which is the same as 212 degrees Fahrenheit.

   a. Use the freezing and boiling points of water to find a formula expressing Celsius temperature C as a linear function of the Fahrenheit temperature F.

   b. What is the slope of the function you found in part a? Explain its meaning in practical terms.

   c. In Example 3.5 we showed that $F = 1.8C + 32$. Solve this equation for C and compare the answer with that obtained in part a.

6. **A Trip to a Science Fair** An elementary school is taking a busload of children to a science fair. It costs $130.00 to drive the bus to the fair and back, and the school pays each student's $2.00 admission fee.

© Layne Kennedy/CORBIS

   a. Use a formula to express the total cost C, in dollars, of the science fair trip as a linear function of the number n of children who make the trip.

   b. Identify the slope and initial value of C, and explain in practical terms what they mean.

   c. Explain in practical terms what C(5) means, and then calculate that value.

   d. Solve the equation $C(n) = 146$ for n. Explain what the answer you get represents.

7. **Digitized Pictures on a Disk Drive** The hard disk drive on a computer holds 800 gigabytes of information. That is 800,000 megabytes. The formatting information, operating system, and applications software take up 6000 megabytes of disk space. The operator wants to store on his computer a collection of digitized pictures, each of which requires 2 megabytes of storage space.

   a. We think of the total amount of storage space used on the disk drive as a function of the number of pictures that are stored on the drive. Explain why this function is linear.

   b. Find a formula to express the total amount of storage space used on the disk drive as a linear function of the number of pictures that are stored on the drive. (Be sure to identify what the letters you use mean.) Explain in practical terms what the slope of this function is.

   c. Express using functional notation the total amount of storage space used on the disk drive if there are 350 pictures stored on the drive, and then calculate that value.

   d. After putting a number of pictures on the disk drive, the operator executes a *directory* command, and at the end of the list the computer displays the message *769,000,000,000 bytes free*. This message means that there are 769,000 megabytes of storage space left on the computer. How many pictures are stored on the disk drive? How many additional pictures can be added before the disk drive is filled?

8. **Speed of Sound** The speed of sound in air changes with the temperature. When the temperature T is 32 degrees Fahrenheit, the speed S of sound is 1087.5 feet per second. For each degree increase in temperature, the speed of sound increases by 1.1 feet per second.

   a. Explain why speed S is a linear function of temperature T. Identify the slope of the function.

   b. Use a formula to express S as a linear function of T.

   c. Solve for T in the equation from part b to obtain a formula for temperature T as a linear function of speed S.

   d. Explain in practical terms the meaning of the slope of the function you found in part c.

9. Total Cost   The *total cost C* for a manufacturer during a given time period is a function of the number N of items produced during that period. To determine a formula for the total cost, we need to know two things. The first is the manufacturer's *fixed costs*. This amount covers expenses such as plant maintenance and insurance, and it is the same no matter how many items are produced. The second thing we need to know is the cost for each unit produced, which is called the *variable cost*.

Suppose that a manufacturer of widgets has fixed costs of $1500 per month and that the variable cost is $20 per widget (so it costs $20 to produce 1 widget).

a. Explain why the function giving the total monthly cost C, in dollars, of this widget manufacturer in terms of the number N of widgets produced in a month is linear. Identify the slope and initial value of this function, and write down a formula.

b. Another widget manufacturer has a variable cost of $12 per widget, and the total cost is $3100 when 150 widgets are produced in a month. What are the fixed costs for this manufacturer?

c. Yet another widget manufacturer has determined the following: The total cost is $2700 when 100 widgets are produced in a month, and the total cost is $3500 when 150 widgets are produced in a month. What are the fixed costs and variable cost for this manufacturer?

10. Total Revenue and Profit   *This is a continuation of Exercise 9.* The *total revenue R* for a manufacturer during a given time period is a function of the number N of items produced during that period. In this exercise we assume that the selling price per unit of the item is a constant, so it does not depend on the number of items produced. The *profit P* for a manufacturer is the total revenue minus the total cost. If the profit is zero, then the manufacturer is at a *break-even point*.

We consider again the manufacturer of widgets in Exercise 9 with fixed costs of $1500 per month and a variable cost of $20 per widget. Suppose the manufacturer sells 100 widgets for $2300 total.

a. Use a formula to express the total monthly revenue R, in dollars, of this manufacturer in a month as a function of the number N of widgets produced in a month.

b. Use a formula to express the monthly profit P, in dollars, of this manufacturer as a function of the number of widgets produced in a month. Explain how the slope and initial value of P are derived from the fixed costs, variable cost, and price per widget.

c. What is the break-even point for this manufacturer?

d. Make graphs of total monthly cost and total monthly revenue. Include monthly production levels up to 1200 widgets. What is the significance of the point where the graphs cross?

11. Slowing Down in a Curve   A study of average driver speed on rural highways by A. Taragin[4] found a linear relationship between average speed S, in miles per hour, and the amount of curvature D, in degrees, of the road. On a straight road ($D = 0$), the average speed was found to be 46.26 miles per hour. This was found to decrease by 0.746 mile per hour for each additional degree of curvature.

a. Find a linear formula relating speed to curvature.

b. Express using functional notation the speed for a road with a curvature of 10 degrees, and then calculate that value.

12. Real Estate Sales   A real estate agency has fixed monthly costs associated with rent, staff salaries, utilities, and supplies. It earns its money by taking a percentage commission on total real estate sales. During the month of July, the agency had total sales of $832,000 and showed a net income (after paying fixed costs) of $15,704. In August total sales were $326,000 with a net income of only $523.

a. Use a formula to express net income as a linear function of total sales. Be sure to identify what the letters that you use mean.

b. Plot the graph of net income and identify the slope and vertical intercept.

c. What are the real estate agency's fixed monthly costs?

d. What percentage commission does the agency take on the sale of a home?

e. Find the horizontal intercept and explain what this number means to the real estate agency.

13. Currency Conversion   The number P of British pounds you can get from a bank is a linear function of the number D of American dollars you pay. An American tourist arriving at Heathrow airport in England went to a banking window at the airport and gave the teller 70 American dollars.

_____
[4]"Driver Performance on Horizontal Curves," *Proceedings* **33** (Washington, DC: Highway Research Board, 1954), 446–466.

(continued)

She received 34 British pounds in exchange. In this exercise, assume there is no service charge for exchanging currency.

   a. What is the rate of change, or slope, of P with respect to D? Explain in practical terms what this number means. (*Note:* You need two values to calculate a slope, but you were given only one. If you think about it, you know one other value. How many British pounds can you get for zero American dollars?)

   b. A few days later, the American tourist went to a bank in Plymouth and exchanged 130 American dollars for British pounds. How many pounds did she receive?

   c. Upon returning to the airport, she found that she still had £12.32 in British currency in her purse. In preparation for the trip home, she exchanged that for American dollars. How much money, in American dollars, did she get?

14. **Growth in Height**   Between the ages of 7 and 11 years, a certain boy grows 2 inches taller each year. At age 9 he is 48 inches tall.

   a. Explain why, during this period, the function giving the height of the boy in terms of his age is linear. Identify the slope of this function.

   b. Use a formula to express the height of the boy as a linear function of his age during this period. Be sure to identify what the letters that you use mean.

   c. What is the initial value of the function you found in part b?

   d. Studying a graph of the boy's height as a function of his age from birth to age 7 reveals that the graph is increasing and concave down. Does this indicate that his actual height (or length) at birth was larger or smaller than your answer to part c? Be sure to explain your reasoning.

15. **Adult Male Height and Weight**   Here is a rule of thumb relating weight to height among adult males: If a man is 1 inch taller than another, then we expect him to be heavier by 5 pounds.

   a. Explain why, according to this rule of thumb, among typical adult males the weight is a *linear* function of the height. Identify the slope of this function.

   b. A related rule of thumb is that a typical man who is 70 inches tall weighs 170 pounds. On the basis of these two rules of thumb, use a formula to express the trend giving weight as a linear function of height. (Be sure to identify the meaning of the letters that you use.)

   c. If a man weighs 152 pounds, how tall would you expect him to be?

   d. An atypical man is 75 inches tall and weighs 190 pounds. In terms of the trend formula you found in part b, is he heavy or light for his height?

16. **Lean Body Weight in Males**   Your *lean body weight* L is the amount you would weigh if all the fat in your body were to disappear. One text gives the following estimate of lean body weight L (in pounds) for young adult males:

$$L = 98.42 + 1.08W - 4.14A,$$

where W is total weight in pounds and A is abdominal circumference in inches.[5]

   a. Consider a group of young adult males who have the same abdominal circumference. If their weight increases but their abdominal circumference remains the same, how does their lean body weight change?

   b. Consider a group of young adult males who have the same weight. If their abdominal circumference decreases but their weight stays the same, how does their lean body weight change?

   c. Suppose a young adult male has a lean body weight of 144 pounds. Over a period of time, he gains 15 pounds in total weight, and his abdominal circumference increases by 2 inches. What is his lean body weight now?

17. **Lean Body Weight in Females**   *This is a continuation of Exercise 16.* The text cited in Exercise 16 gives a more complex method of calculating lean body weight for young adult females:

$$L = 19.81 + 0.73W + 21.2R - 0.88A$$
$$- 1.39H + 2.43F.$$

Here L is lean body weight in pounds, W is weight in pounds, R is wrist diameter in inches, A is abdominal circumference in inches, H is hip circumference in inches, and F is forearm circumference in inches. Assuming the validity of the formulas given here and in Exercise 16, compare increase in

---

[5] D. Kirkendall, J. Gruber, and R. Johnson, *Measurement and Evaluation for Physical Educators*, 2nd ed. (Champaign, IL: Human Kinetics Publishers, 1987).

lean body weight of young adult males and of young adult females if their weight increases but all other factors remain the same.

18. **Horizontal Reach of Straight Streams**   If a fire hose is held horizontally, then the distance the stream will travel depends on the water pressure and on the *horizontal factor* for the nozzle. The horizontal factor H depends on the diameter of the nozzle. For a 0.5-inch nozzle, the horizontal factor is 56. For each $\frac{1}{8}$-inch increase in nozzle diameter, the horizontal factor increases by 6.

   a. Explain why the function giving the horizontal factor H in terms of the nozzle diameter d (measured in inches) is linear.

   b. Use a formula to express H as a linear function of d.

   c. Once the horizontal factor H is known, we can calculate the distance S in feet that a horizontal stream of water can travel by using
   $$S = \sqrt{Hp}.$$
   Here p is pressure in pounds per square inch. How far will a horizontal stream travel if the pressure is 50 pounds per square inch and the nozzle diameter is 1.75 inches?

   d. Firefighters have a nozzle with a diameter of 1.25 inches. The pumper generates a pressure of 70 pounds per square inch. The hose nozzle is 75 feet from a fire. Can a horizontal stream of water reach the fire?

19. **Vertical Reach of Fire Hoses**   If a fire hose is held vertically, then the height the stream will travel depends on water pressure and on the *vertical factor* for the nozzle. The vertical factor V depends on the diameter of the nozzle. For a 0.5-inch nozzle, the vertical factor is 85. For each $\frac{1}{8}$-inch increase in nozzle diameter, the vertical factor increases by 5.

   a. Explain why the function giving the vertical factor V in terms of the nozzle diameter d is linear.

   b. Use a formula to express V as a linear function of d (measured in inches).

   c. Once the vertical factor is known, we can calculate the height S in feet that a vertical stream of water can travel by using
   $$S = \sqrt{Vp}.$$
   Here p is pressure in pounds per square inch. How high will a vertical stream travel if the pressure is 50 pounds per square inch and the nozzle diameter is 1.75 inches?

   d. Firemen have a nozzle with a diameter of 1.25 inches. The pumper generates a pressure of 70 pounds per square inch. From street level, they need to get water on a fire 60 feet overhead. Can they reach the fire with a vertical stream of water?

20. **Budget Constraints**   Your family likes to eat fruit, but because of budget constraints, you spend only $5 each week on fruit. Your two choices are apples and grapes. Apples cost $0.50 per pound, and grapes cost $1 per pound. Let a denote the number of pounds of apples you buy and g the number of pounds of grapes. Because of your budget, it is possible to express g as a linear function of the variable a. To find the linear formula, we need to find its slope and initial value.

   a. If you buy one more pound of apples, how much less money do you have available to spend on grapes? Then how many fewer pounds of grapes can you buy?

   b. Use your answer to part a to find the slope of g as a linear function of a. (*Hint:* Remember that the slope is the change in the function that results from increasing the variable by 1. Should the slope of g be positive or negative?)

   c. To find the initial value of g, determine how many pounds of grapes you can buy if you buy no apples.

   d. Use your answers to parts b and c to find a formula for g as a linear function of a.

21. **More on Budget Constraints**   *This is a continuation of Exercise 20.* Another way to find a linear formula giving the number of pounds g of grapes that you can buy in terms of a, the number of pounds of apples, is to write the budget constraint as an equation and solve it for g.

   a. If you buy 5 pounds of apples, then of course you spend $0.50 \times 5 = 2.50$ dollars on apples. In general, if you buy a pounds of apples, how much money do you spend on apples? Your answer should be an expression involving a.

   b. Write an expression involving g to represent how much money you spend on grapes if you buy g pounds of grapes.

   c. Use your answers to parts a and b to write an equation expressing the budget constraint of $5 in terms of a and g.

   d. Solve the equation you found in part c for g, and thus find a formula expressing g as a linear function of a. Compare this with your answer to part d of Exercise 20.

(continued)

22. Traffic Signals The number of seconds n for the yellow light is critical to safety at a traffic signal. One study[6] recommends the following formula for setting the time:

$$n = t + 0.5\frac{v}{a} + \frac{w + l}{v}.$$

Here t is the perception-reaction time of a driver in seconds, v is the average approach speed in feet per second, a is the deceleration rate in feet per second per second, w is the crossing-street width in feet, and l is the average vehicle length in feet. If we assume that the average perception-reaction time is 1 second, the approach velocity is 40 miles per hour (58.7 feet per second), the deceleration rate is 15 feet per second per second, and the average vehicle length is 20 feet, then n can be expressed as the following linear function of crossing-street width:

$$n = 3.3 + 0.017w.$$

a. Under the given assumptions, what is the minimum time the yellow light should be on, no matter what the width of the crossing street?

b. If the crossing street for one signal is 10 feet wider than the crossing street for another signal, how should the lengths of the yellow light times compare?

c. Calculate $n(70)$ and explain in practical terms what your answer means.

d. What crossing-street width would warrant a 5-second yellow light?

23. Sleeping Longer A certain man observed that each night he was sleeping 15 minutes longer than he had the night before, and he used this observation to predict the day of his death.[7] If he made his observation right after sleeping 8 hours, how long would it be until he slept 24 hours (and so would never again wake)?

24. The Saros Cycle Total solar eclipses occur on a regular and predictable basis. One that could be viewed in parts of the United States occurred on March 7, 1970. The *saros cycle* is a period of 18 years, $11\frac{1}{3}$ days, and solar eclipses occur according to this cycle. For the purposes of this exercise, we ignore leap years and assume that a year is

exactly 365 days. Because the saros cycle is not a whole number of days, but instead is one-third of a day longer than a day, at the end of the saros cycle Earth will have rotated one-third revolution beyond its location at the beginning of the cycle. Thus if a solar eclipse is viewable at a certain location, the next solar eclipse will not be. How long after March 7, 1970, will a solar eclipse again be viewable from the United States?[8]

25. Life on Other Planets The number N of civilizations per galaxy that might be able to communicate with us can be expressed in the formula

$$N = SPZBIL.$$

Here S is the number of stars per galaxy, about 2×10^{11}. The remaining variables, together with their pessimistic and optimistic estimates,[9] are given in the following table.

Variable	Meaning	Pessimistic value	Optimistic value
P	Fraction of stars with planets	0.01	0.5
Z	Planets per star in life zone for at least 4 billion years	0.01	1
B	Fraction of suitable planets on which life begins	0.01	1
I	Fraction of life forms that evolve to intelligence	0.01	1
L	Fraction of star's life during which a technological society survives	10^{-8}	10^{-4}

a. Calculate the most pessimistic and most optimistic values for N.

b. The single variable with the greatest range of values is L, the fraction of a star's life during which a technological society, once

---

[6]P. L. Olson and R. W. Rothery, "Driver response to the amber phase of traffic signals," *Traffic Engineering* **XXXII (5)** (1962), 17–20, 29.
[7]This is a legend about the mathematician De Moivre, who died in 1754.
[8]The actual date is April 8, 2024, and the eclipse should be viewable through much of the central United States.
[9]Data taken from Michael A. Seeds, *Foundations of Astronomy* (Belmont, CA: Wadsworth, 1994).

established, can survive. The pessimistic estimate corresponds to 100 years, the optimistic estimate to 1,000,000 years. Using pessimistic estimates for other variables, what value of L will give 1 communicating civilization per galaxy?

c. Is this formula helpful in determining whether there is life on other planets? Explain your reasoning.

3.2 SKILL BUILDING EXERCISES

S-1. Slope from Two Values Suppose f is a linear function such that $f(2) = 7$ and $f(5) = 19$. What is the slope of f?

S-2. Slope from Two Values Suppose f is a linear function such that $f(3) = 9$ and $f(8) = 5$. What is the slope of f?

S-3. Function Value from Slope and Run Suppose f is a linear function such that $f(3) = 7$. If the slope of f is 2.7, then what is $f(5)$?

S-4. Function Value from Slope and Run Suppose f is a linear function such that $f(5) = 2$. If the slope of f is 3.1, then what is $f(12)$?

S-5. Run from Slope and Rise Suppose that f is a linear function with slope -3.4 and that $f(1) = 6$. What value of x gives $f(x) = 0$?

S-6. Run from Slope and Rise Suppose that f is a linear function with slope 2.6 and that $f(5) = -3$. What value of x gives $f(x) = 0$?

S-7. Linear Equation from Slope and Point Suppose that f is a linear function with slope 4 and that $f(3) = 5$. Find the equation for f.

S-8. Linear Equation from Slope and Point Suppose that f is a linear function with slope -3 and that $f(2) = 8$. Find the equation for f.

S-9. Linear Equation from Two Points Suppose f is a linear function such that $f(4) = 8$ and $f(9) = 2$. Find the equation for f.

S-10. Linear Equation from Two Points Suppose f is a linear function such that $f(3) = 5$ and $f(7) = -4$. Find the equation for f.

Properties of Linear Functions Exercises S-11 through S-23 explore elementary properties of linear functions.

S-11. Suppose y is a linear function of x. The slope is 3.3. An increase of 5 units in x causes what change in y?

S-12. Suppose y is a linear function of x. The slope is 3.3. A decrease of 4 units in x causes what change in y?

S-13. Suppose y is a linear function of x. The slope is -2.6. An increase of 6.3 units in x causes what change in y?

S-14. Suppose y is a linear function of x. The slope is -2.6. A decrease of 5.4 units in x causes what change in y?

S-15. Suppose y is a linear function of x. The slope is 3.1. An increase of 3 units in y causes what change in x?

S-16. Suppose y is a linear function of x. The slope is 4.3. A decrease of 2.7 units in y causes what change in x?

S-17. Suppose y is a linear function of x. The slope is -3.2. An increase of 4.6 units in y causes what change in x?

S-18. Suppose y is a linear function of x. Increasing x by 2.7 units causes a 9.4-unit increase in y. What is the slope?

S-19. Suppose y is a linear function of x. Increasing x by 3.3 units decreases y by 1.1 units. What is the slope?

S-20. Suppose y is a linear function of x. Decreasing x by 2.3 units causes an increase in y of 3.7 units. What is the slope?

S-21. If x is increased by 7 units, y is increased by 9 units. If x is increased by an additional 7 units, y is increased by only 8 units. Is y a linear function of x?

S-22. Suppose y is a linear function of x. Increasing x by 7 units causes a 5-unit increase in y. What additional increase in y is caused by an additional 3-unit increase in x?

S-23. Suppose y is a linear function of x. Increasing x by 4.4 decreases y by 3.6 units. What change in y is caused by a 3.3-unit decrease in x?

3.3 MODELING DATA WITH LINEAR FUNCTIONS

Information about physical and social phenomena is frequently obtained by gathering data or sampling. For example, collecting data is the basic tool that census takers use to get information about the population of the United States. Once data are gathered, an important key to further analysis is to produce a *mathematical model* describing the data. A model is a function that (1) represents the data either exactly or approximately and (2) incorporates patterns in the data. In many cases, such a model takes the form of a linear function.

Georgios Kollidas/Shutterstock.com

Testing Data for Linearity

Let's look at a hypothetical example. One of the most important events in the development of modern physics was Galileo's description of how objects fall. In about 1590 he conducted experiments in which he dropped objects and attempted to measure their downward velocities $V = V(t)$ as they fell. Here we measure velocity V in feet per second and time t as the number of seconds after release. If Galileo had been able to nullify air resistance, and if he had been able to measure velocity without any experimental error at all, he might have recorded the following table of values.

t = seconds	0	1	2	3	4	5
V = feet per second	0	32	64	96	128	160

What conclusions can be drawn from this table of data? In this case, the key is to look at how velocity changes as the rock falls. From $t = 0$ to $t = 1$, the velocity changes from 0 feet per second to 32 feet per second. That is a change of $32 - 0 = 32$ feet per second. From $t = 1$ to $t = 2$, the velocity changes from 32 feet per second to 64 feet per second, a change in velocity of $64 - 32 = 32$ feet per second. Thus the change in velocity is the same, 32 feet per second, during the first and second seconds of the fall. If we continue this, we obtain the following table.

Change in t	From 0 to 1	From 1 to 2	From 2 to 3	From 3 to 4	From 4 to 5
Change in V	32	32	32	32	32

Evenly spaced data are exactly linear when they show constant differences.

We see that the rate of change in velocity, the *acceleration*, is always the same, 32 feet per second per second. We know this is characteristic of linear functions; their rate of change is always the same. Thus it is reasonable to describe these data with a linear formula. That is, we want a formula of the form $V = mt + b$, where m is the slope and b is the initial value. Since V changes by 32 when t changes by 1, the slope of the linear function is $m = 32$ feet per second per second. We also know that $b = 0$ because the initial velocity is zero. We conclude that the velocity of the rock t seconds after it is dropped is given by

$$V = 32t + 0 = 32t.$$

Linear Models

This linear function serves as a *mathematical model* for the experimentally gathered data, and it gives us more information than is apparent from the data table alone. For example, we can use the linear formula to calculate the velocity of the rock 10.8 seconds after it is released, information that is not given by the table:

$$V(10.8) = 10.8 \times 32$$

$$= 345.6 \text{ feet per second.}$$

The key physical observation that Galileo made was that falling objects have constant acceleration, 32 feet per second per second. He did additional experiments to show that if air resistance is ignored, then this acceleration does not depend on the weight or size of the object. The same table of values would result, and the same acceleration would be calculated, whether the experiment was done with a pebble or with a cannonball. According to tradition, Galileo conducted some of his experiments in public, dropping objects from the top of the leaning tower of Pisa. His observations got him into serious trouble with the authorities because they conflicted with the accepted premise of Aristotle that heavier objects would fall faster than lighter ones.

The feature of this data table that allowed us to construct a linear model was that the change in the function, velocity, was always the same. If a data table (with evenly spaced values for the variable) does not exhibit this behavior, then the data cannot be modeled with a linear function, and some more complicated model may be sought. For example, if instead of measuring velocity, we had measured the distance $D = D(t)$ that the rock fell, we would have obtained the following data.

t = seconds	0	1	2	3	4	5
D = feet	0	16	64	144	256	400

The corresponding table of differences is

Change in t	From 0 to 1	From 1 to 2	From 2 to 3	From 3 to 4	From 4 to 5
Change in D	16	48	80	112	144

We see that the change in D is not constant, so D is not a linear function. A deeper analysis can show that $D(t) = 16t^2$.

KEY IDEA 3.5 **WHEN DATA ARE LINEAR**

A data table for $y = y(x)$ with evenly spaced values for x can be modeled with a linear function if it shows a constant change in y.

If the change in y is not constant, then the data cannot be modeled exactly by a linear function.

Michael Dwyer/Alamy

EXAMPLE 3.6 **SAMPLING VOTER REGISTRATION**

In this hypothetical experiment, a political analyst compiled data on the number of registered voters in Payne County, Oklahoma, each year from 2006 through 2011. For each of the following possible data tables that the analyst might have obtained, determine whether the data can be modeled with a linear function. If so, find such a formula and predict the number of registered voters in Payne County in the year 2013.

Hypothetical Data Table 1:

Date	2006	2007	2008	2009	2010	2011
Registered voters	28,321	28,542	29,466	30,381	30,397	31,144

Hypothetical Data Table 2:

Date	2006	2007	2008	2009	2010	2011
Registered voters	28,321	28,783	29,245	29,707	30,169	30,631

Solution for Data Table 1 First we choose variable and function names: Let d be the number of years since 2006, and let N be the number of registered voters. The table for N as a function of d is then

d	0	1	2	3	4	5
N	28,321	28,542	29,466	30,381	30,397	31,144

Choosing the variable to be years since 2006 simplifies our calculations.

To determine whether the data can be modeled by a linear function, we need to look at the change in the number of registered voters from year to year.

Change in d	0 to 1	1 to 2	2 to 3	3 to 4	4 to 5
Change in N	221	924	915	16	747

This table shows clearly that the change in registered voters is not constant from year to year, and we conclude that it is not appropriate to model these data with a linear function.

Solution for Data Table 2 Again, we let d be the number of years since 2006 and N the number of registered voters. The table for N as a function of d is then

d	0	1	2	3	4	5
N	28,321	28,783	29,245	29,707	30,169	30,631

We construct a table of differences using Data Table 2.

Change in d	0 to 1	1 to 2	2 to 3	3 to 4	4 to 5
Change in N	462	462	462	462	462

The rate of change for a linear function is the amount of change in the function for each unit change in the variable.

Here we see that the number of registered voters increases by 462 each year, so we know that we can model the data with a linear function. To determine the formula, we need to know the slope and the initial value. Since 462 is the change in N corresponding to a change in d of 1, the slope of our linear function is 462 registered voters per year. The initial value is given by the first entry in the table for N: $N(0) = 28,321$. Thus the formula is $N = 462d + 28,321$. To predict the number of registered voters in the year 2013, we use this formula, putting in 7 for d:

$$N(7) = 462 \times 7 + 28,321 = 31,555.$$

We can get this same prediction using only the slope rather than the formula. The change in d from 5 (the year 2011) to 7 (the year 2013) is 2 years, so we find

$$\text{Change in } N = m \times \text{Change in } d$$

$$\text{Change in } N = 462 \times 2$$

$$\text{Change in } N = 924.$$

Thus from $d = 5$ (the year 2011) to $d = 7$ (the year 2013), we expect an increase of 924 registered voters. That gives $30,631 + 924 = 31,555$ registered voters in the year 2013, the same answer we got using the formula.

TEST YOUR UNDERSTANDING | **FOR EXAMPLE 3.6**

The following table shows hypothetical enrollment E at a college t years after 2007.

t = years since 2007	0	1	2	3	4
E = enrollment	8641	8975	9309	9643	9977

Show that the data are linear, and find a linear model for the data. ■

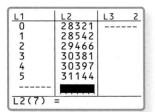

FIGURE 3.31 Voter registration data entered in a calculator table

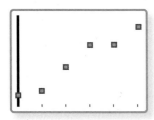

FIGURE 3.32 A view of nonlinear data

When linear data are graphed, the points fall on a straight line.

We note that the choice of the variable d as the number of years since 2006 greatly simplifies finding the linear formula, since the initial value can then be found as the first entry in the table. We could choose the variable to be the actual date. This would not affect the slope, but we would have to calculate the initial value using the slope and one of the data points in the original table. The resulting initial value would be different from the one we found above, since it would represent the number of registered voters present in A.D. 0! That is of course a nonsensical figure, and for this reason, as well as for the purpose of simplifying the calculation, it is better to choose the variable as we did in the example. This is an important point to remember for subsequent examples and exercises in which the variable is time.

We also note that Data Table 2 in Example 3.6 is not very realistic. It almost never happens that statistics of this sort turn out to be exactly linear. But many times such data can be closely approximated by a linear function. We will see how to deal with that in the next section.

Graphing Discrete Data

Calculating differences can always tell you whether data are linear, but many times it is advantageous to view such data graphically. Most graphing calculators will allow you to enter data tables and to display them graphically. In Figure 3.31 we have entered the data for Data Table 1 from Example 3.6. They are graphed in Figure 3.32. This picture clearly verifies our earlier contention that these are not linear data; the points do not fall on a straight line.

Figure 3.32 contrasts nicely with the plot of linear data from Data Table 2 that is shown in Figure 3.33. We see that the points do indeed fall on a straight line, clearly showing the linear nature of the data. Finally, we can add the graph of the linear function $N = 462d + 28,321$ that we found in part 2 of Example 3.6 to the screen as shown in Figure 3.34.

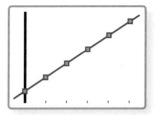

FIGURE 3.33 Linear data points from Data Table 2

FIGURE 3.34 Adding the linear model

Figure 3.33 and Figure 3.34 illustrate clearly what it means to model linear data. The data are displayed in Figure 3.33. The model is the line in Figure 3.34 that passes through the data points but also fills in the gaps and extends beyond the data points.

EXAMPLE 3.7 NEWTON'S SECOND LAW OF MOTION

Newton's second law of motion shows how force on an object, measured in *newtons*,[10] is related to acceleration of the object, measured in meters per second per second. The following experiment might be conducted in order to discover Newton's second law. Objects of various masses, measured in kilograms, were given an acceleration of 5 meters per second per second, and the associated forces were measured and recorded in the table below.

Mass	1	1.3	1.6	1.9	2.2
Force	5	6.5	8	9.5	11

Part 1 Check differences to show that these are linear data.

Part 2 Find the slope of a linear model for the data, and explain in practical terms what the slope means.

Part 3 Construct a linear model for the data.

Part 4 What force does your model show for an object of mass 1.43 kilograms that is accelerating at 5 meters per second?

Part 5 Make a graph showing the data, and overlay it with the graph of the linear model you made in part 3.

Part 6 Newton's second law of motion says that

$$\text{Force} = \text{Mass} \times \text{Acceleration}.$$

Do the results of our hypothetical experiment provide support for the validity of the second law? What additional experiments might be appropriate to provide further verification?

Solution to Part 1 First we choose variable and function names: Let m be the mass and F the force. Now we make a table of differences.

Change in m	1 to 1.3	1.3 to 1.6	1.6 to 1.9	1.9 to 2.2
Change in F	1.5	1.5	1.5	1.5

Since the change in F is always the same, 1.5 newtons, we conclude that the data are linear.

---

[10]One newton is about a quarter of a pound.

Solution to Part 2 Part 1 shows that it is appropriate to model the data with a linear function. In finding the slope we must be careful. For Data Table 2 of Example 3.6, the common difference in *N*, the number of registered voters, and the slope were the same. This will occur exactly when the data for the variable are given in steps of 1 unit, as was the case for the variable *d* in Example 3.6. Here the change in variable is not 1; rather, it is 0.3. Now we get the slope with a familiar formula:

> When the data increment is not 1, we must divide the common difference by the data increment to find the slope.

$$\text{Slope} = \frac{\text{Change in function}}{\text{Change in variable}} = \frac{\text{Change in } F}{\text{Change in } m} = \frac{1.5}{0.3} = 5 \text{ newtons per kilogram.}$$

To explain the slope in practical terms, we recall that the slope is the change in the function that results from a unit increase in the variable. Thus the slope of 5 newtons per kilogram means that when the mass increases by 1 kilogram, the associated force increases by 5 newtons. (For future reference, it is worth noting that the slope 5 turned out to be exactly the same as the acceleration. This is not an accident!)

Solution to Part 3 Since we know the slope, we need only get the initial value *b*. We cannot find this directly from the table. Instead, we use the slope together with any of the data points we wish. We will make the calculation with the first data point, $F(1) = 5$:

$$F = 5m + b$$
$$5 = 5 \times 1 + b$$
$$0 = b.$$

Thus the initial value *b* is 0, and we arrive at the linear model $F = 5m$.

> The linear model provides information not shown in the table.

In this case we could also have found the initial value by reasoning as follows: If the object has a mass of zero kilograms, the force will surely be 0, so the initial value is 0. Although this method gives the correct answer here, in general we should be cautious about going outside of the table to get data points, since the linear model may be appropriate only over the range of values in the table. See Exercise 13 at the end of this section.

Solution to Part 4 To get the force for a mass of 1.43 kilograms, we use this in place of the mass *m* in our linear model:

$$F = 5m = 5 \times 1.43 = 7.15 \text{ newtons.}$$

Solution to Part 5 The first step is to enter the data as shown in Figure 3.35. Once the data are entered, we can plot them to get Figure 3.36. As expected, the points in Figure 3.36 fall on a straight line, giving striking visual verification for our conclusion in part 1 that it is appropriate to make a linear model for the data. To overlay the data with the model we made, we enter the function as shown in Figure 3.37 and then make the graph. We see in Figure 3.38 that our model overlays the data exactly. This provides a good way for us to check our work. If the line had missed some of the points, we would know that we had made an error somewhere.

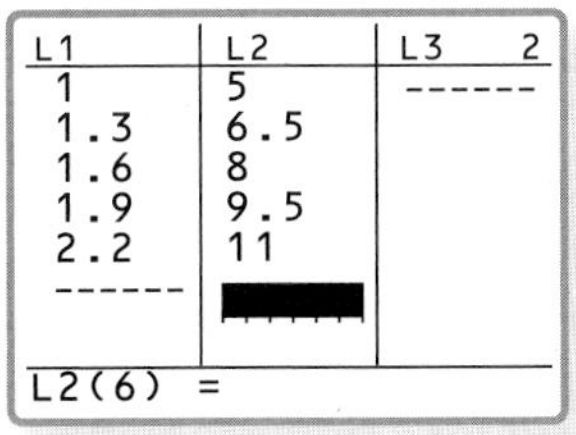

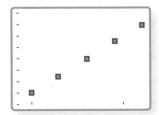

FIGURE 3.35 Data entered for force versus mass

FIGURE 3.36 A plot of data for force versus mass

FIGURE 3.37 Entering the linear model

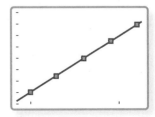

FIGURE 3.38 The linear model overlaying the data

Solution to Part 6 Our model verifies Newton's second law for objects with an acceleration of 5 meters per second per second. There are many other experiments that might be appropriate to give further evidence. One important experiment would be to take an object of fixed mass and measure what happens to the force when acceleration is varied. See Exercise 10 at the end of this section.

TEST YOUR UNDERSTANDING | **FOR EXAMPLE 3.7**

Objects of various masses, measured in kilograms, were given an acceleration of 10 meters per second per second. The associated forces, in newtons, are recorded in the table below.

m = mass	1	1.6	2.2	2.8
F = force	10	16	22	28

Show that the data are linear, and find a linear model for the data. ■

ANSWERS FOR TEST YOUR UNDERSTANDING

3.6 The data show a constant change of 334 students per year. The equation is $E = 334t + 8641$.

3.7 The data show a constant difference of 6 newtons in F for a difference of 0.6 in m. The formula is $F = 10m$.

3.3 EXERCISES

Reminder Round all answers to two decimal places unless otherwise indicated.

Note If no variable and function names are given in an exercise, you are of course expected to choose them and give the appropriate units. In your choice, when the variable is time, you should keep in mind the discussion following Example 3.6. Also, some of the data tables in this exercise set have been altered to allow for exact rather than approximate linear modeling.

1. **Making Ice** Our ice machine is making ice in preparation for the game that starts at 7:00 P.M. The machine is monitored, and the amount of ice is recorded at the end of each hour. The results are in the table below.

Time	12:00 P.M.	1:00 P.M.	2:00 P.M.	3:00 P.M.
Pounds of ice	200	273	346	419

   a. Show that the data are linear.

   b. Let t denote the time in hours since noon, and let I denote the pounds of ice made. Find a linear model for I as a function of t.

   c. If 675 pounds of ice will be needed for the game tonight, will the ice machine produce enough ice by game time?

2. **Adjusted Gross Income** An individual's *adjusted gross income* is the amount of income that is subject to federal income tax. The following table shows the total adjusted gross income (AGI), in trillions of dollars, reported to the IRS in the given year.

Year	2003	2004	2005	2006
AGI (trillions)	6.2	6.8	7.4	8.0

   a. Show that the data are linear.

   b. Let t denote the time in years since 2003, and let A denote the total adjusted gross income. Find a linear model for A as a function of t.

   c. Identify the slope of the linear model you found in part b, and explain its meaning in practical terms.

3. **Price of Amazon's Kindle** The following table shows the price of Amazon's Kindle 2 e-book reader. It is adapted from data available on the web.[11] Here

---

[11]www.pcworld.com/article/220980/will_amazons_kindle_be_free_by_november.html

time is measured in months since February 2009, when the Kindle was launched.

Time	Price
0	$349
5	$299
10	$249
15	$199

a. By calculating differences, show that these data can be modeled using a linear function.

b. Find a linear formula that models these data. Be careful about the sign of the slope.

c. What price does your formula from part b project for January 2012 (35 months after the Kindle was launched)? *Note:* The web data have been the basis for speculation that some day the Kindle would be free.

4. **Tuition at American Private Universities** The following table shows the average yearly tuition and required fees, in dollars, charged by four-year American private universities in the school year ending in the given year.

Date	Average tuition
2005	$25,540
2006	$27,085
2007	$28,630
2008	$30,175
2009	$31,720

a. Show that these data can be modeled by a linear function, and find its formula.

b. Plot the data points and add the graph of the linear formula you found in part a.

c. What prediction does this formula give for average tuition and fees at four-year American private universities for the academic year ending in 2014?

5. **Tuition at American Public Universities** *This is a continuation of Exercise 4.* The following table shows the average yearly in-state tuition and

required fees, in dollars, charged by four-year American public universities in the school year ending in the given year.

Date	Average tuition
2005	$5965
2006	$6381
2007	$6797
2008	$7213
2009	$7629

a. Show that these data can be modeled by a linear function, and find its formula.

b. What is the slope for the linear function modeling tuition and required fees for public universities?

c. What is the slope of the linear function modeling tuition and required fees for private universities? (*Note:* See Exercise 4.)

d. Explain what the information in parts b and c tells you about the rate of increase in tuition in public versus private institutions.

e. Which shows the larger percentage increase from 2005 to 2006?

6. **Total Cost** *The background for this exercise can be found in Exercises 9 and 10 in Section 3.2.* The following table gives the total cost C, in dollars, for a widget manufacturer as a function of the number N of widgets produced during a month.

Number N	Total cost C
200	7900
250	9650
300	11,400
350	13,150

a. What are the fixed costs and variable cost for this manufacturer?

b. The manufacturer wants to reduce the fixed costs so that the total cost at a monthly production level of 350 will be $12,975. What will the new fixed costs be?

(*continued*)

c. Instead of reducing the fixed costs as in part b, the manufacturer wants to reduce the variable cost so that the total cost at a monthly production level of 350 will be $12,975. What will the new variable cost be?

7. Total Revenue and Profit *This is a continuation of Exercise 6.* In general, the highest price p per unit of an item at which a manufacturer can sell N items is not constant but is, rather, a function of N. Suppose the manufacturer of widgets in Exercise 6 has developed the following table showing the highest price p, in dollars, of a widget at which N widgets can be sold.

Number N	Price p
200	43.00
250	42.50
300	42.00
350	41.50

a. Find a formula for p in terms of N modeling the data in the table.

b. Use a formula to express the total monthly revenue R, in dollars, of this manufacturer in a month as a function of the number N of widgets produced in a month. Is R a linear function of N?

c. On the basis of the tables in this exercise and the preceding one, use a formula to express the monthly profit P, in dollars, of this manufacturer as a function of the number of widgets produced in a month. Is P a linear function of N?

8. Dropping Rocks on Mars The behavior of objects falling near Earth's surface depends on the mass of Earth. On Mars, a much smaller planet than Earth, things are different. If Galileo had performed his experiment on Mars, he would have obtained the following table of data.

t = seconds	V = feet per second
0	0
1	12.16
2	24.32
3	36.48
4	48.64
5	60.8

a. Show that these data can be modeled by a linear function, and find a formula for the function.

b. Calculate $V(10)$ and explain in practical terms what your answer means.

c. Galileo found that the acceleration due to gravity of an object falling near Earth's surface was 32 feet per second per second. Physicists normally denote this number by the letter g. If Galileo had lived on Mars, what value would he have found for g?

9. The Kelvin Temperature Scale Physicists and chemists often use the *Kelvin* temperature scale. In order to determine the relationship between the Fahrenheit and Kelvin temperature scales, a lab assistant put Fahrenheit and Kelvin thermometers side by side and took readings at various temperatures. The following data were recorded.

K = kelvins	F = degrees Fahrenheit
200	−99.67
220	−63.67
240	−27.67
260	8.33
280	44.33
300	80.33

a. Show that the temperature F in degrees Fahrenheit is a linear function of the temperature K in kelvins.

b. What is the slope of this linear function? (*Note:* Be sure to take into account that the table lists kelvins in jumps of 20 rather than in jumps of 1.)

c. Find a formula for the linear function.

d. Normal body temperature is 98.6 degrees Fahrenheit. What is that temperature in kelvins?

e. If temperature increases by 1 kelvin, by how many degrees Fahrenheit does it increase? If temperature increases by 1 degree Fahrenheit, by how many kelvins does it increase?

f. The temperature of 0 kelvins is known as *absolute zero.* It is not quite accurate to say that all molecular motion ceases at absolute zero, but at that temperature the system has its minimum possible total energy. It is thought that absolute zero cannot be attained experimentally, although temperatures lower than 0.0000001 kelvin have been attained. Find the temperature of absolute zero in degrees Fahrenheit.

10. **Further Verification of Newton's Second Law**   This exercise represents a hypothetical implementation of the experiment suggested in the solution of part 6 of Example 3.7. A mass of 15 kilograms was subjected to varying accelerations, and the resulting force was measured. In the following table, acceleration is in meters per second per second, and force is in newtons.

Acceleration	Force
8	120
11	165
14	210
17	255
20	300

a. Construct a table of differences and explain how it shows that these data are linear.

b. Find a linear model for the data.

c. Explain in practical terms what the slope of this linear model is.

d. Express, using functional notation, the force resulting from an acceleration of 15 meters per second per second, and then calculate that value.

e. Explain how this experiment provides further evidence for Newton's second law of motion.

11. **Market Supply**   The following table shows the quantity S of wheat, in billions of bushels, that wheat suppliers are willing to produce in a year and offer for sale at a price P, in dollars per bushel.

S = quantity of wheat	P = price
1.0	$1.35
1.5	$2.40
2.0	$3.45
2.5	$4.50

In economics, it is customary to plot S on the horizontal axis and P on the vertical axis, so we will think of S as a variable and of P as a function of S.

a. Show that these data can be modeled by a linear function, and find its formula.

b. Make a graph of the linear formula you found in part a. This is called the *market supply curve*.

c. Explain why the market supply curve should be increasing. (*Hint:* Think about what should happen when the price increases.)

d. How much wheat would suppliers be willing to produce in a year and offer for sale at a price of $3.90 per bushel?

12. **Market Demand**   *This is a continuation of Exercise 11.* The following table shows the quantity D of wheat, in billions of bushels, that wheat consumers are willing to purchase in a year at a price P, in dollars per bushel.

D = quantity of wheat	P = price
1.0	$2.05
1.5	$1.75
2.0	$1.45
2.5	$1.15

In economics, it is customary to plot D on the horizontal axis and P on the vertical axis, so we will think of D as a variable and of P as a function of D.

a. Show that these data can be modeled by a linear function, and find its formula.

b. Add the graph of the linear formula you found in part a, which is called the *market demand curve*, to your graph of the market supply curve from Exercise 11.

c. Explain why the market demand curve should be decreasing.

d. The *equilibrium price* is the price determined by the intersection of the market demand curve and the market supply curve. Find the equilibrium price determined by your graph in part b.

13. **Sports Car**   An automotive engineer is testing how quickly a newly designed sports car accelerates from rest. He has collected data giving the velocity, in miles per hour, as a function of the time, in

(continued)

seconds, since the car was at rest. Here is a table giving a portion of his data:

Time	Velocity
2.0	27.9
2.5	33.8
3.0	39.7
3.5	45.6

a. By calculating differences, show that the data in this table can be modeled by a linear function.

b. What is the slope for the linear function modeling velocity as a function of time? Explain in practical terms the meaning of the slope.

c. Use the data in the table to find the formula for velocity as a linear function of time that is valid over the time period covered in the table.

d. What would your formula from part c give for the velocity of the car at time 0? What does this say about the validity of the linear formula over the initial segment of the experiment? Explain your answer in practical terms.

e. Assume that the linear formula you found in part c is valid from 2 seconds through 5 seconds. The marketing department wants to know from the engineer how to complete the following statement: "This car goes from 0 to 60 mph in _____ seconds." How should they fill in the blank?

14. High School Graduates The following table shows the number, in millions, graduating from high school in the United States in the given year.[12]

Year	Number graduating (in millions)
1985	2.83
1987	2.65
1989	2.47
1991	2.29

a. By calculating differences, show that these data can be modeled using a linear function.

b. What is the slope for the linear function modeling high school graduations? Explain in practical terms the meaning of the slope.

c. Find a formula for a linear function that models these data.

d. Express, using functional notation, the number graduating from high school in 1994, and then use your formula from part c to calculate that value.

15. Later High School Graduates *This is a continuation of Exercise 14.* The following table shows the number, in millions, graduating from high school in the United States in the given year.

Year	Number graduating (in millions)
2001	2.85
2003	2.98
2005	3.11
2007	3.24

a. Find the slope of the linear function modeling high school graduations, and explain in practical terms the meaning of the slope.

b. Find a formula for a linear function that models these data.

c. Express, using functional notation, the number graduating from high school in 2008, and then calculate that value.

d. The actual number graduating from high school in 1994 was about 2.52 million. Compare this with the value given by the formula in part b and with your answer to part d of Exercise 14. Which is closer to the actual value? In general terms, what was the trend in high school graduations from 1985 to 2007?

16. Tax Table Here are selected entries from the 2010 tax table that show the federal income tax owed by those married and filing jointly. The taxable income and the tax are both in dollars.

---

[12]The tables in this exercise and the next are adapted from the *Digest of Education Statistics,* U.S. Department of Education.

Taxable income	Tax
67,500	9291
67,600	9306
67,700	9321
67,800	9336
67,900	9351
68,000	9369
68,100	9394
68,200	9419
68,300	9444
68,400	9469
68,500	9494
68,600	9519

Over what two parts of this table is the tax a linear function of the taxable income? Find formulas for both linear functions, and explain in practical terms what the slopes mean.

17. **Sound Speed in Oceans** Marine scientists use a linear model for the speed c of sound in the oceans as a function of the salinity S at a fixed depth and temperature.[13] In the following table, one scientist recorded data for c and S at a depth of 500 meters and a temperature of 15 degrees Celsius. Here c is measured in meters per second and S in parts per thousand.

Salinity S	Sound speed c
35.0	1515.36
35.6	1516.08
36.2	1516.80
36.8	1517.62
37.4	1518.24

a. Looking over the table, the scientist realizes that one of the entries for sound speed is in error. Which entry is it, and what is the correct speed?

b. Explain in practical terms the meaning of the slope of the linear model for c in terms of S at the given depth and temperature.

c. Calculate $c(39)$ and explain in practical terms what it means.

18. **Focal Length** A refracting telescope has a main lens, or *objective lens*, and a second lens, the *eyepiece* (see Figure 3.39). For a given magnification M of the telescope, the focal length F_o of the objective lens is a linear function of the focal length F_e of the eyepiece. For example, a telescope with magnification $M = 80$ times can be constructed using various combinations of lenses. The following table gives some samples of focal lengths for telescopes with magnification $M = 80$. Here focal lengths are in centimeters.

F_e	0.3	0.5	0.7	0.9
F_o	24	40	56	72

a. Construct a linear model for the data.

b. In this example the magnification M is 80. In general F_o is proportional to F_e, and the constant of proportionality is M. Use this relation to write a formula for F_o in terms of F_e and M.

c. Solve the equation you obtained in part b for M and thus obtain a formula for magnification as a function of objective lens focal length and eyepiece focal length.

d. To achieve a large magnification, how should objective and eyepiece lenses be selected?

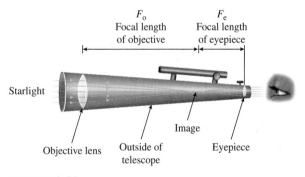

F_o
Focal length of objective

F_e
Focal length of eyepiece

Starlight

Objective lens   Outside of telescope   Image   Eyepiece

FIGURE 3.39

19. **Measuring the Circumference of the Earth** Eratosthenes, who lived in Alexandria around 200 B.C., learned that at noon on the summer solstice the sun

[13]The model here is based on the formula of Kenneth V. Mackenzie, "Nine-term equation for sound speed in the oceans," *J. Acoust. Soc. Am.* **70** (1981), 807–812.

(continued)

shined vertically into a well in Syene (modern Aswan, due south of Alexandria). He found that on the same day in Alexandria the sun was about 7 degrees short of being directly overhead. Since 7 degrees is about 1/50 of a full circle of 360 degrees, he concluded that the distance from Syene to Alexandria was about 1/50 of the Earth's circumference. He knew from travelers that it was a 50-day trip and that camels could travel 100 *stades*[14] per day. What was Eratosthenes' measure, in stades, of the circumference of the Earth? One estimate is that the stade is 0.104 mile. Using this estimate, what was Eratosthenes' measurement of the circumference of the Earth in miles?

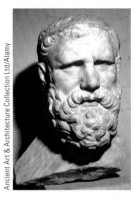

Ancient Art & Architecture Collection Ltd/Alamy

20. **A Research Project on Tax Tables** *This is a continuation of Exercise 16.* Here are selected entries from the 2009 tax table that show the federal income tax owed by those married and filing jointly. The taxable income and the tax are both in dollars.

Taxable income	Tax
67,500	9294
67,600	9309
67,700	9324
67,800	9339
67,900	9356
68,000	9381
68,100	9406
68,200	9431
68,300	9456
68,400	9481
68,500	9506
68,600	9531

a. Over what two parts of this table is the tax a linear function of the taxable income? Find formulas for both linear functions.

b. Compare your answer to part a with your answer to Exercise 16. Consider especially the income level where the transition from one linear function to the next takes place.

c. As a research project, examine tax tables from earlier years and investigate tax law to discover how the transition level is determined. Find out what the phrase *bracket creep* means.

When Data Are Unevenly Spaced   If data are evenly spaced, we need only calculate differences to see whether the data are linear. But if data are not evenly spaced, then we must calculate the average rate of change over each interval to see whether the data are linear. If the average rate of change is constant, it is the slope of the linear function. This fact is used in Exercises 21 and 22.

21. In the following table, show that the average rate of change from 2 to 5 is not the same as the average rate of change from 5 to 6. This shows that the data are not linear even though the differences in *y* are constant.

x	1	2	5	6
y	3	6	9	12

22. In the following table, show that the average rate of change is the same over each interval. This shows the data are linear even though the differences in *y* are not constant. Then find a linear model for the data.

x	1	2	5	6
y	7	10	19	22

3.3 **SKILL BUILDING EXERCISES**

S-1. **Testing Data for Linearity**   Test the following data to see whether they are linear.

x	2	4	6	8
y	12	17	22	27

---

[14]There is some uncertainty about just how long the stade was, so we know only that his estimate is too large by between 4 and 14 percent. In any case, it is considerably better than Columbus's estimate (which was about 40% too small) some 1700 years later. The mathematicians of the day knew full well that Columbus's estimate was nowhere near correct, and that is one of the reasons why he had so much trouble getting financing for his expedition to Cathay.

S-2. Testing Data for Linearity Test the following data to see whether they are linear.

x	2	4	6	8
y	12	17	21	25

S-3. Making a Linear Model Make a linear model for the data in Exercise S-1.

S-4. Making a Linear Model The data in Exercise S-2 are not linear, but if we omit the first point, the remaining data are linear. Make a linear model for these last three points.

S-5. Graphing Discrete Data Plot the data from the table in Exercise S-2.

S-6. Adding a Graph to a Data Plot Add the graph of $y = 2.5x + 7$ to the data plot from Exercise S-5.

S-7. Entering and Graphing Data Enter the following data and plot f against x.

x	1	2	3	4	5
f	8	6	5	3	1

S-8. Adding a Graph Add to the picture in Exercise S-7 the graph of $y = -1.7x + 9.7$.

S-9. Editing Data Plot the squares of the data points in the table from Exercise S-7.

S-10. Data That Are Linear Plot the data from Exercise S-1 along with the linear model from Exercise S-3.

Plotting Data and Functions In Exercises S-11 through S-21, you are asked to produce various graphs.

S-11. Plot the following data points.

x	2	3	5	7	8
y	-2	0	10	28	40

S-12. Add the graph of $y = x^2 - 3x$ to the plot from Exercise S-11.

S-13. Plot the following data points.

x	-2	2	5	6	7
y	2.25	2	27	58	121

S-14. Add the graph of $y = 2^x - x$ to the plot from Exercise S-13.

S-15. Plot the following data points.

x	1	2	5	7	9
y	0.5	0.4	0.19	0.16	0.14

S-16. Add the graph of $y = \dfrac{1}{x + 1}$ to the plot from Exercise S-15.

S-17. Plot the following data points.

x	3	7	10	12	15
y	3.73	5.47	6.48	7.07	7.83

S-18. Add the graph of $y = \sqrt{x} + \sqrt{x + 1}$ to the plot from Exercise S-17.

S-19. Add the graph of $y = \dfrac{x}{x^2 + 1}$ to the plot from Exercise S-15.

S-20. Plot the following data points.

x	2	5	6	7	9
y	-0.5	4	5.17	6.29	8.44

S-21. Add the graph of $y = x - \dfrac{5}{x}$ to the plot from Exercise S-20.

3.4 LINEAR REGRESSION

In real life, rarely is information gathered that perfectly fits any simple formula. In cases such as government spending, many factors influence the budget, including the political make-up of the legislature. In the case of scientific experiments, variations may be due to *experimental error*, the inability of the data gatherer to obtain exact measurements; there may also be elements of chance involved. Under these circumstances, it may be necessary to obtain an approximate rather than an exact mathematical model.

The Regression Lines

To illustrate this idea, let's look at federal Medicare expenditures, in billions of dollars, by the United States as reported by the *Statistical Abstract of the United States* and recorded in the following table.

Date	2004	2005	2006	2007	2008
Expenditures in billions	269.4	298.6	329.9	375.4	390.8

To study these data, we assign variable and function names: Let t be the number of years since 2004, and let E be the expenditures, in billions of dollars. The table for E as a function of t is then

t	0	1	2	3	4
E	269.4	298.6	329.9	375.4	390.8

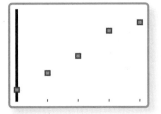

FIGURE 3.40 Data points that are almost on a straight line

It is difficult to discern, by looking at the table, the pattern of spending on Medicare. We can check to see whether these data can be modeled by a linear function by making a table of changes.

Change in t	From 0 to 1	From 1 to 2	From 2 to 3	From 3 to 4
Change in E	29.2	31.3	45.5	15.4

When data points almost fall on a straight line, it may be appropriate to model them with a linear function.

The change in E is not constant, so we know that these data cannot be modeled exactly by a linear function, but let's explore further. If we plot the data points as we have done in Figure 3.40, we see that they *almost* fall on a straight line. To show this better, we have added, in Figure 3.41, a straight line that passes *close* to all the data points. This line is known as the *least-squares fit* or *linear regression* line, and it appears that the data follow this line fairly closely. Thus, although we cannot model the data *exactly* with a linear function, it seems reasonable to use the line in Figure 3.41 as an *approximate* model for Medicare spending.

There are many ways of approximating data that are nearly linear, but the regression line is the one that is most often used. It is the line that gives the least possible total of the squares of the vertical distances from the line to the data points. This is why the regression line is sometimes referred to as the least-squares fit. The mathematical concepts needed to derive the formula for regression lines are beyond the scope of this course. Furthermore, although it is possible to calculate a formula for the regression line by hand, the procedure is cumbersome. Instead, this calculation is nearly always done with a calculator or computer, and that is how we will do it here. You should refer to Appendices B and C to see the exact keystrokes needed to execute this important procedure.

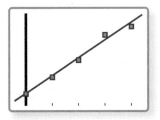

FIGURE 3.41 A line that almost fits the data

FIGURE 3.42 Data for Medicare spending

```
LinReg
 y=ax+b
 a=31.96
 b=268.9
```

FIGURE 3.43 The regression line parameters

The regression line provides an approximate model for data rather than an exact fit.

With the data properly entered in the calculator as shown in Figure 3.42, we can get the regression line information shown in Figure 3.43. As usual, the calculator chooses its own letter[15] names, and we need to record the appropriate correspondences:

$$X = t, \text{ the variable}$$

$$Y = E, \text{ the function name}$$

$$a = 31.96 = \text{slope of the regression line}$$

$$b = 268.90 = \text{vertical intercept of the regression line.}$$

We use this information to make the regression line model for E as a function of t:

$$E = 31.96t + 268.90. \tag{3.8}$$

This is the line we added to the data plot in Figure 3.40 to get the picture in Figure 3.41.

It is important to remember that even though we have written an equality, Equation (3.8) in fact is a model that only approximates the relationship between t and E given by the data table. For example, the initial value according to the linear model in Equation (3.8) is 268.9, whereas the entry in the table for $t = 0$ is $E = 269.4$. We will use the equals sign as above, but you should be aware that in this setting many would prefer to replace it by an approximation symbol, $\approx$, or to use different letters for the regression equation. If you do use different letters for function and variable, it is crucial that you state clearly what they represent.

Uses of the Regression Line: Slope and Trends

The most useful feature of the regression line is its slope, which in many cases provides the key to understanding data. For Medicare spending, the slope, 31.96 billion dollars per year, of the regression line tells us that during the period from 2004 to 2008, Medicare spending grew by about 31.96 billion dollars per year. It tells how the data are changing.

A plot that shows the regression line with the data can be useful in analyzing trends. For example, in Figure 3.41 the fourth data point (corresponding to 2007) lies above the regression line, whereas the fifth data point (corresponding to 2008) lies below it. Certainly spending was higher in 2008 than in 2007, but it could be argued, on the basis of the position of the data points relative to the line, that in 2007 spending was ahead of the trend, whereas in 2008 it was slightly behind.

It is tempting to use the regression line to predict the future. We will show how to do this and then discuss the pitfalls in such practice. What level of spending does the regression line in Equation (3.8) predict for 2009? Since t is the number of years since 2004, we want to get the value of E when $t = 5$. This information is not shown in the table, so we approximate the value of $E(5)$ using the regression line with $t = 5$:

Projected spending in 2009 $= 31.96 \times 5 + 268.9 = 428.7$ billion dollars.

Thus we project 2009 spending to be about 428.7 billion dollars. Consulting the source for our data, the *Statistical Abstract of the United States*, we find that Medicare expenditures in 2009 were in fact $E = 430.1$ billion dollars. In this case, the projection given by the regression line was not exact, but was within 1% of the actual value.

Let's try to use the regression line to go the other way—that is, to estimate money spent before 2004. In particular, we want to determine the level of Medicare expenditures

---

[15]An additional number, often denoted r, is shown at the bottom of the screen display for some calculators. This is a statistical measure of how closely the line fits the data, and we will not make use of it.

in 2000. That is 4 years before 2004, so we want the value of E when $t = -4$. We use the regression line with $t = -4$:

Estimated spending in 2000 $= 31.96 \times (-4) + 268.9 = 141.06$ billion dollars.

Thus we estimate 2000 spending to have been about 141.1 billion dollars. Once again, we refer to the *Statistical Abstract of the United States* and see that the actual federal expenditures for Medicare in 2000 were 197.1 billion dollars. In this case, the value given by the regression line is a bad estimate of the real value.

KEY IDEA 3.6 **USING THE REGRESSION LINE**

The slope is the most important piece of information about the regression line because it estimates the rate of change of the data. The regression line is useful for analyzing trends, but it may not be accurate beyond the limits of the data.

The regression line is a powerful tool for analysis of certain kinds of data, but caution in its use is essential. We saw above that the regression line gave a decent estimate for 2009 expenditures but a bad one for 2000. In general, using the regression line to extrapolate beyond the limits of the data is risky, and the risk increases dramatically for long-range extrapolations. What the regression line really shows is the *linear trend* established by the data. Thus, for our 2009 projection, it would be appropriate to say, "If the trend established in the mid-2000s had persisted, then federal Medicare spending in 2009 would have been about 428.7 billion dollars." A check of the 2009 data showed that this trend did indeed persist into 2009. An appropriate statement for our 2000 analysis might be "If the trend established in the mid-2000s had been valid since 2000, then federal expenditures in 2000 would have been about 141.1 billion dollars." Because the actual expenditure was much more, we might proceed by gathering more data to see exactly how spending changed and then conduct historical, political, and economic investigations into why it changed.

The regression line must be used with care.

Newspaper and magazine articles frequently use data gathered today to make predictions about what might be expected in the future, and in many cases this is done with the regression line. When such data are handled by professionals who have sophisticated tools and insights available to help them, valuable information can be gained. But sometimes predictions based on gathered data are made by people who know a great deal less about handling data than you will by the time you complete this course. As a citizen, it is important that you be able to bring your own insight to bear on such analyses. It is risky to use the regression line to make projections unless you have good reason to believe that the data you are looking at are nearly linear and that the linear nature of the data will persist into the future. On the other hand, the regression line can clearly show the trend of almost linear data and can appropriately be used to determine whether trends persist. If information is available that indicates that linear trends are persisting, then the regression line can be used to make forecasts.

As an example, suppose a legislator who ran for office in 2008 ran on a platform calling for a "significant" change in federal Medicare spending in 2009. Does the record show that the legislator was able to fulfill his campaign promise? If so, then we should be able to detect a "significant" change in 2009 from the trend established from 2004 through 2008. We have already seen that the trend of the mid-2000s leads us to project an expenditure of about 428.7 billion dollars in 2009. Checking the record, we find that actual federal spending, 430.1 billion in 2009, was slightly more than the trend of the mid-2000s would suggest. It is easy to conceive of this legislator taking credit for such a change in 2009. We leave it to the reader to decide whether a re-election vote would be merited.

EXAMPLE 3.8 MILITARY EXPENDITURES

The following table shows the amount $M = M(t)$ of money, in billions of dollars, spent by the United States on national defense.[16] In the time row, $t = 0$ corresponds to 2004, $t = 1$ refers to 2005, and so on.

$t =$ years since 2004	0	1	2	3	4
$M =$ billions of dollars	455.8	495.3	521.8	551.3	616.1

Part 1 Plot the data points. Does it appear that it is appropriate to approximate these data with a straight line?

Part 2 Find the equation of the regression line for M as a function of t, and add the graph of this line to your data plot.

Part 3 Explain in practical terms the meaning of the slope of the regression line model we found in part 2.

Part 4 Compare the slope of the regression line for defense spending with the slope of the regression line for spending on Medicare that we made earlier. What conclusions do you draw from this comparison?

Part 5 Use the regression equation to estimate military spending by the United States in 2009. The actual military expenditures in 2009 were $661.0 billion. Did the trend established in the mid-2000s persist until 2009?

Solution to Part 1 First we enter the data as shown in Figure 3.44. Next we plot the data as shown in Figure 3.45. Note that the horizontal axis corresponds to years since 2004, the vertical axis to billions of dollars of military expenditures. From Figure 3.45, it is clear that the data points do not lie exactly on a line, but they do almost line up, and it is not unreasonable to model these data with a straight line.

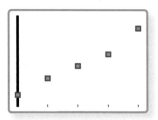

FIGURE 3.44 National defense spending data

FIGURE 3.45 A plot of national defense spending

---

[16]From the *Statistical Abstract of the United States.*

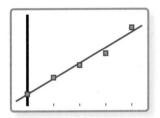

```
LinReg
 y=ax+b
 a=37.66
 b=452.74
```

FIGURE 3.46 Regression line parameters for defense spending

FIGURE 3.47 The regression line added to the data plot

The slope of the regression line yields important information.

Solution to Part 2 We use the calculator to get the regression line parameters shown in Figure 3.46. We record the appropriate associations:

$$X = t, \text{ the variable, years since 2004}$$

$$Y = M, \text{ the function name, billions spent}$$

$$a = 37.66 = \text{slope of the regression line}$$

$$b = 452.74 = \text{vertical intercept of the regression line.}$$

Thus the regression line model we want is $M = 37.66t + 452.74$. We emphasize once more that even though we have written an equality, this does not establish an exact relationship between t and M. Rather, it is the approximation provided by the regression model. We add this equation to the function list and get the graph in Figure 3.47. It is important to note here that the regression line we found passes near the data points. If it didn't, then we would look for a mistake in our work; if we found that no error had been made, we would conclude that it might not be appropriate to model the data using the regression line.

Solution to Part 3 The slope of the regression line shows the rate at which military expenditures were growing during the mid-2000s. We can conclude that in this period, defense spending was growing at a rate of about 37.66 billion dollars per year.

Solution to Part 4 In part 3 we noted that during the mid-2000s, defense spending was growing at a rate of 37.66 billion dollars per year. Earlier we found that the slope of the regression line for Medicare spending was 31.91 billion dollars per year. Thus not only was military spending much greater during the mid-2000s, it was growing faster than federal Medicare spending. (There are other comparisons that may be made that do not show such a sharp distinction. You may note, for example, that the percentage increase in Medicare spending was larger.)

Solution to Part 5 The year 2009 corresponds to $t = 5$. We use these values in the regression equation to make the projection:

$$\text{Projected 2009 spending} = 37.66 \times 5 + 452.74 = 641.04 \text{ billion dollars.}$$

The projection of 641.0 billion dollars for defense spending in 2009 is about 3% lower than the actual figure of 661.0 billion dollars. The trend established in the mid-2000s persisted until 2009, with an indication that the rate of growth may have been increasing slightly.

TEST YOUR UNDERSTANDING │ **FOR EXAMPLE 3.8**

The following table shows the average price P, in dollars, of a gallon of regular gas t years after 2000.

t = years since 2000	0	2	4	7	8
P = price in dollars	1.51	1.36	1.88	2.80	3.27

Find the equation of the regression line for P as a function of t. What price does your model give for gas in 2009? Compare your answer with the actual value of $2.35. ■

ANSWERS FOR TEST YOUR UNDERSTANDING

3.8 $P = 0.24t + 1.17$. The model gives $3.33, which is much higher than the actual value.

3.4 EXERCISES

Reminder Round all answers to two decimal places unless otherwise indicated.

1. **Motor Vehicle Accidents** The table below shows the number A, in millions, of motor vehicle accidents t years after 2004.[17] Find the equation of the regression line for A as a function of t.

t = years since 2004	A = millions of accidents
0	10.9
1	10.7
2	10.4
3	10.6
4	10.2

2. **Federal Methamphetamine Arrests** The table below shows the number A, in thousands, of federal arrests for methamphetamine t years after 2006.[18]

t = years since 2006	A = thousands of arrests
0	5.85
1	5.54
2	4.72
3	4.70

Find the equation of the regression line for A as a function of t.

3. **DirecTV Subscribers** The table below shows the number S, in millions, of subscribers to DirecTV t years after 1995.

t = years since 1995	S = subscribers, in millions
0	1.20
4	6.68
7	11.18
9	13.00
12	16.83
14	18.08

a. Find the equation of the regression line for S as a function of t.

b. What number does this equation give for DirecTV subscribers in 2005? (The actual number was 15.00 million.)

c. Explain in practical terms the meaning of the slope of the line you found in part a.

d. Plot the data points and the regression line.

4. **Crayola Colors** The table below shows the number C of Crayola colors available t years after 1900.

t = years since 1900	3	49	58	72	90	98	103
C = number of colors	8	48	64	72	80	120	120

a. Find the equation of the regression line for C as a function of t.

b. How many Crayola colors does the regression line indicate for 1993? (Round your answer to the nearest whole number. Note that the actual number is 96.)

c. Plot the data points and the regression line.

5. **Is a Linear Model Appropriate?** The number, in thousands, of bacteria in a petri dish is given by the table below. Time is measured in hours.

Time in hours since experiment began	Number of bacteria in thousands
0	1.2
1	2.4
2	4.8
3	9.6
4	19.2
5	38.4
6	76.8

[17]These data are from the *Statistical Abstract of the United States*.
[18]These data are from the *Statistical Abstract of the United States*.

(continued)

The table below shows enrollment,[19] in millions of people, in private colleges in the United States during the years from 2004 through 2008.

Date	Enrollment in millions
2004	4.29
2005	4.47
2006	4.58
2007	4.76
2008	5.13

   a. Plot the data points for number of bacteria. Does it look reasonable to approximate these data with a straight line?

   b. Plot the data points for college enrollment. Does it look reasonable to approximate these data with a straight line?

6. College Enrollment *This is a continuation of Exercise 5.* We use the data in the college enrollment table that appears in Exercise 5.

   a. Find the equation of the regression line model for college enrollment as a function of time, and add its graph to the data plot made in Exercise 5.

   b. Explain the meaning of the slope of the line you found in part a.

   c. Express, using functional notation, the enrollment in American private colleges in 2010, and then estimate that value.

   d. Enrollment in American private colleges in 1990 was 2.97 million. Does it appear that the trend established in the mid-2000s was valid as early as 1990?

7. Tourism The number, in millions, of international tourists who visited the United States is given in the following table.[20]

Date	2005	2006	2007	2008
Millions of tourists	49.21	51.06	55.98	57.94

   a. Plot the data.

   b. Find the equation of the regression line and add its graph to your data plot. (Round the regression line parameters to two decimal places.)

   c. Explain in practical terms the meaning of the slope.

   d. Express, using functional notation, the number of tourists who visited the United States in 2009, and then estimate that value. The actual number was 54.95 million. Can you explain the difference in terms of the global recession?

8. Cell Phones The following table gives the amount spent on cellular service.

Date	Cellular service revenue (in billions)
2005	113.5
2006	125.5
2007	138.9
2008	148.1

   a. Plot the data points.

   b. Find the equation of the regression line and add its graph to the plotted data.

   c. In 2009, $152.6 billion was spent on cellular service. If you had been a financial strategist in 2008 with only the data in the table above available, what would have been your prediction for the amount spent on cellular service in 2009?

9. Long Jump The following table shows the length, in meters, of the winning long jump in the Olympic Games for the indicated year. (One meter is 39.37 inches.)

Year	1900	1904	1908	1912
Length	7.19	7.34	7.48	7.60

   a. Find the equation of the regression line that gives the length as a function of time. (Round the regression line parameters to three decimal places.)

---

[19]This is taken from the *Statistical Abstract of the United States* and includes projections.
[20]These data are from the *Statistical Abstract of the United States*.

b. Explain in practical terms the meaning of the slope of the regression line.

c. Plot the data points and the regression line.

d. Would you expect the regression line formula to be a good model of the winning length over a long period of time? Be sure to explain your reasoning.

Diego Barbieri//Shutterstock.com

e. There were no Olympic Games in 1916 because of World War I, but the winning long jump in the 1920 Olympic Games was 7.15 meters. Compare this with the value that the regression line model gives. Is the result consistent with your answer to part d?

10. Driving You are driving on a highway. The following table gives your speed S, in miles per hour, as a function of the time t, in seconds, since you started making your observations.

Time t	0	15	30	45	60
Speed S	54	59	63	66	68

a. Find the equation of the regression line that expresses S as a linear function of t.

b. Explain in practical terms the meaning of the slope of the regression line.

c. On the basis of the regression line model, when do you predict that your speed will reach 70 miles per hour? (Round your answer to the nearest second.)

d. Plot the data points and the regression line.

e. Use your plot in part d to answer the following: Is your prediction in part c likely to give a time earlier or later than the actual time when your speed reaches 70 miles per hour?

11. The Effect of Sampling Error on Linear Regression A stream that feeds a lake is flooding, and during this flooding period the depth of water in the lake is increasing. The actual depth of the water at a certain point in the lake is given by the linear function $D = 0.8t + 52$ feet, where t is measured in hours since the flooding began. A hydrologist

does not have this function available and is trying to determine experimentally how the water level is rising. She sits in a boat and, each half-hour, drops a weighted line into the water to measure the depth to the bottom. The motion of the boat and the waves at the surface make exact measurement impossible. Her compiled data are given in the following table.

t = hours since flooding began	D = measured depth in feet
0	51.9
0.5	52.5
1	52.9
1.5	53.3
2	53.7

a. Plot the data points.

b. Find the equation of the regression line for D as a function of t, and explain in practical terms the meaning of the slope.

c. Add the graph of the regression line to the plot of the data points.

d. Add the graph of the depth function $D = 0.8t + 52$ to the picture. Does it appear that the hydrologist was able to use her data to make a close approximation of the depth function?

e. What was the actual depth of the water at $t = 3$ hours?

f. What prediction would the hydrologist's regression line give for the depth of the water at $t = 3$?

12. Gross National Product The United States gross national product, in trillions of dollars, is given in the table below.

Date	Gross national product
2005	12.7
2006	13.5
2007	14.2
2008	14.6

a. Find the equation of the regression line, and explain the meaning of its slope. (Round regression line parameters to two decimal places.)

(continued)

b. Plot the data points and the regression line.

c. When would you predict that a gross national product of 15.3 trillion dollars would be reached? The actual gross national product in 2009 was 14.4 trillion dollars. What does that say about your prediction?

13. **Domestic Auto Sales in the United States**   For 2005 through 2008, the following table shows the total U.S. sales, in millions, of domestic automobiles (excluding light trucks).[21]

Date	Domestic cars sold
2005	5.53
2006	5.48
2007	5.25
2008	4.54

a. Get the equation of the regression line (rounding parameters to two decimal places), and explain in practical terms the meaning of the slope. In particular, comment on the meaning of the sign of the slope.

b. Plot the data points and the regression line.

c. In 2009, 3.62 million domestic cars were sold in the United States. How does the forecast obtained from the regression line compare with this figure?

14. **Running Speed Versus Length**   The following table gives the length L, in inches, of an animal and its maximum speed R, in feet per second, when it runs.[22] (For comparison, 10 feet per second is about 6.8 miles per hour.)

Animal	Length L	Speed R
Deermouse	3.5	8.2
Chipmunk	6.3	15.7
Desert crested lizard	9.4	24.0
Grey squirrel	9.8	24.9
Red fox	24.0	65.6
Cheetah	47.0	95.1

a. Does this table support the generalization that larger animals run faster?

b. Plot the data points.

c. Find the equation of the regression line for R as a function of L, and explain in practical terms the meaning of its slope. (Round regression line parameters to two decimal places.) Add the plot of the regression line to the data plot in part b.

d. Judging on the basis of the plot in part c, which is faster *for its size*, the red fox or the cheetah?

15. **Antimasonic Voting**   In *A Guide to Quantitative History*,[23] R. Darcy and Richard C. Rohrs use mathematics to investigate the influence of religious zeal, as evidenced by the number C of churches in a given township, on the percent M of voting that was antimasonic in Genesee County, New York, from 1828 to 1832. The data used there are partially reproduced in the accompanying table.

Township	C = number of church buildings	M = percent antimasonic voting
Alabama	2	60.0
Attica	4	65.0
Stafford	5	74.0
Covington	6	81.7
Elbe	7	88.3

Solely on the basis of the data above, analyze the premise that "The percent of antimasonic voting in Genesee County around 1830 had a direct dependence on the number of churches in the township." Your analysis should include a statement of why you believe there is a relationship and how that relationship might appropriately be described. You are encouraged to consult the book cited for a deeper and more authoritative analysis.

16. **Running Ants**   A scientist collected the following data on the speed, in centimeters per second, at which ants ran at the given ambient temperature, in degrees Celsius.[24]

---

[21]From the *Statistical Abstract of the United States.*
[22]The table is adapted from J. T. Bonner, *Size and Cycle* (Princeton, NJ: Princeton University Press, 1965).
[23]Published by Praeger Publishers (Westport, CT) in 1995.
[24]The table is taken from the data of H. Shapley, "Note on the thermokinetics of Dolichoderine ants," *Proc. Nat. Acad. Sci.* **10** (1924), 436–439.

Temperature	Speed
25.6	2.62
27.5	3.03
30.3	3.57
30.4	3.56
32.2	4.03
33.0	4.17
33.8	4.32

a. Find the equation of the regression line, giving the speed as a function of the temperature.

b. Explain in practical terms the meaning of the slope of the regression line.

c. Express, using functional notation, the speed at which the ants run when the ambient temperature is 29 degrees Celsius, and then estimate that value.

d. The scientist observes the ants running at a speed of 2.5 centimeters per second. What is the ambient temperature?

17. Expansion of Steam When water changes to steam, its volume increases rapidly. At a normal atmospheric pressure of 14.7 pounds per square inch, water boils at 212 degrees Fahrenheit and expands in volume by a factor of 1700 to 1. But when water is sprayed into hotter areas, the expansion ratio is much greater. This principle can be applied to good effect in fire fighting. The steam can occupy such a large volume that oxygen is expelled from the area and the fire may be smothered. The table below shows the approximate volume, in cubic feet, of 50 gallons of water converted to steam at the given temperatures, in degrees Fahrenheit.

T = temperature	V = cubic feet of steam
212	10,000
400	12,500
500	14,100
800	17,500
1000	20,000

a. Make a linear model of volume V as a function of T.

b. If one fire is 100 degrees hotter than another, what is the increase in the volume of steam produced by 50 gallons of water?

c. Calculate $V(420)$ and explain in practical terms what your answer means. Round your answer to the nearest whole number.

d. At a certain fire, 50 gallons of water expanded to 14,200 cubic feet of steam. What was the temperature of the fire?

18. Technological Maturity Versus Use Maturity There are a number of processes used industrially for separation (of salt from water, for example). Such a process has a *technological maturity*, which is the percentage of perfection that the process has reached. A separation process with a high technological maturity is a very well-developed process. It also has a *use maturity*, which is the percentage of total reasonable use. High use maturity indicates that the process is being used to near capacity. In 1987 Keller collected from a number of experts their estimation of technological and use maturities. The table below is adapted from those data.

Process	Technological maturity (%)	Use maturity (%)
Distillation	87	87
Gas absorption	81	76
Ion exchange	60	60
Crystallization	64	62
Electrical separation	24	13

a. Construct a linear model for use maturity as a function of technological maturity.

b. Explain in practical terms the meaning of the slope of the regression line.

c. Express, using functional notation, the use maturity of a process that has a technological maturity of 89%, and then estimate that value.

d. Solvent extraction has a technological maturity of 73% and a use maturity of 61%. Is solvent extraction being used more or less than would be expected from its technological development? How might this information affect an entrepreneur's

(continued)

decision whether to get into the business of selling solvent equipment to industry?

e. Construct a linear model for technological maturity as a function of use maturity.

19. **Whole Crop Weight Versus Rice Weight**   A study by Horie[25] compared the dry weight B of brown rice with the dry weight W of the whole crop (including stems and roots). These data, in tons per hectare,[26] for the variety *Nipponbare* grown in various environmental conditions are partially presented in the table below.

Whole crop weight W	Rice weight B
6	1.8
11.1	3.2
13.7	3.7
14.9	4.3
17.6	5.2

a. Find an approximate linear model for B as a function of W.

b. Which sample might be considered to have a lower rice weight than expected from the whole crop weight?

c. How much additional rice weight can be expected from 1 ton per hectare of additional whole crop weight?

20. **Rice Production in Asia**   Improved agricultural practices, including better utilization of fertilizers

huyangshu/Shutterstock.com

and use of improved plant varieties, have resulted in increased rice yields in Asia. The accompanying table shows the average yield[27] Y, in tons per hectare, as a function of the number of years t since 1980.

t = years since 1980	Y = average yield
5	3.32
10	3.61
15	3.73
20	3.95
25	4.11

a. One study used a linear function to approximate yield as a function of time. Find an approximate linear model for Y as a function of t.

b. Explain what the slope of the linear model tells you.

c. Use the model to calculate $Y(30)$, and explain in practical terms what your answer means.

d. What yield would the model estimate for 2025?

e. The 1993 report of the International Rice Research Institute predicted that rice requirements in Asia would grow exponentially. If the rice requirements in 2025 are 6.44 tons per hectare, will rice production in 2025 meet those requirements?

21. **Energy Cost of Running**   Physiologists have studied the steady-state oxygen consumption (measured per unit of mass) in a running animal as a function of its velocity (i.e., its speed). They have determined that the relationship is approximately linear, at least over an appropriate range of velocities. The table on the following page gives the velocity v, in kilometers per hour, and the oxygen consumption E, in milliliters of oxygen per gram per hour, for the rhea, a large, flightless South American bird.[28]

---

[25]T. Horie, "The effects of climatic variations on agriculture in Japan. 5: The effects on rice yields in Hokkaido." In M. L. Parry, T. R. Carter, and N. T. Konijn, eds. *The Impact of Climatic Variations on Agriculture*, Vol. 1: *Assessment in Cool Temperate and Cold Regions* (Dordrecht, The Netherlands: Kluwer Academic Publishers, 1988), 809–826.

[26]A hectare is about $2\frac{1}{2}$ acres.

[27]IRRI, *World Rice Statistics*. International Rice Research Institute, http://www.irri.org/science.

[28]The table is based on C. R. Taylor, R. Dmi'el, M. Fedak, and K. Schmidt-Nielsen, "Energetic cost of running and hear balance in a large bird, the rhea," *Am. J. Physiol*, **221** (1971), 597–601.

(For comparison, 10 kilometers per hour is about 6.2 miles per hour.)

Velocity v	Oxygen consumption E
2	1.0
5	2.1
10	4.0
12	4.3

a. Find the equation of the regression line for E in terms of v.

b. The slope of the linear function giving oxygen consumption in terms of velocity is called the *cost of transport* for the animal, since it measures the energy required to move a unit mass by 1 unit distance. What is the cost of transport for the rhea?

c. Physiologists have determined the general approximate formula $C = 8.5W^{-0.40}$ for the cost of transport C of an animal weighing W grams. If the rhea weighs 22,000 grams, is its cost of transport from part b higher or lower than what the general formula would predict? Is the rhea a more or a less efficient runner than a typical animal its size?

d. What would your equation from part a lead you to estimate for the oxygen consumption of a rhea at rest?

3.4 SKILL BUILDING EXERCISES

Getting Regression Lines

S-1. For the following data set: (a) Plot the data. (b) Find the equation of the regression line. (c) Add the graph of the regression line to the plot of the data points.

x	1	2	3	4	5
y	2.3	2	1.8	1.4	1.3

S-2. For the following data set: (a) Plot the data. (b) Find the equation of the regression line. (c) Add the graph of the regression line to the plot of the data points.

x	1.3	2.5	3.3	4.2	5.1
y	2.6	2.6	2	1.8	1.5

S-3. For the following data set: (a) Plot the data. (b) Find the equation of the regression line. (c) Add the graph of the regression line to the plot of the data points.

x	2.3	3.7	5.1	6.4	8.2
y	4.8	5.3	7.2	9.6	10.3

S-4. For the following data set: (a) Plot the data. (b) Find the equation of the regression line. (c) Add the graph of the regression line to the plot of the data points.

x	4.1	5.7	7.3	8.9	10.5
y	7.7	8.3	8.4	8.9	9.1

S-5. For the following data set: (a) Plot the data. (b) Find the equation of the regression line. (c) Add the graph of the regression line to the plot of the data points.

x	5.2	8.9	12.6	16.3	20
y	-3.1	-4.8	-5.3	-7.1	-7.9

S-6. For the following data set: (a) Plot the data. (b) Find the equation of the regression line. (c) Add the graph of the regression line to the plot of the data points.

x	16.3	20	23.7	27.4	31.1
y	51.1	68.8	80.3	86.2	99.6

Getting Regression Lines Only

S-7. Find the equation of the regression line for the following data set.

x	1	2	3
y	3	2	1

(continued)

S-8. Find the equation of the regression line for the following data set.

x	1	2	3
y	0	3	4

S-9. Find the equation of the regression line for the following data set.

x	1	2	3
y	3	3	4

S-10. Find the equation of the regression line for the following data set.

x	1	2	3
y	3	0	3

S-11. Find the equation of the regression line for the following data set.

x	1	2	3
y	3	0	0

Correlation Coefficient The *correlation coefficient r* is a measure of how well a regression line fits data. The value of r is always between -1 and 1. If the fit is exact, the value of r is 1 or -1. Values of r near zero indicate a poor fit. Most calculators can be set to display the correlation coefficient.[29]

S-12. For the following data set, get the regression line, plot the data and the regression line on the same screen, and report the value of the correlation coefficient.

x	2	4	6	8	10
y	5	9	13	17	21

S-13. For the following data set, get the regression line, plot the data and the regression line on the same screen, and report the value of the correlation coefficient.

x	2	4	6	8	10
y	5	9	15	17	21

S-14. For the following data set, get the regression line, plot the data and the regression line on the same screen, and report the value of the correlation coefficient.

x	2	4	6	8	10
y	5	9	20	17	21

S-15. For the following data set, get the regression line, plot the data and the regression line on the same screen, and report the value of the correlation coefficient.

x	2	4	6	8	10
y	5	3	0	1	0

S-16. For the following data set, get the regression line, plot the data and the regression line on the same screen, and report the value of the correlation coefficient.

x	2	4	6	8	10
y	1	0	-1	3	2

S-17. For the following data set, get the regression line, plot the data and the regression line on the same screen, and report the value of the correlation coefficient.

x	2	4	6	8	10
y	-2	0	2	-2	0

S-18. For the following data set, get the regression line, plot the data and the regression line on the same screen, and report the value of the correlation coefficient.

x	2	4	6	8	10
y	2	4	6	8	10

[29]See Appendix B.

3.5 SYSTEMS OF EQUATIONS

Many physical problems can be described by a system of two equations in two unknowns, and often the desired information is found by *solving the system of equations.* As we shall see, this involves nothing more than finding the intersection of two lines, and we already know how to do that since we can find the intersection of *any* two graphs.

Graphical Solutions of Systems of Equations

To show the method, we look at a simple example. We have $900 to spend on the repair of a gravel drive. We want to make the repairs using a mix of coarse gravel priced at $28 per ton and fine gravel priced at $32 per ton. To make a good driving surface, we need 3 times as much fine gravel as coarse gravel. How much of each will our budget allow us to buy?

As in any *story problem*, the first step is to convert the words and sentences into symbols and equations. Let's begin by writing equations using the words *tons of coarse gravel* and *tons of fine gravel*. We have a $900 budget, and we spend $28 per ton for coarse gravel and $32 per ton for fine gravel:

$$28 \times \text{Tons of coarse gravel} + 32 \times \text{Tons of fine gravel} = 900. \tag{3.9}$$

We also know that we must use three times as much fine gravel as coarse. In other words, we multiply the amount of coarse gravel by 3 to get the amount of fine gravel:

$$\text{Tons of fine gravel} = 3 \times \text{Tons of coarse gravel.} \tag{3.10}$$

Now let's choose variable names:

$$c = \text{Amount of coarse gravel, in tons}$$

$$f = \text{Amount of fine gravel, in tons.}$$

We complete the process by putting these letter names into Equation (3.9) and Equation (3.10), and we arrive at a *system of equations* that we need to solve:

$$28c + 32f = 900$$

$$f = 3c.$$

To *solve the system* just means to find values of c and f that make *both equations true at the same time*. There are many ways to do this, but we can easily change this into a problem that we already know how to do with the calculator. The key step is first to solve each of the equations for one of the variables. Since the second equation is already solved for f, we do the same for the first equation, $28c + 32f = 900$:

$$28c + 32f = 900$$

$$32f = 900 - 28c$$

$$f = \frac{900 - 28c}{32}.$$

Thus we can replace the original system of equations by

$$f = \frac{900 - 28c}{32} = 3c.$$

The two equations are both satisfied where their graphs cross.

We want to find where *both* of these equations are true. That is, we want to find where their graphs cross. Thus we can proceed using the crossing-graphs method for solving equations that we learned in Chapter 2.

Let's recall the procedure. The first step is to enter both equations as shown in Figure 3.48. We record the appropriate variable correspondences:

$$Y_1 = \frac{900 - 28c}{32}, \text{ first expression for } f \text{ on vertical axis}$$

$$Y_2 = 3c, \text{ second expression for } f \text{ on vertical axis}$$

$$X = c, \text{ coarse gravel on horizontal axis.}$$

A commonsense estimate can help us choose our graphing range. We are buying $900 worth of gravel that costs about $30 per ton, so we will certainly buy no more than about 30 tons of either type. Thus we choose a viewing window with a horizontal span of $c = 0$ to $c = 30$ and a vertical span of $f = 0$ to $f = 30$. The properly configured window is in Figure 3.49.

Now when we graph, we get the display in Figure 3.50, where the thin line is the graph of $f = (900 - 28c)/32$ and the thick line is the graph of $f = 3c$. The solution we seek is the crossing point, which we find as we did in Chapter 2. We see from Figure 3.51 that, rounded to two decimal places, we should buy $c = 7.26$ tons of coarse gravel and $f = 21.77$ tons of fine gravel.

```
Plot1 Plot2 Plot3
\Y₁≡(900−28X)/32

\Y₂≡3X
\Y₃=
\Y₄=
\Y₅=
\Y₆=
```

FIGURE 3.48 Entering functions for the purchase of gravel

```
WINDOW
 Xmin=0
 Xmax=30
 Xscl=1
 Ymin=0
 Ymax=30
 Yscl=1
 Xres=1
```

FIGURE 3.49 Configuring the graphing window

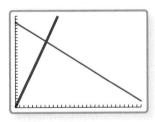

FIGURE 3.50 Graphing a system of equations for gravel

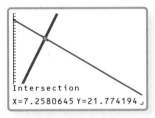

FIGURE 3.51 The solution of the system

KEY IDEA 3.7   **HOW TO SOLVE A SYSTEM OF TWO EQUATIONS IN TWO UNKNOWNS**

Step 1: Solve both equations for one of the variables.
Step 2: The solution of the system of equations is the intersection point of the two graphs. This is found using the crossing-graphs method from Chapter 2.

EXAMPLE 3.9   **A PICNIC**

We have $56 to spend on pizzas and drinks for a picnic. Pizzas cost $12 each and drinks cost $0.50 each. Four times as many drinks as pizzas are needed. How many pizzas and how many drinks will our budget allow us to buy?

Solution The cost of the picnic can be written as

$$12 \times \text{Number of pizzas} + 0.50 \times \text{Number of drinks} = \text{Cost of picnic.}$$

Since the picnic is to cost $56, we can rewrite this as

$$12 \times \text{Number of pizzas} + 0.50 \times \text{Number of drinks} = 56 \text{ dollars}.$$

If we use D for the number of drinks and P for the number of pizzas, this is

$$12P + 0.5D = 56.$$

We also know that we need four times as many drinks as pizzas. In other words, to get the number of drinks, we multiply the number of pizzas by 4:

$$\text{Number of drinks} = 4 \times \text{Number of pizzas}$$

In terms of the letters D and P, this is

$$D = 4P.$$

Thus we need to solve the system of equations

$$12P + 0.5D = 56$$
$$D = 4P.$$

We begin by solving the first equation for D. (Alternatively, it would be correct to solve each equation for P. We chose to solve for D since the work is already done for the second equation.) We have

$$0.5D = 56 - 12P$$
$$D = \frac{56 - 12P}{0.5}.$$

Now we need to use the crossing-graphs method to solve the system of equations

$$D = \frac{56 - 12P}{0.5}$$
$$D = 4P.$$

```
Plot1 Plot2 Plot3
\Y₁☰(56-12X)/0.5

\Y₂☰4X
\Y₃=
\Y₄=
\Y₅=
\Y₆=
```

FIGURE 3.52 Entering a system of equations for a picnic

We enter both functions as shown in Figure 3.52 and record the appropriate variable correspondences:

$$Y_1 = D, \text{ drinks on vertical axis}$$

$$X = P, \text{ pizzas on horizontal axis}.$$

Since pizzas cost $12 each, we can't buy more than 5 of them. And that means we will buy no more than 20 drinks. Thus, to make the graph, we set the horizontal span of $P = 0$ to $P = 5$ and the vertical span of $D = 0$ to $D = 20$. This configuration produces the graphs shown in Figure 3.53, where the thin line is the graph of $D = (56 - 12P)/0.5$, and the thick line is the graph of $D = 4P$.

We find the intersection point using the calculator, and we see from Figure 3.53 that the solution is $X = 4$ and $Y = 16$. Since X corresponds to P and Y corresponds to D, we conclude that we can buy 4 pizzas and 16 drinks.

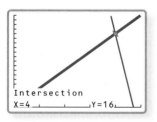

```
Intersection
X=4        Y=16
```

FIGURE 3.53 How many pizzas and drinks to buy

TEST YOUR UNDERSTANDING | **FOR EXAMPLE 3.9**

For dessert we need to buy apples and oranges. Apples cost $0.40 each, and oranges cost $0.60 each. We need 3 times as many apples as oranges, and we have $7.20 to spend on fruit. How many of each do we buy? ∎

An alternative algebraic solution to this system of equations is provided below.

◼ Algebraic Solutions

We can, if we wish, solve systems of equations without using the calculator. Let's see how to do that with the system of equations for the picnic in Example 3.9:

$$12P + 0.5D = 56$$

$$D = 4P.$$

The first step is to solve one equation for one of the variables. In general, we can use either equation and either variable. Sometimes, as in this case, a wise choice can save work. Accordingly, we avoid unnecessary work by choosing to solve the second equation for D:

$$D = 4P.$$

Now we put the expression $4P$ in place of D in the first equation:

$$12P + 0.5(4P) = 56.$$

The result is that the variable D has been *eliminated*, and we are left with a linear equation that involves only the single variable P. We learned in Section 2.3 how to solve such equations. We get

$$12P + 0.5(4P) = 56$$

$$12P + 2P = 56$$

$$14P = 56$$

$$P = \frac{56}{14}$$

$$P = 4.$$

This gives us the solution for P. We put this value for P back into the equation $D = 4P$ to get the value for D:

$$D = 4P = 4 \times 4 = 16.$$

We get the solution $P = 4$ and $D = 16$, and this agrees with our earlier solution using the calculator.

KEY IDEA 3.8   **HOW TO SOLVE A SYSTEM OF TWO EQUATIONS IN TWO UNKNOWNS ALGEBRAICALLY**

Step 1: Solve one of the equations for one of the variables.
Step 2: Put the expression you got in step 1 into the other equation.
Step 3: Solve the resulting linear equation in one variable.
Step 4: To get the other variable, put the value you got from step 3 into the expression you found in step 1.

ANSWERS FOR TEST YOUR UNDERSTANDING
3.9 We buy 12 apples and 4 oranges.

3.5 EXERCISES

Reminder Round all answers to two decimal places unless otherwise indicated.

1. **A Party** You have $36 to spend on refreshments for a party. Large bags of chips cost $2.00 and drinks cost $0.50. You need to buy five times as many drinks as bags of chips. How many bags of chips and how many drinks can you buy?

2. **Mixing Feed** A milling company wants to mix alfalfa, which contain 20% protein, and *wheat mids*,[30] which contain 15% protein, to make cattle feed.

   a. If you make a mixture of 30 pounds of alfalfa and 40 pounds of wheat mids, how many pounds of protein are in the mixture? (*Hint:* In the 30 pounds of alfalfa there are $30 \times 0.2 = 6$ pounds of protein.)

   b. Write a formula that gives the amount of protein in a mixture of a pounds of alfalfa and w pounds of wheat mids.

   c. Suppose the milling company wants to make 1000 pounds of cattle feed that contains 17% protein. How many pounds of alfalfa and how many pounds of wheat mids must be used?

3. **An Order for Bulbs** You have space in your garden for 55 small flowering bulbs. Crocus bulbs cost $0.35 each and daffodil bulbs cost $0.75 each. Your budget allows you to spend $25.65 on bulbs. How many crocus bulbs and how many daffodil bulbs can you buy?

Nicole Gordine/Shutterstock.com

4. **American Dollars and British Pounds** Assume that at the current exchange rate, the British pound is worth $1.65 in American dollars. You have some dollar bills and several British pound coins. There are 17 items altogether, which have a total value of $20.25 in American dollars. How many American dollars and how many British pound coins do you have?

5. **Population Growth** There are originally 255 foxes and 104 rabbits on a particular game reserve. The fox population grows at a rate of 33 foxes per year, and the rabbits increase at a rate of 53 rabbits per year. Under these conditions, how long does it take for the number of rabbits to catch up with the number of foxes? How many of each animal will be present at that time?

6. **Telecommunications** The accompanying table shows the operating revenue, in millions of dollars, of wired and wireless telecommunications carriers in the United States in the given year.[31]

Year	Wired	Wireless
2005	205.7	140.0
2006	195.6	157.5
2007	197.0	173.8
2008	194.8	184.8

Calculate the regression line for each type of carrier, and determine the revenue level at which the two lines cross. Round your answer for the revenue level to one decimal place.

7. **Competition Between Populations** In this exercise we consider the problem of competition between two populations that vie for resources but do not prey on each other. Let m be the size of the first population, let n be the size of the second (both measured in thousands of animals), and assume that the populations coexist eventually. An example of one common model for the interaction is

Per capita growth rate for m is $3(1 - m - n)$

Per capita growth rate for n is $2(1 - 0.7m - 1.1n)$.

At an *equilibrium point* the per capita growth rates for m and for n are both zero. If the populations reach such a point, then they will continue at that size indefinitely. Find the equilibrium point in the example above.

---

[30]When the wheat kernel is removed to produce flour, some parts of the whole wheat, including the bran, are often pressed into pellets known as *wheat mids* and sold as a feed ingredient.
[31]From the *Statistical Abstract of the United States.*

(continued)

8. **Market Supply and Demand** The quantity of wheat, in billions of bushels, that wheat suppliers are willing to produce in a year and offer for sale is called the *quantity supplied* and is denoted by S. The quantity supplied is determined by the price P of wheat, in dollars per bushel, and the relation is $P = 2.13S - 0.75$.

   The quantity of wheat, in billions of bushels, that wheat consumers are willing to purchase in a year is called the *quantity demanded* and is denoted by D. The quantity demanded is also determined by the price P of wheat, and the relation is $P = 2.65 - 0.55D$.

   At the *equilibrium price,* the quantity supplied and the quantity demanded are the same. Find the equilibrium price for wheat.

9. **Boron Uptake** Many factors influence a plant's uptake of boron from the soil, but one key factor is soil type. One experiment[32] compared plant content C of boron, in parts per million, with the amount B, in parts per million, of water-soluble boron in the soil. In *Decatur silty clay* the relation is given by $C = 33.78 + 37.5B$.

   In *Hartsells fine sandy loam* the relation is given by $C = 31.22 + 71.17B$.

   a. What amount of water-soluble boron available will result in the same plant content of boron for Decatur silty clay and Hartsells fine sandy loam? (If you choose to solve this problem graphically, we suggest a horizontal span of 0 to 0.5 for B.)

   b. For available boron amounts larger than that found in part a, which of the two soil types results in the larger plant content of boron?

10. **Male and Female High School Graduates** The table below shows the percentage of male and female high school graduates who enrolled in college within 12 months of graduation.[33]

Year	1960	1965	1970	1975
Males	54%	57.3%	55.2%	52.6%
Females	37.9%	45.3%	48.5%	49%

   a. Find the equation of the regression line for percentage of male high school graduates entering college as a function of time.

   b. Find the equation of the regression line for percentage of female high school graduates entering college as a function of time.

   c. Assume that the regression lines you found in part a and part b represent trends in the data. If the trends persisted, when would you expect first to have seen the same percentage of female and male graduates entering college? (You may be interested to know that this actually occurred for the first time in 1980. The percentages fluctuated but remained very close during the 1980s and 1990s. In the 2000s more female graduates entered college than did males. In 2008, for example, the rate for males was 66% compared with 72% for females.)

11. **Fahrenheit and Celsius** If you know the temperature C in degrees Celsius, you can find the temperature F in degrees Fahrenheit from the formula

$$F = \frac{9}{5}C + 32.$$

   At what temperature will a Fahrenheit thermometer read exactly twice as much as a Celsius thermometer?

12. **A Bag of Coins** A bag contains 30 coins, some dimes and some quarters. The total amount of money in the bag is $3.45. How many dimes and how many quarters are in the bag?

13. **Parabolic Mirrors** Reflector telescopes use *parabolic mirrors* because they have the shape that will reflect incoming light to a single point, the *focal point* (see Figure 3.54). The graph of $y = x^2$ has the shape of a parabola, for example. Light traveling vertically down (parallel to the y-axis) encounters the parabola and is reflected according to the law that the angle of incidence equals the angle of reflection. This law implies that vertical light rays encountering the parabola at a point (a, a^2) will be reflected along the line

$$4ay + (1 - 4a^2)x = a.$$

   a. Find where the light rays reflected from the points $(2, 4)$ and $(3, 9)$ meet.

---

[32]J. I. Wear and R. M. Patterson, "Effect of soil pH and texture on the availability of water-soluble boron in the soil," *SSSA Proc.* **26** (1962), 334–335.
[33]From the *Statistical Abstract of the United States.*

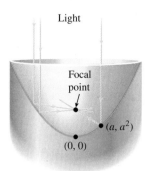

Light

Focal point

(a, a^2)

$(0, 0)$

FIGURE 3.54

b. The point you found in part a is the focal point. Show that every vertical light ray that is reflected from the mirror passes through this point.

14. **Airspeed** The *airspeed* of a plane is its speed in the absence of wind. With a headwind, *ground speed* (the actual speed in relation to the ground) is decreased by the speed of the wind. With a tailwind, ground speed is increased by the speed of the wind. Let A denote the airspeed of a plane and W the speed of the wind, both in miles per hour. Suppose it takes the plane 9 hours to travel the 720 miles from one town to another facing a headwind of W. The return trip, now with a tailwind of W, takes only 6 hours.

   a. Express the ground speed on the trip out in terms of A and W.

   b. Use the information from part a and the fact that distance equals rate times time to find an equation involving A and W for the trip out.

   c. Express the ground speed on the return trip in terms of A and W.

   d. Use the information from part c and the fact that distance equals rate times time to find an equation involving A and W for the return trip.

   e. Find the airspeed and the speed of the wind.

15. **Figuring Taxes** A business has a gross income of $1.47 million. It has promised to pay the CEO a bonus of 20% of net income, which is income after taxes. But the bonus is not subject to taxes because the bonus is an operating expense. The total tax owed is 10% of gross income less the bonus. Our goal is to find the company profit after the bonus and taxes are paid. Let B be the amount on which the bonus is based, and let T be the amount on which taxes are calculated, both in millions of dollars.

   a. Express the taxes paid in terms of the variable T.

   b. The gross income of $1.47 million equals the amount B on which the bonus is based plus the taxes paid. Express this as an equation involving T and B. (Part a may be helpful.)

   c. Express the bonus paid in terms of the variable B.

   d. The gross income also equals the amount T on which taxes are based plus the bonus paid. Express this as an equation involving T and B. (Part c may be helpful.)

   e. Solve the system of two equations in two unknowns from parts b and d for the variables T and B.

   f. How much is the bonus, how much is paid in taxes, and how much profit is left over?

16. **An Interesting System of Equations** What happens when you try to solve the following system of equations? Can you explain what is going on?
$$x + y = 1$$
$$x + y = 2$$

17. **Another Interesting System of Equations** What happens when you try to solve the following system of equations? Can you explain what is going on?
$$x + 2y = 3$$
$$-2x - 4y = -6$$

18. **A System of Three Equations in Three Unknowns** Consider the following system of three equations in three unknowns.
$$2x - y + z = 3$$
$$x + y + 2z = 9$$
$$3x + 2y - z = 4$$

   a. Solve the first equation for z.

   b. Put the solution you got in part a for z into *both* the second and third equations.

   c. Solve the system of two equations in two unknowns that you found in part b.

   d. Write the solution of the original system of three equations in three unknowns.

19. **An Application of Three Equations in Three Unknowns** A bag of coins contains nickels, dimes, and quarters. There are a total of 21 coins in the bag, and the total amount of money in the bag is $3.35. There is one more dime than there are nickels. How many dimes, nickels, and quarters are in the bag?

(continued)

3.5 SKILL BUILDING EXERCISES

S-1. An Explanation Explain why the solution to a system of two equations in two unknowns is the intersection point of the two graphs.

S-2. What Is the Solution? For a certain system of two linear equations in two unknowns, the graphs are distinct parallel lines. What can you conclude about the solution of the system of equations?

Systems That May Be Solved by Hand or with Technology The systems of equations in Exercises S-3 through S-18 may be solved by hand calculation or by using the crossing graphs method. Solve by the method requested.

S-3. Solve the following system using the crossing-graphs method.

$$x + y = 5$$
$$x - y = 1$$

S-4. Solve the system of equations from Exercise S-3 by hand calculation.

S-5. Solve the following system using the crossing-graphs method.

$$2x - y = 0$$
$$3x + 2y = 14$$

S-6. Solve the system of equations from Exercise S-5 by hand calculation.

S-7. Solve the following system using the crossing-graphs method.

$$3x + 2y = 6$$
$$4x - 3y = 8$$

S-8. Solve the system of equations from Exercise S-7 by hand calculation.

S-9. Solve the following system using the crossing-graphs method.

$$3x + y = 0$$
$$x - y = 0$$

S-10. Solve the system of equations from Exercise S-9 by hand calculation.

S-11. Solve the following system using the crossing-graphs method.

$$x - y = 1$$
$$3x + 2y = 8$$

S-12. Solve the system of equations from Exercise S-11 by hand calculation.

S-13. Solve the following system using the crossing-graphs method.

$$x + y = 1$$
$$2x + 3y = 0$$

S-14. Solve the system of equations from Exercise S-13 by hand calculation.

S-15. Solve the following system using the crossing-graphs method.

$$3x + 4y = 6$$
$$2x - 6y = 5$$

S-16. Solve the system of equations from Exercise S-15 by hand calculation.

S-17. Solve the following system using the crossing-graphs method.

$$-7x + 21y = 79$$
$$13x + 17y = 6$$

S-18. Solve the system of equations from Exercise S-17 by hand calculation.

Systems Not Easily Solved by Hand Calculation Exercises S-19 through S-23 involve systems of equations that are not easily solved by hand calculation.

S-19. Solve the following system using the crossing-graphs method.

$$0.7x + 5.3y = 6.6$$
$$5.2x + 2.2y = 1.7$$

S-20. Solve the following system using the crossing-graphs method.

$$-6.6x - 26.5y = 17.1$$
$$6.9x + 5.5y = 8.4$$

S-21. Solve the following system using the crossing-graphs method.

$$2.1x + 3.3y = -1.43$$
$$2.2x - 1.1y = 3.78$$

S-22. Solve the following system using the crossing-graphs method.

$$3.7x - 4.3y = 9.88$$
$$6.5x + 7.3y = 64$$

S-23. Solve the following system using the crossing-graphs method.

$$-4.1x + 5.2y = 29.74$$
$$6.8x - 3.9y = -36.23$$

CHAPTER 3   SUMMARY

One of the simplest and most important geometric objects is the straight line, and the line is intimately tied to the idea of a *linear function*. This is a function with a constant rate of change, and the constant is known as the *slope*.

3.1   THE GEOMETRY OF LINES

An important characterization of a straight line such as that determined by a roof is that it rises or falls at the same rate everywhere. This rate is known as the *slope* of the line and is commonly denoted by the letter *m*. Critical relationships involving the slope are summarized below.

Calculating the slope: The slope *m* can be calculated using

$$m = \frac{\text{Vertical change}}{\text{Horizontal change}}.$$

Fundamental property of the slope: The slope of a line tells how steeply it is increasing or decreasing.

- If *m* is positive, the line is rising from left to right.
- If $m = 0$, the line is horizontal.
- If *m* is negative, the line is falling from left to right.

Using the slope: The slope *m* can be used to calculate vertical change.

$$\text{Vertical change} = m \times \text{Horizontal change}.$$

3.2   LINEAR FUNCTIONS

A linear function is one with a constant rate of change. As with lines, this is known as the slope and is denoted by *m*. Linear functions can be written in a special form:

$$y = mx + b,$$

where *m* is the slope and *b* is the *initial value*. The graph of a linear function is a straight line. The slope of the linear function is the same as the slope of the line, and the initial value of the function is the same as the vertical intercept of the line.

The slope of a linear function is used much like the slope of a line.

- The slope *m* can be calculated from the change in *y* corresponding to a given change in *x*:

$$m = \frac{\text{Change in function}}{\text{Change in variable}} = \frac{\text{Change in } y}{\text{Change in } x}.$$

- The slope can be used to calculate the change in *y* resulting from a given change in *x*:

$$\text{Change in } y = m \times \text{Change in } x.$$

For example, the income tax liability *T* is, over a suitably restricted range of incomes, a linear function of the adjusted gross income *I*. Suppose the tax table shows a tax liability of $T = 780$ dollars when the adjusted gross income is $I = 15{,}000$ dollars and shows a tax liability of $825 for an adjusted gross income of $15,500.

- We can calculate the slope of the linear function $T = T(I)$ as follows:

$$m = \frac{\text{Change in tax}}{\text{Change in income}} = \frac{825 - 780}{15{,}500 - 15{,}000} = \frac{45}{500} = 0.09.$$

This means that a taxpayer in this range could expect to pay 9 cents tax on each additional dollar earned. Economists would term this the *marginal tax rate*.

- Once the slope, or marginal tax rate, is known, one can calculate the tax liability for other incomes. If the adjusted gross income is \$15,350, the additional tax over that owed on \$15,000 is given by

Additional tax $= m \times$ Additional income $= 0.09 \times 350 = 31.50$ dollars.

Thus the total tax liability is $780 + 31.50 = 811.50$ dollars.

3.3 MODELING DATA WITH LINEAR FUNCTIONS

It is appropriate to model observed data with a linear function, provided that the data show a constant rate of change. When the data for the variable are evenly spaced, this is always evidenced by a constant change in function values. For example, if a rock is dropped and its velocity is recorded each second, one might collect the following data:

t = seconds	0	1	2	3	4	5
V = feet per second	0	32	64	96	128	160

To see whether it is appropriate to model velocity with a linear function, we look at successive differences:

Change in t	From 0 to 1	From 1 to 2	From 2 to 3	From 3 to 4	From 4 to 5
Change in V	32	32	32	32	32

Since the change in V is always the same, we conclude that V is a linear function of t. We can get a formula for this linear function by first calculating the slope:

$$m = \frac{\text{Change in } V}{\text{Change in } t} = \frac{32}{1} = 32,$$

so then

$$V = \text{Slope} \times t + \text{Initial value} = 32t + 0 = 32t.$$

3.4 LINEAR REGRESSION

Experimentally gathered data are always subject to error, and rarely will data points show an exactly constant rate of change. When appropriate, we can still model the data with a linear function by using *linear regression*. Linear regression gives the line with the least possible total of the squares of the vertical distances from the line to the data points. It is by far the most commonly used method of approximating linear data, and it is nearly always accomplished with a computer or calculator.

The Medicare expenditures $E = E(t)$ of the federal government, in billions of dollars, are given by the following table. Here t is time in years since 2004.

t = years since 2004	0	1	2	3	4
E = billions spent	269.4	298.6	329.9	375.4	390.8

The data table does not show a constant rate of change, but the plot of the data in Figure 3.55 shows that the data points nearly fall on a straight line. If we wish to model this with a linear function, we use the calculator to get the regression line $E = 31.96t + 268.90$. In Figure 3.56 we have added the regression line to the data points to show how the approximate model works.

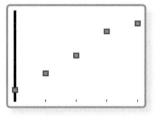

FIGURE 3.55 Medicare spending

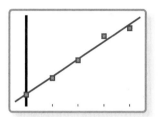

FIGURE 3.56 Regression line model for Medicare spending

The regression line shows the linear trend, if any exists, exhibited by observed data. Extreme caution should be exercised in any use of the regression line beyond the range of observed data points.

3.5 SYSTEMS OF EQUATIONS

Many physical problems are described by a system of two linear equations in two unknowns, and often the needed information is obtained by *solving the system of equations*. This can be accomplished either graphically or by hand calculation. The graphical solution is the easier way and fits nicely with the spirit of this course. The procedure is as follows.

Step 1: Solve both equations for one of the variables.

Step 2: The solution of the system is the intersection point of the two graphs. This is found using the crossing-graphs method.

For example, the purchase of c tons of coarse gravel and f tons of fine gravel might be described by the following system of equations:

$$28c + 32f = 900$$

$$f = 3c.$$

We solve the first equation for f to obtain the equivalent system

$$f = \frac{900 - 28c}{32}$$

$$f = 3c.$$

If we graph each of these functions and find the crossing point, we discover that we should buy $c = 7.26$ tons of coarse gravel and $f = 21.77$ tons of fine gravel.

CHAPTER 3   REVIEW EXERCISES

Reminder   Round all answers to two decimal places unless otherwise indicated.

1. **Drainage Pipe Slope**   Suppose a drainage pipe slopes downward 6 inches for each 8-foot horizontal stretch. What is the slope of the pipe?

2. **Height from Slope and Horizontal Distance**   A large ladder leans against a house with the base of the ladder 4.4 horizontal feet from the base of the house. If the ladder has a slope of 3.7 feet per foot, at what height does the top of the ladder meet the house?

3. **A Ramp into a Van**   Suppose a ramp is used to move a heavy object into a van. Suppose also that the van door is 18 inches off the ground.

   a. If the bottom of the ramp rests on the ground 4 feet from the van, what is the slope of the ramp?

   b. If the ramp cannot have a slope of more than 1.2 inches per foot, how far from the van should the ramp rest?

4. **A Reception Tent**   A circular reception tent has a center pole 25 feet high, and the poles along the outside are 10 feet high. Assume that the distance from the outside poles to the center pole is 30 feet.

   a. What is the slope of the line that follows the roof of the reception tent?

   b. How high is the tent 5 feet in from the outside poles?

   c. Ropes are used to stabilize the tent following the line of the roof of the tent to the ground. How far away from the outside poles are the ropes attached to the ground?

5. **Function Value from Slope and Run**   Suppose g is a linear function such that $g(3.7) = 5.1$. If the slope of g is -2.2, what is $g(6.8)$?

6. **Lanes on a Curved Track**   On a curved track, the lanes are arcs of circles, but each with a different radius. For a typical 100-meter curved track with several lanes, the inner radius of the nth lane is

$$R(n) = \frac{100}{\pi} + 1.22(n - 1)$$

   in meters.

   a. What is the radius of the first lane?

   b. What is the width of each lane?

   c. It is more difficult to run in a lane with a small radius. If you wish to run in a lane with a radius of at least 35 meters, which lane should you pick?

7. **Linear Equation from Two Points**   Suppose g is a linear function such that $g(1.1) = 3.3$ and $g(-2) = 7.2$. Find the equation for g.

8. **Working on a Commission**   A certain man works in sales and earns a base salary of $1000 per month plus 5% of his total sales for the month.

   a. Explain why his total monthly income I is a linear function of total sales S, both measured in dollars.

   b. How much does he earn if he sells $1600 in merchandise in a month?

   c. Write a formula that gives total monthly income as a linear function of sales in a month.

   d. What should his monthly total sales be if he wishes to earn $1350 this month?

9. **Testing Data for Linearity**   Consider the following data.

x	3	3.3	3.6	3.9
f	8	7.4	6.8	6.2

   a. Test the data to see whether they are linear.

   b. Make a linear model for the data.

10. **An Epidemic**   During a certain flu epidemic, the total number T of patients diagnosed with the flu after d days is given in the table below.

d	0	5	10	15	20
T	35	41	47	53	59

    a. Check differences to show that these are linear data.

    b. What is the practical meaning of the slope?

    c. Find the slope.

    d. Use a formula to express T as a linear function of d.

    e. What would you expect to be the total number of diagnosed flu cases after 17 days?

11. **Testing and Plotting Linear Data** Consider the following data.

x	f
−1	−1.12
1	−0.88
3	−0.64
5	−0.40

a. Test the data to see whether they are linear.
b. Make a linear model for the data.
c. Plot the data together with the linear model.

12. **Marginal Tax Rate** The following table shows tax due for the given taxable income level for a single taxpayer.

Taxable income	Tax due
$97,000	$21,913
$97,050	$21,927
$97,100	$21,941
$97,150	$21,955
$97,200	$21,969

a. Show that the data in the table are linear.
b. How much additional tax is due on each dollar over $97,000?
c. What would you expect for your tax due if you had a taxable income of $97,000? of $98,000?
d. Find a linear formula that gives tax due if your income is A dollars over $97,000.

13. **Plotting Data and Regression Lines** For the following data set: (a) Plot the data. (b) Find the equation of the regression line. (c) Add the regression line to the plot of the data points.

x	y
1.1	19.3
1.2	17.2
1.3	16.4
1.4	14.7
1.5	12.4

14. **Meaning of Slope of Regression Line** The slope of the regression line of the number of births per thousand to unmarried 18- to 19-year-old women for various years as a function of year is −1.13. Explain in practical terms what this means.

15. **Life Expectancy** The following table shows the average life expectancy, in years, of a child born in the given year.[34]

Year	Life expectancy
2003	77.1
2004	77.5
2005	77.4
2006	77.7
2007	77.9

a. Find the equation of the regression line, and explain the meaning of its slope.
b. Plot the data points and the regression line.
c. Explain in practical terms the meaning of the slope of the regression line.
d. Based on the trend of the regression line, what do you predict as the life expectancy of a child born in 2013?
e. Based on the trend of the regression line, what do you predict as the life expectancy of a child born in 1580? 2300?

16. **XYZ Corporation Stock Prices** The following table shows the average stock price, in dollars, of XYZ Corporation in the given month.

Month	Stock price
January 2011	$43.71
February 2011	$44.22
March 2011	$44.44
April 2011	$45.17
May 2011	$45.97

a. Find the equation of the regression line. (Round the regression coefficients to three decimal places.)

[34]From the *Statistical Abstract of the United States.*

b. Plot the data points and the regression line.

c. Explain in practical terms the meaning of the slope of the regression line.

d. Based on the trend of the regression line, what do you predict the stock price to be in January 2012? January 2013?

17. Crossing Graphs   Solve using crossing graphs.

$$4x - 2y = 9$$
$$x + y = 0$$

18. Hand Calculation   Solve the system of equations in Exercise 17 by hand calculation.

19. Bills   A stack contains $400 worth of paper money consisting of $20 bills and $50 bills. There are 11 bills altogether. How many $20 bills and how many $50 bills are in the stack?

20. Mixing   You're mixing blue paint with yellow paint to get a total of 10 gallons of the mixture. You want to use 3 times as much yellow paint as blue paint. How many gallons of each should you use?

A FURTHER LOOK

Parallel and Perpendicular Lines

Recall that the slope of the line $y = mx + b$ is m. The slope of a line determines the direction in which it points. It is then not surprising to note that lines with the same slope are parallel or coincident. In Figure 3.57 we show the graphs of $y = 2x + 1$ and $y = 2x + 4$. They have the same slope, 2, and indeed appear to be parallel. Note that we get the upper graph by shifting the lower graph up 3 units.

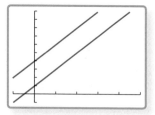

FIGURE 3.57 Parallel lines
resulting from the same slope

For perpendicular lines, the relationship is a bit more complicated. The notion of *similarity* from high school geometry can be used to show that lines $y = m_1x + b$ and $y = m_2x + b$ are perpendicular if and only if

$$m_1 = -\frac{1}{m_2}.$$

In words, lines are perpendicular exactly when their slopes are negative reciprocals.

For example, the line $y = 4x + 5$ is perpendicular to the line $y = -\frac{1}{4}x + 7$ because the slope of the second line, $-\frac{1}{4}$, is the negative reciprocal of the slope, 4, of the first line.

EXAMPLE 3.10 **FINDING PARALLEL AND PERPENDICULAR LINES**

Consider the line L whose equation is $y = 2x + 4$.

Part 1 Find the equation of the line that is parallel to L and passes through the point (3, 11).

Part 2 Find the equation of the line that is perpendicular to L and passes through the point (6, 4).

Solution to Part 1 Because the line we want is parallel to L, it has the same slope as L. That slope is 2. Hence the desired equation has the form $y = 2x + b$. We need to find b.

The fact that the line passes through the point (3, 11) tells us that when $x = 3$, y is equal to 11. We put these two values into the equation to find b:

$$y = 2x + b$$

$$11 = 2 \times 3 + b$$

$$11 = 6 + b$$

$$5 = b.$$

Thus $b = 5$, and the equation of the line is $y = 2x + 5$.

Solution of Part 2 The line L has slope 2. Hence the slope of the line we want is the negative reciprocal of 2. That is $-1/2$. So the equation of the line we seek has the form $y = -\frac{1}{2}x + b$. We need to find b. We proceed as we did in part 1. We know that when $x = 6$, y is 4:

$$y = -\frac{1}{2}x + b$$

$$4 = -\frac{1}{2} \times 6 + b$$

$$4 = -3 + b$$

$$7 = b.$$

Thus $b = 7$, and the equation of the perpendicular line is $y = -\frac{1}{2}x + 7$.

EXERCISES

Reminder Round all answers to two decimal places unless otherwise indicated.

Parallel and Perpendicular Lines In Exercises 1 through 4, calculate the slopes of the given lines to determine whether they are parallel, perpendicular, or neither.

1. Line l_1 passes through (1, 2) and (5, 10); line l_2 passes through (1, 3) and (7, 15). Determine whether l_1 and l_2 are parallel, perpendicular, or neither.

2. Line l_1 passes through (1, 2) and (5, 10); line l_2 passes through (1, 3) and (3, 2). Determine whether l_1 and l_2 are parallel, perpendicular, or neither.

3. Line l_1 passes through (1, 2) and (5, 10); line l_2 passes through (1, 3) and (3, 4). Determine whether l_1 and l_2 are parallel, perpendicular, or neither.

4. Line l_1 passes through (1, 2) and (5, 10); line l_2 passes through (1, 3) and (−5, 15). Determine whether l_1 and l_2 are parallel, perpendicular, or neither.

5. **Slope of Perpendicular Lines** The lines l_1 and l_2 are perpendicular. The slope of l_1 is $-\frac{5}{8}$. Find the slope of l_2.

6. **Finding Parallel Lines** Find the equation of the line parallel to $y = 3x + 4$ and passing through the point (1, 5).

7. **Finding Parallel Lines** Find the equation of the line parallel to $y = 5x + 3$ and passing through the point (−2, 15).

8. **Finding Perpendicular Lines** Find the equation of the line perpendicular to $y = 3x + 1$ that passes through the point (9, 1).

9. **Finding Perpendicular Lines** Find the equation of the line perpendicular to $y = -4x + 5$ that passes through the point (12, 6).

A FURTHER LOOK

Secant Lines

For graphs, the average rate of change has an important geometric meaning that we explore here. The straight line joining two points on a graph is known as a *secant line* to the graph. In Figure 3.58 we have marked two points on a graph, and in Figure 3.59 we have added the corresponding secant line.

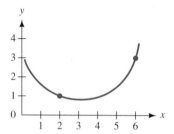

FIGURE 3.58 A graph with two points singled out

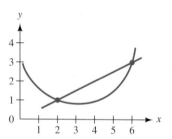

FIGURE 3.59 The secant line added

The secant line in Figure 3.59 passes through the points (2, 1) and (6, 3). We can use these two points to find the slope of the secant line:

$$\text{Slope} = \frac{\text{Vertical change}}{\text{Horizontal change}}$$

$$= \frac{3 - 1}{6 - 2}$$

$$= \frac{1}{2}.$$

It is important to observe that this slope is also the average rate of change from $x = 2$ to $x = 6$.

Thus the equation of the line is $y = \frac{1}{2}x + b$. We can use the fact that the line passes through (2, 1) to find b. When $x = 2$, $y = 1$:

$$y = \frac{1}{2}x + b$$

$$1 = \frac{1}{2} \times 2 + b$$

$$1 = 1 + b$$

$$0 = b.$$

Hence the equation of the secant line is $y = \frac{1}{2}x$.

Once more, it is important to note that the slope of a secant line is also an average rate of change for the underlying function.

EXAMPLE 3.11 FINDING SLOPES OF SECANT LINES

Find the equation of the secant line for the graph of $y = x^2$ from $x = 2$ to $x = 4$. See Figure 3.60.

Solution The slope of the secant line is the same as the average rate of change for $f(x) = x^2$ from $x = 2$ to $x = 4$:

$$\text{Slope} = \text{Average rate of change}$$

$$= \frac{4^2 - 2^2}{4 - 2}$$

$$= 6.$$

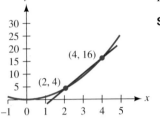

FIGURE 3.60 Figure for Example 3.11

Thus $y = 6x + b$. When $x = 2$, $y = 2^2 = 4$. We can use these values to find the equation of the secant line:

$$y = 6x + b$$

$$4 = 6 \times 2 + b$$

$$4 = 12 + b$$

$$-8 = b.$$

The equation of the secant line is $y = 6x - 8$.

TEST YOUR UNDERSTANDING │ FOR EXAMPLE 3.11

Show that the equation of the secant line for the graph of $y = \sqrt{x}$ from $x = 100$ to $x = 121$ is $y = \frac{1}{21}x + \frac{110}{21}$. See Figure 3.61.

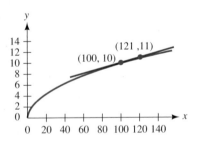

FIGURE 3.61 Figure for TEST YOUR UNDERSTANDING 3.11

Using average rates of change to approximate function values as we did in Section 1.2 is the same as approximating the graph by its secant line. For example, in TEST YOUR UNDERSTANDING 3.11, we saw that the equation of the secant line for $f(x) = \sqrt{x}$ from $x = 100$ to $x = 121$ is

$$y = \frac{1}{21}x + \frac{110}{21}.$$

Note how closely the secant line follows the graph of $y = \sqrt{x}$, at least from $x = 100$ to $x = 121$. We can use this fact to good effect. Suppose we needed to calculate $\sqrt{110}$ but had no calculator available. We could approximate the value using the secant line. To do so, we put 110 in place of x in the equation for the secant line, $y = \frac{1}{21}x + \frac{110}{21}$:

$$\sqrt{110} \approx \frac{1}{21} \times 110 + \frac{110}{21}$$

$$= \frac{220}{21},$$

or about 10.476. If we use a calculator to compute $\sqrt{110}$, we find 10.488 to three decimal places. Our approximation using the secant line is very close in this case. Using secant lines for approximations is the same as using average rates of change. Hence the results are best when a relatively short interval is used.

The graph in Figure 3.59 is concave up, and it is easily seen that for such a graph, the secant line lies above the graph between the points (2, 1) and (6, 3) and below the graph outside those points. Thus, when a graph is concave up, approximating function values using average rates of change gives an answer that is too big on the interval between the points used and is too small outside that interval. You will be asked in the exercises to analyze what happens for a graph that is concave down.

The graph in Figure 3.59 also shows why approximation using the average rate of change works best when we are dealing with points that are relatively close. The secant line between $x = 2$ and $x = 3$ lies much closer to the graph (at least between those two points) than does the secant line between $x = 2$ and $x = 6$.

If we put our finger on the mark at (6, 3) and move the marker along the graph back toward the point (2, 1), the secant line changes to follow our movement. If we push all the way back to (2, 1), the resulting line is known as the *tangent line,* and the slope of the tangent line is the *instantaneous rate of change,* or *derivative,* of the function at (2, 1). This is a critical idea in calculus. ∎

ANSWERS FOR TEST YOUR UNDERSTANDING

3.11 The slope is $\frac{11 - 10}{121 - 100} = \frac{1}{21}$. To show that the initial value is $b = 110/21$, put in $x = 100, y = 10$.

EXERCISES

Reminder Round all answers to two decimal places unless otherwise indicated.

1. Secant Lines Find the equation of the secant line for the graph of $y = x^3$ from $x = 1$ to $x = 3$. See Figure 3.62.

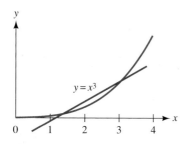

FIGURE 3.62 Figure for Exercise 1

2. Secant Lines Find the equation of the secant line for the graph of $y = \sqrt{x}$ from $x = 4$ to $x = 9$. See Figure 3.63.

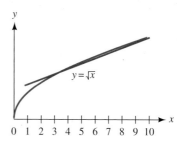

FIGURE 3.63 Figure for Exercise 2

3. Secant Lines Find the equation of the secant line for the graph of $y = x^2 + 1$ from $x = 2$ to $x = 3$.

(continued)

4. **Secant Lines**   Find the equation of the secant line for the graph of $y = \dfrac{15}{x}$ from $x = 3$ to $x = 5$.

5. **Secant Lines**   Find the equation of the secant line for the graph of $y = x^2 + x$ from $x = 4$ to $x = 6$.

6. **Approximations:**

   a. Find the equation of the secant line to the graph of $f(x) = x^2/2$ from $x = 2$ to $x = 6$.

   b. Use the secant line you found in part a to approximate the value of $f(3)$.

7. **Approximations:**

   a. Find the equation of the secant line to the graph of $f(x) = 1 - x^3$ from $x = 2$ to $x = 4$.

   b. Use the secant line you found in part a to approximate the value of $f(3)$.

8. **Secant Lines for Graphs That Are Concave Down**   Sketch a graph that is concave down, and add a secant line.

   a. Where is the secant line above the graph, and where is it below the graph?

   b. If you use an average rate of change to estimate function values for a graph that is concave down, where will the estimate be too large, and where will it be too small?

9. **Which Is Best?**   Figure 3.64 shows secant lines marked 1, 2, and 3 and a function value marked

$f(a)$. Which of the three secant lines would give the best approximation for $f(a)$?

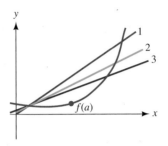

FIGURE 3.64 Figure for Exercise 9

10. **Difference Quotient**   Calculate the slope of the secant line from a to x for $f(x) = x^2$. *Note:* This is often called a *difference quotient*.

11. **Derivative**   *This exercise uses the result of Exercise 10.* We find the derivative of f at a by taking the slope of the secant line from a to x and then letting x go to a. What is the derivative of $f = x^2$ at a?

12. **Difference Quotient**   Calculate the slope of the secant line from $x = a$ to $x = a + h$ for $g(x) = x^2 + 1$. *Note:* This is often called a *difference quotient*.

13. **Derivative**   *This exercise uses the result of Exercise 12.* We find the derivative of g at a by taking the slope of the secant line from $x = a$ to $x = a + h$ and then letting h go to 0. What is the derivative of $g = x^2 + 1$ at a?

EXPONENTIAL FUNCTIONS 4

Len DeLessio/Getty Images

Live 8/Getty Images Entertainment/
Getty Images

When a band such as U2 performs, the loudness is often measured in decibels, which are defined using exponential and logarithmic functions. See Example 4.13 on page 297.

EXPONENTIAL FUNCTIONS occur almost as often as linear functions in nature, science, and mathematics. We have already encountered examples of exponential functions in looking at population growth, radioactive decay, free fall subject to air resistance, and interest on bank loans. We will now look more closely at the structure of such functions and show where the formulas we presented in those applications originated.

Student resources are available on the website
www.cengagebrain.com

4.1 EXPONENTIAL GROWTH AND DECAY

Recall that a linear function $y = y(x)$ with slope m is one that has a constant rate of change m. Another way of saying this is that a linear function changes by constant sums of m. That is, when x is increased by 1, we get the new value of y by adding m to the old value of y. In contrast, an *exponential function* $N = N(t)$ with *base a* is one that changes by constant *multiples* of a. That is, when t is increased by 1, we get the new value of N by multiplying the current N value by a.

 Exponential Growth

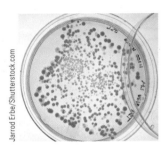

Exponential functions occur naturally in some population growth analyses, and we consider the following biology experiment to provide an illustration. There are initially 3000 bacteria in a petri dish, and the population doubles each hour. We can calculate the number of bacteria present after each hour.

Hour	Number of bacteria
0	3000
1	$2 \times 3000 = 6000$
2	$2 \times 6000 = 12{,}000$
3	$2 \times 12{,}000 = 24{,}000$

Hour "0" is when the → experiment begins. The number of bacteria at this time is the *initial value*.

After each hour, the number of bacteria in the petri dish is multiplied by 2.

In general, we let $N = N(t)$ denote the number of bacteria present after t hours. We can get the population N at 1 hour in the future by multiplying the current N value by 2. This means that N is an exponential function with base 2. When the base of an exponential function is greater than 1, we sometimes refer to the base as the *growth factor*. In this case, we would say that the bacteria exhibit *exponential growth* with an hourly growth factor of 2 and an initial value of 3000.

Change by constant multiples always results in an exponential function.

If we calculate several values of the function N for bacteria that was described above and write the answer in a special form, we can see how to get a formula for N:

$$N(0) = 3000 = 3000 \times 2^0$$

$$N(1) = 3000 \times 2 = 3000 \times 2^1$$

$$N(2) = 3000 \times 2 \times 2 = 3000 \times 2^2$$

$$N(3) = 3000 \times 2 \times 2 \times 2 = 3000 \times 2^3$$

$$N(4) = 3000 \times 2 \times 2 \times 2 \times 2 = 3000 \times 2^4.$$

If we get the next hour's amount by multiplying by 2, then after t hours the amount will be the initial amount times 2^t.

The pattern should be evident. To get the population N after t hours, we multiply the initial amount by the hourly growth factor 2 a total of t times, and that is the same as multiplying the initial population by 2^t:

$$N = 3000 \times 2^t.$$

This formula makes it easy to calculate the number of bacteria present at any time t. For example, after 6 hours there are $N(6) = 3000 \times 2^6 = 192{,}000$ bacteria present.

Jarrod Erbe/Shutterstock.com

Or after $1\frac{1}{2}$ hours there are $N(1.5) = 3000 \times 2^{1.5} = 8485$ bacteria present. (Here we have rounded our answer to the nearest integer because we don't expect to see parts of bacteria in the petri dish.)

The formula $N = 3000 \times 2^t$ that we obtained is typical of exponential functions in general. The formula for an exponential function $N = N(t)$ with base a and initial value P is

$$N = Pa^t.$$

Exponential Decay

Let's look at the bacteria experiment again, but this time let's suppose that an antibiotic has been introduced into the petri dish, so that rather than growing in number, each hour half of the bacteria die. Under these conditions there will be only 1500 bacteria left after 1 hour. After another hour, there will be 750 bacteria left. In general, we can get the number of bacteria present 1 hour in the future by multiplying the present population by $\frac{1}{2}$. Thus, under these conditions, N is an exponential function with base $\frac{1}{2}$. Because the population is decreasing rather than increasing, we would say that the population exhibits *exponential decay* with an hourly *decay factor* of $\frac{1}{2}$. Just as we did in the case where the population was growing, we can show a pattern that will lead us to a formula for $N = N(t)$:

$$N(0) = 3000 = 3000 \times \left(\frac{1}{2}\right)^0$$

$$N(1) = 3000 \times \frac{1}{2} = 3000 \times \left(\frac{1}{2}\right)^1$$

$$N(2) = 3000 \times \frac{1}{2} \times \frac{1}{2} = 3000 \times \left(\frac{1}{2}\right)^2$$

$$N(3) = 3000 \times \frac{1}{2} \times \frac{1}{2} \times \frac{1}{2} = 3000 \times \left(\frac{1}{2}\right)^3$$

$$N(4) = 3000 \times \frac{1}{2} \times \frac{1}{2} \times \frac{1}{2} \times \frac{1}{2} = 3000 \times \left(\frac{1}{2}\right)^4.$$

In general, we see that $N = 3000 \times \left(\frac{1}{2}\right)^t$. This is an exponential function with initial value 3000 and base, or hourly decay factor, $\frac{1}{2}$.

In Figure 4.1 we have made the graph of the exponential function 3000×2^t. A table of values led us to choose a viewing window with a horizontal span of $t = 0$ to $t = 5$ and a vertical span of $N = 0$ to $N = 100{,}000$. The graph is increasing and concave up, and this is typical of exponential growth. Growth is slow at first, but it later becomes more rapid. In Figure 4.2 we have graphed $3000 \times \left(\frac{1}{2}\right)^t$ using a horizontal span of $t = 0$

> Exponential growth may initially be slow, but eventually it is quite rapid.

> Exponential decay may be rapid at first but eventually slows to near zero.

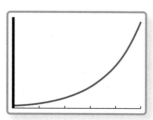

FIGURE 4.1 Exponential growth

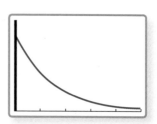

FIGURE 4.2 Exponential decay

to $t = 5$ and a vertical viewing span of $N = 0$ to $N = 3500$. This graph is decreasing and concave up, and it shows typical behavior for exponential decay. Decay is initially rapid, but the decay rate later slows.

KEY IDEA 4.1   **EXPONENTIAL FUNCTIONS**

A function $N = N(t)$ is exponential with base a if N changes in constant multiples of a. That is, if t is increased by 1, the new value of N is found by multiplying by a.

1. The formula for an exponential function with base a and initial value P is

$$N = Pa^t.$$

2. If $a > 1$, then N shows exponential growth with *growth factor a*. The graph of N will be similar in shape to that in Figure 4.1.

3. If $a < 1$, then N shows exponential decay with *decay factor a*. The graph of N will be similar in shape to that in Figure 4.2. The limiting value of such a function is 0.

EXAMPLE 4.1   **RADIOACTIVE DECAY**

Radioactive substances decay over time, and the rate of decay depends on the element. If, for example, there are G grams of tritium (a radioactive form of hydrogen) in a container, then as a result of radioactive decay, 1 year later there will be $0.945G$ grams of tritium left. Suppose we begin with 50 grams of tritium.

Part 1 Use a formula to express the number G of grams left after t years as an exponential function of t. Identify the base, or yearly decay factor, and the initial value.

Part 2 How much tritium is left after 5 years?

Part 3 Plot the graph of $G = G(t)$ over the first 30 years and describe in words how the amount of tritium that is present changes with time.

Part 4 How long will it take for half of the tritium to decay?

Solution to Part 1 The initial value, 50 grams is given. We are also told that to find the amount G of tritium 1 year in the future we multiply the present amount by 0.945. Thus $G = G(t)$ is an exponential function with initial value 50 grams, and base, or yearly decay factor, 0.945. We conclude that the formula for G is given by

$$G = \text{Initial value} \times (\text{Yearly decay factor})^t$$
$$G = 50 \times 0.945^t.$$

Solution to Part 2 To find the amount of tritium left after 5 years, we use the formula above to calculate $G(5)$:

$$G(5) = 50 \times 0.945^5 = 37.68 \,\text{grams}.$$

Solution to Part 3 We made the graph shown in Figure 4.3 using a horizontal span of $t = 0$ to $t = 30$ and a vertical span of $G = 0$ to $G = 60$. You may wish to look at a table of values to see why we chose these settings. We note that in this graph the correspondences are

$$\mathsf{X} = t, \text{ time in years on horizontal axis}$$
$$\mathsf{Y} = G, \text{ amount left in grams on vertical axis.}$$

To find next year's amount from this year's amount we multiply by 0.945, so after t years the amount is 0.945^t times the initial value.

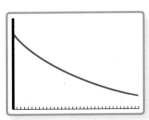

FIGURE 4.3 The decay of tritium

USGS

The features of this graph are typical of exponential decay. The amount of tritium decreases rapidly over the first few years, but after that the rate of decrease slows.

Solution to Part 4 We want to know when there will be 25 grams of tritium left. That is, we want to solve the equation

$$50 \times 0.945^t = 25$$

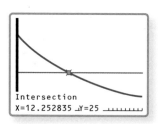

Intersection
X=12.252835 Y=25

FIGURE 4.4 The half-life of tritium

for t. We can do this using the crossing-graphs method. In Figure 4.4 we have added the graph of $G = 25$ and used the calculator to find the intersection point shown in Figure 4.4. We see that there will be 25 grams of tritium left after about 12.25 years. This number is known as the *half-life* of tritium because it tells how long it takes for half of the amount to decay. It is a fact that the half-life of a radioactive substance does not depend on the initial amount. Thus we would have found the same answer, 12.25 years, if we had started with 1000 grams of tritium and asked how long it took until only 500 grams were left. See Exercise 7 at the end of this section.

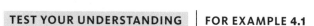

TEST YOUR UNDERSTANDING | **FOR EXAMPLE 4.1**

Polonium-210 is another radioactive substance. To find next week's amount of polonium-210 we multiply this week's amount by 0.97. Suppose there are initially 10 grams of polonium-210. Find a formula that gives the amount P, in grams, of polonium-210 present after t weeks. How long will it be before there are only 5 grams remaining?   ■

Now we apply these ideas to a study of the growth of the world population.

EXAMPLE 4.2   **WORLD POPULATION GROWTH**

In 1900 the world population was about 1.6 billion. On average to find next year's population we multiply this year's population by 1.014.

Part 1 Explain why these assumptions tell us that the world population is an exponential function of time.

Part 2 Find a formula that gives the world population W, in billions, as an exponential function of t, the time in years since 1900.

Part 3 Make a graph of the world population from 1900 to 2010.

Part 4 According to this model, when did the world population reach 5 billion?

Solution to Part 1 The world population is an exponential function of time because our assumptions tell us that the population changes by constant multiples.

To find next year's population, we multiply this year's population by 1.014, so the population after t years is the initial value times 1.014^t.

Solution to Part 2 To find next year's population we multiply this year's population by 1.014, so this number is the yearly growth factor. The initial value is $W(0) = 1.6$. Therefore,

$$W = \text{Initial value} \times (\text{Yearly growth factor})^t$$

$$W = 1.6 \times 1.014^t.$$

Solution to Part 3 We use the following correspondences to make the graph:

$$X = t, \text{time in years since 1900 on the horizontal axis}$$

$$Y_1 = W, \text{population in billions on the vertical axis.}$$

The time interval from 1900 to 2010 corresponds to the range $t = 0$ to $t = 110$. Hence we choose a horizontal span from 0 to 110. The table of values shown in Figure 4.5 leads us to choose a vertical span of 0 to 10. The graph is shown in Figure 4.6.

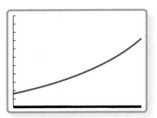

X	Y₁
0	1.6
20	2.1129
40	2.7902
60	3.6847
80	4.8658
100	6.4256
120	8.4854

X=0

FIGURE 4.5 A table of values for world population

FIGURE 4.6 A graph of world population

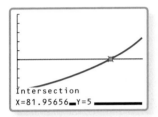

Intersection
X=81.95656 Y=5

FIGURE 4.7 Finding when world population reached 5 billion

Solution to Part 4 We want to know when the population reached 5 billion. That is, we want to solve the equation

$$1.6 \times 1.014^t = 5.$$

We solve this equation using the crossing-graphs method as shown in Figure 4.7. The time $t = 81.96$ corresponds to about 1982.

TEST YOUR UNDERSTANDING │ **FOR EXAMPLE 4.2**

In a small country, the population is initially 1.4 million. To find next year's population from this year's population we multiply by 1.02. Find an exponential formula that gives the population N, in millions, after t years. When does the population reach 6 million?

Unit Conversion

Sometimes it is important to represent exponential functions in units that are not the same as those that are given. For example, from 1790 to 1860, the U.S. population grew exponentially with yearly growth factor 1.03. If we are interested in modeling the population as it changed each decade, we want to know the *decade* growth factor. Since the yearly growth factor is 1.03, in 10 years we would multiply by 1.03 a total of 10 times—that is, by 1.03^{10}:

Decade growth factor = (Yearly growth factor)^{10}

Decade growth factor = $1.03^{10} = 1.344$, rounded to three decimal places.

Thus if we want the population at the end of a decade, we multiply the population at the beginning of the decade by 1.344. This is the decade growth factor, and if we use d to denote decades since 1790, we have the information we need to write a formula for our exponential function:

U = Initial value × (Decade growth factor)^d

$U = 3.93 \times 1.344^d.$

Let's turn the problem around. Census data are normally collected each decade, and rather than being given information on a yearly basis, we might have been told that from 1790 to 1860 the decade growth factor for the U.S. population was 1.344. If we wanted to express the exponential function in terms of years, we would need to recover the yearly growth factor from the decade growth factor. Since the population grows by a factor of 1.344 each decade, in d decades it will grow by a factor of 1.344^d. In particular, in 1 year, which is one-tenth of a decade, it will grow by a factor of

Yearly growth factor = (Decade growth factor)$^{1/10}$

Yearly growth factor = $1.344^{1/10} = 1.03$, rounded to two decimal places.

Thus we have recovered the yearly growth factor of 1.03.
The reasoning we have used above applies in general.

KEY IDEA 4.2 **UNIT CONVERSION FOR GROWTH FACTORS**

If the growth or decay factor for 1 period of time is a, then the growth or decay factor A for k periods of time is given by

$$A = a^k.$$

This relationship may be illustrated with the following diagram.

GROWTH FACTOR FOR ONE PERIOD OF TIME	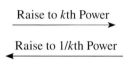 Raise to kth Power ⟶ Raise to $1/k$th Power ⟵	GROWTH FACTOR FOR k PERIODS OF TIME

EXAMPLE 4.3 **GETTING THE DECAY FACTOR FROM THE HALF-LIFE**

It is standard practice to give the rate at which a radioactive substance decays in terms of its *half-life*. That is the amount of time it takes for half of the substance to decay. The half-life of carbon-14 is 5770 years.

Part 1 If you start with 1 gram of carbon-14, how long will it take for only 1/4 gram to remain?

Part 2 What is the yearly decay factor rounded to five decimal places?

Part 3 Assuming that we start with 25 grams of carbon-14, find a formula that gives the amount of carbon-14 left after t years.

Solution to Part 1 To say that the half-life is 5770 years means that no matter how much carbon-14 is originally in place, at the end of 5770 years, half of that amount will remain. If we start with 1 gram, in 5770 years 1/2 gram will remain. In 5770 more years, there will be half of that amount, 1/4 gram, remaining. It takes two half-lives, or $5770 + 5770 = 11{,}540$ years, for 1 gram of carbon-14 to decay to 1/4 gram of carbon-14.

Solution to Part 2 In one half-life, carbon-14 decays by a factor of 1/2. Thus in h half-lives, it will decay by a factor of

$$\left(\frac{1}{2}\right)^h.$$

Since 1 year is 1/5770 half-life, in 1 year carbon-14 will decay by a factor of

Yearly decay factor = (Half-life decay factor)$^{1/5770}$

Yearly decay factor = $\left(\frac{1}{2}\right)^{1/5770} = 0.99988$, rounded to five decimal places.

Solution to Part 3 The initial value is 25, and from part 2 we know that the yearly decay factor is 0.99988. Thus if we let $C = C(t)$ be the amount, in grams, of carbon-14 left after t years, then

$$C = \text{Initial value} \times (\text{Yearly decay factor})^t$$

$$C = 25 \times 0.99988^t.$$

| **TEST YOUR UNDERSTANDING** │ **FOR EXAMPLE 4.3** |

Cesium-137 is a common by-product of nuclear reactors. It has a half-life of 30 years. What is the yearly decay factor for cesium-137? (Report your answer to 3 decimal places.)   ■

ANSWERS FOR TEST YOUR UNDERSTANDING

4.1 $P = 10 \times 0.97^t$. After 22.76 weeks there will be 5 grams.

4.2 $N = 1.4 \times 1.02^t$. The population reaches 6 million after $t = 73.49$ years.

4.3 0.977

4.1 EXERCISES

Reminder   Round all answers to two decimal places unless otherwise indicated.

1. **Exponential Growth with Given Initial Value and Growth Factor**   Write the formula for an exponential function with initial value 23 and growth factor 1.4. Plot its graph.

2. **Exponential Decay with Given Initial Value and Decay Factor**   Write the formula for an exponential function with initial value 200 and decay factor 0.73. Plot its graph.

3. **A Weight-Gain Program**   A woman, having recovered from a serious illness, begins a diet regimen designed to get her back to a healthy weight. She currently weighs 104 pounds. She hopes each week to multiply her weight by 1.03.

   a. Find a formula for an exponential function that gives the woman's weight w, in pounds, after t weeks on the regimen.

   b. How long will it be before she reaches her normal weight of 135 pounds?

4. **Unit Conversion with Exponential Growth**   The exponential function $N = 3.93 \times 1.34^d$ gives the approximate U.S. population, in millions, d decades after 1790. (The formula is valid only up to 1860.)

   a. What is the yearly growth factor? Find a formula that gives the population y years after 1790.

   b. What is the century growth factor? Find a formula that gives the U.S. population c centuries after 1790. (Assume that the original formula is valid over several centuries.)

5. **Unit Conversion with Exponential Decay**   The exponential function $P = 300 \times 0.86^m$ gives the amount in parts per million of PCBs in a contaminated site m months after a cleanup process begins.[1]

   a. What is the weekly decay factor? Find a formula that gives the amount P in parts per million w weeks after cleanup begins. Assume there are four weeks in each month.

   b. What is the yearly decay factor? Find a formula that gives the amount P in parts per million y years after cleanup begins.

6. **Radioactive Iodine**   Iodine-131 is a radioactive form of iodine. After the crisis at a Japanese nuclear power plant in March 2011, elevated levels of this substance were detected thousands of miles away from Japan. Iodine-131 has a half-life of 8 days. What is the daily decay factor for this substance?

7. **Half-Life of Tritium**   *This is a continuation of Example 4.1.* We stated in Example 4.1 that the half-life of a radioactive substance does not depend on the initial amount. Using the information from Example 4.1, show that it takes the same amount of time for 100 grams of tritium to decay to 50 grams as for 50 grams to decay to 25 grams. How long will it take for 100 grams of tritium to decay to 6.25 grams? (*Note:* You can do this without your calculator!)

8. **Doubling Time**   The current world population is about 7 billion. Under current conditions the population is growing exponentially with a yearly growth factor of 1.014. In parts b and c, round your answers to the nearest year.

   a. Find a formula that gives the world population N, in billions, after t years.

   b. How long will it take the population to double?

   c. How long after doubling will it take for the population to double again?

[1]The chemical polychlorinated biphenyl is a serious health hazard that is present in some industrial waste sites.

9. **The MacArthur–Wilson Theory of Biogeography**
Consider an island separated from the mainland, which contains a pool of potential colonizer species. The MacArthur–Wilson theory of biogeography[2] hypothesizes that some species from the mainland will migrate to the island but that increasing competition on the island will lead to species extinction. It further hypothesizes that both the rate of migration and the rate of extinction of species are exponential functions, and that an *equilibrium* occurs when the rate of extinction matches the rate of immigration. This equilibrium point is thought to be the point at which immigration and extinction stabilize. Suppose that, for a certain island near the mainland, the rate of immigration of new species is given by

$$I = 4.2 \times 0.93^t \text{ species per year}$$

and that the rate of species extinction on the island is given by

$$E = 1.5 \times 1.1^t \text{ species per year.}$$

According to the MacArthur–Wilson theory, how long will be required for stabilization to occur, and what are the immigration and extinction rates at that time?

10. **Long-Term Population Growth** Although exponential growth can often be used to model population growth accurately for some periods of time, there are inevitably, in the long term, limiting factors that make purely exponential models inaccurate. From 1790 to 1860 the U.S. population could be modeled by $N = 3.93 \times 1.03^t$ million people, where t is the time in years since 1790. If this exponential growth rate had continued until today, what would the population of the United States have been in 2010? Compare your answer with the actual population of the United States in 2010, which was about 309 million.

11. **Gold Prices** During the period from 2003 through 2010, gold prices doubled every 3 years (approximately).
    a. What was the yearly growth factor for the price of gold during this period?
    b. Explain in practical terms the meaning of the growth factor you found in part a.

12. **Growth of Bacteria** The organism *E. coli* is a common bacterium. Under certain conditions it undergoes cell division approximately each 20 minutes. During cell division, each cell divides into two cells.
    a. Explain why the number of *E. coli* cells present is an exponential function of time.
    b. What is the hourly growth factor for *E. coli*?
    c. Express the population N of *E. coli* as an exponential function of time t measured in hours. (Use N_0 to denote the initial population.)
    d. How long will it take a population of *E. coli* to triple in size?

13. **Grains of Wheat on a Chess Board** A children's fairy tale tells of a clever elf who extracted from a king the promise to give him one grain of wheat on a chess board square today, two grains on an adjacent square tomorrow, four grains on an adjacent square the next day, and so on, doubling the number of grains each day until all 64 squares of the chess board were used. How many grains of wheat did the hapless king contract to place on the 64th square? There are about 1.1 million grains of wheat in a bushel. Assume that a bushel of wheat sells for $4.25. What was the value of the wheat on the 64th square?

14. **An Expensive Book** A scientist visited Amazon.com to purchase a book.[3] To his amazement he found the book priced at $23,698,655.93. He discovered that there were two vendors selling this book on Amazon, and each was using a computer to adjust the price based on the other seller's price. The result was that each day the first vendor's price was multiplied by 1.27. Assume that the initial price was $134.27, and determine how many days it took for the first vendor's price to reach the exorbitant figure above. Round your answer to the nearest day.

15. **Mobile Phones** According to one source, the amount of data passing through mobile phone networks doubles each year.[4]
    a. Explain why the amount of data passing through mobile phone networks is an exponential function of time.
    b. Use D_0 for the initial amount of data, and find a formula that gives the data D as an exponential function of the time t in years.
    c. If this trend continues, how long will it be before the amount of data is 100 times its initial value?

(continued)

---

[2]R. H. MacArthur and E. O. Wilson, *The Theory of Island Biogeography* (Princeton, NJ: Princeton University Press, 1967); by the same authors: "An equilibrium theory of insular zoogeography," *Evolution* **17** (1963), 373–387.
[3]See the article at http://www.michaeleisen.org/blog/?p=358
[4]See http://marketplace.publicradio.org/display/web/2010/11/29/am-carriers-fight-to-meet-bandwidth-demands-/?refid=0

16. Internet Domain Hosts Often new technology spreads exponentially. Between 1995 and 2005, each year the number of Internet domain hosts was 1.43 times the number of hosts in the preceding year. In 1995 the number of hosts was 8.2 million.

   a. Explain why the number of hosts is an exponential function of time.

   b. Find a formula for the exponential function that gives the number N of hosts, in millions, as a function of the time t in years since 1995.

   c. According to this model, in what year did the number of hosts reach 24 million?

17. Headway on Four-Lane Highways When traffic is flowing on a highway, the *headway* is the average time between vehicles. On four-lane highways, the probability P that the headway is at least t seconds is given to a good degree of accuracy[5] by
$$P = e^{-qt},$$
where q is the average number of vehicles per second traveling one way on the highway.

   a. On a four-lane highway carrying an average of 500 vehicles per hour in one direction, what is the probability that the headway is at least 15 seconds? (*Note:* 500 vehicles per hour is $500/3600 = 0.14$ vehicle per second.)

   b. On a four-lane highway carrying an average of 500 vehicles per hour, what is the decay factor for the probability that headways are at least t seconds? *Reminder:* An important law of exponents tells us that $a^{bc} = (a^b)^c$.

4.1 SKILL BUILDING EXERCISES

S-1. Exponential Growth An amount A is initially 10. To find next year's amount we multiply this year's amount by 3. Find a formula that gives the amount A after t years.

S-2. Exponential Growth An amount A is initially 8. To find next week's amount we multiply this week's amount by 4. Find a formula that gives the amount A after t weeks.

S-3. Exponential Decay An amount A is initially 7. To find tomorrow's amount we multiply today's amount by 0.6. Find a formula that gives the amount A after t days.

S-4. Exponential Decay An amount A is initially 25. To find the next decade's amount we multiply this decade's amount by 0.7. Find a formula that gives the amount A after t decades.

S-5. Exponential Change An amount A is initially 8. To find next week's amount we divide this week's amount by 7. Find a formula that gives the amount A after t weeks.

S-6. Exponential Change The initial amount is 4. To find next week's amount we divide this week's amount by 3. Find a formula that gives the amount A after t weeks.

S-7. Function Value from Initial Value and Growth Factor Suppose that f is an exponential function with growth factor 2.4 and that $f(0) = 3$. Find $f(2)$. Find a formula for $f(x)$.

S-8. Function Value from Initial Value and Decay Factor Suppose that f is an exponential function with decay factor 0.094 and that $f(0) = 400$. Find $f(2)$. Find a formula for $f(x)$.

S-9. Finding the Growth Factor Suppose that f is an exponential function with $f(4) = 8$ and $f(5) = 10$. What is the growth factor for f?

S-10. Exponential Decay Is the graph of exponential decay versus time increasing or decreasing?

S-11. Exponential Decay What is the concavity of a graph showing exponential decay?

S-12. Exponential Growth What is the concavity of a graph showing exponential growth?

S-13. Rate of Change What can be said about the rate of change of an exponential function?

S-14. Exponential Growth Is the graph of exponential growth versus time increasing or decreasing?

S-15. Changing Units A certain quantity has a yearly growth factor of 1.17. What is its monthly growth factor?

S-16. Changing Units A certain quantity has a yearly growth factor of 1.17. What is its decade growth factor?

S-17. Changing Units A certain quantity has a decade growth factor of 3.4. What is its yearly growth factor?

S-18. Changing Units A certain quantity has a yearly decay factor of 0.99. What is its century decay factor?

S-19. Changing Units A certain quantity has a century decay factor of 0.2. What is its yearly decay factor?

[5]See Institute of Traffic Engineers, *Transportation and Traffic Engineering Handbook*, ed. John E. Baerwald (Englewood Cliffs, NJ: Prentice-Hall, 1976), p. 102.

4.2 CONSTANT PERCENTAGE CHANGE

One of the most common ways we encounter exponential functions in daily life is in the form of constant percentage change. When a news report says that we expect 3% inflation, or that an endangered population is declining by 10% per year, or that an investment is growing by 5% each year, exponential functions are at work behind the scenes. Constant percentage change always gives rise to an exponential function.

Percentage Growth Rate and Growth Factor

Let's focus for the moment on how a 5% raise affects our salary. The raise is 5% of our current salary, and that amount is added to our current salary. So our new salary will be our current salary plus 5% more. That is, the new salary is 100% plus 5%, or 105%, of our current salary. The result is that to find our new salary we multiply our current salary by 1.05.

In general if our raise is r percent as a decimal, then to find our new salary we multiply our current salary by $1 + r$. Some may find an algebraic derivation helpful. Let C denote our current salary and N denote our new salary. Our raise is rC. To find our new salary N we add this raise to our current salary:

$$N = C + rC$$

$$N = C(1 + r). \qquad \text{Factor out the } C.$$

This is an algebraic justification for the fact that a raise of r percent as a decimal multiplies our current salary by $1 + r$.

If we get a 5% raise every year, then each year we multiply by 1.05. That means our salary is an exponential function of time with yearly growth factor 1.05. We say that our salary has a *yearly percentage growth rate* of 5%.

More generally, if a function shows constant percentage growth of r as a decimal, then it is an exponential function with growth factor $1 + r$.

A classic example of exponential growth is the balance of an account that earns compound interest. If the account earns 6% interest each year and interest is compounded annually, then each year the balance is increased by 6%. Thus, the balance is an exponential function with yearly growth factor 1.06.

> Constant percentage growth always gives an exponential function.

EXAMPLE 4.4   COMPOUND INTEREST

A credit card holder begins the year owing $395.00 to a bank card for credit card purchases. The bank charges 1.2% interest on the outstanding balance each month. For purposes of this exercise, assume that no additional payments or charges are made and that no additional service charges are levied. Let $B = B(t)$ denote the balance of the account t months after January 1.

Part 1 Explain why B is an exponential function of t. Identify the monthly percentage growth rate, the monthly growth factor, and the initial value. Write a formula for $B = B(t)$.

Part 2 How much is owed after 7 months?

Part 3 Assume that there is a limit of $450 on the card and that the bank will demand a payment the first month this limit is exceeded. How long is it before a payment is required?

> A percentage growth rate of 1.2% corresponds to a growth factor of 1.012.

Solution to Part 1 For each dollar that is owed at the beginning of the month, an additional 0.012 dollar will be owed at the end of the month. Thus for each dollar owed at the beginning of the month, $1.012 will be owed at the end of the month. That means the value of B at the end of the month can be found by multiplying the balance at the beginning of the month by 1.012. Therefore, B is an exponential function with monthly growth factor 1.012. The monthly percentage growth rate is the monthly interest rate, 1.2%. The initial value is $395. Thus B is given by the formula

$$B = \text{Initial value} \times (\text{Monthly growth factor})^t$$

$$B = 395 \times 1.012^t.$$

X	Y$_1$
6	424.31
7	429.4
8	434.55
9	439.77
10	445.04
11	450.38
12	455.79
Y$_1$=450.383771297	

FIGURE 4.8 A table of values for a credit card

Solution to Part 2 To calculate the balance after 7 months, we put $t = 7$ into the formula we found in part 1:

$$B(7) = 395 \times 1.012^7 = 429.40 \text{ dollars.}$$

Solution to Part 3 To find when the credit limit is reached, we need to solve

$$395 \times 1.012^t = 450$$

for t. We could solve this using the crossing-graphs method, but in this case it is easier to make a table of values for $B(t)$ as we have done in Figure 4.8. The table shows that the credit limit will be exceeded in 11 months.

TEST YOUR UNDERSTANDING | **FOR EXAMPLE 4.4**

Suppose $900 is invested in a savings account that earns interest of 4% per year compounded annually. Thus, each year the balance is increased by 4%. Find a formula that gives the account balance B, in dollars, after t years. What is the value of the account after 10 years? ∎

Percentage Decay Rate and Decay Factor

The percentage growth rate is not the same as the growth factor, and the percentage decay rate is not the same as the decay factor.

Suppose a population declines by 10% this year. That means 90% of the population remains. Thus, to find next year's population from this year's population we multiply by 0.90. If the population declines by 10% each year, then to find the new population we always multiply the old population by 0.90. That is, a 10% per year decrease indicates an exponential function with yearly decay factor 0.90. More generally, if a function shows a constant percentage decrease of r as a decimal, then it is an exponential function with decay factor $1 - r$.

KEY IDEA 4.3 **ALTERNATIVE CHARACTERIZATION OF EXPONENTIAL FUNCTIONS**

A function is exponential if it shows constant percentage growth or decay.

1. For an exponential function with discrete (yearly, monthly, etc.) percentage growth rate r as a decimal, the growth factor is $a = 1 + r$.

2. For an exponential function with discrete percentage decay rate r as a decimal, the decay factor is $a = 1 - r$.

EXAMPLE 4.5 AN OIL SPILL

A 1000-gallon oil spill has contaminated a reservoir. The company hired to clean the reservoir can remove 10% of the remaining oil each week.

Part 1 Explain why the amount O, in gallons, of oil remaining in the reservoir is an exponential function of t, the time in weeks since cleanup began.

Part 2 Find an exponential model for the amount of oil in the reservoir through the cleanup period.

Part 3 How long will it be until there are only 100 gallons of oil in the reservoir? Round your answer to the nearest week.

Part 4 The cleanup company charges $7600 per week for its services. How much will it cost to remove the first 900 gallons of oil from the reservoir?

Part 5 After there are 100 gallons of oil left, how much will it cost to remove 90 more gallons of oil from the reservoir?

Solution to Part 1 The amount of oil remaining is an exponential function of time because it is decreasing by the same percentage each week.

Constant percentage decrease always gives an exponential function.

Solution to Part 2 Now 10% as a decimal is $r = 0.10$. Hence the weekly decay factor is

$$1 - r = 1 - 0.10 = 0.90.$$

A percentage decrease of 10% corresponds to a decay factor of 0.90.

The initial amount is 1000 gallons. Therefore,

$$O = \text{Initial value} \times \text{Decay factor}^t$$

$$O = 1000 \times 0.90^t.$$

X	Y₁	Y₂
0	1000	100
5	590.49	100
10	348.68	100
15	205.89	100
20	121.58	100
25	71.79	100
30	42.391	100

X=0

FIGURE 4.9 A table of values for an oil spill

Solution to Part 3 We need to know when there will be 100 gallons of oil remaining. That is, we need to solve the equation

$$1000 \times 0.90^t = 100.$$

We use the crossing-graphs method to do this with the following variable assignments:

$$X = t, \text{ time in weeks on the horizontal axis}$$

$$Y_1 = O, \text{ gallons of oil remaining on the vertical axis}$$

$$Y_2 = 100, \text{ target value on the vertical axis.}$$

The table of values in Figure 4.9 leads us to choose a horizontal span of 0 to 25 and a vertical span of 0 to 1000. The solution $t = 21.85$ is shown in Figure 4.10. We round this result to 22 weeks.

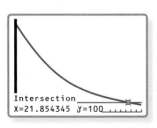

Intersection
X=21.854345 Y=100

FIGURE 4.10 Time to remove 900 gallons

Solution to Part 4 From part 3 we know that it took 22 weeks to remove the first 900 gallons of oil from the reservoir. It costs $7600 each week. So the total cost is

$$\text{Cost} = \text{Cost per week} \times \text{Number of weeks}$$

$$= 7600 \times 22$$

$$= 167,200 \text{ dollars.}$$

Solution to Part 5 Removal of an additional 90 gallons of oil leaves only 10 gallons in the reservoir. Thus we need to solve the equation

$$1000 \times 0.90^t = 10.$$

We solve this equation using the crossing-graphs method just as we did in part 3. The reader should verify that there are 10 gallons left after 43.71 weeks. That is $43.71 - 21.85 = 21.86$ or about 22 weeks after we reached the 100-gallon level. So the additional cost is

$$\text{Cost for next 90 gallons} = 7600 \times 22$$

$$= 167,200 \text{ dollars.}$$

Note that the cost for removal of the later 90 gallons of oil is the same as the cost for removal of the first 900 gallons (because the time required is the same). This outcome is typical of exponential decay. Decline may be rapid at first, but the rate of decline always slows to near zero.

TEST YOUR UNDERSTANDING | **FOR EXAMPLE 4.5**

There are 2000 pounds of a pollutant in a reservoir. The cleanup operation can remove 12% of the pollutant each month. Find a formula that gives the amount P, in pounds, of pollutant remaining after t months. How long will it be until half the pollutant is gone? ■

 ## Percentage Change and Unit Conversion

Often we are interested in the long-term effect of percentage increases. For example, suppose we received a 3.5% raise each year for 10 years. We want to determine the percentage increase in our wages over this decade. To do this we first find the yearly growth factor. The yearly percentage growth rate is $r = 0.035$ as a decimal, so the yearly growth factor is $a = 1 + r = 1.035$. Now we perform unit conversion. There are 10 years in a decade, so the decade growth factor is $1.035^{10} = 1.411$ rounded to three decimal places. The last step is to go back to a percentage growth rate. A decade growth factor of 1.411 gives a decade percentage growth rate of $1.411 - 1 = 0.411$ as a decimal. We conclude that our wages increased by 41.1% over the decade. To find the cumulative effect of a 3.5% annual increase in our wages over 10 years, it is tempting simply to multiply 3.5% by 10, but doing so gives the wrong answer of 35%. The correct answer of 41.1% is higher because it takes into account the effects of compounding.

EXAMPLE 4.6 **GRAY WOLVES**

A recent report[6] states that the gray wolf population that ranges through Idaho, Montana, Wyoming, Oregon, and Washington declined by 5% from 2009 to 2010. Suppose that the wolf population continues to decline at a rate of 5% per year.

Part 1 Use W_0 for the initial number of wolves, and find an exponential function that gives the wolf population W after t years.

Part 2 If the 5% decline continues for 5 years, what will be the percentage decrease in the wolf population over the five-year period? Report your answer to the nearest tenth of a percent.

Solution to Part 1 The yearly percentage decay rate as a decimal is $r = 0.05$. Hence the yearly decay factor is $a = 1 - r = 1 - 0.05 = 0.95$. We get the formula

$$W = W_0 \times 0.95^t.$$

[6]See http://howlingforjustice.wordpress.com/tag/endangered-species-act

Solution to Part 2 Because the yearly decay factor is 0.95, the 5-year decay factor is $0.95^5 = 0.774$. Therefore, the 5-year percentage decay rate is $1 - 0.774 = 0.226$. We conclude that the population will have decreased by 22.6%. In other words, a 5-year decay factor of 0.774 means that after 5 years 77.4% of the initial population remains, so the population has decreased by 22.6%.

TEST YOUR UNDERSTANDING | **FOR EXAMPLE 4.6**

You get a 5% raise each year for 5 years. What is your percentage salary increase over the 5-year period? Report your answer to the nearest tenth of a percent. ∎

FIGURE 4.11 5% inflation (thin graph) versus 4% inflation (thick graph)

Exponential Functions and Daily Experience

Exponential functions are common in daily news reports, and it is important that citizens spot them and understand their significance. Economic reports are often driven by exponential functions, and inflation is a prime example. When it is reported that the inflation rate is 3% per year, the meaning is that prices are increasing at a rate of 3% each year, and therefore prices are growing exponentially. It might be reported as good economic news that "Last year's 5% inflation rate is now down to 4%." Although this should indeed be encouraging, the lower rate does not change the fact that we are still dealing with an exponential function whose graph will eventually get very steep. The reduction from 5% to 4% changes the growth factor from 1.05 to 1.04 and so delays our reaching the steep part of the curve, but, as is shown in Figure 4.11, if the price function remains exponential in nature, then over the long term, catastrophic price increases will occur.

Short-term population growth is another phenomenon that is, in many settings, exponential in nature. It is the shape of the exponential curve, not necessarily the exact exponential formula, that causes concerns about eventual overcrowding. Many environmental issues related to human and animal populations are inextricably tied to exponential functions. It may be relatively inexpensive to dispose properly of large amounts of materials at a waste cleanup site, but further cleanup may be much more expensive. That is, as we begin the cleanup process, the amount of objectionable material remaining to be dealt with may be a decreasing exponential function. Since certain toxic substances are dangerous even in minute quantities, it can be very expensive to reduce them to safe levels. Once again, it is the nature of exponential decay, not the exact formula, that contributes to the astronomical expense of environmental cleanup.

Whenever phenomena are described in terms of percentage change, they may be modeled by exponential functions, and understanding how exponential functions behave is the key to understanding their true behavior. One of the most important features of exponential phenomena is that they may change at one rate over a period of time but change at a much different rate later on. Exponential growth may be modest for a time, but eventually the function will increase at a dramatic rate. Similarly, exponential decay may show encouragingly rapid progress at first, but such rates of decrease cannot continue indefinitely.

ANSWERS FOR TEST YOUR UNDERSTANDING

4.4 $B = 900 \times 1.04^t$. After 10 years the balance is $1332.22.

4.5 $P = 2000 \times 0.88^t$. Half the pollutant will be removed after $t = 5.42$ months.

4.6 27.6%

4.2 EXERCISES

Reminder   Round all answers to two decimal places unless otherwise indicated.

1. **Theater Productions**   Data from the *Statistical Abstract of the United States* show that in 1995 there were 56.61 thousand performances in the United States by nonprofit professional theaters. From 1995 through 2007, this number increased on average by about 10% each year.

   a. Let P denote the number of performances, in thousands, and let t denote the time in years since 1995. Make an exponential model for P versus t.

   b. How many performances by nonprofit professional theaters does your model give for 2007? (The actual number was 197 thousand.)

2. **Marriage Length**   In this exercise we consider data from the *Statistical Abstract of the United States* on the fraction of women married for the first time in 1960 whose marriage reached a given anniversary number. The data show that the fraction of women who reached their fifth anniversary was 0.928. After that, for each one-year increase in the anniversary number, the fraction reaching that number drops by about 2%. These data describe constant percentage change, so it is reasonable to model the fraction M as an exponential function of the number n of anniversaries since the fifth.

   a. What is the yearly decay factor for the exponential model?

   b. Find an exponential model for M as a function of n.

   c. According to your model, what fraction of women married for the first time in 1960 celebrated their 40th anniversary? (Take $n = 35$.) Round your answer to three decimal places. (The actual fraction is 0.449 or 44.9%.)

3. **A Population with Given Per Capita Growth Rate**   A certain population has a yearly per capita growth rate of 2.3%, and the initial value is 3 million.

   a. Use a formula to express the population as an exponential function.

   b. Express using functional notation the population after 4 years, and then calculate that value.

4. **Unit Conversion with Exponential Growth**   The exponential function $N = 3500 \times 1.77^d$, where d is measured in decades, gives the number of individuals in a certain population.

   a. Calculate $N(1.5)$ and explain what your answer means.

   b. What is the percentage growth rate per decade?

   c. What is the yearly growth factor rounded to three decimal places? What is the yearly percentage growth rate?

   d. What is the growth factor (rounded to two decimal places) for a century? What is the percentage growth rate per century?

5. **Unit Conversion with Exponential Decay**   The exponential function $N = 500 \times 0.68^t$, where t is measured in years, shows the amount, in grams, of a certain radioactive substance present.

   a. Calculate $N(2)$ and explain what your answer means.

   b. What is the yearly percentage decay rate?

   c. What is the monthly decay factor rounded to three decimal places? What is the monthly percentage decay rate?

   d. What is the percentage decay rate per second? (*Note:* For this calculation, you will need to use all the decimal places that your calculator can show.)

6. **A Savings Account**   You initially invest $500 in a savings account that pays a yearly interest rate of 4%.

   a. Write a formula for an exponential function giving the balance in your account as a function of the time since your initial investment.

   b. What monthly interest rate best represents this account? Round your answer to three decimal places.

   c. Calculate the decade growth factor.

   d. Use the formula you found in part a to determine how long it will take for the account to reach $740. Explain how this is consistent with your answer to part c.

7. **Gray Wolves in Idaho**   The report cited in Example 4.6 tells us that in 2009 there were 870 gray wolves in Idaho, but that the population declined by 19% that year. For purposes of this problem we assume that this 19% annual rate of decrease continues.

   a. Find an exponential model that gives the wolf population W as a function of the time t in years since 2009.

   b. It is expected that the wolf population cannot recover if there are fewer than 20 individuals.

How long must this rate of decline continue for the wolf population to reach 20?

8. **How Fast Do Exponential Functions Grow?** At age 25 you start to work for a company and are offered two rather fanciful retirement options.

    Retirement Option 1 When you retire, you will be paid a lump sum of $25,000 for each year of service.

    Retirement Option 2 When you start to work, the company will deposit $10,000 into an account that pays a monthly interest rate of 1%. When you retire, the account will be closed and the balance given to you.

    Which retirement option is more favorable to you if you retire at age 65? Which retirement option is more favorable if you retire at age 55?

9. **Inflation** The yearly *inflation rate* tells the percentage by which prices increase. For example, from 1990 through 2000 the inflation rate in the United States remained stable at about 3% each year. In 1990 an individual retired on a fixed income of $36,000 per year. Assuming that the inflation rate remains at 3%, determine how long it will take for the retirement income to deflate to half its 1990 value. (*Note:* To say that retirement income has deflated to half its 1990 value means that prices have doubled.)

10. **Rice Production in Asia** In Exercise 20 of Section 3.4, it was estimated that annual rice yield in Asia was increasing by about 0.04 ton per hectare each year. The International Rice Research Institute[7] predicts an increase in demand for Asian rice of about 2.1% per year. Discuss the future of rice production in Asia if these estimates remain valid.

11. **The Population of Mexico[8]** In 1980 the population of Mexico was about 67.38 million. For the years 1980 through 1985, the population grew at a rate of about 2.6% per year.

    a. Find a formula for an exponential function that gives the population of Mexico.

    b. Express using functional notation the population of Mexico in 1983, and then calculate that value.

    c. Use the function you found in part a to predict when the population of Mexico will reach 90 million.

12. **Cleaning Contaminated Water** A tank of water is contaminated with 60 pounds of salt. In order to bring the salt concentration down to a level consistent with EPA standards, clean water is being piped into the tank, and the well-mixed overflow is being collected for removal to a toxic-waste site. The result is that at the end of each hour there is 22% less salt in the tank than at the beginning of the hour. Let $S = S(t)$ denote the number of pounds of salt in the tank t hours after the flushing process begins.

    a. Explain why S is an exponential function and find its hourly decay factor.

    b. Give a formula for S.

    c. Make a graph of S that shows the flushing process during the first 15 hours, and describe in words how the salt removal process progresses.

    d. In order to meet EPA standards, there can be no more than 3 pounds of salt in the tank. How long must the process continue before EPA standards are met?

    e. Suppose this cleanup procedure costs $8000 per hour to operate. How much does it cost to reduce the amount of salt from 60 pounds to 3 pounds? How much does it cost to reduce the amount of salt from 3 pounds to 0.1 pound?

13. **Tsunami Waves in Crescent City** Crescent City, California, has historically been at risk for tsunami waves.[9] The probability P (as a decimal) of no tsunami wave of height 15 feet or more striking Crescent City over a period of Y years decreases as the time interval increases. Increasing the time interval by 1 year decreases the probability[10] by about 2%.

    a. Explain why P is an exponential function of Y.

    b. What is the decay factor for P?

    c. Recall that the probability of a certainty is 1. What is the initial value of P?

    d. Find a formula for P as a function of Y.

    e. What is the probability of no tsunami waves 15 feet or higher striking Crescent City over a 10-year period? Over a 100-year period?

    f. Find a formula for the probability Q that at least one wave 15 feet or higher will strike Crescent City over a period of Y years. (*Suggestion:* It should be helpful to note that the probability of the

(continued)

[7]IRRI, *IRRI Rice Almanac 1993–1995* (Los Banos, the Philippines: International Rice Research Institute, 1993).
[8]Adapted from D. Hughes-Hallett, A. M. Gleason, et al., *Calculus*, 3rd ed. (New York: John Wiley, 2002).
[9]Its harbor was severely damaged by the tsunami originating off the coast of Japan on March 11, 2011.
[10]This value is taken from data provided on page 294 of Robert L. Wiegel, ed., *Earthquake Engineering* (Englewood Cliffs, NJ: Prentice-Hall, 1970).

occurrence of an event plus the probability that it will not occur is the probability of a certainty, 1.)

14. The Beer–Lambert–Bouguer Law When light strikes the surface of a medium such as water or glass, its intensity decreases with depth. The *Beer–Lambert–Bouguer law* states that the percentage of decrease is the same for each additional unit of depth. In a certain lake, intensity decreases about 75% for each additional meter of depth.

a. Explain why intensity I is an exponential function of depth d in meters.

b. Use a formula to express intensity I as an exponential function of d. (Use I_0 to denote the initial intensity.)

c. Explain in practical terms the meaning of I_0.

d. At what depth will the intensity of light be one-tenth of the intensity of light striking the surface?

15. The Photic Zone *This is a continuation of Exercise 14.* In the ocean, the *photic zone* is the region where there is sufficient light for photosynthesis to occur. (See Figure 4.12.) For marine phytoplankton, the photic zone extends from the surface of the ocean to a depth where the light intensity is about 1% of surface light. Near Cape Cod, Massachusetts, the depth of the photic zone is about 16 meters. In waters near Cape Cod, by what percentage does light intensity decrease for each additional meter of depth?

16 m

FIGURE 4.12

16. Decibels Sound exerts a pressure P on the human ear. This pressure increases as the loudness of the sound increases. It is convenient to measure the loudness D in decibels and the pressure P in dynes per square centimeter. It has been found that each increase of 1 decibel in loudness causes a 12.2% increase in pressure. Furthermore, a sound of loudness 97 decibels produces a pressure of 15 dynes per square centimeter.

a. Explain why P is an exponential function of D and find the growth factor.

b. Find $P(0)$ and explain in practical terms what your answer means.

c. Find an exponential model for P as a function of D.

d. When pressure on the ear reaches a level of about 200 dynes per square centimeter, physical damage can occur. What decibel level should be considered dangerous?

17. A Diet An overweight man makes lifestyle changes in order to lose weight. He currently weighs 260 pounds and has set a target weight of 200 pounds. Each month the difference D, in pounds, between his current weight and his target weight decreases by 10%.

a. Make an exponential model of D versus the time t in months since the diet began.

b. How long will it take for his weight to reach 210 pounds?

18. APR and APY Recall that financial institutions sometimes report the annual interest rate that they offer on investments as the APR, often called the *nominal* interest rate. To indicate how an investment will actually grow, they advertise the *annual percentage yield*, or APY.[11] In mathematical terms, this is the yearly percentage growth rate for the exponential function that models the account balance. In this exercise and the next, we study the relationship between the APR and the APY. We assume that the APR is 10%, or 0.1 as a decimal.

To determine the APY when we know the APR, we need to know how often interest is compounded. For example, suppose for the moment that interest is compounded twice a year. Then to say that the APR is 10% means that in half a year, the balance grows by $\frac{10}{2}$% (or 5%). In other words, the $\frac{1}{2}$-year percentage growth rate is $\frac{0.1}{2}$ (as a decimal). Thus the $\frac{1}{2}$-year growth factor is $1 + \frac{0.1}{2}$. To find the *yearly* growth factor, we need to perform a unit conversion: One year is 2 half-year periods, so the yearly growth factor is $\left(1 + \frac{0.1}{2}\right)^2$, or 1.1025.

a. What is the yearly growth factor if interest is compounded four times a year?

b. Assume that interest is compounded n times each year. Explain why the formula for the yearly growth factor is

$$\left(1 + \frac{0.1}{n}\right)^n.$$

c. What is the yearly growth factor if interest is compounded daily? Give your answer to four decimal places.

[11]For loans, financial institutions often refer to this as the effective annual rate, or EAR.

19. Continuous Compounding *This is a continuation of Exercise 18.* In this exercise we examine the relationship between the APR and the APY when interest is compounded continuously—in other words, at every instant. We will see by means of an example that the relationship is

$$\text{Yearly growth factor} = e^{\text{APR}}, \qquad (4.1)$$

and so

$$\text{APY} = e^{\text{APR}} - 1 \qquad (4.2)$$

if both the APR and the APY are in decimal form and interest is compounded continuously. Assume that the APR is 10%, or 0.1 as a decimal.

a. The yearly growth factor for continuous compounding is just the limiting value of the function given by the formula in part b of Exercise 18. Find that limiting value to four decimal places.

b. Compute e^{APR} with an APR of 0.1 (as a decimal).

c. Use your answers to parts a and b to verify that Equation (4.1) holds in the case where the APR is 10%.

Note: On the basis of part a, one conclusion is that there is a limit to the increase in the yearly growth factor (and hence in the APY) as the number of compounding periods increases. We might have expected the APY to increase without limit for more and more frequent compounding.

4.2 SKILL BUILDING EXERCISES

Round each percentage increase or decrease that you calculate to the nearest tenth of a percent.

S-1. Percentage Growth Rate from Growth Factor An exponential function has a growth factor of 1.06. What is the percentage growth rate?

S-2. Percentage Growth Rate from Growth Factor An exponential function has a growth factor of 3.61. What is the percentage growth rate?

S-3. Percentage Decay Rate from Decay Factor An exponential function has a decay factor of 0.70. What is the percentage decay rate?

S-4. Percentage Decay Rate from Decay Factor An exponential function has a decay factor of 0.06. What is the percentage decay rate?

S-5. Function from Initial Value and Percentage Growth Rate Suppose that f is an exponential function with percentage growth rate of 5% and that $f(0) = 8$. Find a formula for $f(x)$.

S-6. Function from Initial Value and Percentage Growth Rate Suppose that f is an exponential function with percentage growth rate of 1% and that $f(0) = 7$. Find a formula for $f(x)$.

S-7. Function from Initial Value and Percentage Decay Rate Suppose that f is an exponential function with percentage decay rate of 5% and that $f(0) = 6$. Find a formula for $f(x)$.

S-8. Percentage Growth A certain quantity has an initial value of 10 and grows at a rate of 7% per year. Give an exponential function that describes this quantity.

S-9. Percentage Growth A certain quantity has an initial value of 30 and grows at a rate of 4% per year. Give an exponential function that describes this quantity.

S-10. Percentage Growth A certain quantity has an initial value of 50 and grows at a rate of 9% per month. Give an exponential function that describes this quantity.

S-11. Percentage Decay A certain quantity has initial value 10 and decays by 4% each year. Give an exponential function that describes this quantity.

S-12. Percentage Decay A certain quantity has initial value 13 and decays by 6% each year. Give an exponential function that describes this quantity.

S-13. Percentage Decay A certain quantity has initial value 5 and decays by 3% each day. Give an exponential function that describes this quantity.

S-14. Percentage Decay A certain quantity has initial value 100 and decays by 20% each year. Give an exponential function that describes this quantity.

S-15. Percentage Change A quantity increases by 5% for each of 10 years. What is its percentage increase over the 10-year period?

S-16. Percentage Change A quantity increases by 10% for each of 10 years. What is its percentage increase over the 10-year period?

S-17. Percentage Change A quantity decreases by 5% for each of 10 years. What is its percentage decrease over the 10-year period?

S-18. Percentage Change A quantity decreases by 10% for each of 10 years. What is its percentage decrease over the 10-year period?

S-19. Percentage Change A quantity decreases by 3% for each of 5 years. What is its percentage decrease over the 5-year period?

(continued)

S-20. Percentage Change A bank account grows by 9% each year. By what percentage does it grow each month?

S-21. Percentage Change A certain population grows by 15% per decade. What is its percentage growth each year?

S-22. Percentage Change A certain population declines by 5% each year. What is the percentage decline each month?

S-23. Percentage Change An investment declines in value by 2% each month. By what percentage does its value decline in a year?

S-24. Percentage Change A certain radioactive substance decays at a rate of 17% each year. By what percentage does it decay each month?

S-25. Percentage Change An investment grows by 1% each year. What is the percentage growth after 50 years?

4.3 MODELING EXPONENTIAL DATA

Just as we did with linear data, we will show how to recognize exponential data and develop the appropriate tools for constructing exponential models.

Recognizing Exponential Data

The table below shows how the balance in a savings account grows over time since the initial investment.

Time in months	0	1	2	3	4
Savings balance	$3500.00	$3542.00	$3584.50	$3627.52	$3671.05

Let t be the time in months since the initial investment, and let B be the balance in dollars. Certainly a linear model for the data is not appropriate, since the balance is growing more and more quickly over time. For example, over the first month the balance grows by $3542.00 - 3500.00 = 42.00$ dollars, whereas over the second month it grows by $3584.50 - 3542.00 = 42.50$ dollars. In fact, we expect that the balance will grow as an exponential function if the interest rate is constant. How can we test the data to see whether they really are exponential in nature? If the data are growing exponentially, then each month we should get the new balance by multiplying the old balance by the monthly growth factor:

$$\text{New } B = \text{Monthly growth factor} \times \text{Old } B.$$

Using division to rewrite this yields

$$\frac{\text{New } B}{\text{Old } B} = \text{Monthly growth factor}. \tag{4.3}$$

Thus a data table representing an exponential function should show common quotients if we divide each function entry by the one preceding it, assuming that we have evenly spaced values for the variable. In the following table, we have calculated this quotient for each of the data entries, rounding our answers to three decimal places.

Time increment	From 0 to 1	From 1 to 2	From 2 to 3	From 3 to 4
Ratios of B	$\dfrac{3542.00}{3500.00} = 1.012$	$\dfrac{3584.50}{3542.00} = 1.012$	$\dfrac{3627.52}{3584.50} = 1.012$	$\dfrac{3671.05}{3627.52} = 1.012$

According to Equation (4.3), this table tells us two things. First, the ratios are always the same, so the data are exponential. Second, the common quotient 1.012 is the monthly growth factor. This means that the monthly interest rate is $1.012 - 1 = 0.012$ as a decimal, or 1.2%.

Comparing this with how we handled linear data in Chapter 3, we notice an important analogy. Linear functions are functions that change by *constant sums*, so we detect linear data by looking for a common *difference* in function values when we use successive data points, assuming evenly spaced values for the variable. When our data set is given in variable increments of 1, this common difference is the slope of the linear model. Exponential functions are functions that change by *constant multiples*, so we detect exponential data by looking for a common *quotient* when we use successive data points, again assuming evenly spaced values for the variable. When our data set is given in increments of 1, this common quotient is the growth (or decay) factor of the exponential model.

> Linear data show constant differences, but exponential data show constant ratios.

Constructing an Exponential Model

In our discussion of the balance in a savings account, we saw that the data were exponential. Since we know the growth factor, we need only one further bit of information to make an exponential model: the initial value. But that also appears in the table as $B(0) = 3500.00$ dollars. Thus the formula, which will serve as our exponential model for B as a function of time t in months, is

$$B = \text{Initial value} \times (\text{Monthly growth factor})^t$$

$$B = 3500.00 \times 1.012^t \text{ dollars.}$$

In some situations the data themselves are not exponential, but the difference from a limiting value can be modeled by an exponential function.

EXAMPLE 4.7 CONTAMINATED WELL

Underground water seepage, such as that from a toxic-waste site, can contaminate water wells many miles away. Suppose that water seeping from a toxic-waste site is polluted with a certain contaminant at a level of 64 milligrams per liter. Several miles away, monthly tests are made on a water well to monitor the level of this contaminant in the drinking water. In the table on the next page, we record the difference between the contaminant level of the waste site (64 milligrams per liter) and the contaminant level of the water well (in milligrams per liter).

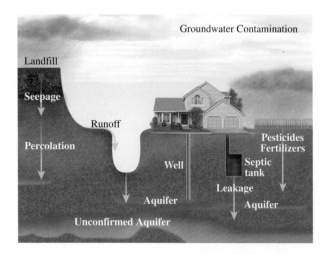

Time in months	0	1	2	3	4	5
Contaminant level difference	64	45.44	32.26	22.91	16.26	11.55

Part 1 Explain why the nature of the data suggests that an exponential model may be appropriate.

Part 2 Test to see that the data are exponential.

Part 3 Find an exponential model for the difference in contaminant level.

Part 4 Use your answer in part 3 to make a model for the contaminant level as a function of time.

Part 5 This contaminant is considered dangerous in drinking water when it reaches a level of 57 milligrams per liter. When will this dangerous level be reached?

We are modeling the contaminant level difference—not the contaminant level itself.

Solution to Part 1 Let t be the time, in months, since the leakage began, and let D be the difference, in milligrams per liter, between the contaminant level of the waste site and that of the water well. The table shows that the difference D is narrowing, so the water well is becoming more heavily polluted. We note further that D decreases quickly at first but that the rate of decrease slows later on. This is a feature of exponential decay, and it leads us to suspect that these data may be appropriately modeled by an exponential function.

Solution to Part 2 In the following table, we have calculated successive quotients for the data entries, rounding our answers to two decimal places.

Time increment	From 0 to 1	From 1 to 2	From 2 to 3	From 3 to 4	From 4 to 5
Ratios of D	$\dfrac{45.44}{64} = 0.71$	$\dfrac{32.26}{45.44} = 0.71$	$\dfrac{22.91}{32.26} = 0.71$	$\dfrac{16.26}{22.91} = 0.71$	$\dfrac{11.55}{16.26} = 0.71$

Since the common quotient is 0.71, the data can be modeled by an exponential function.

Solution to Part 3 The decay factor for D is the common quotient 0.71 that we found in part 2. The initial value of D is in the table: $D(0) = 64$ milligrams per liter. Thus we have

$$D = \text{Initial value} \times (\text{Decay factor})^t$$

$$D = 64 \times 0.71^t.$$

We find the contaminant level from the difference by simple algebraic manipulation.

Solution to Part 4 Let $C = C(t)$ be the contaminant level of the water well in milligrams per liter. Since D represents the contaminant level of the waste site, 64 milligrams per liter, minus the contaminant level C of the water well, we find a formula for C as follows:

$$D = 64 - C$$

$$C + D = 64$$

$$C = 64 - D$$

$$C = 64 - 64 \times 0.71^t \text{ milligrams per liter.}$$

Solution to Part 5 We want to solve the equation

$$64 - 64 \times 0.71^t = 57$$

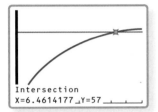

X	Y₁	Y₂
3	41.094	57
4	47.737	57
5	52.453	57
6	55.802	57
7	58.179	57
8	59.867	57
9	61.066	57

X = 7

FIGURE 4.13 A table of values for contaminant levels

Intersection
X=6.4614177 Y=57

FIGURE 4.14 When the contaminant level reaches 57 milligrams per liter

for t. We use the crossing-graphs method to do that. We entered the contaminant function $C = 64 - 64 \times 0.71^t$ and the target level, 57 milligrams per liter, into the calculator. We record the appropriate variable correspondences:

$Y_1 = C$, contaminant level, in milligrams per liter, on vertical axis

$X = t$, months on horizontal axis.

We made the table of values in Figure 4.13 to find out how to set up the graphing window. This table shows that the critical level of contamination will occur sometime before 7 months. This led us to choose a horizontal span of $t = 0$ to $t = 8$. Since the contaminant level starts at 0 and will never be more than 64, we add a little extra room, using a vertical span of $C = 0$ to $C = 75$. The resulting graphs are in Figure 4.14. We have used the calculator to get the crossing point, and from the prompt at the bottom of Figure 4.14, we see that the contaminant level of the water well will reach 57 milligrams per liter 6.46 months after the leakage began.

TEST YOUR UNDERSTANDING | **FOR EXAMPLE 4.7**

The salinity of a saltwater aquarium is being raised to a target value of 32 parts per thousand of dissolved solids in the tank. The table below shows the difference D between the target salinity and the current salinity S after t hours. Both D and S are measured in parts per thousand.

t = time in hours	0	1	2	3	4
D = salinity difference	32	25.60	20.48	16.38	13.11

Make a model of D as an exponential function of t. Find a formula that gives the salinity S as a function of t. ∎

The graph in Figure 4.14 not only answers for us the question we asked but also provides a display that shows the classic exponential behavior for this phenomenon. As we noted earlier, the difference D between the contaminant level of the waste site and that of the water well decreases quickly at first, but the rate of decrease slows later on. As Figure 4.14 shows, this means that the contamination itself increases rapidly at first, but as the level nears that of the polluted source, the rate of increase slows.

EXAMPLE 4.8 **MAKING ICE**

A freezer maintains a constant temperature of 6 degrees Fahrenheit. An ice tray is filled with tap water and placed in the refrigerator to make ice. The difference between the temperature of the water and that of the freezer was sampled each minute and recorded in the table below.

Time in minutes	0	1	2	3	4	5
Temperature difference	69.0	66.3	63.7	61.2	58.8	56.5

Part 1 Test to see that the data are exponential.

Part 2 Find an exponential model for temperature difference.

Part 3 Use your answer in part 2 to make a model for the temperature of the cooling water as a function of time.

Part 4 When will the temperature of the water reach 32 degrees?

The ratio we check is
(Next entry)/(Current entry).

Solution to Part 1 Let t be the time in minutes and D the temperature difference. To test whether the data are exponential, we make a table of successive quotients. (We rounded our answers to two decimal places.)

Time increment	From 0 to 1	From 1 to 2	From 2 to 3	From 3 to 4	From 4 to 5
Ratios of D	$\dfrac{66.3}{69.0} = 0.96$	$\dfrac{63.7}{66.3} = 0.96$	$\dfrac{61.2}{63.7} = 0.96$	$\dfrac{58.8}{61.2} = 0.96$	$\dfrac{56.5}{58.8} = 0.96$

The table of values shows a common quotient of 0.96, and we conclude that the data can be modeled by an exponential function.

The common ratio gives the growth or decay factor.

Solution to Part 2 We know that the decay factor for D is the common ratio 0.96 that we calculated in part 1. We also know that when $t = 0$, $D = 69.0$, and that is the initial value of D. Thus we have

$$D = \text{Initial value} \times (\text{Decay factor})^t$$

$$D = 69 \times 0.96^t.$$

Solution to Part 3 Let W be the temperature of the water, in degrees Fahrenheit. The function D is the temperature W of the water minus 6, the temperature of the freezer:

$$W - 6 = D$$

$$W = 6 + D$$

$$W = 6 + 69 \times 0.96^t \text{ degrees Fahrenheit.}$$

Solution to Part 4 We want to solve the equation

$$6 + 69 \times 0.96^t = 32.$$

We enter the water temperature function W and the target temperature, 32 degrees Fahrenheit. We record the variable correspondences:

$$Y_1 = W, \text{ water temperature, on vertical axis}$$

$$X = t, \text{ minutes, on horizontal axis.}$$

We made the table of values in Figure 4.15 using an increment value of 10 minutes to get an estimate of when the temperature will reach 32 degrees. The table shows that this will happen within 30 minutes. Thus we set up our graphing window using a horizontal span of $t = 0$ to $t = 30$ and a vertical span of $W = 0$ to $W = 75$. The graphs with the crossing point that was calculated appear in Figure 4.16. We see that the temperature will reach 32 degrees in 23.91 minutes.

X	Y$_1$	Y$_2$
0	75	32
10	51.873	32
20	36.498	32
30	26.276	32
40	19.48	32
50	14.962	32
60	11.958	32
X=0		

FIGURE 4.15 A table of values for water temperature

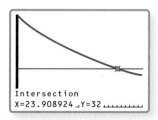

Intersection
X=23.908924 Y=32

FIGURE 4.16 When the temperature is 32 degrees

TEST YOUR UNDERSTANDING | **FOR EXAMPLE 4.8**

A snake is growing toward its maximum length of 46 inches. The table below shows the difference D between its maximum length of 46 inches and its current length L. Both D and L are measured in inches.

t = time in years	0	1	2	3	4
D = difference in length	25.00	18.75	14.06	10.55	7.91

Find a model of the difference D as an exponential function of the time t in years. Find a function that gives the length L in terms of t. ∎

Growth and Decay Factor Units in Exponential Modeling

In the contaminated well example, the decay factor was the common quotient we calculated. This will always occur when time measurements are given in 1-unit increments. But when data are measured in different increments, adjustments to account for units must be made to get the right decay factor. We can show what we mean by looking at the contaminated water well example again, but this time we suppose that measurements were taken every 3 months. Under those conditions, the data table would have been as follows:

t = months	0	3	6	9	12	15
D = difference in contaminant level	64	22.91	8.2	2.93	1.05	0.38

We test to see whether the data are exponential by calculating successive ratios, rounding to two decimal places.

Time increment	From 0 to 3	From 3 to 6	From 6 to 9	From 9 to 12	From 12 to 15
Ratios of D	$\dfrac{22.91}{64} = 0.36$	$\dfrac{8.2}{22.91} = 0.36$	$\dfrac{2.93}{8.2} = 0.36$	$\dfrac{1.05}{2.93} = 0.36$	$\dfrac{0.38}{1.05} = 0.36$

Since the successive ratios are the same, 0.36, we conclude once again that the data are exponential. But the common ratio 0.36 is not the monthly decay factor. Rather, it is the *3-month decay factor*. To get the monthly decay factor, we need to change the units. Since 1 month is one-third of the 3-month recording time, we get

Monthly decay factor $= $ (3-month decay factor)$^{1/3}$

Monthly decay factor $= 0.36^{1/3} = 0.71$ rounded to two decimal places.

The initial value is 64, so we arrive at the same exponential model, 64×0.71^t, as we did when we modeled the data given in 1-month intervals.

> **KEY IDEA 4.4** **MODELING EXPONENTIAL DATA**
>
> **1.** To test whether data with evenly spaced values for the variable show exponential growth or decay, calculate successive quotients. If the quotients are all the same, it is appropriate to model the data with an exponential function. If the quotients are not all the same, some other model will be needed.
>
> **2.** To model exponential data, use the common successive quotient for the growth (or decay) factor, but if the data are not measured in 1-unit increments, it will be necessary to make an adjustment for units. One of the data points can be used to determine the initial value.

EXAMPLE 4.9 **FINDING THE TIME OF DEATH**

One important topic of forensic medicine is the determination of the time of death. A method that is sometimes used involves temperature. Suppose at 6:00 P.M. a body is discovered in a basement of a building where the ambient air temperature is maintained at 72 degrees.[12] At the moment of death, the body temperature was 98.6 degrees, but after death the body cools, and eventually its temperature matches the ambient air temperature. Beginning at 6:00 P.M., the body temperature is measured and the difference $D = D(t)$ between body temperature and ambient air temperature is recorded. The measurement is repeated every 2 hours.

t = hours since 6:00 P.M.	0	2	4	6	8
D = temperature difference	12.02	8.08	5.44	3.65	2.45

Part 1 Show that the data can be modeled by an exponential function.

Part 2 Find an exponential model for the data that shows temperature difference as a function of hours.

Part 3 Find a formula for a function $T = T(t)$ that gives the temperature of the body at time t.

Part 4 What was the time of death?

Solution to Part 1 To see whether the data are exponential, we calculate successive quotients, rounding to two decimal places.

Time increment	From 0 to 2	From 2 to 4	From 4 to 6	From 6 to 8
Ratios of D	$\dfrac{8.08}{12.02} = 0.67$	$\dfrac{5.44}{8.08} = 0.67$	$\dfrac{3.65}{5.44} = 0.67$	$\dfrac{2.45}{3.65} = 0.67$

Since the successive quotients are the same, we conclude that the data show exponential decay.

---

[12]If the air temperature fluctuates, establishing time of death using body temperature is more difficult. It is noted in the book *Helter Skelter*, by Vincent Bugliosi, that body temperature was used to help establish the time of death in the murders of Sharon Tate and others by the Charles Manson group.

When the increment is not 1, we must adjust the common ratio to find the growth or decay factor.

Solution to Part 2 To give an exponential model, we need to know the initial value and the hourly decay factor. From the data table, the initial value is $D(0) = 12.02$. Our calculation of successive quotients in part 1 gave us the 2-hour decay factor, 0.67. We want the 1-hour decay factor. Since 1 hour is half of 2 hours, we get

$$\text{Hourly decay factor} = (\text{2-hour decay factor})^{1/2}$$

$$\text{Hourly decay factor} = 0.67^{1/2} = 0.82, \text{rounded to two decimal places.}$$

Thus we have

$$D = 12.02 \times 0.82^t.$$

Solution to Part 3 The function $D = 12.02 \times 0.82^t$ gives the difference between the temperature of the body and that of the air. Since we are assuming that the air has a temperature of 72 degrees, we add that to D to get the temperature of the body:

$$T = 72 + 12.02 \times 0.82^t.$$

Solution to Part 4 To find the time of death, we need to know when the temperature was that of a living person, 98.6 degrees. Thus we need to solve the equation

$$72 + 12.02 \times 0.82^t = 98.6$$

for t. Before we begin, we note that since $t = 0$ corresponds to some time several hours after death, the value of t that we are looking for is surely negative. We solve the equation by using the crossing-graphs method. In Figure 4.17 we have made the graph of T versus t using a horizontal span of $t = -5$ to $t = 5$ hours and a vertical span of $T = 70$ to $T = 120$ degrees. You may wish to consult a table of values to see why we chose these window settings. We then added the graph of the target temperature $T = 98.6$ and found the intersection point. From Figure 4.17, we see that death occurred about 4 hours before the body was found. Since $t = 0$ corresponds to 6:00 P.M., we set the time of death at about 2:00 P.M.

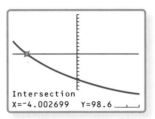

```
Intersection
X=-4.002699    Y=98.6
```

FIGURE 4.17 Finding the time of death

TEST YOUR UNDERSTANDING | **FOR EXAMPLE 4.9**

A radioactive substance is decaying. The amount A, in grams, present after t years is given in the table below.

t = years	0	10	20	30	40
A = amount	500.00	300.00	180.00	108.00	64.80

Find a formula that gives A as an exponential function of t. ■

ANSWERS FOR TEST YOUR UNDERSTANDING

4.7 $D = 32 \times 0.80^t$. $S = 32 - 32 \times 0.80^t$

4.8 $D = 25 \times 0.75^t$. $L = 46 - 25 \times 0.75^t$

4.9 $A = 500 \times 0.95^t$

4.3 EXERCISES

Reminder   Round all answers to two decimal places unless otherwise indicated.

1. **Making an Exponential Model**   Show that the accompanying data are exponential and find a formula for an exponential model.

t	$f(t)$	t	$f(t)$
0	3.80	3	4.27
1	3.95	4	4.45
2	4.11	5	4.62

2. **An Exponential Model with Unit Adjustment**   Show that the following data are exponential and find a formula for an exponential model. (*Note:* It will be necessary to make a unit adjustment. For this exercise, round your answers to three decimal places.)

t	$g(t)$	t	$g(t)$
0	38.30	12	16.04
4	28.65	16	11.99
8	21.43	20	8.97

3. **Data That Are Not Exponential**   Show that the following data are not exponential.

t	$h(t)$	t	$h(t)$
0	4.9	3	200.2
1	26.6	4	352.1
2	91.7	5	547.4

4. **Linear and Exponential Data**   One of the following two tables shows linear data, and the other shows exponential data. Identify which is which, and find models for both.

t	$f(t)$	t	$f(t)$
0	6.70	3	10.46
1	7.77	4	12.13
2	9.02	5	14.07

Table A

t	$g(t)$	t	$g(t)$
0	5.80	3	10.99
1	7.53	4	12.72
2	9.26	5	14.45

Table B

5. **A Coin Collection**   The value of a coin collection increases as new coins are added and the value of some rare coins in the collection increases. The value V, in dollars, of the collection t years after the collection was started is given by the following table.

t = time in years	V = value in dollars
0	130.00
1	156.00
2	187.20
3	224.64
4	269.57

   a. Show that these data are exponential.

   b. Find an exponential model for the data.

   c. According to the model, when will the collection have a value of $500?

6. **Insect Control**   DDT (dichlorodiphenyltrichloroethane) was used extensively from 1940 to 1970 as an insecticide. It still sees limited use for control of disease. But DDT was found to be harmful to plants and animals, including humans, and its effects were found to be lasting. The amount of time that DDT remains in the environment depends on many factors, but the following table shows what can be expected of 100 kilograms of DDT that has seeped into the soil.

t = time in years since application	D = DDT remaining, kilograms
0	100.00
1	95.00
2	90.25
3	85.74

a. Show that the data are exponential.

b. Make a model of D as an exponential function of t.

c. What is the half-life of DDT in the soil? That is, how long will it be before only 50 kilograms of DDT remain?

7. **Magazine Sales**   The following table shows the income from sales of a certain magazine, measured in thousands of dollars, at the start of the given year.

Year	Income
2005	7.76
2006	8.82
2007	9.88
2008	10.94
2009	12.00
2010	13.08
2011	14.26
2012	15.54

Over an initial period the sales grew at a constant rate, and over the rest of the time the sales grew at a constant percentage rate. Calculate differences and ratios to determine what these time periods are, and find the growth rate or percentage growth rate, as appropriate.

8. **Population Growth**   A population of animals is growing exponentially, and an ecologist has made the following table of the population size, in thousands, at the start of the given year.

Year	Population in thousands
2007	5.25
2008	5.51
2009	5.79
2010	6.04
2011	6.38
2012	6.70

Looking over the table, the ecologist realizes that one of the entries for population size is in error. Which entry is it, and what is the correct population? (Round the ratios to two decimal places.)

9. **An Investment**   You have invested money in a savings account that pays a fixed monthly interest on the account balance. The following table shows the account balance over the first 5 months.

Time in months	Savings balance
0	$1750.00
1	$1771.00
2	$1792.25
3	$1813.76
4	$1835.52
5	$1857.55

a. How much money was originally invested?

b. Show that the data are exponential and find an exponential model for the account balance.

c. What is the monthly interest rate?

d. What is the yearly interest rate?

e. Suppose that you made this investment on the occasion of the birth of your daughter. Your plan is to leave the money in the account until she starts college at age 18. How large a college fund will she have?

f. How long does it take your money to double in value? How much longer does it take it to double in value again?

10. **A Bald Eagle Murder Mystery**   At 3:00 P.M. a park ranger discovered a dead bald eagle that had been impaled by an arrow. Only two archers were found in the region. The first archer is able to establish that between 11:00 A.M. and 1:00 P.M., he was in a nearby diner having lunch. The second archer can show that he was in camp with friends between 9:00 A.M. and 11:00 A.M. The air temperature in the park has remained at a constant 62 degrees. Beginning at 3:00 P.M., the difference $D = D(t)$ between the temperature of the dead eagle and that of the air was measured and recorded in the following table. (Here t is the time in hours since 3:00 P.M.)

(continued)

t = hours since 3:00 P.M.	D = temperature difference
0	26.83
1	24.42
2	22.22
3	20.22
4	18.40
5	16.74

This table, together with the fact that the body temperature of a living bald eagle is 105 degrees, exonerates one of the archers, but the other may remain a suspect. Which archer's innocence is established?

11. **A Skydiver** When a skydiver jumps from an airplane, his downward velocity increases until the force of gravity matches air resistance. The velocity at which this occurs is known as the *terminal velocity*. It is the upper limit on the velocity a skydiver in free fall will attain (in a stable, spread position), and for a man of average size, its value is about 176 feet per second (or 120 miles per hour). A skydiver jumped from an airplane, and the difference $D = D(t)$ between the terminal velocity and his downward velocity in feet per second was measured at 5-second intervals and recorded in the following table.

t = seconds into free fall	D = velocity difference
0	176.00
5	73.61
10	30.78
15	12.87
20	5.38
25	2.25

a. Show that the data are exponential and find an exponential model for D. (Round all your answers to two decimal places.)

b. What is the percentage decay rate per second for the velocity difference of the skydiver? Explain in practical terms what this number means.

c. Let $V = V(t)$ be the skydiver's velocity t seconds into free fall. Find a formula for V.

d. How long would it take the skydiver to reach 99% of terminal velocity?

12. **The Half-Life of ^{239}U** Uranium-239 is an unstable isotope of uranium that decays rapidly. In order to determine the rate of decay, 1 gram of ^{239}U was placed in a container, and the amount remaining was measured at 1-minute intervals and recorded in the table below.

Time in minutes	Grams remaining
0	1
1	0.971
2	0.943
3	0.916
4	0.889
5	0.863

a. Show that these are exponential data and find an exponential model. (For this problem, round all your answers to three decimal places.)

b. What is the percentage decay rate each minute? What does this number mean in practical terms?

c. Use functional notation to express the amount remaining after 10 minutes and then calculate that value.

d. What is the half-life of ^{239}U?

13. **An Inappropriate Linear Model for Radioactive Decay** *This is a continuation of Exercise 12.* Physicists have established that radioactive substances display constant percentage decay, and thus radioactive decay is appropriately modeled exponentially. This exercise is designed to show how using data without an understanding of the phenomenon that generated them can lead to inaccurate conclusions.

a. Plot the data points from Exercise 12. Do they appear almost to fall on a straight line?

b. Find the equation of the regression line and add its graph to the one you made in part a.

c. If you used the regression line as a model for decay of ^{239}U, how long would it take for the initial 1 gram to decay to half that amount? Compare this with your answer to part d of Exercise 12.

d. The linear model represented by the regression line makes an absurd prediction concerning the amount of uranium-239 remaining after 1 hour. What is this prediction?

14. **Account Growth**   The table below shows the balance B in a savings account, in dollars, in terms of time t, measured as the number of years since the initial deposit was made.

Time t	Balance B
0	125.00
1	131.25
2	137.81
3	144.70
4	151.94

a. Was the yearly interest rate constant over the first 4 years? If so, explain why and find that rate. If not, explain why not. (Round the ratios to two decimal places.)

b. Estimate $B(2.75)$ and explain in practical terms what your answer means. (Assume that interest is compounded and deposited continuously.)

15. **Rates Vary**   The following table shows the balance in a savings account, in dollars, in terms of time, measured as the number of years since the initial deposit was made.

Time	Balance	Time	Balance
0	250.00	4	302.43
1	262.50	5	316.04
2	275.63	6	330.26
3	289.41		

Explain in terms of interest rates how the account grew. (Round the ratios to three decimal places.)

16. **Wages**   A worker is reviewing his pay increases over the past several years. The table below shows the hourly wage W, in dollars, that he earned as a function of time t, measured in years since the beginning of 2000.

Time t	Wage W
1	15.30
2	15.60
3	15.90
4	16.25

a. By calculating ratios, show that the data in this table are exponential. (Round the quotients to two decimal places.)

b. What is the yearly growth factor for the data?

c. The worker can't remember what hourly wage he earned at the beginning of 2000. Assuming that W is indeed an exponential function, determine what that hourly wage was.

d. Find a formula giving an exponential model for W as a function of t.

e. What percentage raise did the worker receive each year?

f. Given that prices increased by 26% over the decade of the 2000s, use your model to determine whether the worker's wage increases kept pace with inflation.

17. **Stochastic Population Growth**   Many populations are appropriately modeled by an exponential function, at least for a limited period of time. But there are many factors contributing to the growth of any population, and many of them depend on chance. There are a number of ways to produce stochastic models. We consider one that is an illustration of a simple *Monte Carlo method*. We want to model a population that is initially 500 and grows at an average rate of 2% per year. To do this we make a table of values for population according to the following procedure. The first entry in the table is for time $t = 0$, and it records the initial value 500. To get the entry corresponding to $t = 1$, we roll a die and change the population according to the face that appears, using the following rule.

Face appearing	Population change
1	Down 2%
2	Down 1%
3	No change
4	Up 2%
5	Up 4%
6	Up 9%

To get the entry corresponding to $t = 2$, we roll a die and change the population from $t = 1$, again using the above rule. This procedure is then followed for $t = 3$, and so on.

(continued)

a. Using this procedure, make a table recording the population values for years 0 through 10.

b. Plot the data points from your table and the exponential model on the same screen. Comment on the level of agreement.

18. **Growth Rate of a Tubeworm**   A study of the growth rate and life span of a marine tubeworm concludes that it is the longest-lived noncolonial marine invertebrate known.[13] Since tubeworms live on the ocean floor and have a long life span, scientists do not measure their age directly. Instead, scientists measure their growth rate at various lengths and then construct a model for growth rate in terms of length. On the basis of that model, scientists can find a relationship between *age* and length. This is a good example of how rates of change can be used to determine a relationship when direct measurement is difficult or impossible. The following table shows for a tubeworm the rate of growth in length, measured in meters per year, at the given length, in meters.

Length in meters	Growth rate in meters per year
0	0.0510
0.5	0.0255
1.0	0.0128
1.5	0.0064
2.0	0.0032

a. Often in biology the growth rate is modeled as a decreasing linear function of length. For some organisms, however, it may be appropriate to model the growth rate as a decreasing exponential function of length. Use the data in the table to decide which model is more appropriate for the tubeworm, and find that model. Give a practical explanation of the slope or percentage decay rate, whichever is applicable.

b. Use functional notation to express the growth rate at a length of 0.64 meter, and then calculate that value using your model from part a.

4.3 SKILL BUILDING EXERCISES

Finding an Exponential Formula   In Exercises S-1 through S-7, use the given information to find a formula for the exponential function $N = N(t)$.

S-1. The initial value of N is 7. If t is increased by 1, N is multiplied by 8.

S-2. The initial value of N is 6. If t is increased by 1, N is divided by 13.

S-3. The initial value of N is 12. If t is increased by 7, the effect is to multiply N by 62.

S-4. The initial value of N is 8. Increasing t by 2 divides N by 9.

S-5. $N(1) = 4$ and $N(3) = 16$.

S-6. $N(2) = 8$ and $N(5) = 1$.

S-7. The initial value of N is 4. If t is increased by 3, N is divided by 8.

Testing Exponential Data   In Exercises S-8 through S-16, test the given data to see whether it is exponential. For those that are exponential, find an exponential model.

S-8.

x	0	1	2	3
y	2.6	7.8	23.4	70.2

S-9.

x	0	2	4	6
y	5	10	20	40

S-10.

x	0	2	4	6
y	5	9	21	43

S-11.

x	0	2	4	6
y	6	18	54	162

S-12.

x	0	2	4	6
y	5	15	25	35

---

[13]D. Bergquist, F. Williams, and C. Fisher, "Longevity record for deep-sea invertebrate," *Nature* **403** (2000), 499–500. Their conservative estimate for the life span is between 170 and 250 years.

S-13.

x	0	2	4	6
y	100	90	80	70

S-14.

x	0	2	4	6
y	1000	100	10	1

S-15.

x	3	6	9	12
y	5	10	20	40

S-16.

x	3	6	9	12
y	10	5	2.5	1.25

Growth Rate In Exercises S-17 through S-21, you are asked about the growth factor of an exponential function $N = N(t)$.

S-17. Increasing t by 2 units multiplies N by a, where $a > 0$. Find the growth factor for N.

S-18. Increasing t by 2 units divides N by a, where $a > 0$. Find the growth factor for N.

S-19. Increasing t by 1 unit multiplies N by a^4, where $a > 0$. How does an increase of t by 5 units affect N?

S-20. One entry in a data table for N gives $t = 5$ and $N = 9$. Another entry gives $t = 7$ and $N = 45$. What value of N should be given for $t = 8$?

S-21. One entry in a data table for N gives $t = 4$ and $N = 6$. Another gives $t = 7$ and $N = 48$. What value of N should be given for $t = 8$?

4.4 MODELING NEARLY EXPONENTIAL DATA

As in the case of linear data, rarely can experimentally gathered data be modeled exactly by an exponential function, but such data can in many cases be closely approximated by an exponential model. This process is called *exponential regression.*

Exponential Regression

The following data, taken from the *Information Please Almanac,* show the U.S. population from 1800 to 1860, just prior to the Civil War.

Date	1800	1810	1820	1830	1840	1850	1860
Population in millions	5.31	7.24	9.64	12.87	17.07	23.19	31.44

Let t be the time in years since 1800 and N the population in millions. The table for N as a function of t is then

t	0	10	20	30	40	50	60
N	5.31	7.24	9.64	12.87	17.07	23.19	31.44

Since the data table shows population growth, it is not unreasonable to suspect that the data might be exponential in nature. If you calculate successive quotients of N, you will find that the data are not exactly exponential, but as with linear regression, it may still be appropriate to make an exponential model that approximates the data. That is the first question we want to answer:

"Is it reasonable to approximate the data with an exponential model?"

First we enter the data in the calculator as we have done in Figure 4.18. Next we plot them to get a look at the overall trend. Figure 4.19 shows the classic shape of exponential growth, and the idea that the data may be exponential is reinforced.

L1	L2	L3 1
0	5.31	------
10	7.24	
20	9.64	
30	12.87	
40	17.07	
50	23.19	
60	31.44	
L1(1)=0		

FIGURE 4.18 Entering pre–Civil War population data

The way we analyze nearly exponential data reminds us of the way nearly linear data are analyzed.

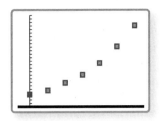

FIGURE 4.19 A plot of pre–Civil War data

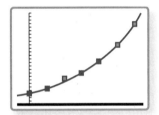

FIGURE 4.20 Regression data

However, a caution is in order here. It is very difficult in general to be sure that data are exponential by looking at their plot, because there are other types of data that give a similar appearance. For the current example, it is enough to know that an exponential model is often appropriate for population growth and that a plot of the data seems to confirm this.

Now we find an exponential function to model the data. We obtain the information for the exponential model shown in Figure 4.20. Let's write down what all the letters there mean. We have rounded to three decimal places.[14]

$$X = t, \text{years since 1800}$$

$$Y = N, \text{the population in millions}$$

$$a = 5.337 = \text{initial value of regression model}$$

$$b = 1.030 = \text{growth factor of regression model.}$$

Thus we can approximate N with the exponential function

$$N = 5.337 \times 1.030^t.$$

In Figure 4.21 we have added the graph of this function to the original data plot of N in Figure 4.19. The level of agreement between the exponential curve and the data points is striking.[15]

We use the resulting formula in the same ways in which we used the regression line formula for linear data. For example, the growth factor 1.030 for N tells us immediately that from 1800 to 1860, U.S. population grew at a rate of 3.0% per year, or by about 34% per decade. We might also use the formula to estimate the population in 1870. To do this, we would put 70 in for t:

Exponential regression estimate for 1870 $= 5.337 \times 1.030^{70} = 42.26$ million.

There is some uncertainty about the actual U.S. population in 1870. The 1870 census reported the number as 38.56 million. This figure was later revised to 39.82 million because it was thought that the southern population had been undercounted. Whether we use the original or the revised 1870 census estimate, it is clear that U.S. population grew a good deal less than we would have expected from the exponential trend established from 1800 to 1860. Such a discrepancy leads us to seek a historical explanation, and the most obvious culprit is the Civil War. In fact the death and disruption of the Civil War may have had long-lasting effects on U.S. population growth. From 1790 to 1860, population grew at a steady rate of around 34% per decade. From 1860 to 1870, population grew by only 27%, and this rate steadily declined to its historical low of 7.2% from 1930 to 1940, the decade of the Great Depression.

FIGURE 4.21 Exponential model for U.S. population data

KEY IDEA 4.5 **EXPONENTIAL REGRESSION**

When a plot of data appears to be exponential in shape, using exponential regression to find a model may be appropriate. The real-world context can provide important confirmation that an exponential model is appropriate.

---

[14]In exponential regression, it is often necessary to use more digits than our standard convention of two.
[15]Although exponential models are often used to study population growth, such close agreement between the model and real data is unusual.

EXAMPLE 4.10   **COLONIAL POPULATION GROWTH**

The following data table shows colonial population from 1610 to 1670, where t is the number of years since 1610 and $C = C(t)$ is the population in thousands.[16] The source for the data, the *Information Please Almanac*, cautions that records from this period are spotty, so the numbers should be considered estimates.

t = years since 1610	0	10	20	30	40	50	60
C = population in thousands	0.35	2.3	4.6	26.6	50.4	75.1	111.9

Part 1 Plot the data to get an overall view of their nature. Does the plot indicate that the data might be modeled by an exponential function?

Part 2 Use exponential regression to construct a model for C. (Round regression parameters to three decimal places.) Add the graph of the exponential model to the plot of population data.

Part 3 Compare the population growth rate as given by exponential regression during colonial times with that in the 60 years preceding the Civil War.

Solution to Part 1 We enter the data in the calculator as shown in Figure 4.22 and then plot them as in Figure 4.23. The points in Figure 4.23 show some features of exponential growth; it is slow at first and faster later on, but the picture certainly leaves room for doubt.

Solution to Part 2 We get the exponential regression parameters for C as shown in Figure 4.24. Let's record the appropriate correspondences:

$$Y_1 = C, \text{population in thousands}$$

$$X = t, \text{years since 1610}$$

$$a = 0.704 = \text{initial value of regression model}$$

$$b = 1.100 = \text{growth factor of regression model.}$$

Using these values, we get the exponential model for C:

$$C = 0.704 \times 1.100^t.$$

Finally, in Figure 4.25, we add the graph of the exponential model to the plot of data points for C. The fit is not very good.

L1	L2	L3	1
0	.35	------	
10	2.3		
20	4.6		
30	26.6		
40	50.4		
50	75.1		
60	111.9		

L1(1)=0

FIGURE 4.22 Entering colonial data

FIGURE 4.23 Plot of colonial data

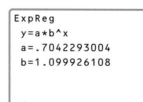

```
ExpReg
y=a*b^x
a=.7042293004
b=1.099926108
```

FIGURE 4.24 Regression data for C

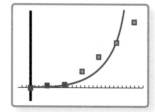

FIGURE 4.25 Exponential model added to plot of C versus t

A growth factor of 1.100 corresponds to a percentage growth rate of 10%.

Solution to Part 3 According to the exponential model we have constructed, colonial population grew at a rate of about 10% per year as compared with 3% per year from 1800 to 1860. But since the colonial model is so inaccurate, the estimate of 10% is unreliable.

[16]Historical records from these periods ignored Native Americans.

TEST YOUR UNDERSTANDING | **FOR EXAMPLE 4.10**

The following table shows the (estimated) world population in millions t years after 1750.

t = time in years since 1750	0	50	100	150	200	250
W = population in millions	791	978	1262	1650	2519	6070

Make an exponential model for W as a function of t. One date from the table shows a population much larger than that given by the exponential model. Which date is that? ∎

In the following example the data themselves are not approximately exponential, but taking the difference with a fixed value gives nearly exponential data.

EXAMPLE 4.11 **LIFE EXPECTANCY**

The table below shows the at-birth life expectancy E, in years, of a white male born in the United States t years after 1900. The bottom line of the table shows $D = 80 - E$, the difference between 80 years and life expectancy.

t = time in years since 1900	10	20	40	50	70	90	100
E = life expectancy	50.2	56.3	62.8	66.3	67.9	72.7	74.8
$D = 80 - E$	29.8	23.7	17.2	13.7	12.1	7.3	5.2

Part 1 Use exponential regression to model the difference D as an exponential function of t.

Part 2 Use your answer to part 1 to find a model for the life expectancy E in terms of t.

Part 3 What does your model yield for the life expectancy of a white male born in 2010? (The official projected value is 76.5 years.)

Part 4 Plot the data for E and the graph of the model you found in part 2.

Part 5 According to this model, when would the life expectancy be 45 years?

Solution to Part 1 The regression parameters are shown in Figure 4.26. Rounding to three decimal places gives the model

$$D = 35.441 \times 0.982^t.$$

```
ExpReg
 y=a*b^x
 a=35.4413213
 b=.9822046574
```

FIGURE 4.26 Exponential regression parameters

Solution to Part 2 Because D is the difference between 80 years and life expectancy, $D = 80 - E$. We use this formula to find an expression for E:

$$D = 80 - E$$
$$E + D = 80$$
$$E = 80 - D$$
$$E = 80 - 35.441 \times 0.982^t.$$

Solution to Part 3 Now 2010 is $t = 110$ years since 1900, so that is the value we use for t:

$$E = 80 - 35.441 \times 0.982^t$$
$$E(110) = 80 - 35.441 \times 0.982^{110}$$
$$= 75.19.$$

The model gives a life expectancy of about 75.2 years for a child born in 2010.

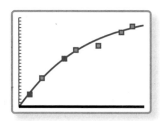

FIGURE 4.27 Model and data

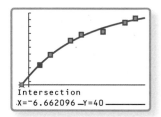

FIGURE 4.28 When life expectancy was 40 years

Solution to Part 4 We record the following correspondences for the graph in Figure 4.27:

$$X = t, \text{time in years since 1900 on horizontal axis}$$

$$Y_1 = E, \text{Life expectancy on vertical axis.}$$

Solution to Part 5 We need to solve the equation

$$40 = 80 - 35.441 \times 0.982^{110}.$$

We use the crossing-graphs method to do this. In Figure 4.28 we have enlarged the viewing window a bit and added the target value of 40 to the picture. We find a life expectancy of 40 years when $t = -6.66$. That is 6.66 years *before* 1900, or about 1893.

TEST YOUR UNDERSTANDING | **FOR EXAMPLE 4.11**

A small dog will weigh 20 pounds at adulthood. The following table shows the weight w, in pounds, of the dog t months after birth. The difference D between the maximum weight of 20 pounds and the current weight W, in pounds, is recorded in the third row of the table.

$t = $ time in months	2	4	7	8	10
$W = $ weight in pounds	5.0	10.5	15.5	17.0	19.0
$D = 20 - W$	15.0	9.5	4.5	3.0	1.0

Find an exponential model for D as a function of t. Use your model for D to give a formula for W in terms of t. ∎

 Choosing a Model

Once we learn about a new type of regression model, it is tempting to apply the model without much reflection. But we should keep in mind that the calculator can perform linear regression on any data—regardless of whether use of a linear model for the data is fitting. A similar statement is true for exponential regression. That is why it is crucial to consider carefully whether a particular type of model (say, linear or exponential) is appropriate given the real-world context. This is especially important when we realize that there are several different models for nonlinear data.

Linear models are appropriate when the context leads us to expect the data to exhibit a nearly constant rate of change. A plot of the data should then reflect the linear nature. Exponential models are appropriate when we expect a nearly constant percentage rate of change. In this case the plot should be curved. But there are other important models whose graphs have a shape similar to that of exponential functions. So one must be wary of relying solely on the perceived shape of a graph or even on how closely a graph fits given data points. It is always possible to choose a model that fits given data points *exactly*. Often it is appropriate to rely on experts in a given field to tell us which type of model may be best applied.

ANSWERS FOR TEST YOUR UNDERSTANDING

4.10 $W = 659.892 \times 1.008^t$. For 2000 the population shown in the table was much larger than that given by the model.

4.11 $D = 33.450 \times 0.725^t$; $W = 20 - 33.450 \times 0.725^t$

4.4 EXERCISES

Special Rounding Instructions   For this exercise set, round all regression parameters to three decimal places, but round all other answers to two decimal places unless otherwise indicated.

1. **Choose the Model**   A car travels down a straight highway at a constant speed. Which is a better model for the distance traveled as a function of time: linear or exponential?

2. **Postal Rates**   The table below shows the cost s, in cents, of a domestic first-class postage stamp in the United States t years after 1900.

t = time in years since 1900	s = cost of stamp
19	2
32	3
58	4
71	8
78	15
85	22
95	32
102	37
109	44

   a. Use exponential regression to model s as an exponential function of t.

   b. What cost does your model give for a 1988 stamp? Report your answer to the nearest cent. (The actual cost was 25 cents.)

   c. Plot the data and the exponential model.

3. **Rare Coins**   The table below shows the estimated value C, in dollars, of an 1877 Indian Head Cent (Philadelphia mint mark) in *very fine* condition t years after 1950.

t = time in years since 1950	C = value in dollars
0	25
30	400
45	625
54	1750
60	2000

   a. Use exponential regression to model C as an exponential function of t.

   b. According to your exponential model, by what percentage does the value of the 1877 cent increase from year to year?

4. **Cost of Scientific Periodicals**   The table below shows the average cost C, in dollars, of chemistry and physics periodicals t years after 1980.[17]

t = years since 1980	C = cost in dollars
0	140
5	250
10	410
15	780
20	1300
22	1520

   a. Make an exponential model of C as a function of t.

   b. Plot the data and the exponential model.

   c. What was the yearly percentage growth rate of the cost of chemistry and physics periodicals?

   d. If this exponential trend continues, what will be the expected average cost of physics and chemistry periodicals in 2015? Round your answer to the nearest dollar.

5. **Population Growth**   The following table shows the size, in thousands, of an animal population at the start of the given year.

---

[17]Data taken from "History and future of scholarly publishing" by Robert B. Townsend, American Historical Association, *Perspectives,* (October 2003).

Find an exponential model for the population.

Year	Population (thousands)
2008	2.30
2009	2.51
2010	2.73
2011	2.98
2012	3.25

6. **Magazine Sales**   The following table shows the income, measured in thousands of dollars, from sales of a certain magazine at the start of the given year.

Year	2008	2009	2010	2011	2012
Income	8.10	8.59	9.10	9.65	10.23

Find an exponential model for the income.

7. **Cell Phones**   The following table shows the number, in millions, of cell phone subscribers in the United States at the end of the given year.

Year	Subscribers (millions)
2005	207.9
2006	233.0
2007	255.4
2008	270.3
2009	285.6

a. Plot the data points.

b. Use exponential regression to construct an exponential model for the subscriber data.

c. Add the graph of the exponential model to the plot in part a.

d. What was the yearly percentage growth rate from the end of 2005 through the end of 2009 for cell phone subscribership?

e. In 2009 an executive had a plan that could make money for the company, provided that there would be at least 320 million cell phone subscribers by the end of 2011. Solely on the basis of an exponential model for the data in the table, would it be reasonable for the executive to implement the plan?

8. **Household Income**   The following table shows the median income, in thousands of dollars, of American families for 2003 through 2008.

Year	Income (thousands of dollars)
2003	52.68
2004	54.06
2005	56.19
2006	58.41
2007	61.36
2008	61.52

a. Plot the data.

b. Use exponential regression to construct an exponential model for the income data.

c. What was the yearly percentage growth rate in median family income during this period?

d. From 2003 through 2008, inflation was about 3% per year. Did median family income keep pace with inflation during this period?

9. **National Health Care Spending**   The following table shows national health care costs, measured in billions of dollars.

Date	1970	1980	1990	2000	2010
Costs in billions	75	253	714	1353	2570

Dmitriy Shironosov/Shutterstock.com

a. Plot the data.

b. Find an exponential function that approximates the data for health care costs.

c. By what percent per year were national health care costs increasing during the period from 1970 through 2010?

(continued)

d. Use functional notation to express how much money was spent on health care in the year 2011, and then estimate that value.

10. A Bad Data Point A scientist sampled data for a natural phenomenon that she has good reason to believe is appropriately modeled by an exponential function $N = N(t)$. A laboratory assistant reported that there may have been an error in recording one of the data points but is not certain which one. Which is the suspect data point? Explain your reasoning.

t	N	t	N
0	21.3	4	204.4
1	37.5	5	359.7
2	66.0	6	633.1
3	95.3		

11. Grazing Rabbits

stock_shot/Shutterstock.com

The amount A of vegetation (measured in pounds) eaten in a day by a grazing animal is a function of the amount V of food available (measured in pounds per acre).[18] Even if vegetation is abundant, there is a limit, called the *satiation level,* to the amount the animal will eat. The following table shows, for rabbits, the difference D between the satiation level and the amount A of food eaten for a variety of values of V.

V = vegetation level	D = satiation level − A
27	0.16
36	0.12
89	0.07
134	0.05
245	0.01

a. Draw a plot of D versus V.

b. Find an exponential function that approximates D.

c. The satiation level of a rabbit is 0.18 pound per day. Use this, together with your work in part b, to find a formula for A.

d. Find the vegetation level V for which the amount of food eaten by the rabbit will be 90% of its satiation level of 0.18 pound per day.

12. Growth in Length In the fishery sciences it is important to determine the length of a fish as a function of its age. One common approach, the von Bertalanffy model, uses a decreasing exponential function of age to describe the growth in length yet to be attained; in other words, the difference between the maximum length and the current length is supposed to decay exponentially with age. The following table shows the length L, in inches, at age t, in years, of the North Sea sole.[19]

t = age	L = length	t = age	L = length
1	3.7	5	12.7
2	7.5	6	13.5
3	10.0	7	14.0
4	11.5	8	14.4

The maximum length attained by the sole is 14.8 inches.

a. Make a table showing, for each age, the difference D between the maximum length and the actual length L of the sole.

b. Find the exponential function that approximates D.

c. Find a formula expressing the length L of a sole as a function of its age t.

d. Draw a graph of L versus t.

e. If a sole is 11 inches long, how old is it?

f. Calculate $L(9)$ and explain in practical terms what your answer means.

[18]This exercise is based on the work of J. Short, "The functional response of kangaroos, sheep and rabbits in an arid grazing system," *Journal of Applied Ecology* **22** (1985), 435–447.
[19]The table is from the work of A. Bückmann, as described by R. J. H. Beverton and S. J. Holt, *On the Dynamics of Exploited Fish Populations,* Fishery Investigations, Series 2, Volume 19 (London: Ministry of Agriculture, Fisheries and Food, 1957).

13. **Nearly Linear or Exponential Data** One of the two tables below shows data that are better approximated with a linear function, and the other shows data that are better approximated with an exponential function. Make plots to identify which is which, and then use the appropriate regression to find models for both.

t	f(t)
1	3.62
2	23.01
3	44.26
4	62.17
5	83.25

Table A

t	g(t)
1	3.62
2	5.63
3	8.83
4	13.62
5	21.22

Table B

14. **Atmospheric Pressure** The table below gives a measurement of atmospheric pressure, in grams per square centimeter, at the given altitude, in kilometers.[20]

Altitude	Atmospheric pressure
5	569
10	313
15	172
20	95
25	52

Rob kemp/Shutterstock.com

(For comparison, 1 kilometer is about 0.6 mile, and 1 gram per square centimeter is about 2 pounds per square foot.)

   a. Plot the data on atmospheric pressure.

   b. Make an exponential model for the data on atmospheric pressure.

   c. What is the atmospheric pressure at an altitude of 30 kilometers?

   d. Find the atmospheric pressure on Earth's surface. This is termed *standard atmospheric pressure*.

   e. At what altitude is the atmospheric pressure equal to 25% of standard atmospheric pressure?

15. **Sound Pressure** Sound exerts pressure on the human ear. Increasing loudness corresponds to greater pressure. The table below shows the pressure P, in dynes per square centimeter, exerted on the ear by sound with loudness D, measured in decibels.

Loudness D	Pressure P
65	0.36
85	3.6
90	6.4
105	30
110	50

   a. Plot the data.

   b. Find an exponential model of P as a function of D.

   c. How is pressure on the ear affected when loudness is increased by 1 decibel?

16. **Magnitude and Distance** Astronomers measure brightness of stars using both the *absolute magnitude*, a measure of the true brightness of the star, and the *apparent magnitude*, a measure of the brightness of a star as it appears from Earth.[21] The difference between apparent and absolute magnitude should yield information about the distance to the star. The table on the following page gives magnitude difference m and distance d, measured in light-years, for several stars.

---

[20]This exercise is based on *Space Mathematics* by B. Kastner, published in 1985 by NASA.
[21]Larger apparent magnitudes correspond to dimmer stars.

(continued)

Star	Magnitude difference m	Distance d
Algol	2.56	105
Aldebaran	1.56	68
Capella	0.66	45
Canopus	2.38	98
Pollux	0.13	35
Regulus	2.05	84

a. Plot d versus m.

b. Give an exponential model for the data.

c. If one star shows a magnitude difference 1 greater than the magnitude difference that a second star shows, how do their distances from Earth compare?

d. Alphecca shows a magnitude difference of 1.83. How far is Alphecca from Earth?

e. Alderamin is 52 light-years from Earth and has an apparent magnitude of 2.47. Find the absolute magnitude of Alderamin.

17. **Growth in Length of Haddock**   A study by Riatt showed that the maximum length a haddock could be expected to grow is about 53 centimeters. Let $D = D(t)$ denote the difference between 53 centimeters and the length at age t years. The table below gives experimentally collected values for D.

Age t	Difference D
2	28.2
5	16.1
7	9.5
13	3.3
19	1.0

a. Find an exponential model of D as a function of t.

b. Let $L = L(t)$ denote the length in centimeters of a haddock at age t years. Find a model for L as a function of t.

c. Plot the graph of the experimentally gathered data for the length L at ages 2, 5, 7, 13, and 19 years along with the graph of the model you made for L. Does this graph show that the 5-year-old haddock is a bit shorter or a bit longer than would be expected?

d. A fisherman has caught a haddock that measures 41 centimeters. What is the approximate age of the haddock?

18. **Caloric Content Versus Shell Length**   In 1965 Robert T. Paine[22] gathered data on the length L, in millimeters, of the shell and the caloric content C, in calories, for a certain mollusk. The table below is adapted from those data.

L = length	C = calories
7.5	92
13	210
20	625
24	1035
31	1480

a. Find an exponential model of calories as a function of length.

b. Plot the graph of the data and the exponential model. Which of the data points show a good deal less caloric content than the model would predict for the given length?

c. If length is increased by 1 millimeter, how is caloric content affected?

19. **Injury Versus Speed**   The following data are adapted from a report[23] by D. Solomon that relates the number N of persons injured per 100 accident-involved vehicles to the travel speed s in miles per hour.

s	20	30	40	50	60
N	25	32	38	43	60

---

[22]See "Natural history, limiting factors and energetics of the opisthobranch *Navanax inermis*," *Ecology* **46** (1965), 603–619.
[23]*Accidents on Main Rural Highways Related to Speed, Driver, and Vehicle* (Washington, DC: Federal Highway Administration, July 1964), p. 11.

a. Find an exponential model for N as a function of s.

b. Calculate $N(70)$ and explain in practical terms what your answer means.

c. How does an increase in 1 mile per hour of speed affect the number of people injured per 100 accident-involved vehicles?

20. **Gray Wolves in Wisconsin** Gray wolves were among the first mammals protected under the Endangered Species Act in the 1970s. Wolves recolonized in Wisconsin beginning in 1980. Their population grew reliably after 1985 as follows:[24]

Year	Wolves	Year	Wolves
1985	15	1993	40
1986	16	1994	57
1987	18	1995	83
1988	28	1996	99
1989	31	1997	145
1990	34	1998	178
1991	40	1999	197
1992	45	2000	266

a. Explain why an exponential model may be appropriate.

b. Are these data exactly exponential? Explain.

c. Find an exponential model for these data.

d. Plot the data and the exponential model.

e. Comment on your graph in part d. Which data points are below or above the number predicted by the exponential model?

21. **Walking in Seattle** It is common in large cities for people to travel to the center city and then walk to their final destination. In Seattle the percent P of pedestrians who walk at least D feet from parking facilities in the center city is given partially in the accompanying table.[25]

Distance D	Percent P walking at least D feet
300	60
500	40
1000	18
1500	9
2000	5

a. Make a model of P as an exponential function of D.

b. What percentage of pedestrians walk at least 200 feet from parking facilities?

c. When models of this sort are made, it is important to remember that many times they are only rough indicators of reality. Often they apply only to parts of the data and may tell very little about extremes of the data. Explain why it is not appropriate to view this as an accurate model for very short distances walked.

22. **Traffic in the Lincoln Tunnel** Characteristics of traffic flow include density D, which is the number of cars per mile, and average speed s in miles per hour. Traffic system engineers have investigated several methods for relating density to average speed. One study[26] considered traffic flow in the north tube of the Lincoln Tunnel and fitted an exponential function to observed data. Those data are partially presented in the table on the following page.

Caro/Alamy

(continued)

---

[24]Data courtesy of the International Wolf Center, Ely, Minnesota. For more information about wolves and their recovery, go to http://www.wolf.org

[25]Adapted from Institute of Traffic Engineers, *Transportation and Traffic Engineering Handbook,* ed. John E. Baerwald (Englewood Cliffs, NJ: Prentice Hall, 1976, 180).

[26]H. Greenberg, *A Mathematical Analysis of Traffic Flow,* Tunnel Traffic Capacity Study, the Port of New York Authority, New York, 1958.

Speed s	Density D
32	34
25	53
20	74
17	88
13	102

a. Make an approximate exponential model of D as a function of s.

b. Express using functional notation the density of traffic flow when the average speed is 28 miles per hour, and then calculate that density.

c. If average speed increases by 1 mile per hour, what can be said about density?

23. **Frequency of Earthquakes** The table below gives the average number N of earthquakes[27] of magnitude at least M that occur each year worldwide.

Magnitude M	Number N with magnitude at least M
6	95.8
6.1	77.8
6.6	27.6
7	12.1
7.3	6.5
8	1.5

a. Gutenberg and Richter fitted these data with an exponential function. Find an approximate exponential model for the data.

b. How many earthquakes per year of magnitude at least 5.5 can be expected?

c. Gutenberg and Richter found that the model fit observed data well up to a magnitude of about 8, but for magnitudes above 8 there were consider-

ably fewer quakes than predicted by the model. How many quakes per year of magnitude 8.5 or greater does the model predict?

d. What is the limiting value for N? Explain in practical terms what this means.

e. How does the number of earthquakes per year of magnitude M or greater compare with the number of earthquakes of magnitude $M + 1$ or greater?

24. **Medicare Disbursements** The following table is taken from the *Statistical Abstract of the United States*. It shows Medicare disbursements M in billions of dollars t years since 1990.

t = time in years since 1990	M = disbursements in billions
0	109.71
5	180.10
10	219.28
15	336.88
16	380.46
19	499.84

a. Make an exponential model of M versus t.

b. Plot the data and the exponential model.

c. When does the exponential model indicate that disbursements will reach one trillion dollars (that is, $M = 1000$)?

25. **Economic Growth of the United States** This exercise and the next refer to the *gross domestic product* (GDP), which is the market value of the goods and services produced in a country in a given year. The data in these exercises are adapted from a report that predicted rates of growth for the world economy over a 15-year period.[28]

The following table shows the GDP of the United States, in trillions of dollars. The data are based on figures and projections in the report.

---

[27]Data adapted from B. Gutenberg and C. F. Richter, "Earthquake magnitude, intensity, energy, and acceleration," *Bull. Seism. Soc. Am.* **46** (1956).

[28]"Foresight 2020: Economic, industry and corporate trends," a report from the Economist Intelligence Unit, sponsored by Cisco Systems, March 2006, available at http://newsroom.cisco.com/dlls/tln/research.studies/2020foresight/index.html. The numbers for GDP are measured in U.S. dollars and have been adjusted to eliminate the difference in price levels between countries.

Year	GDP in trillions of dollars
2005	12.46
2008	13.62
2011	14.85
2014	16.13
2017	17.52

Use exponential regression to model the GDP of the United States as a function of time in years since 2005.

26. Economic Growth of China *This is a continuation of Exercise 25.* The economy of China is smaller than that of the United States but is growing more quickly. Here is a projection for the GDP of China:

$$C = 8.2 \times 1.06^t.$$

In this projection t is the number of years since 2005 and C is the GDP of China, measured in trillions of U.S. dollars.

a. Compare the precentage growth rates for the GDP of the United States and China.

b. Find the year when the GDP of China will overtake the U.S. GDP.

27. Research Project For this project, you should collect and analyze data for a population of M&M's™. Start with four candies, toss them on a plate, and add one for each candy that has the M side up; record the data. Repeat this seven times and see how close the data are to being exponential.

4.4 SKILL BUILDING EXERCISES

Round all answers to two decimal places.

S-1. Population Growth A population of deer is introduced into a region with abundant resources. Which regression model should be used to approximate the population data over time: linear or exponential?

S-2. Inflation During a certain period, the price of a bag of groceries grows by about 3% each year. Which regression model should be used to approximate the price of a bag of groceries as a function of time: linear or exponential?

S-3. Running Speed Versus Length We have gathered data on running speed versus length of small animals. We want to use either a linear model or

an exponential model to fit the data. A biologist tells us that, for the group of small animals in our study, a 1-unit increase in length always results in the same increase in running speed. Should we use a linear model or an exponential model?

S-4. Educational Spending We have gathered data on state educational spending during the past 10 years. A state auditor tells us that state spending on education has increased by the same percentage each year during the past 10 years. Should we use a linear model or an exponential model for our data?

S-5. Exponential Regression Use exponential regression to fit the following data set.

x	1	2	3	4	5
y	53.0	55.2	57.4	59.7	62.0

S-6. Exponential Regression Use exponential regression to fit the following data set.

x	1	2	3	4	5
y	46.0	49.2	52.7	56.4	60.3

S-7. Exponential Regression Use exponential regression to fit the following data set. Give the exponential model, and plot the data along with the model.

x	1	2	3	4	5
y	4.1	8.7	19.2	28.6	64.7

S-8. Exponential Regression Use exponential regression to fit the following data set. Give the exponential model, and plot the data along with the model.

x	1	2	3	4	5
y	0.7	0.3	0.1	0.05	0.01

S-9. Exponential Regression Use exponential regression to fit the following data set. Give the exponential model, and plot the data along with the model.

x	4.2	7.9	10.8	15.5	20.2
y	7.5	8.1	8.5	10.2	12.3

(continued)

S-10. Exponential Regression   Use exponential regression to fit the following data set. Give the exponential model, and plot the data along with the model.

x	22.4	27.3	29.4	34.1	38.6
y	0.053	0.025	0.011	0.005	0.002

S-11. Exponential Regression   Use exponential regression to fit the following data set. Give the exponential model, and plot the data along with the model.

x	1	2	3	4	5
y	3.7	4.3	6.1	9.1	13.6

S-12. Exponential Regression   Use exponential regression to fit the following data set. Give the exponential model, and plot the data along with the model.

x	3	7	9	10	15
y	33.5	988.8	5470.8	12,830	893,442

S-13. Exponential Regression   Use exponential regression to fit the following data set. Give the exponential model, and plot the data along with the model.

x	2	5	6	9	10
y	4.2	6.4	7.4	11.3	12.9

S-14. Exponential Regression   Use exponential regression to fit the following data set. Give the exponential model, and plot the data along with the model.

x	4	11	15	17	18
y	1.0	17.3	87.6	197.1	295.6

Using Exponential Models   In Exercises S-15 through S-24, we use exponential models from Exercises S-5 through S-14 to obtain further information.

S-15. Use the model found in Exercise S-5 to calculate $y(3.3)$ and to solve $y = 65$ for x.

S-16. Use the model found in Exercise S-6 to calculate $y(4.4)$ and to solve $y = 70$ for x.

S-17. Use the model found in Exercise S-7 to calculate $y(7)$ and to solve $y = 21$ for x.

S-18. Use the model found in Exercise S-8 to calculate $y(2.5)$ and to solve $y = 0.25$ for x.

S-19. Use the model found in Exercise S-9 to calculate $y(10)$ and to solve $y = 9$ for x.

S-20. Use the model found in Exercise S-10 to calculate $y(25)$ and to solve $y = 0.04$ for x.

S-21. Use the model found in Exercise S-11 to calculate $y(2.2)$ and to solve $y = 8$ for x.

S-22. Use the model found in Exercise S-12 to calculate $y(9.5)$ and to solve $y = 900$ for x.

S-23. Use the model found in Exercise S-13 to calculate $y(8.8)$ and to solve $y = 10$ for x.

S-24. Use the model found in Exercise S-14 to calculate $y(12)$ and to solve $y = 220$ for x.

4.5 LOGARITHMIC FUNCTIONS

Logarithmic functions have important applications to the world around us. The most common applications, such as the Richter scale for measuring seismic disturbances, make use of the common, or base-10, logarithm. The natural logarithm also occurs in applications. We will mention some of its properties here and will make use of its connection with the exponential function in the next section.

The Richter Scale

The properties of the logarithm are easier to understand if we first look at a familiar scale that is logarithmic in nature. The seismograph measures ground movement during an earthquake, and that movement is normally reported using the *Richter scale*.

The Richter scale was developed in 1935 by Charles F. Richter as a device for comparing the sizes of earthquakes. A moderate earthquake may measure 5.3 on the Richter scale, and a strong earthquake may measure 6.3. The largest shocks ever recorded have a magnitude of 9.5. Thus a small change in the Richter number indicates a large change in the severity of the earthquake. In fact, an earthquake 6.3 in magnitude is 10 times as powerful as a 5.3 earthquake, and a 7.3 earthquake is 10 times 10, or 100, times as powerful as a 5.3 earthquake. In general, an increase of t units on the Richter scale indicates an earthquake 10^t times as strong. This is the key factor in understanding how the Richter scale works.

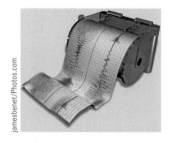

EXAMPLE 4.12 COMPARING SOME FAMOUS EARTHQUAKES

On December 16, 1811, an earthquake occurred near New Madrid, Missouri, that temporarily reversed the course of the Mississippi River. This was actually one of a series of earthquakes in the area, one of which is estimated to have had a Richter magnitude of 8.8. The area was sparsely populated at the time, and there were thought to be few fatalities. On October 17, 1989, a calamitous earthquake measuring 7.1 on the Richter scale occurred in the San Francisco Bay area. The earthquake killed 67 and injured over 3000.

Part 1 How much more powerful was the New Madrid quake than the 1989 San Francisco quake?

Part 2 If an earthquake 1000 times as powerful as the 1989 San Francisco earthquake occurred, what would its Richter scale measurement be?

Part 3 On December 26, 2004, an earthquake 80 times as powerful as the 1989 San Francisco quake struck the Indian Ocean near Indonesia. It caused a tsunami that resulted in the deaths of hundreds of thousands. What was the Richter scale reading for this quake?

Solution to Part 1 The New Madrid quake registered $8.8 - 7.1 = 1.7$ points higher on the Richter scale than the San Francisco quake. Thus the New Madrid quake was $10^{1.7} = 50.12$ times as strong as the San Francisco quake.

Solution to Part 2 A quake 10^t times as powerful registers t points higher on the Richter scale. Since $1000 = 10^3$, the proposed quake would register 3 points higher, or $7.1 + 3 = 10.1$, on the Richter scale.

Solution to Part 3 This problem is similar to part 2, except that it is easier to write 1000 as a power of 10 than it is to write 80 as a power of 10. We need to solve the equation

$$10^t = 80.$$

The crossing-graphs method can be used to do this. (We will discover a more direct method shortly.) In Figure 4.29 we have plotted 10^t, and we see from the prompt at the bottom of the screen that $10^t = 80$ when $t = 1.90$. That means that the quake was $10^{1.9}$ times as powerful as the San Francisco quake. Thus the quake registered $7.1 + 1.9 = 9.0$ on the Richter scale. At the time, this was the most powerful quake in 40 years.

```
Intersection
X=1.90309  Y=80
```

FIGURE 4.29 Solving $10^t = 80$

TEST YOUR UNDERSTANDING | **FOR EXAMPLE 4.12**

On January 12, 2010, a catastrophic earthquake measuring 7.0 on the Richter scale struck Haiti. On March 11, 2011, a quake of magnitude 9.0 struck off the coast of Japan. It caused a tsunami with devastating effects. How did the power of the two quakes compare? ∎

How the Common Logarithm Works

The logarithm of x is the exponent of 10 that gives x.

The common logarithm $\log x$ is defined as the power of 10 that gives x. For example, $\log 100$ is 2 because 2 is the power of 10 that gives 100. Alternatively, $\log x$ is the solution for t of the equation $10^t = x$. For example, with the calculator we find that $\log 5 = 0.7$. You should check using the calculator that $10^{0.7} = 5.0$, which is what the definition of the logarithm guarantees. This definition makes common logarithms of integral powers of 10 easy to calculate.

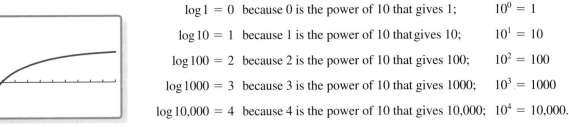

$$\log 1 = 0 \quad \text{because 0 is the power of 10 that gives 1;} \qquad 10^0 = 1$$
$$\log 10 = 1 \quad \text{because 1 is the power of 10 that gives 10;} \qquad 10^1 = 10$$
$$\log 100 = 2 \quad \text{because 2 is the power of 10 that gives 100;} \qquad 10^2 = 100$$
$$\log 1000 = 3 \quad \text{because 3 is the power of 10 that gives 1000;} \qquad 10^3 = 1000$$
$$\log 10{,}000 = 4 \quad \text{because 4 is the power of 10 that gives 10,000;} \quad 10^4 = 10{,}000.$$

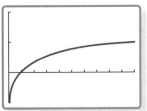

FIGURE 4.30 $\log x$ from 0 to 10

These calculations suggest two important features of the logarithm. The first is that, although the logarithm function increases without bound, it increases very slowly. One million is 10 to the sixth power, so $\log 1{,}000{,}000 = 6$. Similarly, the logarithm of 1 billion is only 9. We can see how the logarithm function grows slowly by looking at its graph. In Figure 4.30 we have graphed $\log x$ with a horizontal span of 0 to 10 and a vertical span of -1 to 2. Note that the graph crosses the horizontal axis at $x = 1$ since $\log 1 = 0$. In Figure 4.31 we use a horizontal span of 10 to 1,000,000 and a vertical span of 0 to 7. Note that the graph is indeed increasing slowly and that it is concave down. That is, it increases at a decreasing rate.

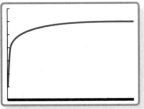

FIGURE 4.31 $\log x$ up to 1 million

The second feature of the logarithm that is apparent from these calculations is that increasing x by a factor of 10 increases the logarithm by 1 unit. This is precisely what happens with the Richter scale, which indicates that indeed it is a logarithmic scale.

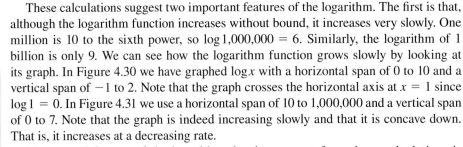

KEY IDEA 4.6 **THE COMMON LOGARITHM**

- The common logarithm, $\log x$, is the power of 10 that gives x.
  Alternatively, $\log x$ is the solution for t of the equation $10^t = x$.
- If we multiply a number by 10^t, the logarithm is increased by t units.
- The function $\log x$ increases very slowly, and its graph is concave down.

Note that the second property in Key Idea 4.6 is precisely what happens with the Richter scale. This property is critical when we are working with the Richter scale and with the logarithm in general.

The alternative definition of the logarithm also gives us an easy way to solve problems such as part 3 of Example 4.12. There we solved the equation $10^t = 80$ using the crossing-graphs method. The alternative definition of the logarithm says that the solution of this equation is $t = \log 80 = 1.90$. One of the reasons why the logarithm is so useful is that it enables us to solve such equations directly.

There are other important and familiar scales that are logarithmic in nature. Here is an example.

Sound	Decibels
Raindrops	40
Normal conversation	60
Busy city traffic	85
Hair dryers	90
Rock concerts	105
Chainsaws	110
iPod (at peak volume)	115
Jackhammers	120
Gunshot, fireworks	140

Sources: dangerousdecibels.org; WSJ Research.

EXAMPLE 4.13 **DECIBEL AS A MEASURE OF LOUDNESS**

The intensity of sound is measured in watts per square meter. Normally this is given as a *relative intensity*.[29] Loudness of sound is measured in *decibels*, and the number of decibels is 10 times the logarithm of the relative intensity:

$$\text{Decibels} = 10 \log (\text{Relative intensity}).$$

Part 1 If one sound has a relative intensity 100 times that of another sound, how do their decibel readings compare?

Part 2 A stereo speaker is producing music at 60 decibels. If a second stereo speaker producing music at 60 decibels is placed next to the first, the relative intensity doubles. What is the decibel reading of the combined speakers?

The solution of $10^t = 2$ is $t = \log 2$.

Solution to Part 1 Relative intensity is increased by a factor of $100 = 10^2$. Thus the logarithm is increased by 2 units. Since the decibel is 10 times the logarithm, the decibel level is increased by $10 \times 2 = 20$ units.

Solution to Part 2 We know that if relative intensity is increased by a factor of 10^t, then t is added to the logarithm. The relative intensity in this case is doubled. Thus we need first to solve $10^t = 2$. The solution is $t = \log 2 = 0.30$. Thus, doubling the relative intensity is the same as increasing the relative intensity by a factor of $10^{0.3}$. This means that the logarithm is increased by 0.3 unit. Since the decibel is 10 times the logarithm, we need to add $10 \times 0.3 = 3$ units on the decibel scale. The pair of speakers produces a reading of 63 decibels.

TEST YOUR UNDERSTANDING | **FOR EXAMPLE 4.13**

The noise from a lawn mower may reach 100 decibels. What would be the decibel reading of 3 such lawn mowers mowing side-by-side?   ■

---

[29]Physicists have adopted a base intensity of 10^{-12} watts per square meter. The relative intensity is intensity divided by the base intensity.

It is important to note that the decibel scale is logarithmic in nature because that is how we actually hear. As sound increases in intensity, perceived loudness increases logarithmically. The same principle applies to vision. Increasing intensity of light produces, to the human eye, a logarithmic increase in perceived brightness. Fechner's law, proposed in 1859, states that *if a stimulus is increased in geometrical progression, its resulting sensation increases in an arithmetic progression.* This means that sensation is a logarithmic function of stimulus. Although it should be considered a rule of thumb rather than a physical law, Fechner's law is at least correct for human vision and hearing.

 ## The Logarithm as the Inverse of the Exponential Function

We can think of the Richter scale in two different ways. We might record how powerful the quake was and then calculate the Richter magnitude. To examine this, let R be the Richter magnitude for an earthquake of relative strength P.[30] We have already noted that the Richter magnitude is a logarithmic function of P, and in fact

$$R = \log P.$$

But we can also look at things in the opposite or *inverse* way, thinking of the relative strength P as a function of the Richter magnitude R. In this setting, we look at the Richter reading and determine how powerful the quake was. How do we calculate P from R? We know that increasing P by a factor of 10 adds 1 unit to R. Thus, adding 1 unit to R causes P to be multiplied by 10. This means that P is an exponential function of R with growth factor 10, and

$$P = 10^R.$$

In general, there is a close relationship between the exponential and logarithmic functions: One is the inverse of the other.

KEY IDEA 4.7   **THE LOGARITHM AS AN INVERSE**

- The function $\log x$ is the inverse of the function 10^t:

$$t = \log x \text{ means that } x = 10^t.$$

- For the logarithmic function $\log x$, increasing x by a factor of 10^t adds t units to the logarithm.

- For the exponential function 10^x, increasing x by t units increases the function by a factor of 10^t.

EXAMPLE 4.14   **APPARENT MAGNITUDE AND BRIGHTNESS**

The apparent brightness of a star observed from Earth is a logarithmic function of the intensity of the light arriving from the star. Astronomers use the *magnitude* scale to measure apparent brightness. The *relative intensity* I is calculated by forming the ratio of the intensity of light from the star Vega to the intensity of light from the star of interest:

$$I = \text{Relative intensity of star} = \frac{\text{Intensity of Vega}}{\text{Intensity of star}}.$$

---

[30] By relative strength we mean the energy released at the source of the quake divided by the energy released at the source of a quake with magnitude 0 on the Richter scale.

The magnitude m is then given by

$$m = 2.5 \log I.$$

On the magnitude scale, higher magnitudes indicate dimmer stars.

Part 1 Since Vega is used as a base scale, the relative intensity of light coming from Vega is 1. What is the magnitude of Vega?

Part 2 Light arriving at Earth from the star Phecda has a relative intensity of 9.46. That is, light from Vega is 9.46 times as intense as light from Phecda. What is the magnitude of Phecda?

Part 3 Since magnitude is a logarithmic function of relative intensity, relative intensity is an exponential function of magnitude. Use a formula to express relative intensity as an exponential function of magnitude.

Part 4 Higher magnitudes indicate dimmer stars, and very bright stars can even have negative magnitudes. Sirius is the star that appears brightest of all in the night sky. It has an apparent magnitude of -1.45. How does the intensity of light arriving from Sirius compare with that of light arriving from Vega?

Solution to Part 1 For Vega, $I = 1$. Thus its magnitude is given by $m = 2.5 \log 1$. Since $10^0 = 1$, we have $\log 1 = 0$. Thus the magnitude of Vega is 0.

Solution to Part 2 The magnitude of Phecda is calculated using $m = 2.5 \log 9.46 = 2.44$.

Solution to Part 3 We first rearrange the formula for magnitude by dividing both sides by 2.5:

$$m = 2.5 \log I$$

$$\frac{m}{2.5} = \log I.$$

Now $m/2.5 = \log I$ means that $I = 10^{m/2.5}$.

Solution to Part 4 We can calculate the relative intensity of light arriving from Sirius by using the relationship we found in part 3:

$$I = 10^{m/2.5}$$

$$I = 10^{-1.45/2.5} = 0.26.$$

Thus light from Vega is 0.26 times as intense as light from Sirius.

TEST YOUR UNDERSTANDING | **FOR EXAMPLE 4.14**

If light arriving from a star is twice the intensity of light from Vega, what is the magnitude of the star? (*Caution:* Is the relative intensity 2 or 1/2?) ∎

The Natural Logarithm

Recall from page 5 of Chapter P that e is the special number whose approximate value is 2.71828.

The *natural logarithm*, denoted by $\ln x$, has properties similar to those of the common logarithm, and mathematicians generally prefer the natural logarithm to the common logarithm. The number e plays the role for the natural logarithm that the number 10 plays for the common logarithm. Just as $\log x$ is the power of 10 that gives x, $\ln x$ is the power of e that gives x. This gives immediately the following:

$$\ln 1 = 0 \quad \text{because} \quad 0 \text{ is the power of } e \text{ that gives } 1: e^0 = 1$$

$$\ln e = 1 \quad \text{because} \quad 1 \text{ is the power of } e \text{ that gives } e: e^1 = e$$

$$\ln e^2 = 2 \quad \text{because} \quad 2 \text{ is the power of } e \text{ that gives } e^2$$

$$\ln e^t = t \quad \text{because} \quad t \text{ is the power of } e \text{ that gives } e^t.$$

Just as with the common logarithm, we find $\ln x$ using a calculator when x is not an obvious power of e. For example, with the calculator we find that $\ln 7 = 1.95$. Key Idea 4.8 summarizes and compares the properties of the common and natural logarithms.

KEY IDEA 4.8   **COMMON AND NATURAL LOGARITHMS**

Common Logarithm	Natural Logarithm
$\log x$ is the power of 10 that gives x. $t = \log x$ is the solution for t of $10^t = x$. $$\log 10^t = t$$ $$10^{\log t} = t$$ The common logarithm is the inverse of the exponential function $y = 10^x$.	$\ln x$ is the power of e that gives x. $t = \ln x$ is the solution for t of $e^t = x$. $$\ln e^t = t$$ $$e^{\ln t} = t$$ The natural logarithm is the inverse of the exponential function $y = e^x$.

ANSWERS FOR TEST YOUR UNDERSTANDING

4.12 The quake in Japan was 100 times as powerful as the quake in Haiti.

4.13 104.77 or about 105 decibels.

4.14 -0.75

4.5 EXERCISES

Reminder   Round all answers to two decimal places unless otherwise indicated.

1. **Earthquakes in Alaska and Chile**   In 1964 an earthquake measuring 9.2 on the Richter scale occurred in Alaska.

   Jerome Scholler/Shutterstock.com

   a. How did the power of the New Madrid earthquake described in Example 4.12 compare with that of the 1964 Alaska earthquake?

   b. In 1960 an earthquake occurred in Chile that was 2 times as powerful as the Alaska quake. What was the Richter scale reading for the Chilean earthquake?

2. **State Quakes**   The largest recorded earthquake centered in Idaho measured 7.2 on the Richter scale.

   a. The largest recorded earthquake centered in Montana was 3.16 times as powerful as the Idaho earthquake. What was the Richter scale reading for the Montana earthquake?

   b. The largest recorded earthquake centered in Arizona measured 5.6 on the Richter scale. How did the power of the Idaho quake compare with that of the Arizona quake?

3. **Recent Earthquakes** On April 7, 2011, an earthquake of magnitude 6.5 on the Richter scale struck Veracruz, Mexico. On August 23, 2011, an earthquake of magnitude 5.8 on the Richter scale shook the East Coast of the United States.

   a. How did the power of the East Coast and Veracruz quakes compare?

   b. What would be the magnitude of a quake twice as powerful as the Veracruz quake?

4. **Magnitude and Energy** The magnitude M of an earthquake is related to the energy E that it releases. The formula is

$$M = \frac{\log E + 7.5}{1.5}.$$

Here the unit of energy is the *terajoule*, which is one trillion joules. One joule is approximately the energy expended in lifting 3/4 of a pound one foot. One terajoule is approximately the energy produced by burning one million cubic feet of natural gas.

   a. What is the magnitude of a 10-terajoule earthquake? Round your answer to one decimal place.

   b. On January 18, 2011, a quake of magnitude 7.2 struck Pakistan. How much energy was released by that quake?

   c. The atomic bomb dropped on Hiroshima released about 60 terajoules of energy. How many such bombs would be needed to produce the energy released by the Pakistan quake from part b? Round your answer to the nearest whole number.

5. **Moore's Law** The speed of a computer chip is closely related to the number of transistors on the chip, and the number of transistors on a chip has increased with time in a remarkably consistent way. In fact, in the year 1965 Dr. Gordon E. Moore (now chairman emeritus of Intel Corporation) observed a trend and predicted that it would continue for a time. His observation, now known as *Moore's law*, is that every two years or so a chip is introduced with double the number of transistors of its fastest predecessor. This law can be restated in the following way: If time increases by

1 year, then the number of transistors is multiplied by $10^{0.15}$. More generally, the rule is that if time increases by t years, then the number of transistors is multiplied by $10^{0.15t}$. For example, after 8 years the number of transistors is multiplied by $10^{0.15 \times 8}$, or about 16. The Itanium 9300 Series processor was released by Intel Corporation in the year 2010.

   a. If a chip were introduced in the year 2017, how many times the transistors of the Itanium 9300 Series would you expect it to have? Round your answer to the nearest whole number.

   b. The limit of conventional computing will be reached when the size of a transistor is on the scale of an atom. At that point the number of transistors on a chip will be 1000 times that of the Itanium 9300 Series. When, according to Moore's law, will that limit be reached?

   c. Even for unconventional computing, the laws of physics impose a limit on the speed of computation.[31] The fastest speed possible corresponds to having about 10^{40} times the number of transistors as on the Itanium 9300 Series. Assume that Moore's law will continue to be valid even for unconventional computing, and determine when this limit will be reached. Round your answer to the nearest century.

6. **Brightness of Stars** *This is a continuation of Example 4.14.* Refer to Example 4.14 for the relationship between relative intensity of light and apparent magnitude of stars.

   a. The light striking Earth from Vega is 2.9 times as bright as that from the star Fomalhaut. What is the apparent magnitude of Fomalhaut?

   b. The star Antares has an apparent magnitude of 0.92. How does the intensity of light reaching Earth from Antares compare with that of light from Vega?

   c. If the intensity of light striking Earth from one star is twice that of light from another, how do the stars' magnitudes compare?

7. **The pH Scale** Acidity of a solution is determined by the concentration H of hydrogen ions in the solution (measured in moles per liter of solution). Chemists use the negative of the logarithm of the

(continued)

---

[31]S. Lloyd, "Ultimate physical limits to computation," *Nature* **406** (2000), 1047–1054.

concentration of hydrogen ions to define the pH scale:

$$pH = -\log H.$$

Lower pH values indicate a more acidic solution.

a. Normal rain has a pH value of 5.6. Rain in the eastern United States often has a pH level of 3.8. How much more acidic is this than normal rain?

b. If the pH of water in a lake falls below a value of 5, fish often fail to reproduce. How much more acidic is this than normal water with a pH of 5.6?

8. Gross Tonnage The *gross tonnage G* is a standardized measure of a ship's capacity. It is calculated in terms of the volume V, in cubic meters, of the ship. There are no units associated with gross tonnage. It is calculated using the formula

$$G = V(0.2 + 0.02\log V).$$

In this exercise round your answers to the nearest whole number.

a. Find the gross tonnage of a ship with a volume of 13,000 cubic meters.

b. Use the crossing-graphs method to find the volume of a ship with a gross tonnage of 3000.

9. Whispers A whisper in a quiet library is about 30 decibels. What decibel level would be recorded by a table of 10 library patrons each whispering at 30 decibels?

10. Dispersion Models Animal populations move about and disperse. A number of models for this dispersion have been proposed, and many of them involve the logarithm. For example, in 1965 O. H. Paris[32] released a large number of pill bugs and after 12 hours recorded the number n of individuals that could be found within r meters from the point of release. He reported that the most satisfactory model for this dispersion was

$$n = -0.772 + 0.297\log r + \frac{6.991}{r}.$$

a. Make a graph of n versus r for the circle around the release point with radius 15 meters.

b. How many pill bugs were to be found within 2 meters from the release point?

c. How far from the release point would you expect to find only a single individual?

11. Weight Gain Zoologists have studied the daily rate of gain in weight G as a function of daily milk-energy intake M during the first month of life in several ungulate (that is, hoofed mammal) species.[33] (Both M and G are measured per unit of mean body weight.) They developed the model

$$G = 0.067 + 0.052\log M,$$

with appropriate units for M and G.

a. Draw a graph of G versus M. Include values of M up to 0.4 unit.

b. If the daily milk-energy intake M is 0.3 unit, what is the daily rate of gain in weight?

c. A zookeeper wants to bottle-feed an elk calf so as to maintain a daily rate of gain in weight G of 0.03 unit. What must the daily milk-energy intake be?

d. The study cited above noted that "the higher levels of milk ingested per unit of body weight are used with reduced efficiency." Explain how the shape of the graph supports this statement.

12. Reaction Time For certain decisions, the time it takes to respond is a logarithmic function of the number of choices faced.[34] One model is

$$R = 0.17 + 0.44\log N,$$

where R is the reaction time in seconds and N is the number of choices.

a. Draw a graph of R versus N. Include values of N from 1 to 10 choices.

b. Express using functional notation the reaction time if there are seven choices, and then calculate that time.

c. If the reaction time is to be at most 0.5 second, how many choices can there be?

d. If the number of choices increases by a factor of 10, what happens to the reaction time?

e. Explain in practical terms what the concavity of the graph means.

---

[32]"Vagility of P[32]-labeled isopods in grassland," *Ecology* **46** (1965), 635–648.
[33]C. Robbins et al., "Growth and nutrient consumption of elk calves compared to other ungulate species," *J. Wildlife Management* **45** (1981), 172–186.
[34]The model in this exercise is based on *Space Mathematics* by B. Kastner, published in 1985 by NASA.

13. **Age of Haddock** The age T, in years, of a haddock can be thought of as a function of its length L, in centimeters. One common model uses the natural logarithm:

$$T = 19 - 5\ln(53 - L).$$

   a. Draw a graph of age versus length. Include lengths between 25 and 50 centimeters.

   b. Express using functional notation the age of a haddock that is 35 centimeters long, and then calculate that value.

   c. How long is a haddock that is 10 years old?

14. **Growth Rate** An animal grows according to the formula

$$L = 0.6\log(2 + 5T).$$

   Here L is the length in feet and T is the age in years.

   a. Draw a graph of length versus age. Include ages up to 20 years.

   b. Explain in practical terms what $L(15)$ means, and then calculate that value.

   c. How old is the animal when it is 1 foot long?

   d. Explain in practical terms what the concavity of the graph means.

   e. Use a formula to express the age as a function of the length.

15. **Stand Density** Forest managers are interested in measures of how crowded a given forest stand is.[35] One measurement used is the *stand-density index,* or SDI. It can be related to the number N of trees per acre and the diameter D, in inches, of a tree of average size (in terms of cross-sectional area at breast height) for the stand. The relation is

$$\log SDI = \log N + 1.605\log D - 1.605.$$

   a. A stand has 500 trees per acre, and the diameter of a tree of average size is 7 inches. What is the stand-density index? (Round your answer to the nearest whole number.)

   b. What is the effect on the stand-density index of increasing the number of trees per acre by a factor of 10, assuming that the average size of a tree remains the same?

   c. What is the relationship between the stand-density index and the number of trees per acre if the diameter of a tree of average size is 10 inches?

16. **Fully Stocked Stands** *This is a continuation of Exercise 15.* In this exercise we study one of the ingredients used in formulating the relation given in the preceding exercise among the stand-density index SDI, the number N of trees per acre, and the diameter D, in inches, of a tree of average size.[36] This ingredient is an empirical relationship between N and D for *fully stocked* stands—that is, stands for which the tree density is in some sense optimal for the given size of the trees. This relationship, which was observed by L. H. Reineke in 1933, is

$$\log N = -1.605\log D + k,$$

   where k is a constant that depends on the species in question.

   a. Assume that for loblolly pines in an area the constant k is 4.1. If in a fully stocked stand the diameter of a tree of average size is 8 inches, how many trees per acre are there? (Round your answer to the nearest whole number.)

   b. For fully stocked stands, what effect does multiplying the average size of a tree by a factor of 2 have on the number of trees per acre?

   c. What is the effect on N of increasing the constant k by 1 if D remains the same?

17. **Spectroscopic Parallax** Stars have an apparent magnitude m, which is the brightness of light reaching Earth. They also have an *absolute magnitude* M, which is the intrinsic brightness and does not depend on the distance from Earth. The difference $S = m - M$ is the *spectroscopic parallax.* Spectroscopic parallax is related to the distance D from Earth, in *parsecs,*[37] by

$$S = 5\log D - 5.$$

   a. The distance to the star Kaus Australis is 38.04 parsecs. What is its spectroscopic parallax?

   b. The spectroscopic parallax for the star Rasalhague is 1.27. How far away is Rasalhague?

   c. How is spectroscopic parallax affected when distance is multiplied by 10?

(continued)

---

[35]See Thomas E. Avery and Harold E. Burkhart, *Forest Measurements*, 4th ed. (New York: McGraw-Hill, 1994).
[36]The other main ingredient is implicit in part c of Exercise 15.
[37]One parsec is the distance from Earth that would show a 1-second parallax angle. That is the apparent angle produced by movement of Earth from one extreme of its orbit to the other. One parsec is about 3.26 light-years.

d. The star Shaula is 3.78 times as far away as the star Atria. How does the spectroscopic parallax of Shaula compare to that of Atria?

18. Rocket Flight   The velocity v attained by a launch vehicle during launch is a function of c, the exhaust velocity of the engine, and R, the *mass ratio* of the spacecraft.[38] The mass ratio is the vehicle's takeoff weight divided by the weight remaining after all the fuel has been burned, so the ratio is always greater than 1. It is close to 1 when there is room for only a little fuel relative to the size of the vehicle, and one goal in improving the design of spacecraft is to increase the mass ratio. The formula for v uses the natural logarithm:

$$v = c \ln R.$$

Here we measure the velocities in kilometers per second, and we assume that $c = 4.6$ (which can be attained with a propellant that is a mixture of liquid hydrogen and liquid oxygen).

a. Draw a graph of v versus R. Include mass ratios from 1 to 20.

b. Is the graph in part a increasing or decreasing? In light of your answer, explain why increasing the mass ratio is desirable.

c. To achieve a stable orbit, spacecraft must attain a velocity of 7.8 kilometers per second. With $c = 4.6$, what is the smallest mass ratio that allows this to happen? (*Note:* For such a propellant, the mass ratio needed for orbit is usually too high, and that is why the launch vehicle is divided into stages. The next exercise shows the advantage of this.)

19. Rocket Staging   *This is a continuation of Exercise 18.* One way to raise the effective mass ratio is to divide the launch vehicle into different stages. See Figure 4.32. After one stage has burned its fuel, it drops away, and the next stage begins to fire. We assume that there are two stages and that each stage has the same exhaust velocity c. Then the total velocity v attained is the sum of the velocities attained by each stage:

$$v = c \ln R_1 + c \ln R_2,$$

where R_1 and R_2 are the mass ratios for stages 1 and 2, respectively. In this exercise we assume that $c = 3.7$ kilometers per second, and we assume that each stage has a mass ratio of 3.4.

FIGURE 4.32

a. What is the total velocity attained by this two-stage craft?

b. Can this craft achieve a stable orbit, as described in part c of Exercise 18?

20. Stereo Speakers   See Example 4.13 for the relationship between the decibel scale and relative intensity of sound.

a. In this part assume that four stereo speakers, each producing 40 decibels, are placed side by side, multiplying by a factor of 4 the relative intensity of sound produced by only one of the speakers. What is the decibel level of the four speakers together?

b. Find an exponential formula giving relative intensity I as a function of decibel level D.

c. What is the growth factor for relative intensity as an exponential function of decibels? *Reminder:* $a^{b/c} = (a^{1/c})^b$.

d. If the decibel level increases by 3 units, what happens to relative intensity?

21. Relative Abundance of Species   A collection of animals may contain a number of species. Some rare species may be represented by only 1 individual in the collection. Others may be represented by more. In 1943 Fisher, Corbert, and Williams[39] proposed a model for finding the number of species

[38]This exercise and the next are based on *Space Mathematics* by B. Kastner, published in 1985 by NASA.
[39]"The relation between the number of species and the number of individuals in a random sample of an animal population," *J. Anim. Ecol.* **12** (1943), 42–58.

Unless otherwise noted, all art on this page is © Cengage Learning.

in certain types of collections represented by a fixed number of individuals. They associate with an animal collection two constants, α and x, with the following property: The number of species in the sample represented by a single individual is αx; the number in the sample represented by 2 individuals is $\alpha(x^2/2)$; and, in general, the number of species represented by n individuals is $\alpha(x^n/n)$. If we add all these numbers up, we get the total number S of species in the collection. It is an important fact from advanced mathematics that this sum also yields a natural logarithm:

$$S = \alpha x + \alpha \frac{x^2}{2} + \alpha \frac{x^3}{3} + \cdots = -\alpha \ln(1-x).$$

If N is the total number of individuals in the sample, then it turns out that we can find x by solving the equation

$$\frac{x-1}{x} \ln(1-x) = \frac{S}{N}.$$

We can then find the value of α using

$$\alpha = \frac{N(1-x)}{x}.$$

In 1935 at the Rothamsted Experimental Station in England, 6814 moths representing 197 species were collected and catalogued.

a. What is the value of S/N for this collection? (Keep four digits beyond the decimal point.)

b. Draw a graph of the function

$$\frac{x-1}{x} \ln(1-x)$$

using a horizontal span of 0 to 1.

c. Use your answer to part a and your graph in part b to determine the value of x for this collection. (Keep four digits beyond the decimal point.)

d. What is the value of α for this collection? (Keep two digits beyond the decimal point.)

e. How many species of moths in the collection were represented by 5 individuals?

f. For this collection, plot the graph of the number of species represented by n individuals versus n. Include values of n up to 20.

Using the Logarithm to Test for Exponential Data It can be shown that if y is an exponential function of x, then $\ln y$ is a linear function of x. For example, consider the following data.

x	0	1	2	3	4
y	1	2	4	8	16
$\ln y$	0	0.69	1.39	2.08	2.77

Clearly y is an exponential function of x with growth factor 2. A plot of $\ln y$ versus x is shown in Figure 4.33, and we see that $\ln y$ is a linear function of x. Conversely, the fact that the data for $\ln y$ versus x fall on a straight line indicates that y is an exponential function of x. For data that are almost exponential, the natural logarithm of data points will nearly fall on a straight line. Exercises 22 and 23 make use of these observations.

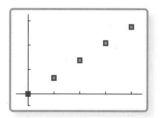

FIGURE 4.33 The logarithm of exponential data is linear.

22. Population Growth The following table shows the size, in thousands, of an animal population at the start of the given year.

Year	2008	2009	2010	2011	2012
Population in thousands	1.56	1.62	1.69	1.75	1.83

Plot the natural logarithm of the population versus time in years since 2008. Does this plot support the use of an exponential function to model the population as a function of time?

23. Magazine Circulation The following table shows the circulation, in thousands, of a magazine at the start of the given year.

Year	2008	2009	2010	2011	2012
Circulation in thousands	2.64	2.78	2.93	3.08	3.25

Plot the natural logarithm of the circulation versus time in years since 2008. Does this plot support the use of an exponential function to model the circulation as a function of time?

4.5 SKILL BUILDING EXERCISES

Richter Scale   Exercises S-1 through S-6 use elementary properties of the Richter Scale.

S-1. One earthquake reads 4.2 on the Richter scale, and another reads 7.2. How do the two quakes compare?

S-2. One earthquake measures 5.5 on the Richter scale, and another measures 4.5. How do the two quakes compare?

S-3. One earthquake has a Richter scale reading of 6.5. A second is 100 times as strong. What is its Richter scale reading?

S-4. One earthquake has a Richter scale reading of 5.4. Another is one-thousandth as strong. What is its Richter scale reading?

S-5. If the relative intensity of a quake is multiplied by 10^t, how is the Richter scale reading affected?

S-6. If the Richter scale reading of one quake is t units larger than that of a second quake, how do the relative intensities of the quakes compare?

The Decibel Scale   Exercises S-7 through S-10 refer to the decibel scale.

S-7. If one sound has a relative intensity 1000 times that of another, how do their decibel levels compare?

S-8. If one sound has a relative intensity one-tenth that of another, how do their decibel levels compare?

S-9. One sound has a decibel reading of 5. Another has a decibel reading of 7. How do the relative intensities of the sounds compare?

S-10. One sound has a decibel reading of 6. Another has a decibel reading of 5. How do their relative intensities of the sounds compare?

Calculating Common Logarithms   In Exercises S-11 through S-14, you are asked to find common logarithms without using your calculator.

S-11. $\log 1000$

S-12. $\log 1$

S-13. $\log \dfrac{1}{10}$ $\left(Suggestion: \text{Recall that } \dfrac{1}{10} = 10^{-1}. \right)$

S-14. $\log 10^{643.77}$

Calculating Natural Logarithms by Hand   In Exercises S-15 through S-19, calculate the given natural logarithm without using your calculator.

S-15. $\ln e$

S-16. $\ln 1$

S-17. $\ln \dfrac{1}{e}$ $\left(Suggestion: \text{Recall that } \dfrac{1}{e} = e^{-1}. \right)$

S-18. $\ln e^{10}$

S-19. $\ln e^e$

S-20. **How the Common Logarithm Changes**   How is $\log x$ affected if x is multiplied by 10^t?

S-21. **How the Natural Logarithm Changes**   How is $\ln x$ affected if x is multiplied by e^t?

Solving Logarithmic Equations   In Exercises S-22 through S-27, solve the given equation by hand.

S-22. $\log x = 2$

S-23. $\log x = 1$

S-24. $\log x = -2$

S-25. $\ln x = 3$ (*Note:* Your answer will involve e.)

S-26. $\ln x = 1$ (*Note:* Your answer will involve e.)

S-27. $\ln x = 0$

S-28. **How the Logarithm Increases**   Suppose we know that $\log x = 6.6$. Find the following:

a. $\log(10x)$

b. $\log(1000x)$

c. $\log \dfrac{x}{10}$

S-29. **How the Logarithm Increases**   If $\log x = 8.3$ and $\log y = 10.3$, how do x and y compare?

Solving Exponential Equations   In Exercises S-30 through S-33, solve by hand the given exponential equation. Your answer will involve either the common logarithm or the natural logarithm.

S-30. $10^t = 5$

S-31. $e^t = 5$

S-32. $10^t = a$

S-33. $e^t = a$

CHAPTER 4 | SUMMARY

Exponential functions are almost as pervasive as linear functions. They are commonly used to describe population growth, radioactive decay, free fall subject to air resistance, bank loans, inflation, and many other familiar phenomena. Their defining property is similar to that of linear functions. Linear functions are those with a constant rate of change, whereas exponential functions have a constant percentage or proportional rate of change. Any phenomenon that can be described in terms of yearly (monthly, daily, etc.) percentage growth (or decay) is properly modeled with an exponential function.

4.1 EXPONENTIAL GROWTH AND DECAY

A linear function with slope m changes by constant sums of m. That is, if the variable is increased by 1 unit, the function is increased by adding m units. By contrast, an exponential function with base a changes by constant multiples of a. That is, when the variable is increased by 1, the function value is multiplied by a. Exponential functions are of the form

$$N = Pa^t,$$

where a is the *base*, or *growth/decay factor*, and P is the initial value. When a is larger than 1, the exponential function grows rapidly, eventually becoming exceptionally large. When the base a is less than 1, the exponential function decays rapidly toward zero.

The growth factor for an exponential function is tied to a time period. It is sometimes important to know the growth factor for a different time period. We do this with unit conversion.

> **Unit conversion for exponential growth factors:** If the growth factor for one period of time is a, then the growth factor A for k periods of time is given by $A = a^k$.

As an example, U.S. population in the early 1800s grew with yearly growth factor 1.029. The appropriate exponential function is of the form $N = P \times 1.029^t$, where t is measured in years. If we want to express population growth in terms of decades, we need a new growth factor:

$$1.029^{10} = 1.331.$$

Thus $N = P \times 1.331^d$, where d is measured in decades.

4.2 CONSTANT PERCENTAGE CHANGE

Many times, exponential functions are described in terms of constant percentage growth or decay. If r is a decimal showing a percentage increase, then the growth factor is $1 + r$; if r shows a percentage decrease, then the decay factor is $1 - r$. To convert units when we are given percentage growth or decay rates, first we find the corresponding growth or decay factor.

4.3 MODELING EXPONENTIAL DATA

Data that are evenly spaced for the variable are linear if the function values show constant successive differences. They are exponential if the function values show constant *successive quotients*. When the increment for the data is 1 unit, the common successive quotient is the growth (or decay) factor for the exponential function.

A simple example is provided by a hypothetical well that is being contaminated by seepage of water containing a pollutant at a level of 64 milligrams per liter. Each month the difference D between 64 and the contaminant level of the well is recorded.

t = months	0	1	2	3	4	5
D = contaminant level difference	64	45.44	32.26	22.91	16.26	11.55

The first quotient is calculated as $45.44/64 = 0.71$. It can be verified that each successive quotient in the table gives this same value. We conclude that the data are exponential with decay factor 0.71. Since the initial value is 64, we obtain $D = 64 \times 0.71^t$.

4.4 MODELING NEARLY EXPONENTIAL DATA

As with linear data, sampling error and other factors make it rare that experimentally gathered data can be modeled exactly by an exponential function. But as with linear functions, it may be appropriate to use the exponential model that most closely approximates the data. This process is *exponential regression*. The real-world context can provide important confirmation that an exponential model is appropriate.

4.5 LOGARITHMIC FUNCTIONS

The common logarithm $\log x$ is the power of 10 that gives x. Thus, if we can express x as a power of 10, we get the logarithm immediately using

$$\log 10^t = t.$$

The formula above tells us that the common logarithm is the *inverse* of the exponential function 10^t, and we can think of the logarithm of x as the solution for t of the equation

$$x = 10^t.$$

This in turn provides us with a direct way of solving certain exponential equations. For example, the solution of the equation

$$10^t = 7.6$$

is $\log 7.6$, which is about 0.88.

Increasing x by a factor of 10 increases the logarithm by 1 unit, and in general if x is multiplied by 10^t, then t units are added to the logarithm. This can be expressed using the formula

$$\log(10^t x) = t + \log x.$$

Many common scales, such as the Richter scale and the decibel scale, are logarithmic in nature.

CHAPTER 4   REVIEW EXERCISES

Reminder   Round all answers to two decimal places unless otherwise indicated.

1. **Percentage Growth**   A population grows by 4.5% each year. By what percentage does it grow each month?

2. **Percentage Decline**   A population declines by 0.5% each year. By what percentage does it decline each decade?

3. **Credit Cards**   Suppose you make a charge of $1000 to a new credit card but then charge nothing else

and make the minimum payment each month. If the finance charges amount to a monthly rate of 2% and the minimum payment is 5% of the new balance, then your balance B, in dollars, after n payments is an exponential function given by the formula

$$B = 1000 \times a^n.$$

Here the decay factor a is obtained from the finance charge and the minimum payment by the formula

$$a = (1 + 0.02)(1 - 0.05).$$

a. Find the exact value of the decay factor a.

b. By what percentage does your balance decrease each month?

c. What will your balance be after you have made payments for 3 years?

4. **More on Credit Cards**   *This is a continuation of Exercise 3.* Suppose again that you make a charge of $1000 to a new credit card but then make no other charges. Assume that the finance charges amount to a monthly rate of r as a decimal. Let the minimum payment as a percentage of the new balance be given by the number m when expressed as a decimal. Then your balance B, in dollars, is an exponential function of the number of payments n, with initial value $1000 and base

$$a = (1 + r)(1 - m).$$

a. Find the exact value of the base a if the monthly finance charge is 1.5% and the minimum payment is 4% of the new balance.

b. In the situation of part a, by what percentage does your balance decrease each month?

c. What is the exact value of the base a if the monthly finance charge is 1.5% and the minimum payment is 1% of the new balance? Is this a realistic situation?

5. **Testing Exponential Data**   Determine whether the following table shows exponential data.

x	0	1	2	3
y	25.0	30.0	36.0	43.2

6. **Modeling Exponential Data**   Make an exponential model for the data from the preceding exercise.

7. **Credit Card Balance**   You get a new credit card and make an initial charge but then charge nothing else. The following table shows your balance B, in dollars, after you have made n monthly payments.

Payment n	0	1	2	3
Balance B	500.00	485.00	470.45	456.34

a. Find an exponential model for the data.

b. What was your initial charge?

c. What will your balance be after you have made payments for 2 years?

8. **Inflation**   An economist tracks the price of a certain item at the beginning of several years and compiles the following table.

Year	2006	2007	2008	2009
Price in dollars	265.50	273.47	281.67	290.12

a. Show that the price is growing as an exponential function.

b. Find an exponential model for the data.

c. At the beginning of some year the price will surpass $325. Use your model to determine which year.

9. **Exponential Regression**   Use exponential regression to fit the following data set.

x	1	5	7	10	11
y	28.1	90.7	162.9	392.0	525.3

10. **Exponential Regression**   Use exponential regression to fit the following data set. Give the exponential model, and plot the data along with the model.

x	1	2	3	4	5
y	8.8	7.0	5.6	4.5	3.6

11. **Sales**   A market analyst is studying sales figures for two competing products, Alpha and Beta. In the accompanying table, A represents the annual sales of Alpha, in thousands of dollars, and B represents the annual sales of Beta, in thousands of dollars. Time t is measured in years since 2005.

t	0	1	2	3
A	8.90	9.08	9.26	9.44
B	2.30	2.92	3.71	4.71

(continued)

a. Use exponential regression to find models for A and B. (Round regression parameters to two decimal places.)

b. By what annual percentage are sales for Alpha growing? By what annual percentage are sales for Beta growing?

c. Use your models to predict when sales for Beta will overtake sales for Alpha.

12. Credit Card Payments You make one charge to a new credit card but then charge nothing else and make the minimum payment each month. You can't find all of your statements, but the accompanying table shows, for those you *do* have, your balance B, in dollars, after you make n payments.

Payment n	2	4	7	11
Balance B	478.73	440.74	389.33	329.99

a. Use regression to find an exponential model for the data in the table. (Round the decay factor to four decimal places.)

b. What was your initial charge?

c. For such a payment scheme, the decay factor equals $(1 + r)(1 - m)$. Here r is the monthly finance charge as a decimal, and m is the minimum payment as a percentage of the new balance when expressed as a decimal. Assume that your minimum payment is 5%, so $m = 0.05$. Use the decay factor in your model to determine your monthly finance charge.

13. Solving Logarithmic Equations Solve by hand the equation $\log x = -1$.

14. Comparing Logarithms If $\log x = 3.5$ and $\log y = 2.5$, how do x and y compare?

15. Comparing Earthquakes In 1992 there was an earthquake at Little Skull Mountain, Nevada, measuring 5.5 on the Richter scale. In 1994 there was an earthquake near Double Spring Flat, Nevada, measuring 6.0 on the Richter scale. How did the power of the Double Spring Flat quake compare with that of the Little Skull Mountain quake?

16. Population Growth A biologist studying the growth of an animal population has obtained the model

$$T = 25 \log N - 50$$

for the time T, in years, at which the population has reached the size N.

a. At what time does the population reach a size of 200?

b. What was the initial population size?

A FURTHER LOOK

Solving Exponential Equations

We noted earlier that the solution for t of the equation $10^t = x$ is $t = \log x$ and that the solution for t of $e^t = x$ is $t = \ln x$. In fact, we can use the logarithm (either common or natural) to solve more complicated exponential equations. To do so we need the basic laws of logarithms listed in Key Idea 4.9. Note the similarity in the behavior of the natural and common logarithms.

KEY IDEA 4.9 **LAWS OF LOGARITHMS**

For A and B positive:

	Common Logarithm	**Natural Logarithm**
Product law	$\log AB = \log A + \log B$	$\ln AB = \ln A + \ln B$
Quotient law	$\log \dfrac{A}{B} = \log A - \log B$	$\ln \dfrac{A}{B} = \ln A - \ln B$
Power law	$\log A^p = p \log A$	$\ln A^p = p \ln A$

In this section we will focus on how these laws are used.

Here is a simple example of how to apply the laws of logarithms: Suppose we know that $\log A = 2.8$ and $\log B = 1.5$. Then we can calculate the following.

$$\log AB = \log A + \log B = 2.8 + 1.5 = 4.3$$

$$\log \frac{A}{B} = \log A - \log B = 2.8 - 1.5 = 1.3$$

$$\log \sqrt{A} = \log A^{1/2} = \frac{1}{2} \log A = \frac{1}{2} \times 2.8 = 1.4.$$

Now let's use logarithms to solve the equation $5 \times 3^t = 8$. First we divide each side by 5:

$$3^t = \frac{8}{5} = 1.6.$$

Next we take the logarithm of each side:

$$\log 3^t = \log 1.6.$$

Now we apply the power law to bring the t outside the logarithm:

$$t \log 3 = \log 1.6.$$

Finally we divide each side by log 3:

$$t = \frac{\log 1.6}{\log 3} = 0.43.$$

The same answer can be found using the natural logarithm rather than the common logarithm. The steps are virtually identical and are shown below.

$$5 \times 3^t = 8$$

$$3^t = \frac{8}{5} = 1.6$$

$$\ln 3^t = \ln 1.6$$

$$t \ln 3 = \ln 1.6 \qquad \text{Power law}$$

$$t = \frac{\ln 1.6}{\ln 3} = 0.43.$$

As evidenced by the example above, only one logarithm is really needed. But both the common and natural logarithm occur in applications of mathematics, and familiarity with both is important. Their behaviors are so similar that it is not difficult to employ either one or both.

We can use the laws of logarithms to solve more complicated exponential equations. For example, let's solve $15^t = 4 \times 2^{3t}$. We use natural logarithms. The reader is encouraged to show that the use of the common logarithm yields the same answer.

$$15^t = 4 \times 2^{3t}$$

$$\ln 15^t = \ln (4 \times 2^{3t})$$

$$\ln 15^t = \ln 4 + \ln 2^{3t} \qquad \text{Product law}$$

$$t \ln 15 = \ln 4 + 3t \ln 2 \qquad \text{Power law}$$

$$t \ln 15 - 3t \ln 2 = \ln 4$$

$$t(\ln 15 - 3 \ln 2) = \ln 4$$

$$t = \frac{\ln 4}{\ln 15 - 3 \ln 2} = 2.21.$$

Solving Logarithmic Equations

Just as we use logarithms to solve exponential equations, we can use exponents to solve logarithmic equations. The keys to doing so are the equations

$$10^{\log A} = A$$

$$e^{\ln A} = A.$$

For example, let's solve the equation $\log (2t + 1) = 2$. We use 10 to exponentiate each side of the equation:

$$10^{\log (2t+1)} = 10^2 = 100.$$

Now, because $10^{\log (2t+1)} = 2t + 1$, this simplifies to $2t + 1 = 100$. We solve this equation to find $t = 49.5$.

Note that to solve $\ln (2t + 1) = 2$ we would use e rather than 10 to exponentiate each side. The reader should verify that the solution of this equation is

$$t = \frac{e^2 - 1}{2} = 3.19.$$

Finally we note that the solutions of some equations involve the use of both logarithms and exponentiation. To illustrate this fact we solve $\ln(1 + a^x) = b$ for x. Here we assume that a and b are positive and that $a \neq 1$.

$$\ln(1 + a^x) = b$$
$$e^{\ln(1 + a^x)} = e^b$$
$$1 + a^x = e^b \qquad \qquad e^{\ln A} = A$$
$$a^x = e^b - 1$$
$$\ln a^x = \ln(e^b - 1)$$
$$x \ln a = \ln(e^b - 1) \qquad \text{Power law}$$
$$x = \frac{\ln(e^b - 1)}{\ln a}.$$

EXERCISES

Reminder   Round all answers to two decimal places unless otherwise indicated.

Using the Laws of Logarithms   For Exercises 1 through 6, suppose that $\ln A = 3$, $\ln B = 4$, and $\ln C = 5$. Evaluate the given expression.

1. $\ln \dfrac{AB}{C}$

2. $\ln \dfrac{A}{BC}$

3. $\ln \sqrt{C}$

4. $\ln A^2 B^3$

5. $\ln \dfrac{1}{A}$

6. $\ln \dfrac{A^2}{BC}$

Solving Exponential Equations   In Exercises 7 through 16, solve the exponential equation for t. You may use either the common logarithm or the natural logarithm.

7. $3 \times 4^t = 21$

8. $\dfrac{5^t}{4} = 9$

9. $7^{t-1} = 4^{t+1}$

10. $2 \times 5^t = 3^{2t}$

11. $5 \times 4^t = 7$

12. $2^{t+3} = 7^{t-1}$

13. $\dfrac{12^t}{9^t} = 7 \times 6^t$

14. $a^{2t} = 3a^{4t}$ if $a > 0$ and $a \neq 1$

15. $ab^t = c$ if a, b, and c are positive and $b \neq 1$

16. $a^{t+1} = b^{t-1}$ if a and b are positive and $a \neq b$

Solving Logarithmic Equations   In Exercises 17 through 22, solve the logarithmic equations for t.

17. $\ln(3t - 1) = 4$

18. $\log(6t + 4) = -2$

19. $\ln(at + b) = c$ if $a \neq 0$

20. $\log(at - b) = c$ if $a \neq 0$

21. $\ln(2^t + 1) = 3$

22. $\log(3^t - 2) = 2$

Illustrative Applications   Exercises 23 through 26 illustrate applications of the logarithm.

23. **Decibels**   For this exercise we use the relation

$$\text{Decibels} = 10 \log(\text{Relative intensity}).$$

  a. Use the product law for logarithms to show that doubling the relative intensity adds 3.01 to the decibel rating of the sound.

  b. Using D for decibels and R for relative intensity, solve the equation $D = 10 \log R$ for R in order

(continued)

to express relative intensity as a function of decibels.

24. **Spectroscopic Parallax Preview**   In Exercise 17 at the end of Section 4.5, we saw how to find the *spectroscopic parallax S* of a star in terms of its distance D from Earth in *parsecs*. The relationship is $S = 5 \log D - 5$.

   a. Use the laws of logarithms to determine what happens to S if D is doubled.

   b. Solve the equation above for D to express distance as a function of spectroscopic parallax.

25. **Doubling Time**   The logarithm can be used to determine how long it takes an exponential function to double the initial value.

   a. The balance after t months of a certain investment is given by $B = 5000 \times 1.005^t$ dollars. The original investment was $5000. Use the logarithm to determine how long is

required for the account to double the original investment.

   b. Use the logarithm to find the doubling time for the exponential function $y = a \times b^t$.

26. **Moore's Law**   According to *Moore's law,* the number of transistors on a computer chip doubles about every 2 years.[40]

   a. Use the common logarithm to solve the equation $10^k = 2^{1/2}$ for k.

   b. Use the laws of exponents to show that an exponential function with yearly growth factor $2^{1/2}$ doubles every 2 years.

   c. Explain using part b why Moore's law says that every year the number of transistors on a computer chip increases by a factor of 10^k, where k is the solution to the equation in part a.

---

[40]See Exercise 5 at the end of Section 4.5.

A SURVEY OF OTHER COMMON FUNCTIONS

5

Len DeLessio/Getty Images

© Rob Howard/Corbis

The number of species on an island is modeled by a power function of the area. See Example 5.6, p. 333.

ALTHOUGH LINEAR and exponential functions probably are the mathematical functions that are most commonly found in applications, other important functions occur as well. We look first at *logistic functions*, which have features similar to exponential functions but always have limiting values. We then look closely at *power functions* in the next two sections of this chapter. The fourth section treats composition of functions and piecewise-defined functions. The fifth section is devoted to *polynomials* and *rational functions*.

Student resources are available on the website **www.cengagebrain.com**

5.1 LOGISTIC FUNCTIONS

Population growth eventually encounters limiting factors.

In many settings, it is reasonable to assume that growth is exponential, at least over limited time periods. For population growth, an exponential model is a direct consequence of the assumption that percentage change (birth rate minus death rate) is constant. In reality, however, a population cannot undergo such rapid growth indefinitely. The species will begin to exhaust local resources, and we expect that, instead of remaining constant, the birth rate will begin to decrease and the death rate to increase. Thus, there will be a reduced rate of growth as the population increases in size. One important model that takes into account the limited potential for growth is the *logistic growth model,* which is the topic of this section.

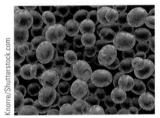

Logistic Growth Model

One of the best known examples of logistic growth is the classic study of the growth of a yeast culture.[1] The data gathered in that study are plotted in Figure 5.1, and the data has been fitted with a curve in Figure 5.2. This S-shaped curve (sometimes called a sigmoid shape) is the trademark of logistic growth.

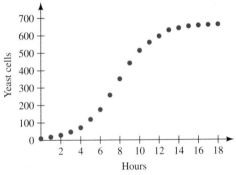

FIGURE 5.1 Population growth data for yeast

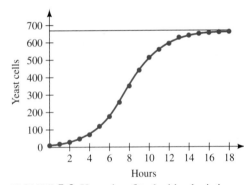

FIGURE 5.2 Yeast data fitted with a logistic curve

Here are some of the key features of logistic growth displayed by the curve for the yeast culture:

Initial rapid growth rate: Note that if we restrict our attention to the curve over about the first 8 hours, it has the classic shape of exponential growth. Over this period the yeast have abundant resources, and population growth is nearly exponential.

Point of inflection: The graph has a point of inflection at approximately $t = 8$ hours. This is the steepest part of the graph, so the inflection point represents the time of most rapid growth. Before the inflection point, growth is almost exponential. After the inflection point the growth rate declines.

---

[1] The data are taken from an early study of yeast population growth by Tor Carlson, as described by R. Pearl in "The growth of populations," *Quart. Rev. Biol.* **2** (1927), 532–548.

Eventual declining growth rate and limiting value: After the inflection point, the graph is increasing and concave down. The growth rate declines, and the population appears to have a limiting value of just under 700 cells. The horizontal line corresponding to this value has been added to the graph in Figure 5.2. This limiting value is known as the *carrying capacity*. It is the maximum population that available resources can support.

We remark that logistic models apply to many populations other than yeast cultures. The logistic model was introduced by P. F. Verhulst in 1838 to describe the growth of human populations. In 1920, R. Pearl and L. J. Reed derived the same model to describe the growth of the population of the United States since 1790 and to attempt predictions of that population at future times.[2] At the end of this section, we will return to their predictions.

Many settings have the characteristics of a logistic model: an initial rapid growth rate that is eventually slowed and then capped. For example, if we introduce a new newspaper in a small city, the increase in circulation may be quite rapid over some initial time period. But eventually circulation is limited by the population of the city, so circulation growth must eventually slow. One common phenomenon displaying logistic growth is the spread of a new technology (see Example 5.3 later in this section).

Our discussion of the logistic model so far has been qualitative—we have given no formulas. We only know the shape and some of the general features of the logistic curve. But even at this level the logistic model yields interesting results, as the following application shows.

Application to Harvesting Renewable Resources

In the management of a harvested population (as found in a marine fishery), an important problem is to determine at what level to maintain the population in order to sustain a maximum harvest. This is the problem of optimum yield. The logistic model for population growth gives a theoretical basis for solving this problem, leading to the theory of *maximum sustainable yield*.

If we harvest when the population is small, then the population level will be reduced so far that it will take a long time to recover and, in extreme cases, the population may be driven to extinction. In Figure 5.3 we have illustrated harvesting at the lower part of the curve for a species undergoing logistic growth. The unbroken curve shows how the population would grow if it were left alone, and the broken graph shows the result of periodic harvesting. It shows that if we harvest when the population reaches a level of about 125, we will be able to harvest about 100 individuals every 3 days. At the other extreme, we could maintain the population near its maximum level. But if the harvested population grows according to the logistic model, this means maintaining the population near the carrying capacity of the environment, where the rate of population growth is *slow* because the graph flattens out there. After a small harvest we will have to wait a relatively long time for the population to recover. This is illustrated in Figure 5.4, and we see that in the case shown, we are able to harvest about 100 individuals every 2 days.

---

[2]"On the rate of growth of the population of the United States since 1790 and its mathematical representation," *Proc. Nat. Acad. Sci.* **6** (1920), 275–288.

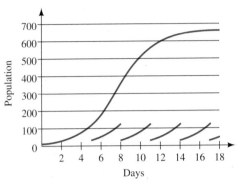

FIGURE 5.3 Harvesting at the lower part of the curve for a population undergoing logistic growth

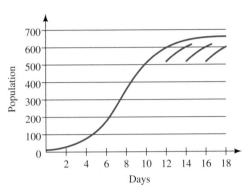

FIGURE 5.4 Harvesting near the carrying capacity for a population undergoing logistic growth

To get the best harvest, it makes sense to determine the population size at which the rate of growth is the largest, since the population will recover most quickly near that size. This scenario is illustrated in Figure 5.5, and we see that under these conditions, we are able to harvest 100 individuals each day. That population size, known as the *optimum yield level*, corresponds to the steepest point on the graph, which is near the middle of the curve. This suggests that the optimum yield level is half of the carrying capacity. It is marked in Figure 5.6. Before the population reaches this steepest point, the rate of growth is increasing; after this point, the rate of growth is decreasing. In other words, the optimum yield level corresponds to the inflection point on the graph. Our conclusion is that optimum yields come from populations maintained not at maximum size but at maximum growth rates, and this occurs at half of the carrying capacity.

Optimum yield occurs at half of the carrying capacity.

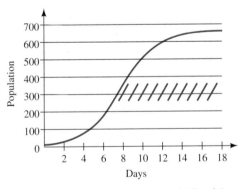

FIGURE 5.5 Harvesting near the middle of the logistic curve

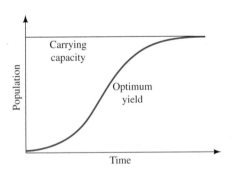

FIGURE 5.6 The level of optimum yield is half the carrying capacity.

Note: The importance of wise management is demonstrated by the history of the Peruvian anchovy fishery, which flourished off the coast of Peru from the mid-1950s until the early 1970s. In 1970 it accounted for 18% of the total world harvest of fish. But in the early 1970s, a change in environmental conditions, combined with overfishing, caused the collapse of the fishery. This had a dramatic impact on food prices worldwide. The fishery has never recovered from this collapse.

> **KEY IDEA 5.1** **LOGISTIC CURVE**
>
> The logistic growth curve has the following properties:
>
> - Initially the growth is rapid, nearly exponential.
> - The inflection point represents the time of most rapid growth.
> - After the inflection point, the growth rate declines. The function has a limiting value, known as the *carrying capacity.*
> - The point of inflection occurs at half of the carrying capacity. This is the level of maximum growth. In the case of harvested populations, this level is known as the *optimum yield level.*

Formula for the Logistic Model

Deriving the formula for logistic growth requires techniques beyond the scope of this text, and we simply state the result. The formula for a logistic model is

$$N = \frac{K}{1 + be^{-rt}}.$$

The parameter K in the formula for logistic growth is the carrying capacity.

Here K is the carrying capacity. It is the limiting value of N, so the point of inflection occurs at the level $N = K/2$. The constant b is determined by

$$b = \frac{K}{N(0)} - 1.$$

The constant r is the intrinsic exponential growth rate. In population studies, ecologists often refer to the r value of a species in an environment. In the absence of limiting factors, growth would be exponential according to the formula $N = N(0)e^{rt}$. The r value is measured in units such as per year or per day. The corresponding growth factor for this exponential function is $a = e^r$. Thus, we can find r from a using the natural logarithm: $r = \ln a$.

The following example illustrates the use of the logistic formula.

EXAMPLE 5.1 **THE PACIFIC SARDINE**

Studies to fit a logistic model to the Pacific sardine population have yielded

$$N = \frac{2.4}{1 + 239e^{-0.338t}},$$

where t is measured in years and N is measured in millions of tons of fish.[3]

Part 1 What is r for the Pacific sardine?

Part 2 According to the logistic model, in the absence of limiting factors, what would be the annual percentage growth rate for the Pacific sardine?

[3]From G. I. Murphy, "Vital statistics of the Pacific sardine (*Sardinops caerulea*) and the population consequences," *Ecology* **48** (1967), 731–736.

liveostockimages/Shutterstock.com

Part 3 What is the environmental carrying capacity K?

Part 4 What is the optimum yield level?

Part 5 Make a graph of N versus t.

Part 6 At what time t should the population be harvested?

Part 7 What portion of the graph is concave up? What portion is concave down?

Solution to Part 1 The formula for N is written in the standard form for a logistic function. Since the coefficient of t in the exponential is -0.338, it must be that $r = 0.338$ per year.

Solution to Part 2 In the absence of limiting factors, the growth would be exponential, with a yearly growth factor of $a = e^r = e^{0.338} = 1.40$. That is a yearly percentage rate of $1.40 - 1 = 0.40$ or 40%.

Solution to Part 3 The constant in the numerator is 2.4, so $K = 2.4$ million tons of fish.

Solution to Part 4 The optimum yield level is half of the carrying capacity, and by part 3 this is $\frac{1}{2} \times 2.4 = 1.2$ million tons of fish.

Solution to Part 5 Using a table of values, we determine a horizontal span of 0 to 40 and a vertical span of 0 to 2.5. The graph is shown in Figure 5.7.

Solution to Part 6 The population should be harvested at the optimum yield level, and according to part 4 this is at the population level $N = 1.2$, so we want to solve the equation $N(t) = 1.2$ for t. We do this using the crossing-graphs method, as shown in Figure 5.8. We find that the time t is about 16.2 years.

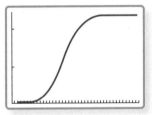

FIGURE 5.7 Population growth for the Pacific sardine

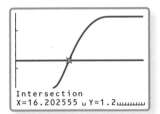

FIGURE 5.8 Time of optimum yield

Solution to Part 7 By examining Figure 5.7 we see that the graph is concave up until the population reaches the level for optimum yield and is concave down after that. Using part 6, we see that the graph is concave up over the first 16.2 years and concave down thereafter.

Note: The Pacific sardine fishery along the California coast expanded rapidly from about 1920 until the 1940s, with an annual catch of around 800 thousand tons at its height. In the late 1940s to early 1950s the fishery collapsed. The economic consequences of the collapse were severe. The major factors that contributed to the collapse were heavy fishing and environmental changes that seem to have favored a competing fish, the anchovy.[4] This example illustrates that the model of maximum sustainable yield which we have developed is only a first step in the study of managing renewable resources. A variety of factors must be considered.

---

[4]For more information on the Pacific sardine fishery, see the account by Michael Culley in *The Pilchard* (Oxford, England: Pergamon Press, 1971).

Another fish population follows the logistic function

$$N = \frac{6.2}{1 + 188e^{-0.44t}} \text{ million tons.}$$

What is the carrying capacity? What is the initial population? In the absence of limiting factors, what would be the annual percentage growth rate? ∎

KEY IDEA 5.2 **LOGISTIC MODEL**

The formula for a logistic model is

$$N = \frac{K}{1 + be^{-rt}}.$$

- The constant K is the carrying capacity. It is the limiting value of N. The point of inflection occurs at $N = K/2$.

- The constant b is determined by $b = \frac{K}{N(0)} - 1$.

- The r value is the intrinsic exponential growth rate. In the absence of limiting factors, growth would be exponential according to the formula $N = N(0)e^{rt}$. The corresponding growth factor for this exponential function is $a = e^r$, so we can find r from a using $r = \ln a$.

The next example shows how to construct a logistic model.

EXAMPLE 5.2 **CONSTRUCTING A LOGISTIC FUNCTION**

We begin selling a new magazine in our hometown. Initial sales are 200 magazines per month. We believe that in the absence of limiting factors, our sales will increase by 6% per month, but the size of the town limits our total circulation to 800 magazines per month.

Part 1 Make a logistic model for magazine sales under these conditions. Round r to three decimal places.

Part 2 Make a graph of sales during the first 60 months, and determine when sales can be expected to reach 600 magazines per week.

Solution to Part 1 Let N denote the sales per month after t months. Then $N(0) = 200$. The limiting value is $K = 800$. We get b using

$$b = \frac{K}{N(0)} - 1 = \frac{800}{200} - 1 = 3.$$

We need the r value to construct the model. Because sales would grow by 6% per month in the absence of limiting factors, the monthly growth factor is $a = 1 + 0.06 = 1.06$. To find r we use the formula $r = \ln a$. The result is $r = \ln 1.06 = 0.058$ per month. This gives the logistic model

$$N = \frac{800}{1 + 3e^{-0.058t}}.$$

The r value is the natural logarithm of the growth factor.

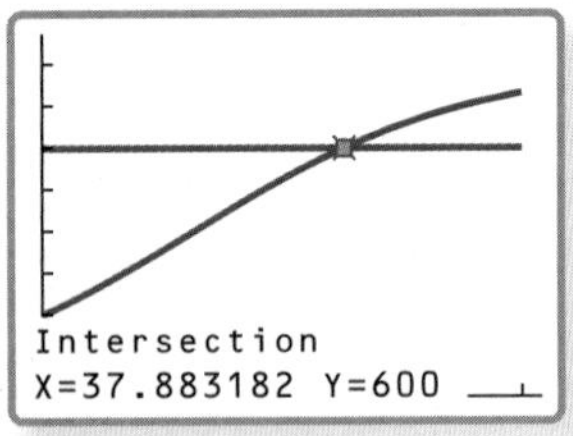

FIGURE 5.9 Magazine sales

Solution to Part 2 We use a horizontal span of 0 to 60 and a vertical span of 0 to 850. The graph is shown in Figure 5.9. Using the crossing-graphs method, we find that sales reach 600 per month at $t = 37.88$ or about 38 months.

TEST YOUR UNDERSTANDING | **FOR EXAMPLE 5.2**

Initial sales of a newspaper are 500 copies per week. Sales would increase by 10% each week if no limiting factors were involved, but the size of the town limits circulation to 2500 copies per week. Make a logistic model for newspaper circulation N as a function of time t in weeks. ∎

Fitting Logistic Data Using Regression

The calculator can be used to fit a logistic model to data by means of *logistic regression*. It is best to employ regression when the real-world context for the data suggests that a logistic model is appropriate. A plot of the data can confirm that regression gives a good fit.

We noted earlier that often the spread of a new technology displays logistic growth. In the next example we see how to fit a logistic model to the growth of Internet access.

EXAMPLE 5.3 **INTERNET ACCESS**

The accompanying table shows the number, in millions, of adults in the United States who had Internet access (home, work, or other) in the given year. The data are taken from the *Statistical Abstract of the United States*.

Year	1997	2000	2003	2006	2009
Number, in millions	46	113	166	177	196

Part 1 Plot the data and discuss whether a logistic model is appropriate.

Part 2 Use regression to find a logistic model for the data. Round the r value to three decimal places and the other parameters to two decimal places.

Part 3 Add the graph of the model you found in part 2 to the plot of the data.

Part 4 Find the carrying capacity for the logistic model and explain its meaning.

Part 5 At what level of Internet access was the rate of growth of access the largest?

Solution to Part 1 We let t denote the time in years since 1997 and I the number, in millions, of adults with Internet access. The properly entered data for I in terms of t are shown in Figure 5.10. A plot of the data is shown in Figure 5.11. We expect data showing the spread of a new technology to exhibit logistic growth, and the plot confirms that an S-shaped curve should fit the data.

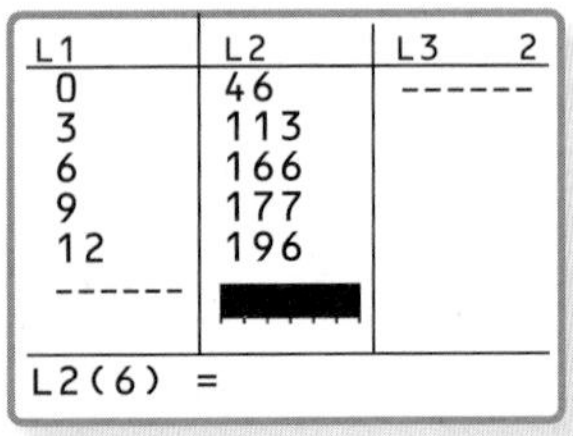

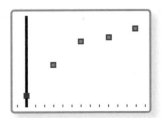

FIGURE 5.10 Data for Internet access

FIGURE 5.11 Plot of the data

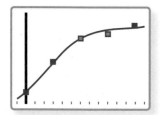

```
Logistic
 y=c/(1+ae^(¯bx))
 a=3.067775924
 b=.4859912791
 c=191.8466405
```

FIGURE 5.12 Regression parameters

Solution to Part 2 We use the calculator to find a logistic model. The regression parameters are shown in Figure 5.12. The corresponding logistic model is

$$I = \frac{191.85}{1 + 3.07e^{-0.486t}}.$$

Solution to Part 3 The graph is shown in Figure 5.13.

Solution to Part 4 From the model in part 2 we see that the carrying capacity is 191.85 million. This means that in the long run about 192 million adults will have Internet access. This estimate is less than the number in 2009, which shows the limitations of the logistic model.

Solution to Part 5 For a logistic function the rate of growth is the largest at half of the carrying capacity. Thus, the largest rate of growth in Internet access occurred at the level $I = 191.85/2$ or about 96 million adults.

FIGURE 5.13 Graph of model with data

TEST YOUR UNDERSTANDING | **FOR EXAMPLE 5.3**

The following table shows the number C of concerts, in thousands, presented by symphony orchestras in the United States t years after 1990. These data are from the *Statistical Abstract of the United States*.

t = time in years since 1990	0	5	10	12	15	17
C = concerts (thousands)	18.93	29.33	33.15	37.12	37.20	37.17

Make a logistic model of C as a function of t. ∎

The Value of the Logistic Model

The logistic model for population growth, like the exponential model, has limitations. As we have noted, the original article by Pearl and Reed in 1920 used this model to predict population growth in the United States. According to their data, the population was to stabilize at about 197 million. Of course, this level has been far surpassed. At a more fundamental level, the basic assumptions of the logistic model have been called into question. Various other models have been suggested, but they involve more complicated mathematics.

The main value of the models we have examined lies in their qualitative form. They enable us to discuss, in general terms, population trends and the reasons for such trends. Students should concentrate on understanding how the S-shaped graph reflects the underlying assumptions of this model.

ANSWERS FOR TEST YOUR UNDERSTANDING

5.1 The carrying capacity is 6.2 million tons of fish. The initial population is 0.03 million tons of fish. In the absence of limiting factors, the population would grow by 55% per year.

5.2 $N = \dfrac{2500}{1 + 4e^{-0.095t}}$

5.3 $C = \dfrac{38.26}{1 + 1.01e^{-0.227t}}$

5.1 EXERCISES

Special Rounding Instructions   When you perform logistic regression, round the r value to three decimal places and the other parameters to two decimal places. Round all answers to two decimal places unless otherwise indicated.

1. **Estimating Optimum Yield**   In Figure 5.14 a logistic growth curve is sketched. Estimate the optimum yield level and the time when this population should be harvested.

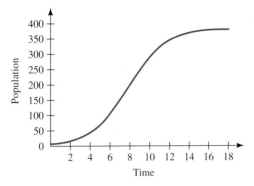

FIGURE 5.14 A logistic growth curve

2. **Estimating Carrying Capacity**   In Figure 5.15 a portion of a logistic growth curve is sketched. Estimate the optimum yield level and the carrying capacity.

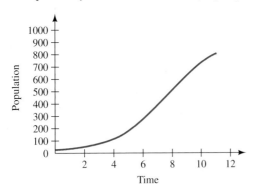

FIGURE 5.15 A portion of a logistic growth curve

3. **Magazine Sales**   Our new magazine initially sells 300 copies per month. Research indicates that a vigorous advertising campaign could increase sales by 20% each month if our market were unlimited. But research also indicates that magazine sales in our area are unlikely to exceed 1200 per month. Make a logistic model of projected magazine sales.

4. **Fluorescent Bulbs**   Compact fluorescent light bulbs save energy when compared with traditional incandescent bulbs. Our green energy campaign includes efforts to get local residents to exchange their incandescent bulbs for fluorescent bulbs. Initially 200 households make the change. Market studies suggest that, in the absence of limiting factors, we could increase that number by 25% each month. In our target area there are 250,000 households, which we take as the limiting value. Make a logistic model that gives the number of households converting to fluorescent bulbs after t months.

5. **African Bees**   There are 3600 commercial bee hives in a region threatened by African bees. Today African bees have taken over 50 hives. Experience in other areas shows that, in the absence of limiting factors, the African bees will increase the number of hives they take over by 30% each year. Make a logistic model that shows the number of hives taken over by African bees after t years, and determine how long it will be before 1800 hives are affected.

6. **An Epidemic**   In a city of half a million, there are initially 800 cases of a particularly virulent strain of flu. The Centers for Disease Control and Prevention in Atlanta claims that the cumulative number of infections of this flu strain will increase by 40% per week if there are no limiting factors. Make a logistic model of the potential cumulative number of cases of flu as a function of weeks from initial outbreak, and determine how long it will be before 100,000 people are infected.

7. **PTA Participation**   A school board has a plan to increase participation in the PTA. Currently only about 25 parents attend meetings. Suppose the school board plan results in logistic growth of attendance. The school board believes their plan can eventually lead to an attendance level of 50 parents. In the absence of limiting factors the school board

believes its plan can increase participation by 10% each month. Let m denote the number of months since the participation plan was put in place, and let P be the number of parents attending PTA meetings.

a. What is the carrying capacity K for a logistic model of P versus m?

b. Find the constant b for a logistic model.

c. Find the r value for a logistic model. Round your answer to three decimal places.

d. Find a logistic model for P versus m.

8. Mortality Versus Age   One study[5] of mortality versus age used the following model to give the probability P of death from measles if contracted at age t years:

$$P = \frac{1}{1 + 77.39e^{-0.08t}}.$$

Here we assume that t is at least 2.

a. What is the limiting value for this logistic function? *Note*: In other contexts this would be known as the carrying capacity.

b. Explain in practical terms the meaning of the limiting value you found in part a.

c. At what age does the model predict that mortality due to measles is 70%? (That is a value of 0.7 for P.)

9. Reliability Curves   A company is developing a new computer chip. Each month a collection of prototypes is tested and the percentage P of chips that operate successfully is recorded. Here are the results.

m = month	1	2	3	4	5	6	7
P = % successful	22	29	37	45	54	63	71

a. Find a logistic model for P versus m.

b. A *reliability curve* is a graph of percentage of successes versus time. Make a reliability curve for this computer chip. Include both the data and the model in your graph, and cover months from 0 to 15.

c. The chip is ready for production when reliability reaches 95%. When does the logistic model predict the chip will be ready for production? Round your answer to the nearest month.

10. Natural Gas Production   The following table shows natural gas production N in trillions of cubic feet in the United States t years after 1940.[6]

t = years since 1940	N = cubic ft. in trillions
0	3.75
10	8.48
20	15.09
30	23.79
40	21.87
50	21.52
60	24.15

a. Make a logistic model for N as a function of t.

b. Graph the data and the logistic model.

c. Which year's production was farthest from the prediction of the logistic model?

d. What does the logistic model predict for the amount of natural gas that will be produced in the long run? *Note*: In other contexts this would be known as the carrying capacity.

11. Paramecium Cells   The following table is adapted from a paramecium culture experiment conducted by Gause in 1934. The data show the paramecium population N as a function of time t in days.

T	2	3	5	6	8	9	10	11
N	14	34	94	189	330	416	507	580

a. Use regression to find a logistic model for this population.

b. Make a graph of the model you found in part a.

c. According to the model you made in part a, when would the population reach 450?

12. Long-Term Data and the Carrying Capacity   *This is a continuation of Exercise 11.* Ideally logistic data grow toward the carrying capacity but never go beyond this limiting value. The following table shows additional data on paramecium cells.

t	12	13	14	15	16	17	18	19	20
N	610	513	593	557	560	522	565	517	500

(continued)

---

[5]See http://ije.oxfordjournals.org/content/33/3/566.full
[6]Data taken from the U.S. Energy Information Administration.

a. Add these data to the graph in part b of Exercise 11.

b. Comment on the relationship of the data to the carrying capacity.

13. **Northern Yellowstone Elk**   The northern Yellowstone elk winter in the northern range of Yellowstone National Park.[7] A moratorium on elk hunting was imposed in 1969, and after that the growth of the elk population was approximately logistic for a time. The following table gives data on the growth.

Terry W Ryder/Shutterstock.com

Year	N
1968	3172
1969	4305
1970	5543
1971	7281
1972	8215
1973	9981
1974	10,529

a. Use regression to find a logistic model for this elk population.

b. According to the model you made in part a, when would the elk population reach half of the carrying capacity?

Note: At one time the gray wolf was a leading predator of the elk, but it was not a factor during this study period. The level at which the elk population stabilized suggests that food supply (and not just predators) can effectively regulate population size in this setting.

14. **Cable TV**   The following table shows the number C, in millions, of basic subscribers to cable TV in the indicated year. These data are from the *Statistical Abstract of the United States*.

Year	1975	1980	1985	1990	1995	2000
C	9.8	17.5	35.4	50.5	60.6	66.3

a. Use regression to find a logistic model for these data.

b. By what annual percentage would you expect the number of cable subscribers to grow in the absence of limiting factors?

c. The estimated number of subscribers in 2005 was 65.3 million. What light does this shed on the model you found in part a?

15. **World Population**   The following table shows world population N, in billions, in the given year.

Year	1950	1960	1970	1980	1990	2000	2010
N	2.56	3.04	3.71	4.45	5.29	6.09	6.85

a. Use regression to find a logistic model for world population.

b. What r value do these data yield for humans on planet Earth?

c. According to the logistic model using these data, what is the carrying capacity of planet Earth for humans?

d. According to this model, when will world population reach 90% of carrying capacity? Round to the nearest year.

Note: This represents a rather naive analysis of world population.

16. **Using New Starting Values to Make Logistic Models**   Suppose a population is growing according to the logistic formula $N = \dfrac{500}{1 + 3e^{-0.41t}}$, where t is measured in years.

a. Suppose that today there are 300 individuals in the population. Find a new logistic formula for the population using the same K and r values as the formula above but with initial value 300.

b. How long does it take the population to grow from 300 to 400 using the formula in part a?

17. **More on the Pacific Sardine**   *This is a continuation of Example 5.1. In this exercise we explore further the Pacific sardine population using the model in Example 5.1.*

a. If the current level of the Pacific sardine population is 50,000 tons, how long will it take for the population to recover to the optimum growth level of 1.2 million tons? (*Suggestion:* One way

---

[7]This exercise is based on the study by Douglas B. Houston in *The Northern Yellowstone Elk* (New York: Macmillan, 1982). Houston uses a different method to fit a logistic model.

to solve this is to make a new logistic formula using $K = 2.4$, $r = 0.338$, and $N(0) = 0.05$.)

b. The value of r used in Example 5.1 ignores the effects of fishing. If fishing mortality is taken into account, then r drops to 0.215 per year (with the carrying capacity still at 2.4 million tons). Answer the question in part a using this lower value of r.

Note: The population estimate of 50,000 tons and the adjusted value of r are given in the paper by Murphy (see footnote 3 on page 319). Murphy points out that factoring in growth of the competing anchovy population makes the recovery times even longer, and he adds, "It is disconcerting to realize how slowly the population will recover to its level of maximum productivity . . . even if fishing stops."

18. **Modeling Human Height with a Logistic Function**
A male child is 21 inches long at birth and grows to an adult height of 73 inches. In this exercise we make a logistic model of his height as a function of age.

a. Use the given information to find K and b for the logistic model.

b. Suppose he reaches 95% of his adult height at age 16. Use this information and that from part a to find r. (*Suggestion:* You will need to use either the crossing-graphs method or some algebra involving the logarithm.)

c. Make a logistic model for the height H, in inches, as a function of the age t, in years.

d. According to the logistic model, at what age is he growing the fastest?

e. Is your answer to part d consistent with your knowledge of how humans grow?

19. **Eastern Pacific Yellowfin Tuna**
Studies to fit a logistic model to the Eastern Pacific yellowfin tuna population have yielded

louise murray/Alamy

$$N = \frac{148}{1 + 3.6e^{-2.61t}},$$

where t is measured in years and N is measured in thousands of tons of fish.[8]

a. What is the r value for the Eastern Pacific yellowfin tuna?

b. What is the carrying capacity K for the Eastern Pacific yellowfin tuna?

c. What is the optimum yield level?

d. Use your calculator to graph N versus t.

e. At what time was the population growing the most rapidly?

20. **Pacific Halibut** For a Pacific halibut population, the value of r is 0.71 per year, and the carrying capacity is 89 thousand tons of fish.[9]

a. Find the time it takes the population to grow from the optimum yield level to 90% of carrying capacity if the number b in the logistic model is 1.5.

b. Find the time it takes the population to grow from the optimum yield level to 90% of carrying capacity if the number b in the logistic model is 4.8.

c. Compare the times in parts a and b. Why could this result have been expected?

21. **An Inverted Logistic Curve** When the parameter r in the logistic formula is negative (resulting in a positive power of e), the logistic curve has a different shape. For example, the following formula[10] gives the approximate number of deaths due to tuberculosis as a fraction of all deaths in the United States t years after 1875:

$$T = \frac{0.13}{1 + 0.07e^{0.05t}}.$$

a. Make a graph of T versus t. Include dates up to 2000.

b. What is the limiting value of T? Explain in practical terms what this limiting value means.

c. Estimate when the fraction of deaths due to tuberculosis was decreasing most rapidly. Give your answer accurate to the nearest decade (for example, 1920).

22. **Gompertz Model** One possible substitute for the logistic model of population growth is the Gompertz model, according to which

$$\text{Rate of growth} = r\, N\, \ln\!\left(\frac{K}{N}\right).$$

(continued)

[8]From a study by M. B. Schaefer, as described by Colin W. Clark in *Mathematical Bioeconomics*, 2nd ed. (New York: Wiley, 1990).
[9]From a study by H. S. Mohring, as described by Colin W. Clark, ibid.
[10]Adapted from http://phe.rockefeller.edu/death/

For simplicity in this problem we take $r = 1$, so this reduces to

$$\text{Rate of growth} = N \ln\left(\frac{K}{N}\right).$$

a. Let $K = 10$, and make a graph of the rate of growth versus N for the Gompertz model.

b. Use the graph you obtained in part a to determine for what value of N the growth rate reaches its maximum. This is the optimum yield level under the Gompertz model with $K = 10$.

c. Under the logistic model the optimum yield level is $K/2$. What do you think is the optimum yield level in terms of K under the Gompertz model? (*Hint:* Repeat the procedure in parts a and b using different values of K, such as $K = 1$ and $K = 100$. Try to find a pattern.)

5.1 SKILL BUILDING EXERCISES

S-1. Logistic Growth When we add the notion of *carrying capacity* to the basic assumptions leading to the exponential model, we get a different model. What is the name of this model?

S-2. Percentage Rate of Change for Logistic Model For exponential growth we assume that the percentage rate of change is constant. What basic assumption regarding the percentage growth rate underlies the logistic model?

S-3. Harvesting What is the name of the theory that says that a renewable resource growing logistically should be harvested at half of the carrying capacity?

S-4. Harvesting Suppose a renewable population grows logistically according to

$$N = \frac{778}{1 + 5e^{-0.02t}}.$$

According to the theory of maximum sustainable yield, what is the optimum harvesting level?

S-5. Harvesting Continued The theory of maximum sustainable yield says that a renewable resource that grows logistically should be harvested at half of the carrying capacity. What is the significance of the corresponding point on the graph of population versus time?

Finding Logistic Parameters In Exercises S-6 through S-19, you are asked to find the values of certain numbers associated with the logistic formula

$$N = \frac{K}{1 + be^{-rt}}.$$

S-6. Find b if the carrying capacity is 600 and the initial population is 50.

S-7. Find b if the optimum yield level is 400 and the initial population is 200.

S-8. Find the carrying capacity if $b = 7$ and the initial population is 300.

S-9. Find the optimum yield level if $b = 5$ and the initial population is 150.

S-10. Find the initial population if the carrying capacity is 900 and $b = 9$.

S-11. Find the initial population if the carrying capacity is 1000 and $b = 3$.

S-12. The r value is 0.3 per year. What would be the growth factor in the absence of constraints?

S-13. The r value is 0.3 per year. What percentage growth rate would the population show in the absence of constraints?

S-14. The r value is 0.7 per year. What would be the growth factor in the absence of constraints?

S-15. The r value is 0.7 per year. What percentage growth rate would the population show in the absence of constraints?

S-16. In the absence of constraints, a population would have a yearly growth factor of 1.11. What is its r value?

S-17. In the absence of constraints, a population would have an 11% increase each year. What is its r value?

S-18. In the absence of constraints, a population would have a yearly growth factor of 1.64. What is its r value?

S-19. In the absence of constraints, a population would have a 64% increase each year. What is its r value?

Finding the Logistic Formula In Exercises S-20 through S-26, find the formula for logistic growth using the given information.

S-20. The carrying capacity is 1800, the r value is 0.21 per year, and $b = 26$.

S-21. The carrying capacity is 2300, the r value is 0.77 per year, and $b = 7$.

S-22. The carrying capacity is 800, the r value is 0.06 per year, and the initial population is 200.

S-23. The carrying capacity is 400, the r value is 0.44 per year, and the initial population is 10.

S-24. The r value is 0.015 per year, the carrying capacity is 2390, and the initial population is 120.

S-25. The optimum yield is 400, the initial value is 100, and the population would, in the absence of constraints, grow by 22% per year.

S-26. The optimum yield is 750, the initial value is 120, and the population would, in the absence of constraints, grow by 42% per year.

Using Logistic Regression   Exercises S-27 through S-31 require a calculator that can perform logistic regression. Round the r value to three decimal places and the other parameters to two decimal places.

S-27. Use logistic regression to model the following data.

t	2	3	5	6	7	9
N	120	155	207	210	215	220

S-28. Use logistic regression to model the following data.

t	1	3	4	7	8	10
N	60	110	155	280	290	313

S-29. *This is a continuation of Exercise S-28.*
   a. What is the carrying capacity for this population?
   b. When will the population reach 125?

S-30. Use logistic regression to model the following data.

t	5	10	15	20	25	30
N	300	710	1260	1830	2220	2261

S-31. *This is a continuation of Exercise S-30.*
   a. What is the carrying capacity for this population?
   b. When will the population reach 1625?

5.2 POWER FUNCTIONS

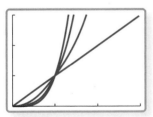

FIGURE 5.16 Power functions with positive powers are increasing functions.

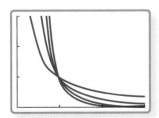

FIGURE 5.17 Power functions with negative powers decrease toward 0.

Recall that an exponential function has the form $f(x) = Pa^x$, where the base a is fixed and the exponent x varies. For a power function these properties are reversed—the base varies and the exponent remains constant—so a power function has the form $f(x) = cx^k$. The number k is called the *power*, the most significant part of a power function, and the coefficient c is equal to $f(1)$. For most applications, we are interested only in positive values of the variable x and of c, but we allow k to be any number.

We can use the graphing calculator to illustrate how power functions work and the role of k. In the exercises at the end of this section, you will have an opportunity to explore the role of c graphically. We look first at what happens when the power k is positive. In Figure 5.16 we show the graphs of x, x^2, x^3, and x^4. For the viewing window, we have used both a horizontal and a vertical span of 0 to 3. The first thing we note is that when the power is positive, the graph of the power function is increasing. We see also that when $k = 1$, the graph is a straight line. This tells us that a power function with power 1 is a linear function. Finally, we observe that, for values of x larger than 1 (the common crossing point in Figure 5.16), larger powers cause the power function to grow faster.

In Figure 5.17, we look at what happens when the power k is negative. Here we show the graphs of x^{-1}, x^{-2}, x^{-3}, and x^{-4} using both a horizontal and a vertical span of 0 to 3. We see that for negative powers, power functions decrease toward 0. Furthermore, negative powers that are larger in size cause the graph to approach the horizontal axis more rapidly than do negative powers that are smaller in size.

> **KEY IDEA 5.3** **POWER FUNCTIONS**
>
> For a power function $f(x) = cx^k$ with c and x positive:
>
> **1.** If k is positive, then f is increasing. Larger positive values of k cause f to increase more rapidly.
>
> **2.** If k is negative, then f decreases toward zero. Negative values of k that are larger in size cause f to decrease more rapidly.

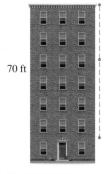

70 ft

FIGURE 5.18

EXAMPLE 5.4 **DISTANCE FALLEN AS A FUNCTION OF TIME**

When a rock is dropped from a tall structure, it will fall $D = 16t^2$ feet in t seconds. (See Figure 5.18.)

Part 1 Make a graph that shows the distance the rock falls versus time if the building is 70 feet tall.

Part 2 How long does it take the rock to strike the ground?

Solution to Part 1 The first step is to enter the function in the calculator function list and to record the variable associations:

$$Y_1 = D, \text{ distance, in feet, on vertical axis}$$

$$X = t, \text{ time, in seconds, on horizontal axis.}$$

Since the building is 70 feet tall, we allow a bit of extra room and use the vertical span of 0 to 100. Since our everyday experience tells us that it will take only a few seconds for the rock to reach the ground, we use the horizontal span of 0 to 5. The graph appears in Figure 5.19.

Solution to Part 2 We want to know the value of t when the rock strikes the ground—that is, when $D = 70$ feet. Thus we need to solve the equation

$$16t^2 = 70.$$

We proceed with the crossing-graphs method. We enter the target distance, 70, on the function list and plot to see the picture in Figure 5.20. We use the calculator to find the crossing point at $t = 2.09$ seconds as shown in Figure 5.20.

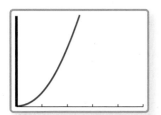

FIGURE 5.19 Graph of distance fallen versus time

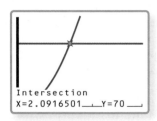

Intersection
X=2.0916501 Y=70

FIGURE 5.20 When the rock strikes the ground

TEST YOUR UNDERSTANDING | **FOR EXAMPLE 5.4**

In the situation of the example, if the structure is 90 feet tall, how long does it take for the rock to hit the ground? ∎

Homogeneity Property of Power Functions

Many times, the qualitative nature of a function is as important as the exact formula that describes it. It is, for example, crucial to physicists to understand that distance fallen is proportional to the square of the time (as opposed to an exponential, linear, or some other type of function). This is an important qualitative observation about how gravity acts on objects near the surface of the Earth. To make it clear what we mean, we look at an important mathematical property of power functions that is known as *homogeneity*.

Suppose that in a power function $f(x) = cx^k$, the value of x is increased by a factor of t. What happens to the value of f? To help answer this question, let's first look at some specific examples.

How tripling the side of a square affects its area: The area A of a square of side with length s is equal to the square of s. Thus $A = s^2$. Suppose a square initially has sides of length 4 feet. If the length of the sides of the square is tripled, how is the area affected? To answer this, we calculate the original area and compare it to the area after the sides of the square are tripled:

$$\text{Original area} = 4^2 = 16 \text{ square feet.}$$

To get the new area, we use $s = 12$:

$$\text{New area} = 12^2 = 144 \text{ square feet.}$$

Thus, if the side of the square is tripled, the area increases from 16 square feet to 144 square feet; that is, it increases by a factor of 9. The key thing to note here is that 9 is 3^2. Increasing the length of the side by a factor of 3 increased the area by a factor of 3^2— that is, 3 raised to the same power as the function.

How doubling the radius of a sphere affects its volume: A slightly more complicated calculation will show the same phenomenon. The volume V inside a sphere, such as a tennis ball or basketball, depends on its radius r and is proportional to the cube of the radius. Specifically, from elementary geometry we know that $V = \frac{4\pi}{3}r^3$. Suppose a balloon initially has a radius of 5 inches. If air is pumped into the balloon until the radius doubles, what is the effect on the volume? We proceed as before, calculating the original volume and comparing it with the volume after the radius is doubled. Initially the radius is $r = 5$. Accuracy is important in this calculation, so in this case we report all the digits given by the calculator:

$$\text{Original volume} = \frac{4\pi}{3}5^3 = 523.5987756.$$

To get the new volume, we use $r = 10$:

$$\text{New volume} = \frac{4\pi}{3}10^3 = 4188.790205.$$

To understand how the volume has changed, we divide:

$$\frac{\text{New volume}}{\text{Old volume}} = \frac{4188.790205}{523.5987756} = 8.$$

Increasing the radius by a factor of 2 results in an increase in volume by a factor of 8. Once again, the key thing to observe is that 8 is 2^3. Summarizing, we see that if the radius is increased by a factor of 2, then the volume is increased by a factor of 2^3—that is, 2 raised to the power of the function.

The phenomenon we observed in these two examples is characteristic of power functions, and we can show this using some elementary properties of exponents. Let's return to our original question. Suppose $f = cx^k$. If x is increased by a factor of t, what is the effect on f? We have

$$\text{Old value} = cx^k.$$

To get the new value, we replace x by tx:

$$\text{New value} = c(tx)^k = ct^k x^k = t^k(cx^k) = t^k \times \text{Old value}.$$

Thus, just as we observed in our examples, increasing x by a factor of t increases f by a factor of t^k.

Linear: $y = mx + b$. Adding t to x adds mt to y.

Exponential: $y = ca^x$. Adding t to x multiplies y by a^t.

Power function: $y = cx^k$. Multiplying x by t multiplies y by t^k.

KEY IDEA 5.4 **HOMOGENEITY PROPERTY OF POWER FUNCTIONS**

For a power function $f = cx^k$, if x is increased by a factor of t, then f is increased by a factor of t^k.

EXAMPLE 5.5 **EXPANSION OF A SHOCK WAVE**

The shock wave produced by a large explosion expands rapidly. (See Figure 5.21.) The radius R of the wave 1 second after the explosion is a power function of the energy E released by the explosion, and the formula is

$$R = 4.16E^{0.2}.$$

Here R is measured in centimeters and E in ergs.

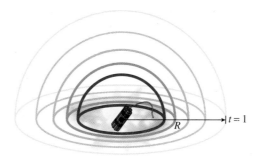

FIGURE 5.21

Part 1 If one explosion releases 1000 times as much energy as another, how much larger will the radius of the shock wave be after 1 second?

Part 2 If one explosion releases half as much energy as another, how much smaller will the radius of the shock wave be after 1 second?

Solution to Part 1 We use the homogeneity property of power functions. Here R is a power function of E with power $k = 0.2$. Thus when E is increased by a factor of 1000, the function R is increased by a factor of $1000^{0.2}$. Since $1000^{0.2} = 3.98$, after 1 second the radius of the shock wave for the larger explosion will be 3.98 times as large as that for the smaller explosion.

Multiplying E by 1000 causes R to be multiplied by $1000^{0.2}$.

Solution to Part 2 When E is multiplied by 0.5, the function R is multiplied by $0.5^{0.2}$, or about 0.87. Thus after 1 second the radius of the shock wave for the smaller explosion will be 0.87 times as large as that for the larger explosion.

TEST YOUR UNDERSTANDING | **FOR EXAMPLE 5.5**

In the situation of the example, if one explosion releases triple the energy of another, how do the radii of their shock waves compare after one second? ■

Note that in this example the coefficient $c = 4.16$ did not enter into the computations. The next example further illustrates that the homogeneity property for power functions can be useful in situations where only the power is known.

EXAMPLE 5.6 **NUMBER OF SPECIES VERSUS AVAILABLE AREA**

Ecologists have studied how the number S of species that make up a given group existing in a closed environment (often an island) varies with the area A that is available.[11] They use the approximate *species-area relation*

$$S = cA^k$$

to estimate, among similar habitats, the number of species as a function of available area. For birds on islands in the Bismarck Archipelago near New Guinea, the value of k is estimated to be about $k = 0.18$.

Part 1 If one island in the Bismarck Archipelago were twice as large as another, how many more species of birds would it have? Interpret your answer in terms of percentages.

---

[11]See J. Diamond and R. M. May, "Island biogeography and the design of natural reserves," in R. M. May (ed.), *Theoretical Ecology*, 2nd ed. (Sunderland, MA: Sinauer Associates, 1981). See also the references therein.

Part 2 If there are 50 species of birds on an island in the Bismarck Archipelago of area 100 square kilometers, find the value of c, and then make a graph of the number of species as a function of available area for islands in the range of 50 to 200 square kilometers. (Here we measure A in square kilometers.)

Part 3 It is thought that a power relation $S = cA^k$ also applies to the Amazon rain forest, whose size is being reduced by (among other things) burning in preparation for farming. Ecologists use the estimate $k = 0.30$ for the power. If the rain forest were reduced in area by 20%, by what percentage would the number of surviving species be expected to decrease?

Part 4 Experience has shown that in a certain chain of islands, a 10% reduction in area leads to a 4% reduction in the number of species. Find the value of k in the species-area relation. What would be the result of a 25% reduction in the usable area of one of these islands?

Solution to Part 1 The number of species S is a power function of the area A, with power $k = 0.18$. If the variable A is doubled, then, by the homogeneity property of power functions, the function S increases by a factor of $2^{0.18} = 1.13$. Thus the number of species of birds increases by a factor of 1.13. In terms of percentages, doubling A is a 100% increase, and changing S by a factor of 1.13 represents a 13% increase. Thus increasing the area by 100% has the effect of increasing the number of species by 13%.

Solution to Part 2 We know that $S = cA^{0.18}$ and that an area of 100 square kilometers supports 50 species. That is, $S = 50$ when $A = 100$:

$$50 = c \times 100^{0.18}$$

$$50 = c \times 2.29.$$

This is a linear equation, and we can solve for c by hand calculation:

$$c = \frac{50}{2.29} = 21.83.$$

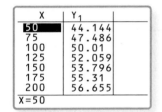

X	Y$_1$
50	44.144
75	47.486
100	50.01
125	52.059
150	53.796
175	55.31
200	56.655
X=50	

FIGURE 5.22 A table of values for number of species

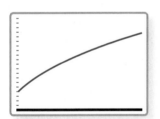

FIGURE 5.23 Number of species versus area in the Bismarck Archipelago

We conclude that in the Bismarck Archipelago, the number of species is related to area by $S = 21.83A^{0.18}$. We first enter this on our calculator function list and record variable correspondences:

$$Y_1 = S, \text{ number of species on vertical axis}$$

$$X = A, \text{ area, in square kilometers, on horizontal axis.}$$

We are asked to make the graph for islands from 50 to 200 square kilometers in area. Thus we use a horizontal span of 50 to 200. For the vertical span, we look at the table of values in Figure 5.22. Allowing a little extra room, we use a vertical span of 40 to 60. The completed graph is shown in Figure 5.23.

A 20 percent reduction means multiplying the area by 0.8, which in turn multiplies the number of species by $0.8^{0.30}$.

Solution to Part 3 To say that there is a 20% decrease in area means that the new area is 80% of its original value. Thus A has changed by a factor of 0.8. Using the homogeneity property of power functions, we conclude that S, the number of species, changes by a factor of $0.8^{0.30} = 0.94$. Thus 94% of the original species can be expected to survive, so the number of species has been reduced by 6%.

Solution to Part 4 We know that a 10% reduction in the area of an island will result in a 4% reduction in the number of species. In other words, if A is changed by a factor of 0.9, then S is changed by a factor of 0.96. The homogeneity property of power functions tells us that $0.96 = 0.9^k$. We can use the crossing-graphs method to solve this equation as is shown in Figure 5.24, where we have used a horizontal span of 0 to 1 and a vertical span

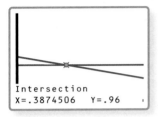

FIGURE 5.24 Finding the power in the species-area relation

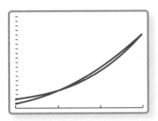

FIGURE 5.25 A limited span where a power function and an exponential function appear similar

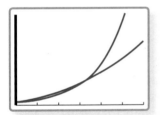

FIGURE 5.26 The characteristic dominance of exponential functions over power functions in the long term

of 0.8 to 1.2. We read from the prompt at the bottom of the figure that $k = 0.39$, rounded to two decimal places.

If the area is reduced by 25%, we are changing A by a factor of 0.75. Thus S changes by a factor of $0.75^{0.39} = 0.89$. That is an 11% reduction in the number of species.

TEST YOUR UNDERSTANDING | **FOR EXAMPLE 5.6**

Use the information from part 3 of the example to determine what would happen to the number of species in the Amazon rain forest if the area could be doubled. ∎

 | **Comparing Exponential and Power Functions**

Over limited ranges, the graphs of exponential functions and power functions may appear to be similar. We see this in Figure 5.25, where we have graphed the exponential function 2^x and the power function x^2 using a horizontal span of 1 to 4 and a vertical span of 0 to 20. As the figure shows, the two graphs are nearly identical on this span. This sometimes makes it difficult to determine whether observed data should be modeled with an exponential function or with a power function. We want to illustrate the consequences of an inappropriate choice. Figure 5.25 shows the similarity of the functions over the displayed range, but Figure 5.26 shows a dramatic difference if we view the graphs with a horizontal span of 1 to 7 and a vertical span of 0 to 70. The exponential function 2^x is the graph in Figure 5.26 that rises rapidly above the graph of the power function.

If you view these graphs on an even larger horizontal span, the differences will be even more dramatic. This behavior is typical of the comparison of any exponential function with base larger than 1 with any power function, no matter how large the power. Power functions may look similar to exponential functions over a limited range, and they may even grow more rapidly for brief periods. But eventually, exponential functions always grow many times faster than power functions. This makes the consequences of choosing the wrong model quite serious. For example, if we were to choose an exponential model for federal spending when a power model was in fact appropriate, our mistake would lead us to predict federal spending at levels many orders of magnitude too large. Or, if we chose a power model when in fact an exponential model was appropriate, we would be led to predict future federal spending at levels far too low.

The fact that exponential functions eventually grow many times faster than power functions is one of the most important qualitative distinctions between these two types of functions.

> **KEY IDEA 5.5** **EXPONENTIAL FUNCTIONS GROW FASTER THAN POWER FUNCTIONS**
>
> Over a sufficiently large horizontal span, an exponential function (with base larger than 1) will increase much more rapidly than a power function.

ANSWERS FOR TEST YOUR UNDERSTANDING

5.4 After 2.37 seconds

5.5 The radius of the shock wave of the larger explosion is 1.25 times that of the smaller explosion.

5.6 There would be 1.23 times as many species.

5.2 EXERCISES

Reminder   Round all answers to two decimal places unless otherwise indicated.

1. **The Role of the Coefficient in Power Functions with Positive Power**   Consider the family of power functions $f(x) = cx^2$. On the same screen, make graphs of f versus x for $c = 1$, $c = 2$, $c = 3$, and $c = 4$. We suggest a horizontal span of 0 to 5. A table of values will be helpful in choosing a vertical span. On the basis of the plots you make, discuss the effect of the coefficient c on a power function when the power is positive.

2. **The Role of the Coefficient in Power Functions with a Negative Power**   Consider the family of power functions $f(x) = cx^{-2}$. On the same screen, make graphs of f versus x for $c = 1$, $c = 2$, $c = 3$, and $c = 4$. We suggest a horizontal span of 0 to 5. A table of values will be helpful in choosing a vertical span. On the basis of the plots you make, discuss the effect of the coefficient c on a power function when the power is negative.

3. **Speed and Stride Length**   The speed at which certain animals run is a power function of their stride length, and the power is $k = 1.7$. (See Figure 5.27.) If one animal has a stride length three times as long as another, how much faster does it run?

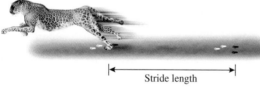

|←──────── Stride length ────────→|

FIGURE 5.27

4. **Weight and Length**   A biologist has discovered that the weight of a certain fish is a power function of its length. He also knows that when the length of the fish is doubled, its weight increases by a factor of 8. What is the power k?

5. **Stevens's Power Law**   In 1957 Stanley Smith Stevens proposed a law that relates the magnitude of a physical stimulus and the perceived intensity.[12] One application of his model relates pressure p felt on the palm as a result of a static force f on the skin. The relationship is

$$p = kf^{1.1},$$

where k is a constant. According to Stevens's law, how would a doubling of force affect perceived pressure on the palm?

6. **Zipf's Law**   The linguist George Kingsley Zipf (1902–1950) proposed a law that bears his name. In a typical English text the most commonly occurring word is "the." We say that its *frequency rank r* is 1. The frequency f of occurrence of "the" is about $f = 7\%$. That is, in a typical English text, the word "the" accounts for about 7 out of 100 occurrences of words. The second most common word is "of," so it has a frequency rank of $r = 2$. Its frequency of occurrence is $f = 3.5\%$. Zipf's law gives a power relationship between frequency of occurrence f, as a percentage, and frequency rank r. (Note that a higher frequency rank means a word that occurs less often.) The relationship is

$$f = cr^{-1},$$

where c is a constant.

   a. Use the frequency information given for "the" to determine the value of c.

   b. The third most common English word is "and." According to Zipf's law, what is the frequency of this word in a typical English text? Round your answer to one decimal place.

   c. If one word's frequency rank is twice that of another, how do their frequencies of occurrence compare?

7. **Mosteller Formula for Body Surface Area**   Body surface area is an important piece of medical information because it is a factor in temperature regulation as well as some drug level determinations. The Mosteller formula[13] gives one way of estimating body surface area B in square meters. The formula uses the weight w in kilograms and the height h in centimeters. The relation is

$$B = \frac{1}{60}h^{1/2}w^{1/2}.$$

   a. Use the Mosteller formula to estimate the body surface area for a man who is 188 centimeters tall and weighs 86 kilograms.

   b. The weight of an adult increases by 10%. How will this change affect his body surface area? Give your answer as a percentage. (*Note*: You may assume that the adult's height will stay the same.)

---

[12]This power model contrasts with the logarithmic model proposed by Fechner and presented in Chapter 4.
[13]R. D. Mosteller, "Simplified calculation of body-surface area," *N. Engl. J. Med.* **317,** No. 17 (1987); 1098 (letter).

8. **Ship Propellers** An ideal diameter d, in feet, of a ship's propeller is given by the formula

$$d = ch^{1/5}r^{-3/5}.$$

Here h is the horsepower of the engine driving the propeller, r is the (maximum) number of revolutions per minute of the propeller, and c is a constant. In both parts give your answer in terms of a percentage.

   a. If the horsepower is increased by 20% while the number of revolutions per minute remains the same, how is the propeller diameter affected?

   b. If the horsepower remains the same while the number of revolutions per minute is increased by 20%, how is the propeller diameter affected?

9. **Length of Skid Marks Versus Speed** When a car skids to a stop, the length L, in feet, of the skid marks is related to the speed S, in miles per hour, of the car by the power function $L = \frac{1}{30h}S^2$. Here the constant h is the *friction coefficient*, which depends on the road surface.[14] For dry concrete pavement, the value of h is about 0.85.

   a. If a driver going 55 miles per hour on dry concrete jams on the brakes and skids to a stop, how long will the skid marks be?

   b. A policeman investigating an accident on dry concrete pavement finds skid marks 230 feet long. The speed limit in the area is 60 miles per hour. Is the driver in danger of getting a speeding ticket?

   c. This part of the problem applies to any road surface, so the value of h is not known. Suppose you are driving at 60 miles per hour but, because of approaching darkness, you wish to slow to a speed that will cut your emergency stopping distance in half. What should your new speed be? (*Hint:* You should use the homogeneity property of power functions here. By what factor should you change your speed to ensure that L changes by a factor of 0.5?)

10. **Binary Stars** *Binary stars* are pairs of stars that orbit each other. The *period p* of such a pair is the time, in years, required for a single orbit. The separation s between such a pair is measured in seconds of arc. The *parallax* angle a (also in seconds of arc) for any stellar object is the angle of its apparent movement as the Earth moves through one half of its orbit around the sun. Astronomers can calculate the total mass M of a binary system using

$$M = s^3a^{-3}p^{-2}.$$

Here M is the number of *solar masses*.

   a. Alpha Centauri, the nearest star to the sun, is in fact a binary star. The separation of the pair is $s = 17.6$ seconds of arc, its parallax angle is $a = 0.76$ second of arc, and the period of the pair is 80.1 years. What is the mass of the Alpha Centauri pair?

   b. How would the mass change if the separation angle were doubled, but parallax and period remained the same as for the Alpha Centauri system?

   c. How would the mass change if the parallax angle were doubled but separation and period remained the same?

   d. How would the mass change if the period doubled but parallax angle and separation remained the same?

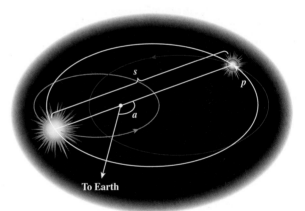

11. **Life Expectancy of Stars** The life expectancy E of a main-sequence star[15] depends on its mass M. The relation is given by

$$E = M^{-2.5},$$

where M is solar masses and E is solar lifetimes. The sun is thought to be at the middle of its life, with a total life expectancy of about 10 billion years. Thus the value $E = 1$ corresponds to a life expectancy of 10 billion years.

   a. Does a more massive star have a longer or a shorter life expectancy than a less massive star?

   b. Spica is a main-sequence star that is about 7.3 solar masses. What is the life expectancy of Spica?

(*continued*)

[14]See J. C. Collins, *Accident Reconstruction* (Springfield, IL: Charles C. Thomas, 1979). Collins notes that the friction coefficient is actually reduced at higher speeds.
[15]About 90% of all stars are main-sequence stars.

c. Express using functional notation the life expectancy of a main-sequence star with mass equal to 0.5 solar mass, and then calculate that value.

d. Vega is a main-sequence star that is expected to live about 6.36 billion years. What is the mass of Vega?

e. If one main-sequence star is twice as massive as another, how do their life expectancies compare?

12. Height of Tsunami Waves   When waves generated by tsunamis approach shore, the height of the waves generally increases. Understanding the factors that contribute to this increase can aid in controlling potential damage to areas at risk.

   Green's law tells how water depth affects the height of a tsunami wave. If a tsunami wave has height H at an ocean depth D, and the wave travels to a location of water depth d, then the new height h of the wave is given by $h = HR^{0.25}$, where R is the water depth ratio given by $R = D/d$.

   a. Calculate the height of a tsunami wave in water 25 feet deep if its height is 3 feet at its point of origin in water 15,000 feet deep.

   b. If water depth decreases by half, the depth ratio R is doubled. How is the height of the tsunami wave affected?

13. Tsunami Waves and Breakwaters   *This is a continuation of Exercise 12.* Breakwaters affect wave height by reducing energy. (See Figure 5.28.) If a tsunami wave of height H in a channel of width W encounters a breakwater that narrows the channel to a width w, then the height h of the wave beyond the breakwater is given by $h = HR^{0.5}$, where R is the width ratio $R = w/W$.

   a. Suppose a wave of height 8 feet in a channel of width 5000 feet encounters a breakwater that narrows the channel to 3000 feet. What is the height of the wave beyond the breakwater?

   b. If a channel width is cut in half by a breakwater, what is the effect on wave height?

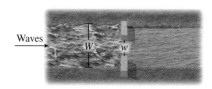

FIGURE 5.28

14. Bores   Under certain conditions, tsunami waves encountering land will develop into *bores*. A bore is a surge of water much like what would be expected if a dam failed suddenly and emptied a reservoir into a river bed. In the case of a bore traveling from the ocean into a dry river bed, one study[16] shows that the velocity V of the tip of the bore is proportional to the square root of its height h. Expressed in a formula, this is

$$V = kh^{0.5},$$

where k is a constant.

   a. A bore travels up a dry river bed. How does the velocity of the tip compare with its initial velocity when its height is reduced to half of its initial height?

   b. How does the height of the bore compare with its initial height when the velocity of the tip is reduced to half its initial velocity?

   c. If the tip of one bore surging up a dry river bed is three times the height of another, how do their velocities compare?

15. Terminal Velocity   By comparing the surface area of a sphere with its volume and assuming that air resistance is proportional to the square of velocity, it is possible to make a heuristic argument to support the following premise: *For similarly shaped objects, terminal velocity varies in proportion to the square root of length.* Expressed in a formula, this is

$$T = kL^{0.5},$$

where L is length, T is terminal velocity, and k is a constant that depends on shape, among other things. This relation can be used to help explain why small mammals easily survive falls that would seriously injure or kill a human.

   a. A 6-foot man is 36 times as long as a 2-inch mouse (neglecting the tail). How does the terminal velocity of a man compare with that of a mouse?

   b. If the 6-foot man has a terminal velocity of 120 miles per hour, what is the terminal velocity of the 2-inch mouse?

   c. Neglecting the tail, a squirrel is about 7 inches long. Again assuming that a 6-foot man has a terminal velocity of 120 miles per hour, what is the terminal velocity of a squirrel?

[16]Y. M. Fukui, H. S. Nakamura, and Y. Sasaki, "Hydraulic study on tsunami," *Coastal Engr. Japan* **6** (1963), 67–82.

16. **Dropping Rocks on Other Planets** It is a consequence of Newton's law of gravitation that near the surface of any planet, the distance D fallen by a rock in time t is given by $D = ct^2$. That is, distance fallen is proportional to the square of the time, no matter what planet one may be on. But the value of c depends on the mass of the planet. For Earth, if time is measured in seconds and distance in feet, the value of c is 16.

    a. Suppose a rock is falling near the surface of a planet. What is the comparison in distance fallen from 2 seconds to 6 seconds into the drop? (*Hint:* This question may be rephrased as follows: "If time increases by a factor of 3, by what factor will distance increase?")

    b. For objects falling near the surface of Mars, if time is measured in seconds and distance in feet, the value of c is 6.4. If a rock is dropped from 70 feet above the surface of Mars, how long will it take for the rock to strike the ground?

    c. On Venus, a rock dropped from 70 feet above the surface takes 2.2 seconds to strike the ground. What is the value of c for Venus?

17. **Newton's Law of Gravitation** According to Newton's law of gravitation, the gravitational attraction between two massive objects such as planets or asteroids is proportional to d^{-2}, where d is the distance between the centers of the objects. Specifically, the gravitational force F between such objects is given by $F = cd^{-2}$, where d is the distance between their centers. The value of the constant c depends on the masses of the two objects and on the *universal gravitational constant*.

    a. Suppose the force of gravity is causing two large asteroids to move toward each other. What is the effect on the gravitational force if the distance between their centers is halved? What is the effect on the gravitational force if the distance between their centers is reduced to one-quarter of its original value?

    b. Suppose that for a certain pair of asteroids whose centers are 300 kilometers apart, the gravitational force is 2,000,000 newtons. (One newton is about one-quarter of a pound.) What is the value of c? Find the gravitational force if the distance between the centers of these asteroids is 800 kilometers.

    c. Using the value of c you found in part b, make a graph of gravitational force versus distance between the centers of the asteroids for distances from 0 to 1000 kilometers. What happens to the gravitational force when the asteroids are close together? What happens to the gravitational force when the asteroids are far apart?

18. **Geostationary Orbits** For communications satellites to work properly, they should appear from the surface of the Earth to remain stationary. That is, they should orbit the Earth exactly once each day. For *any* satellite, the *period P* (the length of time required to complete an orbit) is determined by its mean distance A from the center of the Earth. For a satellite of negligible mass, P and A are related by a power function $A = cP^{2/3}$.

    a. The moon is 239,000 miles from the center of the Earth and has a period of about 28 days. How high above the center of the Earth should a geostationary satellite be? (*Hint:* You want the distance A for a satellite with period $1/28$ that of the moon. The homogeneity property of power functions is applicable.)

    b. The radius of the Earth is about 3963 miles. How high above the surface of the Earth should a geostationary satellite be?[17]

19. **Giant Ants and Spiders** Many science fiction movies feature animals such as ants, spiders, or apes growing to monstrous sizes and threatening defenseless
Earthlings. (Of course, they are in the end defeated by the hero and heroine.) Biologists use power functions as a rough guide to relate body weight and cross-sectional area of limbs to length or height. Generally, weight is thought to be proportional to the cube of length, whereas cross-sectional area of limbs is proportional to the square of length. Suppose an ant, having been exposed to "radiation," is enlarged to 500 times its normal length. (Such an event can occur only

(continued)

---

[17]The actual height is 22,300 miles above the surface of the Earth. You will get a slightly different answer because we have neglected the mass of the moon.

in Hollywood fantasy. Radiation is utterly incapable of causing such a reaction.)

a. By how much will its weight be increased?

b. By how much will the cross-sectional area of its legs be increased?

c. Pressure on a limb is weight divided by cross-sectional area. By how much has the pressure on a leg of the giant ant increased? What do you think is likely to happen to the unfortunate ant?[18] *Note:* The factor by which pressure increases is given by

$$\frac{\text{Factor of increase in weight}}{\text{Factor of increase in area}}.$$

20. Kepler's Third Law By 1619 Johannes Kepler had completed the first accurate mathematical model describing the motion of planets around the sun. His model consisted of three laws that, for the first time in history, made possible the accurate prediction of future locations of planets. Kepler's third law related the period (the length of time required for a planet to complete a single trip around the sun) to the mean distance D from the planet to the sun. In particular, he stated that the period P is proportional to $D^{1.5}$.

a. Neptune is about 30 times as far from the sun as is the Earth. How long does it take Neptune to complete an orbit around the sun? (*Hint:* The period for the Earth is 1 year. If the distance is increased by a factor of 30, by what factor will the period be increased?)

b. The period of Mercury is about 88 days. The Earth is about 93 million miles from the sun. How far is Mercury from the sun? (*Hint:* The period of Mercury is different from that of the Earth by a factor of $88/365$.)

21. Time to Failure A building that is subjected to shaking (caused, for example, by an earthquake) may collapse. Failure depends both on intensity and on duration of the shaking. If an intensity I_1 causes a building to collapse in t_1 seconds, then an intensity I_2 will cause the collapse in t_2 seconds, where

$$\frac{t_1}{t_2} = \left(\frac{I_2}{I_1}\right)^2.$$

If a certain building collapses in 30 seconds at one intensity, how long would it take the building to collapse at triple that intensity?

5.2 SKILL BUILDING EXERCISES

Graphs of Power Functions Exercises S-1 through S-6 deal with graphs of power functions.

S-1. Sketch the graph of cx^k for $x \geq 0$ where c is positive and $k > 1$.

S-2. Sketch the graph of cx^k for $x \geq 0$ where c is negative and $k > 1$.

S-3. Sketch the graph of cx^k for $x \geq 0$ where c is positive and $0 < k < 1$.

S-4. Sketch the graph of cx^k for $x \geq 0$ where c is negative and $0 < k < 1$.

S-5. Sketch the graph of cx^k for $x \geq 0$ where c is positive and $k < 0$.

S-6. Sketch the graph of cx^k for $x \geq 0$ where c is negative and $k < 0$.

Homogeneity Exercises S-7 through S-13 deal with the homogeneity property.

S-7. Let $f(x) = cx^4$. If x is doubled, by what factor is f increased?

S-8. Let $f(x) = cx^{1.47}$. If x is tripled, by what factor is f increased?

S-9. Let $f(x) = cx^{2.53}$. By what factor must x be increased in order to triple the value of f?

S-10. Let $f(x) = cx^{3.11}$. How do the function values $f(3.6)$ and $f(5.5)$ compare?

S-11. Let $f(x) = cx^{3.11}$. Suppose that $f(y)$ is 9 times as large as $f(z)$. How do y and z compare?

S-12. Let $f(x) = cx^k$. If x is multiplied by 7, by what factor is f increased?

S-13. Let $f(x) = cx^k$. Suppose that $f(6.6)$ is 6.2 times as large as $f(1.76)$. What is the value of k?

S-14. Constant Term Let $f(x) = cx^{4.2}$ and suppose that $f(4) = 8$. Find the value of c.

S-15. Constant Term Let $f(x) = cx^{-1.32}$ and suppose that $f(5) = 11$. Find the value of c.

S-16. Exponential Versus Power Functions Which function is eventually larger: x^{1023} or 1.0002^x?

---

[18]Similar arguments about creatures of extraordinary size go back at least to Galileo. See his famous book *Two New Sciences*, published in 1638.

5.3 MODELING DATA WITH POWER FUNCTIONS

In this section, we will learn how to construct power function models much as we did linear and exponential models in the preceding chapters.

Getting a Power Model from Data

The steps needed to execute power regression are much like those needed to execute linear or exponential regression.

To illustrate the procedure of finding a power model, we look at the power function $f(x) = 3x^2$. In Figure 5.29 we have entered $x = 1, 2, \ldots, 7$ in the first column and $f(x) = 3 \times 1^2, 3 \times 2^2, \ldots, 3 \times 7^2$ in the second column. Our goal is to use the calculator to recover the power function $f(x) = 3x^2$ from these data. We use *power regression* to do this. The power regression parameters are shown in Figure 5.30. They tell us immediately that the data were produced by the function $f(x) = 3x^2$.

```
L1        L2       L3     2
1         3        ------
2         12
3         27
4         48
5         75
6         108
7         147

L2(1)=3
```

FIGURE 5.29 Data for x and $3x^2$

```
PwrReg
 y=a*x^b
 a=3
 b=2
```

FIGURE 5.30 Regression parameters

EXAMPLE 5.7 **THE VOLUME INSIDE A SPHERE**

FIGURE 5.31

The following table of values gives the volume V in cubic inches inside a sphere of radius r inches. (See Figure 5.31.)

Radius r	1	2	3	4	5
Volume V	4.19	33.51	113.10	268.08	523.60

Part 1 Use regression to find a formula for the volume of a sphere as a function of the radius.

Part 2 To check your work, plot the graph of the original data points together with the function you found in part 1.

Solution to Part 1 The first step is to enter the data into the calculator as shown in Figure 5.32. Then we use power regression to find the parameters in Figure 5.33. Rounding these parameters to two decimal places gives the power function $V = 4.19 \times r^3$.

```
L1       L2        L3     2
1        4.19      ------
2        33.51
3        113.1
4        268.08
5        523.6
-----

L2(6) =
```

FIGURE 5.32 Radius and volume of spheres

```
PwrReg
 y=a*x^b
 a=4.189716194
 b=2.999830868
```

FIGURE 5.33 Power regression parameters

Solution to Part 2 To check our work, we plotted the original data for volume versus radius in Figure 5.34 and added the graph of $4.19r^3$ in Figure 5.35. This shows excellent agreement between the given data and the power model we constructed.

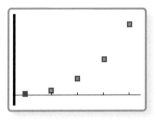

FIGURE 5.34 Plot of data for volume versus radius

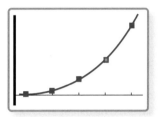

FIGURE 5.35 Adding the power function model

TEST YOUR UNDERSTANDING │ **FOR EXAMPLE 5.7**

The following data show the surface area S of a sphere of radius r. Find a power function that gives the surface area as a function of the radius.

Radius r	1	2	3	4	5
Surface area S	12.57	50.27	113.10	201.06	314.16

EXAMPLE 5.8 KEPLER'S THIRD LAW

Johannes Kepler was the first to give an accurate description of the motion of the planets about the sun. He presented his model in the form of three laws. His third law states, "It is absolutely certain and exact that the ratio which exists between the periodic times of any two planets is precisely the ratio of the $\frac{3}{2}$th power of the mean distances [of the planet to the sun]."[19] Kepler formulated his third law by examining carefully the recorded measurements of the astronomer Tycho Brahe, but he never gave any derivation of it from other principles. That is to say, Kepler believed there was a power relation between the period P of a planet (the time required to complete a revolution about the sun) and its mean distance D from the sun, and he found the correct power by looking at data. We will use more accurate data and much more powerful calculation techniques than were available to Kepler to arrive at similar conclusions. The following table gives distances measured in millions of miles and periods measured in years for the nine planets.

Planet	Mercury	Venus	Earth	Mars	Jupiter	Saturn	Uranus	Neptune	Pluto
Distance D	36.0	67.1	92.9	141.7	483.4	886.1	1782.7	2793.1	3666.1
Period P	0.24	0.62	1	1.88	11.87	29.48	84.07	164.90	249

---

[19]From *The Harmonies of the World*, translated by Charles Glenn Wallis in the *Great Books*. Cited by Victor J. Katz in *A History of Mathematics* (New York: HarperCollins, 1993).

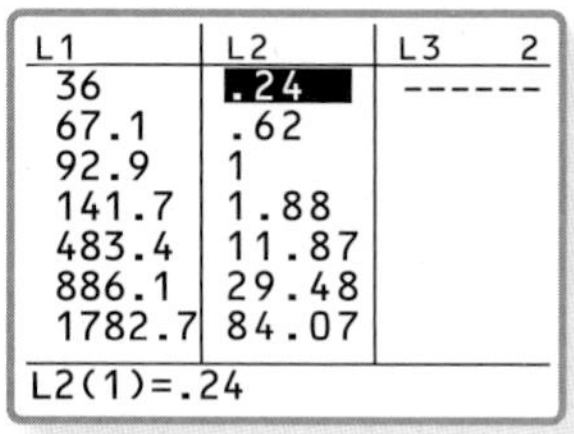

```
L1        L2      L3    2
 36       .24    ------
 67.1     .62
 92.9    1
141.7    1.88
483.4    11.87
886.1    29.48
1782.7   84.07

L2(1)=.24
```

FIGURE 5.36 Data for orbital period versus distance from the sun

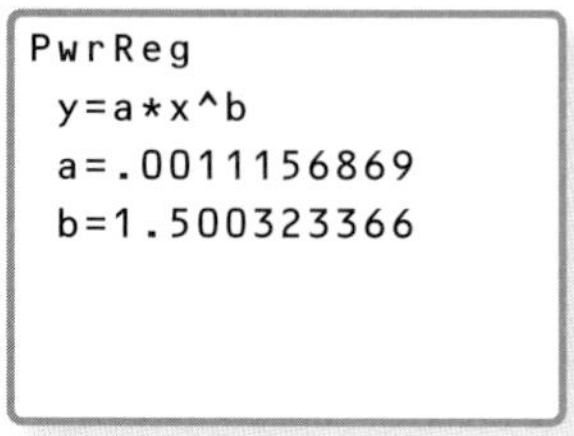

```
PwrReg
 y=a*x^b
 a=.0011156869
 b=1.500323366
```

FIGURE 5.37 Power regression parameters

Part 1 Use regression to find a formula expressing P as a power function of D. Round the regression parameters to four decimal places.

Part 2 If one planet were twice as far from the sun as another, how would their periods compare?

Solution to Part 1 The first step is to enter the data as shown in Figure 5.36. The power regression parameters are shown in Figure 5.37. This gives the power model $P = 0.0011D^{1.5}$.

Solution to Part 2 Here we use the homogeneity property of power functions. If the distance is increased by a factor of 2, then the period will be increased by a factor of $2^{1.5} = 2.83$. Thus, if one planet is twice as far away from the sun as another, its period will be 2.83 times as long.

TEST YOUR UNDERSTANDING | **FOR EXAMPLE 5.8**

We can use the same planetary data to calculate period P as a power function of distance D. What formula do the data give? Round the regression parameters to two decimal places. ∎

Almost Power Data

As with linear and exponential models, many times observed data cannot be modeled exactly by a power function, but regression enables us to make power models that approximately fit the data.

EXAMPLE 5.9 GENERATION TIME AS A POWER FUNCTION OF LENGTH

The *generation time* for an organism is the time it takes to reach reproductive maturity. Biologists have observed that generation time depends on size, and in particular on the length of an organism. Table 5.1 gives the length L measured in feet and generation time T measured in years for various organisms.[20]

TABLE 5.1 Length and Generation Time

Organism	Length L	Generation time T
House fly	0.023	0.055 (20 days)
Cotton deermouse	0.295	0.192 (70 days)
Tiger salamander	0.673	1
Beaver	2.23	2.8
Grizzly bear	5.91	4
African elephant	11.5	12.3
Yellow birch	72.2	40
Giant sequoia	262	60

---

[20]These data are adapted from J. T. Bonner, *Size and Cycle* (Princeton, NJ: Princeton University Press, 1965). We have presented only a sample of the 46 organisms for which Bonner gives data.

Part 1 Use regression to find a formula that models T as a power function of L, and plot the function along with the given data. Round the regression parameters to one decimal place.

Part 2 If one organism is 5 times as long as another, what would be the expected comparison of generation times? (*Suggestion:* Use the homogeneity property of power functions.)

Solution to Part 1 The data are entered in Figure 5.38, and the power regression parameters are shown in Figure 5.39.

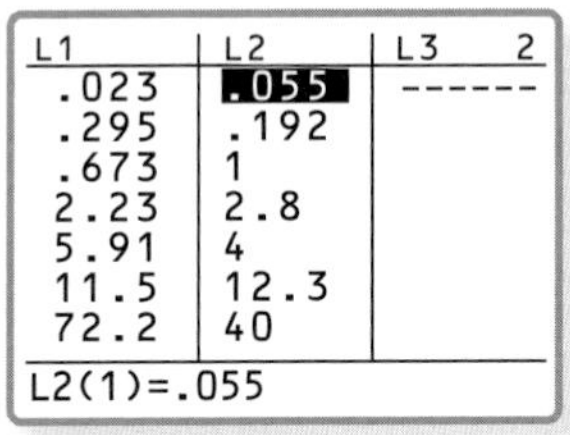

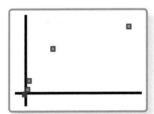

FIGURE 5.38 Data for L and T

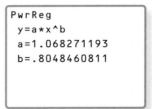

FIGURE 5.39 Finding the power regression parameters

Rounding gives the power formula $T = 1.1L^{0.8}$. In Figure 5.40 we have plotted the original data, and we have added the power model in Figure 5.41.

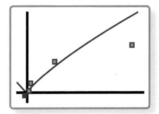

FIGURE 5.40 Plot of data for generation time versus length

FIGURE 5.41 Adding the power model

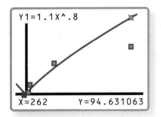

FIGURE 5.42 A closer look at the sequoia

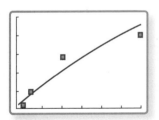

FIGURE 5.43 Changing the window to a smaller range

We note that the generation time for the giant sequoia, the last data point in Figure 5.41, seems to lie well below the power function model. In Figure 5.42 we have traced the graph and set the cursor to $X = 262$. We see that the power model shows a generation time of 94 years, in contrast to the actual time of 60 years. In some settings, this might cause us to question the validity of the model. Such questions are always appropriate, but in this case we are happy with a model that can be a starting point for further analysis rather than an exact relationship, such as that which we had for Kepler's third law. We would be more concerned if the value given by the model were off by a factor of 10. In fact, one reason for the difference is the extreme variation in both length and generation time that is to some degree characteristic of data to which power function models are applied.[21] The 262-foot sequoia is over 10,000 times longer than the house fly, and its 60-year generation time is over 1000 times longer than that of the house fly. This is the cause of the bunching of data in Figure 5.41. In Figure 5.43 the plot is shown with a horizontal span of 0 to 6 and a vertical span of 0 to 5. It shows more of the smaller data points but leaves out the elephant, birch, and sequoia.

[21] In contrast, data that should be modeled exponentially may show relatively small variation in the horizontal direction but much larger variation in the vertical direction.

Solution to Part 2 The homogeneity property of a power function such as $T = 1.1L^{0.8}$ tells us that if length L is increased by a factor of 5, then generation time T will be increased by a factor of $5^{0.8} = 3.62$. Thus we would expect the longer organism to have a generation time 3.62 times as long as the shorter organism.

TEST YOUR UNDERSTANDING | **FOR EXAMPLE 5.9**

We can use the same data to model length L as a power function of generation time T. Find such a model. (Round the regression parameters to one decimal place.) According to the model, how does doubling generation time affect length? ■

ANSWERS FOR TEST YOUR UNDERSTANDING

5.7 $S = 12.57r^2$

5.8 $P = 92.87 \times D^{0.67}$

5.9 $L = 1.0T^{1.2}$. If generation time doubles, length is multiplied by 2.30.

5.3 EXERCISES

Reminder Round all answers to two decimal places unless otherwise indicated.

1. **Stopping Distance** The table below shows the average stopping distance D, in feet, for a car on dry pavement versus the speed S of the car, in miles per hour.

S = speed (mph)	15	25	35	40	60	75
D = stopping distance (feet)	44	85	136	164	304	433

   a. Find a model of stopping distance as a power function of speed.

   b. If speed is doubled, how is stopping distance affected?

   c. Plot the data and power model on the same screen.

Paul Gibson/Alamy

2. **Distance to the Horizon** A sailor records the distances D, in miles, to the visible horizon at several heights h, in feet, above the surface of a calm ocean.

h = height	6	8	12	16	19
D = distance to horizon	3.3	3.6	4.7	5.4	5.9

   a. Make a model of D as a power function of h.

   b. If height above sea level is increased by 10%, by what percentage is distance to the horizon increased? Round your answer to the nearest whole number.

3. **Hydroplaning** On wet roads, under certain conditions the front tires of a car will *hydroplane*, or run along the surface of the water. The critical speed V at which hydroplaning occurs is a function of p, the tire inflation pressure.[22] The following table shows hypothetical data for p, in pounds per square inch, and V, in miles per hour.

Tire inflation pressure p	Critical speed V for hydroplaning
20	46.3
25	51.8
30	56.7
35	61.2

(continued)

---

[22] See J. C. Collins, *Accident Reconstruction* (Springfield, IL: Charles C. Thomas, 1979). The critical speed for hydroplaning also increases with the amount of tire tread available.

a. Find a formula that models V as a power function of p.

b. In the rain a car (with tires inflated to 35 pounds per square inch) is traveling behind a bus (with tires inflated to 60 pounds per square inch), and both are moving at 65 miles per hour. If they both hit their brakes, what might happen?

4. **Urban Travel Times** Population of cities and driving times are related, as shown in the accompanying table, which shows the 1960 population N, in thousands, for several cities, together with the average time T, in minutes, spent by residents driving to work.

City	Population N	Driving time T
Los Angeles	6489	16.8
Pittsburgh	1804	12.6
Washington	1808	14.3
Hutchinson	38	6.1
Nashville	347	10.8
Tallahassee	48	7.3

An analysis[23] of these data, along with data from 17 other cities in the United States and Canada, led to a power model of average driving time as a function of population.

a. Construct a power model of driving time in minutes as a function of population measured in thousands.

b. Is average driving time in Pittsburgh more or less than would be expected from its population?

c. If you wish to move to a smaller city to reduce your average driving time to work by 25%, how much smaller should the city be?

5. **Mass-Luminosity Relation** Roughly 90% of all stars are *main-sequence stars*. Exceptions include supergiants, giants, and dwarfs. For main-sequence stars (including the sun) there is an important relationship called the *mass-luminosity relation* between the relative luminosity[24] L and the mass M in terms of solar masses. Relative masses and luminosities of several main-sequence stars are reported in the accompanying table.

Star	Solar mass M	Luminosity L
Spica	7.3	1050
Vega	3.1	55
Altair	1	1.1
The Sun	1	1
61 Cygni A	0.17	0.002

a. Find a power model for the data in this table. (Round the power and the coefficient to one decimal place.) The function you find is known to astronomers as the mass-luminosity relation.

b. Kruger 60 is a main-sequence star that is about 0.11 solar mass. Use functional notation to express the relative luminosity of Kruger 60, and then calculate that value.

c. Wolf 359 has a relative luminosity of about 0.0001. How massive is Wolf 359?

d. If one star is 3 times as massive as another, how do their luminosities compare?

6. **Growth Rate Versus Weight** Ecologists have studied how a population's intrinsic exponential growth rate r is related to the body weight W for herbivorous mammals.[25] In Table 5.2, W is the adult weight measured in pounds, and r is growth rate per year.

TABLE 5.2 Weight and Exponential Growth Rate

Animal	Weight W	r
Short-tailed vole	0.07	4.56
Norway rat	0.7	3.91
Roe deer	55	0.23
White-tailed deer	165	0.55
American elk	595	0.27
African elephant	8160	0.06

Find a formula that models r as a power function of W, and draw a graph of this function.

[23]A. M. Voorhees, S. J. Bellomo, J. L. Schofer, and D. E. Cleveland, "Factors in work trip lengths," Highway Research Record No. 141 (Washington, DC: Highway Research Board, 1966), 24–46. The analysis is based on data reported by Alan M. Voorhees & Associates.
[24]Relative luminosity is the ratio of the luminosity of a star to that of the sun.
[25]G. Caughley and C. J. Krebs, "Are big mammals simply little mammals writ large?" *Oecologia* **59** (1983), 7–17.

7. **Speed in Flight Versus Length** Table 5.3 gives the length L, in inches, of a flying animal and its maximum speed F, in feet per second, when it flies.[26] (For comparison, 10 feet per second is about 6.8 miles per hour.)

TABLE 5.3 Length and Flying Speed

Animal	Length L	Flying speed F
Fruit fly	0.08	6.2
Horse fly	0.51	21.7
Ruby-throated hummingbird	3.2	36.7
Willow warbler	4.3	39.4
Flying fish	13	51.2
Bewick's swan	47	61.7
White pelican	62	74.8

a. Judging on the basis of this table, is it generally true that larger animals fly faster?

b. Find a formula that models F as a power function of L.

c. Make the graph of the function in part b.

d. Is the graph you found in part c concave up or concave down? Explain in practical terms what your answer means.

e. If one bird is 10 times longer than another, how much faster would you expect it to fly? (Use the homogeneity property of power functions.)

8. **Speed Swimming Versus Length** *This is a continuation of Exercise 7.* Table 5.4 gives the length L, in inches, of a swimming animal and its maximum speed S, in feet per second, when it swims.

a. Find a formula that models S as a power function of L. (Round the power to one decimal place.) On the basis of the power you found, what special type of power function is this?

TABLE 5.4 Length and Swimming Speed

Animal	Length L	Swimming speed S
Bacillus	9.8×10^{-5}	4.9×10^{-5}
Paramecium	0.0087	0.0033
Water mite	0.051	0.013
Flatfish larva	0.37	0.38
Goldfish	2.8	2.5
Dace	5.9	5.7
Adélie penguin	30	12.5
Dolphin	87	33.8

b. Add the graph of S against L to the graph of F that you drew in Exercise 7.

c. Is flying a significant improvement over swimming if an animal is 1 foot long (the approximate length of a flying fish)?

d. Would flying be a significant improvement over swimming for an animal 20 feet long?

e. A blue whale is about 85 feet long, and its maximum speed swimming is about 34 feet per second. Judging on the basis of these facts, do you think the trend you found in part a continues indefinitely as the length increases?

9. **Metabolism** Physiologists who study warm-blooded animals are interested in the *basal metabolic rate*, which is one measure of the energy needed for survival. It can be measured from the volume and composition of expired air. Table 5.5 on the next page gives the weight W, in pounds, of an animal and its basal metabolic rate B, in kilocalories per day.[27]

a. Find a formula that models B as a power function of W.

b. Define the *metabolic weight* of a warm-blooded animal to be $W^{0.75}$ if W is its weight in pounds. Direct comparison of food among animals of

(continued)

---

[26]The tables in this exercise and the next are adapted from J. T. Bonner, op. cit.

[27]The table is adapted from M. Kleiber, "Body size and metabolism," *Hilgardia* **6** (1932), 315–353. See also his book *The Fire of Life*, rev. ed. (Huntington, New York: Robert E. Krieger, 1975).

different sizes is not easy. Clearly, large animals will consume more food than small ones, and to compensate for this, we should take account of the energy needed for survival—that is, the basal metabolic rate B. To compare maximum daily food intake among animals of different sizes, we divide this food intake by the metabolic weight.

TABLE 5.5 Weight and Basal Metabolic Rate

Animal	Weight W	Basal metabolic rate B
Rat	0.38	20.2
Pigeon	0.66	30.8
Hen	4.3	106
Dog	25.6	443
Sheep	101	1220
Cow	855	6421

i. Explain why dividing by the metabolic weight is more meaningful than dividing by the weight in comparing maximum daily food intake.

ii. Find the metabolic weight of a 2.76-pound animal and that of a 126.8-pound animal.

iii. An ecologist found the maximum daily food intake of a 2.76-pound rabbit to be about 0.18 pound and that of a 126.8-pound merino sheep to be about 2.8 pounds.[28] Divide each intake by the corresponding metabolic weight. How do the daily consumption levels compare on this basis?

10. Metabolism and Surface Area *This is a continuation of Exercise 9.* Table 5.6 gives the weight W, in pounds, the basal metabolic rate B, in kilocalories per day, and the surface area A, in square inches, of a variety of marsupials.[29]

a. Find a formula that models B as a power function of W for this group of marsupials.

TABLE 5.6 Weight, Basal Metabolic Rate, and Surface Area

Animal	Weight W	Basal metabolic rate B	Surface area A
Fat-tailed marsupial mouse	0.03	2.16	12.4
Brown marsupial mouse	0.08	4.18	20.5
Long-nosed bandicoot	1.5	37.0	122
Brush-tailed possum	4.4	71.8	260
Tammar wallaby	10.6	159	465
Red kangaroo	71.6	643	1796

b. How does the formula you found in part a compare with the one you found in part a of the preceding exercise? What does this tell you about the metabolic rates of marsupials?

c. Find a formula that models A as a power function of W.

d. The surface area of a 0.26-pound sugar glider (a marsupial similar to a flying squirrel) is about 95 square inches. Use your model in part c to compare this with the trend for surface area versus weight in the table. Can you explain the deviation?

e. Often the power $k = 2/3$ is used to model A as a power function of W. Compare this with the power you found in part c.

f. In early studies of metabolic rates, scientists often assumed that the basal metabolic rate of an animal was proportional to its surface area. Is this assumption supported by the data in the table?

11. Proportions of Trees Table 5.7 gives the diameter d and height h, both in feet, of some "champion"

---

[28]J. Short, "Factors affecting food intake of rangelands herbivores." In G. Caughley, N. Shepherd, and J. Short, eds., *Kangaroos* (Cambridge, England: Cambridge University Press, 1987).
[29]The data in this table are adapted from T. J. Dawson and A. J. Hulbert, "Standard metabolism, body temperature, and surface areas of Australian marsupials," *Am. J. Physiol.* **218** (1970), 1233–1238.

trees (largest American specimens) of a variety of shapes.[30] (See Figure 5.44.)

TABLE 5.7 Diameter and Height of Champion Trees

Tree	Diameter d	Height h
Plains cottonwood	2.9	80
Hackberry	5.7	113
Weeping willow	6.2	95
Ponderosa pine	8.6	162
Douglas fir	14.4	221

FIGURE 5.44

a. Find a formula that models h as a power function of d.

b. Which is taller *for its diameter*: the plains cottonwood or the weeping willow?

c. It has been determined that the critical height at which a column made from green wood of diameter d, in feet, would buckle under its own weight is $140d^{2/3}$ feet.

  i. How does your answer to part a compare with this formula?

  ii. Are any of the trees in the table taller than their critical buckling height?

12. **Weight Versus Length** The accompanying table shows the relationship between the length L, in centimeters, and the weight W, in grams, of the North Sea plaice (a type of flatfish).[31]

  a. Find a formula that models W as a power function of L. (Round the power to one decimal place.)

  b. Explain in practical terms what $W(50)$ means, and then calculate that value.

  c. If one plaice were twice as long as another, how much heavier than the other should it be?

L	W
28.5	213
30.5	259
32.5	308
34.5	363
36.5	419
38.5	500
40.5	574
42.5	674
44.5	808
46.5	909
48.5	1124

13. **Self-Thinning** When seeds of a plant are sown at high density in a plot, the seedlings must compete with each other. As time passes, individual plants grow in size, but the density of the plants that survive decreases.[32] This is the process of *self-thinning*. In one experiment, horseweed seeds were sown on October 21, and the plot was sampled on successive dates. The results are summarized in Table 5.8, which gives for each date the density p, in number per square meter, of surviving

(continued)

---

[30]This exercise is based on the work of T. McMahon, "Size and shape in biology," *Science* **179** (1973), 1201–1204. He considers data for 576 trees, primarily from the American Forestry Association lists of champions.

[31]The table is from Lowestoft market samples in 1946, as described by R. J. H. Beverton and S. J. Holt, *On the Dynamics of Exploited Fish Populations*, Fishery Investigations, Series 2, Volume 19 (London: Ministry of Agriculture, Fisheries and Food, 1957).

[32]This exercise is based on the work of K. Yoda, T. Kira, H. Ogawa, and K. Hozumi, "Self-thinning in overcrowded pure stands under cultivated and natural conditions," *J. Biol. Osaka City Univ.* **14** (1963), 107–129. See also Chapter 6 of John L. Harper, *Population Biology of Plants* (London: Academic Press, 1977).

TABLE 5.8 Density and Weight of Surviving Plants

Date	Density p	Weight w
November 7	140,400	1.6×10^{-4}
December 16	36,250	7.7×10^{-4}
January 30	22,500	0.0012
April 2	9100	0.0049
May 13	4510	0.018
June 25	2060	0.085

TABLE 5.9 Area and Number of Species of Amphibians and Reptiles

Island	Area A	Number S of species
Cuba	44,000	76
Hispaniola	29,000	84
Jamaica	4200	39
Puerto Rico	3500	40
Montserrat	40	9
Saba	5	5

plants and the average dry weight w, in grams, per plant.

a. Explain how the table illustrates the phenomenon of self-thinning.

b. Find a formula that models w as a power function of p.

c. If the density decreases by a factor of $1/2$, what happens to the weight?

d. The *total plant yield y* per unit area is defined to be the product of the average weight per plant and the density of the plants: $y = w \times p$. As time goes on, the average weight per plant increases while the density decreases, so it's unclear whether the total yield will increase or decrease. Use the power function you found in part b to determine whether the total yield increases or decreases *with time*. Check your answer using the table.

14. Species-Area Relation Ecologists have studied the relationship between the number S of species of a given taxonomic group within a given habitat (often an island) and the area A of the habitat.[33] They have discovered a consistent relationship: Over similar habitats, S is approximately a power function of A, and for islands the powers fall within the range 0.2 to 0.4. Table 5.9 gives, for some islands in the West Indies, the area in square miles and the number of species of amphibians and reptiles.

a. Find a formula that models S as a power function of A.

b. Is the graph of S against A concave up or concave down? Explain in practical terms what your answer means.

c. The species-area relation for the West Indies islands can be expressed as a rule of thumb: If one island is 10 times larger than another, then it will have ___ times as many species. Use the homogeneity property of the power function you found in part a to fill in the blank in this rule of thumb.

d. In general, if the species-area relation for a group of islands is given by a power function, the relation can be expressed as a rule of thumb: If one island is 10 times larger than another, then it will have ___ times as many species. How would you fill in the blank? (*Hint:* The answer depends only on the power.)

15. Cost of Transport Physiologists have discovered that steady-state oxygen consumption (measured per unit of mass) in a running animal increases linearly with increasing velocity. The slope of this line is called the *cost of transport* of the animal, since it measures the energy required to move a unit mass 1 unit of distance. Table 5.10 gives the weight W, in grams, and the cost of transport C, in milliliters of oxygen per gram per kilometer, of seven animals.[34]

a. Judging on the basis of the table, does the cost of transport generally increase or decrease with increasing weight? Are there any exceptions to this trend?

b. Find a formula that models C as a power function of W.

---

[33]See the discussion in Example 5.6 of Section 5.2. The data in this exercise are adapted from Philip J. Darlington, Jr., *Zoogeography* (Huntington, NY: Robert E. Krieger, 1980 reprint).
[34]The data in this table are taken from C. R. Taylor, K. Schmidt-Nielsen, and J. L. Raab, "Scaling of energetic cost of running to body size in mammals," *Am. J. Physiol.* **219** (1970), 1104–1107. For an extensive collection of data with references, see M. A. Fedak and H. J. Seeherman, "Reappraisal of energetics of locomotion shows identical cost in bipeds and quadrupeds including ostrich and horse," *Nature* **282** (1979), 713–716.

TABLE 5.10 Weight and Cost of Transport

Animal	Weight W	Cost of transport C
White mouse	21	2.83
Kangaroo rat	41	2.01
Kangaroo rat	100	1.13
Ground squirrel	236	0.66
White rat	384	1.09
Dog	2600	0.34
Dog	18,000	0.17

c. The cost of transport for a 20,790-gram emperor penguin is about 0.43 milliliter of oxygen per gram per kilometer. Use your model in part b to compare this with the trend for cost of transport versus weight in the table. Does this confirm the stereotype of penguins as awkward waddlers?

16. **Parallax Angle**   If we view a star now, and then view it again 6 months later, our position will have changed by the diameter of the Earth's orbit around the sun. (See Figure 5.45.) For stars within about 100 light-years of Earth, the change in viewing location is sufficient to make the star appear to be in a different location in the sky. Half of the angle from one location to the next is known as the *parallax angle*. Even for nearby stars, the parallax angle is very small[35] and is normally measured in seconds of arc. The distance to a star can be determined from the parallax angle. The table below gives parallax angle p measured in seconds of arc and the distance d from the sun measured in light-years.

Star	Parallax angle	Distance
Markab	0.030	109
Al Na'ir	0.051	64
Alderamin	0.063	52
Altair	0.198	16.5
Vega	0.123	26.5
Rasalhague	0.056	58

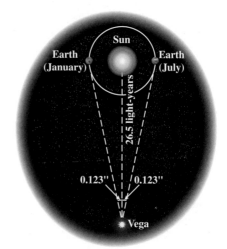

FIGURE 5.45

a. Make a power function model of the data for d in terms of p.

b. If one star has a parallax angle twice that of a second, how do their distances compare?

c. The star Mergez has a parallax angle of 0.052 second of arc. Use functional notation to express how far away Mergez is, and then calculate that value.

d. The star Sabik is 69 light-years from the sun. What is its parallax angle?

17. **Exponential Growth Rate and Generation Time**   In this exercise, we will examine the relationship between a population's intrinsic exponential growth rate r and the generation time T—that is, the time it takes an organism to reach reproductive maturity. The following table gives values of the generation time and the exponential growth rate for a selection of lower organisms.[36] The basic unit of time is a day.

Organism	Generation time T	r
Bacterium *E. coli*	0.014	60
Protozoan *Paramecium aurelia*	0.5	1.24
Spider beetle *Eurostus hilleri*	110	0.01
Golden spider beetle	154	0.006
Spider beetle *Ptinus sexpunctatus*	215	0.006

(continued)

---

[35]Early astronomers were caught up in the problem of determining whether the Earth or the sun should be thought of as the center of motion of the solar system. They were well aware that if the Earth orbited the sun, then stars would show a parallax angle. But the angle is so small that they were unable to detect it. This was presented as justification for considering Earth a stationary object.
[36]The table is adapted from E. R. Pianka, *Evolutionary Ecology*, 4th ed. (New York: Harper & Row, 1988).

a. Find a formula that models r as a power function of T, and make a graph of this function. (Round the regression parameters to one decimal place, and use a horizontal span of 0 to 10.)

b. In Example 5.9 we found a power relation between generation time T, in years, and length L. Use this relation and the results of part a to find a formula for r as a function of length L for this group of lower organisms. (Convert the power relation from Example 5.9 to measure time in days, and remember that $T^{-1} = 1/T$.)

18. Weight Versus Height   The following data show the height h, in inches, and weight w, in pounds, of an average adult male.

h	61	62	66	68	70	72	74	75
w	131	133	143	149	155	162	170	175

a. Make a power model for weight versus height.

b. According to the model from part a, what percentage increase in weight can be expected if height is increased by 10%?

Testing for Power Data   The natural logarithm function from Section 4.5 can be used to test for power data. It can be shown that y is a power function of x when $\ln y$ is a linear function of $\ln x$. As an example, we have taken the data from Example 5.8 and added rows for $\ln D$ and $\ln P$.

A plot of $\ln P$ versus $\ln D$ is shown in Figure 5.46.

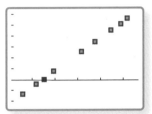

FIGURE 5.46

The fact that the data points for $\ln P$ versus $\ln D$ fall on a straight line indicates that P is a power function of D. This provides a justification for making the power model as we did in Example 5.8. For data that are *almost* power data, the logarithm of data points will *nearly* fall on a straight line. Exercises 19 and 20 make use of this observation.

19. Generation Time as a Function of Length   For this exercise use the data from Example 5.9. Make a plot of $\ln T$ versus $\ln L$. Does this plot support the use of a power model for T as a function of L?

20. Hydroplaning Again   This exercise uses the hydroplaning data from Exercise 3. Make a graph of $\ln V$ versus $\ln p$. Does this graph support the use of a power model for V as a function of p?

Planet	Mercury	Venus	Earth	Mars	Jupiter	Saturn	Uranus	Neptune	Pluto
Distance D	36.0	67.1	92.9	141.7	483.4	886.1	1782.7	2793.1	3666.1
Period P	0.24	0.62	1	1.88	11.87	29.48	84.07	164.90	249
$\ln D$	3.58	4.21	4.53	4.95	6.18	6.79	7.49	7.93	8.21
$\ln P$	−1.43	−0.48	0	0.63	2.47	3.38	4.43	5.11	5.52

5.3 SKILL BUILDING EXERCISES

Choosing a Model In Exercises S-1 through S-4, you are given information and asked to choose the appropriate model.

S-1. Broad Jump Record In examining broad jump records at your university, you find that over each 5-year period the school record increases by 2 inches. Does this information suggest a linear, exponential, or power function model for the broad jump record at your university?

S-2. A Rumor A rumor is spreading, and you find that each day the number of people who have heard the rumor is 50% larger than the day before. Does this information suggest a linear, exponential, or power function model for the spread of the rumor?

S-3. Further Reflection Regarding the Rumor *This is a continuation of Exercise S-2.* Suppose the rumor from Exercise S-2 is spreading across your college campus but will never spread beyond the limits of the campus. How might this new information alter your choice of model?

S-4. Stopping Distance Records show that if speed of an automobile is multiplied by t, then stopping distance is multiplied by a fixed power of t. Does this information suggest a linear, exponential, or power function model for stopping distance versus speed?

S-5. An Easy Power Formula Model the following data with a power formula. You should be able to do this exercise quickly and easily without using your calculator.

x	1	2	3	4	5
y	1	8	27	64	125

S-6. An Easy Power Formula Model the following data with a power formula. You should be able to do this exercise quickly and easily without using your calculator.

x	1	2	3	4	5
y	1	$\frac{1}{4}$	$\frac{1}{9}$	$\frac{1}{16}$	$\frac{1}{25}$

S-7. Modeling Power Data The following data table was generated by a power function f. Find a formula for f and plot the data points along with the graph of the formula.

x	1	2	3	4	5
f	3.6	8.86	15.02	21.83	29.17

S-8. Modeling Almost Power Data Model the following data with a power function. Give the formula and plot the data points along with the model.

x	1	2	3	4	5
f	6.3	1.9	0.6	0.2	0.07

S-9. Modeling Almost Power Data Model the following data with a power function. Give the formula and plot the data points along with the model.

x	1	2	3	4	5
f	2.2	51.7	338.9	1236.5	3188.8

S-10. Modeling Almost Power Data Model the following data with a power function. Give the formula and plot the data points along with the model.

x	1	2	3	4	5
f	5.5	15.1	28.8	63.1	84.2

S-11. Modeling Almost Power Data Model the following data with a power function. Give the formula and plot the data points along with the model.

x	0.3	1.3	2.2	3.3	4.1
f	5.6	2	0.92	0.77	0.51

S-12. Modeling Almost Power Data Find a formula for a power function that models the following data.

x	1	2	3	4	5
f	0.5	2.0	4.5	8.0	12.5

S-13. Modeling Almost Power Data Find a formula for a power function that models the following data.

x	1	2	3	4	5
f	2.30	0.76	0.40	0.25	0.18

5.4 COMBINING AND DECOMPOSING FUNCTIONS

We have up to now looked most closely at linear, exponential, and power functions. But we have also encountered various combinations of these and other functions. In this section we look more carefully at how functions are combined to make new ones and at the meanings of the individual pieces.

 ## Sums, Products, and Limiting Values

In looking at linear and exponential functions, we have already emphasized the meanings of various parts of functions. Suppose, for example, that the cost c in dollars of joining a music club and purchasing n CDs is given by the linear function

$$c(n) = 20 + 12n.$$

The pieces of the formula have important meanings with which we are familiar. The first piece, 20, is the initial value, and it tells us that it costs $20 to join the club. The second piece, $12n$, shows the slope, 12, and it tells us that once we join the music club, we can buy CDs for $12 each. Thus the total cost function c is made up of two pieces, the price of joining the club and the cost of buying n CDs. Furthermore, we can glean this information by looking at pieces of the formula.

In the previous chapter we used the exponential function

$$N(t) = 5.34 \times 1.03^t$$

to model the U.S. population from 1800 through 1860. Here t is the number of years since 1800, and N is the population in millions. Once again, the pieces of the formula have familiar meanings. The number 5.34 is the initial value of the function, and it tells us that the U.S. population in 1800 was about 5.34 million. The second part of the exponential function, 1.03^t, gives the growth factor 1.03, and it tells us that population grew by about 3% per year. The total population N is thus made up of two pieces, the initial population and the effect of t years of growth.

Many models are obtained by putting together linear, exponential, or power functions in various ways, and often we can get desired information by looking at these pieces.

If a yam, initially at room temperature, is placed in a preheated oven, its temperature P after t minutes may be given by $P = 400 - D$, where $D = 325 \times 0.98^t$. Note that P is made up of two pieces: the constant 400 and the exponential function D. Examining the role that each of these plays will help us better understand how the temperature of the yam changes. You should check by looking at a table of values that D has a limiting value of 0. Thus $P = 400 - D$ has a limiting value of 400. Since we expect the temperature of the yam eventually to match that of the oven, we conclude that the oven temperature is 400 degrees.

To understand the function D better, we rearrange the formula for P to obtain

$$D = 400 - P.$$

Thus D represents the difference between the temperature of the oven, which is 400 degrees, and that of the yam. Hence the pieces of the temperature function represent

Sums, products, differences, and quotients are the simplest way of combining functions.

the limiting value (oven temperature) and the difference between oven temperature and that of the yam:

$$P = \text{Limiting value} - \text{Temperature difference}$$
$$P = \text{Oven temperature} - \text{Temperature difference.}$$

We noted that the limiting value of the exponential function $D = 325 \times 0.98^t$ is 0. Indeed, as we noted in Section 4.1, the limiting value of any function that represents exponential decay is 0.

EXAMPLE 5.10 BAKING A CAKE

A pan of cake batter is initially at a room temperature of 75 degrees. The pan is placed in a 350-degree oven to bake. Let $C = C(t)$ denote the temperature of the cake batter t minutes after it is placed in the oven. In this example, all temperatures are measured in degrees Fahrenheit.

Part 1 The temperature of the cake batter is given by

$$C = \text{Limiting value} - \text{Temperature difference.}$$

What is the limiting value of C?

Part 2 Let D denote the difference between oven temperature and that of the batter. Then *Newton's law of cooling* tells us that D is an exponential function. What is the initial value of D?

Part 3 After 10 minutes, we find that the cake batter has heated to 165 degrees. Find a formula for D.

Part 4 Find a model for the temperature of the cake batter t minutes after it is placed in the oven.

Solution to Part 1 If the cake batter is left in the oven for a long time, its temperature will match that of the oven. Thus the limiting value of C is 350 degrees.

Solution to Part 2 Since the initial temperature of the cake batter is room temperature, or 75 degrees, the initial value of D is $350 - 75 = 275$ degrees.

Solution to Part 3 We know from part 2 that D can be written as

$$D = 275a^t,$$

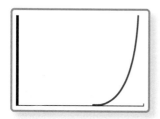

FIGURE 5.47 Temperature difference versus decay factor

where a is the decay factor for D. After 10 minutes the temperature of the batter is 165 degrees, so the temperature difference after 10 minutes is $350 - 165 = 185$ degrees. That is, $D = 185$ when $t = 10$. We can find the decay factor a by solving the equation

$$275a^{10} = 185.$$

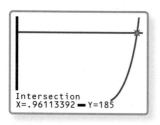

Intersection
X=.96113392 ▬ Y=185

FIGURE 5.48 Finding the decay factor

We use the crossing-graphs method to get the solution. In Figure 5.47 we have graphed the temperature difference $D = 275a^{10}$ using a horizontal span of 0 to 1 and a vertical span of 0 to 225. Note that in this figure the horizontal axis represents the decay factor, not time. In Figure 5.48 we have added the horizontal line 185, and we see that the graphs cross at $a = 0.96$. We conclude that

$$D = 275 \times 0.96^t.$$

Solution to Part 4 To get a model for C, we use the information found in parts 1 and 3:

$$C = \text{Limiting value} - \text{Temperature difference}$$

$$C = 350 - 275 \times 0.96^t.$$

TEST YOUR UNDERSTANDING │ **FOR EXAMPLE 5.10**

When the cake in the example reaches a temperature of 300 degrees, it is baked. The cake is taken from the oven and placed on a counter to cool. The temperature of the room is 75 degrees. The difference between the cake's temperature and room temperature is an exponential function of time, in minutes, with decay factor 0.85. What is the initial value of this exponential function? What is the limiting value of the cake's temperature? Find a formula that gives the cake's temperature after t minutes.   ■

■ │ Composition of Functions

Suppose we know that the weight w of a certain young boy from ages 6 through 16 depends on his height h in the following way:

$$w = 4.4h - 150.$$

Here weight is in pounds and height in inches. Suppose we also know that the boy's height in inches depends on his age t in years according to

$$h = 1.9t + 40.$$

Thus we have weight $w = w(h)$ expressed as a function of height, and in turn height $h = h(t)$ is expressed as a function of age. Suppose we wish to find the boy's weight at age 11. We can do this by first calculating the height at age 11:

$$h = h(11) = 60.9 \text{ inches.}$$

Next we use this value to calculate the weight:

$$w = w(60.9) = 117.96 \text{ pounds.}$$

Function composition is an important method of combining functions.

An alternative method for making the same calculation is to *compose* the two functions that we have to make a new function that enables us to calculate the weight directly as a function of the age. We get the formula for this new function by simply replacing h in the formula for w by its expression in terms of t as shown below:

$$w = w(h) = 4.4h - 1.50$$

$$w = w(h(t)) = 4.4(1.9t + 40) - 150.$$

This new function is known as w *composed with* h, and it enables us to calculate the weight at age 11 directly in a single calculation:

$$w = w(h(11)) = 4.4(1.9 \times 11 + 40) - 150 = 117.96.$$

Putting together, or composing, functions in this way is important not only for making direct calculations but also for clearer analysis. After composition we can, for example, view the graph of weight versus time or make a table of values. A partial table of values is shown in Figure 5.49, and the graph, where we have used a horizontal span of 10 to 15 years and a vertical span of 100 to 200 pounds, is shown in Figure 5.50.

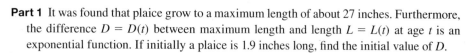

X	Y_1
9	101.24
10	109.6
11	117.96
12	126.32
13	134.68
14	143.04
15	151.4
X=9	

FIGURE 5.49 Table of values for w composed with h

FIGURE 5.50 Graph of $w(h(t))$

EXAMPLE 5.11 NORTH SEA PLAICE

In biological studies a *cohort* is a collection of individuals all of the same age. A study by Beverton and Holt[37] contains information on a cohort of North Sea plaice, which is a type of flatfish. Some of that information is adapted for this exercise. We measure length in inches and weight in pounds. Also, the variable t measures the so-called recruitment age in years, which we refer to simply as the age.

Part 1 It was found that plaice grow to a maximum length of about 27 inches. Furthermore, the difference $D = D(t)$ between maximum length and length $L = L(t)$ at age t is an exponential function. If initially a plaice is 1.9 inches long, find the initial value of D.

Part 2 It was found that a 5-year-old plaice is about 11.3 inches long. Find a formula that gives L in terms of t.

Part 3 The weight W of a plaice can be calculated from its length L using

$$W = 0.000322L^3.$$

Use function composition to find a formula for weight W as a function of age t.

Part 4 After t years, the total number of fish in the cohort is given by the exponential function

$$N = 1000e^{-0.1t}.$$

The total *biomass* of the cohort is the number of fish times the weight per fish. Find a formula for the total biomass B of the cohort as a function of the age of the cohort.

Part 5 How old is the cohort when biomass is at a maximum?

Solution to Part 1 Since the initial length is 1.9 inches, the initial difference is $27 - 1.9 = 25.1$ inches. This is the initial value of D.

Solution to Part 2 We first find a formula for length difference $D(t)$. From part 1 we know that D is an exponential function with initial value 25.1, so it can be written as

$$D = 25.1a^t,$$

where a is the decay factor. A 5-year-old plaice is about 11.3 inches long, and this means the length difference is $27 - 11.3 = 15.7$ inches. That is, $D = 15.7$ when $t = 5$. Hence we can find the decay factor a by solving the equation

$$15.7 = 25.1a^5.$$

[37] R. J. H. Beverton and S. J. Holt, *On the Dynamics of Exploited Fish Populations*, Fishery Investigations, Series 2, Volume 19 (London: Ministry of Agriculture, Fisheries and Food, 1957).

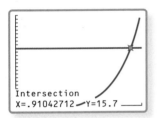

FIGURE 5.51 Finding the decay factor for length difference

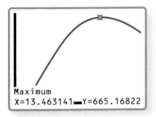

FIGURE 5.52 Finding maximum biomass

In Figure 5.51 we have used the crossing-graphs method with a horizontal span of 0 to 1 and a vertical span of 0 to 25. We find that $a = 0.91$.

We conclude that $D = 25.1 \times 0.91^t$. Finally, from rearranging $D = 27 - L$ we obtain

$$L = 27 - D = 27 - 25.1 \times 0.91^t.$$

Solution to Part 3 We simply need to replace L in the formula $W = 0.000322L^3$ by its expression in terms of t, which is $L = 27 - 25.1 \times 0.91^t$. We obtain

$$W = 0.000322(27 - 25.1 \times 0.91^t)^3.$$

Solution to Part 4 The total biomass is the number of fish times the weight of each. That is, $B = NW$. Replacing N and W by their expressions in terms of t gives

$$B = 1000e^{-0.1t}[0.000322(27 - 25.1 \times 0.91^t)^3].$$

Solution to Part 5 In Figure 5.52 we have graphed B versus t using a horizontal span of 0 to 20 years and a vertical span of 0 to 700 pounds. We read from the prompt at the bottom of the screen that a maximum biomass of about 665 pounds occurs at an age of about 13.5 years.

TEST YOUR UNDERSTANDING | **FOR EXAMPLE 5.11**

The *generation time T* for an organism is the time it takes to reach reproductive maturity. For many species there is a relationship between the generation time, in years, and the length L, in feet: $T = 1.1L^{0.8}$. For a group of related species, the length is related to the weight W, in pounds, by $L = 1.2W^{0.3}$. Express generation time in terms of weight for this group. ∎

Piecewise-Defined Functions

Piecewise-defined functions are functions defined by different formulas for different values of the variable. A simple example is the cost of first-class U.S. postage for a letter as a function of the weight of the letter. For a letter weighing 1 ounce or less, the postage at the time of the printing of this text is 45 cents; for a letter weighing more than 1 ounce but no more than 2 ounces, the postage is 65 cents; and for a letter weighing more than 2 ounces but no more than 3 ounces, the postage is 85 cents. Thus the expression for postage P in cents as a function of weight w in ounces involves several different formulas: $P(w) = 45$ for $w \leq 1$, $P(w) = 65$ for $1 < w \leq 2$, and $P(w) = 85$ for $2 < w \leq 3$. This can be written somewhat less awkwardly by using a brace:

$$P = \begin{cases} 45 & \text{for} \quad w \leq 1 \\ 65 & \text{for} \quad 1 < w \leq 2 \\ 85 & \text{for} \quad 2 < w \leq 3 \end{cases}$$

Piecewise-defined functions are used to describe functions that are best described, or most naturally described, by different formulas for different intervals of variable values. Here are a couple of examples showing how this idea can be used.

EXAMPLE 5.12   **ROOF LINES**

The outside wall of a house is 10 feet high. From the western wall, the roof rises 4 feet over a horizontal span of 16 feet and then descends at the same rate over another horizontal span of 16 feet to the eastern wall (see Figure 5.53).

FIGURE 5.53 Sketch of a roof line

Part 1 Write a formula for the line that describes the ascending part of the roof.

Part 2 Write a formula for the line that describes the descending part of the roof.

Part 3 Write a piecewise-defined function describing the entire roof line.

Part 4 Graph the piecewise-defined function from part 3.

Solution to Part 1 If we measure the horizontal distance from the western wall, then the line has an initial value of 10 feet. The slope is the rise divided by the run, which is $4/16 = 0.25$, so a formula for the line describing the ascending part is $R = 0.25h + 10$. Here h is the horizontal distance from the western wall, R is the height of the roof (both in feet), and the formula is valid for h up to 16 feet.

Solution to Part 2 This line descends at the same rate at the ascending line from part 1, so this time the slope is -0.25. To find the initial value, we imagine this roof line extending back west beyond the peak. We can see that it would lie 8 feet over the top of the western wall, or 18 feet above ground. Thus a formula for this part of the roof line is $R = -0.25h + 18$, with h and R as in part 1. This formula is valid for h between 16 and 32 feet.

Solution to Part 3 The roof is ascending for $h \leq 16$ and it is descending for $16 \leq h \leq 32$, so the roof line itself can be described as

$$R = \begin{cases} 0.25h + 10 & \text{for} \quad h \leq 16 \\ -0.25h + 18 & \text{for} \quad 16 \leq h \leq 32. \end{cases}$$

Solution to Part 4 Graphing the parts of the two lines gives Figure 5.54.[38] The horizontal span is from 0 to 32, and the vertical span is from 5 to 20.

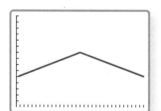

FIGURE 5.54 Roof line graph

TEST YOUR UNDERSTANDING | **FOR EXAMPLE 5.12**

Find a formula for the roof line in the example if the center height is 8 feet above the outside wall rather than 4 feet.   ■

In the next example, we will need to weave together two formulas to make a single piecewise-defined function.

EXAMPLE 5.13   **PUBLIC HIGH SCHOOL ENROLLMENTS 1987–2010**

Here are two different models for U.S. public high school enrollments. One model is $N = 0.021t^2 - 0.09t + 11.74$, where N is enrollment in millions of students, t is

[38]See Appendices B and C for guidance in graphing functions that are defined piecewise.

time in years since 1987, and the formula is valid from 1987 to 1999. Another model is $N = -0.033t^2 + 0.46t + 13.37$, where N is enrollment in millions of students, t is time in years since 2000, and the formula is valid from 2000 to 2010.

Part 1 Determine whether these two different formulas give contradictory information.

Part 2 Find a way to write these formulas in terms of the same variable.

Part 3 Write N as a single piecewise-defined function.

Part 4 Graph N for the years 1987 through 2010.

Solution to Part 1 Each formula is valid only for a specified interval of years: 1987 through 1999 for one and 2000 through 2010 for the other. Since neither is valid in the other's years, there can be no contradiction, despite the different formulas that could give different values for the same year.

Solution to Part 2 As written, each formula uses t as the variable, but t means something different for each formula: For the first model, t is years since 1987, whereas for the second model, t is years since 2000. We need a new variable. Because the final model will cover the years from 1987 through 2010, it makes sense to use years since 1987 as the variable, but we need a new name. Let's denote the years since 1987 by T. Then $T = t$ for the 1987–1999 model, but we need to relate T to t for the 2000–2010 model. A table comparing the two for a few years is helpful:

Year	2000	2001	2002	2003
t years since 2000	0	1	2	3
T years since 1987	13	14	15	16

Clearly $T = t + 13$; equivalently, $t = T - 13$. Substituting $t = T - 13$ into the 2000–2010 model—that is, writing N as a composition of functions—enables us to put both models in terms of the same variable T, years since 1987.

Solution to Part 3 Using the variable T and substituting, we find that

$$N = \begin{cases} 0.021T^2 - 0.09T + 11.74 & \text{for} \quad 0 \le T \le 12 \\ -0.033(T - 13)^2 + 0.46(T - 13) + 13.37 & \text{for} \quad 13 \le T \le 23. \end{cases}$$

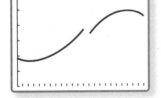

FIGURE 5.55 High school enrollments 1987–2010

Solution to Part 4 Graphing the two functions gives Figure 5.55. Here the horizontal span is from 0 to 23, and the vertical span is from 10 to 16.[39]

TEST YOUR UNDERSTANDING | **FOR EXAMPLE 5.13**

Write a formula for N using the variable s, defined to be the time in years since 1980. ■

ANSWERS FOR TEST YOUR UNDERSTANDING

5.10 The initial value of the difference is 225 degrees. The limiting value of the cake's temperature is 75 degrees. The temperature C of the cake after t minutes is $C = 75 + 225 \times 0.85^t$ degrees.

5.11 $T = 1.1(1.2W^{0.3})^{0.8}$. This can also be written as $T = 1.27W^{0.24}$.

5.12 $R = \begin{cases} 0.5h + 10 & \text{for} \quad h \le 16 \\ -0.5h + 26 & \text{for} \quad 16 \le h \le 32 \end{cases}$

5.13 $N = \begin{cases} 0.021(s - 7)^2 - 0.09(s - 7) + 11.74 & \text{for} \quad 7 \le s \le 19 \\ -0.033(s - 20)^2 + 0.46(s - 20) + 13.37 & \text{for} \quad 20 \le s \le 30 \end{cases}$

---

[39]For functions defined piecewise, it is often wise to set the graphing mode so that the graph is not connected. On your calculator this mode may be called "dot" or something similar.

5.4 EXERCISES

Reminder   Round all answers to two decimal places unless otherwise indicated.

1. **River Flow**   The *cross-sectional area C*, in square feet, of a river is given by the formula

$$C = wd,$$

where w is the width of the river and d is the depth, both in feet. (See Figure 5.56.) The speed s of the flow, in feet per minute, is measured using a meter. The *flow rate F*, in cubic feet per minute, of the river is given by

$$F = sC.$$

   a. Find a formula that gives F in terms of the variables w, d and s.

   b. Use the formula you found in part a to find the flow rate of a river that has a width of 23 feet, a depth of 2 feet, and a flow speed of 24 feet per minute.

FIGURE 5.56 Cross-sectional area of a river

2. **Net Profit Margin**   The *net profit margin* tells how much profit a company makes for every dollar it generates in revenue. If N is the net income (the income after taxes have been paid) and R is the total revenue, then the net profit margin M is given by

$$M(N, R) = \frac{N}{R}.$$

   A certain company pays a tax rate of 20% on its income.

   a. Use I for the income before taxes, and express the net income N in terms of I. (Be careful: N is the part of I left after taxes—not the part you pay in taxes.)

   b. Use a formula to express the net profit margin in terms of the variables I and R.

3. **A Skydiver**   If a skydiver jumps from an airplane, his velocity v, in feet per second, starts at 0 and increases toward terminal velocity. An average-sized man has a terminal velocity T of about 176 feet per second. The difference $D = T - v$ is an exponential function of time.

   a. What is the initial value of D?

   b. Two seconds into the fall, the velocity is 54.75 feet per second. Find an exponential formula for D.

   c. Find a formula for v.

   d. Express using functional notation the velocity 4 seconds into the fall, and then calculate that value.

4. **Present Value**   If you invest P dollars (the *present value* of your investment) in a fund that pays an interest rate of r, as a decimal, compounded yearly, then after t years your investment will have a value F dollars, which is known as the *future value*. The *discount rate D* for such an investment is given by

$$D = \frac{1}{(1 + r)^t},$$

where t is the life, in years, of the investment. The present value of an investment is the product of the future value and the discount rate. Find a formula that gives the present value in terms of the future value, the interest rate, and the life of the investment.

5. **Immigration**   Suppose a certain population is initially absent from a certain area but begins migrating there at a rate of v individuals per day. Suppose further that this is an animal group that would normally grow at an exponential rate. Then the population after t days in the new area is given by

$$N = \frac{v}{r}(e^{rt} - 1),$$

where r is a constant that depends on the species and the environment. If the new location proves unfavorable, then the value of r may be negative. In such a case, we can rewrite the population function as

$$N = \frac{v}{r}(a^t - 1),$$

where a is less than 1. Under these conditions, what is the limiting value of the population?

(continued)

6. **Correcting Respiration Rate for Temperature** In a study by R. I. Van Hook[40] of a grassland ecosystem in Tennessee, the rate O of energy loss to respiration of consumers and predators was initially modeled using

$$O = aW^B,$$

where W is weight and a and B are constants. The model was then corrected for temperature by multiplying O by $C = 1.07^{T-20}$, where T is temperature in degrees Celsius.

   a. What effect does the correction factor have on energy loss due to respiration if the temperature is larger than 20 degrees Celsius?

   b. What effect does the correction factor have on energy loss due to respiration if the temperature is exactly 20 degrees Celsius?

   c. What effect does the correction factor have on energy loss due to respiration if the temperature is less than 20 degrees Celsius?

7. **Biomass of Haddock** In the study by Beverton and Holt[41] described in Example 5.11, information is also provided on haddock. We have adapted it for this exercise. We measure length in inches, weight in pounds, and age t in years.

   a. It was found that haddock grow to a maximum length of about 21 inches. Furthermore, the difference $D = D(t)$ between maximum length and length $L = L(t)$ at age t is an exponential function. If initially a haddock is 4 inches long, find the initial value of D.

   b. It was found that a 6-year-old haddock is about 15.8 inches long. Find a formula that gives L in terms of t.

   c. The weight W in pounds of a haddock can be calculated from its length L using

$$W = 0.000293L^3.$$

   Use function composition to find a formula for weight W as a function of age t.

   d. After t years, the total number of fish in the cohort is given by the exponential function

$$N = 1000e^{-0.2t}.$$

The total biomass of the cohort is the number of fish times the weight per fish. Find a formula for the total biomass B of the cohort as a function of the age of the cohort.

   e. How old is the cohort when biomass is at a maximum? (Consider ages up to 10 years.)

8. **Traffic Flow** For traffic moving along a highway, we use q to denote the *mean flow rate*. That is, the average number of vehicles per hour passing a certain point. We let q_m denote the maximum flow rate, k the *mean traffic density* (that is, the average number of vehicles per mile), and k_m the density at which flow rate is a maximum (that is, the value of k when $q = q_m$).

   a. An important measurement of traffic on a highway is the *relative density R*, which is defined as

$$R = \frac{k}{k_m}.$$

   i. What does a value of $R < 1$ indicate about traffic on a highway?

   ii. What does a value of $R > 1$ indicate about traffic on a highway?

   b. Let u denote the mean speed of vehicles on the road and u_f the *free speed*—that is, the speed when there is no traffic congestion at all. One study[42] proposes the following relation between density and speed:

$$u = u_f e^{-0.5R^2}.$$

   Use function composition to find a formula that directly relates mean speed to mean traffic density.

   c. Make a graph of mean speed versus mean traffic density, assuming that k_m is 122 cars per mile and u_f is 75 miles per hour. (Include values of mean traffic density up to 250 vehicles per mile.) Paying particular attention to concavity, explain the significance of the point $k = 122$ on the graph.

   d. Traffic is considered to be seriously congested if the mean speed drops to 35 miles per hour. Use the graph from part c to determine what density will result in serious congestion.

---

[40]"Energy and nutrient dynamics of spider and orthopteran populations in a grassland ecosystem," *Ecol. Monogr.* **41** (1971), 1–26.
[41]R. J. H. Beverton and S. J. Holt, *On the Dynamics of Exploited Fish Populations*, Fishery Investigations, Series 2, Volume 19 (London: Ministry of Agriculture, Fisheries and Food, 1957).
[42]J. Drake, A. D. May, and J. L. Schofer, "A statistical analysis of speed density hypotheses," *Traffic Flow Characteristics*, Highway Research Record No. 154 (Washington, DC: Highway Research Board, 1967), 53–87.

9. **Waiting at a Stop Sign**
Consider a side road
connecting to a major
highway at a stop sign.
According to a study
by D. R. Drew,[43] the
average delay D, in
seconds, for a car
waiting at the stop sign
to enter the highway is
given by

$$D = \frac{e^{qT} - 1 - qT}{q},$$

where q is the *flow rate*, or the number of cars per
second passing the stop sign on the highway, and T
is the *critical headway*, or the minimum length of
time in seconds between cars on the highway that
will allow for safe entry. We assume that the critical
headway is $T = 5$ seconds.

a. What is the average delay time if the flow rate is
500 cars per hour (0.14 car per second)?

b. The *service rate s* for a stop sign is the number
of cars per second that can leave the stop sign. It
is related to the delay by

$$s = D^{-1}.$$

Use function composition to represent the
service rate as a function of flow rate.
Reminder: $(a/b)^{-1} = b/a$.

c. What flow rate will permit a stop sign service
rate of 5 cars per minute (0.083 car per second)?

10. **Average Traffic Spacing** The *headway h* is the average time between vehicles. On a highway carrying
an average of 500 vehicles per hour, the probability
P that the headway is at least t seconds is given[44] by

$$P = 0.87^t.$$

a. What is the limiting value of P? Explain what
this means in practical terms.

b. The headway h can be calculated as the quotient
of the spacing f, in feet, which is the average
distance between vehicles, and the average speed
v, in feet per second, of traffic. Thus the probability that spacing is at least f feet is the same
as the probability that the headway is at least

f/v seconds. Use function composition to find
a formula for the probability Q that the spacing
is at least f feet. *Note:* Your formula will involve
both f and v.

c. If the average speed is 88 feet per second
(60 miles per hour), what is the probability that
the spacing between two vehicles is at least
40 feet?

11. **Probability of Extinction** An exponential model of
growth follows from the assumption that the yearly
rate of change in a population is $(b - d)N$, where
b is births per year, d is deaths per year, and N is
current population. The increase is in fact to some
degree probabilistic in nature. If we assume that
population increase is *normally distributed* around
rN, where $r = b - d$, then we can discuss the
probability of extinction of a population.

a. If the population begins with a single individual,
then the probability of extinction by time t is
given by

$$P(t) = \frac{d(e^{rt} - 1)}{be^{rt} - d}.$$

If $d = 0.24$ and $b = 0.72$, what is the probability that this population will eventually become
extinct? (*Hint:* The probability that the population will eventually become extinct is the limiting value for P.)

b. If the population starts with k individuals, then
the probability of extinction by time t is

$$Q = P^k,$$

where P is the function in part a. Use function
composition to obtain a formula for Q in terms
of t, b, d, r, and k.

c. If $b > d$ (births greater than deaths), so that
$r > 0$, then the formula obtained in part b can
be rewritten as

$$Q = \left(\frac{d(1 - a^t)}{b - da^t} \right)^k,$$

where $a < 1$. What is the probability that a
population starting with k individuals will
eventually become extinct?

(continued)

---

[43]*Traffic Flow Theory and Control* (New York: McGraw-Hill, 1968).
[44]See Institute of Traffic Engineers, *Transportation and Traffic Engineering Handbook*, ed. by John E. Baerwald (Englewood Cliffs, NJ:
Prentice-Hall, 1976), 102.

Mark Aplet/Shutterstock.com

d. If b is twice as large as d, what is the probability of eventual extinction if the population starts with k individuals?

e. What is the limiting value of the expression you found in part d as a function of k? Explain what this means in practical terms.

12. Decay of Litter Litter such as leaves falls to the forest floor, where the action of insects and bacteria initiates the decay process. Let A be the amount of litter present, in grams per square meter, as a function of time t in years. If the litter falls at a constant rate of L grams per square meter per year, and if it decays at a constant proportional rate of k per year, then the limiting value of A is $R = L/k$. For this exercise and the next, we suppose that at time $t = 0$, the forest floor is clear of litter.

a. If D is the difference between the limiting value and A, so that $D = R - A$, then D is an exponential function of time. Find the initial value of D in terms of R.

b. The yearly decay factor for D is e^{-k}. Find a formula for D in terms of R and k. *Reminder:* $(a^b)^c = a^{bc}$.

c. Explain why $A = R - Re^{-kt}$.

13. Applications of the Litter Formula *This is a continuation of Exercise 12.* Here we make use of the formula $A = R - Re^{-kt}$ obtained in the previous exercise.

a. Show that 95% of the limiting value is reached when $t = 3/k$.

b. A study[45] of sand dune soils near Lake Michigan found the decay rate to be $k = 0.003$ per year. How long is required to reach 95% of the limiting value?

c. The decay rate for forest litter in the Sierra Nevada mountains in California is about 0.063 per year, whereas the decay rate in the Congo[46] is about 4 per year. How does the time required to reach 95% of the limiting value compare for the Congo and the Sierra Nevada mountains?

14. High School Graduates The following table shows the number, in millions, graduating from high school in the United States in the given year.[47]

Year	Number
1975	3.13
1980	3.04
1985	2.68
1990	2.57
1994	2.46
1999	2.76
2004	3.05
2009	3.32

a. Make a plot of the data and explain why a linear model is not appropriate.

b. Use regression to find a linear model for the years 1975 through 1990. (In this part and the next, round regression line parameters to three decimal places.)

c. Use regression to find a linear model for the years 1994 through 2009.

d. Write a formula for a model of the number, in millions, graduating as a piecewise-defined function using the linear models from part b and part c.

e. Make a graph of the formula you found in part d.

f. The number graduating in 1995 was 2.52 million. On the basis of your graph in part e, determine how this compares with what would be expected from your formula.

15. Gray Wolves in Michigan Gray wolves recolonized in the Upper Peninsula of Michigan beginning in 1990. Their population has been documented as shown in the accompanying table.[48]

a. Explain why one would expect an exponential model to be appropriate for these data.

b. Find an exponential model for the data given.

---

[45] J. S. Olson, "Rates of succession and soil changes on southern Lake Michigan sand dunes," *Bot. Gaz.* **119** (1958), 125–170.
[46] J. S. Olson, "Energy storage and the balance of producers and decomposers in ecological systems," *Ecology* **44** (1963), 322–331.
[47] The data in this exercise are from the *Digest of Education Statistics*.
[48] Data courtesy of the International Wolf Center, Ely, Minnesota. For more information about wolves and their recovery, go to http://www.wolf.org

Year	Wolves
1990	6
1991	17
1992	21
1993	30
1994	57
1995	80
1996	116
1997	112
1998	140
1999	174
2000	216

c. Graph the data and the exponential model. Would it be better to use a piecewise-defined function?

d. Find an exponential model for 1990 through 1996 and another for 1997 through 2000.

e. Write a formula for the number of wolves as a piecewise-defined function using the two exponential models. Is the combined model a better fit?

16. **Mother-to-Child Transmission of HIV** The retrovirus HIV can be transmitted from mother to child during breastfeeding. But under conditions of poverty or poor hygiene, the alternative to breastfeeding carries its own risk for infant mortality. These risks differ in that the risk of HIV transmission is the same regardless of the age of the child, whereas the risk of mortality due to artificial feeding is high for newborns but decreases with the age of the child.[49] Figure 5.57 illustrates the additional risk of death due to artificial feeding and due to HIV transmission from breastfeeding.

Here the red curve is additional risk of death from artificial feeding, and the blue curve is additional risk of death from breastfeeding.

FIGURE 5.57 Additional risk of death from breastfeeding and from artificial feeding as a function of the age of the child.

For the purposes of this exercise, assume that the additional risk of death from artificial feeding is $A = 5 \times 0.785^t$ and that the additional risk from breastfeeding is $B = 2$, where t is the child's age in months.

a. At what age is the additional risk of HIV transmission from breastfeeding the same as the additional risk from artificial feeding?

b. For which ages of the child does breastfeeding carry the smaller additional risk?

c. What would be the optimal plan for feeding a child to minimize additional risk of death?

d. Write a formula for a piecewise-defined function R of t giving the additional risk of death under the optimal plan from part c.

17. **Quarterly Pine Pulpwood Prices** In southwest Georgia, the average pine pulpwood prices vary predictably over the course of the year, primarily because of weather. Prices in 2009 followed this pattern. At the beginning of the first quarter, the average price P was $9 per ton. During the first quarter, prices declined steadily to $8 per ton, then remained steady at $8 per ton through the end of the third quarter. During the fourth quarter, prices increased steadily from $8 to $10 per ton.[50]

a. Sketch a graph of pulpwood prices as a function of the quarter in the year.

b. What formula for price P as a function of t, the quarter, describes the price from the beginning of the year through the first quarter?

(continued)

---

[49]Adapted from J. Ross, "A spreadsheet model to estimate the effects of different infant feeding strategies on mother-to-child transmission of HIV and on overall infant mortality," preprint, 1999, Academy for Educational Development, Washington, DC. For simplicity, we have suppressed the units for additional risk.

[50]The data in this exercise and the next are adapted from the *Forestry Report* (Winter 2010) published by F&W Forestry Services and available at https://www.fwforestry.com/public/pdf/2010_Winter.pdf

c. What formula for price P as a function of t, the quarter, describes the price from the first to the third quarter?

d. What formula for price P as a function of t, the quarter, describes the price from the third through the fourth quarter?

e. Write a formula for price P throughout the year as a piecewise-defined function of t, the quarter.

18. Quarterly Sawtimber Prices *This is a continuation of Exercise 17.* In southwest Georgia, the average pine small sawtimber prices vary predictably over the course of the year, primarily because of weather. Prices in 2009 followed this pattern. At the beginning of the year, the average price P was $18 per ton. During the first two quarters of the year, prices declined steadily to $16 per ton. During the last two quarters of the year, prices decreased down to $15 per ton and then back up to $16 per ton.

a. Assuming small sawtimber prices are a linear function of the quarter in the year for the first two quarters, what formula for price P describes the price during the first two quarters?

b. Assume that t^2 accurately describes the shape of the curve during the third and fourth quarters. Compose the function with an appropriate translation $t - a$ for some a, and add a constant to the function to obtain a function P that describes the price throughout the third and fourth quarters.

c. Make a graph of the price over the course of a year.

d. Write a formula for price P throughout the year as a piecewise-defined function of t, the quarter.

5.4 SKILL BUILDING EXERCISES

Formulas for Composed Functions In Exercises S-1 through S-4, use a formula to express w as a function of t.

S-1. Use a formula to express w as a function of t if $w = s^2 + 1$ and $s = t - 3$.

S-2. Use a formula to express w as a function of t if $w = \dfrac{s}{s + 1}$ and $s = t^2 + 2$.

S-3. Use a formula to express w as a function of t if $w = \sqrt{2s + 3}$ and $s = e^t - 1$.

S-4. Use a formula to express w as a function of t if $w = \dfrac{s}{s - 1}$ and $s = \dfrac{t}{t - 1}$.

Formulas for Composed Functions In Exercises S-5 through S-8, find $f(g(x))$ and $g(f(x))$.

S-5. If $f(x) = 3x + 1$ and $g(x) = \dfrac{2}{x}$, find $f(g(x))$ and $g(f(x))$.

S-6. If $f(x) = x^2 + x$ and $g(x) = x - 1$, find $f(g(x))$ and $g(f(x))$.

S-7. If $f(x) = \dfrac{1}{x}$ and $g(x) = \dfrac{1}{x}$, find $f(g(x))$ and $g(f(x))$.

S-8. If $f(x) = (1 - x)^2$ and $g(x) = 2 - x$, find $f(g(x))$ and $g(f(x))$.

S-9. Limiting Values Find the limiting value of $7 + a \times 0.6^t$.

S-10. Multiplying Functions A certain function f is the product of weight, given by $t + \dfrac{1}{t}$, and height, given by t^2. Find a formula for f in terms of t.

S-11. Adding Functions A certain function f is the sum of two temperatures, one given by $t^2 + 3$, and the other given by $\dfrac{t}{t^2 + 1}$. Find a formula for f in terms of t.

S-12. Decomposing Functions Let $f(x) = x^2$ and $g(x) = x + 1$. Express $(x + 1)^2 + 1$ in terms of compositions of f and g.

S-13. Decomposing Functions If $f(x) = \sqrt{x^2 + 3}$, express f as the composition of two functions.

S-14. Composing a Function with Itself Let $f(x) = x^2 + 1$. Find formulas for $f(f(x))$ and $f(f(f(x)))$.

S-15. Decomposing Functions To join a book club, you pay an initial fee and then a fixed price each month for a book. The total cost in dollars of joining the club and buying n books is given by $C = 30 + 17n$. What is the initial fee? What is the cost of each book after you are a club member?

S-16. Decomposing Functions The population of a certain species is given by $N = 128 \times 1.07^t$, where t is measured in years. What is the initial population? By what percentage does the population grow each year?

S-17. Combining Functions Let $f(x) = x^2 - 1$ and $g(x) = 1 - x$. Find a formula for $f(g(x)) + g(f(x))$ in terms of x.

5.5 POLYNOMIALS AND RATIONAL FUNCTIONS

In addition to linear, exponential, logarithmic, and power functions, many other types of functions occur in mathematics and its applications. In this section, we will look at polynomials and rational functions. In particular, we will consider quadratics and higher-degree polynomials, and we will use regression to find models. We will also consider the use of rational functions as models.

A *polynomial* function is a function whose formula can be written as a sum of power functions where each of the powers is a non-negative whole number. For example, $5x^3 + 6x + 4$ is a polynomial because it is a sum of the power functions $5x^3$, $6x$, and $4 = 4x^0$. Another example of a polynomial is $(x + 3)^2$ since it can be written as $x^2 + 6x + 9$. On the other hand, $\sqrt{x}$ and $1/x$ are not polynomials since the powers of x are not non-negative whole numbers. Also 2^x, e^x, and $\ln x$ cannot be written as sums of power functions and so are not polynomials either.

The graphs of polynomial functions usually have important geometric properties needed in models. For example, polynomials are the simplest functions having a minimum, or having a point of inflection and a maximum. The simplicity of polynomials makes them attractive models for data that have a minimum or a maximum or other geometric features.

Quadratics

Quadratic functions are polynomials for which the highest power that occurs is 2. Quadratics have the form $ax^2 + bx + c$. For example, $x^2 + 3x + 5$ is a quadratic. We have already encountered quadratics in a number of contexts. A refresher on their algebraic properties is provided in the appendices.

Quadratic functions are useful in many applications of mathematics when a single power function is not sufficient. For example, when a rock is dropped, the distance D, in feet, that it falls in t seconds is given by the power function $D = 16t^2$. But if the rock is thrown downward with an initial velocity of 5 feet per second, then the distance it falls is no longer governed by a power function. Instead, the quadratic function $D = 16t^2 + 5t$ is needed to describe the distance traveled by the rock.

The graph of a quadratic function is known as a *parabola* and has a distinctive shape that occurs often in nature. In Figure 5.58 we show the parabola that is the graph of $x^2 - 3x + 7$, and in Figure 5.59 we show the parabola that is the graph of $-x^2 + 3x + 7$. A positive sign on x^2 makes the parabola open upward, as in Figure 5.58, and a negative sign on x^2 makes the parabola open downward, as in Figure 5.59.

If air resistance is not a factor, then a projectile such as a bullet or cannonball follows a parabolic path.

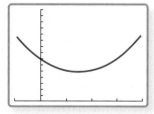

FIGURE 5.58 The graph of $x^2 - 3x + 7$, an upward-opening parabola

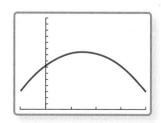

FIGURE 5.59 The graph of $-x^2 + 3x + 7$, a downward-opening parabola

EXAMPLE 5.14 **FLIGHT OF A CANNONBALL**

A cannonball fired from a cannon will follow the path of the parabola that is the graph of $y = y(x)$, where $y = -16(1 + s^2)(x/v_0)^2 + sx$ feet. Here s is the slope of inclination of the cannon barrel, v_0 is the initial velocity in feet per second, and the variable x is the distance downrange in feet. Suppose that a cannon is elevated with a slope of 0.5 (corresponding to an angle of inclination of about 27 degrees) and the cannonball is given an initial velocity of 250 feet per second. (See Figure 5.60).

$v_0 = 250$ ft/sec

FIGURE 5.60 Canonball's inclination and initial velocity

Part 1 How far will the cannonball travel?

Part 2 What is the maximum height that the cannonball will reach?

Part 3 Suppose the cannon is firing at a fort whose 20-foot-high wall is 1540 feet downrange. Show that the cannonball will not clear the fort wall. There are two possible adjustments that might be made so that the cannonball will clear the wall: Increase the initial velocity (add powder) or change the slope of inclination.

   a. How would you adjust the initial velocity so that the wall will be cleared?

   b. How would you adjust the slope of inclination so that the wall will be cleared?

Solution to Part 1 Since the slope of inclination is 0.5 and the initial velocity is 250 feet per second, the function we wish to graph is

$$-16(1 + 0.5^2)\left(\frac{x}{250}\right)^2 + 0.5x = -20\left(\frac{x}{250}\right)^2 + 0.5x.$$

After consulting a table of values, we chose a horizontal span of 0 to 1800 and a vertical span of -50 to 220 to make the graph shown in Figure 5.61. We want to know how far downrange the cannonball will land, so we use the calculator to get the crossing point $x = 1562.5$ feet shown in Figure 5.61.

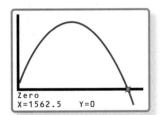

Zero
X=1562.5   Y=0

FIGURE 5.61 Where the cannonball will land

Solution to Part 2 To get the maximum height of the cannonball, we use the calculator to locate the peak of the parabola as shown in Figure 5.62. We see that the cannonball will reach a maximum height of 195.31 feet at 781.25 feet downrange.

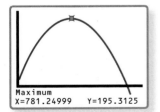

Maximum
X=781.24999   Y=195.3125

FIGURE 5.62 The maximum height of the cannonball

Solution to Part 3 In Figure 5.63 we have added the horizontal line $y = 20$ to the picture and calculated the intersection point. We see that the cannonball will be 20 feet high at 1521.42 feet downrange. Since the 20-foot-high wall is several feet farther downrange, the cannonball will not clear it.

For part a, we want to choose an initial velocity v_0 so that the height

$$-16(1 + 0.5^2)\left(\frac{x}{v_0}\right)^2 + 0.5x$$

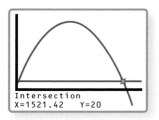

FIGURE 5.63 The cannonball won't clear the 20-foot wall.

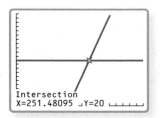

FIGURE 5.64 Adjusting initial velocity

will be 20 when the distance x is 1540. That is, we want to solve the equation

$$-20\left(\frac{1540}{v_0}\right)^2 + 0.5 \times 1540 = 20$$

for v_0. We can do this by using the crossing-graphs method as shown in Figure 5.64. Since the cannonball doesn't need much increase in initial velocity to clear the wall, we use a horizontal span of 240 to 260 and a vertical span of 0 to 40. We see that an initial velocity of 251.48 feet per second is needed to clear the wall.

For part b, we want to leave the initial velocity at 250 feet per second and change the slope of inclination, so we need to solve the equation

$$-16(1 + s^2)\left(\frac{1540}{250}\right)^2 + s \times 1540 = 20$$

for s. Once again, we don't need to change the slope much, so we use a horizontal span of 0.4 to 0.6 and a vertical span of 0 to 40. When we calculate the intersection, we find that a slope of 0.51 is required.

TEST YOUR UNDERSTANDING | **FOR EXAMPLE 5.14**

Find a formula for the path of the cannonball if the slope of the angle of elevation is $s = 0.7$ and the initial velocity is $v_0 = 300$ feet per second. What is the maximum height the cannonball reaches? How far downrange does the maximum height occur? ■

Quadratic Regression

In many practical settings it is necessary to model data that show a maximum or minimum. Such data cannot be accurately modeled by linear, exponential, or logistic functions. But quadratic functions often serve the purpose well. The process of fitting a quadratic function is similar to other forms of regression that we have studied in this text (see Appendices B and C). It's important to consider whether a quadratic model is appropriate before we perform regression. Good reasons might include an expectation that the data ought to be quadratic—for example, in the case of the flight of a projectile.

EXAMPLE 5.15 **OSPREYS IN THE TWIN CITIES**

According to the Raptor Center 1995 Osprey Report,[51] the numbers of introduced ospreys fledged in the Twin Cities, Minnesota, from 1992 through 1995 are as given in the accompanying table.

Year	1992	1993	1994	1995
Ospreys fledged	8	18	18	7

Part 1 Plot the number of ospreys fledged versus time in years since 1992.

Part 2 Assume that a quadratic model is deemed appropriate. Use quadratic regression to find a model for the data.

Part 3 Add the graph of the model to the plot of data points from part 1. Discuss the fit of the model.

---

[51]See http://www.raptor.cvm.umn.edu

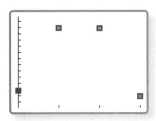

FIGURE 5.65 Plot of data

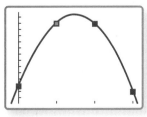

FIGURE 5.66 Quadratic fit

Part 4 What does this model predict for the number of ospreys fledged in 1996?

Solution to Part 1 The plot of the data appears in Figure 5.65. Here the variable is t, the time in years since 1992.

Solution to Part 2 Quadratic regression yields the formula

$$I = -5.25t^2 + 15.45t + 7.95,$$

where I is the number of introduced ospreys fledged.

Solution to Part 3 The plot of the data points and the quadratic model is shown in Figure 5.66. The fit appears to be excellent.

Solution to Part 4 Now 1996 corresponds to $t = 4$, and the formula for the quadratic regression model gives $I(4) = -14.25$. The number, being negative, certainly could not be a reasonable prediction for the number of introduced ospreys. This is an example of the dangers of using models to extrapolate beyond the range of the data collected.

TEST YOUR UNDERSTANDING │ **FOR EXAMPLE 5.15**

In the study "Economics of Scale in High School Operation" by J. Riew,[52] the author studied data from the early 1960s on expenditures for high schools ranging from 150 to 2400 enrollment. The data he observed were similar to those in the table below.

n = enrollment	200	600	1000	1400	1800
C = cost per student, in dollars	667.4	545.0	461.0	415.4	408.2

Model these data using a quadratic function. (Round regression parameters to four decimal places.) According to this model, what enrollment produces the minimum cost per student? ∎

The Quadratic Formula

We have taken the point of view that linear functions should be solved by hand calculation and that others should be solved using the calculator. However, there is a well-known formula, the *quadratic formula*, that makes possible the solution of quadratic equations by hand calculation, and some instructors may prefer that their students use it. In general, a quadratic equation $ax^2 + bx + c = 0$ has two solutions, which differ only by a sign in one place:

$$x = \frac{-b + \sqrt{b^2 - 4ac}}{2a}$$

and

$$x = \frac{-b - \sqrt{b^2 - 4ac}}{2a}.$$

To illustrate the use of the quadratic formula, we consider a rock tossed upward from a 135-foot-high building with an initial velocity of 38 feet per second. By elementary physics the distance D up from the ground, in feet, is given by

$$D = -16t^2 + 38t + 135,$$

[52]Published in *Review of Economics and Statistics* **48** (August 1966), 280–287. In our presentations of Riew's results, we have suppressed variables such as teachers' salaries.

where t is the time in seconds after the rock is thrown. Suppose we want to know when the rock will be 145 feet above the ground. That is, we wish to solve the equation

$$-16t^2 + 38t + 135 = 145.$$

To solve this equation using the quadratic formula, we first subtract 145 from each side:

$$-16t^2 + 38t - 10 = 0.$$

Next we identify the values of a, b, and c:

$$a = -16$$
$$b = 38$$
$$c = -10.$$

Plugging these values into the quadratic formula yields

$$t = \frac{-b + \sqrt{b^2 - 4ac}}{2a} = \frac{-38 + \sqrt{38^2 - 4 \times (-16) \times (-10)}}{2 \times (-16)} = 0.30$$

$$t = \frac{-b - \sqrt{b^2 - 4ac}}{2a} = \frac{-38 - \sqrt{38^2 - 4 \times (-16) \times (-10)}}{2 \times (-16)} = 2.07.$$

Thus we see that at 0.30 second and 2.07 seconds after the rock is thrown it is 145 feet above the ground. (There are two such times because the rock reaches this height on the way up and on the way down.) You may wish to check to see that you get the same answers using the crossing-graphs method with the calculator.

Higher-Degree Polynomials

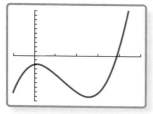

FIGURE 5.67 The cubic function $y = x^3 - 4x^2 + x - 2$

Higher-degree polynomials are named for the highest power that occurs in the polynomials. For example, *cubic* polynomials have a highest term of degree 3, *quartic* polynomials have a highest term of degree 4, and so on. These polynomials often have interesting graphs. The cubic graph in Figure 5.67 has a horizontal span of -1 to 5 and a vertical span of -10 to 10. It displays both a maximum and a minimum. The quartic graph in Figure 5.68 has a horizontal span of -5 to 5 and a vertical span of -10 to 10. It shows two maxima and a minimum.

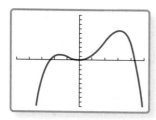

FIGURE 5.68 The quartic function $y = -0.1x^4 + 0.2x^3 + x^2$

Cubic, quartic, and higher-degree polynomials have some desirable features that make them attractive as models. Typically cubics have one maximum and one minimum,[53] and quartics may have two maxima and one minimum or two minima and one maximum. Higher-degree polynomials may change direction many times. The keystrokes required to perform regression to find cubic and quartic models are presented in Appendices B and C.

EXAMPLE 5.16 **SALES OF CONDOS AND CO-OPS**

The following table shows the number of units sold, in thousands, of existing apartment condos and co-ops in various years. The data are from the *Statistical Abstract of the United States*.

Year	2003	2004	2005	2006	2007	2008	2009	2010
Units sold (thousands)	732	820	896	801	713	563	590	599

---

[53]Some cubics have no maxima or minima.

Let A denote the number of units sold and t the time in years since 2003.

Part 1 Plot the data points for A versus t.

Part 2 Find the equation of a cubic model for A in terms of t, and comment on its fit with the data. Round the regression parameters to three decimal places.

Part 3 What does the cubic model predict for future sales?

Solution to Part 1 The plot of the data points is in Figure 5.69. We see that the points increase, then decrease, then increase again. As we noted earlier, the property of having one maximum and one minimum is typical of a cubic polynomial.

Having one maximum and one minimum is typical of a cubic polynomial.

Solution to Part 2 Using cubic regression, we obtain the model

$$A = 7.106t^3 - 81.876t^2 + 208.536t + 720.803.$$

The graph in Figure 5.70 shows that the model fits the data points fairly well.

Solution to Part 3 In Figure 5.71 we have enlarged the horizontal span to 12 (corresponding to the year 2015), and we have enlarged the vertical span to 3800. We see in Figure 5.71 that the model predicts that sales will climb very steeply (so steeply that the data points are bunched at the lower part of the screen).

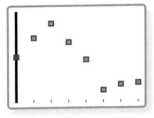

FIGURE 5.69 Plot of units sold

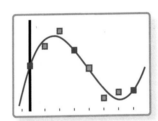

FIGURE 5.70 Cubic regression curve added to plot

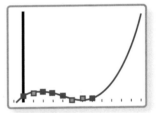

FIGURE 5.71 Using a cubic model to predict future sales

TEST YOUR UNDERSTANDING | **FOR EXAMPLE 5.16**

The following fictitious table shows kryptonite prices, in dollars per gram, t years after 2000.

t = years since 2000	0	1	2	3	4	5	6	7	8	9	10
K = price	56	51	50	55	58	52	45	43	44	48	51

Make a quartic model for these data. Round the regression parameters to two decimal places.

It is worthwhile to note part 3 of the example. The model predicts a steep rise in units sold, but such a rise seems unwarranted by the plot of the data in Figure 5.69. Polynomial functions can grow quickly, and they may fit data and give reasonable predictions for only a very small set of values. In addition, rounding of regression coefficients can radically affect fit.

⬛ | Rational Functions

Rational functions are functions that can be written as a ratio of polynomial functions—that is, as a fraction whose numerator and denominator are both polynomials. Simple examples include any polynomial, $1/x$, $(x^{12} - x)/(0.7x^6 + 3)$, and $x + (1/x^2)$. By way

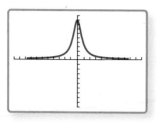

FIGURE 5.72 A graph of the rational function $y = 8/(x^2 + 1)$

of contrast, e^x, $\log x$ and $\sqrt{x}$ are not rational functions. The graphs of rational functions can be quite interesting and often exhibit behavior not found in other functions.

One rational function is $8/(x^2 + 1)$, whose graph is displayed in Figure 5.72. Unlike polynomial functions, for larger values of x this function gets closer to zero, which is similar to the behavior of decreasing exponential functions. Unlike exponential functions, this function gets close to zero both for large positive and for large (in size) negative values of x.

A more interesting rational function is $x/(x^2 - 1)$, whose graph is displayed in Figure 5.73. The graph looks quite ragged, and it is unclear what is happening near $x = -1$ and $x = 1$. The reason for this becomes apparent when we consider that when $x = -1$ or 1, then $x^2 - 1 = 0$; thus the function is trying to evaluate $\frac{1}{0}$, which is not possible. In Figure 5.74, we graph $y = x/(x^2 - 1)$, looking more closely near $x = 1$. We use a horizontal span of 0 to 2 and a vertical span of -10 to 10.

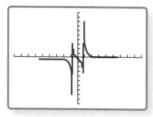

FIGURE 5.73 A graph of the rational function $y = x/(x^2 - 1)$

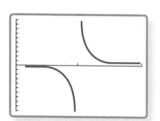

FIGURE 5.74 Looking more closely at a pole

Poles typically occur when there is danger of division by zero.

The function $x/(x^2 - 1)$ has no value at $x = 1$. Note that for x to the left of 1, as x tends toward 1 the function values are negative and become larger and larger in size; by contrast, as x tends toward 1 from the right, the function values are positive and become larger and larger positive numbers. In general, a function has a *pole* at $x = a$ if the function is not defined at $x = a$ and the values of the function become larger and larger in size as x gets near a. Not all rational functions have poles, as we saw earlier in considering $y = 8/(x^2 + 1)$. In general, poles occur when the denominator of a rational function has a zero but the numerator does not have the same zero.[54] If a function has

Near poles, functions typically get very large.

a pole at the point $x = a$, then we say that the line $x = a$ is a *vertical asymptote* of the graph of the function, since the graph gets closer and closer to this line as x tends toward a. For example, the graph of the function $x/(x^2 - 1)$ in Figure 5.74 has the line $x = 1$ as a vertical asymptote.

In addition to vertical asymptotes, rational functions may also have *horizontal asymptotes*. Like vertical asymptotes, a horizontal asymptote is a line to which the graph gets closer and closer. Since the line is horizontal, it has slope 0, so its equation is of the form $y = b$ for some b. Since the graph gets closer and closer to $y = b$, the number b is a limiting value of the function. Thus the idea of "horizontal asymptote" is nothing more than a geometric version of "limiting value." For example, the function $x/(x^2 - 1)$ from the previous paragraph can be seen to have a horizontal asymptote $y = 0$ since the function values get closer and closer to 0 for large values of x.

Horizontal asymptotes correspond to limiting values.

In physical applications, rational functions are far less common than the other functions discussed in this text. When they have poles, the poles usually occur at numbers to which the physical situation does not apply. For example, Ohm's law says that if electric current is flowing across a resistor, then $i = v/R$, where i is current, v is voltage, and

---

[54]We are assuming that factors common to the numerator and denominator have been canceled.

R is resistance. For fixed v, the function i is a rational function of R with a pole when $R = 0$. Note that if $R = 0$, then there is no resistance, so we are no longer in the situation of current flowing across a resistor. On the other hand, the behavior of the function i for values of R close to zero does have meaning: As R gets smaller and smaller, i gets larger and larger.

EXAMPLE 5.17 FUNCTIONAL RESPONSE

Holling's functional response curve[55] describes the feeding habits of a predator in terms of the density of the prey. An example of such a curve is given by the rational function

$$y = \frac{16x}{1 + 2x}.$$

Here x is the density of the prey—that is, the number of prey per unit area—and y is the number of prey eaten per day by a certain predator.

Part 1 Make a graph of y as a function of x.

Part 2 In terms of the predator and its prey, why is it reasonable for the graph to be increasing and concave down?

Part 3 Does the function have a limiting value? What is the equation of the horizontal asymptote?

Part 4 Explain the significance of the horizontal asymptote from part 3 in terms of the predator and prey.

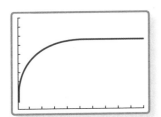

FIGURE 5.75 Prey eaten as a function of prey density

Solution to Part 1 The graph in Figure 5.75 has a horizontal span of 0 to 10 and a vertical span of 0 to 10.

Solution to Part 2 It is reasonable for the graph to increase, because the greater the density of the prey, the easier it is for the predator to catch and eat the prey. On the other hand, the predator can only eat so much prey in one day, so it is reasonable to expect the graph to have the shape you found in part 1. That is, we expect that the rate of increase in prey eaten will decrease as x increases.

Solution to Part 3 Using a table or a graph, it is easy to see that 8 is a limiting value of the function. Thus the equation of the horizontal asymptote is $y = 8$.

Solution to Part 4 The horizontal asymptote indicates that for large density of prey, the predator will eat about 8 per day.

TEST YOUR UNDERSTANDING | FOR EXAMPLE 5.17

The time t, in minutes, required to drive 1 mile in central London is given approximately by the formula

$$t = \frac{600}{280 - 0.08q},$$

where q is the traffic flow measured in vehicles per hour.[56] The formula is valid as long as $280 - 0.08q$ is positive. What value of q corresponds to a pole for this function? Explain in practical terms the meaning of the pole. ∎

---

[55]Based on the work of C. S. Holling, "Some characteristics of simple types of predation and parasitism," *Can. Ent.* **91** (1959), 385–398.

[56]Adapted from J.M. Thomson, *Road Pricing in Central London*, PRP 18, 1962.

ANSWERS FOR TEST YOUR UNDERSTANDING

5.14 $y = -16(1 + 0.7^2)(x/300)^2 + 0.7x$. It reaches a maximum height of 462.46 feet at 1321.31 feet downrange.

5.15 The model is $C = 0.00012n^2 - 0.402n + 743$. The model shows a minimum cost per student when enrollment is 1675 students.

5.16 $K = 0.03t^4 - 0.55t^3 + 2.75t^2 - 4.58t + 54.78$

5.17 The pole occurs when q is 3500 vehicles per hour. When traffic flow is near 3500 per hour, it takes a very long time to travel one mile. So traffic is almost stopped.

5.5 EXERCISES

Reminder Round all answers to two decimal places unless otherwise indicated.

1. **Sales Growth**   The rate of growth G, in thousands of dollars per year, in sales of a certain product is a function of the current sales level s, in thousands of dollars, and the model uses a quadratic function:

$$G = 1.2s - 0.3s^2.$$

The model is valid up to a sales level of 4 thousand dollars.

   a. Draw a graph of G versus s.

   b. Express using functional notation the rate of growth in sales at a sales level of $2260, and then estimate that value.

   c. At what sales level is the rate of growth in sales maximized?

2. **Marine Fishery**   A class of models for population growth rates in marine fisheries assumes that the harvest from fishing is proportional to the population size. One such model uses a quadratic function:

$$G = 0.3n - 0.2n^2.$$

Here G is the growth rate of the population, in millions of tons of fish per year, and n is the population size, in millions of tons of fish.

   a. Make a graph of G versus n. Include values of n up to 1.7 million tons.

   b. Calculate $G(1.62)$ and explain what your answer means in practical terms.

   c. At what population size is the growth rate the largest?

3. **Cox's Formula**   Assume that a long horizontal pipe connects the bottom of a reservoir with a drainage area. Cox's formula provides a way of determining the velocity v of the water flowing through the pipe:

$$\frac{Hd}{L} = \frac{4v^2 + 5v - 2}{1200}.$$

Here H is the depth of the reservoir in feet, d is the pipe diameter in inches, L is the length of the pipe in feet, and the velocity v of the water is in feet per second. (See Figure 5.76.)

FIGURE 5.76

   a. Graph the quadratic function $4v^2 + 5v - 2$ using a horizontal span from 0 to 10.

   b. Judging on the basis of Cox's formula, is it possible to have a velocity of 0.25 foot per second?

   c. Find the velocity of the water in the pipe if its diameter is 4 inches, its length is 1000 feet, and the reservoir is 50 feet deep.

   d. If the water velocity is too high, there will be erosion problems. Assuming that the pipe length is 1000 feet and the reservoir is 50 feet deep, determine the largest pipe diameter that will ensure that the water velocity does not exceed 10 feet per second.

4. **Estimating Wave Height**   Sailors use the following quadratic function to estimate wave height h, in feet, from wind speed w, in miles per hour:

$$h = 0.02w^2.$$

   a. What wave height does the formula give for a wind speed of 25 miles per hour?

   b. A sailor observes that the wave height is 4 feet. According to the formula above, what is the speed of the wind?

(continued)

5. **Surveying Vertical Curves** When a road is being built, it usually has straight sections, all with the same grade, that must be linked to each other by curves. (By this we mean curves up and down rather than side to side, which would be another matter.) It's important that as the road changes from one grade to another, the rate of change of grade between the two be constant.[57] The curve linking one grade to another grade is called a *vertical curve.*

Surveyors mark distances by means of *stations* that are 100 feet apart. To link a straight grade of g_1 to a straight grade of g_2, the elevations of the stations are given by

$$y = \frac{g_2 - g_1}{2L}x^2 + g_1x + E - \frac{g_1L}{2}.$$

Here y is the elevation of the vertical curve in feet, g_1 and g_2 are percents, L is the length of the vertical curve in hundreds of feet, x is the number of the station, and E is the elevation in feet of the intersection where the two grades would meet. (See Figure 5.77.) The station $x = 0$ is the very beginning of the vertical curve, so the station $x = 0$ lies where the straight section with grade g_1 meets the vertical curve. The last station of the vertical curve is $x = L$, which lies where the vertical curve meets the straight section with grade g_2.

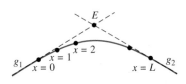

FIGURE 5.77

Assume that the vertical curve you want to design goes over a slight rise, joining a straight section of grade 1.35% to a straight section of grade −1.75%. Assume that the length of the curve is to be 500 feet (so $L = 5$) and that the elevation of the intersection is 1040.63 feet.

a. What is the equation for the vertical curve described above? Don't round the coefficients.

b. What are the elevations of the stations for the vertical curve?

c. Where is the highest point of the road on the vertical curve? (Give the distance along the vertical curve and the elevation.)

6. **A Leaking Can** The side of a cylindrical can full of water springs a leak, and the water begins to stream out. (See Figure 5.78.) The depth H, in inches, of water remaining in the can is a function of the distance D in inches (measured from the base of the can) at which the stream of water strikes the ground. Here is a table of values of D and H:

Distance D in inches	Depth H in inches
0	1.00
1	1.25
2	2.00
3	3.25
4	5.00

a. Use regression to find a formula for H as a quadratic function of D.

b. When the depth is 4 inches, how far from the base of the can will the water stream strike the ground?

c. When the water stream strikes the ground 5 inches from the base of the can, what is the depth of water in the can?

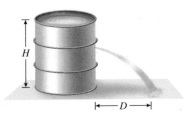

FIGURE 5.78

7. **Traffic Accidents** The following table shows the rate R of vehicular involvement in traffic accidents (per 100,000,000 vehicle-miles) as a function of vehicular speed s, in miles per hour, for commercial vehicles driving at night on urban streets.[58]

[57]Adapted from F. H. Moffitt and H. Bouchard, *Surveying*, 9th ed. (New York: HarperCollins, 1992).
[58]The data are adapted from J. C. Marcellis, "An economic evaluation of traffic movement at various speeds," *Travel Time and Vehicle Speed*, Record 35 (Washington, DC: Highway Research Board, 1963).

Speed s	Accident rate R
20	1600
25	700
30	250
35	300
40	700
45	1300

a. Use regression to find a quadratic model for the data.

b. Calculate $R(50)$ and explain what your answer means in practical terms.

c. At what speed is vehicular involvement in traffic accidents (for commercial vehicles driving at night on urban streets) at a minimum?

8. **Vehicles Parked**   The table below shows the number, in thousands, of vehicles parked in the central business district of a certain city on a typical Friday as a function of the hour of the day.[59]

Hour of the day	Vehicles parked (thousands)
9 A.M.	6.2
11 A.M.	7.5
1 P.M.	7.6
3 P.M.	6.6
5 P.M.	3.9

a. Use regression to find a quadratic model for the data. (Round the regression parameters to three decimal places.)

b. Express using functional notation the number of vehicles parked on a typical Friday at 2 P.M., and then estimate that value.

c. At what time of day is the number of vehicles parked at its greatest?

9. **Women Employed Outside the Home**   The following table shows the number, in millions, of women employed outside the home in the given year.[60]

Year	Number in millions
1942	16.11
1943	18.70
1944	19.17
1945	19.03
1946	16.78

a. Use regression to find a quadratic model for the data. (Round the regression parameters to three decimal places.)

b. Express using functional notation the number of women working outside the home in 1947, and then estimate that value.

c. The actual number of women working outside the home in 1947 was 16.90 million, whereas in 1948 the number was 17.58 million. In light of this, is a quadratic model appropriate for the period from 1942 through 1948?

10. **Resistance in Copper Wire**   Electric resistance in copper wire changes with the temperature of the wire. If $C(t)$ is the electric resistance at temperature t, in degrees Fahrenheit, then the resistance ratio $C(t)/C(0)$ can be measured.

Temperature t in degrees	$\dfrac{C(t)}{C(0)}$ ratio
0	1
10	1.0393
20	1.0798
30	1.1215
40	1.1644

(continued)

---

[59]The data are adapted from *Nashville Metropolitan Area Transportation Study—Downtown Parking* (Nashville, TN: Nashville Parking Board, 1960), 57.
[60]The data are taken from the *1950 Statistical Abstract of the United States*.

a. Use regression to find a quadratic formula for the ratio $C(t)/C(0)$ as a function of temperature t. Do not round the regression parameters.

b. At what temperature is the electric resistance double that at 0 degrees?

c. Suppose that you have designed a household appliance to be used at room temperature (72 degrees) and you need to have the wire resistance inside the appliance accurate to plus or minus 10% of the predicted resistance at 72 degrees.

  i. What resistance ratio do you predict at 72 degrees? (Use four decimal places.)

  ii. What range of resistance ratios represents plus or minus 10% of the resistance ratio for 72 degrees?

  iii. What temperature range for the appliance will ensure that your appliance operates within the 10% tolerance? Is this range reasonable for use inside a home?

11. A Falling Rock   A rock is thrown downward, and the distance D, in feet, that it falls in t seconds is given by $D = 16t^2 + 3t$. Find how long it takes for the rock to fall 400 feet by using

a. the quadratic formula.

b. the crossing-graphs method.

12. Linear and Quadratic Data   One of the two tables below shows data that can best be modeled by a linear function, and the other shows data that can best be modeled by a quadratic function. Identify which table shows the linear data and which table shows the quadratic data, and find a formula for each model.

x	0	1	2	3	4
$f(x)$	10	17	26	37	50

Table A

x	0	1	2	3	4
$g(x)$	10	17	24	31	38

Table B

13. Builder's Old Measurement   The *Builder's Old Measurement* was instituted by law in England in 1773 as the way to estimate the total tonnage T of a wooden ship from its beam width W and length L, both measured in feet. The formula is

$$T = \frac{(L - 0.6W)W^2}{188}.$$

In this exercise we consider wooden ships of length 120 feet.

a. What tonnage is given by the Builder's Old Measurement for a ship of beam width 25 feet?

b. What beam width will give a tonnage of 400?

14. Speed of Sound in the North Atlantic   The speed of sound in ocean water is 1448.94 meters per second, provided that the ocean water has a salinity of 35 parts per thousand, the temperature is 0 degrees Celsius, and the measurement is taken at the surface. If any one of these three factors varies, the speed of sound also changes. Different oceans often differ in salinity. In the North Atlantic Central Water (the main body of water for the northern half of the Atlantic Ocean), the salinity can be determined from the temperature, so the speed of sound depends only on temperature and depth. A simplified polynomial formula for velocity in this body of water is[61]

$$V = 1447.733 + 4.7713T - 0.05435T^2$$
$$+ 0.0002374T^3 + 0.0163D + 1.675$$
$$\times 10^{-7}D^2 - 7.139 \times 10^{-13}T\,D^3.$$

Here V is the speed of sound in meters per second, T is water temperature in degrees Celsius, and D is depth in meters. This formula is valid for depths up to 8000 meters and for temperatures between 0 and 30 degrees Celsius.

a. What type of polynomial is V as a function of T alone? Of D alone?

b. For a fixed depth of 1000 meters, write the formula for V in terms of T alone.

c. Graph V as a function of T for the fixed depth of 1000 meters.

d. What is the concavity of the graph from part c? What does this imply about the speed of sound at that depth as temperature increases?

---

[61]Adapted from P. C. Etter, *Underwater Acoustic Modeling* (London: E & FN Spon, 1996), 15–16.

15. **Traffic Accidents**   The following table shows the cost C of traffic accidents, in cents per vehicle-mile, as a function of vehicular speed s, in miles per hour, for commercial vehicles driving at night on urban streets.[62]

Speed s	20	25	30	35	40	45	50
Cost C	1.3	0.4	0.1	0.3	0.9	2.2	5.8

The *rate* of vehicular involvement in traffic accidents (per vehicle-mile) can be modeled[63] as a quadratic function of vehicular speed s, and the cost per vehicular involvement is roughly a linear function of s, so we expect that C (the product of these two functions) can be modeled as a cubic function of s.

   a. Use regression to find a cubic model for the data. (Keep two decimal places for the regression parameters written in scientific notation.)

   b. Calculate $C(42)$ and explain what your answer means in practical terms.

   c. At what speed is the cost of traffic accidents (for commercial vehicles driving at night on urban streets) at a minimum? (Consider speeds between 20 and 50 miles per hour.)

16. **Poiseuillé's Law for Rate of Fluid Flow**   Poiseuillé's law for the rate of flow of a fluid through a tube is a fourth-order polynomial that is also a power function:

$$F = cR^4,$$

where F is the flow rate (measured as a volume per unit time), c is a constant, and R is the radius of the tube.

   a. Assume that R increases by 10%. Explain why F increases by 46.41%. (*Hint:* Consider 1.10 raised to the fourth power.)

   b. What is the flow rate through a 3/4-inch pipe compared with that through a 1/2-inch pipe?

   c. Suppose that an artery supplying blood to the heart muscle (see Figure 5.79) is partially blocked and is only half its normal radius. What percentage of the usual blood flow will flow through the partially blocked artery?

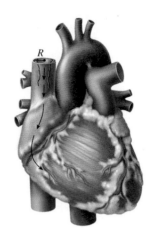

FIGURE 5.79

17. **Population Genetics**   In the study of population genetics, an important measure of inbreeding is the proportion of *homozygous genotypes*—that is, instances in which the two alleles carried at a particular site on an individual's chromosomes are both the same. For populations in which blood-related individuals mate, there is a higher than expected frequency of homozygous individuals. Examples of such populations include endangered or rare species, selectively bred breeds, and isolated populations. In general, the frequency of homozygous children from matings of blood-related parents is greater than that for children from unrelated parents.[64]

   Measured over a large number of generations, the proportion of heterozygous genotypes—that is, nonhomozygous genotypes—changes by a constant factor λ_1 from generation to generation. The factor λ_1 is a number between 0 and 1. If $\lambda_1 = 0.75$, for example, then the proportion of heterozygous individuals in the population decreases by 25% in each generation. In this case, after 10 generations the proportion of heterozygous individuals in the population decreases by 94.37%, since $0.75^{10} = 0.0563$, or 5.63%. In other words, 94.37% of the population is homozygous.

   For specific types of matings, the proportion of heterozygous genotypes can be related to that of previous generations and is found from an equation.

(continued)

---

[62]The data are adapted from J. C. Marcellis, "An economic evaluation of traffic movement at various speeds," *Travel Time and Vehicle Speed*, Record 35 (Washington, DC: Highway Research Board, 1963).
[63]See Exercise 7.
[64]This exercise and the following exercises are adapted from Chapter 9 of A. Jacquard's *The Genetic Structure of Populations* (New York: Springer-Verlag, 1974).

For matings between siblings, λ_1 can be determined as the largest value of λ for which

$$\lambda^2 = \frac{1}{2}\lambda + \frac{1}{4}.$$

This equation comes from carefully accounting for the genotypes for the present generation (the λ^2 term) in terms of those of the previous two generations (represented by λ for the parents' generation and by the constant term for the grandparents' generation).

a. Find both solutions to the quadratic equation above and identify which is λ_1. (Use a horizontal span of -1 to 1 in this exercise and the following exercise.)

b. After 5 generations what proportion of the population will be homozygous?

c. After 20 generations what proportion of the population will be homozygous?

18. **Population Genetics—First Cousins** *This is a continuation of Exercise 17.* Double first cousins are first cousins in two ways—that is, each of the parents of one is a sibling of a parent of the other. First cousins in general have six different grandparents, whereas double first cousins have only four different grandparents. For matings between double first cousins, the proportions of the genotypes of the children can be accounted for in terms of those of parents, grandparents, and great-grandparents:

$$\lambda^3 = \frac{1}{2}\lambda^2 + \frac{1}{4}\lambda + \frac{1}{8}.$$

a. Find all solutions to the cubic equation above and identify which is λ_1.

b. After 5 generations what proportion of the population will be homozygous?

c. After 20 generations what proportion of the population will be homozygous?

19. **An Epidemic Model** A certain disease is contracted in a population at an average age of A_0 if no one is immunized at birth. If q percent of the population is immunized at birth, then the average age at which the disease is contracted is $A(q)$. One model gives the relation

$$A(q) = \frac{100A_0}{100 - q}.$$

a. What value of q gives a pole for this rational function?

b. Explain in practical terms what happens to $A(q)$ when q is near the pole.

20. **Forming a Pen** You want to form a rectangular pen of area 80 square feet. (See Figure 5.80.) One side of the pen is to be formed by an existing building and the other three sides by a fence. If w is the length, in feet, of the sides of the rectangle perpendicular to the building, then the length of the side parallel to the building is $80/w$, so the total amount $F = F(w)$, in feet, of fence required is the rational function

$$F = 2w + \frac{80}{w}.$$

a. Make a graph of F versus w.

b. Explain in practical terms the behavior of the graph near the pole at $w = 0$.

c. Determine the dimensions of the rectangle that requires a minimum amount of fence.

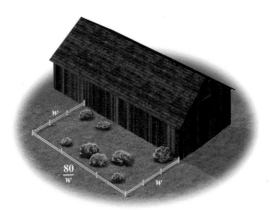

FIGURE 5.80

21. **Traffic Signals** The number of seconds n for the yellow light is critical to safety at a traffic signal. One study recommends a formula for setting the time that permits a driver who sees the yellow light shortly before entering the intersection either to stop the vehicle safely or to cross the intersection at the current approach speed before the end of the yellow light.[65] For a street of width 70 feet under standard conditions, the formula is

$$n = 1 + \frac{v}{30} + \frac{90}{v}.$$

[65]P. L. Olson and R. W. Rothery, "Driver response to the amber phase of traffic signals," *Traffic Engineering* **XXXII,** No. 5 (1962), 17–20, 29.

Here v is the approach speed in feet per second. (See Figure 5.81.)

|←——— 70 ft ———→|

FIGURE 5.81

a. Make a graph of n as a function of v. Include speeds from 30 to 80 feet per second (roughly 20 to 55 miles per hour).

b. Express using functional notation the length of the yellow light when the approach speed is 45 feet per second, and then calculate that value.

c. Explain in practical terms the behavior of the graph near the pole at $v = 0$.

d. What is the minimum length of time for a yellow light?

22. Catch Equation   If a lake is stocked with fish of the same age, then the total number C of these fish caught by fishing over the life span of the fish is given by the catch equation

$$C = \frac{F}{M + F} N_0,$$

where F is the proportion of fish caught annually, M is the proportion of fish that die of natural causes annually, and N_0 is the original number of fish stocked in the lake. Assume that a lake is stocked with 1000 channel catfish and that the natural mortality rate is $M = 0.1$ (this means that the annual mortality rate is 10%).

a. Write the catch equation for the lake stocked with channel catfish.

b. Assuming that $F = 1$ means that 100% of the fish are caught within 1 year, what might $F = 2$ mean?

c. What is the horizontal asymptote for the catch equation?

d. Explain what the horizontal asymptote means in terms of the original 1000 fish.

e. What is the vertical asymptote for the catch equation? Does the pole of the equation have any meaning in terms of the fish population?

23. Hill's Law   When a force is applied to muscle tissue, the muscle contracts. Hill's law is an equation that relates speed of muscle contraction with force applied to the muscle.[66] The equation is given by the rational function

$$S = \frac{(F_\ell - F)b}{F + a},$$

where S is the speed at which the muscle contracts, F_ℓ is the maximum force of the muscle at the given length ℓ, F is the force against which the muscle is contracting, and a and b are constants that depend on the muscle tissue itself. This is valid for non-negative F no larger than F_ℓ. For a fast-twitch vertebrate muscle—for example, the leg muscle of a sprinter— we may take $F_\ell = 300$ kPa, $a = 81$, and $b = 6.75$. These are the values we use in this exercise.

a. Write the equation for Hill's law using the numbers above for fast-twitch vertebrate muscles.

b. Graph S versus F for forces up to 300 kPa.

c. Describe how the muscle's contraction speed changes as the force applied increases.

d. When is S equal to zero? What does this mean in terms of the muscle?

e. Does the rational function for S have a horizontal asymptote? What meaning, if any, does the asymptote have in terms of the muscle?

f. Does the rational function for S have a vertical asymptote? What meaning, if any, does the asymptote have in terms of the muscle?

24. The Michaelis–Menten Relation   Enzymes are proteins that act as catalysts converting one type of substance, the *substrate*, into another type. An example of an enzyme is invertase, an enzyme in your body, which converts sucrose into fructose and glucose. Enzymes can act very rapidly; under the right circumstances, a single molecule of an enzyme can convert millions of molecules of the substrate per minute. The Michaelis–Menten relation expresses the initial speed of the reaction as a rational function of the initial concentration of the substrate:

$$v = \frac{Vs}{s + K_m},$$

where v is the initial speed of the reaction (in moles per liter per second), s is the initial concentration of

(continued)

[66]Adapted from R. McN. Alexander, *Animal Mechanics*, 2nd ed. (London: Blackwell, 1983).

the substrate (in moles per liter), and V and K_m are constants that are important measures of the kinetic properties of the enzyme.[67]

For this exercise, graph the Michaelis–Menten relation giving v as a function of s for two different values of V and of K_m.

a. On the basis of your graphs, what is the horizontal asymptote of v?

b. On the basis of your graphs, what value of s makes $v(s) = V/2$? How is that value related to K_m?

c. In practice you don't know the values of V or K_m. Instead, you take measurements and find the graph of v as a function s. Then you use the graph to determine V and K_m. If you have the graph, how will that enable you to determine V? How will that enable you to determine K_m?

5.5 SKILL BUILDING EXERCISES

S-1. Quadratic Formula   Use the quadratic formula to solve $-2x^2 + 2x + 5 = 0$.

S-2. Quadratic Formula   Use the quadratic formula to solve $5x^2 - 8 = 0$.

S-3. Quadratic Regression   Use quadratic regression to find a model for the following data set.

x	1	3	5	6	8
$f(x)$	2.2	9.7	27.7	35.2	62.1

S-4. Quadratic Regression   Use quadratic regression to find a model for the following data set.

x	1	3	5	6	8
$f(x)$	1	3	7	5	2

S-5. Quadratic Regression   Use quadratic regression to find a model for the following data set.

x	-2.4	-1.1	5.7	16.4	18.3
$f(x)$	17.3	4.1	93.6	743.6	864.5

Recognizing Polynomials   In Exercises S-6 through S-9, you are asked to determine whether the given function is a polynomial.

S-6. Is $x^8 - 17x + 1$ a polynomial? If it is a polynomial, give its degree.

S-7. Is $\sqrt{x} + 8$ a polynomial? If it is a polynomial, give its degree.

S-8. Is $9.7x - 53.1x^4$ a polynomial? If it is a polynomial, give its degree.

S-9. Is $x^{3.2} - x^{2.3}$ a polynomial? If it is a polynomial, give its degree.

S-10. Rational Function?   Is $\dfrac{x - \sqrt{x}}{1 + x}$ a rational function?

S-11. Cubic Regression   Use cubic regression to model the following data set.

x	1	3	4	6	7	8	10
y	1	3	5	2	1	4	8

S-12. Cubic Regression   Use cubic regression to model the following data set.

x	1	3	4	6	7	8	10
y	1.5	3.8	7.4	4.6	8.8	9.1	12.4

S-13. Quartic Regression   Use quartic regression to model the data from Exercise S-11.

S-14. Quartic Regression   Use quartic regression to model the data from Exercise S-12.

S-15. Finding Poles   Find the poles of
$$\frac{x}{x^2 - 3x + 2}.$$

S-16. Finding Poles   Find the poles of
$$\frac{x + 1}{x^2 + 7x}.$$

S-17. Horizontal Asymptotes   Find all the horizontal asymptotes of
$$\frac{2x^2 + 1}{x^2 - 1}.$$

S-18. Horizontal Asymptotes   Find all the horizontal asymptotes of
$$\frac{6x^4 - 5}{2x^4 + x}.$$

---

[67]See S. I. Rabinow, *Introduction to Mathematical Biology* (New York: Wiley, 1975), 46–51.

CHAPTER 5   SUMMARY

Linear and exponential functions are among the most common in mathematical applications, but other functions occur often as well. These include *logistic functions*, *power functions*, *polynomial functions*, and *rational functions*. Also, these functions can be combined to form new functions.

5.1   LOGISTIC FUNCTIONS

The formula for a logistic function is $N = \dfrac{K}{1 + be^{-rt}}$ where K, r, and b are all positive.

Logistic functions are useful not only in describing population growth but also in many other situations where growth is initially rapid but later slows. Important features of the logistic function include:

- The constant K is known as the carrying capacity and is the limiting value of N.
- A logistic function has a point of inflection at $N = \dfrac{K}{2}$. This is the point of maximum increase. In the case of a renewable resource, it is the *optimum yield*. The value of N is always increasing. It is concave up before the inflection point and concave down after.

- The constant r is a characteristic of the population. In the absence of limiting factors, growth would be exponential according to the formula $N = N_0 e^{rt}$.

5.2   POWER FUNCTIONS

An exponential function is of the form Pa^x, where a is fixed and x is the variable. For *power functions*, these roles are reversed. Power functions are of the form $f(x) = cx^k$. The number k is the *power*, and c is $f(1)$. For a power function with c and x positive:

- If k is positive, then f is increasing.
- If k is negative, then f decreases toward zero.

One common example of a power function is the distance D, in feet, that a rock that is dropped will fall in t seconds if air resistance is ignored: $D = 16t^2$.

A key property of power functions that is often important in applications is the *homogeneity property*. It says that for a power function $f = cx^k$, if x is increased by a factor of t, then f is increased by a factor of t^k. For example, the volume V of a sphere is a power function of its radius: $V = \frac{4\pi}{3}r^3$. The homogeneity property tells us that if the radius of a sphere is doubled, the volume will be increased by a factor of 2^3. That is, the volume will be 8 times larger.

Although power functions with positive power increase, in the long term they do not increase as rapidly as exponential functions. The classic dominance of exponential functions over power functions is sometimes key to the choice of an appropriate model. Exponential functions should be used when eventual dramatic growth is anticipated.

5.3   MODELING DATA WITH POWER FUNCTIONS

We can use regression to model data with a power function just as we did for linear, exponential, and logistic models.

5.4 COMBINING AND DECOMPOSING FUNCTIONS

Often the individual pieces of a function have important meanings that help us understand the behavior of natural phenomena. This idea has been exploited in several forms throughout this text. For example, if a certain population is modeled by $N = Pa^t$, then P is the initial size of the population, and a is the growth or decay factor.

One of the most useful ways to combine functions is through *function composition*. This may involve substituting an expression for a variable into another formula. If the weight W of a certain animal is proportional to the cube of its length L, then $W = kL^3$. If the length is in turn a linear function of age t, then $L = at + b$. We can use function composition to express weight as a function of age:

$$W = kL^3 = k(at + b)^3.$$

This enables us to make a direct investigation of how age affects weight.

Sometimes functions are defined in a piecewise fashion. United States postal rates provide a prime example. At the time of printing of this text, first-class letters required 45 cents for the first ounce and 20 cents for each additional ounce up to 3 ounces total, the weight being rounded upward to the next whole number of ounces. Thus a 1.2-ounce letter cost $45 + 20 = 65$ cents to mail.

5.5 POLYNOMIALS AND RATIONAL FUNCTIONS

A *polynomial* function is one whose formula can be written as a sum of power functions where each of the powers is a non-negative whole number. Higher-degree[68] polynomials are interesting and useful. Their graphs may have many peaks and valleys, and they are useful in modeling phenomena with one or more maximum or minimum values.

A quadratic function is a polynomial where the highest power that occurs is 2. The formula for a quadratic function is $ax^2 + bx + c$. For example, if a rock is thrown downward with an initial velocity of 5 feet per second, the distance D in feet that it travels in t seconds is given by the quadratic function $D = 16t^2 + 5t$.

The graph of a quadratic function is a *parabola*, an important geometric shape because it is the path followed by any object ejected near the surface of the Earth. Quadratic regression may be used to fit a quadratic model to data.

In this text, we have made a practice of solving nonlinear equations using the graphing calculator. But there is a formula, known as the *quadratic formula*, that allows for hand solution of quadratic equations. For the quadratic equation $ax^2 + bx + c = 0$, the solution is given by

$$x = \frac{-b \pm \sqrt{b^2 - 4ac}}{2a}.$$

Cubic and quartic polynomials are useful models and can be fit to data using cubic or quartic regression.

Rational functions are quotients of polynomials. They occur, for example, in describing the flow of electric current across a resistor. Many rational functions have *poles*, which may occur where the denominator is zero. Rational functions grow rapidly near poles. The vertical and horizontal *asymptotes* of rational functions often have practical meaning.

---

[68]The degree is the largest power that occurs in the polynomial.

CHAPTER 5 REVIEW EXERCISES

Reminder Round all answers to two decimal places unless otherwise indicated.

1. **Logistic Model** A population grows according to the logistic model. The r value is 0.02, and the environmental carrying capacity is 2500. Write the logistic equation satisfied by the population if $N(0) = 100$.

2. **Logistic Formula** A population grows according to the logistic model

$$N = \frac{25}{1 + 0.5e^{-1.4t}},$$

where t is measured in years and N is measured in thousands.

   a. What is r for this population?

   b. What is the environmental carrying capacity K?

   c. This population is subject to harvesting. What is the optimum yield level?

3. **Inflection Point** A population is growing logistically, and its graph has an inflection point when the time is 3 years and the population level is 320. What is the carrying capacity?

4. **Homogeneity** Let $f(x) = cx^{3.2}$. If x is doubled, what happens to f?

5. **Power** Let $f(x) = cx^k$ be a power function such that $f(10)$ is twice the size of $f(1)$. What is the power k?

6. **Flow Rate** When fluid flows at a moderate velocity through a tube, the radius R of the tube is related to the flow rate F (measured as volume per unit time) by the formula

$$R = cF^{0.25}$$

for some constant c.

   a. A technician adjusts the radius of a tube and finds that the flow rate has doubled. By what factor has the radius changed?

   b. If the radius is tripled, what happens to the flow rate?

7. **Falling Object** An object is dropped near the surface of a planet. The time T, in seconds, it takes for the object to fall s feet is given by the formula

$$T = cs^{0.5}$$

for some constant c that depends on the planet.

   a. On Earth the constant c is 0.25. How long does it take an object to fall 16 feet on Earth?

   b. On a certain planet an object takes 5 seconds to fall 20 feet. What is the value of the constant c for this planet?

   c. If it takes 1 second for an object to fall 10 feet on a planet, how long does it take an object to fall 40 feet?

8. **Modeling Almost Power Data** Find a formula for a power function that models the following data.

x	1	2	3	4	5
f	3.50	1.52	0.94	0.66	0.51

9. **Gas Cost** The following table shows the distance D, in miles, that you can drive on \$1 worth of gas as a function of the cost g, in dollars, of a gallon of gas.

Cost g	1.65	1.75	1.85	1.95	2.05
Distance D	13.3	12.6	11.9	11.3	10.7

   a. Find a power model for the data.

   b. If the price of gas increases by 50%, what happens to the distance you can drive on \$1 worth of gas?

   c. What is your gas mileage (measured in miles per gallon)? (*Hint:* What distance can you drive if a gallon of gas costs \$1?)

10. **Falling Rocks** The following table shows the time T, in seconds, it takes for a rock to fall s feet on a certain planet.

Distance s	10	20	30	40	50
Time T	0.63	0.89	1.10	1.26	1.41

   a. Find a power model for the data.

   b. How long does it take for a rock to fall 70 feet?

   c. How far does a rock fall in 2 seconds?

   d. What effect does doubling the distance fallen have on the time?

11. **Composing Functions** Use a formula to express y as a function of t if $y = 3x^2 + 5x$ and $x = t - 1$.

(*continued*)

12. Decomposing Functions   The balance in a bank account is given by the formula $B = 120 \times 1.02^t$. Here B is the balance in dollars t years since the start of 2012. What was the balance at the start of 2012? What is the annual interest rate?

13. Population Growth   Here is one model of the per capita growth rate R, measured per year, as a function of population size N: For values of N between 0 and 100, R takes the constant value 0.01. For values of N greater than 100 and less than 200, R is given by the expression

$$R = 0.02\left(1 - \frac{N}{200}\right).$$

a. Make a graph of R as a function of N. Use a horizontal span of 0 to 200.

b. What is the per capita growth rate at a population level of 150?

14. Volume   A spherical balloon is being inflated with air. The radius R, in inches, of the balloon is a function of the volume V, in cubic inches, of air in the balloon. The formula is

$$R = \left(\frac{3V}{4\pi}\right)^{1/3}.$$

The volume V is a function of time t, in seconds, and the formula is

$$V = 1.5 + 0.1t.$$

a. Use function composition to find a formula for R as a function of t.

b. What is the radius at time $t = 2$ seconds?

15. Quadratic Formula   Use the quadratic formula to solve $2x^2 - x - 1 = 0$.

16. Quadratic Regression   Use quadratic regression to find a model for the following data set.

x	1	4	5	7	8
$f(x)$	0	-30	-52	-114	-154

17. Chemical Reaction   The following table shows, for a certain chemical reaction, the rate of reaction R, in moles per cubic meter per second, as a function of the concentration x, in moles per cubic meter, of the product.

Concentration x	10	20	30	40	50
Reaction rate R	18	12	7	3	0

a. Use quadratic regression to find a model for the data. Round regression parameters to three decimal places.

b. Use your model to estimate $R(24)$, and explain what your answer means.

c. Estimate the concentration at which the reaction rate is 6 moles per cubic meter per second. (Consider concentrations only up to a level of 50 moles per cubic meter.)

18. Cubic Regression   Use cubic regression to model the following data.

x	-2	-1	0	1	2
y	1.4	2.7	3.0	2.9	3.0

19. Finding Poles   Find the poles of

$$\frac{2x - 5}{x^2 + 4x + 3}.$$

20. Expanding Balloon   A spherical balloon is being inflated. The following table shows how the volume V, in cubic inches, depends on time t, in seconds.

Time t	0	1	2	3	4
Volume V	4.19	7.24	11.49	17.16	24.43

a. It turns out that the radius is a linear function of time. Because the volume of a sphere is a cubic function of the radius, a cubic model for the data is appropriate. Use cubic regression to find a model for the data.

b. Express, using functional notation, the volume after 5 seconds, and use your model to estimate that value.

21. Travel Time   The time T, in hours, required to drive 100 miles is a function of the average speed s, in miles per hour. The formula is

$$T = \frac{100}{s}.$$

a. Make a graph T versus s covering speeds up to 70 miles per hour.

b. Calculate $T(25)$ and explain in practical terms what your answer means.

c. Explain in practical terms the behavior of the graph near the pole at $s = 0$.

A FURTHER LOOK

Fitting Logistic Data Using Rates of Change

We discussed earlier the process of fitting logistic data by means of logistic regression. Here we present an alternative approach that is readily explained in terms of rates of change for logistic functions. For ease of description we consider the case of population growth.

For exponential growth we assume that the population changes by a constant percentage r. For logistic growth the percentage growth rate starts at the r value. But as the population increases toward the carrying capacity K, the percentage growth rate decreases to 0. For logistic growth, the graph of percentage growth rate as a function of population is a straight line, as shown in Figure 5.82. This model assumes that each individual added to the population decreases the percentage growth rate by an equal amount, until this rate is 0 at the carrying capacity.

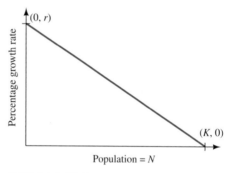

FIGURE 5.82 Percentage growth rate as a function of population size

The line in Figure 5.82 has vertical intercept r and horizontal intercept K, so the slope is

$$\frac{r - 0}{0 - K} = -\frac{r}{K}.$$

Thus, if we write G for the percentage growth rate as a decimal, the equation of the line in Figure 5.82 is

$$G = -\frac{r}{K}N + r.$$

This equation gives us a way of fitting data with a logistic model when populations and percentage growth rates are given. We use linear regression to fit the data for percentage growth rates in terms of population. The r value is the vertical intercept of this line, and if we call its slope m, then we solve the equation $m = -r/K$ for K to find $K = -r/m$. The method is shown in the following example.

387

EXAMPLE 5.18   **LOGISTIC GROWTH FOR THE WATER FLEA**

The following table gives population sizes and corresponding percentage growth rates G per day (in decimal form) for the water flea with limited resources.[69] Here, N is measured as number per cubic centimeter.

N	1	2	4	8	16
G	0.230	0.216	0.208	0.145	0.057

Part 1 Plot the given data points to show that a linear model for G versus N is appropriate for this population.

Part 2 Use the data to estimate the r value and carrying capacity K for the water flea in this environment. Round the slope of the regression line for G versus N to three decimal places and the vertical intercept to two decimal places.

Part 3 Make a logistic model for the water flea under the assumption that $N(0) = 1$.

Solution to Part 1 The plot is shown in Figure 5.83. Because this population is growing with limited resources, we expect it to exhibit logistic growth. This means that G should be modeled as a linear function of N, and the plot confirms that a linear model is appropriate.

Solution to Part 2 We calculate the regression line, with the result shown in Figure 5.84. This corresponds to the equation

$$G = -0.012N + 0.24.$$

We can see that the vertical intercept of the regression line is about 0.24, our estimate for r. Also the slope of the regression line is about -0.012, so by the preceding discussion this gives the estimate $K = -\frac{0.24}{-0.012} = 20$. Thus, for the water flea in this environment, r is about 0.24 per day, and the carrying capacity K is about 20 individuals per cubic centimeter.

Solution to Part 3 To find the logistic formula, we recall that

$$b = \frac{K}{N(0)} - 1 = \frac{20}{1} - 1 = 19.$$

Thus if t denotes time in days, we have the logistic model

$$N = \frac{K}{1 + be^{-rt}} = \frac{20}{1 + 19e^{-0.24t}}.$$

Note: To use this procedure we need to know the per capita growth rates for a sample of population sizes. In practice, these rates can be difficult to obtain. One way to estimate the percentage growth rate as a decimal is

$$\text{Percentage growth rate} = \frac{\text{Average rate of change}}{\text{Population}}.$$

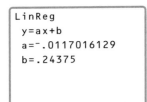

FIGURE 5.83 Data for percentage growth rate as a function of population size

```
LinReg
 y=ax+b
 a=⁻.0117016129
 b=.24375
```

FIGURE 5.84 Equation of the regression line

[69]The data are taken from P. W. Frank, C. D. Boll, and R. W. Kelly, in "Vital statistics of laboratory cultures of *Daphnia pulex* DeGeer as related to density," *Physiol. Zool.* **30** (1957), 287–305.

> **KEY IDEA 5.6** **LOGISTIC RATE OF CHANGE**
>
> Assume that population growth is logistic. If N denotes the population size and G the percentage growth rate (as a decimal), then a linear model for G in terms of N is appropriate. The linear model is
>
> $$G = -\frac{r}{K}N + r.$$
>
> Thus the r value is the vertical intercept of this line, and we can find the carrying capacity K using $K = -r/m$. Here m is the slope of the line.

EXERCISES

Reminder Round all answers to two decimal places unless otherwise indicated.

Using Logistic Regression In Exercises 1 and 2, you are given data for the percentage growth rate G per year (in decimal form) and the population size N. Use linear regression to find a model for G in terms of N, and then find the r value and the carrying capacity for the population. Round the slope of the regression line to three decimal places and the vertical intercept to two decimal places.

1.

N	2	3	5	6	7	9
G	0.206	0.204	0.201	0.197	0.195	0.192

2.

N	1	3	4	7	8	10
G	0.408	0.405	0.401	0.395	0.394	0.390

3. Negative Growth Rates The original water flea study described in Example 5.18 also included growth from population sizes of $N = 24$ and $N = 32$.

  a. Use the regression line computed in Example 5.18 to estimate the percentage growth rates (in decimal form) for these two values of N.

  b. Explain in terms of the carrying capacity why the growth rates from part a should both be negative.

4. Logistic Rate of Change Suppose that in a logistic model the percentage growth rate G per year (in decimal form) as a function of N declines from $G = 0.13$ when $N = 0$ to $G = 0$ when $N = 378$.

  a. What is the r value?

  b. What is the carrying capacity?

  c. At what value of N will there be a maximum growth rate?

5. World Population from Two Points One source states that the world population growth rate has steadily declined from 2.19% per year in 1963 to 1.14% per year in 2000. The population in 1963 was 3.21 billion, and the population in 2000 was 6.08 billion.

  a. Assume that the world population exhibits logistic growth, and use the information given above to find a linear formula for the percentage growth rate G per year (in decimal form) as a function of N, the population in billions. Round the slope of the line to four decimal places and the vertical intercept to three decimal places. [*Suggestion:* The information above says that the points $(3.21, 0.0219)$ and $(6.08, 0.0114)$ are on the line.]

  b. Use part a to find the r value for humans on planet Earth.

  c. Use part a to find the carrying capacity for humans on planet Earth. *Note:* This represents a rather naive analysis of world population.

(*continued*)

6. **Maximum Growth Rate for the Logistic Model** We have stated that when logistic growth is occurring, the maximum growth rate occurs at half of the carrying capacity K. The goal of this exercise is to justify that statement.

   a. When logistic growth is occurring, the percentage growth rate G (in decimal form) of the population size N is given by

   $$G = -\frac{r}{K}N + r.$$

   (See Key Idea 5.6.) In general, we calculate the percentage growth rate of N as a decimal using

   $$\text{Percentage growth rate} = \frac{\text{Growth rate}}{N}.$$

   Combine these results to show that

   $$\text{Growth rate} = -\frac{r}{K}N^2 + rN.$$

   b. At what values of N is the growth rate 0?

   c. It is a fact that the vertex of the parabola $y = ax^2 + bx + c$ occurs at $x = -b/(2a)$. Also, when a is negative the vertex represents a maximum of the function. Apply these facts in part a to show that the maximum growth rate occurs at $N = K/2$.

7. **Maximum Growth Rate for Tuna** *This is a continuation of Exercise 6.* For Eastern Pacific yellowfin tuna, it is found that $r = 2.61$ per year and $K = 148$ thousand tons.

   a. Use the given values of r and K and the formula from part a of Exercise 6 to find a formula for the growth rate for yellowfin tuna. Use five decimal places.

   b. Use your calculator to make a graph of the growth rate in terms of N.

   c. Use your graph in part b to find the value of N at which the growth rate is a maximum.

   d. How does your answer from part c compare with $K/2$?

8. **Maximum Growth Rate** *This is a continuation of Exercise 6.* In Exercise 6 we have seen that under the logistic model

   $$\text{Growth rate} = -\frac{r}{K}N^2 + rN$$

   and that the growth rate is at its maximum when $N = K/2$.

   a. Use these formulas to express this maximum growth rate in terms of r and K.

   b. Use your answer to part a to find the maximum growth rate for the Pacific sardine population of Example 5.1 in Section 5.1.

A FURTHER LOOK

Factoring Polynomials, Behavior at Infinity

The *Fundamental Theorem of Algebra* is one of the most important theorems in elementary mathematics, and it tells us that every polynomial has at least one (possibly complex) zero. This theorem was proved in about 1800 by C. F. Gauss, and it enables us to draw important conclusions about the structure of polynomials. Another important (though much easier) theorem is the *factor theorem*, which says that a is a zero of a polynomial if and only if $x - a$ is a factor. For example, it is easy to check that $x = 1$ is a zero of $x^3 - x^2 - 4x + 4$. The factor theorem tells us then that $x - 1$ is a factor of $x^3 - x^2 - 4x + 4$. Indeed, you can check by direct multiplication that

$$x^3 - x^2 - 4x + 4 = (x - 1)(x^2 - 4).$$

Using this factorization, we can find the remaining zeros. Note that $x^2 - 4 = 0$ when $x = 2$ or $x = -2$. Thus the zeros of $x^3 - x^2 - 4x + 4$ are 1, 2, and -2, and we can check that $x^3 - x^2 - 4x + 4 = (x - 1)(x - 2)(x + 2)$.

This example suggests a method for finding all the zeros of a polynomial and for writing it in factored form.

Step 1: Find a zero, say a_1, of $P(x)$.
Step 2: Factor $P(x)$ as $P(x) = (x - a_1)P_1(x)$.
Step 3: Find a zero a_2 of $P_1(x)$.
Step 4: Factor $P_1(x)$ as $P_1(x) = (x - a_2)P_2(x)$.

We continue this process until we have $P(x)$ fully factored:

$$P(x) = k(x - a_1)(x - a_2) \cdots$$

Unfortunately, for high-degree polynomials there is no practical way of carrying out this procedure. In general, one can find approximate zeros but not exact zeros. Nonetheless, this procedure reveals important information about polynomials. We see immediately, for example, that a polynomial of degree n has exactly n linear factors (possibly repeated) and at most n zeros. At first glance, it also tells us that a polynomial of degree n has exactly n real zeros, but brief reflection shows that there are difficulties with this. Remember that some of the zeros a_i may be complex numbers, and it is possible that they are not distinct. For example, let's apply this process to $x^2 - 2x + 1$. Now $x = 1$ is a zero, and the polynomial factors as $(x - 1)(x - 1)$. In this case, 1 is said to be a *repeated zero* of $x^2 - 2x + 1$, and even though this is a second-degree polynomial, it has only one zero. In the labeling procedure above, we would have $a_1 = 1$ and $a_2 = 1$. When there are repeated zeros, a polynomial of degree n will have fewer than n distinct zeros. In the case of $x^2 + 1$, the two linear factors involve complex numbers: $x^2 + 1 = (x - i)(x + i)$. Thus this is an example of a polynomial with no real zeros. It turns out that for real polynomials, the complex zeros always occur just like this—that is, as zeros of a quadratic factor with real coefficients. We summarize our discussion with the following result.

Decomposition theorem: If $P(x)$ is a polynomial with real coefficients, then $P(x)$ can be factored into a product of linear and quadratic factors with real coefficients.

Once again, although the decomposition theorem is extremely important, there is no practical way of actually finding the factors for high-degree polynomials.

In general, we expect the graph of a polynomial of degree n to cross the horizontal axis at most n times and to have at most $n - 1$ maxima or minima. Figure 5.85 shows the graph of a typical cubic, and Figure 5.86 shows the graph of a typical quartic.

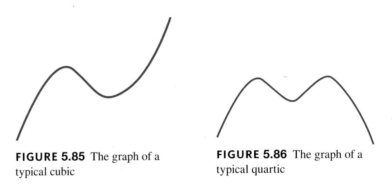

FIGURE 5.85 The graph of a typical cubic

FIGURE 5.86 The graph of a typical quartic

The graph of a polynomial may go up and down for a while, but eventually it goes up to infinity or down to minus infinity. In fact, for long-term behavior, only the leading, or highest-degree, term matters. To explain better what we mean, let's look at $x^3 + 10x^2 + 1$. It has leading term x^3. In Figure 5.87 we have graphed both x^3 and $x^3 + 10x^2 + 1$. The lighter graph is x^3. The horizontal span is -10 to 5, and the vertical span is -150 to 200. The graphs appear very different. In Figure 5.88 we have changed the horizontal span to 0 to 30 and the vertical span to 0 to 10,000. In this window the graphs have the same shape, but the graph of $x^3 + 10x^2 + 1$ is above the graph of x^3.

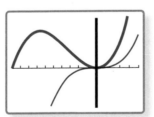

FIGURE 5.87 Comparing graphs on a small span

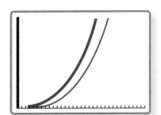

FIGURE 5.88 Comparing graphs on a wider span

In Figure 5.89 on the next page, we have changed the window to a horizontal span of 0 to 100 and a vertical span of 0 to 1,000,000. The graphs appear to be practically the same. As another comparison, we have in Figure 5.90 plotted the graph of the quotient $(x^3 + 10x^2 + 1)/x^3$. The horizontal span is 0 to 2100, and the vertical span is 0 to 2. We have traced the graph, and we see from the prompt at the bottom of the screen that the limiting value of the quotient appears to be 1. This is indeed the case (recalling the notation from "A Further Look: Limits" in Chapter 2):

$$\lim_{x \to \infty} \frac{x^3 + 10x^2 + 1}{x^3} = 1.$$

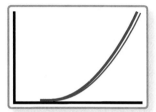

FIGURE 5.89 Comparing graphs on a much wider span

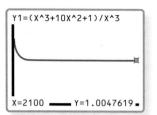

FIGURE 5.90 Examining the quotient

It is a fact that the limiting behavior of polynomials depends so strongly on the leading coefficient that when we are calculating the limiting value of a rational function, we need consider only leading coefficients: If $a_n, b_k \neq 0$, then

$$\lim_{x \to \infty} \frac{a_n x^n + a_{n-1} x^{n-1} + \cdots}{b_k x^k + b_{k-1} x^{k-1} + \cdots} = \lim_{x \to \infty} \frac{a_n x^n}{b_k x^k}.$$

This makes limiting values of rational functions very easy to calculate. We simply delete all but the leading terms. Let's look at some examples.

$$\lim_{x \to \infty} \frac{4x^3 - 7x^2 + 13}{2x^3 + x - 8} = \lim_{x \to \infty} \frac{4x^3}{2x^3} = \lim_{x \to \infty} 2 = 2.$$

$$\lim_{x \to \infty} \frac{4x + 13}{x^3 + 1} = \lim_{x \to \infty} \frac{4x}{x^3} = \lim_{x \to \infty} \frac{4}{x^2} = 0.$$

EXERCISES

Reminder   Round all answers to two decimal places unless otherwise indicated.

1. **Finding the Degree of a Polynomial**   Figure 5.91 shows the graph of a polynomial that has no complex zeros and no repeated zeros. What is the degree of the polynomial?

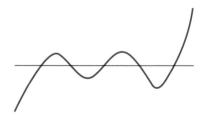

FIGURE 5.91 A polynomial with no complex zeros and no repeated zeros

2. **Choosing a Model**   Suppose you have a data set that shows three maxima and two minima. What is the degree of the smallest-degree polynomial that could possibly fit the data exactly?

Finding Polynomials   In Exercises 3 through 6, you are to find a polynomial that meets the given criteria.

3. A polynomial of degree 3 has zeros 2, 3, and 4. The leading coefficient is 1. Find the polynomial.

4. A polynomial of degree 3 has zeros 2, -2, and 0. The leading coefficient is 3. Find the polynomial.

5. $x = 1$ is the only zero of a polynomial of degree 5, and the leading coefficient is 1. Find such a polynomial.

6. $x = 0$ is the only zero of a polynomial of degree 10, and the leading coefficient is 5. Find such a polynomial.

Calculating Limits   In Exercises 7 through 10, calculate the given limit.

7. Calculate $\lim\limits_{x \to \infty} \dfrac{5x^4 + 4x^3 + 7}{2x^4 - x^2 + 4}$.

8. Calculate $\lim\limits_{x \to \infty} \dfrac{6x^3 + 4x^2 + 5}{3x^3 - 5x + 14}$.

(continued)

9. Calculate $\lim\limits_{x \to \infty} \dfrac{3x^4 + 4x^2 - 9}{2x^5 + 4x^3 - 8}$.

10. Calculate $\lim\limits_{x \to \infty} \dfrac{ax^5 + bx^2 + c}{dx^5 - ex^2 + f}$, $d \neq 0$.

Getting a Polynomial from Points   A polynomial of degree n is determined by $n + 1$ points. For example, a quadratic is determined by three points. In Exercises 11 through 14, find the quadratics that pass through the given points.

11. Find the quadratic determined by $(1, 5)$, $(-1, 3)$, and $(2, 9)$. (*Suggestion:* We are looking for a quadratic $ax^2 + bx + c$. We need to find a, b, and c.) The fact that the quadratic passes through the three points gives the following:

$$a + b + c = 5$$
$$a - b + c = 3$$
$$4a + 2b + c = 9.$$

Solve this system of equations for a, b, and c.

12. Find the quadratic determined by $(1, 5)$, $(2, 4)$, and $(3, 19)$. (See the suggestion for Exercise 11.)

13. Find the quadratic determined by $(1, -2)$, $(0, -5)$, and $(-1, -2)$. (See the suggestion for Exercise 11.)

14. Find the quadratic determined by $(-1, 1)$, $(0, 0)$, and $(1, 1)$. (See the suggestion for Exercise 11.)

15. Lagrangian Polynomials   The polynomials

$$P_1(x) = \frac{(x - x_2)(x - x_3)}{(x_1 - x_2)(x_1 - x_3)}$$

$$P_2(x) = \frac{(x - x_1)(x - x_3)}{(x_2 - x_1)(x_2 - x_3)}$$

$$P_3(x) = \frac{(x - x_1)(x - x_2)}{(x_3 - x_1)(x_3 - x_2)}$$

are known as *Lagrangian polynomials*. Here x_1, x_2, and x_3 are given numbers.

a. Show that if $i \neq j$, then $P_i(x_j) = 0$.

b. Show that $P_i(x_i) = 1$.

c. Show that $y_1P_1 + y_2P_2 + y_3P_3$ passes through the points (x_1, y_1), (x_2, y_2), and (x_3, y_3).

d. Use Lagrangian polynomials to find the quadratic that passes through the points $(1, 2)$, $(2, 5)$, and $(3, 6)$.

RATES OF CHANGE

Rates of change can be used to model sardine populations subject to fishing. See Example 6.10 on page 432.

6.1 VELOCITY

6.2 RATES OF CHANGE FOR OTHER FUNCTIONS

6.3 ESTIMATING RATES OF CHANGE

6.4 EQUATIONS OF CHANGE: LINEAR AND EXPONENTIAL FUNCTIONS

6.5 EQUATIONS OF CHANGE: GRAPHICAL SOLUTIONS

▪ Summary

▪ Chapter Review Exercises

A KEY IDEA in analyzing natural phenomena as well as the functions that may describe them is the *rate of change*, and it is a familiar idea from everyday experience. When you are driving your car, the rate of change in your location is *velocity*. Discussions of cars or airplanes commonly involve velocity because it is virtually impossible to convey key ideas about motion without reference to velocity. The rate of change is no less descriptive for other events. If you step on the gas pedal to pass a slowly moving truck, then your velocity itself changes, and the rate of change in velocity is *acceleration*. Rates of change occur in other contexts as well. In fact the idea is pervasive in mathematics, science, engineering, social science, and daily life because it is such a powerful tool for description and analysis.

Student resources are available on the website **www.cengagebrain.com**

6.1 VELOCITY

We look at velocity first because it is a familiar rate of change. Consider a rock tossed upward from ground level. As the rock rises and then falls back to Earth, we locate its position as the distance up from the ground. In Figure 6.1 we have sketched a possible graph of distance up versus time. It shows the height of the rock increasing as it rises and then decreasing after it reaches its peak and begins to fall.

■ | **Getting Velocity from Directed Distance**

The *velocity* of the rock is the rate of change in distance up from the ground. That is, at any point in the flight of the rock, the velocity measures how fast the rock is rising or falling. The rock gets some initial velocity at the moment of the toss, but the effect of gravity makes it slow down as it rises toward its peak. After the rock reaches its peak, gravity causes it to accelerate toward the ground, and its *speed* increases. In everyday language, the terms *speed* and *velocity* are often used interchangeably, but there is an important, if subtle, difference. Speed is always a positive number—the number you might read on the speedometer of your car, for example. But velocity has an additional component; it has a sign (positive or negative) attached that indicates the direction of movement. The key to understanding this is to remember that velocity is the rate of change in *directed distance*. As the rock moves upward, its distance up from the ground is increasing. Thus the rate of change in distance up, the velocity, is positive. But when the rock starts to fall back to Earth, its distance up from the Earth is decreasing. Thus the rate of change in distance up, the velocity, is negative.

The relationships between the graph of distance up and the graph of velocity are crucial to understanding velocity and, indeed, rates of change in general. In Figure 6.2 we have sketched a graph of the velocity versus time for the rock. During the period when the rock is moving upward, the graph of distance up is increasing. Since distance up is increasing, velocity is positive. This is shown in Figure 6.2 by the fact that the graph of velocity is above the horizontal axis until the rock reaches the peak of its flight. We also note that during this period, the graph of velocity is decreasing toward zero, indicating that the rock is slowing as it rises. At the point where the rock reaches its peak, the graph of velocity crosses the horizontal axis, indicating that the velocity is momentarily zero. When the rock is moving downward, the graph of distance up is decreasing. During this period, velocity is negative, and this is shown in Figure 6.2 by the fact that the graph of velocity is below the horizontal axis. Note that the *speed* of

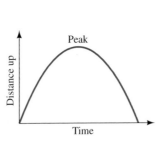

FIGURE 6.1 Distance up of a rock versus time

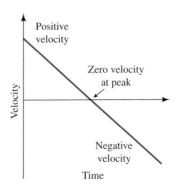

FIGURE 6.2 Velocity of a rock versus time

the rock is increasing as it falls, so its velocity (the negative of its speed) is decreasing. Velocity continues to decrease until the rock strikes the ground.

In Figure 6.2 we have represented the graph of velocity as a straight line. This is not an apparent consequence of the graph in Figure 6.1, but as we shall discover later (see Section 6.4 and also Exercise 9 of Section 6.2), it is a consequence of the fact that near the surface of the Earth, gravity imparts a constant acceleration. Finally, we note that the sign of velocity depends on the perspective chosen to measure position. In this case, we located the rock using its distance up from the surface of the Earth. In Exercise 2 at the end of this section, you will be asked to analyze the velocity of the rock when the event is viewed from the top of a tall building and the rock's position is considered as its distance down from the top of the building.

The relationships we observed between the graph of distance up for the rock and the graph of its velocity are fundamental, and they are true in a general setting. When directed distance is increasing, velocity is positive. When directed distance is decreasing, velocity is negative. When directed distance is not changing, even momentarily, velocity is zero. In particular, when directed distance reaches a peak (maximum) or a valley (minimum), velocity is zero.

> **KEY IDEA 6.1** | **VELOCITY AND DIRECTED DISTANCE: THE FUNDAMENTAL RELATIONSHIP**
>
> **1.** Velocity is the rate of change in directed distance.
>
> **2.** **When directed distance is increasing, velocity is positive.** (The graph of velocity is above the horizontal axis.)
>
> **3.** **When directed distance is decreasing, velocity is negative.** (The graph of velocity is below the horizontal axis.)
>
> **4.** **When directed distance is not changing, velocity is zero.** (The graph of velocity is on the horizontal axis.)

Constant Velocity Means Linear Directed Distance

In the case of the rock we just looked at, velocity is always changing, but in many situations there are periods when the velocity does not change. A familiar example is that of a car that might accelerate from a yield sign onto a freeway and travel at the same speed for a time before exiting the freeway and parking at its destination. If the car is traveling westward on the freeway, and we locate its position as distance west from the yield sign, a possible graph of its velocity, the rate of change in distance west, is shown in Figure 6.3. We want to consider what this means for the function $L = L(t)$, which gives the location of the car at time t as distance west of the yield sign. Our interest is in determining the nature of L during the period when the car is traveling at a constant velocity on the freeway—say, at 65 miles per hour. During this period, each hour the car moves 65 additional miles. This means that the distance L increases by 65 miles each hour, so it is a linear function with slope 65.

In more general terms, since velocity of the car is the rate of change in L, during this period the rate of change in L is constant. But we know that a function with a constant rate of change is linear, and this is the key observation we wish to make. Whenever velocity, the rate of change in directed distance, is constant, directed distance must be a

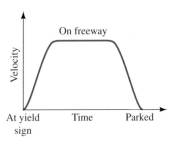

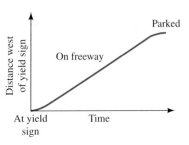

FIGURE 6.3 The velocity of a car traveling on a freeway

FIGURE 6.4 Constant velocity means linear directed distance

linear function. Furthermore, the slope of L is the velocity. The linearity of L is reflected in Figure 6.4 by the fact that for the portion of the graph that represents the time when the car was on the freeway, we drew the graph of directed distance as a straight line.

> **KEY IDEA 6.2 DIRECTED DISTANCE FOR CONSTANT VELOCITY**
>
> When velocity is constant, the rate of change in directed distance is constant. Thus directed distance is a linear function with slope equal to the constant velocity, so its graph is a straight line.

EXAMPLE 6.1 FROM NEW YORK TO MIAMI

An airplane leaves Kennedy Airport in New York and flies to Miami, where it is serviced and receives new passengers before returning to New York. Assume that the trip is uneventful and that after each takeoff the airplane accelerates to its standard cruising speed, which it maintains until it decelerates prior to landing.

Part 1 Describe what the graph of distance south of New York looks like during the period when the airplane is maintaining its standard cruising speed on the way to Miami.

Part 2 Say we locate the airplane in terms of its distance south of New York. Make possible graphs of its distance south of New York versus time and of the velocity of the airplane versus time.

Part 3 Say we locate the airplane in terms of its distance north of Miami. Make possible graphs of its distance north of Miami versus time and of the velocity of the airplane versus time.

Solution to Part 1 To say that the airplane is maintaining its standard cruising speed means that its velocity is not changing. In other words, the rate of change in distance south is constant. As we noted above, any function with a constant rate of change is linear. We conclude that during this period, the graph of distance south is a straight line whose slope is the standard cruising speed.

Solution to Part 2 We will make the graph of distance south in several steps, creating a template for solving problems of this type.

Step 1 *Locate and mark the places on the graph where directed distance is zero and where it reaches its extremes.* The distance south of New York is zero at the beginning and end of the trip. These points will lie on the horizontal axis, and we have marked and labeled them in Figure 6.5. The graph of distance south will be at its maximum while the airplane is being serviced in Miami. This is also marked and labeled in Figure 6.5.

Step 2 *Label on the graph the regions where directed distance is increasing and the regions where it is decreasing.* Directed distance is increasing on the trip from New York to Miami and is decreasing on the return leg of the trip. These regions are labeled in italics in Figure 6.6.

Step 3 *Complete the graph, incorporating any additional information known about directed distance.* In this case, the important additional information is that for most of the trip, the airplane is flying at its standard cruising speed. As we observed in part 1, this means that the graph of distance during these periods is a straight line. Our graph of distance south of New York for the airplane is shown in Figure 6.7. Note that in the completed graph, we have labeled the horizontal and vertical axes as well as other important features. You are encouraged to follow this practice.

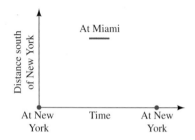

FIGURE 6.5 Zeros and extremities of distance south

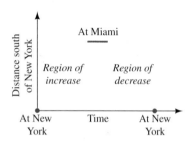

FIGURE 6.6 Regions of increase and decrease for distance south

FIGURE 6.7 The completed graph of distance south of New York

Now let's consider velocity for the same trip. To make the graph, we use steps similar to those we used to graph distance south.

Step 1 *Locate and label the points on the graph where the velocity is zero.* The velocity is zero when the airplane is on the ground at the beginning and end of the trip and during the period when it is being serviced in Miami. These times must lie on the horizontal axis and are so marked and labeled in Figure 6.8. For the graph of velocity, we put the horizontal axis in the middle since we expect to graph both positive and negative values.

Step 2 *Label the regions where velocity is positive and where it is negative.* On the flight from New York to Miami, distance south of New York is increasing, so its rate of change, or velocity, is positive. On the return leg, distance south of New York is decreasing, so its rate of change, or velocity, is negative. These regions are labeled in italics in Figure 6.9. At this point, it is important to check to be sure that the places in Figure 6.6 we marked as *increasing* match the places in Figure 6.9 that we marked as *positive*. Similarly, the *decreasing* labels in Figure 6.6 must match the *negative* labels in Figure 6.9.

Step 3 *Complete the graph, incorporating any other known features of the graph.* For velocity, the important additional feature is that during most of both legs of the trip, the airplane is maintaining a constant cruising speed. That means that the graph of velocity must be horizontal in these regions. This is shown in our completed graph in Figure 6.10. Note once again that we have labeled the axes as well as other important features of the graph.

We want to emphasize that Figure 6.7 and Figure 6.10 show the fundamental relationship between graphs of directed distance and velocity that we noted earlier. Observe in particular that during the flight from New York to Miami, the graph of directed distance

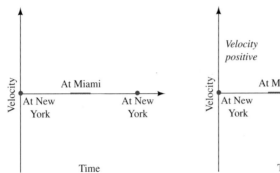

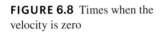

FIGURE 6.8 Times when the velocity is zero

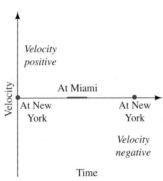

FIGURE 6.9 Labeling regions of positive and negative velocity

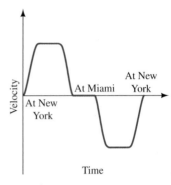

FIGURE 6.10 The completed graph of velocity

in Figure 6.7 is increasing, and the corresponding part of the graph of velocity in Figure 6.10 is above the horizontal axis. Also, during the return leg of the trip, the graph of directed distance in Figure 6.7 is decreasing, and the corresponding graph of velocity in Figure 6.10 is below the horizontal axis.

Solution to Part 3 We want to look now at the airplane flight from the perspective of a Miami resident. We will follow the same steps as we did in part 2 to get the graphs. If you are waiting at the Miami airport, the distance north to the airplane is the same large positive number at the beginning and end of the trip, when the plane is in New York. These points are marked and labeled in Figure 6.11. The distance north is zero while the airplane is in Miami, and this region is also marked in Figure 6.11.

As the plane flies from New York toward Miami, the distance north from Miami decreases, and this is noted in italics in Figure 6.11. On the return leg, distance north increases, and this is also noted in italics in Figure 6.11. Following the notes in Figure 6.11, we complete the graph in Figure 6.12, showing (as before) the regions where the graph is a straight line.

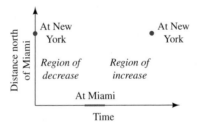

FIGURE 6.11 Interesting points and regions of increase and decrease for distance north

FIGURE 6.12 The completed graph of location from the Miami perspective

We proceed similarly to get the graph of velocity. We first mark the places, New York and Miami, where the velocity is zero. Our change in perspective does not affect this, so the appropriate picture is the same as Figure 6.8. But the change in perspective does affect the sign of velocity. On the first leg of the flight, distance north of Miami is decreasing, so the velocity is negative. This is marked in italics in Figure 6.13, and we note that it corresponds to the *decreasing* label in Figure 6.11. Similarly, on the return trip, distance north of Miami is increasing, so the velocity is positive. This is marked in italics in Figure 6.13, and we note (as before) that it corresponds to the *increasing* label in Figure 6.11. We complete the picture in Figure 6.14, being careful to incorporate constant cruising speed.

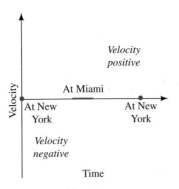

FIGURE 6.13 Labeling velocity features from the Miami perspective

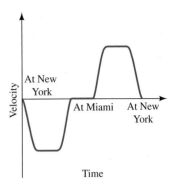

FIGURE 6.14 The completed graph of velocity from the Miami perspective

---

EXAMPLE 6.2   **A FORGOTTEN WALLET**

A man leaves home driving west on a straight road. We locate the position of the car as the distance west from home. Suppose the man begins his trip and sets his cruise control, but after a few minutes he notices that he has forgotten his wallet. He returns home, gets his wallet, and then resumes his journey. He figures that he may be late, so he sets the cruise control higher than before. Make graphs of distance west versus time and of velocity versus time for the forgetful man's car.

Solution We first make the graph showing the location of the car by following the steps outlined in Example 6.1. The distance west from home is zero when the trip starts and when the man is back home retrieving his wallet. In Figure 6.15 we have marked these places, as well as the point where the man remembered the forgotten wallet and the car turned around and headed home. Distance west is increasing when the car is moving away from home. This occurs at the beginning of the trip and also later, after the wallet has been retrieved. The only period when distance west is decreasing is when the man is returning home to get the wallet. These regions are marked in italics in Figure 6.15.

To get the completed graph in Figure 6.16, we made use of the notes from Figure 6.15 and also incorporated the information we have about cruise control settings: On the first leg of the trip, the cruise control is set, so the velocity is constant. This means that for this period, distance west is a linear function, so we have drawn the graph there as a straight

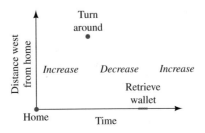

FIGURE 6.15 Important features in the location of a car

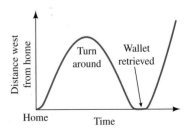

FIGURE 6.16 The completed graph for a car trip

line. We have also made the graph for the trip back home a straight line, although that information is not provided in the problem. For the final leg of the trip, we know once again that the graph is a straight line, but we also know a little more. The cruise control is set higher than it was on the first leg of the trip, and therefore the velocity is greater. Thus we must draw a line with a larger slope (a steeper line) than we drew for the first part of the trip.

We proceed with a similar analysis for velocity. We know that the velocity is zero when the trip starts and when the wallet is being retrieved. But the velocity is also zero when the car turns around. At this point, just as we saw at the peak of the rock's flight, the rate of change in distance west is momentarily zero. We have marked and labeled all of these places in Figure 6.17.

We can get the right regions for positive and negative velocity from the increasing and decreasing labels in Figure 6.15. Each *increase* label in Figure 6.15 corresponds to a *positive velocity* label in Figure 6.17, and the *decrease* label corresponds to the *negative velocity* label in Figure 6.17. We use the notes in Figure 6.17 to get the completed graph of velocity in Figure 6.18, keeping in mind that since the cruise control is set for most of the trip, the velocity is constant most of the time. This means that the graph of velocity is horizontal in these regions. Finally, we note that since the car went faster on the last part of the trip, the graph of velocity is higher there.

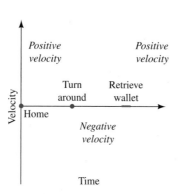

FIGURE 6.17 Important features for velocity

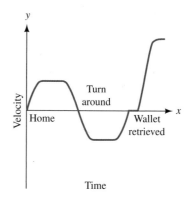

FIGURE 6.18 The completed graph for the velocity of the car

When Distance Is Given by a Formula

Many times, directed distance, or velocity, is given by a formula. For example, if we stand atop a building 30 feet high and toss a rock upward with an initial velocity of 18 feet per second, then elementary physics can be used to show that the distance up $D = D(t)$ from the ground of the rock t seconds after it is tossed is given by

$$D = 30 + 18t - 16t^2 \text{ feet.}$$

We are assuming that when the rock comes back down, it does not hit the top of the building where we are standing but, rather, falls all the way to the ground. Let's begin our analysis by making a graph of D versus t. We can use our everyday experience to choose a window setting. The rock surely won't go over 50 feet high, and it will take only a few seconds for the rock to hit the ground. Thus in Figure 6.19, which shows the flight of the rock, we used a horizontal span of $t = 0$ to $t = 5$ and a vertical span of $D = 0$ to $D = 50$.

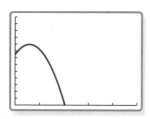

FIGURE 6.19 A rock tossed upward from the top of a building

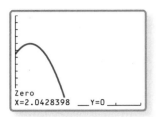

FIGURE 6.20 When the rock strikes the ground

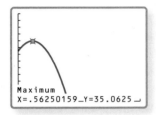

FIGURE 6.21 The peak of the rock's flight

How long did it take the rock to hit the ground? That happens when the distance up is zero. Thus we want to solve for t in the equation

$$30 + 18t - 16t^2 = 0.$$

We have used the single-graph method to do that in Figure 6.20, and we see that it takes about 2.04 seconds for the rock to complete its flight.

We might also want to know how high the rock went and when it reached its peak. In Figure 6.21 we got that information by using the calculator to locate the maximum. We see that the rock reached its highest point of 35.06 feet just over half a second after it was tossed.

Let's look now at the velocity $V = V(t)$ of the rock. We proceed as in earlier examples, first marking important features and then noting in particular where velocity is positive, where it is negative, and where it is zero. This is shown in Figure 6.22. We use this information to complete the graph of velocity in Figure 6.23.

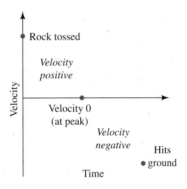

FIGURE 6.22 Interesting features of velocity

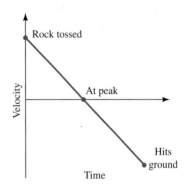

FIGURE 6.23 The graph of velocity versus time

6.1 EXERCISES

Reminder Round all answers to two decimal places unless otherwise indicated.

1. **From New York to Miami Again** The city of Richmond, Virginia, is about halfway between New York and Miami. A Richmond resident might locate the airplane in Example 6.1 using distance north of Richmond. Make the graphs of location and velocity of the airplane from this perspective.

2. **The Rock with a Changed Reference Point** Make graphs of position and velocity for a rock tossed upward from ground level as it might be viewed by someone standing atop a tall building. Thus the location of the rock is measured by its distance down from the top of the building.

3. **The Rock with a Formula** If from ground level we toss a rock upward with a velocity of 30 feet per second, we can use elementary physics to show that the height in feet of the rock above the ground t seconds after the toss is given by $S = 30t - 16t^2$.

   a. Use your calculator to plot the graph of S versus t.
   b. How high does the rock go?
   c. When does it strike the ground?
   d. Sketch the graph of the velocity of the rock versus time.

4. **Getting Velocity from a Formula** When a man jumps from an airplane with an opening parachute, the distance $S = S(t)$, in feet, that he falls in t seconds is given by

$$S = 20\left(t + \frac{e^{-1.6t} - 1}{1.6}\right).$$

   a. Use your calculator to make a graph of S versus t for the first 5 seconds of the fall.
   b. Sketch a graph of velocity for the first 5 seconds of the fall.

5. **Walking and Running** You live east of campus, and you are walking from campus toward your home at

(continued)

a constant speed. When you get there, you rest for 5 minutes and then run back west at a rapid speed. After a few minutes you reach your destination, and then you rest for 10 minutes. Measure your location as your distance west of your home, and make graphs of your location and velocity.

6. **A Rubber Ball**   A rubber ball is dropped from the top of a building. The ball lands on concrete and bounces once before coming to rest on the grass. Measure the location of the ball as its distance up from the ground. Make graphs of the location and velocity of the ball.

7. **Gravity on Earth and on Mars**   The acceleration due to gravity near the surface of a planet depends on the mass of the planet; larger planets impart greater acceleration than smaller ones. Mars is much smaller than Earth. A rock is dropped from the top of a cliff on each planet. Give its location as the distance down from the top of the cliff.

   a. On the same coordinate axes, make a graph of distance down for each of the rocks.

   b. On the same coordinate axes, make a graph of velocity for each of the rocks.

8. **Traveling in a Car**   Make graphs of location and velocity for each of the following driving events. In each case, assume that the car leaves from home moving west down a straight road and that position is given as the distance west from home.

   a. *A vacation:* Being eager to begin your overdue vacation, you set your cruise control and drive faster than you should to the airport. You park your car there and get on an airplane to Spain. When you fly back 2 weeks later, you are tired and drive at a leisurely pace back home. (*Note:* Here we are talking about location of your car, not of the airplane.)

   b. *On a country road:* A car driving down a country road encounters a deer. The driver slams on the brakes and the deer runs away. The journey is cautiously resumed.

   c. *At the movies:* In a movie chase scene, our hero is driving his car rapidly toward the bad guys. When the danger is spotted, he does a Hollywood 180-degree turn and speeds off in the opposite direction.

9. **Making Up a Story about a Car Trip**   You begin from home on a car trip. Initially your velocity is a small positive number. Shortly after you leave,

velocity decreases momentarily to zero. Then it increases rapidly to a large positive number and remains constant for this part of the trip. After a time, velocity decreases to zero and then changes to a large negative number.

   a. Make a graph of velocity for this trip.

   b. Discuss your distance from home during this driving event, and make a graph.

   c. Make up a driving story that matches this description.

10. **Car Trips with Given Graphs**

   a. The graph in Figure 6.24 shows your distance west of home on a car trip. Make a graph of velocity.

   b. The graph in Figure 6.25 shows your velocity on a different car trip. Assuming you start at home, make a graph of your distance west of home.

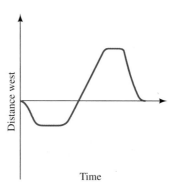

FIGURE 6.24 A graph of distance west of home

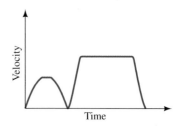

FIGURE 6.25 A graph of velocity for a different car trip

11. **A Car Moving in an Unusual Way**   In making graphs for location and velocity, the authors of this text have tried to avoid sharp corners. This problem is designed to show you why. Suppose a car's distance west of home is given by the graph in Figure 6.26.

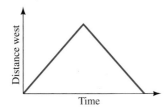

FIGURE 6.26 Position for an unusual driving event

a. Make a graph of velocity versus time. In making your graph, take extra care near the peak shown in Figure 6.26.

b. Carefully describe the motion of the car near the peak in Figure 6.26. Do you think it is possible to drive in a way that matches the graph in Figure 6.26?

12. Sporting Events Analyze the following sporting activities according to the instructions.

a. *Skiing:* A skier is going down a ski slope at Park City, Utah. The hill is initially gentle, but then it gets steep about halfway down before flattening out at the bottom. Locate position on the slope as the distance from the top, and make graphs of location and velocity.

b. *Practicing for the NCAA basketball tournament:* You are bouncing a basketball. Locate the basketball by its distance up from the floor. Make graphs of location and velocity.

c. *Hiking:* The graph of the distance west of base camp for a hiker in Colorado is given in Figure 6.27. Make a graph of velocity versus time, and then make up a story that matches this description.

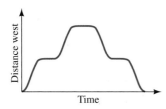

FIGURE 6.27 A hiking trip

6.1 SKILL BUILDING EXERCISES

S-1. Velocity What is the rate of change in directed distance?

S-2. Sign of Velocity When directed distance is decreasing, is velocity positive or negative? What is the velocity when directed distance is not changing?

S-3. Sign of Velocity When the graph of directed distance is decreasing, is the graph of velocity above or below the horizontal axis?

S-4. Constant Velocity When velocity is constant, what kind of function is directed distance?

S-5. Constant Velocity When the graph of directed distance is a straight line, what can be said about the graph of velocity?

S-6. At a Valley When the graph of directed distance reaches a minimum (at a valley), what is the velocity at this minimum?

S-7. A Car A car is driving at a constant velocity of 60 miles per hour. A perspective has been chosen so that directed distance is increasing. Since velocity is constant, we know that directed distance is a linear function. What is the slope of that linear function?

S-8. A Trip A car is driving on a highway that leads west from home. We locate its position as distance west from home. Determine whether velocity is positive, negative, or zero in each of the following situations.

a. The car is driving west.

b. The car is stopped at a traffic light.

c. The car is driving east.

S-9. A Rock A rock is tossed upward and reaches its peak 2 seconds after the toss. Its location is determined by its distance up from the ground. What is the sign of velocity at each of the following times?

a. 1 second after the toss

b. 2 seconds after the toss

c. 3 seconds after the toss

S-10. Graph of Velocity If the graph of velocity lies on the horizontal axis, what can be said about the graph of directed distance?

S-11. Change in Direction A graph of directed distance switches from increasing to decreasing. What happens to velocity?

S-12. Zero Velocity When velocity is zero for a time, what is true about directed distance?

6.2 RATES OF CHANGE FOR OTHER FUNCTIONS

Velocity is the rate of change in directed distance, and this same idea applies to any function. For a function $f = f(x)$, we can look at the rate of change in f with respect to x, which tells how f changes for a given change in x. Rates of change are so pervasive in mathematics, science, and engineering that they are given a special notation, most commonly $\frac{df}{dx}$ or $f'(x)$, and a special name, the *derivative of f with respect to x*. We will use the notation $\frac{df}{dx}$ but will consistently refer to it as the *rate of change in f with respect to x*. We should note that some applications texts use the notation $\frac{\Delta f}{\Delta x}$ for the rate of change.[1]

Examples of Rates of Change

The notation $\frac{df}{dx}$ means the rate of change in f with respect to x. Specifically, $\frac{df}{dx}$ tells how much f is expected to change if x increases by 1 unit. We note that $\frac{df}{dx}$ is a function of x; in general, the rate of change varies as x varies. Some examples may help clarify things.

1. If $S = S(t)$ gives directed distance for an object as a function of time t, then $\frac{dS}{dt}$ is the rate of change in directed distance with respect to time. This is *velocity*. It tells the additional distance we expect to travel in 1 unit of time. For example, if we are currently located $S = 100$ miles south of Dallas, Texas, and if we are traveling south with a velocity $\frac{dS}{dt}$ of 50 miles per hour, then in 1 additional hour we would expect to be 150 miles south of Dallas.

2. If $V = V(t)$ is the velocity of an object as a function of time t, then $\frac{dV}{dt}$ is the rate of change in velocity with respect to time. This is *acceleration*. It tells the additional velocity we expect to attain in 1 unit of time. For example, if we are traveling with a velocity $V = 50$ miles per hour, and if we start to pass a truck, our acceleration $\frac{dV}{dt}$ might be 2 miles per hour each second. Then 1 second in the future, we would expect our velocity to be 52 miles per hour.

3. If $T = T(D)$ denotes the amount of income tax, in dollars, that you pay on an income of D dollars, then $\frac{dT}{dD}$ is the rate of change in tax with respect to the money you earn. It is known as the *marginal tax rate*. If you have already accumulated D dollars, then $\frac{dT}{dD}$ is the additional tax you expect to pay if you earn 1 additional dollar. For example, suppose we have a tax liability of $3000 and our marginal tax rate $\frac{dT}{dD}$ is 0.2, or 20 cents per dollar. Then if we earn an additional $1, we would expect our tax liability to increase by 20 cents to a total of $3000.20; if instead we earn an additional $100, we would expect our tax liability to increase by $20 to a total of $3020. The marginal tax rate is a crucial bit of information for financial planning.

4. If $P = P(i)$ is the profit, in dollars, that you expect to earn on an investment of i dollars, then $\frac{dP}{di}$ is the rate of change in profit with respect to dollars invested. It tells how much additional profit is to be expected if 1 additional dollar is invested. In economics this is known as *marginal profit*. For example,

---

[1]This is seen most often in business and agricultural applications but appears in other places as well. There is a technical difference between the meanings of $\frac{df}{dx}$ and $\frac{\Delta f}{\Delta x}$. The notation $\frac{\Delta f}{\Delta x}$ means an *average rate of change over a given change in x*, whereas $\frac{df}{dx}$ is the *instantaneous rate of change*. The distinction is important in advanced mathematics but less so in many applications.

if our current investment in a project gives a profit of $1000, and if our marginal profit $\frac{dP}{di}$ is 0.2, or 20 cents per dollar, then we would expect that an additional investment of $100 would yield an additional profit of $20 for a total profit of $P = 1020$ dollars.

Properties That All Rates of Change Share

Fortunately, all rates of change exhibit the properties that we have already seen in our study of velocity. If $S = S(t)$ denotes directed distance at time t, then we know that when S is increasing, the velocity $\frac{dS}{dt}$ is positive; when S is decreasing, the velocity $\frac{dS}{dt}$ is negative; and when S is not changing, the velocity $\frac{dS}{dt}$ is zero. This fundamental relationship holds true for all rates of change.

KEY IDEA 6.3   **FUNDAMENTAL PROPERTIES OF RATES OF CHANGE**

For a function $f = f(x)$ we will use the notation $\frac{df}{dx}$ to denote the rate of change in f with respect to x.

1. The expression $\frac{df}{dx}$ tells how f changes in relation to x. It gives the additional value that is expected to be added to f if x increases by 1 unit.
2. When f is increasing, $\frac{df}{dx}$ is positive.
3. When f is decreasing, $\frac{df}{dx}$ is negative.
4. When f is not changing, $\frac{df}{dx}$ is zero.

EXAMPLE 6.3   **PASSING A TRUCK**

You are driving with your cruise control set when you encounter a slowly moving truck. You speed up to pass the truck. When you have overtaken the truck, you slow down and resume your previous speed. Let $V = V(t)$ denote your velocity during this event as a function of time t.

Part 1 Explain in practical terms the meaning of $\frac{dV}{dt}$.

Part 2 Make a graph of $V = V(t)$, marking important points on the graph.

Part 3 Make a graph of acceleration $A = \frac{dV}{dt}$.

Solutiont to Part 1 The function $\frac{dV}{dt}$ is the rate of change in velocity. This is acceleration. It tells what additional velocity you expect to attain over 1 unit of time.

Solution to Part 2 Since the cruise control is set, velocity is positive and constant at the beginning. That means that the graph of V versus t starts above the horizontal axis and is horizontal. When you start around the truck, you speed up, making the graph of $V(t)$ go up. After you overtake the truck, you slow down and resume your original velocity. This makes the graph of velocity go back down to its original level and flatten out again. Our graph is shown in Figure 6.28.

Solution to Part 3 We will get the graph of acceleration directly from the graph of velocity in Figure 6.28 and then verify that it makes sense. The basic tools we use are the fundamental properties of rates of change: When V is increasing, $\frac{dV}{dt}$ is positive; when V is decreasing, $\frac{dV}{dt}$ is negative; and when V is not changing, $\frac{dV}{dt}$ is zero.

Before and after passing, the graph of velocity in Figure 6.28 is horizontal and so V is not changing at all. This means that the rate of change in velocity $A = \frac{dV}{dt}$ is zero during these periods. Thus the graph of A lies on the horizontal axis, as we have shown in Figure 6.29. From the time the pass begins until the truck is overtaken, the graph of velocity in Figure 6.28 is increasing. This means that the rate of change in velocity $\frac{dV}{dt}$ is positive, and we have marked that in italics in Figure 6.29. At the peak of the graph in Figure 6.28, the acceleration is zero; however, from there to the time the pass is completed, velocity is decreasing, so its rate of change—the acceleration—is negative, as is marked in italics in Figure 6.29.

We used the information from Figure 6.29 to draw the completed graph of $\frac{dV}{dt}$ in Figure 6.30. We note that the graph makes sense. At the beginning and end of the graph, the cruise control is set, so velocity is not changing. That is, acceleration is zero. As you begin the pass, you accelerate. Thus the graph of acceleration is above the horizontal axis for this period, as is shown in Figure 6.30. Once the truck is overtaken, you ease off the gas pedal, and the car *decelerates* back to your original cruising speed. Deceleration is the same as negative acceleration, and so from the time when you overtake the truck until the pass is completed, the graph of acceleration is below the horizontal axis, as is represented in Figure 6.30. Thus the graph we made using the fundamental properties of rates of change agrees with our intuitive analysis.

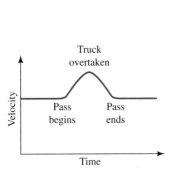

FIGURE 6.28 Velocity when passing a truck

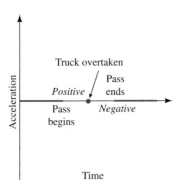

FIGURE 6.29 Important features for acceleration

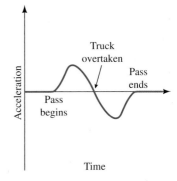

FIGURE 6.30 Completed graph of acceleration when passing a truck

When the Rate of Change Is Zero

In Example 6.3 we used the graph of velocity V in Figure 6.28 to get the graph of acceleration $\frac{dV}{dt}$ in Figure 6.30. In the process, we noted two situations in which the acceleration was zero: when velocity was constant while the cruise control was set (here acceleration was zero over a span of time) and when velocity reached its maximum value at the point when the truck was overtaken (here acceleration was zero for an instant in time). Once again, these observations remain true for rates of change in general. When f is constant, its rate of change $\frac{df}{dx}$ is zero over a span of x values. This is intuitively clear, since stating that $\frac{df}{dx}$ is zero is the same as saying that f is not changing and therefore remains at a constant value. At a peak or valley of f, the rate of change $\frac{df}{dx}$ will be zero for that single value of x. These important facts are illustrated in Figure 6.31 and Figure 6.32.

Let's look at an example to see how we use the fact that the rate of change is zero at extrema.

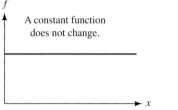

FIGURE 6.31 A constant function has persistent zero rate of change.

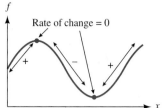

FIGURE 6.32 At extrema, the rate of change is momentarily zero.

EXAMPLE 6.4 MARGINAL PROFIT FOR A TIRE COMPANY

The CEO of a tire company has kept records of the profit $P = P(n)$ that the company makes when it produces n tires per day.

Part 1 The rate of change $\frac{dP}{dn}$ is known as the *marginal profit*. Explain the meaning of marginal profit in practical terms.

Part 2 What action should the CEO take if the marginal profit is positive?

Part 3 What action should the CEO take if the marginal profit is negative?

Part 4 The CEO has used her records to make the graph of marginal profit shown in Figure 6.33. According to this graph, how many tires per day should the company be producing?

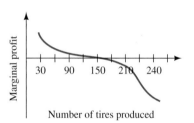

FIGURE 6.33 Marginal profit for a tire company

Solution to Part 1 The marginal profit $\frac{dP}{dn}$ is the rate of change in profit with respect to the number of items produced. It is the change in profit that the CEO can expect to earn by producing 1 more tire.

Solution to Part 2 Marginal profit tells the CEO how much profit can be expected to change if 1 additional tire is produced. Since marginal profit is positive, additional profit can be made by increasing production. In other words, since the rate of change $\frac{dP}{dn}$ is positive, the profit P is an increasing function of the production level n. This means that the CEO should increase production, since doing so will cause the profit to increase. This situation might occur if, for example, the demand for tires were larger than current production.

Solution to Part 3 In this case the marginal profit is negative, which means that if production is increased, profits will change by a negative amount. That is, profits will decrease. In other words, since the rate of change $\frac{dP}{dn}$ is negative, the profit P is a *decreasing* function of n. In this scenario, it would be wise for the CEO to cut back on production. This might happen if the number of tires currently being produced exceeded consumer demand.

Solution to Part 4 In part 2 we noted that if the marginal profit is positive, then the profit P is increasing with n, and production should be increased. In part 3, we saw that if the marginal profit is negative, then the profit P is decreasing with n, and production should be decreased. Thus it makes sense that for the CEO, the ideal value of marginal profit is zero. In other words, to get a maximum profit P, we want to find the point where P switches from increasing to decreasing (since at this point the profit is at its highest value), and that is where the marginal profit $\frac{dP}{dn}$ switches from positive to negative. This is where the graph in Figure 6.33 crosses the horizontal axis—that is, where the marginal profit is zero. The crossing point occurs at about $n = 160$, so the company should be producing about 160 tires per day.

When the Rate of Change Is Constant

We observed in Section 6.1 that when velocity $\frac{dS}{dt}$ is constant, we can conclude that directed distance S is a linear function whose slope is the constant velocity. Virtually all the observations we made about the relationships between velocity and directed distance are true for rates of change in general, and this one is no exception: If the rate of change in f, namely $\frac{df}{dx}$, is constant, then f is a linear function with slope $\frac{df}{dx}$. We would emphasize that this is not really a new observation. The characterization of linear functions as those with constant rate of change has been used repeatedly in this course. The only thing new about our expression here of this fact is the use of $\frac{df}{dx}$ to denote the rate of change.

EXAMPLE 6.5 **A LEAKY BALLOON**

A balloon is initially full of air, but then it springs a leak. Let $B = B(t)$ represent the volume, in liters, of air in the balloon at time t, measured in minutes.

Part 1 Explain what $\frac{dB}{dt}$ means in practical terms. As air is leaking out of the balloon, is $\frac{dB}{dt}$ positive or negative?

Part 2 Assume that air is leaking from the balloon at a constant rate of 0.5 liter per minute.[2] What does this information tell you about $\frac{dB}{dt}$? What does it tell you about B?

Part 3 A little later the leak is patched, a pump that outputs 2 liters of air per minute is attached to the balloon, and it is inflated back to its original size. As the balloon inflates, what can we conclude about $\frac{dB}{dt}$? What can we conclude about B?

Part 4 Make a graph of B versus t as the balloon leaks air and then is reinflated to its original size.

Part 5 Make a graph of $\frac{dB}{dt}$ versus t as the balloon leaks air and then is reinflated to its original size.

Solution to Part 1 The function $\frac{dB}{dt}$ is the rate of change in B with respect to t. This is the rate of change in the volume of air in the balloon with respect to time. In practical terms, $\frac{dB}{dt}$ is the change in the volume of air in the balloon that we expect to occur in 1 minute of time. Since the volume B of air in the balloon is decreasing, $\frac{dB}{dt}$ is negative.

Solution to Part 2 To say that air is leaking from the balloon at a rate of 0.5 liter per minute means that the volume B of air in the balloon is decreasing by 0.5 liter per minute. In other words, B is changing by -0.5 liter per minute. We conclude that $\frac{dB}{dt} = -0.5$. Since B is a function with a constant rate of change of -0.5, we know that it is a linear

jocic/Shutterstock.com

---

[2]This is not, in fact, how we would expect a balloon to leak. Rather, we would expect it to leak rapidly when it is almost full and more slowly when it is nearly empty. We will examine this more realistic description of a leaky balloon in Section 6.4.

function with slope −0.5. We were not asked to do so, but with this information we have almost everything we need to write a formula for B, namely B = −0.5t + b. In this formula, what is the practical meaning of b?

Solution to Part 3 As the balloon is inflated, the volume B of air in the balloon is increasing by 2 liters per minute. That is, $\frac{dB}{dt}$ = 2 liters per minute. Since B has a constant rate of change of 2, we conclude that B is a linear function with slope 2 during the period of reinflation.

As in part 2, we were not asked to write a formula for B, but we have almost everything we need to do that. The formula is B = 2t + c. Explain the practical meaning of c in this formula.

Solution to Part 4 We know that while the balloon is losing air, B is linear with slope −0.5 and that after the pump is attached, B is linear with slope 2. Thus the graph of B should start as a straight line with slope −0.5 and then change to a straight line with slope 2. These features are reflected in Figure 6.34.

Note in Figure 6.34 that we have rounded the graph at the bottom where the pump is attached. Alternatively, one might draw this graph with a sharp corner at the bottom. Which do you think is correct?

Solution to Part 5 We know that while the balloon is leaking, $\frac{dB}{dt}$ has a constant value of −0.5. Thus its graph during this period is a horizontal line located 0.5 unit below the horizontal axis. After the pump is attached, $\frac{dB}{dt}$ has a constant value of 2. Thus during this period, its graph is a horizontal line 2 units above the horizontal axis. How the two line segments are connected near the time when the pump is attached depends on how we drew the graph of B near that point. If you think there should be a sharp corner there, how would this affect the graph of $\frac{dB}{dt}$ that we have drawn in Figure 6.35?

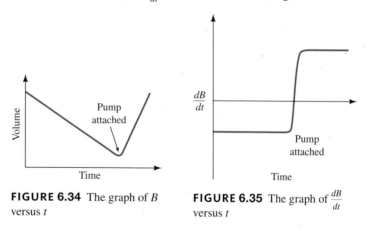

FIGURE 6.34 The graph of B versus t

FIGURE 6.35 The graph of $\frac{dB}{dt}$ versus t

6.2 EXERCISES

Reminder Round all answers to two decimal places unless otherwise indicated.

1. **Estimating Rates of Change** Use your calculator to make the graph of $f(x) = x^3 − 5x$.
   a. Is $\frac{df}{dx}$ positive or negative at x = 2?
   b. Identify a point on the graph of f where $\frac{df}{dx}$ is negative.

2. **The Spread of AIDS** This table shows the cumulative number N = N(t) of AIDS cases in the United States that have been reported to the Centers for Disease Control and Prevention by the end of the year given. (The source for these data, the U.S. Centers for Disease Control and Prevention in Atlanta, cautions that they are subject to retrospective change.)

(continued)

a. What does $\frac{dN}{dt}$ mean in practical terms?

b. From 2002 to 2009 was $\frac{dN}{dt}$ ever negative?

t = year	N = total cases reported
2002	849,780
2003	892,875
2004	908,905
2005	946,578
2006	983,343
2007	1,021,242
2009	1,132,836

3. **Mileage for an Old Car** The gas mileage M that you get on your car depends on its age t in years.

   a. Explain the meaning of $\frac{dM}{dt}$ in practical terms.

   b. As your car ages and performance degrades, do you expect $\frac{dM}{dt}$ to be positive or negative?

4. **Investing in the Stock Market** You are considering buying three stocks whose prices at time t are given by $P_1(t)$, $P_2(t)$, and $P_3(t)$. You know that $\frac{dP_1}{dt}$ is a large positive number, $\frac{dP_2}{dt}$ is near zero, and $\frac{dP_3}{dt}$ is a large negative number. Which stock will you buy? Explain your answer.

5. **Hiking** You are hiking in a hilly region, and $E = E(t)$ is your elevation at time t.

   a. Explain the meaning of $\frac{dE}{dt}$ in practical terms.

   b. Where might you be when $\frac{dE}{dt}$ is a large positive number?

   c. You reach a point where $\frac{dE}{dt}$ is briefly zero. Where might you be?

   d. Where might you be when $\frac{dE}{dt}$ is a large negative number?

6. **Marginal Profit** A small firm produces at most 15 widgets in a week. Its profit P, in dollars, is a function of n, the number of widgets manufactured in a week. The marginal profit $\frac{dP}{dn}$ for the firm is given by the formula

   $$\frac{dP}{dn} = 72 + 6n - n^2.$$

   a. Use your calculator to make a graph of $\frac{dP}{dn}$ versus n.

   b. For what values of n is the profit P decreasing?

   c. How many widgets should the firm produce in a week to maximize profit P?

7. **Health Plan** The managers of an employee health plan for a firm have studied the balance B, in millions of dollars, in the plan account as a function of t, the number of years since the plan was instituted. They have determined that the rate of change $\frac{dB}{dt}$ in the account balance is given by the formula

   $$\frac{dB}{dt} = 10e^{0.1t} - 12.$$

   a. Use your calculator to make a graph of $\frac{dB}{dt}$ versus t over the first 5 years of the plan.

   b. During what period is the account balance B decreasing?

   c. At what time is the account balance B at its minimum?

8. **A Race** A man enters a race that involves running and swimming. He lines up at the starting gate and begins running very fast, hoping to take the initial lead before settling into a constant, slower pace for a distance run. To complete the race, he must swim across a river and back before running back to the starting gate. He swims at a constant rate but cannot swim as fast as he runs. Let $S = S(t)$ denote the distance of the contestant from the starting gate.

   a. Make a graph of S versus t. Note on your graph the places where you know that S is linear.

   b. Make a graph of the contestant's velocity.

   c. Make a graph of the contestant's acceleration.

9. **The Acceleration Due to Gravity** From the time of Galileo, physicists have known that near the surface of the Earth, gravity imparts a constant acceleration of 32 feet per second per second. Explain how this shows that if air resistance is ignored, velocity for a falling object is a linear function of time.

10. **Water in a Tank** Water is leaking out of a tank. The amount of water in the tank t minutes after it springs a leak is given by $W(t)$ gallons.

    a. Explain what $\frac{dW}{dt}$ means in practical terms.

    b. As water leaks out of the tank, is $\frac{dW}{dt}$ positive or negative?

    c. For the first 10 minutes, water is leaking from the tank at a rate of 5 gallons per minute. What do you conclude about the nature of the function W during this period?

d. After about 10 minutes, the hole in the tank suddenly gets larger, and water begins to leak out of the tank at 12 gallons per minute.

    i. Make a graph of W versus t. Be sure to incorporate linearity where it is appropriate.

    ii. Make a graph of $\frac{dW}{dt}$ versus t.

11. A Population of Bighorn Sheep   There is an effort in Colorado to restore the population of bighorn sheep. Let $N = N(t)$ denote the number of sheep in a certain protected area at time t.

a. Explain the meaning of $\frac{dN}{dt}$ in practical terms.

b. A small breeding population of bighorn sheep is initially introduced into the protected area. Food is plentiful and conditions are generally favorable for bighorn sheep. What would you expect to be true about the sign of $\frac{dN}{dt}$ during this period?

c. This summer a number of dead sheep were discovered, and all were infected with a disease that is known to spread rapidly among bighorn sheep and is nearly always fatal. How would you expect an unchecked spread of this disease to affect $\frac{dN}{dt}$?

d. If the reintroduction program goes well, then the population of bighorn sheep will grow to the size the available food supply can support and will remain at about that same level. What would you expect to be true of $\frac{dN}{dt}$ when this happens?

12. Eagles   In an effort to restore the population of bald eagles, ecologists introduce a breeding group into a protected area. Let $N = N(t)$ denote the population of bald eagles at time t. Over time, you observe the following information about $\frac{dN}{dt}$.

- Initially, $\frac{dN}{dt}$ is a small positive number.

- A few years later, $\frac{dN}{dt}$ is a much larger positive number.

- Many years later, $\frac{dN}{dt}$ is positive but near zero. Make a possible graph of $N(t)$.

13. Visiting a Friend   I live in the suburbs and my friend lives out in the country. The speed limit between my home and a stop sign 1 mile away is 35 miles per hour. After that the speed limit is 65 miles per hour. Assume that I drove out to visit my friend. As I pulled into his driveway, I saw that he wasn't at home, and so I drove back to my house. Assume that I obeyed all the traffic laws and had an otherwise uneventful trip. Make graphs of $S = S(t)$, my location in relation to my home; the velocity $V = V(t)$; and the acceleration $A = A(t)$. Be sure your graphs show linearity where it is appropriate.

14. A Car Trip   You are moving away from home in your car, and at no time do you turn back. Your acceleration is initially a large positive number, but it decreases slowly to zero, where it remains for a while. Suddenly your acceleration decreases rapidly to a large negative number before returning slowly to zero.

a. Make a graph of acceleration for this event.

b. Make a graph of velocity for this event.

c. Make a graph of the location of the car.

d. Make up a driving story that matches this description.

15. Velocity of an Airplane   An airplane leaves Atlanta flying to Dallas. Because of heavy air traffic, it circles the airport at Dallas for a time before landing. Let $D(t)$ be the distance west from Atlanta to the airplane. Make a graph of $D(t)$. Thinking of velocity $V(t)$ as the rate of change in $D(t)$, make a graph of the velocity of the airplane.

16. Growth in Height   The following graph gives a man's height $H = H(t)$, in inches at age t in years.

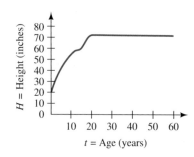

a. Explain what $\frac{dH}{dt}$ means in practical terms.

b. Sketch a graph of $\frac{dH}{dt}$ versus t.

(continued)

6.2 SKILL BUILDING EXERCISES

S-1. Meaning of Rate of Change What is the common term for the rate of change of each of the following phenomena?

a. Directed distance as a function of time

b. Velocity as a function of time

c. Tax due as a function of income

d. Profit as a function of dollars invested

S-2. A Mathematical Term If $f = f(x)$, then we use $\frac{df}{dx}$ to denote the rate of change in f. What is the technical mathematical term for $\frac{df}{dx}$?

S-3. Sign of the Derivative Suppose $f = f(x)$. What is the sign of $\frac{df}{dx}$ in each of the following situations?

a. The function f is increasing.

b. The graph of f has reached a peak.

c. The function f is decreasing.

d. The graph of f is a horizontal line.

S-4. Constant Rate of Change When $\frac{df}{dx}$ is constant, what kind of function is f?

S-5. A Value for the Rate of Change If $\frac{df}{dx}$ has a constant value of 10, we know that f is a linear function. What is the slope of f?

S-6. Elevation If $E(t)$ is your elevation at time t, and you are walking up a steep slope, what can be said about $\frac{dE}{dt}$?

S-7. Marginal Tax Rate You have the opportunity to earn some extra income working at $20 per hour. If your marginal tax rate is 34%, how much of the hourly wage will you get to keep?

S-8. Graph of Rate of Change What can be said about the graph of $\frac{df}{dx}$ in each of the following situations?

a. The graph of f is increasing.

b. The graph of f is at a peak.

c. The graph of f is decreasing.

d. The graph of f is a straight line.

S-9. Graph of f What can be said about the graph of f if the graph of $\frac{df}{dx}$ is below the horizontal axis?

S-10. Advertising Let $s(a)$ denote sales generated by spending a dollars on advertising. My goal is to increase sales. If $\frac{ds}{da}$ is negative, should I spend more or less money on advertising?

S-11. Gasoline Prices The price P of gasoline decreases to a minimum and then starts to increase. What is the rate of change $\frac{dP}{dt}$ of the price with respect to time t at the time when the price reaches a minimum?

S-12. Food Prices Let B denote the price of a bag of groceries at time t, and assume that $\frac{dB}{dt}$ switches from positive to negative at time $t = 1$. What does the graph of B look like at $t = 1$?

6.3 ESTIMATING RATES OF CHANGE

Up to now we have looked at rates of change qualitatively, emphasizing how the sign of a rate of change affects the function. But exact values, or even estimates, for rates of change can provide additional important information. It is easiest to make such estimates for functions given by tables.

Rates of Change for Tabulated Data

The following brief table shows the location S (measured as distance in miles east of Los Angeles) of an airplane flying toward Denver.

Time	1:00 P.M.	1:30 P.M.
Distance from L.A.	360 miles	612 miles

We want to know the velocity $\frac{dS}{dt}$ at 1:00 P.M. We do this using a familiar formula:

$$\text{Average velocity} = \frac{\text{Distance traveled}}{\text{Elapsed time}}.$$

Between 1:00 and 1:30 the airplane traveled $612 - 360 = 252$ miles, and it took half an hour to travel that far:

$$\text{Average velocity from 1:00 to 1:30} = \frac{252}{0.5} = 504 \text{ miles per hour.}$$

It is important to point out that this calculation can be expected to give the exact velocity $\frac{dS}{dt}$ at 1:00 P.M. only if the airplane is traveling at the same speed over the entire 30-minute time interval. If the airplane was speeding up between 1:00 and 1:30, then 504 miles per hour was its average velocity over the 30-minute time interval, but its exact velocity at 1:00 would have been somewhat less than 504 miles per hour. If, on the other hand, the airplane was slowing down, then its exact velocity at 1:00 would have been more than 504 miles per hour. In general, when velocity is calculated from a table of values, as it was here, it should be regarded as an approximation of the exact velocity. Furthermore, it is clear that shorter time intervals yield better approximations, because the moving object doesn't have as much time to change velocity.[3] Suppose, for example, that instead of the table above, we had been given the following table.

Time	1:00 P.M.	1 second after 1:00 P.M.
Distance from L.A.	360 miles	360.140 miles

If we use this table to calculate velocity, we see that the airplane traveled 0.140 mile in 1 second. Since there are 3600 seconds in an hour, we get

$$\text{Average velocity from 1:00 P.M. to 1 second later} = \frac{0.140}{\frac{1}{3600}} = 504 \text{ miles per hour.}$$

The velocity of the airplane can change very little over a 1-second time interval, so we would now feel confident reporting its velocity at 1:00 P.M. as 504 miles per hour.

Exactly the same idea can be used to approximate the rate of change for any function given by tabulated data. For example, the following table shows the percentage $B = B(t)$ of babies born in 2005 and 2009 to unmarried women in the United States.

Year	2005	2009
Percentage born to unmarried women	36.8%	41.0%

The function $\frac{dB}{dt}$ is the rate of change in B. If we were able to calculate it exactly for 2005, it would tell us how fast the percentage of births to unmarried women was increasing in 2005. That would indicate how much growth in this percentage could be expected in 1 year's time. As with the example of the airplane, the table does not give enough information to allow for an exact calculation of $\frac{dB}{dt}$ for 2005, but we can approximate its value using the average rate of change from 2005 to 2009 just as we did for the velocity of the airplane.

During the period from 2005 to 2009, the change in percentage was $41.0 - 36.8 = 4.2$ percentage points. This occurred over a period of 4 years, so we divide to get

---

[3]This is discussed further in Section 6.4 under the topic of instantaneous rates of change.

Average rate of change from 2005 to 2009 $= \dfrac{4.2}{4} = 1.05$ percentage points per year.

We can use this number as an approximation of the value of $\frac{dB}{dt}$.

> **KEY IDEA 6.4** **ESTIMATING RATES OF CHANGE FOR TABULATED DATA**
>
> If $f = f(x)$ is a function given by a table of values, then $\frac{df}{dx}$, the rate of change of f with respect to x, cannot without further information be calculated exactly. But it can be estimated using
>
> $$\text{Average rate of change in } f \text{ with respect to } x = \frac{\text{Change in } f}{\text{Change in } x}.$$

Calculating rates of change in this way for tabulated data is not a new idea at all, and we have already made and used this type of calculation many times in the text. Look back at Example 1.3 in Section 1.2, where we studied tabulated data for the number $W = W(t)$ of women in the United States employed outside the home. In the solution to part 3 of that example, we noted that the number increased by 14.0 million from 1970 to 1980. We concluded that during the decade of the 1970s, the number of women employed outside the home was increasing by an average of $14.0/10 = 1.40$ million per year, and we used that number to estimate the number of women employed outside the home in 1972. This is exactly the calculation presented in Key Idea 6.4. Thus, in the language of rates of change, we would say that $\frac{dW}{dt}$ is approximately the average rate of change from 1970 to 1980—1.40 million per year—and that this number tells us the increase in W that would be expected in 1 year. This is once again an illustration of how mathematics distills ideas from many contexts into a single fundamental concept. Precisely the same mathematical idea, the rate of change, is used to analyze applications from velocity to population growth to numbers of women employed outside the home.

EXAMPLE 6.6 WATER FLOWING FROM A TANK

Consider the following table, which shows the number of gallons W of water left in a tank t hours after it starts to leak.

$t =$ hours	0	3	6	9	12
$W =$ gallons left	860	725	612	515	433

Part 1 Explain the meaning of $\frac{dW}{dt}$ in practical terms, and estimate its value at $t = 6$, using the average rate of change from $t = 6$ to $t = 9$.

Part 2 Use your answer from part 1 to estimate the amount of water in the tank 8 hours after the leak begins.

Solution to Part 1 The function $\frac{dW}{dt}$ is the rate of change in water remaining in the tank. This is the change in volume we expect over an hour. Since the amount of water is decreasing, $\frac{dW}{dt}$ is negative. Its size is the rate at which water is leaking from the tank.

We emphasize that the calculation we make here is the same one we made for women employed outside the home in Example 1.3. We are using a new language but not a new idea. Six hours after the leak began, there were 612 gallons in the tank. When $t = 9$ there are 515 gallons left. Thus the water level changed by $515 - 612 = -97$ gallons over the 3-hour period. Hence

$$\text{Average rate of change from 6 hours to 9 hours} = \frac{-97}{3}$$
$$= -32.33 \text{ gallons per hour.}$$

This is the estimate for $\frac{dW}{dt}$ that we were asked to find.

Solution to Part 2 For each hour after 6, we expect the number of gallons in the tank to decrease by 32.33 gallons. From 6 to 8 hours is a 2-hour span, so we expect $2 \times 32.33 = 64.66$ gallons to leak out. That leaves $612 - 64.66 = 547.34$ gallons in the tank 8 hours after the leak began.

Rates of Change for Functions Given by Formulas

For functions given by formulas, it is possible using calculus to find exact formulas for rates of change. But for many applications, a close approximation to the rate of change is sufficient, and that is what we will use in this text. The idea is first to use the formula to generate a brief table of values and then to compute the rate of change just as we did above. Let's look, for example, at a falling rock. If air resistance is ignored, elementary physics can be used to show that the rock falls $S = 16t^2$ feet during t seconds of fall. Let's estimate the downward velocity $\frac{dS}{dt}$ of the rock 2.5 seconds into the fall. We will first show the steps involved in making this computation and will then show how the graphing calculator offers a shortcut.

We know from the computations we made for the velocity of the airplane that we get more reliable answers if we keep the time interval short. Below we have made a brief table of values for $S = 16t^2$ using only $t = 2.5$ and $t = 2.50001$. You may wish to use your calculator to check that $16 \times 2.5^2 = 100$ and $16 \times 2.50001^2 = 100.0008$.

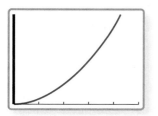

FIGURE 6.36 Graph of distance versus time

t	2.5	2.50001
S	100	100.0008

Now the change in S is $100.0008 - 100 = 0.0008$ foot. This is how far the rock falls from 2.5 to 2.50001 seconds into the fall. It takes $2.50001 - 2.5 = 0.00001$ second to fall this far, so the rate of change in S with respect to t (in other words, the velocity) is approximated as follows:

$$\frac{dS}{dt} = \frac{\text{Change in } S}{\text{Change in } t} = \frac{0.0008}{0.00001} = 80 \text{ feet per second.}$$

We will do this computation again, but this time we use the automated features provided by the calculator. The first step is to make the graph as shown in Figure 6.36. We used a window setup with a horizontal span of $t = 0$ to $t = 5$ and a vertical span of $S = 0$ to $S = 300$. Next we use the calculator to get the rate of change $\frac{dS}{dt}$ at $t = 2.5$. The result (see Figure 6.37) shows that the velocity $\frac{dS}{dt}$ at $t = 2.5$ is 80 feet per second, and this agrees with our earlier computation. We would emphasize that no magic has been performed. The calculator has just internally made a table of values and performed a computation similar to the one we did when we made the computation by hand. The calculator just automates the procedure. For the exact keystrokes[4] needed to do this, consult Appendices B and C.

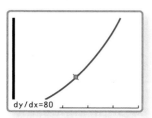

FIGURE 6.37 Getting $\frac{dS}{dt}$, the velocity

---

[4]On some calculators, you will sometimes not be able to get $\frac{dy}{dx}$ for the exact x value you want.

EXAMPLE 6.7 **A CANNONBALL**

A cannonball fired from the origin with a muzzle velocity of 300 feet per second follows the path of the graph of

$$h = x - 32\left(\frac{x}{300}\right)^2,$$

where distances are measured in feet (see Figure 6.38). This simple model ignores air resistance.

Part 1 Plot the graph of the flight of the cannonball.

Part 2 Use the graph to estimate the height h of the cannonball 734 feet downrange.

Part 3 By looking at the graph of h, do you expect $\frac{dh}{dx}$ to be positive or negative at $x = 734$?

Part 4 Calculate $\frac{dh}{dx}$ at 734 feet downrange, and explain in practical terms what this number means.

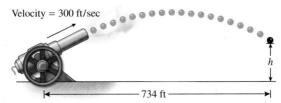

FIGURE 6.38 Path of a cannonball

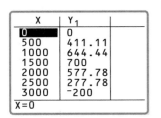

FIGURE 6.39 A table of values for the height of the cannonball

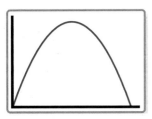

FIGURE 6.40 The flight of the cannonball

Solution to Part 1 To get the graph, we first enter the function and then record the appropriate variable correspondences:

$$Y_1 = h, \text{ height, in feet, on vertical axis}$$

$$X = x, \text{ distance downrange, in feet, on horizontal axis.}$$

Consulting the table of values in Figure 6.39, we choose a window setup with a horizontal span of $x = 0$ to $x = 3000$ and a vertical span of $h = 0$ to $h = 750$. The resulting graph is shown in Figure 6.40.

Solution to Part 2 In Figure 6.41 we have put the cursor at X=734, and we read from the bottom of the screen that the height of the cannonball 734 feet downrange is $h = 542.44$ feet.

Solution to Part 3 From Figure 6.41 we see that at $x = 734$, the graph of h is increasing. That is, at this distance downrange, the cannonball is still rising. Since h is increasing, we know that $\frac{dh}{dx}$ is positive.

Solution to Part 4 In Figure 6.42 we have used the calculator to get the value of $\frac{dh}{dx}$ at $x = 734$ feet downrange. The rate of change, 0.48 foot per foot (rounded to two decimal places), can now be read[5] from the dy/dx= prompt at the bottom of Figure 6.42. This is a measure of how steeply the cannonball is rising. It tells us that if we move 1 more foot downrange to 735 feet, we can expect the cannonball to rise 0.48 foot from 542.44 to $542.44 + 0.48 = 542.92$ feet.

---

[5]Some calculators will give a slightly different answer because they do not calculate the rate of change exactly at $x = 734$.

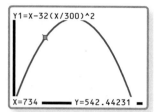

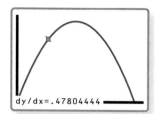

FIGURE 6.41 The height of the cannonball 734 feet downrange

FIGURE 6.42 The rate of change in height 734 feet downrange

6.3 EXERCISES

Reminder Round all answers to two decimal places unless otherwise indicated.

1. **Population Growth** The following table[6] shows the population of reindeer on an island as of the given year.

Date	1945	1950	1955	1960
Population	40	165	678	2793

We let t be the number of years since 1945, so that $t = 0$ corresponds to 1945, and we let $N = N(t)$ denote the population size.

   a. Approximate $\frac{dN}{dt}$ for 1955 using the average rate of change from 1955 to 1960, and explain what this number means in practical terms.

   b. Use your work from part a to estimate the population in 1957.

   c. The number you calculated in part a is an approximation to the actual rate of change. As you will be asked to show in the next exercise, the reindeer population growth can be closely modeled by an exponential function. With this in mind, do you think your answer in part a is too large or too small? Explain your reasoning.

2. **Further Analysis of Population Growth** *This is a continuation of Exercise 1.* Our goal is to make an exponential model of the data and use it to get a more accurate estimate of the rate of change in population in 1955.

   a. Use regression to obtain an exponential model for population growth. (For details on the method here, see Section 4.4.)

   b. Use the formula you found in part a to get $\frac{dN}{dt}$ in 1955.

   c. How does your answer from part b of this exercise compare with your answer from part a of Exercise 1? Does this agree with your answer in part c of Exercise 1?

3. **Deaths from Heart Disease** Tables A and B show the deaths per 100,000 caused by heart disease in the United States for males and females aged 55 to 64 years. The function H_m gives deaths per 100,000 for males, and H_f gives deaths per 100,000 for females.

   a. Approximate the value of $\frac{dH_m}{dt}$ in 2004 using the average rate of change from 2004 to 2007.

   b. Explain the meaning in practical terms of the number you calculated in part a. You should, among other things, tell what the sign means.

TABLE A Heart Disease Deaths per 100,000 for Males Aged 55 to 64 Years

t = year	H_m = deaths per 100,000
1990	537.3
2000	371.7
2003	331.7
2004	312.8
2007	288.8

(continued)

---

[6]The table is based on the study by D. Klein, "The introduction, increase, and crash of reindeer on St. Matthew Island," *J. Wildlife Management* **32** (1968), 350–367.

c. Use your answer from part a to estimate the heart disease death rate for males aged 55 to 64 years in 2006.

d. Approximate the value of $\frac{dH_f}{dt}$ for 2004 using the average rate of change from 2004 to 2007.

e. Explain what your calculations from parts a and d tell you about comparing heart disease deaths for men and women in 2004.

TABLE B Heart Disease Deaths per 100,000 for Females Aged 55 to 64 Years

t = year	H_f = deaths per 100,000
1990	215.7
2000	159.3
2003	141.9
2004	131.5
2007	117.9

4. The Cannon with a Different Muzzle Velocity If the cannonball from Example 6.7 is fired with a muzzle velocity of 370 feet per second, it will follow the graph of

$$h = x - 32\left(\frac{x}{370}\right)^2,$$

where distances are measured in feet.

a. Plot the graph of the flight of the cannonball.

b. Find the height of the cannonball 3000 feet downrange.

c. By looking at the graph of h, determine whether $\frac{dh}{dx}$ is positive or negative at 3000 feet downrange.

d. Calculate $\frac{dh}{dx}$ at 3000 feet downrange and explain what this number means in practical terms.

5. Falling with a Parachute When an average-sized man with a parachute jumps from an airplane, he will fall $S = 12.5(0.2^t - 1) + 20t$ feet in t seconds.

a. Plot the graph of S versus t over at least the first 10 seconds of the fall.

b. How far does the parachutist fall in 2 seconds?

c. Calculate $\frac{dS}{dt}$ at 2 seconds into the fall and explain what the number you calculated means in practical terms.

6. Free Fall Subject to Air Resistance Gravity and air resistance contribute to the characteristics of a falling object. An average-sized man will fall

$$S = 968(e^{-0.18t} - 1) + 176t$$

feet in t seconds after the fall begins.

a. Plot the graph of S versus t over the first 5 seconds of the fall.

b. How far will the man fall in 3 seconds?

c. Calculate $\frac{dS}{dt}$ at 3 seconds into the fall and explain what the number you calculated means in practical terms.

7. A Yam Baking in the Oven A yam is placed in a preheated oven to bake. An application of Newton's law of cooling gives the temperature Y, in degrees, of the yam t minutes after it is placed in the oven as

$$Y = 400 - 325e^{-t/50}.$$

a. Make a graph of the temperature of the yam at time t over 45 minutes of baking time.

b. Calculate $\frac{dY}{dt}$ at the time 10 minutes after the yam is placed in the oven.

c. Calculate $\frac{dY}{dt}$ at the time 30 minutes after the yam is placed in the oven.

d. Explain what your answers in parts b and c tell you about the way the yam heats over time.

8. A Floating Balloon A balloon is floating upward, and its height in feet above the ground t seconds after it is released is given by a function $S = S(t)$.

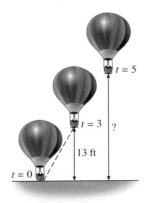

FIGURE 6.43

Suppose you know that $S(3) = 13$ and that $\frac{dS}{dt}$ is 4 when t is 3. Estimate the height of the balloon 5 seconds after it is released. Explain how you got your answer. (See Figure 6.43.)

9. **A Pond**   Water is running out of a pond through a drainpipe. The amount of water, in gallons, in the pond t minutes after the water began draining is given by a function $G = G(t)$.

   a. Explain the meaning in practical terms of $\frac{dG}{dt}$.

   b. While water is running out of the pond, do you expect $\frac{dG}{dt}$ to be positive or negative?

   c. When $t = 30$, water is running out of the drainpipe at a rate of 8000 gallons per minute. What is the value of $\frac{dG}{dt}$?

   d. When $t = 30$, there are 2,000,000 gallons of water in the pond. Using the information from part c, estimate the value of $G(35)$.

10. **Marginal Profit**   *This refers to Example 6.4 of Section 6.2.* A small firm produces at most 20 widgets in a week. Its profit P, in dollars, is a function of n, the number of widgets manufactured in a week, and the formula is

$$P = 26n - n^2.$$

   Recall that the rate of change $\frac{dP}{dn}$ is called the *marginal profit*.

   a. Make a graph of the profit as a function of the number of widgets manufactured in a week.

   b. Determine the marginal profit if the firm produces 10 widgets in a week, and explain what your answer means in practical terms.

   c. Determine the marginal profit if the firm produces 15 widgets in a week, and explain what your answer means in practical terms.

   d. How many widgets should be produced in a week to maximize profit?

   e. Use the calculator to determine the marginal profit at the production level you found in part d. How does your answer compare to what Example 6.4 of Section 6.2 indicates for the marginal profit when the profit is maximized?

6.3 SKILL BUILDING EXERCISES

S-1. **Rate of Change for a Linear Function**   If f is the linear function $f = 7x - 3$, what is the value of $\frac{df}{dx}$?

S-2. **Rate of Change from Data**   Suppose $f = f(x)$ satisfies $f(2) = 5$ and $f(2.005) = 5.012$. Estimate the value of $\frac{df}{dx}$ at $x = 2$.

S-3. **Rate of Change from Data**   Suppose $f = f(x)$ satisfies $f(3) = 8$ and $f(3.005) = 7.972$. Estimate the value of $\frac{df}{dx}$ at $x = 3$.

S-4. **Estimating Rates of Change**   By direct calculation, estimate the value of $\frac{df}{dx}$ for $f(x) = x^2 + 1$ at $x = 3$. Use an increment of 0.0001.

S-5. **Estimating Rates of Change**   By direct calculation, estimate the value of $\frac{df}{dx}$ for $f(x) = 1/x^2$ at $x = 4$. Use an increment of 0.0001.

S-6. **Estimating Rates of Change with the Calculator**   Make a graph of $x^3 - x^2$ and use the calculator to estimate its rate of change at $x = 3$. (We recommend a horizontal span of 0 to 4 and a vertical span of 0 to 30.)

S-7. **Estimating Rates of Change with the Calculator**   Make a graph of $x + (1/x)$ and use the calculator to estimate its rate of change at $x = 3$. (We recommend a horizontal span of 1 to 4 and a vertical span of 0 to 5.)

S-8. **Estimating Rates of Change with the Calculator**   Make a graph of 2^x and use the calculator to estimate its rate of change at $x = 3$. (We recommend a horizontal span of 0 to 4 and a vertical span of 0 to 20.)

S-9. **Estimating Rates of Change with the Calculator**   Make a graph of 3^{-x} and use the calculator to estimate its rate of change at $x = 3$. (We recommend a horizontal span of 0 to 4 and a vertical span of -1 to 1.)

S-10. **Limiting Value of a Rate of Change**   Looking at the graph of 3^{-x} from Exercise S-9, what is $\frac{df}{dx}$ for large values of x?

S-11. **Limiting Value of a Rate of Change**   Look at the graph of $x + (1/x)$ from Exercise S-7. What is $\frac{df}{dx}$ for large values of x?

S-12. **Estimating Rates of Change**   Make a graph of $(x - 1)^2 + 3$ and use the calculator to estimate its rate of change at $x = 1$. (We recommend a horizontal span of 0 to 2 and a vertical span of 0 to 5.) Could you have predicted this rate of change just by looking at the graph?

6.4 EQUATIONS OF CHANGE: LINEAR AND EXPONENTIAL FUNCTIONS

Many natural phenomena can be understood by first analyzing their rates of change, and this analysis can often be used to construct a mathematical model. For example, in 1798 Malthus considered the problem of poverty in terms of the rate of change of population compared to the rate of change of food production: "Population, when unchecked, increases in a geometrical ratio. Subsistence increases only in an arithmetic ratio. A slight acquaintance with numbers will shew the immensity of the first power in comparison with the second."[7] Even earlier, Galileo observed that near the surface of the Earth, downward acceleration due to gravity is constant. Physicists use the letter g to denote this constant, and its value is about 32 feet per second per second. For a freely falling object, the observation that acceleration is constant is actually an observation about $\frac{dV}{dt}$, the rate of change in velocity with respect to time. (We measure downward velocity V in feet per second and time t in seconds.) Furthermore, we can express this in an equation:

$$\text{Acceleration} = \text{Constant}$$
$$\frac{dV}{dt} = g.$$

We will refer to equations of this type as *equations of change* since the equation expresses how the function changes. These equations are a special case of what, in advanced mathematics, science, and engineering, are known more formally as *differential equations*.

Equation of Change for Linear Functions

The key mathematical observation that we need to make about the equation of change $\frac{dV}{dt} = g$ is that it tells us that the rate of change in V is the constant g. But we already know that functions with constant rate of change are linear functions. We conclude that the velocity V is a linear function of time with slope $g = 32$.

Since V is a linear function with slope 32, we know that we can write V as $V = 32t + b$. In order to complete the formula for V, we need only know the initial value b of velocity. For example, if we throw a rock downward with an initial velocity of 10 feet per second, then $b = 10$, and $V = 32t + 10$.

The same analysis holds true for any function with constant rate of change.

> **KEY IDEA 6.5** **EQUATIONS OF CHANGE AND LINEAR FUNCTIONS**
>
> The equation of change $\frac{df}{dx} = m$, where m is a constant, says that f has a constant rate of change m and hence that f is a linear function with slope m. That is, $f = mx + b$. An initial condition is needed to determine the value of b.

We want to emphasize that Key Idea 6.5 is not a new idea. It is just a restatement, in terms of equations of change, of a fundamental property of linear functions that we have already studied in some detail.

---

[7]From page 14 of *First Essay on Population, 1798*, by Thomas Robert Malthus, reprinted in 1926 by Macmillan & Co. Ltd., London.

EXAMPLE 6.8   **DROPPING A ROCK ON MARS**

On Mars, a falling object satisfies the equation of change

$$\frac{dV}{dt} = 12.16,$$

where V is downward velocity in feet per second and t is time in seconds.

Part 1 What is the value of acceleration due to gravity on Mars?

Part 2 Suppose an astronaut stands atop a cliff on Mars and throws a rock downward with an initial velocity of 8 feet per second. What is the velocity of the rock 3 seconds after release?

Solution to Part 1 The equation of change $\frac{dV}{dt} = 12.16$ tells us that the rate of change in velocity for an object falling near the surface of Mars is 12.16 feet per second per second. Since the rate of change in velocity is acceleration, we conclude that acceleration is 12.16 feet per second per second.

Solution to Part 2 The equation of change $\frac{dV}{dt} = 12.16$ tells us that the rate of change in V has a constant value of 12.16. We conclude that V is a linear function with slope 12.16. Since the rock was tossed downward with a velocity of 8 feet per second, the initial value of V is 8. We now have the information we need to write a formula for the function V:

$$V = 12.16t + 8.$$

We want the velocity of the rock when $t = 3$:

$$V = 12.16 \times 3 + 8 = 44.48 \text{ feet per second.}$$

Instantaneous Rates of Change

Linear functions have a constant rate of change, but for other functions this rate varies, and we need to think more carefully about how to interpret the rate of change. When we study equations of change, we are thinking of instantaneous rates of change, not average rates of change.

For example, a car's speedometer reading gives the *instantaneous* rate of change in distance. To understand this, consider the following question: How do we interpret a reading of 60 miles per hour? It would be inaccurate to interpret this reading as saying that we will actually travel 60 miles in the next hour, since the car could slow down or speed up through the hour. It would be more accurate to say that we expect to go 1 mile in the next minute; this is the same average rate of change, but it is computed over a shorter time interval, so there should be less variation in speed. Still more accurate would be the interpretation that we will go one-sixtieth of a mile (88 feet) in a second. In all three interpretations, the average rate of change is 60 miles per hour, but our common sense tells us that the last interpretation is the most accurate, because computing over a time interval of 1 second is the closest to instantaneous change.

We stress that, even though our rates of change are instantaneous, the units used are the same as for average rates of change. For example, our speedometer readings are measured in miles per hour, not in "miles per instant."

Equation of Change for Exponential Functions

We know that the equation of change for a linear function is $\frac{df}{dx} = m$ because that equation tells us that the rate of change is constant. Let's look at the balance in a savings account as a familiar example of an exponential function and find its equation of change. Suppose we open an account by investing \$100 with a financial institution that advertises an APR of 3%, or (as a decimal) 0.03, with interest compounded *continuously*. Our balance $B = B(t)$, with B in dollars and time t in years since opening the account, is an exponential function since it has a constant percentage rate of change. We want to express the fact that B has a constant percentage rate of change using an equation of change. All that is required is to look more closely at what the description of B tells us. The rate of change $\frac{dB}{dt}$ is the amount by which we expect our account balance to increase in a unit of time. This is just the interest. On an annual basis, the interest is 3% of the balance, or $0.03 \times B$. We find that the equation of change is $\frac{dB}{dt} = 0.03B$.

One question that arises is why we use the APR of 0.03 here and not the APY (annual percentage yield), which takes into account the compounding.[8] The reason is that the rate of change is instantaneous, and during an instant we can ignore compounding. We compute on an annual basis, though, because the unit of measurement is a year, not an instant.

Which exponential function does the equation of change $\frac{dB}{dt} = 0.03B$ represent? To answer this we need to find the yearly growth factor for B. We saw in Exercise 19 of Section 4.2 that when interest is compounded continuously, we find the yearly growth factor by exponentiating the APR. Then the growth factor for our exponential function B is $e^{0.03}$. The initial value is 100 dollars (our initial investment), so we have the formula

$$B = 100 \times (e^{0.03})^t,$$

or, since $e^{0.03}$ is about 1.0305, the approximate formula

$$B = 100 \times 1.0305^t.$$

There is an alternative form of B that avoids this last step of computing the exponential of 0.03 (the APR). To find it we use the fact that $(e^{0.03})^t = e^{0.03t}$. Then the alternative form is

$$B = 100e^{0.03t}.$$

This description of the account balance is typical of all exponential functions. If the rate of change in f is a constant multiple of f—that is, if $\frac{df}{dx} = rf$—then f has a constant proportional rate of change and hence is an exponential function. A formula for f in the alternative form is then $f = Pe^{rx}$, where P is the initial value of f. To find the standard form, we first get the growth (or decay) factor by exponentiating:

$$\text{Growth (or decay) factor} = e^r.$$

The formula for f is then $f = P \times (e^r)^x$. Note that finding the alternative form requires one less step, and it is more accurate since it avoids the rounding error in calculating the growth factor.

In this context we refer to the number r as the *exponential growth rate*. This terminology is borrowed from ecology, where the number r is a measure of the per capita growth rate of a population that is assumed to be growing continuously.

---

[8]For loans, financial institutions often refer to the latter rate as the effective annual rate, or EAR.

> **KEY IDEA 6.6** **EQUATIONS OF CHANGE AND EXPONENTIAL FUNCTIONS**
>
> The equation of change $\frac{df}{dx} = rf$, where r is a constant, says that f has a constant proportional (and hence percentage) rate of change and is therefore an exponential function. The exponential growth rate for f is r, so the growth (or decay) factor is e^r. That is,
>
> $$f = Pe^{rx},$$
>
> or
>
> $$f = P \times (e^r)^x,$$
>
> where P is the initial value of f.

As with equations of change for linear functions, we want to emphasize that Key Idea 6.6 is not really new. It is simply a restatement, in terms of equations of change, of the fundamental properties of exponential functions.

EXAMPLE 6.9 NEWTON'S LAW OF COOLING

Several exercises in Chapters 1 and 4 involved the way objects heat or cool. The formulas used in those exercises were all derived from *Newton's law of cooling*. Isaac Newton observed that when a hot object is placed in the open air to cool, the way it cools depends on the difference between the temperature of the object and the temperature of the air. Specifically, if $D = D(t)$ is the difference, in degrees, between the temperature of the object and the temperature of the air, then the rate of change in D is proportional to D. The value of the constant of proportionality depends on the nature of the cooling object and is normally determined experimentally. For a hot cup of coffee placed on the table to cool, if we measure our time t in minutes since the coffee was poured, the value of the cooling constant is about -0.06 per minute. Assume that coffee poured from the pot is at a temperature of 190 degrees and that room temperature is 72 degrees.

Part 1 Write an equation of change that shows how the coffee cools.

Part 2 Describe the function D and give a formula for it.

Part 3 Find a formula for the temperature of the coffee at time t.

Part 4 What will be the temperature of the coffee after 5 minutes? When will the temperature of the coffee be 130 degrees?

Solution to Part 1 The function D is the temperature difference, so the rate of change in temperature difference is $\frac{dD}{dt}$. Newton's law of cooling tells us that the rate of change in temperature difference is -0.06 times D:

Rate of change in temperature difference $= -0.06 \times$ Temperature difference

$$\frac{dD}{dt} = -0.06 \times D.$$

Solution to Part 2 The equation of change we got is that of an exponential function because it gives the rate of change in D as a constant multiple of D.

The constant of proportionality, or cooling constant, is -0.06. Therefore, D is an exponential function with exponential growth rate -0.06. To write a formula for D, we need to know its initial value. When the coffee was poured, its temperature was 190 degrees. Room temperature is 72 degrees. Thus the initial temperature difference, the initial value for D, is $190 - 72 = 118$ degrees, so

$$D = 118e^{-0.06t}.$$

This gives a formula for D in the alternative form. To find the standard form, we first calculate the decay factor: $e^{-0.06}$ is about 0.94. The standard form is then $D = 118 \times 0.94^t$. In the rest of the solution, we will use the alternative form since it is more accurate.

Solution to Part 3 Let $T = T(t)$ denote the temperature, in degrees, of the coffee at time t. We already know a formula for the difference D between coffee temperature and air temperature. To get a formula for T, we need to add the temperature of the air:

$$\text{Temperature difference} = T - \text{Air temperature}$$

$$T = \text{Air temperature} + \text{Temperature difference}$$

$$T = 72 + D$$

$$T = 72 + 118e^{-0.06t}.$$

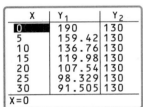

X	Y₁	Y₂
0	190	130
5	159.42	130
10	136.76	130
15	119.98	130
20	107.54	130
25	98.329	130
30	91.505	130
X=0		

FIGURE 6.44 A table of values for a cooling cup of coffee

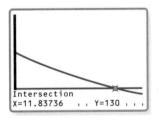

Intersection
X=11.83736 , , Y=130 , , ,

FIGURE 6.45 When the temperature reaches 130 degrees

Solution to Part 4 To get the temperature of the coffee after 5 minutes, we put 5 in place of t in the formula we found in part 3:

$$T(5) = 72 + 118e^{-0.06 \times 5} = 159.42 \text{ degrees.}$$

To answer the second question, we want to know when the temperature of the coffee is 130 degrees. That is, we want to solve for t in the equation

$$72 + 118e^{-0.06t} = 130.$$

We do this using the crossing-graphs method. First we enter the function and the target temperature and record the appropriate correspondences:

$$Y_1 = T, \text{ temperature, in degrees, on vertical axis}$$

$$Y_2 = 130, \text{ target temperature, in degrees}$$

$$X = t, \text{ minutes on horizontal axis.}$$

Next we make the table of values shown in Figure 6.44 to determine how to set the graphing window. The table shows that we can expect the temperature difference to reach 130 degrees between 10 and 15 minutes after the coffee cup is left to cool. Thus we set up the graphing window using a horizontal span of $t = 0$ to $t = 15$ and a vertical span of $T = 100$ to $T = 250$. In Figure 6.45 we made the graphs and then used the calculator to find the intersection point. We see that the coffee reaches a temperature of 130 degrees after about 12 minutes of cooling.

■ Why Equations of Change?

The equation of change that we encountered in Example 6.9 is of the form $\frac{df}{dx} = rf$, and it is a remarkable fact that this same equation actually describes a diverse set of natural phenomena. As you will see in Exercise 6 at the end of this section, this equation describes how a balloon leaks air. It is not apparent that leaky balloons and cooling coffee cups have anything at all in common, but the temperature of the coffee and the volume of air in the balloon actually behave in a very similar fashion. One important power of mathematics is its ability to distill ideas from many areas into a common idea. This is a good case in point. The following table shows a number of phenomena that all obey an equation of change of the form $\frac{df}{dx} = rf$.

Phenomenon	Equation of Change	Comments
Newton's law of cooling	$\dfrac{dD}{dt} = rD$	D is the temperature difference between cooling object and ambient air; r depends on the object.
Exponential population growth	$\dfrac{dN}{dt} = rN$	N is the population size; r is the exponential growth rate.
Compound interest	$\dfrac{dB}{dt} = rB$	B is the account balance; r is the APR with continuous compounding, t in years.
Radioactive decay	$\dfrac{dA}{dt} = rA$	A is the amount of radioactive substance remaining; r depends on the radioactive substance.

These and other phenomena can be modeled by exponential functions because they all obey equations of change of the form $\frac{df}{dx} = rf$. These phenomena also have another important feature, which we will see in the next section: In making a mathematical model, it is easiest first to understand the rate of change and then to make an equation of change to model it. In this section, we have exploited the equation of change to get a formula to serve as our model. In the next section, we will see that there are many other ways to get information from equations of change.

6.4 EXERCISES

Reminder   Round all answers to two decimal places unless otherwise indicated.

1. **Looking Up**   The constant $g = 32$ feet per second per second is the *downward* acceleration due to gravity near the surface of the Earth. If we stand on the surface of the Earth and locate objects using their distance up from the ground, then the positive direction is up, so down is the negative direction. With this perspective, the equation of change in velocity for a freely falling object would be expressed as

$$\frac{dV}{dt} = -g.$$

(We measure upward velocity V in feet per second and time t in seconds.) Consider a rock tossed upward from the surface of the Earth with an initial velocity of 40 feet per second upward.

a. Use a formula to express the velocity $V = V(t)$ as a linear function. (*Hint:* You get the slope of V from the equation of change. The vertical intercept is the initial value.)

b. How many seconds after the toss does the rock reach the peak of its flight? (*Hint:* What is the velocity of the rock when it reaches its peak?)

c. How many seconds after the toss does the rock strike the ground? (*Hint:* How does the time it takes for the rock to rise to its peak compare with the time it takes for it to fall back to the ground?)

2. **An Investment**   You open an account by investing $250 with a financial institution that advertises an APR of 5.25%, with continuous compounding. What account balance would you expect 1 year after making your initial investment?

3. **A Better Investment**   You open an account by investing $250 with a financial institution that advertises an APR of 5.75%, with continuous compounding.

a. Find an exponential formula for the balance in your account as a function of time. In your answer, give both the standard form and the alternative form for an exponential function.

(*continued*)

b. What account balance would you expect 5 years after your initial investment? Answer this question using both of the forms you found in part a. Which do you think gives a more accurate answer? Why?

4. **Baking Muffins** Chocolate muffins are baking in a 350-degree oven. Let $M = M(t)$ be the temperature of the muffins t minutes after they are placed in the oven, and let $D = D(t)$ be the difference between the temperature of the oven and the temperature of the muffins. Then D satisfies Newton's law of cooling, and its equation of change is

$$\frac{dD}{dt} = -0.04D.$$

a. If the initial temperature of the muffins is the temperature of the room, 73 degrees, find a formula for D.

b. Find a formula for M.

c. The muffins will be done when they reach a temperature of 225 degrees. When should we take the muffins out of the oven?

5. **Borrowing Money** Suppose that you borrow $10,000 at 7% APR and that interest is compounded continuously. The equation of change for your account balance $B = B(t)$ is

$$\frac{dB}{dt} = 0.07B.$$

Here t is the number of years since the account was opened, and B is measured in dollars.

a. Explain why B is an exponential function.

b. Find a formula for B using the alternative form for exponential functions.

c. Find a formula for B using the standard form for exponential functions. (Round the growth factor to three decimal places.)

d. Assuming that no payments are made, use your formula from part b to determine how long it would take for your account balance to double.

6. **A Leaky Balloon** In Example 6.5 we used a linear function to describe air leaking from a balloon, but we indicated there that this is not, in fact, an accurate description. Let's look more carefully. Suppose a balloon initially holds 2 liters of air. Let $B = B(t)$ denote the volume, in liters, of air in the balloon t seconds after it starts leaking. When a balloon is almost full, the air pressure inside is large, so it leaks rapidly. When it is almost empty, it leaks more slowly. More formally, the rate of change in B is proportional to B. Assume that for this particular balloon, the constant of proportionality is -0.05.

a. Write an equation of change for B.

b. Find a formula for B.

c. How long will it take for half of the air to leak out of the balloon?

7. **A Population of Bighorn Sheep** A certain group of bighorn sheep live in an area where food is plentiful and conditions are generally favorable to bighorn sheep. Consequently, the population is thriving. There are initially 30 sheep in this group. Let $N = N(t)$ be the population t years later. The population changes each year because of births and deaths. The rate of change in the population is proportional to the number of sheep currently in the population. For this particular group of sheep, the constant of proportionality is 0.04.

a. Express the sentence "The rate of change in the population is proportional to the number of sheep currently in the population" as an equation of change. (Incorporate in your answer the fact that the constant of proportionality is 0.04.)

b. Find a formula for N.

c. How long will it take this group of sheep to grow to a level of 50 individuals?

8. **Water Flow** Water is flowing into a tank at a steady rate of 2 gallons per minute. Let t be the time, in minutes, since the process began, and let V be the volume, in gallons, of water in the tank.

a. Explain why V is a linear function of t, and write an equation of change for V.

b. If initially there were 4 gallons of water in the tank, find a formula for V.

c. The tank can hold 20 gallons. How long will it take to fill the tank?

9. **Growing Child** A certain girl grew steadily between the ages of 3 and 12 years, gaining $5\frac{1}{2}$ pounds each year. Let W be the girl's weight, in pounds, as a function of her age t, in years, between the ages of $t = 3$ and $t = 12$.

a. Is W a linear function or an exponential function? Be sure to explain your reasoning.

b. Write an equation of change for W.

c. Given that the girl weighed 30 pounds at age 3, find a formula for W.

10. Competing Investments You initially invest $500 with a financial institution that offers an APR of 4.5%, with interest compounded continuously. Let B be your account balance, in dollars, as a function of the time t, in years, since you opened the account.

    a. Write an equation of change for B.

    b. Find a formula for B.

    c. If you had invested your money with a competing financial institution, the equation of change for your balance M would have been $\frac{dM}{dt} = 0.04M$. If this competing institution compounded interest continuously, what APR would they offer?

11. Radioactive Decay The amount remaining A, in grams, of a radioactive substance is a function of time t, measured in days since the experiment began. The equation of change for A is

$$\frac{dA}{dt} = -0.05A.$$

    a. What is the exponential growth rate for A?

    b. If initially there are 3 grams of the substance, find a formula for A.

    c. What is the half-life of this radioactive substance? Recall that the half-life of a radioactive substance is the time it takes for half of the substance to change form through radioactive decay.

12. Magazines Two magazines, *Alpha* and *Beta*, were introduced at the same time with the same circulation of 100. The circulation of *Alpha* is given by the function A, which has the equation of change

$$\frac{dA}{dt} = 0.10A.$$

The circulation of *Beta* is given by the function B, which has the equation of change

$$\frac{dB}{dt} = 10.$$

Here t is the time, in years, since the magazines were introduced.

    a. One of these functions is growing in a linear way, whereas the other is growing exponentially. Identify which is which, and find formulas for both functions.

    b. Which magazine is growing more rapidly in circulation? Be sure to explain your reasoning.

6.4 SKILL BUILDING EXERCISES

S-1. Technical Terms What is the common mathematical term for an equation of change?

S-2. Linear Function If f satisfies the equation of change $\frac{df}{dx} = m$, what kind of function is f?

S-3. Slope If f satisfies the equation of change $\frac{df}{dx} = 5$, then f is a linear function. What is the slope of f?

S-4. Exponential Functions If f satisfies the equation of change $\frac{df}{dx} = cf$, what kind of function is f?

S-5. Solving an Equation of Change If f satisfies the equation of change $\frac{df}{dx} = 8f$, then f is an exponential function and hence can be written as $f = Ae^{ct}$. Find the value of c.

S-6. Why Equations of Change List some commonly occurring phenomena that satisfy the equation of change $\frac{df}{dx} = cf$.

S-7. A Leaky Balloon A balloon leaks air (changes volume) at a rate of one-third the volume per minute. Write an equation of change that describes the volume V of air in the balloon at time t in minutes.

S-8. Solving an Equation of Change Solve the equation of change from Exercise S-7 if there are initially 4 liters of air in the balloon.

S-9. Solving an Equation of Change Solve the equation of change $\frac{df}{dx} = 3$ if the initial value of f is 7.

S-10. A Falling Rock Let t be time in seconds, and let $v = v(t)$ be the downward velocity, in feet per second, of a falling rock. The acceleration of the rock has a constant value of 32 feet per second per second. Write an equation of change that is satisfied by v.

S-11. Filling a Tank The water level in a tank rises 4 feet every minute. Write an equation of change that describes the height H, in feet, of the water level at time t in minutes.

S-12. Solving an Equation of Change Solve the equation of change $\frac{df}{dx} = 5f$ if the initial value of f is 2. Use the alternative form for exponential functions.

6.5 EQUATIONS OF CHANGE: GRAPHICAL SOLUTIONS

When an equation of change is of the form $\frac{df}{dx} = c$, we know that f is a linear function with slope c. When it is of the form $\frac{df}{dx} = cf$, we know that f is an exponential function with growth (or decay) factor e^c. Determining which function goes with more complicated equations of change involves calculus and is beyond the scope of this text. Indeed, for many equations of change, even calculus is not powerful enough to allow recovery of the function. Nevertheless, even without being able to find the exact function, we are often able to sketch a graph of it and to find useful information about the phenomena being modeled. The key tool that we will use is the fundamental relationship between a function and its rate of change. When the rate of change is positive, the function is increasing. When the rate of change is negative, the function is decreasing. When the rate of change is momentarily 0, the function may be at a maximum or minimum. When the rate of change is 0 over a period of time, the function is not changing and so is constant over that period.

Equilibrium Solutions

Equilibrium or *steady-state* solutions for equations of change are solutions that never change and hence are constant. Thus they occur where the rate of change is 0 over a period of time. Let's look at an example. The formulas we have used to study falling objects subject to air resistance actually came from an equation of change. When objects fall, they are accelerated downward by gravity. On the other hand, air resistance slows down the fall, which is why a feather falls so slowly. In general, the downward acceleration of a falling object subject to air resistance is given by

$$\text{Acceleration} = \text{Acceleration due to gravity} - \text{Retardation from air resistance.} \quad (6.1)$$

This drag due to air resistance is in practice difficult to determine, and engineers use wind tunnels and other high-tech devices to help in the design of more aerodynamically efficient cars, airplanes, and rockets. The simplest model for air resistance assumes that the drag is proportional to the velocity. That is,

$$\text{Retardation due to drag} = rV,$$

where V is the downward velocity and r is a constant known as the *drag coefficient*. We will show in Exercise 3 one way in which its value can be experimentally determined. Downward acceleration is the rate of change in velocity with respect to time, $\frac{dV}{dt}$. We put these bits of information into Equation (6.1) to get the equation of change:

Acceleration $\dfrac{dV}{dt}$	=	Acceleration due to gravity g	−	Retardation from air resistance rV

Thus the velocity of an object falling subject to air resistance follows the equation of change $\frac{dV}{dt} = g - rV$. The acceleration due to gravity is $g = 32$ feet per second per second, and for an average-sized man, the drag coefficient has been determined to be approximately $r = 0.1818$ per second. Hence a skydiver falls subject to the equation of change

$$\frac{dV}{dt} = 32 - 0.1818V. \quad (6.2)$$

(We measure velocity V in feet per second and time t in seconds.)

Let's find the equilibrium solutions of Equation (6.2) and investigate their physical meaning. Equilibrium solutions are those that never change—that is, those for which the rate of change is 0. To find them, we should put 0 in for $\frac{dV}{dt}$ and solve the resulting equation for V:

$$0 = 32 - 0.1818V.$$

This is a linear equation, which we proceed to solve. (You may if you wish solve it with the calculator.) We get

$$0 = 32 - 0.1818V$$
$$0.1818V = 32$$
$$V = \frac{32}{0.1818} = 176 \text{ feet per second (approximately).}$$

Thus $V = 176$ feet per second is an equilibrium or steady-state solution. When $V = 176$, then $\frac{dV}{dt}$ is zero. Consequently, if the skydiver reaches a velocity of 176 feet per second, he will continue to fall at that same speed until the parachute opens. This is the velocity at which the downward force of gravity matches retardation due to air resistance: the terminal velocity of the skydiver. To emphasize this, we have graphed the steady-state solution $V = 176$ in Figure 6.46. The fact that the graph of velocity versus time is a horizontal line emphasizes that velocity is not changing. Such horizontal lines are typical of equilibrium solutions.

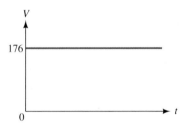

FIGURE 6.46 The equilibrium solution $V = 176$ for the equation of change of a skydiver

There are two important things to note here. One is that the equilibrium solution was easy to find from the equation of change. The second is that the equilibrium solution had an important physical interpretation. It is the velocity we would expect the sky diver to have in the long term. This is typical of steady-state solutions: They yield important information, and that is why we use equations of change to find them.

KEY IDEA 6.7 **EQUILIBRIUM OR STEADY-STATE SOLUTIONS**

Equilibrium or steady-state solutions of an equation of change occur where the rate of change is 0. We get them by setting the rate of change equal to 0 and solving. In many cases they show long-term behavior.

In Section 5.1, we looked at logistic population growth. Logistic growth is one of many growth models that assume there is an upper limit, the *carrying capacity*, beyond which the population cannot grow. In essentially all these models the population grows rapidly at first, but the growth rate slows as the population approaches the carrying capacity. The logistic model is perhaps the simplest model for limited growth. In its original formulation, it was presented as an equation of change.

If a population exhibits logistic growth in an area with a carrying capacity K, then the logistic equation for the population $N = N(t)$ can be written

$$\frac{dN}{dt} = rN\left(1 - \frac{N}{K}\right).$$

The constants K and r depend on the species and on environmental factors, and their values are normally determined experimentally. The number N is sometimes the actual number of individuals, but more commonly it is the combined weight of the population (the *biomass*). In several examples and exercises in this text, we have presented logistic growth with a formula. That formula was obtained by using calculus to solve the equation of change above, a feat we will not reproduce here. Rather, we will gain information directly from the equation of change itself. The logistic model of growth is useful in the study of some species, and we illustrate its use in the case of Pacific sardines (which we earlier considered as Example 5.1 in Section 5.1).

EXAMPLE 6.10 LOGISTIC POPULATION GROWTH: SARDINES

In California in the 1930s and 1940s, a large part of the fishing-related industry was based on the catch of Pacific sardines. Studies have shown that the sardine population grows approximately logistically; moreover, the numbers K and r have been determined experimentally to be $K = 2.4$ million tons and $r = 0.338$ per year.[9] For Pacific sardines, our logistic equation is therefore

$$\frac{dN}{dt} = 0.338N\left(1 - \frac{N}{2.4}\right), \tag{6.3}$$

where N is measured in millions of tons of fish and t is measured in years.

Find the equilibrium solutions of Equation (6.3) and give a physical interpretation of their meaning.

Solution We get the equilibrium solutions where $\frac{dN}{dt}$ is zero. Thus we want to solve the equation

$$0.338N\left(1 - \frac{N}{2.4}\right) = 0.$$

We use the single-graph method to do that. To make and understand the graph, it is crucial that we identify correspondences among the variables. The importance of carefully employing this practice with all graphs made in this section cannot be overemphasized. In this case we have

$Y_1 = \frac{dN}{dt}$, rate of change in biomass, in millions of tons per year, on vertical axis

$X = N$, biomass, in millions of tons, on horizontal axis.

Now we enter the formula for the function $\frac{dN}{dt}$ given in Equation (6.3) and look at the table of values in Figure 6.47 to see how to set up the graphing window. The table shows that $N = 0$ is one equilibrium solution and that another occurs between $N = 2$ and $N = 3$. Allowing a little extra room, we use a window setup with a horizontal span of $N = 0$ to $N = 3$ and a vertical span of $\frac{dN}{dt} = -0.2$ to $\frac{dN}{dt} = 0.3$. This gives the graph in

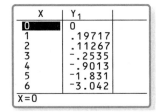

X	Y_1
0	0
1	.19717
2	.11267
3	-.2535
4	-.9013
5	-1.831
6	-3.042

X=0

FIGURE 6.47 A table of values to set up a graphing window

[9]From a study by G. I. Murphy, in "Vital statistics of the Pacific sardine (*Sardinops caerulea*) and the population consequences," *Ecology* **48** (1967), 731–736.

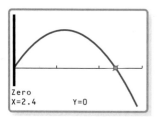

FIGURE 6.48 Getting an equilibrium solution from the graph of $\frac{dN}{dt}$ versus N

Figure 6.48. Using the calculator, we get the second equilibrium solution $N = 2.4$, as shown in Figure 6.48.

We have two equilibrium solutions, $N = 0$ and $N = 2.4$ million tons. If the biomass ever reaches either of these levels, it will remain there and not change. This is certainly obvious when $N = 0$: If there are no fish present, no new ones will be born. The solution $N = 2.4$ million tons is the environmental carrying capacity for sardines in this region. If the biomass ever reaches 2.4 million tons of fish, then environmental limitations match growth tendencies, and the population stays the same.

We emphasize once more that Figure 6.48 does not show the graph of N versus t. It does not directly give the size of the fish biomass. Rather, it is the graph of $\frac{dN}{dt}$ versus N. It shows the growth rate that can be expected for a given biomass. Our next topic is how to use this graph to sketch a graph of the biomass itself.

Sketching Graphs

Equilibrium solutions for equations of change can yield important information about the physical situation that they describe, but the fundamental relationship between a function and its rate of change can tell us even more. To see this, let's continue our analysis of the equation of change $\frac{dN}{dt} = 0.338N\left(1 - \frac{N}{2.4}\right)$ for Pacific sardines. We know that if the biomass today is either 0 or 2.4 million tons, then in the future, the biomass will not change. Can we say what is to be expected to happen in the future if there are today 0.4 million tons of Pacific sardines? The key is the graph of $\frac{dN}{dt}$ versus N that we made in Figure 6.48. Table 6.1 was made by looking at where the graph is above the horizontal axis and where it is below.

TABLE 6.1 The Sign of $\frac{dN}{dt}$

Range for N	From 0 to 2.4	Greater than 2.4
Sign of $\frac{dN}{dt}$	Positive	Negative

The key to understanding what happens to the population N is the fundamental property of rates of change: When $\frac{dN}{dt}$ is positive, N is increasing, and when $\frac{dN}{dt}$ is negative, N is decreasing. In Table 6.2 we have added another row to Table 6.1 to include this information.

TABLE 6.2 The Information We Get from the Graph of $\frac{dN}{dt}$ Versus N

Range for N	From 0 to 2.4	Greater than 2.4
Sign of $\frac{dN}{dt}$	Positive	Negative
Effect on N	Increasing	Decreasing

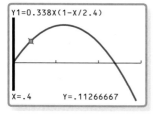

FIGURE 6.49 When $N = 0.4$, the rate of change $\frac{dN}{dt}$ is positive.

The information we get from this table is that whenever N is between 0 and 2.4 million tons, the graph of N versus t will increase; that is, the biomass will grow. When N is greater than 2.4 million tons, the graph of N versus t will decrease; that is, the biomass will decrease over time. Since $N = 0.4$ is between 0 and 2.4, we know that the graph of N versus t will start out increasing. For further evidence of this, in Figure 6.49 we have put the graphing cursor at $N = 0.4$, and we see that the value of $\frac{dN}{dt}$ is 0.11, a positive number. Since $\frac{dN}{dt}$ is positive, N is increasing. Thus if we start with a biomass of 0.4 million tons, we can expect the biomass to increase.

We want to start with this bit of information and make a hand-drawn sketch of the graph of N versus t. We started our picture in Figure 6.50 by first drawing the equilibrium solutions $N = 0$ and $N = 2.4$ and indicating what we already know about how the graph starts out.

To complete the graph of N versus t, we see from Figure 6.48 (or from Table 6.2) that $\frac{dN}{dt}$ remains positive as long as N is less than the equilibrium solution of $N = 2.4$ million tons. This tells us that the graph of N continues to increase toward the horizontal line that represents the equilibrium solution, as we have drawn in Figure 6.51. Note that the horizontal lines representing equilibrium solutions provided us with important guides in sketching the graph of N versus t. Whenever you are asked to sketch a graph from an equation of change, you should first look to see whether there are equilibrium solutions. If so, draw them in before you start.

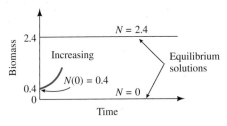

FIGURE 6.50 Equilibrium solutions and beginning the graph of N versus t

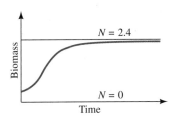

FIGURE 6.51 Completing the graph of biomass versus time

As we expected, the graph in Figure 6.51 looks like the classic logistic curve. But we would emphasize that it is a free-hand drawing and that graphs made in this way cannot be expected to show accurate function values. What is important is to use information about the rate of change properly to get the shape of the graph as nearly correct as is possible.

Let's analyze Pacific sardines in one final situation. What will the biomass be in the future if today it is 2.8 million tons? We proceed as before. In Figure 6.52, on the graph of $\frac{dN}{dt}$ versus N we have put the cursor at $N = 2.8$ million tons. We see that when the biomass is at this level, $\frac{dN}{dt} = -0.16$, a negative number. Thus the graph of N versus t will be decreasing. In Figure 6.53 we have started our graph just as we did in the first case.

We see further from Figure 6.48 (or from Table 6.2) that as long as N is greater than the equilibrium solution $N = 2.4$, $\frac{dN}{dt}$ will remain negative. Thus the graph of N versus t continues to decrease toward the horizontal line that represents the equilibrium solution. Our completed graph is shown in Figure 6.54. The graph in Figure 6.54 is reasonable because the biomass started out greater than the environmental carrying capacity of 2.4 million tons. The environment cannot support that many fish, so we expect the biomass to decrease.

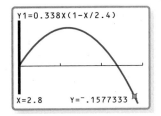

FIGURE 6.52 When biomass is 2.8 million tons, $\frac{dN}{dt}$ is negative.

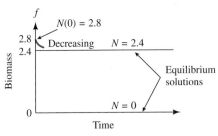

FIGURE 6.53 Starting the graph of N versus t when $N(0) = 2.8$

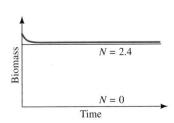

FIGURE 6.54 When population is greater than carrying capacity

> ### KEY IDEA 6.8 | USING THE EQUATION OF CHANGE TO GRAPH FUNCTIONS
>
> To make a hand-drawn sketch of the graph of f versus x from the equation of change for f, first draw horizontal lines representing any obvious equilibrium solutions. Next, use the calculator to graph $\frac{df}{dx}$ versus f. (If the equation of change is of the form
>
> $$\frac{df}{dx} = \text{Right-hand side}$$
>
> you graph the right-hand side.)
>
> 1. Where the graph of $\frac{df}{dx}$ versus f crosses the horizontal axis, we have $\frac{df}{dx} = 0$, so f does not change. This is an equilibrium solution for f.
>
> 2. When the graph of $\frac{df}{dx}$ versus f is above the horizontal axis, the graph of f versus x is increasing.
>
> 3. When the graph of $\frac{df}{dx}$ versus f is below the horizontal axis, the graph of f versus x is decreasing.

EXAMPLE 6.11 FURTHER ANALYSIS OF FALLING OBJECTS

Earlier we found the equilibrium solution $V = 176$ feet per second for the equation of change

$$\frac{dV}{dt} = 32 - 0.1818V$$

which governs the velocity of a skydiver.

Part 1 Sketch the graph of the skydiver's velocity if the initial downward velocity is 0. (This might occur if the skydiver jumped from an airplane that was flying level.)

Part 2 Sketch the graph of the skydiver's velocity if the initial downward velocity is 200 feet per second. (This might occur if a pilot ejected from an airplane that was diving out of control.)

Solution to Part 1 We enter the right-hand side of the equation of change in the calculator and record the important correspondences:

$$Y_1 = \frac{dV}{dt}, \text{ acceleration, in feet per second per second, on vertical axis}$$

$$X = V, \text{ velocity, in feet per second, on horizontal axis.}$$

To get the graph of $\frac{dV}{dt}$ versus V shown in Figure 6.55, we wanted to be sure the value $V = 200$, which will be needed in part 2, was included. We used a horizontal span of $V = 0$ to $V = 250$, and we got the vertical span of $\frac{dV}{dt} = -20$ to $\frac{dV}{dt} = 35$ from a table of values. We know already that the crossing point $V = 176$ shown in Figure 6.55 corresponds to the equilibrium solution that represents terminal velocity.

We have recorded in Table 6.3 the information about V that we get from Figure 6.55. Since the initial velocity for this part is 0, we know from Table 6.3 that the graph of V starts out increasing. We make the graph of V versus t in two steps. In Figure 6.56 we have graphed the equilibrium solution and started our increasing graph. Consulting Table 6.3 once again, we see that velocity will continue to increase toward the terminal velocity of 176 feet per second. Thus the graph will continue to increase toward the horizontal line representing this equilibrium solution. Our graph is shown in Figure 6.57.

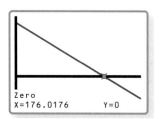

Zero
X=176.0176   Y=0

FIGURE 6.55 The graph of $\frac{dV}{dt}$, acceleration, versus velocity V

TABLE 6.3 The Information We Get from the Graph of $\frac{dV}{dt}$ Versus V

Range for V	Less than 176	Greater than 176
Sign of $\dfrac{dV}{dt}$	Positive	Negative
Effect on V	Increasing	Decreasing

Solution to Part 2 Since the initial velocity is 200 feet per second, we see from Table 6.3 that velocity is decreasing. This bit of information is incorporated into the start of our graph in Figure 6.58. But Table 6.3 also shows that velocity will continue to decrease toward the equilibrium solution of 176 feet per second. Thus, in Figure 6.59 we completed the graph by making it decrease toward the horizontal line corresponding to terminal velocity. We note that the graph in Figure 6.59 makes sense because, at the skydiver's large initial velocity, the retardation due to air resistance is greater than the downward acceleration due to gravity. Thus his speed will decrease toward the terminal velocity.

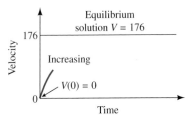

FIGURE 6.56 Starting the graph of V versus t

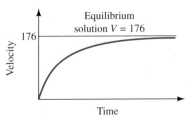

FIGURE 6.57 Completing the graph of V versus t

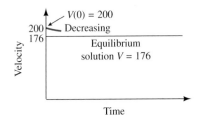

FIGURE 6.58 Starting the graph when $V(0) = 200$

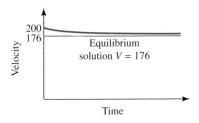

FIGURE 6.59 The completed graph of V versus t for $V(0) = 200$

6.5 EXERCISES

Reminder Round all answers to two decimal places unless otherwise indicated.

1. **Equation of Change for Logistic Growth** The logistic growth formula $N = 6.21/(0.035 + 0.45^t)$ that we used in Chapter 2 for deer on the George Reserve actually came from the following equation of change:

$$\frac{dN}{dt} = 0.8N\left(1 - \frac{N}{177}\right).$$

a. Plot the graph of $\frac{dN}{dt}$ versus N, and use it to find the equilibrium solutions. Explain their physical significance.

b. In one plot, sketch the equilibrium solutions and graphs of N versus t in each of the two cases $N(0) = 10$ and $N(0) = 225$.

c. To what starting value for N does the solution $N = 6.21/(0.035 + 0.45^t)$ correspond?

2. **A Catfish Farm** Catfish in a commercial pond can be expected to exhibit logistic population growth. Consider a pond with a carrying capacity of $K = 4000$ catfish. Take the r value for catfish in this pond to be $r = 0.06$.

   a. Write the equation of change for logistic growth of the catfish population. (*Hint:* If you have difficulty here, refer to Example 6.10.)

   b. Make a graph of $\frac{dN}{dt}$ versus N.

   c. For what values of N would the catfish population be expected to increase?

   d. For what values of N would the catfish population be expected to decrease?

   e. Recall from Section 5.1 that the *maximum sustainable yield* model says that a renewable resource should be maintained at a level where its growth rate is at a maximum, since this allows the population to replenish itself quickly. According to this model, at what level should the catfish population be maintained?

3. **Experimental Determination of the Drag Coefficient** When retardation due to air resistance is proportional to downward velocity V, in feet per second, falling objects obey the equation of change

   $$\frac{dV}{dt} = 32 - rV,$$

   where r is known as the *drag coefficient*. One way to measure the drag coefficient is to measure and record terminal velocity.

   a. We know that an average-sized man has a terminal velocity of 176 feet per second. Use this to show that the value of the drag coefficient is $r = 0.1818$ per second. (*Hint:* To say that the terminal velocity is 176 feet per second means that when the velocity V is 176, velocity will not change. That is, $\frac{dV}{dt} = 0$. Put these bits of information into the equation of change and solve for r.)

   b. An ordinary coffee filter has a terminal velocity of about 4 feet per second. What is the drag coefficient for a coffee filter?

4. **Other Models of Drag Due to Air Resistance** When some objects move at high speeds, air resistance has a more pronounced effect. For such objects, retardation due to air resistance is often modeled as being proportional to the square of velocity or to even higher powers of velocity.

   a. A rifle bullet fired downward has an initial velocity of 2100 feet per second. If we use the model that gives air resistance as the square of velocity,

   then the equation of change for the downward velocity V, in feet per second, of a rifle bullet is

   $$\frac{dV}{dt} = 32 - rV^2.$$

   If the terminal velocity for a bullet is 1600 feet per second, find the drag coefficient r.

   b. A meteor may enter the Earth's atmosphere at a velocity as high as 90,000 feet per second. If the downward velocity V, in feet per second, of a meteor in the Earth's atmosphere is governed by the equation of change

   $$\frac{dV}{dt} = 32 - 2 \times 10^{-18}V^4,$$

   how fast will it be traveling when it strikes the ground, assuming that it has reached terminal velocity?

5. **Fishing for Sardines** *This is a continuation of Example 6.10.* If we take into account an annual fish harvest of F million tons of fish, then the equation of change for Pacific sardines becomes

   $$\frac{dN}{dt} = 0.338N\left(1 - \frac{N}{2.4}\right) - F.$$

   a. Suppose that there are currently 1.8 million tons of Pacific sardines off the California coast and that you are in charge of the commercial fishing fleet. It is your goal to leave the Pacific population of sardines as you found it. That is, you wish to set the fishing level F so that the biomass of Pacific sardines remains stable. What value of F will accomplish this? (*Hint:* You want to choose F so that the current biomass level of 1.8 million tons is an equilibrium solution.)

   b. For the remainder of this exercise, take the value of F to be 0.1 million tons per year. That is, assume the catch is 100,000 tons per year.

   i. Make a graph of $\frac{dN}{dt}$ versus N, and use it to find the equilibrium solutions.

   ii. For what values of N will the biomass be increasing? For what values will it be decreasing?

   iii. On the same graph, sketch all equilibrium solutions and the graphs of N versus t for each of the initial populations $N(0) = 0.3$ million tons, $N(0) = 1.0$ million tons, and $N(0) = 2.3$ million tons.

   iv. Explain in practical terms what the picture you made in part iii tells you. Include in your explanation the significance of the equilibrium solutions.

(continued)

6. **Sprinkler Irrigation in Nebraska**   Logistic growth can be used to model not only population growth but also economic and other types of growth. For example, the total number of acres $A = A(t)$, in millions, in Nebraska that are being irrigated by modern sprinkler systems has shown approximate logistic growth since 1955, closely following the equation of change

$$\frac{dA}{dt} = 0.15A\left(1 - \frac{A}{3}\right).$$

Here time t is measured in years.

   a. According to this model, how many total acres in Nebraska can be expected eventually to be irrigated by sprinkler systems? (*Hint:* This corresponds to the carrying capacity in the logistic model for population growth.)

   b. How many acres of land were under sprinkler irrigation when sprinkler irrigation was expanding at its most rapid rate?

7. **Logistic Growth with a Threshold**   Most species have a survival *threshold* level, and populations of fewer individuals than the threshold cannot sustain themselves. If the carrying capacity is K and the threshold level is S, then the logistic equation of change for the population $N = N(t)$ is

$$\frac{dN}{dt} = -rN\left(1 - \frac{N}{S}\right)\left(1 - \frac{N}{K}\right).$$

For Pacific sardines, we may use $K = 2.4$ million tons and $r = 0.338$ per year, as in Example 6.10. Suppose we also know that the survival threshold level for the sardines is $S = 0.8$ million tons.

   a. Write the equation of change for Pacific sardines under these conditions.

   b. Make a graph of $\frac{dN}{dt}$ versus N and use it to find the equilibrium solutions. How do the equilibrium solutions correspond with S and K?

   c. For what values of N is the graph of N versus t increasing, and for what values is it decreasing?

   d. Explain what can be expected to happen to a population of 0.7 million tons of sardines.

   e. At what population level will the population be growing at its fastest?

8. **Chemical Reactions**   In a *second-order reaction*, one molecule of a substance A collides with one molecule of a substance B to produce a new substance, the *product*. If t denotes time and $x = x(t)$ denotes the concentration of the product, then its rate of change $\frac{dx}{dt}$ is called the *rate of reaction*. Suppose the initial concentration of A is a and the initial concentration of B is b. Then, assuming a constant temperature, x satisfies the equation of change

$$\frac{dx}{dt} = k(a - x)(b - x)$$

for some constant k. This is because the rate of reaction is proportional both to the amount of A that remains untransformed and to the amount of B that remains untransformed. Here we study a reaction between isobutyl bromide and sodium ethoxide in which $k = 0.0055$, $a = 51$, and $b = 76$. The concentrations are in moles per cubic meter, and time is in seconds.[10]

   a. Write the equation of change for the reaction between isobutyl bromide and sodium ethoxide.

   b. Make a graph of $\frac{dx}{dt}$ versus x. Include a span of $x = 0$ to $x = 100$.

   c. Explain what can be expected to happen to the concentration of the product if the initial concentration of the product is 0.

9. **Competition Between Bacteria**   Suppose there are two types of bacteria, type A and type B, in a place with limited resources. If the bacteria had unlimited resources, both types would grow exponentially, with exponential growth rates a for type A and b for type B. Let P be the *proportion* of type A bacteria, so P is a number between 0 and 1. For example, if $P = 0$, then there are no type A bacteria and all are of type B. If $P = 0.5$, then half of the bacteria are of type A and half are of type B. Each population of bacteria grows in competition for the limited resources, so the proportion P changes over time. The function P is subject to the equation of change

$$\frac{dP}{dt} = (a - b)P(1 - P).$$

Suppose $a = 2.3$ and $b = 1.7$.

   a. Does P have a logistic equation of change?

   b. What happens to the populations if initially $P = 0$?

   c. What happens to P in the long run if $P(0)$ is positive (but not zero)? In this case, does it matter what the exact value of $P(0)$ is?

---

[10]From a study by I. Dostrovsky and E. D. Hughes, as described by Gordon M. Barrow in *Physical Chemistry*, 4th ed. (New York: McGraw-Hill, 1979).

10. **Growth of Fish**   Let $w = w(t)$ denote the weight of a fish as a function of its age t. For the North Sea cod, the equation of change

$$\frac{dw}{dt} = 2.1w^{2/3} - 0.6w$$

holds. Here w is measured in pounds and t in years.

   a. Explain what $\frac{dw}{dt}$ means in practical terms.

   b. Make a graph of $\frac{dw}{dt}$ versus w. Include weights up to 45 pounds.

   c. What is the weight of the cod when it is growing at the greatest rate?

   d. To what weight does the cod grow?

11. **Grazing Sheep**   The amount C of food consumed in a day by a merino sheep is a function of the amount V of vegetation available. The equation of change for C is

$$\frac{dC}{dV} = 0.01(2.8 - C).$$

Here C is measured in pounds and V in pounds per acre.

   a. Explain what $\frac{dC}{dV}$ means in practical terms.

   b. Make a graph of $\frac{dC}{dV}$ versus C. Include consumption levels up to 3 pounds.

   c. What is the most you would expect a merino sheep to consume in a day?

6.5 SKILL BUILDING EXERCISES

S-1. Equilibrium Solutions   What is an equilibrium solution of an equation of change?

S-2. More Equilibrium Solutions   If $f = 10$ is an equilibrium solution of an equation of change involving $\frac{df}{dx}$, what is the value of $\frac{df}{dx}$ when f is 10?

S-3. Finding Equilibrium Solutions   Find an equilibrium solution of $\frac{df}{dx} = 2f - 6$.

S-4. Sketching Graphs   Consider the equation of change $\frac{df}{dx} = 2f - 6$.

   a. What is the sign of $\frac{df}{dx}$ when f is less than 3?

   b. What can you say about the graph of f versus x when f is less than 3?

   c. What is the sign of $\frac{df}{dx}$ when f is greater than 3?

   d. What can you say about the graph of f versus x when f is greater than 3?

S-5. Water   Water flows into a tank, and a certain part of it drains out through a valve. The volume v in cubic feet of water in the tank at time t satisfies the equation $\frac{dv}{dt} = 5 - (v/3)$. If the process continues for a long time, how much water will be in the tank?

S-6. Water   At some time in the process described in Exercise S-5, there are 23 cubic feet of water in the tank. Is the volume of water increasing or decreasing?

S-7. Water   At some time in the process described in Exercise S-5, there are 8 cubic feet of water in the tank. Is the volume increasing or decreasing?

S-8. Population   A certain population grows according to $\frac{dN}{dt} = 0.03N\left(1 - \frac{N}{6300}\right)$. What is the carrying capacity of the environment for this particular population?

S-9. Population   At some time, the population described in Exercise S-8 is at the level $N = 4238$. Is the population level increasing or decreasing?

S-10. Population   At some time, the population described in Exercise S-8 is at the level $N = 8716$. Is the population level increasing or decreasing?

S-11. Equation of Change   For the equation of change $\frac{df}{dx} = 5f - 7$, determine whether f is increasing or decreasing when $f = 1$.

S-12. Equation of Change   For the equation of change $\frac{df}{dx} = f(2f - 1)$, determine whether f is increasing or decreasing when $f = 1$.

CHAPTER 6   SUMMARY

This chapter develops the notion of the rate of change for a function. Earlier in the book, rates of change arose in contexts such as studying tabulated functions using average rates of change, analyzing graphs in terms of concavity, and characterizing linear functions as having a constant rate of change. In the current chapter we see the value of a unified approach to rates of change, and we apply this point of view to a variety of important real-world problems.

6.1 VELOCITY

Velocity is the rate of change in location, or directed distance. There are just a few basic rules that enable us to relate directed distance and velocity.

- When directed distance is increasing, velocity is positive.
- When directed distance is decreasing, velocity is negative.
- When directed distance is not changing, even for an instant, velocity is zero.
- When velocity is constant (for example, when the cruise control on a car is set), directed distance is a linear function of time, so its graph is a straight line.

Using these basic rules, we can sketch graphs of directed distance and velocity from a verbal description of how an object moves.

6.2 RATES OF CHANGE FOR OTHER FUNCTIONS

The idea of velocity as the rate of change in directed distance applies to any function. The rate of change of a function $f = f(x)$ is denoted by $\frac{df}{dx}$. It tells how much f is expected to change if x increases by 1 unit. For example, if V is the velocity of an object as a function of time t, then $\frac{dV}{dt}$ is the change in velocity that we expect in 1 unit of time. This is the acceleration of the object. For example, a car might gain 2 miles per hour in velocity each second.

As with velocity, there are just a few basic rules that express the fundamental relationship between a function f and its rate of change $\frac{df}{dx}$.

- When f is increasing, $\frac{df}{dx}$ is positive.
- When f is decreasing, $\frac{df}{dx}$ is negative.
- When f is not changing, $\frac{df}{dx}$ is zero.
- When $\frac{df}{dx}$ is constant, f is a linear function of x.

One important observation is that we expect the rate of change to be zero at a maximum or minimum of a function. For example, say the CEO of a tire company wants to maximize profit by adjusting the production level. Figure 6.60 is a graph of marginal profit, which is the rate of change in profit as a function of the production level. Using the graph, we see that marginal profit is positive if the production level is less than 160 tires per day, so the profit increases up to this level. If the level is more than 160 tires per day, then marginal profit is negative, and the profit is decreasing. The maximum profit occurs where the marginal profit is zero, and that is at a production level of 160 tires per day.

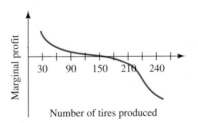

FIGURE 6.60 Marginal profit for a tire company

6.3 ESTIMATING RATES OF CHANGE

The rate of change of a function is again a function, and estimating the rate of change at a point can be of practical importance. The method we use depends on how the function is presented.

For a function $f = f(x)$ given by a table, we can estimate $\frac{df}{dx}$ by using the average rate of change given by

$$\frac{\text{Change in } f}{\text{Change in } x}.$$

For a function given by a formula, we can make a graph and then have the calculator estimate the rate of change at a point.

6.4, 6.5 EQUATIONS OF CHANGE

An equation of change for a function $f = f(x)$ is an equation expressing the rate of change $\frac{df}{dx}$ in terms of f and x. Equations of change arise in many applications where we are given a formula not for the function itself but for its rate of change.

The equation of change $\frac{df}{dx} = m$, where m is a constant, says that f has a constant rate of change m, so f is a linear function with slope m. Once we know the initial value b of f, we can write down the linear formula $f = mx + b$.

The equation of change $\frac{df}{dx} = rf$, where r is a constant, says that f has a constant proportional (or percentage) of change, so f is an exponential function. The exponential growth rate is r, so the growth (or decay) factor is e^r. Once we know the initial value P of f, we can write down the exponential formula $f = Pe^{rx}$, or $f = P \times (e^r)^x$.

Many equations of change that arise in practice do not have the simple characterization we found for linear and exponential functions. But we can still get important information about a function by using the fundamental relationship between the function and its rate of change. The first step is to find the equilibrium solutions of an equation of change. These are the constant solutions, so we find them by setting the rate of change equal to zero. In many cases they show the long-term behavior we can expect. The next step after finding equilibrium solutions is to determine for what values of the function its rate of change is positive or negative, since this will tell us where the function is increasing or decreasing. In many cases of interest, these two steps can be accomplished by analyzing a graph of the rate of change versus the function. After these two steps, we can sketch a graph of the function for a given initial condition.

For example, the population N of deer in a certain area as a function of time t in years has the logistic equation of change

$$\frac{dN}{dt} = 0.5N\left(1 - \frac{N}{150}\right).$$

A graph of $\frac{dN}{dt}$ versus N is shown in Figure 6.61. (We used a horizontal span of 0 to 160 and a vertical span of -5 to 20.) In addition to the equilibrium solution $N = 0$, from the graph we find the equilibrium solution $N = 150$. This is the first step. For the second step we make Table 6.4 on the basis of the graph in Figure 6.61.

The table tells us, for example, that an initial population of 10 deer should increase in time to a size of 150.

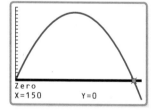

Zero
X=150   Y=0

FIGURE 6.61 A graph of $\frac{dN}{dt}$ versus N

TABLE 6.4 The Information We Get from the Graph of $\frac{dN}{dt}$ Versus N

Range for N	**From 0 to 150**	**Greater than 150**
Sign of $\dfrac{dN}{dt}$	Positive	Negative
Effect on N	Increasing	Decreasing

CHAPTER 6   REVIEW EXERCISES

Reminder   Round all answers to two decimal places unless otherwise indicated.

1. **Maximum Distance**   A graph of directed distance reaches a maximum. What is the velocity there?

2. **Constant Slope**   The graph of directed distance is a straight line with slope -3. What can you say about the velocity?

3. **A Casual Walk**   You leave your home and walk east at a constant speed for 10 minutes until you reach a park. You rest there for 5 minutes and then continue your walk east. Measure your location as your distance east of your home, and make graphs of your location and velocity.

4. **Velocity Given**   A particle is moving in a straight line, and its velocity V at time t is given by the formula $V = 10 + 2t - 3t^2$ for t between 0 and 3.

   a. Use your calculator to plot the graph of V versus t.

   b. Sketch a graph of directed distance for this particle.

5. **Marginal Profit**   The profit P for a firm depends on the number n of widgets made, and the rate of change $\frac{dP}{dn}$ is negative. What is the effect on profit of increasing the number of widgets made?

6. **Water Depth**   A jug is being filled with water, and the depth H of water in the jug is a function of time t. At time $t = 5$ the process stops, and no more water is added. What is the rate of change $\frac{dH}{dt}$ after time $t = 5$?

7. **Helicopter**   The altitude A of a helicopter is a function of time t.

   a. Explain the meaning of $\frac{dA}{dt}$ in practical terms.

   b. What does it mean if $\frac{dA}{dt}$ is zero for a period of time?

   c. The helicopter is coming down for a landing. Is $\frac{dA}{dt}$ positive or negative?

8. **Account Balance**   The balance B, in dollars, of your checking account is a function of t, the number of years since the account was opened. The rate of change $\frac{dB}{dt}$ in the balance is given by the formula

$$\frac{dB}{dt} = 10 + 5t - 3t^2.$$

   a. Use your calculator to make a graph of $\frac{dB}{dt}$ versus t covering the first 3 years.

   b. During what period is the balance B increasing?

   c. At what time is the balance B at its maximum?

9. **Rate of Change from Data**   Suppose $f = f(x)$ satisfies $f(1) = 3$ and $f(1.002) = 3.012$. Estimate the value of $\frac{df}{dt}$ at $x = 1$.

10. **Estimating Rates of Change with the Calculator**   Make a graph of $x^2 - (5/x)$ and use the calculator to estimate its rate of change at $x = 4$. (We recommend a horizontal span of 1 to 5 and a vertical span of -5 to 25.)

11. **SARS**   The following table shows the cumulative number of cases of SARS (severe acute respiratory syndrome) on selected days during the outbreak in 2003. Here t is time in days since the beginning of April, and N is the cumulative number of cases reported by time t.

t	19	22	24	29
N	3547	3947	4439	5642

   a. Approximate the value of $\frac{dN}{dt}$ at $t = 19$ using the average rate of change from $t = 19$ to $t = 22$.

   b. Explain the meaning in practical terms of the number you calculated in part a.

   c. Approximate the value of $\frac{dN}{dt}$ at $t = 24$ using the average rate of change from $t = 24$ to $t = 29$. What do you learn by comparing this result to the result in part a?

12. **Account Balance**   The balance B, in dollars, of your checking account is a function of t, the number of months since the account was opened. The balance is given by the formula

$$B = 1150 + 100t - 9t^2.$$

   a. Use your calculator to make a graph of B versus t covering the first 15 months.

   b. Calculate $\frac{dB}{dt}$ at $t = 8$.

   c. Explain what the number you calculated in part b means in practical terms.

13. **Finding an Equation of Change**   The depth of water in a pool that is being drained decreases by 0.5 foot every minute. Write an equation of change that describes the depth H, in feet, of water in the pool at time t, in minutes.

14. Solving an Equation of Change   Solve the equation of change $\frac{df}{dx} = 6f$ if the initial value of f is 3. Use the alternative form for exponential functions.

15. Traffic Signals   The number n of seconds for the yellow light of a traffic signal is a function of the width w, in feet, of the crossing street.

a. Explain the meaning of $\frac{dn}{dw}$ in practical terms.

b. In certain settings the equation of change for n as a function of w is

$$\frac{dn}{dw} = 0.02,$$

and the number n is 4.7 seconds when the width w is 70 feet. Find a formula for n as a function of w.

16. Atmospheric Pressure   The atmospheric pressure P, in grams per square centimeter, is a function of the altitude h, in kilometers, above the surface of Earth.

a. Explain the meaning of $\frac{dP}{dh}$ in practical terms.

b. The equation of change for P as a function of h is

$$\frac{dP}{dh} = -0.12P.$$

Is P a linear function or an exponential function?

c. The atmospheric pressure on Earth's surface is 1035 grams per square centimeter. Use this and the equation of change in part b to find a formula for P.

17. Equilibrium Solution   Find an equilibrium solution of the equation of change $\frac{df}{dx} = 7f - 14$.

18. Equation of Change   For the equation of change $\frac{df}{dx} = 7f - 14$, determine whether f is increasing or decreasing when $f = 1$.

19. Sales Growth   A study of the sales s, in thousands of dollars, of a product as a function of time t, in years, yields the equation of change

$$\frac{ds}{dt} = 0.3s(4 - s).$$

This is valid for s less than 5.

a. What level of sales will be attained in the long run?

b. What is the largest rate of growth in sales?

20. Critical Threshold   One equation of change that arises in the study of epidemics is

$$\frac{dN}{dt} = -rN\left(1 - \frac{N}{T}\right).$$

Here r and T are positive constants, and T is called the *threshold level*. In this problem assume that $r = 0.5$ and $T = 10$.

a. Write the equation of change for N under these assumptions.

b. Make a graph of $\frac{dN}{dt}$ versus N and use it to find the equilibrium solutions. (Include a span of $N = 0$ to $N = 12$.) Does one of the equilibrium solutions correspond to the threshold level?

c. For what values of N is the graph of N versus t increasing, and for what values is it decreasing?

d. What happens if initially N is close to 10 but smaller than 10? What if initially N is close to 10 but larger than 10? (*Note*: Such an equilibrium solution is said to be *unstable*.)

Section A-1: Order of Operations

Once a mathematical expression has been written in typewriter notation, there is a collection of rules, the *order of operations*, that mathematicians have adopted so that expressions will be unambiguous. Calculators more or less follow these rules, but sometimes they exhibit inconsistencies. These inconsistencies can always be avoided by the proper use of parentheses.

Order of operations rule 1   Expressions inside parentheses are to be evaluated first. If parentheses are nested, innermost pairs are to be evaluated first.

Order of operations rule 2   If there are no parentheses to act as a guide, exponentials are calculated first, multiplications and divisions second, and sums and differences last.

Order of operations rule 3   If two or more from the list of multiplications, divisions, or subtractions occur in sequence with no parentheses to act as a guide, then the calculation should be completed working left to right.

Order of operations rule 4   If two or more exponentials occur in sequence with no parentheses to act as a guide, then the calculation should be completed working from right to left.

The rules lead us correctly even through complicated calculations such as

$$3 + 4 \wedge 2 \times 2 - 12 \div 3 \times 4.$$

$$3 + \boxed{4 \wedge 2} \times 2 - 12 \div 3 \times 4 \quad \text{Exponentials first (rule 2)}$$

$$3 + \boxed{16 \times 2} - 12 \div 3 \times 4 \quad \text{Multiplications and divisions next, from left to right}$$

$$3 + 32 - \boxed{12 \div 3} \times 4$$

$$3 + 32 - \boxed{4 \times 4}$$

$$3 + 32 - 16 = 19 \quad \text{Sums and differences last (rule 2)}$$

If there are parentheses present, we apply rules 2, 3, and 4 as required inside an innermost pair of parentheses. This applies to an expression such as

$$(2 + 3)(12 - (8 \div 2 + 5)).$$

$$(\boxed{2 + 3})(12 - (\boxed{8 \div 2} + 5)) \quad \text{Operations in innermost parentheses (division first)}$$

$$5 \times (12 - (\boxed{4 + 5})) \quad \text{Continue working in innermost parentheses}$$

$$5 \times (\boxed{12 - 9}) \quad \text{Inside parentheses first}$$

$$5 \times 3 = 15$$

Rules 3 and 4 occur less often than do rules 1 and 2. The reader should verify the following calculations using rules 3 and 4:

$$16 \div 4 \div 2 = 2$$

$$2 \wedge 3 \wedge 2 = 512.$$

PRACTICE PROBLEMS

Use the rules for order of operations to calculate the following by hand.

A-1. $9 \div (2 + 1)$ Answer: 3

A-2. $(3 - 5) \div (4 - 2)$ Answer: -1

A-3. $(2 \times 12) \div (4 \times 2)$ Answer: 3

A-4. $24 \div (12 \div 3)$ Answer: 6

A-5. $2 \times 3 + 4$ Answer: 10

A-6. $2 \times 3 + 4 \times 5$ Answer: 26

A-7. $4 \times 2 - 3 \times 3$ Answer: -1

A-8. $6 \div 3 \times 3 \wedge 2$ Answer: 18

A-9. $5 - 4 \times 3 + 8 \div 2 \times 3$ Answer: 5

A-10. $2 \wedge -1 \wedge 2$ Answer: 2

Section A-2: Definition of a Function

At the most elementary level, it is sufficient to think of a function as a rule that tells how one thing depends on another. But in more advanced mathematics, it is important to give precise definitions of the terms we use. The precise definition of a function involves three parts: a set called the *domain* or *input values*, a set called the *range* or *output values*, and a correspondence that assigns to each element of the domain exactly one element of the range.

One of the most common ways we see functions given in elementary mathematics is by formulas. When functions are given by formulas, the domain may not be specified. Normally the domain is assumed to be the largest collection of real numbers for which the formula makes sense. For example, the function f given by the formula $f(x) = 1/x$ has domain all real numbers other than 0. We exclude $x = 0$ because division by 0 is not an allowable operation. For any real number except 0, we find the output value or *function value* by putting the input value in place of x in the formula. So to find $f(3)$ we put 3 in place of x in the formula $1/x$. The result is $f(3) = 1/3$.

In a similar fashion, the function g defined by $g(x) = \sqrt{x}$ has domain all non-negative real numbers. That is because we cannot take the square root of a negative number (without introducing complex numbers). When a function is given by a formula, a good strategy for finding the domain is to rule out the numbers for which the formula does not make sense. The remaining numbers will form the domain.

The range of a function is the set of all outputs of the function. That is, the range is all numbers y such that $f(x) = y$ for some real number x.

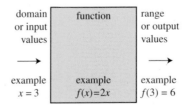

We often use y in place of $f(x)$ and we say that y *is a function of x*. But not all formulas give functions. For example, $y = \pm\sqrt{x}$ does not determine y as a function of x. To see this, consider $x = 4$. The formula assigns two values for y: $y = 2$ and $y = -2$. This violates the rule that a function must assign exactly one y value for each x value.

PRACTICE PROBLEMS

A-1. If $f(x) = \dfrac{x + 3}{x - 1}$, find $f(2)$.

Answer: 5

A-2. If $f(x) = x + \sqrt{x}$, find $f(9)$.

Answer: 12

A-3. If $f(x) = x + 1$, find $f(x + 2)$. *Hint:* Replace x in the formula by $x + 2$.

Answer: $x + 3$

A-4. If $f(x) = x - x^2$, find $f(x^2)$. *Hint:* Replace x in the formula by x^2.

Answer: $x^2 - x^4$

A-5. If $f(x) = \dfrac{x + 1}{x - 1}$, find the domain of f.

Answer: All real numbers except 1.

A-6. If $f(x) = \dfrac{1}{x(x - 3)}$, find the domain of f.

Answer: All real numbers except 0 and 3.

A-7. If $f(x) = \sqrt{x - 4}$, find the domain of f.

Answer: All real numbers greater than or equal to 4.

A-8. Does the equation $y^3 = x$ determine y as a function of x?

Answer: Yes

A-9. Does the equation $y^4 = 2x$ determine y as a function of x?

Answer: No

A-10. If $f(x) = 5 \times 0.5^{x/4}$, calculate $f(0)$ and $f(5)$. [Use two decimal places for $f(5)$.]

Answer: $f(0) = 5$; $f(5) = 2.10$

Section A-3: Geometric Constructions

Many classic problems in mathematics involve geometric figures, and solving these problems often requires expressing one geometric property in terms of another. Some basic formulas are needed.

Perimeter of a rectangle   Perimeter $= 2 \times$ Width $+ 2 \times$ Height. (See Figure A-1.)

Area of a rectangle   Area $=$ Width $\times$ Height. (See Figure A-1.)

Area of a triangle   Area $= \dfrac{1}{2} \times$ Base $\times$ Height. (See Figure A-2.)

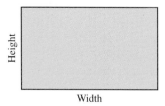

FIGURE A-1 A rectangle

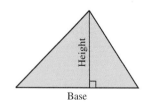

FIGURE A-2 A triangle

Circumference of a circle Circumference $= 2\pi \times$ Radius. (See Figure A-3.)

Area of a circle Area $= \pi \times$ Radius^2. (See Figure A-3.)

Surface area of a sphere Surface area $= 4\pi \times$ Radius^2. (See Figure A-4.)

Volume of a sphere Volume $= \dfrac{4}{3}\pi \times$ Radius^3. (See Figure A-4.)

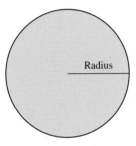
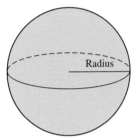

FIGURE A-3 A circle   **FIGURE A-4** A sphere (a ball)

Surface area of a right circular cylinder (a can with no top or bottom)
Surface area $= 2\pi \times$ Radius $\times$ Height. (See Figure A-5.)

Volume of a right circular cylinder Volume $= \pi \times$ Radius$^2 \times$ Height. (See Figure A-5.)

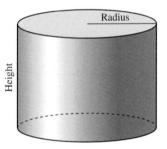

FIGURE A-5 A right circular cylinder

Thus if a rectangle has width 4 and height 2, then its perimeter is $P = 2 \times 4 + 2 \times 2 = 12$, and its area is $A = 2 \times 4 = 8$. A rectangle with width 5 and height 1 has perimeter $P = 2 \times 5 + 2 \times 1 = 12$ and area $A = 1 \times 5 = 5$. Note that these two rectangles have the same perimeter, but their areas are different.

One hundred feet of wire is used to construct a rectangular pen. Suppose the width of the rectangle is w feet. Then we can express both the height h and area A in terms of w. Because the perimeter is 100, $100 = 2w + 2h$. Solving for h gives $h = 50 - w$. We put this value into the formula $A = wh$ for the area of the pen:

$$A = wh$$

$$= w(50 - w).$$

In a similar fashion we can express the area A of a circle in terms of its circumference C. The first step is to solve the circumference formula for the radius r:

$$C = 2\pi r$$

$$\frac{C}{2\pi} = r.$$

Now we put this value of r into the area formula:

$$A = \pi r^2 = \pi \left(\frac{C}{2\pi}\right)^2 = \frac{C^2}{4\pi}.$$

PRACTICE PROBLEMS

A-1. Two hundred yards of fence is to be used to enclose a rectangular area next to a straight river. The river bank acts as one side of the rectangle, and the fence is used to make the other three sides of the rectangle. Suppose the width w in yards of the rectangle is along the river bank.

    a. Express the height of the rectangle in terms of w.

    b. Express the area of the rectangle in terms of w.

        Answer: **a.** Height $h = \dfrac{200 - w}{2}$ yards

                b. Area $A = w\left(\dfrac{200 - w}{2}\right)$ square yards

A-2. A rectangle has an area of 20 square inches. Let w denote the width in inches of the rectangle.

    a. Express the height of the rectangle in terms of w.

    b. Express the perimeter of the rectangle in terms of w.

        Answer: **a.** Height $h = \dfrac{20}{w}$ inches

                b. Perimeter $P = \dfrac{40}{w} + 2w$ inches

A-3. A circle has area A. Combine the formulas for area and circumference of a circle to express the circumference C of the circle in terms of A.

        Answer: $C = 2\sqrt{A\pi}$

A-4. Find the area of half a circle of radius r. Find its perimeter. (The half-circle is made up of a diameter plus half its circumference.)

        Answer: The area is $A = \pi r^2/2$. The perimeter is $P = 2r + \pi r$.

A-5. Find the area of a slice of one-quarter of a pie of radius r. Find its perimeter. (Be sure to include all three edges.)

        Answer: The area is $A = \pi r^2/4$. The perimeter is $P = \pi r/2 + 2r$.

A-6. A ball of radius r is chopped in half. Find its volume V.

        Answer: $V = \dfrac{2}{3}\pi r^3$

A-7. A ball of radius r is chopped in half. Find its surface area S. (*Note*: The ball is solid, so part of the surface of the ball is curved and part is flat.)

        Answer: $S = 3\pi r^2$

A-8. A ball is chopped in half, and that half-ball is chopped in half once more. The result looks like an orange slice that makes up one-quarter of the orange. Find its surface area S in terms of the radius r. (*Suggestion*: The curved part of its surface is a quarter of the surface of the orange. There are two parts to the flat surface. Think about the area of the two flat parts taken together.)

Answer: $S = 2\pi r^2$

A-9. A soda can is made from 40 square inches of aluminum. Let x denote the radius of the top of the can, and let h denote the height, both in inches.

a. Express the total surface area S of the can using x and h. (*Note*: The total surface area is the area of the top plus the area of the bottom plus the area of the cylinder.)

b. Using the fact that the total area is 40 square inches, express h in terms of x.

c. Express the volume V of the can in terms of x.

Answer: **a.** $S = 2\pi xh + 2\pi x^2$ square inches

b. $h = \dfrac{20 - \pi x^2}{\pi x}$

c. $V = x(20 - \pi x^2)$ cubic inches

A-10. A soda can has a volume of 25 cubic inches. Let x denote its radius and h its height, both in inches.

a. Using the fact that the volume of the can is 25 cubic inches, express h in terms of x.

b. Express the total surface area S of the can in terms of x.

Answer: **a.** $h = \dfrac{25}{\pi x^2}$

b. $S = \dfrac{50}{x} + 2\pi x^2$ square inches

 # Section A-4: Inverse Functions

The *inverse* of a function f is a function g such that $f(g(x)) = x$. Intuitively, the function f "undoes" what the inverse function g does—it gets us back where we started. For example, if f has an inverse and $f(4) = 10$, then for the inverse function g of f, we should have $g(10) = 4$. Here is a more interesting example: If $f(x) = x - 1$ and $g(x) = x + 1$, then g is the inverse of f because

$$f(g(x)) = g(x) - 1 = (x + 1) - 1 = x.$$

Determining When Inverses Exist: The Horizontal Line Test

Some functions do not have inverses, and for some of those that do, the formula may be difficult or impossible to find. A typical example of a function that does not have an inverse is the function f defined for all real numbers by $f(x) = x^2$. Note that $f(2) = 4$ and $f(-2) = 4$. So if g were the inverse function of f then we would have both $g(4) = 2$ *and* $g(4) = -2$. Because a function must assign each element of its domain to one value, not two or more, there is no inverse function for f. We should note that a common way of addressing this issue is to restrict the domain of f to make it have an inverse. If we insist that the domain of f is all non-negative real numbers [so that $f(-2) = 4$ is no longer valid], then $g(x) = \sqrt{x}$ is the inverse of f because $f(g(x)) = f(\sqrt{x}) = (\sqrt{x})^2 = x$.

In the preceding example, $f(x) = x^2$ is prevented from having an inverse by the fact that two different values of x ($x = 2$ and $x = -2$) have the same function value. This is shown graphically in Figure A-6, where the horizontal line $y = 4$ crosses the graph of $f(x) = x^2$ at two places, $x = 2$ and $x = -2$. This is a graphical test that can always determine whether a function has an inverse.

The horizontal line test: A function f has an inverse if and only if each horizontal line meets the graph in at most one point.

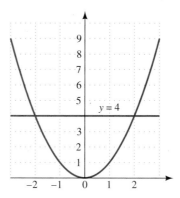

FIGURE A-6 $f(x) = x^2$ does not have an inverse because a horizontal line meets the graph more than once.

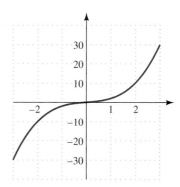

FIGURE A-7 $f(x) = x^3 + x$ does have an inverse because each horizontal line meets the graph at most once.

Notice in Figure A-7 that clearly each horizontal line meets the graph of $f(x) = x^3 + x$ exactly once. This assures us that f has an inverse. But it doesn't tell us what the formula for the inverse might be. For this particular function, the formula will be quite complicated indeed. We will not make an effort to present it here.

Calculating Inverses

For certain functions we can find a formula for the inverse, and the procedure is relatively simple. What we know about solving linear equations is applicable in many cases. Let's look, for example, at $f(x) = 2x + 7$. We are looking for an expression $g(x)$ such that $f(g(x)) = x$; that is, if we let $g(x) = y$, we want to solve the equation $f(y) = 2y + 7 = x$. Subtracting 7 from each side and then dividing by 2, we see that $y = (x - 7)/2$. Thus the inverse function of f is g where $g(x) = (x - 7)/2$. The reader should verify by direct calculation that $f(g(x)) = x$.

A more complex problem is to find the inverse function of $f(x) = \dfrac{2x - 1}{3x + 1}$. We are looking for an expression $g(x) = y$ so that $f(g(x)) = f(y) = x$. That is, we want to solve the equation $\dfrac{2y - 1}{3y + 1} = x$ for y. First we multiply by the denominator of the left-hand side:

$$\frac{2y - 1}{3y + 1} = x$$

$$2y - 1 = x(3y + 1).$$

Then we solve for y:

$$2y - 1 = 3xy + x$$

$$2y - 3xy = x + 1$$

$$y(2 - 3x) = x + 1$$

$$y = \frac{x + 1}{2 - 3x}.$$

Thus the inverse function of f is g where $g(x) = \frac{x + 1}{2 - 3x}$.

PRACTICE PROBLEMS

Using the Horizontal Line Test   In Problems A-1 through A-4, use a graphing calculator to find the graph of f over the horizontal span from $x = -4$ to $x = 4$. Then use the horizontal line test to determine whether f has an inverse. (You are not asked to find a formula for the inverse function.)

A-1. $f(x) = \dfrac{x^5}{100} + x + 1$ Answer: Horizontal, -4 to 4; vertical, -15 to 20

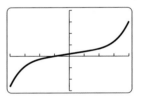

There is an inverse.

A-2. $f(x) = \dfrac{x^5}{100} - x + 1$ Answer: Horizontal, -4 to 4; vertical, -10 to 10

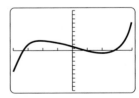

There is no inverse.

A-3. $f(x) = \dfrac{x}{1 + x^2}$ Answer: Horizontal, -4 to 4; vertical, -0.6 to 0.6

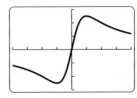

There is no inverse.

A-4. $f(x) = \dfrac{x + x^3}{2 + x^2}$

Answer: Horizontal, -4 to 4; vertical -5 to 5

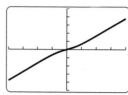

There is an inverse.

Finding Inverse Functions In problems A-5 through A-10, find a formula for the inverse of the given function.

A-5. $f(x) = x^3$ Answer: $y = x^{1/3}$

A-6. $f(x) = 7x$ Answer: $y = \dfrac{x}{7}$

A-7. $f(x) = 7x + 5$ Answer: $y = \dfrac{x - 5}{7}$

A-8. $f(x) = \dfrac{3x + 1}{2x - 5}$ Answer: $y = \dfrac{5x + 1}{2x - 3}$

A-9. $f(x) = \dfrac{x}{4x - 3}$ Answer: $y = \dfrac{3x}{4x - 1}$

A-10. $f(x) = \dfrac{a^3 x}{a^2 x + 1}$ with $a \neq 0$ Answer: $y = \dfrac{-x}{a^2 x - a^3}$

Section A-5: Solving Equations by Factoring

Many nonlinear equations cannot be solved exactly by hand calculation, and some sort of approximation method, such as the crossing-graphs method, may offer the only practical means of solution. But certain special equations can be solved by factoring.

    Recall the following basic factoring formulas.

$$x^2 + (a + b)x + ab = (x + a)(x + b)$$

Binomial: $\quad x^2 + (a - b)x - ab = (x + a)(x - b)$

$$x^2 - (a + b)x + ab = (x - a)(x - b)$$

Special cases:

Perfect square:
$$x^2 + 2ax + a^2 = (x + a)^2$$
$$x^2 - 2ax + a^2 = (x - a)^2$$

Difference of squares: $\quad x^2 - a^2 = (x - a)(x + a)$

Perfect cube:
$$x^3 + 3ax^2 + 3a^2x + a^3 = (x + a)^3$$
$$x^3 - 3ax^2 + 3a^2x - a^3 = (x - a)^3$$

Difference of cubes: $\quad x^3 - a^3 = (x - a)(x^2 + ax + a^2)$

Sum of cubes: $\quad x^3 + a^3 = (x + a)(x^2 - ax + a^2)$

    The key fact that allows us to use factoring to solve equations is that if the product of two numbers is zero, then one factor or the other is zero. That is, if $AB = 0$, then $A = 0$ or $B = 0$.

As a refresher, let's solve the quadratic equation $x^2 + 12 = 7x$:

$$x^2 + 12 = 7x$$

$$x^2 - 7x + 12 = 0$$

$$(x - 4)(x - 3) = 0.$$

We now have a product of two numbers that is zero. Thus, one factor or the other must be 0: $x - 4 = 0$ or $x - 3 = 0$. Thus we get two solutions, $x = 4$ and $x = 3$.

The same idea works with cubic equations (where the highest power is 3) if we can find the factors:

$$x^3 - 7x^2 + 10x = 0$$

$$x(x^2 - 7x + 10) = 0$$

$$x(x - 5)(x - 2) = 0.$$

We get three solutions: $x = 0$, $x = 5$, and $x = 2$.

Theoretically, any equation involving only positive integral powers of x can be solved by factoring. That is, such an equation can always be factored. But from a practical point of view, the factors may be difficult or impossible to find.

PRACTICE PROBLEMS

A-1. Solve by factoring $x^2 - 9 = 0$.   Answer: $x = 3$ or $x = -3$

A-2. Solve by factoring $x^2 - 8x + 16 = 0$.   Answer: $x = 4$

A-3. Solve by factoring $x^2 + 6x + 9 = 0$.   Answer: $x = -3$

A-4. Solve by factoring $x^2 - 8x = 9$.   Answer: $x = 9$ or $x = -1$

A-5. Solve by factoring $x^2 - 5x + 6 = 0$.   Answer: $x = 2$ or $x = 3$

A-6. Solve by factoring $x^3 - 2x^2 = 24x$.   Answer: $x = 0$, $x = 6$, or $x = -4$

A-7. Solve by factoring $x^3 + 12x = 7x^2$.   Answer: $x = 0$, $x = 3$, or $x = 4$

A-8. Solve by factoring $x^3 - 3x^2 + 3x - 1 = 0$. Answer: $x = 1$

A-9. Write $x^3 - 8$ as a product of a linear factor and a quadratic factor.

Answer: $(x - 2)(x^2 + 2x + 4)$

A-10. Write $x^3 + 1$ as a product of a linear factor and a quadratic factor.

Answer: $(x + 1)(x^2 - x + 1)$

Section A-6: Equations of Lines

There are special forms of the equation of a line that enable us to get equations of lines quickly. The first we will look at is the *point–slope* form of a line. This is useful if we know the slope m of a line and a point (a, b) through which it passes. If (x, y) is any point on the line, then we can calculate the slope of the line using (a, b) and (x, y):

$$m = \frac{y - b}{x - a}.$$

This equation can be written as

$$y - b = m(x - a).$$

This is the point–slope form of the line. For example, suppose we know that a line with slope 2 goes through the point (3, 4). Using the point–slope form, we have

$$y - 4 = 2(x - 3)$$
$$y - 4 = 2x - 6$$
$$y = 2x - 2.$$

Another important special form is the *two-point* form. If we know that a line passes through two points (a, b) and (c, d), we can use these two points to calculate the slope:

$$m = \frac{d - b}{c - a}.$$

On the other hand, if (x, y) is any point on the line, we can use (x, y) and (a, b) to find the slope:

$$m = \frac{y - b}{x - a}.$$

Equating these two representations of the slope, we get the two-point form of the equation of a line:

$$\frac{y - b}{x - a} = \frac{d - b}{c - a},$$

which can be written as

$$y - b = \left(\frac{d - b}{c - a}\right)(x - a).$$

For example, if we know that a line passes through the points (1, 3) and (5, 7), we get the equation of the line immediately as follows:

$$y - 3 = \left(\frac{7 - 3}{5 - 1}\right)(x - 1)$$
$$y - 3 = 1(x - 1)$$
$$y = x + 2.$$

PRACTICE PROBLEMS

A-1. Find the equation of the line passing through (2, 1) with slope 3.

Answer: $y = 3x - 5$

A-2. Find the equation of the line passing through (1, 1) with slope -4.

Answer: $y = -4x + 5$

A-3. Find the equation of the line passing through (1, 2) and (5, 10).

Answer: $y = 2x$

A-4. Find the equation of the line passing through (3, 1) and $(-2, 2)$.

Answer: $y = -\frac{1}{5}x + \frac{8}{5}$

A-5. Find the equation of the line passing through (3, 3) with slope 3.

Answer: $y = 3x - 6$

A-6. Find the equation of the line passing through $(8, 2)$ with slope $-1/4$.

Answer: $y = -\frac{1}{4}x + 4$

A-7. Find the equation of the line passing through $(2, 2)$ and $(4, 5)$.

Answer: $y = \frac{3}{2}x - 1$

A-8. Find the equation of the line passing through $(2, 2)$ and $(5, 0)$.

Answer: $y = -\frac{2}{3}x + \frac{10}{3}$

A-9. Find the equation of the line passing through $(1, 2)$ and $(6, 0)$.

Answer: $y = -\frac{2}{5}x + \frac{12}{5}$

A-10. Find the equation of the line passing through $(1, 2)$ and $(3, 7)$.

Answer: $y = \frac{5}{2}x - \frac{1}{2}$

■ | Section A-7: Elementary Properties of Exponents

We will look at exponential functions from a more algebraic point of view here. First we recall the following basic laws of exponents.

Basic Rules of Exponents

For $a > 0$:

Product law: $\quad a^b a^c = a^{b+c}$

Quotient law: $\quad \dfrac{a^b}{a^c} = a^{b-c}$

Power law: $\quad (a^b)^c = a^{bc}$

Negative exponents: $\quad a^{-b} = \dfrac{1}{a^b}$

Zero exponent: $\quad a^0 = 1$

For example, we can use these rules to simplify the expression $a^4(a^3)^2$:

$$a^4(a^3)^2 = a^4 a^{3 \times 2} = a^4 a^6 = a^{4+6} = a^{10}.$$

These rules can also be used to solve certain equations. For example, we will solve the equation $6a^5 = 48a^2$ under the assumption that a is positive. We first divide to put terms involving a on one side and the remaining terms on the other side: Dividing both sides of $6a^5 = 48a^2$ by $6a^2$ gives

$$\frac{6a^5}{6a^2} = \frac{48a^2}{6a^2}$$

or

$$\frac{a^5}{a^2} = 8.$$

Using the rules of exponents to simplify, we find

$$a^{5-2} = 8$$

or

$$a^3 = 8.$$

Therefore, $a = 2$.

Verifying Basic Properties of Exponential Functions

These rules can be used to verify the basic properties of exponential functions stated in Section 4.1. For example, let's verify the fundamental property that exponential functions change by constant multiples. When t is increased by 1, we may consider $N = Pa^t$ as the old value of an exponential function and Pa^{t+1} as the new value. Now

$$\text{New value} = Pa^{t+1} = (Pa^t)a^1 = \text{Old value} \times a.$$

Similarly, we can use these rules to verify how unit conversion works. Let's look first at a specific example. Suppose that in the formula $N = Pa^t$ the variable t represents time measured in years. Then we could more properly write the formula as

$$N = Pa^{t \text{ years}}.$$

Suppose now that we wish to change units to decades. Since d decades is $10d$ years, we have

$$Pa^{10d \text{ years}} = P(a^{10})^{d \text{ decades}}.$$

Thus the decade growth factor is the yearly growth factor raised to the tenth power.

The same idea works in general. Suppose we are measuring t in some unit—let's call it the old unit—for which the growth factor is a. Then

$$N = Pa^{t \text{ old units}}.$$

Suppose we wish to measure using a new unit that is k old units. Then

$$n \text{ new units} = nk \text{ old units.}$$

Hence

$$Pa^{nk \text{ old units}} = P(a^k)^{n \text{ new units}}.$$

Thus we get the new unit growth factor by raising the old unit growth factor to the kth power.

PRACTICE PROBLEMS

Practice with Exponents   Problems A-1 through A-5 provide practice with the basic rules for exponents. In each case use the laws of exponents to simplify the given expression. Write your final answer without using negative exponents.

A-1. $\dfrac{a^3b^2}{a^2b^3}$　　　　　　　　　　　　　　　　　　　Answer: $\dfrac{a}{b}$

A-2. $((a^2)^3)^4$　　　　　　　　　　　　　　　　　　　Answer: a^{24}

A-3. $a^3b^2a^4b^{-1}$　　　　　　　　　　　　　　　　　　Answer: a^7b

A-4. $\dfrac{(ab)^3a^{-2}b^2}{(a^2b^2)^4}$　　　　　　　　　　　　　　　Answer: $\dfrac{1}{a^7b^3}$

A-5. $\dfrac{a^{-4}b^{-3}}{a^{-6}b^{-2}}$　　　　　　　　　　　　　　　　Answer: $\dfrac{a^2}{b}$

Solving Equations   In Problems A-6 through A-10, solve the given equation for a, assuming that all variables are positive.

A-6. $5a^3 = 7$

Answer: $a = \left(\dfrac{7}{5}\right)^{1/3} = 1.12$

A-7. $6a^4 = 2a^2$

Answer: $a = 0.58$

A-8. $a^5 b^2 = ab$

Answer: $a = \left(\dfrac{1}{b}\right)^{1/4} = b^{-1/4}$

A-9. $3a^3 = 7a^5$

Answer: $a = 0.65$

A-10. $\dfrac{3}{a^3} = \dfrac{2}{a^2}$

Answer: $a = \dfrac{3}{2}$

◼ Section A-8: Quadratic Functions and Complex Numbers

A *quadratic function* is a function of the form $f(x) = ax^2 + bx + c$ with $a \neq 0$. Every quadratic function can be written in the form $a(x + p)^2 + q$ by a process known as *completing the square*. The procedure comes from the familiar binomial square formula:

$$(x + p)^2 = x^2 + 2px + p^2.$$

A special case of this formula occurs when $p = \dfrac{b}{2}$:

$$\left(x + \frac{b}{2}\right)^2 = (x^2 + bx) + \frac{b^2}{4}.$$

One way to read this formula is that if we add $\dfrac{b^2}{4}$ to $x^2 + bx$, the result is the perfect square $\left(x + \dfrac{b}{2}\right)^2$. Consider for example $x^2 + 6x$. What can we add to this expression to make it a perfect square? To get the answer, we use $b = 6$ in the formula above. That means we should add $\dfrac{b^2}{4} = \dfrac{6^2}{4} = 9$. The result should be

$$\left(x + \frac{b}{2}\right)^2 = \left(x + \frac{6}{2}\right)^2 = (x + 3)^2.$$

The reader may verify that indeed $(x + 3)^2 = (x^2 + 6x) + 9$.

We can use the process of completing the square to write $3x^2 + 12x + 9$ in the form $a(x + p)^2 + q$ as follows. First factor out the 3:

$$3x^2 + 12x + 9 = 3(x^2 + 4x + 3).$$

The next step is to complete the square for $x^2 + 4x$. Using $b = 4$, we get $x^2 + 4x + 4 = (x + 2)^2$. That is, we need to add 4 to $x^2 + 4x$ in order to make a perfect square, but of course we need to subtract 4 to compensate. Therefore,

$$
\begin{aligned}
3x^2 + 12x + 9 &= 3(x^2 + 4x + 3) \\
&= 3((x^2 + 4x + 4) + 3 - 4) \\
&= 3((x + 2)^2 - 1) \\
&= 3(x + 2)^2 - 3.
\end{aligned}
$$

The Quadratic Formula

The method of completing the square can be used to solve quadratic equations. Let's use this method to solve the quadratic equation $x^2 + 4x = 9$. We want to make the left side into a perfect square. To do that, we need to add $4^2/4 = 4$. We add 4 to each side of the equation to get

$$x^2 + 4x + 4 = 13$$
$$(x + 2)^2 = 13.$$

Now we can take the square root of each side of the equation to get the solution:

$$x + 2 = \pm\sqrt{13}$$
$$x = -2 \pm \sqrt{13}.$$

Thus we get the two solutions $x = -2 + \sqrt{13}$ and $x = -2 - \sqrt{13}$.

Applying the method of completing the square to a general quadratic equation leads to the *quadratic formula* stated in Section 5.5: The solution of the quadratic equation $ax^2 + bx + c = 0$ is

$$x = \frac{-b \pm \sqrt{b^2 - 4ac}}{2a}.$$

This formula is useful when it is not easy to find the factors of a quadratic function.

Complex Numbers

For some quadratic equations, application of the quadratic formula results in taking the square root of a negative number. To accommodate this, we introduce a number i whose square is -1:

$$i^2 = -1.$$

The introduction of this new number enables us to find the square roots of any negative number: If a is positive, then the square roots of $-a$ are $\pm\sqrt{a}\,i$, since

$$(\pm\sqrt{a}\,i)^2 = ai^2 = -a.$$

Thus, for example, $\sqrt{-4} = \pm 2i$. Now we know how to interpret the result of applying the quadratic formula when it involves the square root of a negative number. For example, when we solve $x^2 + 3x + 5 = 0$ using the quadratic formula, we get

$$x = \frac{-3 \pm \sqrt{9 - 20}}{2} = \frac{-3 \pm \sqrt{11}i}{2}.$$

In general, *complex* numbers are numbers of the form $a + bi$, with a and b real numbers. We add complex numbers in the obvious way. For example, $(6 + 4i) + (2 + 5i) = 8 + 9i$. We perform multiplication of complex numbers such as $(6 + 4i)(2 + 5i)$ in the same way we would compute $(6 + 4x)(2 + 5x)$, except that in the end we replace each occurrence of i^2 by -1:

$$(6 + 4i)(2 + 5i) = 12 + 30i + 8i + 20i^2 = 12 + 38i - 20 = -8 + 38i.$$

PRACTICE PROBLEMS

A-1. Complete the square for $x^2 + 8x$ and write the result as a perfect square.

Answer: $(x^2 + 8x) + 16 = (x + 4)^2$

A-2. Complete the square for $x^2 - 10x$ and write the result as a perfect square.

Answer: $(x^2 - 10x) + 25 = (x - 5)^2$

A-3. Write $x^2 - 4x + 5$ in the form $a(x - p)^2 + q$.   Answer: $(x - 2)^2 + 1$

A-4. Write $x^2 + 12x + 1$ in the form $a(x + p)^2 + q$.   Answer: $(x + 6)^2 - 35$

A-5. Use the quadratic equation to solve $x^2 + 2x + 3 = 0$.   Answer: $-1 \pm \sqrt{2}\,i$

A-6. Use the quadratic equation to solve $2x^2 + 3x + 4 = 0$.   Answer: $\dfrac{-3 \pm \sqrt{23}\,i}{4}$

A-7. Write $(3 + 4i) - (2 + 7i)$ in the form $a + bi$.   Answer: $1 - 3i$

A-8. Write $(2 + i)(3 + 5i)$ in the form $a + bi$.   Answer: $1 + 13i$

A-9. Write $(1 + i)(1 + 2i)$ in the form $a + bi$.   Answer: $-1 + 3i$

A-10. Write $(1 + i)^2 + (1 - 2i)^2$ in the form $a + bi$.   Answer: $-3 - 2i$

This guide is intended to provide basic instruction in the use of graphing calculators. Specific instructions are provided for the Texas Instruments TI-83, TI-83 Plus, and TI-84 Plus. Many other graphing calculators operate in fashions similar to these. All the operations necessary for the text are included, and they are presented in the order in which they are needed for the text.

Effective calculator use is essential for success with the text. A concerted effort is made to keep the instructions here basic and straightforward and to make this manual as brief as is practical. This is made easy by the technology itself, which is designed to provide a maximum of power with a minimum learning curve. As you become familiar with the operation of your calculator, you will find that there are many shortcuts available which are not covered here. Most often we show only what we consider to be the easiest method of producing a desired output. The TI-83 and TI-84 are very flexible instruments that may offer a number of options for getting the same result. You will discover some shortcuts by yourself and others by interacting with your colleagues, but you should also look at the extensive manuals which come with the calculator. You may find methods not presented here that you prefer, and you are encouraged to use them. Three topics are covered in depth: arithmetic operations, making tables and graphs, and the treatment of discrete data, particularly the use of regression to find best-fitting functions.

ARITHMETIC OPERATIONS

The TI-82, TI-83, TI-83 Plus, and TI-84 Plus are designed to make arithmetic calculations easy, and many will find that little if any instruction is really necessary. Often all that is needed is a bit of practice to gain familiarity with the operation of the keyboard, and when difficulties do occur, they may be traceable to problems with parentheses and grouping. These are at least partly mathematical in nature, and you are encouraged to read the Prologue of this textbook. The keystrokes for arithmetic operations on the TI-82 and TI-83/84 are virtually identical, but where they differ we will provide separate instructions. There are no differences in our operation of the TI-83, TI-83 Plus, and TI-84 Plus. All but a few of the figures in this text were made using the TI-83, and the output from the TI-82 is sometimes slightly different. The first time this is encountered, special pictures are made from the TI-82 screen, but in general the output of the two calculators is very similar. Where the difference may cause confusion we have pointed out the differences that are to be expected.

 ### Prologue: Basic Calculations

To perform a simple calculation such as $\dfrac{72}{9} + 3 \times 5$ on the TI-82/83/84, we type in the expression just as we might write it on paper: $72 \div 9 + 3 \times 5$. This is displayed in

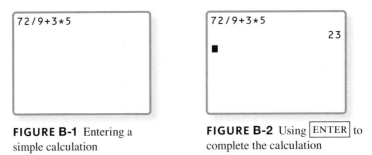

FIGURE B-1 Entering a simple calculation

FIGURE B-2 Using ENTER to complete the calculation

Figure B-1. To get the answer, we press ENTER , and we see 23 as expected, in Figure B-2.

For some calculations you need to access special keys.

The exponent key   To enter an exponent like 2^3 you need to use the black $\land$ key located just below the CLEAR key. Typing 2 $\land$ 3 ENTER will produce the correct result, 8.

The square root key   is at the top-left of the x^2 key.

TI-82

To access these *alternate keys*, you first press the blue 2nd key and the cursor will be replaced by a flashing, upward-pointing arrow. Now press the x^2 key, and the square root symbol will appear on the display. In what follows we will use 2nd [√] to indicate this sequence of keystrokes. So for example, if you want to get $\sqrt{7}$, you type 2nd [√] 7 ENTER . The answer, 2.645751311, is shown in the figure below.

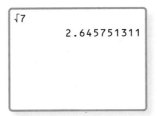

TI-83/84

To access these *alternate keys*, you first press the gold (blue for the TI-84) 2nd key and the cursor will be replaced by a flashing, upward pointing arrow. Now press the x^2 key, and the square root symbol will appear on the display. The TI-83/84 automatically adds a left parenthesis following the square root symbol in anticipation of a longer expression. To complete your entry, you should remember to finish with)) . In what follows we will use 2nd [√] to indicate this sequence of keystrokes. So for example, if you want to get $\sqrt{7}$, you type 2nd [√] 7)) ENTER . The answer is shown in Figure B-3.

The number π The special number π is printed above the $\land$ key. To access it you use 2nd [π]. When you do this, the calculator will display the symbol for π as shown in the first line of Figure B-4. When you press ENTER , the calculator will display the numerical approximation shown in the second line of Figure B-4.

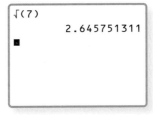

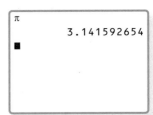

FIGURE B-3 Using the square root key

FIGURE B-4 Accessing π

The number *e*

TI-82

The special number *e* is most often used in the context of an *exponential function*. For this reason, when you press $\boxed{\text{2nd}}$ [e^x] (above the black $\boxed{\text{LN}}$ key) the calculator automatically adds the $\wedge$ in anticipation of an exponent. To get the number $e = e^1$ we provide an exponent of 1. That is, we type $\boxed{\text{2nd}}$ [e^x] 1 and $\boxed{\text{ENTER}}$. The result, 2.718281828, is the numerical approximation of *e* shown below.

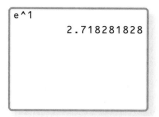

TI-83/84

The special number *e* appears in gold (blue for the TI-84) above the division key $\boxed{\div}$. You access it using $\boxed{\text{2nd}}$ [*e*] and $\boxed{\text{ENTER}}$. The result, 2.718281828, is a numerical approximation of *e*, which is shown in Figure B-6.

We should note, however, that this number is most often used in the context of an *exponential function*, and this provides an alternative way of getting the number *e*. While this method takes more keystrokes, it is the way *e* will be most commonly accessed in what follows. There is a second key $\boxed{\text{2nd}}$ [e^x] (above the black $\boxed{\text{LN}}$ key) which automatically adds the $\wedge$ and a left parenthesis in anticipation of an exponent. To get the number $e = e^1$ using this key we use $\boxed{\text{2nd}}$ [e^x] 1 $\boxed{)}$ and $\boxed{\text{ENTER}}$. The result, which matches our first approximation of *e* in Figure B-6, is in Figure B-5.

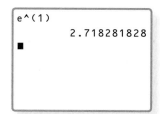

FIGURE B-5 $\wedge$ (and (on the TI-83/84) automatically displayed when you use $\boxed{\text{2nd}}$ [e^x]

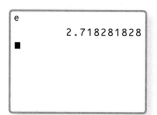

FIGURE B-6 Using $\boxed{\text{2nd}}$ [*e*] to access *e* directly (not available on the TI-82)

▤ Prologue: Parentheses

Two of the most important keys on your calculator are the black parentheses keys $\boxed{(}$ and $\boxed{)}$. When parentheses are indicated in a calculation, you should use them, and sometimes it is necessary to add additional parentheses. To make a calculation such as 2.7(3.6 + 1.8) you type 2.7 $\boxed{(}$ 3.6 $\boxed{+}$ 1.8 $\boxed{)}$ and then $\boxed{\text{ENTER}}$. The result is shown in Figure B-7. To calculate $\dfrac{2.7 - 3.3}{6.1 + 4.7}$, we must use parentheses around 2.7 − 3.3 to indicate that the whole expression goes in the numerator, and we must use parentheses around 6.1 + 4.7 to indicate that the whole expression goes in the denominator. Thus we type

$\boxed{(}$ 2.7 $\boxed{-}$ 3.3 $\boxed{)}$ $\boxed{\div}$ $\boxed{(}$ 6.1 $\boxed{+}$ 4.7 $\boxed{)}$ and $\boxed{\text{ENTER}}$.

The result is in Figure B-8.

```
2.7(3.6+1.8)
              14.58
```

FIGURE B-7 A calculation using parentheses

```
(2.7-3.3)/(6.1+4
.7)
          ⁻.0555555556
```

FIGURE B-8 A calculation where additional parentheses must be supplied

When you are in doubt about whether parentheses are necessary or not, it is good practice to use them. For example, if you calculate $\dfrac{3 \times 4}{7}$ using 3 $\boxed{\times}$ 4 $\boxed{\div}$ 7, you will get the right answer, 1.71. You will get the same answer if you use parentheses to emphasize to the calculator that the entire expression 3×4 goes in the numerator: $\boxed{(}$ $\boxed{(}$ 3 $\boxed{\times}$ 4 $\boxed{)}$ $\boxed{\div}$ 7.

If on the other hand you calculate $\dfrac{7}{3 \times 4}$, you must use parentheses to get the correct answer, 0.58. The correct entry is 7 $\boxed{\div}$ $\boxed{(}$ $\boxed{(}$ 3 $\boxed{\times}$ 4 $\boxed{)}$. If you leave out the parentheses, using 7 $\boxed{\div}$ 3 $\boxed{\times}$ 4, your calculator will think that only the 3 goes in the denominator, and you will get the wrong answer.

The use of parentheses leaves no room for doubt about what goes where, and their correct use is essential to the operation of the calculator.

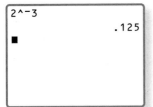

FIGURE B-9 Using $\boxed{(-)}$ to calculate 2^{-3}

FIGURE B-10 Syntax error displayed when $\boxed{-}$ is used in 2^{-3}

Prologue: Minus Signs

If you look on your TI-82/83/84 keyboard, you will find two minus signs. One is the key $\boxed{-}$ located just above the $\boxed{+}$ key. The other is the $\boxed{(-)}$ key that is just to the left of the $\boxed{\text{ENTER}}$ key. The $\boxed{-}$ key is used for subtraction as in 9 $\boxed{-}$ 4, while the $\boxed{(-)}$ key is used to denote a negative number. For example, if you want to get 2^{-3} you must use 2 $\boxed{\wedge}$ $\boxed{(-)}$ 3 as shown in Figure B-9. If you try to use 2 $\boxed{\wedge}$ $\boxed{-}$ 3, you will get a *syntax error* as seen in Figure B-10.

The mathematical distinction between the subtraction operation and the sign of a number is discussed more fully in the Prologue of this textbook. Briefly, you use the subtraction key $\boxed{-}$ when two numbers are involved, and you use the minus key $\boxed{(-)}$ when only one number is involved. With a little practice, the distinction is easy.

Prologue: Chain Calculations

Some calculations are most naturally done in stages. The TI-82/83/84 $\boxed{\text{2nd}}$ [ANS] key is helpful here. This key refers to the results of the last calculation. To show its use, let's look at

$$\left(\frac{2 + \pi}{7}\right)^{\left(3 + \frac{7}{9}\right)}.$$

This can be done in a single calculation, but we will show how to do it in two steps. First we calculate the exponent $3 + \dfrac{7}{9}$ as shown in Figure B-11. To finish, we need

$$\left(\frac{2 + \pi}{7}\right)^{\text{Answer from the last calculation}}.$$

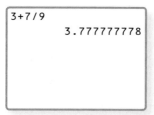

FIGURE B-11 The first step in a chain calculation

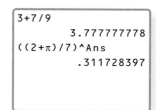

FIGURE B-12 Completing a chain calculation using 2nd [ANS]

FIGURE B-13 Accessing the factorial symbol

FIGURE B-14 Calculating 6!

The answer from the last calculation is obtained using 2nd [ANS]. Thus to complete the calculation, we use (((2 + 2nd [π])) ÷ 7)) ∧ 2nd [ANS].

The completed calculation is in Figure B-12.

Chapter 2: The Factorial Function

The *factorial* function occurs most often in probabilistic and statistical applications of mathematics. It applies only to non-negative integers and is denoted by $n!$. It is defined as follows:

$$0! = 1$$
$$\text{For } n > 0, n! = n \times (n-1) \times (n-2) \times \cdots \times 3 \times 2 \times 1.$$

Thus, for example, $3! = 3 \times 2 \times 1 = 6$ and $5! = 5 \times 4 \times 3 \times 2 \times 1 = 120$. Both the TI-82 and TI-83/84 can calculate the factorial function. Let's see how to use the calculator to get 6!. First enter 6. Next press the MATH key. Then use the right arrow key to highlight PRB as shown in Figure B-13. Finally, use the down arrow to highlight the exclamation point shown in Figure B-13 and press ENTER. The calculator will display 6!. Press ENTER once more to get the answer shown in Figure B-14.

TABLES AND GRAPHS

The feature of modern calculators which makes them different from their forerunners is their ability to make tables and produce graphs. This is a significant technological step, but more importantly it enhances the calculator's utility as a mathematical and scientific problem-solving tool.

Chapter 2: Tables of Values

For a function given by a formula, we can always produce a hand-generated table of values by calculating many individual function values, but the TI-82/83/84 acts as a significant time saver by making such tables automatically.

Three steps are required to make tables of values on a calculator. We will show them in an example. Let's make a table of values for $f = 3x + 1$.

Step 1, Entering the function The first step is to tell the calculator which function we are using. To do this we use Y = to show the *function entry* window. The TI-82/83/84 remembers functions that have been entered before, and so when you do this, you may find leftover formulas already on the screen as shown in Figure B-15.

FIGURE B-15 Clutter from previous work left on the function entry screen

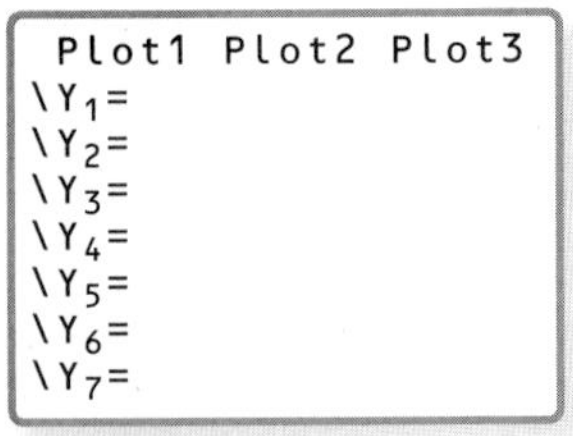

```
Plot1  Plot2  Plot3
\Y₁=
\Y₂=
\Y₃=
\Y₄=
\Y₅=
\Y₆=
\Y₇=
```

FIGURE B-16 Using CLEAR to clean up the function entry window

```
Plot1  Plot2  Plot3
\Y₁≡3X+1
\Y₂=
\Y₃=
\Y₄=
\Y₅=
\Y₆=
\Y₇=
```

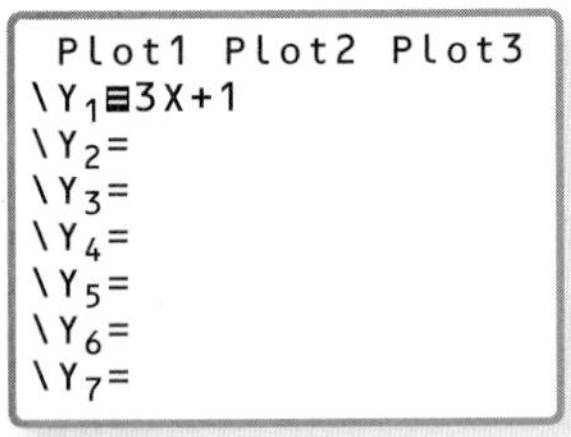

FIGURE B-17 The completed entry of $3x + 1$

```
TABLE SETUP
 TblStart=0
 ΔTbl=1
Indpnt: Auto  Ask
Depend: Auto  Ask
```

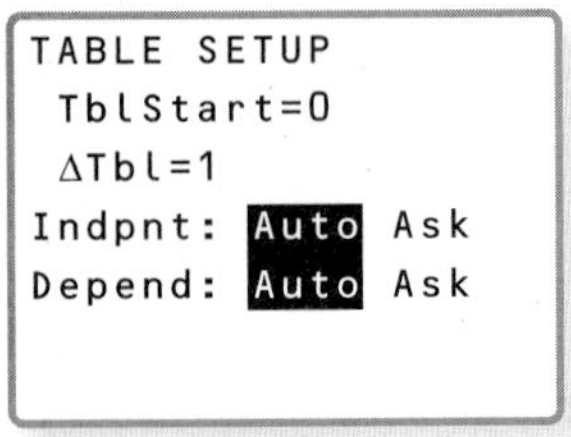

FIGURE B-18 The completed TABLE SETUP menu

If so, move the cursor to each formula that appears and press CLEAR . After the clean-up, you should have a clear function entry window like the one in Figure B-16. (The TI-82 will not show Plot1 Plot2 Plot3 at the top of the screen, nor will it show the marks to the left of the Y's.) The presence of the lines $Y_1 =$ through $Y_7 =$ in Figure B-16 indicates that many functions can be entered. For the moment, we will only use $Y_1 =$. Now we enter the function.[1]

TI-82

To enter the function, we use X, T, Θ for the variable x. (We would use the X, T, Θ key no matter what the name of the variable in our function.) Thus on the $Y_1 =$ line we type 3 X, T, Θ + 1. The properly entered function is in Figure B-17. Notice that the calculator writes the variable x as X. Once again, this is the symbol that the calculator will display no matter what the name of our variable.

TI-83/84

To enter the function, we use X, T, Θ, n for the variable x. (We would use the X, T, Θ, n key no matter what the name of the variable in our function.) Thus on the $Y_1 =$ line we type 3 X, T, Θ, n + 1. The properly entered function is in Figure B-17. Notice that the calculator writes the variable x as X. Once again, this is the symbol that the calculator will display no matter what the name of our variable.

Step 2, Setting up the table The next step is to tell the calculator how we want the table to look. We do this using 2nd [TBLSET]. When you press 2nd [TBLSET] you should see the TABLE SETUP menu shown in Figure B-18. The first option on the screen[2] is TblStart=. This is where the table is to start. We want this table to start at $x = 0$, and so we enter 0 there. The next option is △Tbl=. This gives the increment for the table. For the first view, we will look at the table for $x = 1, 2, 3, \ldots$ That is, we want to increment x by 1 at each step. (We will see later what happens if a different value for △Tbl= is used.) Thus we type 1 here. The properly completed TABLE SETUP menu is in Figure B-18.

Step 3, Viewing the table To view the table we press 2nd [TABLE], and the calcula-tor will present the display shown in Figure B-19. To get information from the table, it is important to remember that the X column corresponds to the variable x, and the Y_1 column gives the corresponding function value $f(x)$. Thus the first line in Figure B-19 tells us that $f(0) = 1$, the second that $f(1) = 4$, and so on. There is more to the table than is shown on the screen. If you use ▽ or △, you will see additional entries. In Figure B-20 we have used ▽ to view further function values.

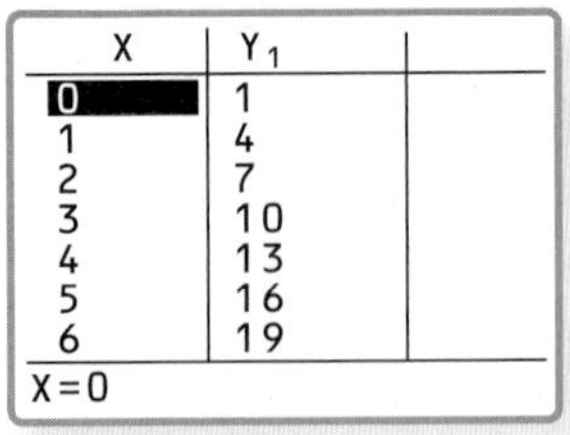

```
 X   | Y₁  |
─────┼─────┼────
 0   | 1   |
 1   | 4   |
 2   | 7   |
 3   | 10  |
 4   | 13  |
 5   | 16  |
 6   | 19  |
─────┴─────┴────
X=0
```

FIGURE B-19 A table of values for $3x + 1$

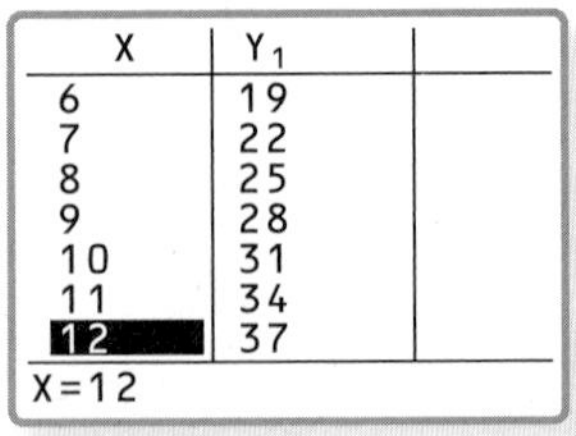

```
 X   | Y₁  |
─────┼─────┼────
 6   | 19  |
 7   | 22  |
 8   | 25  |
 9   | 28  |
 10  | 31  |
 11  | 34  |
 12  | 37  |
─────┴─────┴────
X=12
```

FIGURE B-20 Extending the table using ▽

---

[1]The TI-82 uses X, T, Θ to enter variables, while the TI-83/84 uses X, T, Θ, n. On this first occurrence, we will separate our instructions to accommodate this, but since these keys are so similar, in what follows we will use X, T, Θ, n exclusively without direct reference to the TI-82 X, T, Θ key.

[2]On the TI-82 this appears as TblMin=. As with X, T, Θ and X, T, Θ, n, we will not distinguish between them in what follows.

Adjusting the table Many times we can see what we want from a table by making adjustments with $\triangle$ or $\triangledown$, but for some alterations, this is impractical or impossible. Suppose for example that we wanted to see what happens for $x = 300, 301, 302, \ldots$. It is not very efficient to do this using the $\boxed{\triangledown}$ key. A better strategy is to use $\boxed{2nd}$ [TBLSET] to return to the TABLE SETUP menu and set TblStart= 300. This is shown in Figure B-21. Now when we use $\boxed{2nd}$ [TABLE] to go back to the table, the new TblStart= value causes the entries to start at $x = 300$ as shown in Figure B-22.

```
TABLE SETUP
 TblStart=300
 △Tbl=1
 Indpnt: Auto Ask
 Depend: Auto Ask
```

X	Y₁	
300	901	
301	904	
302	907	
303	910	
304	913	
305	916	
306	919	
X=300		

FIGURE B-21 Changing TblStart= to 300

FIGURE B-22 The table with a new starting value

Let's make another adjustment which will allow us to view the table for $x = 0, 5, 10, 15, \ldots$. That is, we want to view a table for f which starts at $x = 0$ and has an increment of 5. To do this, we use $\boxed{2nd}$ [TBLSET] to return to the TABLE SETUP menu and then use TblStart=0 and $\triangle$Tbl=5. The correctly configured TABLE SETUP menu is in Figure B-23, and when we press $\boxed{2nd}$ [TABLE], we see the new table in Figure B-24.

```
TABLE SETUP
 TblStart=0
 △Tbl=5■
 Indpnt: Auto Ask
 Depend: Auto Ask
```

X	Y₁	
0	1	
5	16	
10	31	
15	46	
20	61	
25	76	
30	91	
X=0		

FIGURE B-23 Changing the value of △Tbl=

FIGURE B-24 The table with an increment of 5

Comparing functions The calculator's ability to deal with more than one function allows us to use tables to make comparisons. For example, let's show how to compare the values of $f = 3x + 1$ with those of $g = 4x - 2$. First we need to use $\boxed{Y=}$ to enter the formula for g. We want to keep f as entered on the $Y_1=$ line, so we don't clear it but move the cursor directly to the $Y_2=$ line. There we enter 4 $\boxed{X, T, \Theta, n}$ $\boxed{-}$ 2 as shown in Figure B-25. Use TABLE SETUP values of TblStart=0 and $\triangle$Tbl=1. When we press $\boxed{2nd}$ [TABLE], we see the table in Figure B-26, which shows function values for f in the Y_1 column and values for g in the Y_2 column. Note that f starts out larger than g. They have the same value when $x = 3$, and after that g has the larger values.

```
Plot1 Plot2 Plot3
\Y₁⊟3X+1
\Y₂⊟4X-2
\Y₃=
\Y₄=
\Y₅=
\Y₆=
\Y₇=
```

X	Y₁	Y₂
0	1	-2
1	4	2
2	7	6
3	10	10
4	13	14
5	16	18
6	19	22
X=0		

FIGURE B-25 Entering a second function

FIGURE B-26 Comparing function values

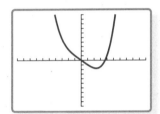

FIGURE B-27 Using $\boxed{Y=}$
to enter $\dfrac{x^4}{50} - x$

Chapter 2: Graphs

The TI-82/83/84 can generate the graph of a function as easily as it makes tables. We will illustrate using $f = \dfrac{x^4}{50} - x$. The first step is to tell the calculator which function we want to graph. We do this using the $\boxed{Y=}$ key exactly as we did in the previous section. Use $\boxed{\text{CLEAR}}$ to delete old functions and enter the new one on the Y_1 line using $\boxed{X, T, \Theta, n}$ $\boxed{\wedge}$ 4 $\boxed{\div}$ 50 $\boxed{-}$ $\boxed{X, T, \Theta, n}$. As with tables, the TI chooses its own names for functions and variables, in this case Y_1, which will appear on the vertical axis for f, and X, which will appear on the horizontal axis for x.

Once the function is properly entered as in Figure B-27, we press $\boxed{\text{ZOOM}}$ and the menu shown in Figure B-28 is shown. From this menu, select 6:ZStandard and the graph in Figure B-29 will be shown.[3]

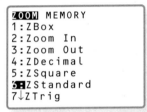

FIGURE B-28 Using $\boxed{\text{ZOOM}}$
to get to the ZOOM menu

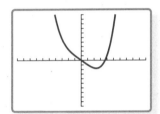

FIGURE B-29 The graph of
$\dfrac{x^4}{50} - x$

There are various ways of adjusting the picture shown by the calculator, and the TI-82/83/84 remembers the settings it used the last time it made a graph. The instructions we gave made what is called the ZStandard view. In this view, the horizontal span is from -10 to 10, and so is the vertical span. If your picture does not match the one in Figure B-29, use $\boxed{Y=}$ to check that your function is properly entered. If this fails, refer to the section Chapter 2: Graphics Configuration Notes (pp. B15 and B16). *It is important that you make your picture match the one in Figure B-29; otherwise, you will have difficulty following the examples and explanations in what follows.*

Once we have a graph on the screen there are several ways to adjust the view or to get information from it.

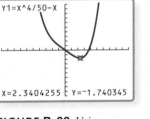

FIGURE B-30 Using
$\boxed{\text{TRACE}}$ to locate the cursor
at the bottom of the graph

Tracing the graph The $\boxed{\text{TRACE}}$ key places a cursor on the graph and enables the left and right arrow keys, $\boxed{\triangleleft}$ and $\boxed{\triangleright}$, to move the cursor along the graph. (If there is more than one graph on the screen, then $\boxed{\triangle}$ or $\boxed{\triangledown}$ moves the cursor to other graphs.) As the cursor moves, its location is recorded at the bottom of the screen. The X= prompt shows where we are on the horizontal axis, and the Y= prompt shows where we are on the vertical axis. In Figure B-30 we have used $\boxed{\text{TRACE}}$ and moved the cursor to the lowest point of the graph, and we see X=2.3404255 and Y= –1.740345 at the bottom of the screen. This tells us that the cursor is located at the point (2.3404255, -1.740345). Since this point lies on the graph, it also tells us that, rounded to two decimal places, $f(2.34) = -1.74$.

---

[3]This assumes the calculator is in its default graphics display mode. As is explained in what follows, this mode may have been altered by any number of calculator operations. Instructions for returning the calculator to its default display mode are given in the section entitled Chapter 2: Graphics Configuration Notes below.

Getting function values If we want to use ⌐TRACE⌐ to get the value of, say, $f(3)$, we might use the arrow keys to try to make the cursor land exactly on X=3. But if we do this, we find that the arrow keys make the X= prompt skip 3, going from 2.9787234 to 3.1914894. Both the TI-82 and the TI-83/84 offer us a way to correct this and make the cursor go exactly to X=3, but the keystrokes to accomplish this are different.

TI-82

We first press ⌐2nd⌐ [CALC]. You will see the CALCULATE menu shown in Figure B-31. We want a function value, so we choose 1:value from the menu. We will be returned to the graph with the prompt Eval X= at the bottom of the screen as shown in Figure B-32. We are interested in $x = 3$, so we type in 3 as seen in Figure B-33. Now when we press ⌐ENTER⌐ we see that the cursor has indeed moved to X=3, and read from Figure B-34 that $f(3) = -1.38$.

TI-83/84

Be sure the ⌐TRACE⌐ option is on. Now type 3, and X=3 will appear at the bottom of the screen as shown in Figure B-35. When we press enter, the cursor will move to X=3 as shown in Figure B-36, and we read from the bottom of the screen that $f(3) = -1.38$.

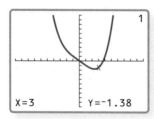

FIGURE B-31 The TI-82 CALCULATE menu

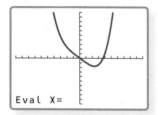

FIGURE B-32 The TI-82 Eval X= prompt

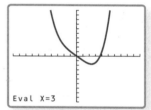

FIGURE B-33 Entering X=3 on the TI-82

FIGURE B-34 Evaluating f at $x = 3$ on the TI-82

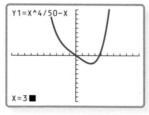

FIGURE B-35 Entering X=3 on the TI-83/84

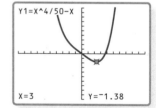

FIGURE B-36 Evaluating f at $x = 3$ on the TI-83/84

We should note that this works only if the X value you are looking for is actually on the viewing screen. If the X value you want is not far to the right or left of the viewing screen, you can use ⌐TRACE⌐ and move toward it. When the tracing cursor gets to the edge of the viewing area, the screen will change to follow it. If the X value is far to the right or left of the viewing area, you will need first to adjust the screen manually as described in **Manual window adjustments** below.

Zooming in and out Figure B-29 doesn't show much detail near the bottom of the graph, but the TI-82/83/84 can provide you with a closer look. Be sure you are using ⌐TRACE⌐ and put the cursor as near as you can to X=2.34. Now press ⌐ZOOM⌐, and you

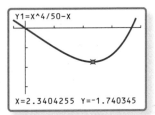

FIGURE B-37 The ZOOM menu

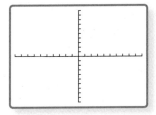

FIGURE B-38 Zooming in near X=2.34

will see the ZOOM menu shown in Figure B-37. Select 2:Zoom In from the menu. You will be put back in the graphing screen with nothing changed. To complete the zoom process, press ENTER, and the graph will be redrawn with a closer view as seen in Figure B-38. Zooming turns off the trace, and so you will need to TRACE again to put the cursor back on the graph. If you ZOOM 2 ENTER again you will see a still closer view of the graph. To back away from the graph, press ZOOM and select 3:Zoom Out from the menu. As before, we press ENTER to complete the zoom process, and then TRACE to put the cursor back on the graph.

Using TRACE ENTER to find lost graphs There are some graphs which will not be shown in the ZStandard view, and adjustments are required to show them. To illustrate this, let's make the graph of $f(x) = x^2 + 1000$. Use Y= and CLEAR out the old functions. On the Y₁= line type X, T, Θ, n ^ 2 + 1000. Now if we ZOOM 6, we will see the axes but no graph at all as shown in Figure B-39. The difficulty is that this graph sits high above the horizontal axis and is out of the viewing area. Press TRACE to put the cursor onto the graph. Now if you press ENTER, the viewing area will move to find the cursor and hence the graph. The resulting graph is shown in Figure B-40.

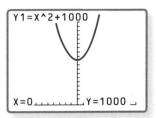

FIGURE B-39 A lost graph

FIGURE B-40 Finding a lost graph with TRACE ENTER

Manual window adjustments You can get satisfactory views of many graphs by tracing and zooming, but some need manual adjustments. To illustrate this, let's look at the graph of $f(x) = \dfrac{x}{50}$. Use Y= and CLEAR out the old functions. Now type X, T, Θ, n ÷ 50 and ZOOM 6 to get the ZStandard view shown in Figure B-41. (We have also pressed TRACE so that the TI-83/84 will identify the graph in the upper-left corner of the screen. The TI-82 identifies graphs by placing a single number n corresponding to Yₙ in the upper-right corner.) This is not a very good picture of the graph. The difficulty is that the graph is a straight line that in the ZStandard view lies so close to the horizontal axis that the calculator has difficulty distinguishing them.

This is a problem that is not easily fixed by zooming. We want to adjust the vertical span of the window to get a better view. We will make a table of values to help us choose an appropriate window size. Since the ZStandard window has a horizontal span of -10 to 10, we make the table start at -10 with an increment of 4. The result is shown in Figure B-42, and it shows that as x ranges from -10 to 10, $f(x)$ ranges from -0.2 to 0.2.

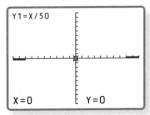

FIGURE B-41 An unsatisfactory view of a line

X	Y₁
-10	-.2
-6	-.12
-2	-.04
2	.04
6	.12
10	.2
14	.28

X=-10

FIGURE B-42 A table of values for $\dfrac{x}{50}$

```
WINDOW
 Xmin=-10
 Xmax=10
 Xscl=1
 Ymin=-10
 Ymax=10
 Yscl=1
 Xres=1
```

FIGURE B-43 The WINDOW menu

Allowing a little extra room, we want to change the vertical span so that it goes from −1 to 1. To accomplish this, we first press the $\boxed{\text{WINDOW}}$ key to get the WINDOW menu shown in Figure B-43. The first two lines of this screen, Xmin=−10 and Xmax=10, make the graphing screen span from −10 to 10 in the horizontal direction. We will leave these settings as they are. The fourth and fifth lines, Ymin=−10 and Ymax=10, make the graphing screen span from −10 to 10 in the vertical direction. The arrow keys, $\boxed{\triangle}$ and $\boxed{\triangledown}$, will allow us to position the cursor so that we can change these to Ymin=$\boxed{(-)}$1, and Ymax=1. (Be sure to use $\boxed{(-)}$ rather than $\boxed{-}$ for these entries.) These new settings, which are shown in Figure B-44, will make the viewing screen span from −10 to 10 in the horizontal direction and from −1 to 1 in the vertical direction. Now $\boxed{\text{GRAPH}}$ to get the better view of the line shown in Figure B-45.

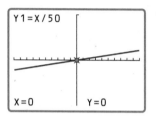

```
WINDOW
 Xmin=-10
 Xmax=10
 Xscl=1
 Ymin=-1
 Ymax=1■
 Yscl=1
 Xres=1
```

FIGURE B-44 Changing the vertical span

FIGURE B-45 A better view of the line

The text gives instructions on how to use a table of values to choose the graphing window in practical settings. See Section 2.2 of this textbook.

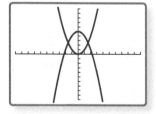

```
Plot1 Plot2 Plot3
\Y₁⧫X^2
\Y₂⧫5-X^2
\Y₃=■
\Y₄=
\Y₅=
\Y₆=
\Y₇=
```

FIGURE B-46 Entering two functions

Showing more than one graph   To show multiple graphs, all you need to do is to enter each function you want to show on the function entry screen. For example, if we want to see the graphs of $f = x^2$ and $g = 5 - x^2$ at the same time, we first use $\boxed{Y=}$ and type $\boxed{X, T, \Theta, n}$ $\boxed{\wedge}$ 2 on the Y₁= line and 5 $\boxed{-}$ $\boxed{X, T, \Theta, n}$ $\boxed{\wedge}$ 2 on the Y₂ = line as shown in Figure B-46. Now if we use $\boxed{\text{ZOOM}}$ 6 we get the ZStandard view of both graphs shown in Figure B-47.

Which graph is which?   We can use $\boxed{\text{TRACE}}$ to help us keep track of which graph goes with which function. When we do this with the TI-83/84, we see in Figure B-48 that the formula for the graph the cursor is on shows in the upper-left corner of the screen. (On the TI-82 a 1 will appear in the upper-right corner indicating that this is the graph of Y₁.) If we use $\boxed{\triangle}$ or $\boxed{\triangledown}$, the cursor switches to the other graph, and as we see in Figure B-49, the corresponding function is shown in the upper-left corner of the screen. (On the TI-82 a 2 will appear in the upper-right corner indicating that this is the graph of Y₂.)

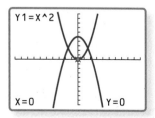

FIGURE B-48 How $\boxed{\text{TRACE}}$ labels graphs on the TI-83/84

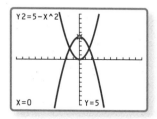

FIGURE B-49 Using $\boxed{\triangle}$ to move to the next graph

FIGURE B-47 Displaying the graphs of two functions

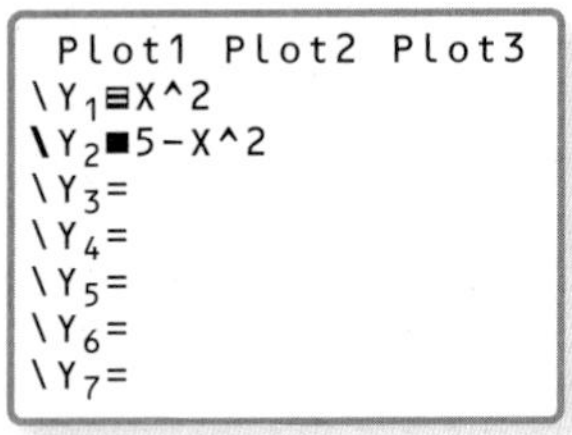

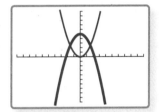

FIGURE B-50 Changing the graphing tag on the TI-83/84

Changing the look of the graph on the TI-83/84   The TI-83/84 allows the display to show different graphs in different styles, providing additional help in keeping track of which graph is which. (This feature is not available on the TI-82.) Let's change the look of the graph of $5 - x^2$. Use $\boxed{Y=}$ and move the highlight to the left of Y$_2$ = and press $\boxed{\text{ENTER}}$. This will cause the slash displayed there to start blinking. Now if we press $\boxed{\text{ENTER}}$, its shape will change as shown in Figure B-50. (If you press $\boxed{\text{ENTER}}$ more times, you will cycle through several display options.) Now when we $\boxed{\text{GRAPH}}$, the graph of $5 - x^2$ is shown with a heavy line as in Figure B-51.

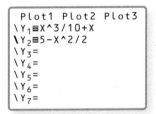

FIGURE B-51 $5 - x^2$ shown with a heavy line on the TI-83/84

Turning off graph displays   There may be occasions when we want to show one graph but not erase the other function from the function entry screen. Let's show how to turn off the display of the graph of x^2 without erasing its formula. Use $\boxed{Y=}$ to return to the function entry screen. Notice that the equal sign on both the Y$_1$ and Y$_2$ lines are highlighted. Move the cursor to the equal sign on the Y$_1$ line and press $\boxed{\text{ENTER}}$. This will turn off the highlight as shown in Figure B-52. Now when we $\boxed{\text{GRAPH}}$ as in Figure B-53, only the graphs of functions with highlighted equal signs will be displayed. To turn the display back on, use $\boxed{Y=}$, move the highlight to the equal sign, and press $\boxed{\text{ENTER}}$ again.

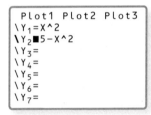

FIGURE B-52 Turning off the highlight on =

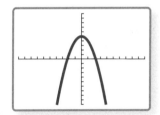

FIGURE B-53 Graph with x^2 turned off

Chapter 2: Solving Equations Using the Crossing-Graphs Method

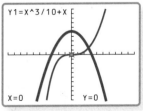

FIGURE B-54 Entering $\frac{x^3}{10} + x$ and $5 - \frac{x^2}{2}$

Let's show how to use the graphing capabilities of the calculator to solve

$$\frac{x^3}{10} + x = 5 - \frac{x^2}{2} \tag{1}$$

using what the text refers to as *the crossing-graphs method*. We want to graph each side of Equation (1) and see where they are the same, that is, where the graphs cross. The first step is to enter the left-hand side of the equation as one function and the right-hand side as a second function. Use $\boxed{Y=}$ and $\boxed{\text{CLEAR}}$ out old functions. On the Y$_1$= line enter $\boxed{X, T, \Theta, n}$ $\boxed{\wedge}$ 3 $\boxed{\div}$ 10 $\boxed{+}$ $\boxed{X, T, \Theta, n}$, and on the Y$_2$= line enter 5 $\boxed{-}$ $\boxed{X, T, \Theta, n}$ $\boxed{\wedge}$ 2 $\boxed{\div}$ 2. The correctly entered functions are in Figure B-54. Use $\boxed{\text{ZOOM}}$ 6 to get the ZStandard view of the graphs shown in the Figure B-55. Note that we have used $\boxed{\text{TRACE}}$ to show which graph goes with which function. The point in Figure B-55 where the graphs cross is where the left-hand and right-hand sides of the equation are the same. To solve Equation (1), we want the X-value of that crossing point.

To locate the crossing point we first press $\boxed{\text{2nd}}$ [CALC] to get the CALCULATE menu shown in Figure B-56. We want to know where the graphs cross, so we choose 5:intersect from the menu. The calculator will return you to the graphing screen with the prompt First curve? at the bottom of the screen. Press $\boxed{\text{ENTER}}$ and you will see the prompt Second curve?. Press $\boxed{\text{ENTER}}$ again. You will now see the graph with Guess? at the bottom. Press $\boxed{\text{ENTER}}$ a final time, and Intersection will appear at the bottom of

FIGURE B-55 The graphs of $\frac{x^3}{10} + x$ and $5 - \frac{x^2}{2}$

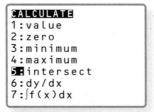

FIGURE B-56 The CALCULATE menu

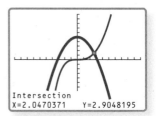

FIGURE B-57 Solving with the crossing-graphs method

the screen together with the coordinates of the crossing point. From Figure B-57 we see that the graphs cross when $x = 2.0470371$, and so, rounding to two decimal places, we report the solution to Equation (1) as $x = 2.05$.

Briefly, to solve with the crossing-graphs method, you use $\boxed{Y=}$ and enter each side of the equation on a separate line. Then $\boxed{\text{GRAPH}}$ and make necessary adjustments so that the crossing point is shown. Press $\boxed{\text{2nd}}$ [CALC] and select 5:intersect from the menu. In general you can get the intersection point from here by pressing $\boxed{\text{ENTER}}$ three times. The intermediate prompts First curve? and Second curve? are there in case you have more than two graphs on the screen. The prompt Guess? is there in case more than one crossing point is on the screen. In that case, move the cursor near the crossing point you want before pressing $\boxed{\text{ENTER}}$. You may also have to move the cursor near the crossing point you want on the rare occasion that the sequence 5:intersect $\boxed{\text{ENTER}}$ $\boxed{\text{ENTER}}$ $\boxed{\text{ENTER}}$ fails to produce an answer. Finally, you should be aware that for this method to work the crossing point you want must appear in the graphing window.

◼ Chapter 2: Solving Equations Using the Single-Graph Method

We will show an alternative way, which we will refer to as *the single-graph method*, for solving Equation (1). First we move everything from the right-hand side of the equation over to the left-hand side, remembering to change the sign of each term:

$$\frac{x^3}{10} + x = 5 - \frac{x^2}{2} \tag{2}$$

$$\frac{x^3}{10} + x - 5 + \frac{x^2}{2} = 0. \tag{3}$$

Now use $\boxed{Y=}$ and $\boxed{\text{CLEAR}}$ the old functions before entering

$$\boxed{X, T, \Theta, n}\ \boxed{\wedge}\ 3\ \boxed{\div}\ 10\ \boxed{+}\ \boxed{X, T, \Theta, n}\ \boxed{-}\ 5\ \boxed{+}\ \boxed{X, T, \Theta, n}\ \boxed{\wedge}\ 2\ \boxed{\div}\ 2$$

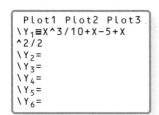

FIGURE B-58 Entering the function in preparation for the single-graph method

as shown in Figure B-58. Use $\boxed{\text{ZOOM}}$ 6 to get the ZStandard view shown Figure B-59. The solution of Equation (3) is where the graph crosses the horizontal axis. This point is referred to as a *root* or *zero*. The TI-82 uses *root* while the TI-83/84 uses *zero*. We will adhere to the TI-83/84 convention, and refer to it as a zero.

To find this point press $\boxed{\text{2nd}}$ [CALC], but this time from the CALCULATE menu, we select 2:zero (or 2:root on the TI-82) as shown in Figure B-60. The graphing screen will appear once more with the prompt Left Bound? (or Lower Bound? on the TI-82) at the bottom as shown in Figure B-61.

The calculator is asking for help to obtain the solution. We provide it by using $\boxed{\triangleleft}$ to move the cursor to any point to the left of the zero and pressing $\boxed{\text{ENTER}}$. We are presented with Figure B-62, where our selection is marked at the top of the screen, and at the bottom we see the new prompt Right Bound? (or Upper Bound? on the

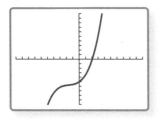

FIGURE B-59 The root or zero

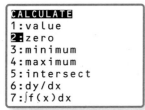

FIGURE B-60 The CALCULATE menu with 2:zero highlighted

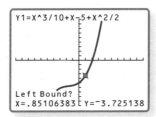

FIGURE B-61 Prompting for the Left Bound

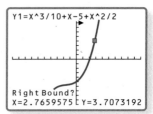

FIGURE B-62 The left bound marked and the Right Bound? prompt

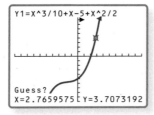

FIGURE B-63 Both bounds marked and the Guess? prompt

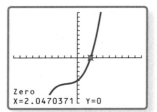

FIGURE B-64 The solution of $\dfrac{x^3}{11} + x - 5 + \dfrac{x^2}{2} = 0$ using the single-graph method

TI-82). This time we use ▷ to move the cursor to the right of the zero and press ENTER again. We see in Figure B-63 that there are now two marks at the top of the screen and a **Guess?** prompt at the bottom. The crossing point must be between the displayed marks. We respond to the **Guess?** prompt by pressing ENTER, and we see the solution $x = 2.0470371$ to Equation (3) in Figure B-64. As we expected, this agrees with the answer we got using the crossing-graphs method.

As with the crossing-graphs method, there are a couple of things that should be noted. Be sure the zero you want actually appears in the graphing window, and that there is only one in the range from the **Left Bound** to the **Right Bound** that you provide. On the rare occasion when you do not get an answer, you may need to move the bounds closer, or at the **Guess?** prompt move the cursor as near the root as you can.

▇ | Chapter 2: Optimization

Maxima and *minima* or *peaks* and *valleys* of a graph can be located on the TI-82/83/84 in much the same way that roots or zeros are found.

To illustrate the method, we will look at the graph of

$$f = 2^x - x^2.$$

Use Y= , CLEAR out any old functions from the list, and enter the new one using 2 ^ X, T, Θ, n − X, T, Θ, n ^ 2. The graph of the function shown in Figure B-65 uses Xmin= (−)1, Xmax=4, Ymin= (−)2, and Ymax=2.

The graph in Figure B-65 shows a maximum, or a peak, and a minimum, or a valley. The keystrokes required to find them are very similar, but we will carefully find both. Let's first find the maximum. Press 2nd [CALC] to get to the CALCULATE menu shown in Figure B-66. We want to find the maximum, so we select 4:maximum from the menu. We are returned to the graphing window with the trace option turned on, and the prompt **Left Bound?** (or **Lower bound?** on the TI-82) appears

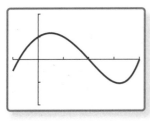

FIGURE B-65 The graph of $2^x - x^2$

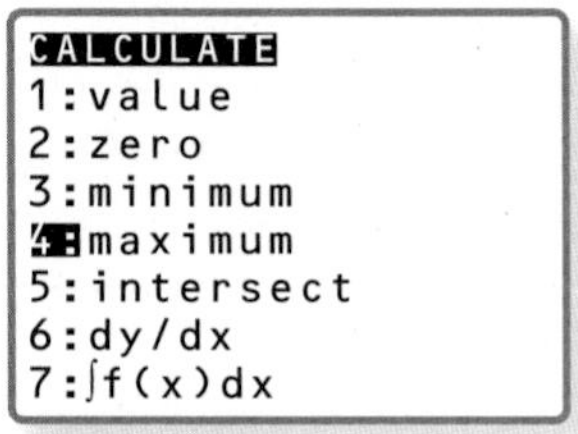

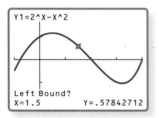

FIGURE B-66 Selecting 4:maximum from the CALCULATE menu

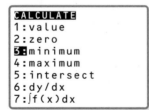

FIGURE B-67 The Left Bound? prompt

at the bottom of the screen as in Figure B-67. The information wanted now by the calculator is similar to what is wanted when we search for zeros or roots. We use the arrow keys to move the cursor to the left of the maximum and press ENTER . A tick mark is placed at the top of the screen showing our selection, and we see the Right Bound? (or Upper bound? on the TI-82) prompt in Figure B-68. We move the cursor to the right of the maximum and press ENTER again. Now we see two tick marks at the top of the screen *which must enclose the maximum we want* and the Guess? prompt in Figure B-69. Rarely is any response other than ENTER required here, and the maximum will be located as in Figure B-70.

The steps for finding the minimum are almost identical with those we used to get the maximum. Press 2nd [CALC], but this time select 3:minimum from the menu as shown in Figure B-71. As with the maximum we are returned to the graphing window with the trace option on and the Left Bound? prompt showing in Figure B-72. We locate the cursor left of the minimum and press ENTER , move the cursor to the right of the minimum and press ENTER , and finally press ENTER at the Guess? prompt. The minimum will be found as in Figure B-73.

We should note that on the rare occasion when the method described here fails to produce the desired result, you should repeat the procedure, but at the Guess? prompt, move the cursor as near the maximum or minimum as you can and press ENTER .

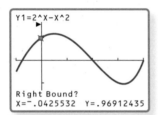

FIGURE B-68 The Right Bound? prompt

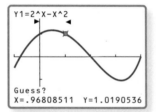

FIGURE B-69 The Guess? prompt

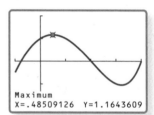

FIGURE B-70 Locating a maximum for $2^x - x^2$

FIGURE B-71 Selecting 3:minimum from the CALCULATE menu

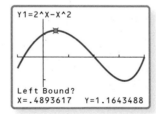

FIGURE B-72 The Left Bound? prompt for the minimum

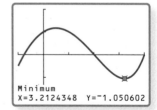

FIGURE B-73 A minimum value for $2^x - x^2$

Chapter 2: Graphics Configuration Notes

There are a number of ways your calculator can be configured to display graphics. If you were able to follow the examples in the earlier section entitled Chapter 2: Graphs, reproducing the screens we made, then you won't need to use these notes. If not, the difficulty may be that your calculator graphics configuration has been altered from the factory-set defaults. These notes address only the two most common graphics settings which may have been changed on your calculator. If this does not solve your graphics configuration difficulties, consult your calculator manual.

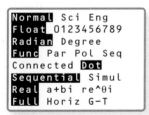

```
ZOOM MEMORY
1:ZPrevious
2:ZoomSto
3:ZoomRc1
4:SetFactors...
```

FIGURE B-74 The ZOOM MEMORY menu

```
ZOOM FACTORS
 XFact=4
 YFact=4
```

FIGURE B-75 The ZOOM FACTORS menu

Problem *My graphs match those in the examples until I zoom in or out. Then I get a different picture.* The probable cause is that the default zoom factors have been changed. To find out and fix the problem, press [ZOOM] and use [▷] to highlight **MEMORY** as shown in Figure B-74. Select 4:Set Factors from the menu. This will take you to the ZOOM FACTORS menu shown in Figure B-75. Your screen should show XFact=4 and YFact=4. If necessary, change these numbers to match Figure B-75.

Problem *When I [GRAPH], I get an error as in Figure B-76, or I get the graphs, but I also get some extra dots as in Figure B-77 or some dots connected by lines as in Figure B-78.* The probable cause is that you have one form or another of statistical plotting turned on. To correct this problem press [2nd] [STAT PLOT] to get to the **STAT PLOTS** menu shown in Figure B-79. We want to turn statistical plotting Off, so we press 4 and then [ENTER].

FIGURE B-76 Error reported when you [GRAPH]

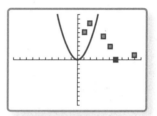

FIGURE B-77 Discrete statistical plot turned on

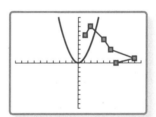

FIGURE B-78 Connected statistical plot turned on

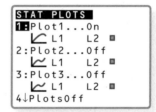

FIGURE B-79 The STAT PLOTS menu

Chapter 5: Piecewise-Defined Functions

Sometimes functions are defined by different formulas over different intervals. Such functions are known as *piecewise-defined* functions. For example, we define

$$f(x) = \begin{cases} x + 1, & \text{if } x < 1 \\ x - 1, & \text{if } x \geq 1. \end{cases}$$

This means that $f(x)$ is $x + 1$ if x is less than 1 but it is $x - 1$ when x is greater than or equal to 1. Thus $f(0) = 0 + 1 = 1$ since $0 < 1$. But $f(3) = 3 - 1 = 2$ since $3 \geq 1$.

The TI-82 and TI-83/84 will graph piecewise-defined functions. Let's see how to graph the function f defined above. The first step is to use the [MODE] key and change from Connected to Dot as shown in Figure B-80. On the function entry screen we want to enter keystrokes that produce the following:

$$(x + 1)(x < 1) + (x - 1)(x \geq 1).$$

```
Normal Sci Eng
Float 0123456789
Radian Degree
Func Par Pol Seq
Connected Dot
Sequential Simul
Real a+bi re^θi
Full Horiz G-T
```

FIGURE B-80 Setting to Dot mode

We do this because the calculator evaluates an expression like $x < 1$ as a function of x that has the value 1 if the relation is true and the value 0 if the relation is false.

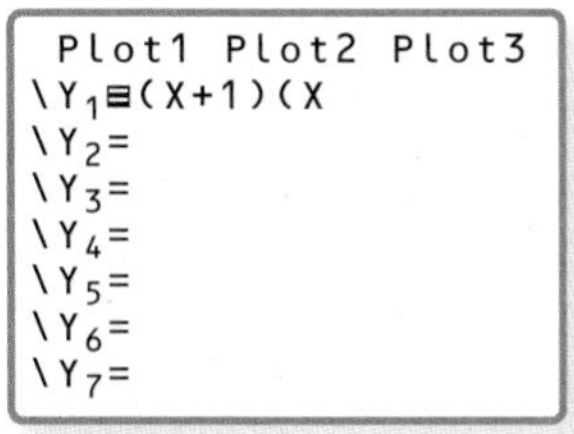

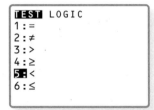

FIGURE B-81 Beginning
the entry

The only difficulty in entering this expression is locating the proper keystrokes for $<$ and $\geq$. We begin using

$$\boxed{(}\ \boxed{X, T, \Theta, n}\ \boxed{+}\ 1\ \boxed{)}\ \boxed{(}\ \boxed{X, T, \Theta, n}$$

as shown in Figure B-81.

Now press $\boxed{\text{2nd}}$ [TEST] and then highlight the $<$ symbol as shown in Figure B-82. Press $\boxed{\text{ENTER}}$, and the $<$ symbol will appear as shown in Figure B-83.

We make the remaining entry up to the $\geq$ symbol using

$$1\ \boxed{)}\ \boxed{+}\ \boxed{(}\ \boxed{X, T, \Theta, n}\ \boxed{-}\ 1\ \boxed{)}\ \boxed{(}\ \boxed{X, T, \Theta, n}.$$

FIGURE B-82 Using $\boxed{\text{2nd}}$
[TEST] to locate the $<$ symbol

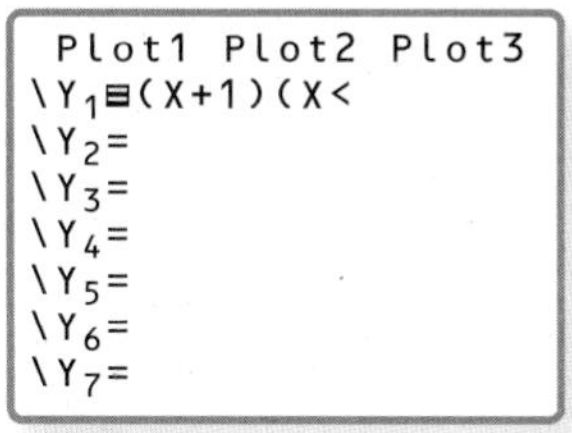

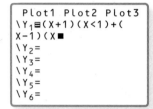

FIGURE B-83 The $<$
symbol entered

FIGURE B-84 Continuing
the entry

This is shown in Figure B-84. To enter the $\geq$ symbol we proceed as before, using $\boxed{\text{2nd}}$ [TEST] and highlighting the appropriate symbol. Press $\boxed{\text{ENTER}}$ to display the symbol as shown in Figure B-85.

Complete the entry using $1\ \boxed{)}$. In Figure B-86 we have used $\boxed{\text{GRAPH}}$ with a horizontal span of 0 to 3 and a vertical span of -1 to 3.

In Figure B-87 we have changed back to Connected mode. This makes the calculator add an extraneous vertical line to the picture, and it illustrates why we prefer the Dot mode.

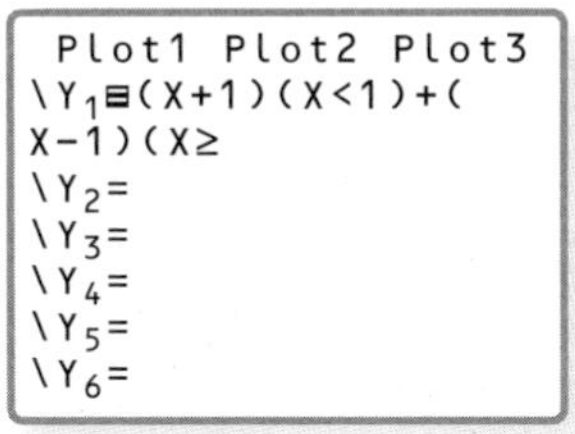

FIGURE B-85 Using $\boxed{\text{2nd}}$
[TEST] to enter the $\geq$ symbol

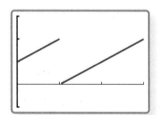

FIGURE B-86 The correct
graph displayed using Dot mode

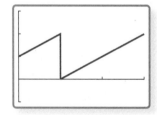

FIGURE B-87 The graph
displayed using Connected mode

DISCRETE DATA

The TI-82/83/84 offers many features for handling discrete data sets. This section shows how to enter data and perform some basic analysis.

Chapter 3: Graphing Discrete Data

Many times information about physical or social phenomena is obtained by gathering individual bits of data and recording them in a table. For example, the following table taken from the *Information Please Almanac* shows median American family income

I by year in terms of 1996 dollars. That means the dollar amounts shown have been adjusted to account for inflation. The variable *d* in the table is years since 1980. Thus, for example, the $d = 2$ column corresponds to 1982.

d = years since 1980	0	1	2	3	4	5
I = median income	21,023	22,388	23,433	24,580	26,433	27,735

Let's show how to enter these data into the calculator and display them graphically.

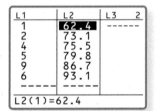

FIGURE B-88 Using $\boxed{\text{STAT}}$ to get the EDIT CALC TESTS menu

Entering data To enter the values from the table above, we press $\boxed{\text{STAT}}$ to get the EDIT CALC TESTS menu shown in Figure B-88. Be sure EDIT is highlighted. If it is not, use $\boxed{\triangleleft}$ to change that. Choose 1:Edit, and you will see a screen similar to the one in Figure B-89. Figure B-89 shows that our calculator has some old data already entered, and we need to clear them out before entering new data. Use $\boxed{\triangle}$ to highlight L1 as shown in Figure B-89 and then press $\boxed{\text{CLEAR}}$ $\boxed{\text{ENTER}}$. Do the same with each column that has unwanted entries. You should now have the clear data entry screen shown in Figure B-90. The L1 column corresponds to *d*, and we type in the *d* values 0 through 5, pressing either $\boxed{\text{ENTER}}$ or $\boxed{\triangledown}$ after each entry. Now use $\boxed{\triangleright}$ to move to the top of the L2 column and enter the median incomes there. The completed data entry is in Figure B-91.

L1	L2	L3	2
1	**62.4**	------	
2	73.1		
4	75.5		
5	79.8		
9	86.7		
6	93.1		
-----	-----		
L2(1)=62.4			

FIGURE B-89 The data entry screen with old data

L1	L2	L3	1
▬▬	------	------	
L1(1) =			

FIGURE B-90 A fresh data entry screen

L1	L2	L3	2
0	21023	------	
1	22388		
2	23433		
3	24580		
4	26433		
5	27735		

L2(7) =			

FIGURE B-91 Completed data entry

Turning on (or off) statistical plots Before we can make a graphical display of data, we need to set up the calculator so that it can display discrete data. That is, we must turn on statistical plotting. This only has to be done the first time we want to make such a picture (or if we ourselves have disabled this feature). Once turned on, the statistical plotting will stay on until we turn it off. To set up statistical plotting, press $\boxed{\text{2nd}}$ [STAT PLOT] to get the menu shown in Figure B-92. Select 1:Plot 1 from that menu, and the TI-83/84 will present you with the menu in Figure B-93. The menu for the TI-82 is similar.

FIGURE B-92 The STAT PLOTS menu

FIGURE B-93 The Plot 1 menu

Highlight On and press $\boxed{\text{ENTER}}$. Your screen should now match the one in Figure B-93. Check the Type:, Xlist, Ylist, and Mark lines to see that they agree with

Figure B-93. If they do not, use the arrow keys to move the highlight where you want it, change the entry, and press $\boxed{\text{ENTER}}$. When you no longer want data points to appear in your graphs, you can repeat this procedure, turning Plot 1 Off.

It is particularly important to note the role of the Xlist and Ylist lines in Figure B-93. The Xlist line determines which of the data columns will be shown on the horizontal axis. Usually, this is L1. To set it on the TI-83/84, use $\boxed{\text{2nd}}$ [L1] (above the 1 key); on the TI-82, highlight L1 and press $\boxed{\text{ENTER}}$. The Ylist line determines which of the data columns will be shown on the vertical axis. Usually, this is L2, which you set on the TI-83/84 by using $\boxed{\text{2nd}}$ [L2] (above the 2 key) and on the TI-82 by highlighting L2 and pressing $\boxed{\text{ENTER}}$.

Plotting the data Now we are ready to view the graph. Generally with discrete data, we don't have to worry with how to set up the window, because the TI-82/83/84 has a special feature which does the job automatically. This feature is accessed as follows. Press $\boxed{\text{ZOOM}}$, and move down the ZOOM menu to 9:ZoomStat as shown in Figure B-94. When you press 9 (or $\boxed{\text{ENTER}}$ with 9:ZoomStat highlighted) you will see the graphical display in Figure B-95. If you got the picture in Figure B-95 with one or more extra graphs added, you need to use $\boxed{Y=}$ and clear out any old functions that are on the function entry screen.

When you use ZoomStat to display data, the calculator decides the window size to use. You can see what its choices were by pressing $\boxed{\text{WINDOW}}$. We see in Figure B-96 that the horizontal span is -0.5 to 5.5 and the vertical span is from $19,881.96$ to $28,876.04$.

Adding regular graphs to data plots You can add any graph you wish to a data plot. For example, using a technique known as *linear regression*, which is discussed in Section 3.4 of this textbook and is reviewed in the next section of this guide, we find that the linear function $y = 1338x + 20919$ closely approximates the median income data. To show their graph, use $\boxed{Y=}$, type 1338 $\boxed{X, T, \Theta, n}$ $\boxed{+}$ 20919, and then $\boxed{\text{GRAPH}}$. This line added to the data plot is in Figure B-97.

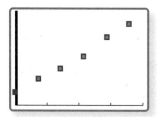

FIGURE B-94 The ZoomStat option

FIGURE B-95 Displaying median income data

```
WINDOW
 Xmin=■.5
 Xmax=5.5
 Xscl=1
 Ymin=19881.96
 Ymax=28876.04
 Yscl=1
 Xres=1
```

FIGURE B-96 The window size selected by ZoomStat

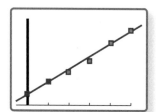

FIGURE B-97 Adding an additional graph

Editing columns There are shortcuts that make it easy to do certain kinds of data edits. For example, suppose we wish to show median income remaining after each family pays a 17% tax. That is, we want to replace each entry in the L2 column by 83% of its current value. Use $\boxed{\text{STAT}}$ and choose 1:Edit from the menu to get to the data we have already entered. Use the arrow keys to highlight L3 as shown in Figure B-98. We want to put in the L3 column the entries in the L2 column times 0.83. We could calculate these values one at a time and enter each individually, but the TI-82/83/84 offers a much easier way to accomplish this. Just type .83 $\boxed{\text{2nd}}$ [L2] as seen at the bottom of the screen in Figure B-99. Now when we press $\boxed{\text{ENTER}}$, the L3 column is filled in automatically as in Figure B-100.

L1	L2	**L3** 3
0	21023	-------
1	22388	
2	23433	
3	24580	
4	26433	
5	27735	
----	----	
L3=		

FIGURE B-98 Preparing for entry in the L3 column

L1	L2	**L3** 3
0	21023	------
1	22388	
2	23433	
3	24580	
4	26433	
5	27735	
----	----	
L3=.83L2		

FIGURE B-99 The formula entered

L1	L2	L3 3
0	21023	17449
1	22388	18582
2	23433	19449
3	24580	20401
4	26433	21939
5	27735	23020
----	----	----
L3(1)=17449.09		

FIGURE B-100 The L3 column filled automatically

Let's show how to plot these new data. Now we want the L1 column on the horizontal axis, but the L3 column rather than the L2 column for the vertical axis. To make this happen, use [2nd] [STAT PLOT] and select 1:Plot1 from the menu. Go to the Ylist: line and type [2nd] [L3] (on the TI-82, highlight L3 and press [ENTER]) as shown in Figure B-101. Now when we [ZOOM] 9 we see the new data displayed in Figure B-102. (Don't forget to use [Y =] and clear out the Y₁ line, which we don't want to see anymore.)

FIGURE B-101 Changing the vertical axis from L2 to L3

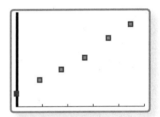

FIGURE B-102 Data after 17% tax

Chapter 3: Linear Regression

Linear regression is a method of getting a linear function which approximates almost linear data. The TI-82/83/84 has built-in features which will do this automatically. Let's show how to do it for the following data table.

x	0	1	2	3	4
y	17	19.3	20.9	22.7	26.1

L1	L2	L3 1
	------	------
L1(1) =		

FIGURE B-103 A clear data entry screen

The first step is to enter the data. This is done exactly as described in the previous section. Use [STAT] and select 1:Edit from the menu. [CLEAR] out old data as necessary to get the clear data entry screen in Figure B-103. Next enter the x data in the L1 column and the y data in the L2 column as shown in Figure B-104.

To get the regression line press [STAT] and use [▷] to move the highlight at the top of the screen to CALC. This will change the menu to the one shown in Figure B-105. From this menu select 4:LinReg(ax+b) (this is 5:LinReg(ax+b) on the TI-82). You will be taken to the calculation screen with LinReg(ax+b) displayed as shown in Figure B-106. Press [ENTER], and you will see the information in Figure B-107. (The TI-82 displays an additional line, which we will not use.)

L1	L2	L3 2
0	17	------
1	19.3	
2	20.9	
3	22.7	
4	26.1	

L2(6) =		

FIGURE B-104 The correctly entered data

This screen tells us that the equation of the regression line is $y = ax + b$, where the slope is $a = 2.16$, and the vertical intercept is $b = 16.88$. (Note here that the TI-82/83/84 uses a for the slope rather than the more familiar m.) Thus the equation of the regression line is

$$y = 2.16x + 16.88.$$

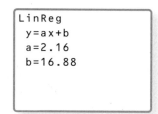

```
EDIT CALC TESTS
1:1-Var Stats
2:2-Var Stats
3:Med-Med
4:LinReg(ax+b)
5:QuadReg
6:CubicReg
7↓QuartReg
```

FIGURE B-105 The statistical calculation menu (LinReg(ax+b) is menu item 5 on the TI-82.)

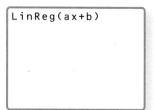

```
LinReg(ax+b)
```

FIGURE B-106 The regression line prompt

```
LinReg
y=ax+b
a=2.16
b=16.88
```

FIGURE B-107 Regression line parameters

We have in fact completed the calculation of the regression line, but it is almost always important to continue and make a display of the data and the regression line together. Among other things, such a display provides a valuable check for our work. To graph the line, we must get the regression line formula into the function entry screen. We will first show how to do this manually and then later show a method for having it automatically entered in the function list at the time it is calculated. For manual entry use $\boxed{Y =}$, $\boxed{\text{CLEAR}}$ out old functions, and type the equation of the regression line as 2.16 $\boxed{X, T, \Theta, n}$ $\boxed{+}$ 16.88.

The correctly entered function is in Figure B-108.

Before we make the graph, we use $\boxed{\text{2nd}}$ [STAT PLOT] and check to see that Plot1 is configured as in Figure B-109. Finally, $\boxed{\text{ZOOM}}$ 9 produces the picture in Figure B-110.

```
Plot1 Plot2 Plot3
\Y₁=2.16X+16.88
\Y₂=
\Y₃=
\Y₄=
\Y₅=
\Y₆=
\Y₇=
```

FIGURE B-108 Manual entry of the regression line formula

```
Plot1 Plot2 Plot3
On Off
Type: ...
Xlist:L1
Ylist:L2
Mark: □  +  ·
```

FIGURE B-109 The properly configured Plot1 window

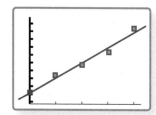

FIGURE B-110 Data and regression line

```
VARS Y-VARS
1:Function...
2:Parametric...
3:Polar...
4:On/Off...
```

FIGURE B-111 The *y*-variables menu

The important feature to note in Figure B-110 is that the regression line closely approximates the data points. If this does not happen, we know either that we have made a mistake or that the data are not properly modeled using a linear function.

Automatic entry of the regression line on the function list Let's calculate the regression line again, this time showing how to make it automatically appear on the function list. This method is available only on the TI-83/84.[4]

Just as before, we use $\boxed{\text{STAT}}$, select the CALC menu, and choose 4:LinReg(ax+b). Now at the regression line prompt shown in Figure B-106 above, press $\boxed{\text{VARS}}$ and highlight Y-VARS at the top of the screen to see the *y*-variables menu. Be sure that 1:Function is highlighted in that menu as shown in Figure B-111. Now when you press $\boxed{\text{ENTER}}$, you will be presented with the list in Figure B-112. Be sure 1:Y₁ is highlighted and press $\boxed{\text{ENTER}}$. You will be returned to the home screen with Y₁ added to the regression

```
FUNCTION
1:Y₁
2:Y₂
3:Y₃
4:Y₄
5:Y₅
6:Y₆
7↓Y₇
```

FIGURE B-112 Function list

[4]There is another method for transferring the equation of the regression line to the function entry screen without typing it, a method that is available on both calculators. Use $\boxed{Y =}$, $\boxed{\text{CLEAR}}$ the Y₁= line, and then press $\boxed{\text{VARS}}$. Select 5:Statistics. Move the highlight at the top of the screen to EQ and then choose 1:RegEQ (on the TI-82 this is 7:RegEQ) from the menu.

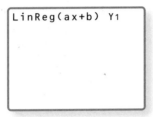

FIGURE B-113 Completing the regression prompt

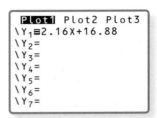

FIGURE B-114 Regression formula automatically entered

prompt, as seen in Figure B-113. Now when you press ⟦ENTER⟧, the regression line parameters will be calculated as usual, but also the regression line formula will appear on the function list. In Figure B-114 we have pressed ⟦Y =⟧ to verify that this has occurred.

We should note that this is a case where the explanation is so long that it may appear to make the procedure more complicated than it really is. If you go through the keystrokes a few times, you may find this quick and easy indeed. This procedure is available to you, and manual function entry is also available. You should choose the method that seems best to you.

■ | Chapter 3: Diagnostics

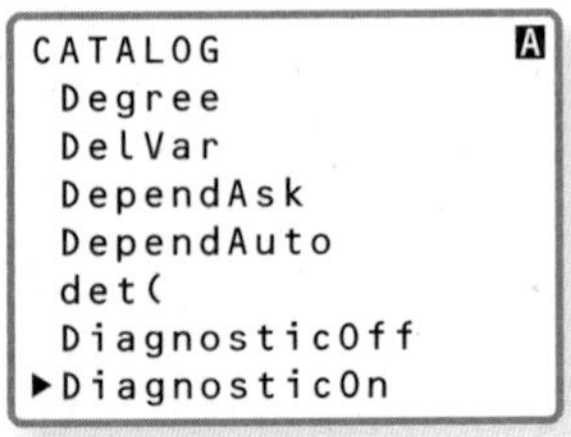

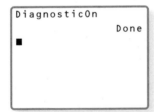

FIGURE B-115 Selecting DiagnosticOn using ⟦2nd⟧ [CATALOG]

Sometimes it is important to get a statistical measure of how well a regression line fits data. One such measure is the *correlation coefficient*, commonly denoted by r. The number r is always between -1 and 1. Values of r near 1 indicate a good fit with a positive slope. Values of r near -1 indicate a good fit with a negative slope. Values of r near 0 indicate a poor fit.

The value of r is automatically displayed by the TI-82 when linear regression is performed. To get the TI-83/84 to display the correlation coefficient, press ⟦2nd⟧ [CATALOG] and select DiagnosticOn as shown in Figure B-115. Press ⟦ENTER⟧ ⟦ENTER⟧, and the diagnostic feature is turned on as shown in Figure B-116.

Now if we perform linear regression the value of r will be displayed when regression coefficients are calculated. This is shown in Figure B-117. (A second statistical measure, denoted by r^2, is also shown on the TI-83/84. We will not discuss this measure here.) Note that the value of r is near 1, indicating a good fit with a positive slope. The data and regression line are plotted in Figure B-118, and the quality of the fit is apparent.

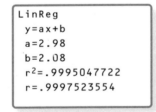

FIGURE B-116 Enabling the diagnostic feature

LinReg
y=ax+b
a=2.98
b=2.08
r²=.9995047722
r=.9997523554

FIGURE B-117 A correlation coefficient near 1

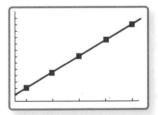

FIGURE B-118 Good fit indicated by r near 1

In Figure B-119 note that the correlation coefficient is near -1, and the good fit with negative slope is shown in Figure B-120.

```
LinReg
 y=ax+b
 a=-2.02
 b=10.08
 r²=.9989228359
 r=-.9994612728
```

FIGURE B-119 A correlation coefficient near −1

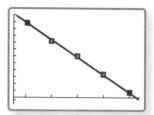

FIGURE B-120 Good fit indicated by r near −1

In Figure B-121 the correlation coefficient is near 0, and the corresponding poor data fit is shown in Figure B-122.

```
LinReg
 y=ax+b
 a=-.5
 b=6.9
 r²=.0323834197
 r=-.1799539377
```

FIGURE B-121 A correlation coefficient near 0

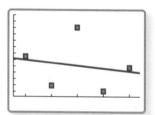

FIGURE B-122 Poor fit indicated by r near 0

Chapter 4: Exponential Regression

Exponential regression can be performed on data. The steps are the same as for linear regression except that a different regression method is selected from the CALC menu: After the data have been entered, press STAT and use ▷ to move the highlight at the top of the screen to CALC. From this menu select 0:ExpReg. You will be taken to the calculation screen with ExpReg displayed. Then press ENTER.

Chapter 5: Logistic Regression

Logistic regression is a method of fitting data with a logistic function, that is, a function of the form $c/(1 + ae^{-bx})$. On the TI calculators, the process is quite similar to linear regression. (Logistic regression is not available on the TI-82.) We will show how to fit the following data set with a logistic function.

x	0	1	2	3	4	5
y	0.40	0.58	0.81	1.06	1.30	1.51

The first step is to use STAT and EDIT to enter the data as usual. The properly entered data are in Figure B-123. Next we use STAT and CALC and highlight B:Logistic as shown in Figure B-124.

```
L1    L2    L3    2
0     .4    ------
1     .58
2     .81
3     1.06
4     1.3
5     1.51
------ ■■■■■
L2(7)=
```

FIGURE B-123 Properly entered data

```
EDIT CALC TESTS
7↑QuartReg
8:LinReg(a+bx)
9:LnReg
0:ExpReg
A:PwrReg
B:Logistic
C:SinReg
```

FIGURE B-124 Selecting logistic regression

Now press ENTER and the regression coefficients will be displayed as in Figure B-125. We read from this screen that the function we want is $2.00/(1 + 4.03e^{-0.50x})$. The data and the logistic fit are displayed in Figure B-126.

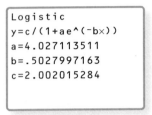

FIGURE B-125 Logistic regression parameters

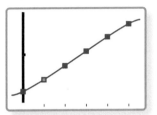

FIGURE B-126 Data and the logistic fit

Chapter 5: Power Regression

Power models can be constructed in much the same way as linear and exponential models. The steps are the same except that a different regression method is selected from the CALC menu: After the data have been entered, press STAT and use ▷ to move the highlight at the top of the screen to CALC. From this menu select A:PwrReg. You will be taken to the calculation screen with PwrReg displayed. Then press ENTER.

Chapter 5: Quadratic, Cubic, and Quartic Regression

Quadratic regression is a method of fitting data with a quadratic polynomial, that is, a function of the form $ax^2 + bx + c$. On the TI calculators, the process is quite similar to linear regression. We will show how to fit the following data set with a quadratic.

FIGURE B-127 Properly entered data

x	0	1	2	3	4	5
y	2.5	1.1	4.3	16.2	47.4	95.3

The first step is to use STAT and EDIT to enter the data as usual. The properly entered data are in Figure B-127. Next we use STAT and CALC and highlight 5:QuadReg (6:QuadReg on the TI-82) as shown in Figure B-128.

Now press ENTER and the regression coefficients will be displayed as in Figure B-129. We read from this screen that the function we want is $6.4x^2 - 14.44x + 5.23$. The data and the quadratic fit are displayed in Figure B-130.

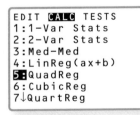

FIGURE B-128 Selecting quadratic regression

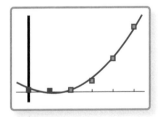

FIGURE B-129 Quadratic regression parameters

FIGURE B-130 Data and the quadratic fit

Cubic regression fits data with a function of the form $ax^3 + bx^2 + cx + d$, and *quartic regression* fits data with a function of the form $ax^4 + bx^3 + cx^2 + dx + e$.

The procedure is the same as for quadratic regression except that for cubic regression we use 6:CubicReg (7:CubicReg on the TI-82) as shown in Figure B-131. The cubic regression coefficients for the data in the table above are displayed in Figure B-132. This tells us that if we wish to fit the data with a cubic, we should use $0.85x^3 - 0.01x^2 - 2.73x + 2.66$.

```
EDIT CALC TESTS
1:1-Var Stats
2:2-Var Stats
3:Med-Med
4:LinReg(ax+b)
5:QuadReg
6:CubicReg
7↓QuartReg
```

FIGURE B-131 Selecting cubic regression

```
CubicReg
 y=ax³+bx²+cx+d
 a=.8546296296
 b=-.0079365079
 c=-2.73478836
 d=2.661111111
```

FIGURE B-132 Cubic regression parameters

If we wish to use quartic regression, we select 7:QuartReg (8:QuartReg on the TI-82) as shown in Figure B-133. The quartic regression parameters in Figure B-134 tell us that the proper quartic to use for the data above is $-0.14x^4 + 2.25x^3 - 4.3x^2 + 1.25x + 2.42$.

```
EDIT CALC TESTS
1:1-Var Stats
2:2-Var Stats
3:Med-Med
4:LinReg(ax+b)
5:QuadReg
6:CubicReg
7↓QuartReg
```

FIGURE B-133 Selecting quartic regression

```
QuarticReg
 y=ax⁴+bx³+...+e
 a=-.1395833333
 b=2.250462963
 c=-4.295138889
 d=1.253306878
 e=2.421825397
```

FIGURE B-134 Quartic regression parameters

Len DeLessio/Getty Images

This guide is intended to provide basic instruction in the use of spreadsheets. Specific instructions are provided for Microsoft Excel®. Many other spreadsheets operate in fashions similar to Excel. All the operations necessary for the text are included, and they are presented in the order in which they are needed for the text. A concerted effort is made to keep the instructions here basic and straightforward and to make this manual as brief as is practical. This is made easy by the technology itself, which is designed to provide a maximum of power with a minimum learning curve. As you become familiar with the operation of your spreadsheet program, you will find that there are many short-cuts available which are not covered here. Most often we show only what we consider to be the easiest method of producing a desired output. Excel, which may on the surface appear to be a simple spreadsheet program, is in fact a very complex and powerful piece of software. You will discover some shortcuts yourself and others by interacting with your colleagues, but you should also look at the extensive manuals which come with Excel. You may find methods not presented here that you prefer, and you are encouraged to use them.

Three topics are covered in depth: arithmetic operations, making tables and graphs, and the treatment of discrete data, particularly the use of regression to find best-fitting functions. Special thanks to Deborah Benton of Wake Technical Community College for her help with the material in this appendix. Whatever errors or shortcomings may be here are the result of our poor execution of her excellent advice.

ARITHMETIC OPERATIONS

Excel is a powerful and complex piece of software. It will do many things, and there is often more than one way to get them done. In general we will present only the way which seems best to the authors, but other methods may be more natural to some. You are encouraged to familiarize yourself with Excel and to choose methods which fit your own style. We will present instructions specifically for Microsoft Excel 2010. For other versions of Excel, the individual steps may vary slightly.

Prologue: Basic Calculations

To perform a simple calculation such as $\frac{72}{9} + 3 \times 5$ we choose any convenient cell and type =72/9+3*5. This is shown in Figure C-1. (If you omit the equal sign, Excel will not perform the calculation.) Note that the symbol * is used to indicate multiplication. Once the expression is correctly typed, we press **Enter** and the calculation is performed as is shown in Figure C-2. Note that with the cell selected, we now read the answer from the cell and the expression from the formula box at the top of the spreadsheet.

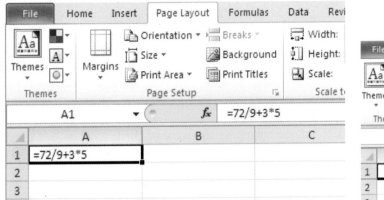

FIGURE C-1 Typing an expression into Excel

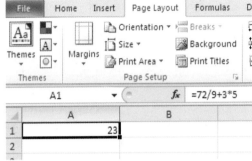

FIGURE C-2 Completing the calculation

Exponentiation To enter an expression like 2^3 we use the caret symbol ^. Typing =2^3 will produce the correct result as shown in Figure C-3. Be sure to begin the expression with an equal sign.

Square roots To get square roots, we use Excel's sqrt function. To calculate $\sqrt{7}$ we type =sqrt(7). This is shown in Figure C-4.

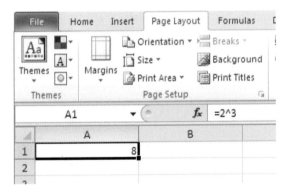

FIGURE C-3 Entering exponents

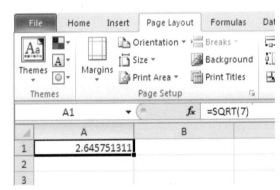

FIGURE C-4 Calculating square roots

The number π As with the square root, we get π by using an Excel function. We type =pi( ). Excel will display a decimal approximation of π as shown in Figure C-5. Note that the parentheses are essential here. If you leave them out, Excel will interpret your input as a word rather than a number.

The number e The Excel function exp(x) is used to denote the exponential function e^x. Thus to get the number e, we use =exp(1). This is shown in Figure C-6. For practice, try calculating $3e^{0.08}$. You should get an answer of 3.24986. If you have trouble, remember that you must begin with an equal sign and that you need to insert the symbol * between the 3 and exp to indicate multiplication.

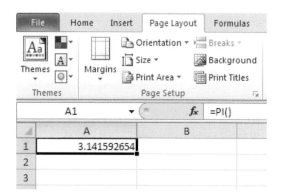

FIGURE C-5 Entering π

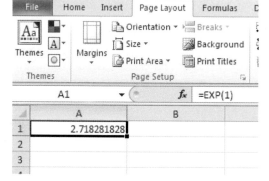

FIGURE C-6 The number e

Prologue: Parentheses

When you are performing arithmetic operations, two of the most important symbols on the keyboard are the parentheses. When parentheses are indicated in a calculation, you should use them, and sometimes it is necessary to supply additional parentheses. To make a calculation such as 2.7(3.6 + 1.8) we type the expression just as shown, being careful to include the parentheses. (Also remember that the symbol * must be used to indicate multiplication.) This is shown in Figure C-7. Some calculations such as

$$\frac{2.7 - 3.3}{6.1 + 4.7}$$

do not have parentheses when we write them on paper with a pencil, but in order to type them correctly we must supply parentheses. We must surround 2.7 − 3.3 with parentheses so that Excel knows that the entire expression goes in the numerator. Similarly, we must enclose 6.1 + 4.7 in parentheses to indicate that the entire expression goes in the denominator. Thus =(2.7-3.3)/(6.1+4.7) is the correctly typed expression. The result is shown in Figure C-8. If the parentheses are not supplied as shown, Excel will misinterpret your input and produce the wrong answer.

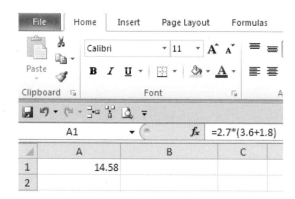

FIGURE C-7 A calculation using parentheses

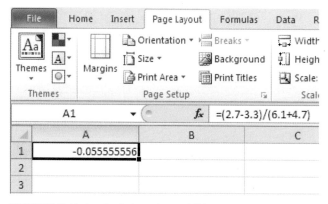

FIGURE C-8 A calculation where additional parentheses must be supplied

When you are in doubt about whether parentheses are necessary or not, it is good practice to use them. For example, if you calculate $\dfrac{3 \times 4}{7}$ using =3*4/7, you will get the right answer, 1.71. You will get the same answer if you use parentheses to emphasize that the entire expression 3×4 goes in the numerator: =(3*4)/7. If on the other hand you calculate $\dfrac{7}{3 \times 4}$, you must use parentheses to get the correct answer of 0.58. The correct entry is =7/(3*4). If you leave out the parentheses and simply type =7/3*4, Excel will think that only the 3 goes in the denominator, and you will get the wrong answer.

The use of parentheses leaves no room for doubt about what goes where, and their correct use is essential to the operation of any calculating device, including Excel.

■ Prologue: Cell Reference

The feature of Excel known as *cell referencing* is quite simple but adds greatly to the power of Excel. In fact, an understanding of cell referencing is fundamental to the operation of Excel. Notice that the columns of Excel are labeled by letters A, B, C, . . . and that the rows are labeled by numbers 1, 2, 3, ... This gives a reference for each cell in the spreadsheet. For example, the reference B3 refers to the third cell down in column B.

We can illustrate one important feature of cell referencing by calculating several values of the function $f(x) = x^2 + 1$. We will show two ways of doing this. Here is the first: Assuming that we start with a clean spreadsheet, get in cell A1 and type 2 (see Figure C-9). Then go to cell B1 and type =A1^2+1. This instruction tells Excel that it should calculate the value of $x^2 + 1$ using whatever is in cell A1 in place of x. In Figure C-10 we see that Excel automatically recalculates the value in cell B1, evaluating $x^2 + 1$ at $x = 2$. Try some other values in A1 so that you have a clear understanding of how Excel handles cell referencing.

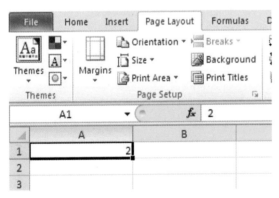

FIGURE C-9 Entering the formula using direct cell reference

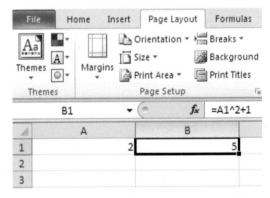

FIGURE C-10 Evaluating the formula at $x = 2$

Employing cell referencing by directly typing in the cell column letter and row number has the advantage of conforming directly to Excel's method of operation. There is an alternative method of cell referencing which allows us to type in formulas exactly as they appear using the variable name rather than the direct cell reference. This is accomplished using Excel's ability to assign names to cells or groups of cells. Let's give the A column the name x. Highlight the A column and click on the name box which is just to the left of the formula box. Replace A1 by x as shown in Figure C-11. Just to remind us that column A now has the name x, we type x in cell A1.

Now we get in cell B2 and type =x^2+1. Note that since we have given the name x to column A, we can type the formula using x as the variable rather than A1. Since we are in cell B2, x refers to the value in A2. (If we were in cell B3, then x would refer to the value in A3.) Note that in Figure C-12 we have typed 2 in cell A2 and the answer, $x^2 + 1$ evaluated at $x = 2$, appears in cell B2.

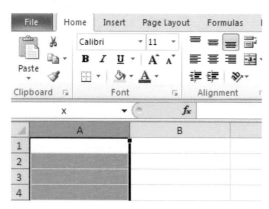

FIGURE C-11 Giving column A the name x

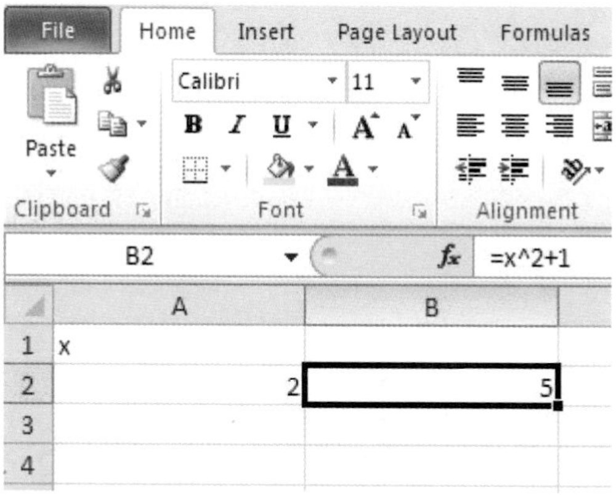

FIGURE C-12 Evaluating $x^2 + 1$ at $x = 2$

Prologue: Chain Calculations

Some calculations are most naturally done in stages. To show how to do this with Excel, let's look at

$$\left(\frac{2 + \pi}{7}\right)^{\left(3+\frac{7}{9}\right)}.$$

This can of course be done in a single calculation, but we want to show how to do it in two steps. First get in cell A1 and calculate the exponent $3 + \dfrac{7}{9}$ as shown in Figure C-13. Now move to cell B1. We now want to calculate

$$\left(\frac{2 + \pi}{7}\right)^{x},$$

where x is the value in cell A1. So we type =((2+pi())/7)^A1. Alternatively, if we have given the A column the name x we can type =((2+pi())/7)^x. The final result (showing the alternative approach) is in Figure C-14.

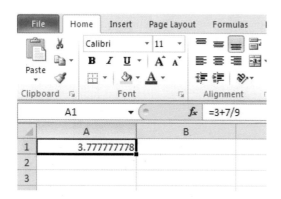

FIGURE C-13 The first step in a chain calculation

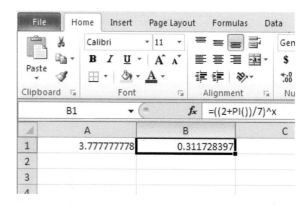

FIGURE C-14 Completing the chain calculation

Chapter 2: The Factorial Function

The *factorial* function occurs most often in probabilistic and statistical applications of mathematics. It applies only to non-negative integers and is denoted by $n!$. It is defined as follows:

$$0! = 1$$
$$\text{For } n > 0, n! = n \times (n-1) \times (n-2) \times \cdots \times 3 \times 2 \times 1.$$

Thus, for example, $3! = 3 \times 2 \times 1 = 6$ and $5! = 5 \times 4 \times 3 \times 2 \times 1 = 120$. Excel can calculate the factorial function. To calculate 6! with Excel, we use Excel's fact function. We simply type =fact(6). Excel will display the answer 720. Try some other examples. If you have trouble, remember to begin with an equal sign and to use parentheses as indicated.

TABLES AND GRAPHS

Excel can make both tables and graphs, and it has a large collection of tools for further analysis.

Chapter 2: Tables of Values

In order to make a table of values in Excel we first need to make a *sequence*. Suppose, for example, that we wish to put the sequence of numbers 1, 2, 3, . . . in column A. We could simply type in the numbers one at a time, but Excel offers a much simpler method. Type 1 in A1 and 2 in A2. Next click on cell A1, hold the mouse button down and drag downward to select both A1 and A2 as shown in Figure C-15. Now move the cursor to the bottom right corner of cell A2. When you do this, the thick cross will change to a thin cross hair. Hold down the left mouse button and drag downward. Excel will look at the first two numbers you typed and try to determine which sequence you had in mind. It is pretty good at getting it right, as Figure C-16 shows in this case.

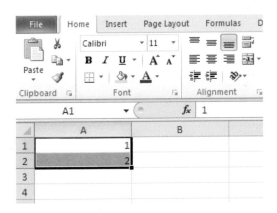

FIGURE C-15 Two sequence cells selected

FIGURE C-16 Dragging the cross hair to extend the sequence

Now let's make a table of values for $f = 3x + 1$. In cell **B1** we type =3*A1+1 as shown in Figure C-17. (If we had given the name x to column **A** we would use 3*x+1.) Don't forget to begin with the equal sign and use the symbol * for multiplication. Notice that in cell **B1** Excel has evaluated f at $x = 1$. We want to evaluate the entire column. To do this, select **B1** and move to the lower right corner until you get a cross hair. Click, hold, and drag downward to extend the sequence as shown in Figure C-18.

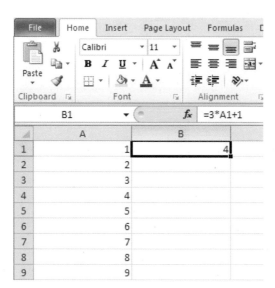

FIGURE C-17 Entering the formula in B1

FIGURE C-18 Dragging the cross hair to complete the table

There are any number of ways in which we may wish to alter the table. If, for example, we decide that we wish to extend the table further, we first use the mouse to select the last two cells of **A** and **B** (four cells in all) as shown in Figure C-19. Move the mouse to the lower right corner of the selection and drag the cross hair downward. The table extends as you drag, as is shown in Figure C-20.

	A	B
1	1	4
2	2	7
3	3	10
4	4	13
5	5	16
6	6	19
7	7	22
8	8	25
9	9	28
10		
11		
12		

FIGURE C-19 Selecting four cells at the bottom

	A	B
1	1	4
2	2	7
3	3	10
4	4	13
5	5	16
6	6	19
7	7	22
8	8	25
9	9	28
10	10	31
11	11	34
12	12	37

FIGURE C-20 Dragging the cross hair to extend the table

Suppose now that we want to see the table begin at 300 rather than at 1. Type 300 in A1 and 301 in A2. Select A1 and A2, and drag them downward just as we did to make our initial sequence. The values in the B column will adjust automatically. The completed table is in Figure C-21.

It is just as easy to change the increment we use in the sequence. Let's start at 0 and move in steps of 0.2. We just enter 0 into cell A1 and 0.2 into cell A2. Now click and drag to extend the sequence as before. The new table is in Figure C-22.

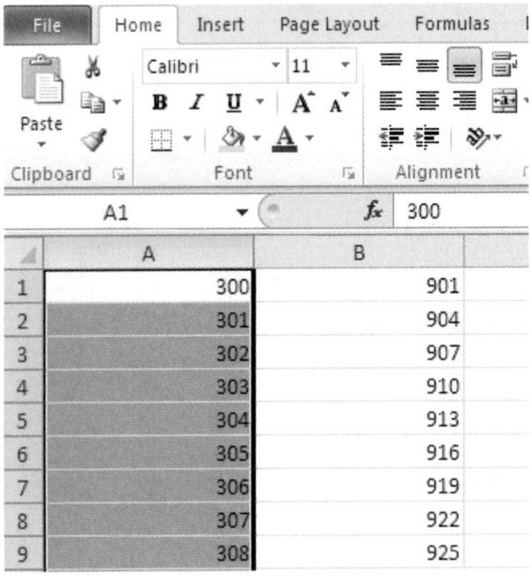

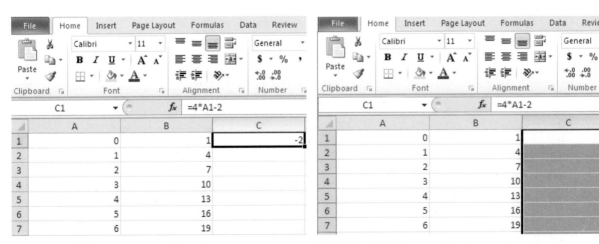

FIGURE C-21 Starting the table at 300

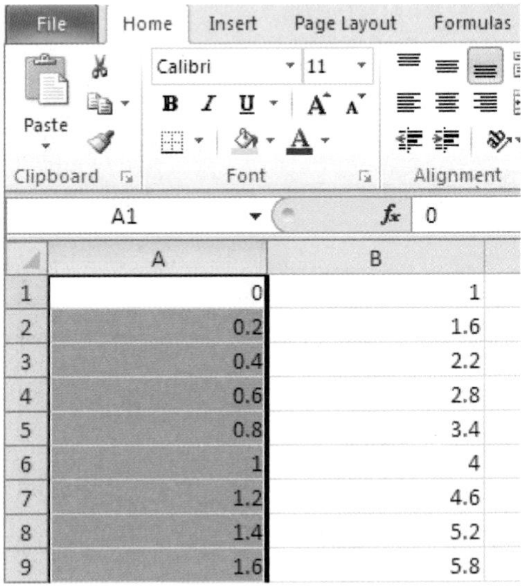

FIGURE C-22 Changing the table increment

Excel easily handles more than one function at a time, and this allows us to use tables to make comparisons. For example, let's show how to compare the values of $f = 3x + 1$ with those of $g = 4x - 2$. We begin with the table of values for $f = 3x + 1$ using x from 0 to 6 in steps of 1. In the C1 cell, we type =4*A1-2 as shown in Figure C-23. (Alternatively, we could type =4*x-2 if column A has been named x.) Now we use the mouse as before to drag downward from the C1 column. The resulting table of values is in Figure C-24. Note that f starts out larger than g. They have the same value when $x = 3$, and after that g has the larger values.

FIGURE C-23 Entering a second function

FIGURE C-24 The completed comparison table

All Excel screen shots are © Microsoft Corporation.

Chapter 2: Graphing

Excel can generate graphs of many types and present them in many ways. We will consider only two types of graphs (traditional function graphing here and discrete graphs in the next chapter). You are invited to explore other types of graphs that Excel can produce. Let's illustrate how to do this using $\dfrac{x^4}{50} - x$. Before beginning, some may wish to name the first column x. The first decision we need to make is what horizontal span to use for the graph. Often functions come from physical situations, and the physical situation might determine the span we want to use.

In this case, let's arbitrarily use the span of -5 to 5. We need to make a table of values running from -5 to 5. Now we have a second decision to make. What step size should we use in our table? A smaller step size gives a more accurate graph, but if we make the step size too small the table may be unnecessarily large. In this case, let's use a step size of 0.25. Thus we enter -5 into A1 and -4.75 into A2 (since $-5 + 0.25 = -4.75$). Now select the two cells and click and drag downward until we reach 5. This is partially shown in Figure C-25. In the B1 cell, we type =A1^4/50-A1. (If you named the first column x, you may type =x^4/50-x.) Click and drag downward to complete the table, which is partially shown in Figure C-26.

FIGURE C-25 A sequence up to 5   **FIGURE C-26** Completing the table

Some will find it convenient before making the graph to give column B the name y. Next click on the Charts menu, which is on the Insert tab. You will be prompted for the type of graph you wish to make. Select Line as shown in Figure C-27. You will see a menu with several possible subtypes. Select the very first. You are now prepared to go on to the next step, by choosing Select Data where you will encounter the Select Data Source screen. On the Chart data range tab of that screen, delete whatever is in the Chart data range box. Now click on the worksheet data and select B1 through B41 as shown in Figure C-28.

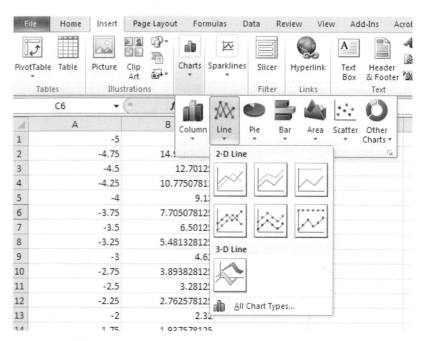

FIGURE C-27 Selecting the chart type

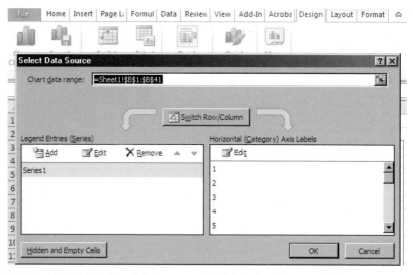

FIGURE C-28 The Chart data range tab of the Chart Source Data screen

Next click on Edit in the Horizontal (Category) Axis Labels box and then use the mouse to select your sequence, A1 through A41. (This particular box will not accept x as the name of the sequence.) The result should appear as in Figure C-29. At this point, you may, if you wish, simply click on OK to get Figure C-30. Excel offers a seemingly unlimited collection of edits and modifications you may choose to make in order to improve the appearance of the graph. You are encouraged to explore these options. Our selection of options, which included clicking on the vertical and horizontal axes and editing them using Format Axis, as well as deleting the Series legend, produced the graph in Figure C-31. Making graphs with Excel may seem a bit complicated at first. That is because Excel offers so many different types and styles of graphs. A bit of practice will make the procedure go quickly and easily.

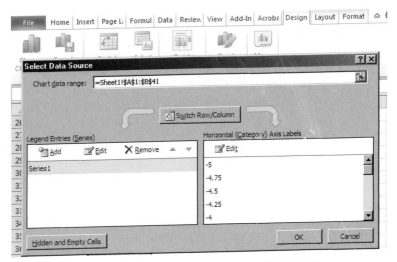

FIGURE C-29 The Select Data Source pop-up of Select Data

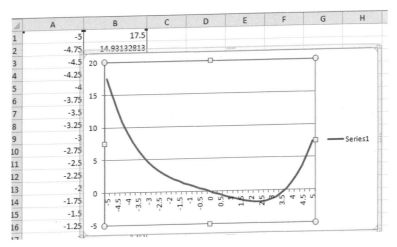

FIGURE C-30 The completed graph

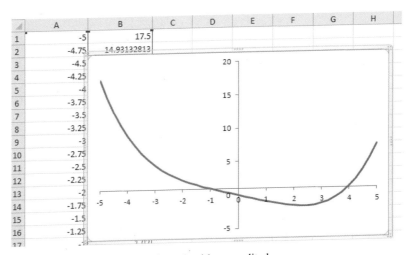

FIGURE C-31 The completed graph with axes edited

Let's see how to add another graph, say that of $5 - \dfrac{x^2}{3}$, to the picture we already have in Figure C-31. You may wish to give column C the name z. Enter the formula in cell C1 as either =5-A1^2/3 or =5-x^2/3, whichever is appropriate. Now click and drag to make a table.

With the chart selected, right click and go to **Select Data**. You will be presented with a **Select Data Source** pop-up. Click on **Chart data range** and replace B with C. The properly configured box is in Figure C-32. Finally click **OK**, and the graph shown in Figure C-33 will appear.

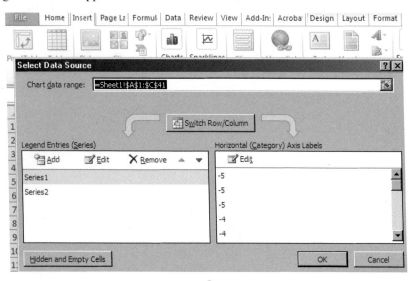

FIGURE C-32 Selecting a new data range

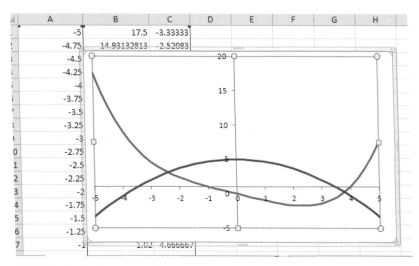

FIGURE C-33 Two graphs displayed

Chapter 2: Solving Equations

On TI calculators, graphs are totally interactive in that they allow one to analyze functions on the graphing screen. Excel does not offer these features, but most can be accomplished outside the graphing environment. It is not necessary in Excel to make a graph in order to solve an equation, but it is often helpful, and it is a much recommended practice.

The methods we use for solving equations using Excel apply to equations of the form $f(t) = 0$ and so correspond to the single-graph method described in Section 2.4 of this text. (More generally, the methods apply to equations of the form $f(t) = c$ for a constant c.) As the text notes, all the equations considered can be put in this form.

We will show two methods for solving equations using Excel. The first method uses the Solver add-in to solve equations. It is accessed from the Tools menu.[1] Let's see how to use Solver to solve $t^2 + t - 3 = 0$. We strongly recommend that you first make a graph as we have done in Figure C-34. The graph shows us that there are two solutions of the equation: one near -2 and another near 1.5.

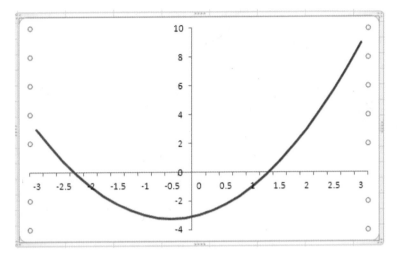

FIGURE C-34 The graph of $t^2 + t - 3$

You may wish to give the cell D1 the name t. To do this, select D1, right click, use Define Name and enter t. Next we want to enter the formula. Go to cell E1 and enter =D1^2+D1-3. Alternatively, if you named D1 t, you can type =t^2+t-3. Assuming there is nothing currently in cell D1, Excel will treat the entry as 0, and -3 will appear in cell E1, indicating that $t^2 + t - 3$ has the value -3 when $t = 0$. Now with cell E1 selected, click on Solver, which is at the far right of the tab Data. Excel will display the screen partially shown in Figure C-35. Be sure the Set Target Cell box reads E1. Be sure the Value of option is selected and the value in the box is set to 0. Finally, put D1 (or t if you named that cell t) in the By Changing Cells box. The properly completed entries are in Figure C-35. The entries can be deciphered to read "make the value of $t^2 + t - 3$ equal 0 by changing the value of t." Now click on Solve, then OK for Solver Resetting, and Excel will present the answer, 1.302776, in D1.

We have found one of the solutions of the equation, but the graph shows us that there is another near -2. We will show two ways to get the other solution. The first is the easiest and is good practice in any case. It is always wise when using Solver to give Excel a little help by starting as near the point you want as you can. In this case, the solution we are looking for is near -2. We make Excel start its work there by first entering -2 into D1. Now select E1 and proceed as before. Be sure that the Solver Parameters box titled Make Unconstrained Variables Non-Negative is unchecked. This time Excel shows the second zero, -2.302775.

---

[1]If Solver is not yet there, go to File, Options, Add-Ins, highlight Solver Add-in and press OK.

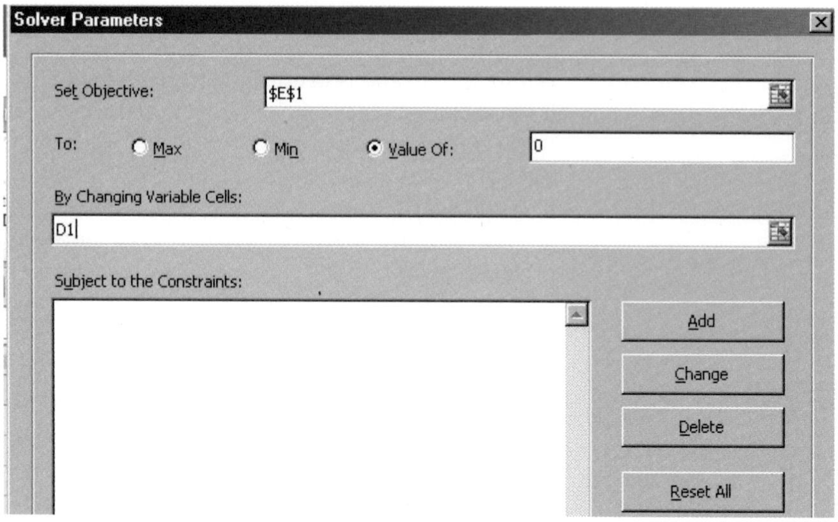

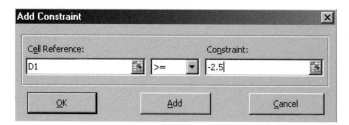

FIGURE C-35 Entering the Solver parameters

For most functions, just forcing Solver to start reasonably near the point you are looking for will get the job done. In some cases it may be necessary to resort to more serious means, which we now present. As before, we select cell E1 and choose Solver from the Tools menu. Again we select the Value of option and put 0 in the Value of box. Type t or D1 in the By Changing Cells box. We want to constrain the solution by $t \geq -2.5$ and $t \leq -2$. To do this go to the Subject to the Constraints section and click on Add. You will see the Add Constraint page shown in Figure C-36. Enter t or D1 in the Cell reference box. There is a pull-down menu in between the two boxes which lets us select >=. Now enter -2.5 in the Constraint box. The completed Add Constraint page is in Figure C-36. Click on Add in order to enter the second constraint, $t \leq -2$. The properly configured Add Constraint box is in Figure C-37. Click on OK and finally on Solve and OK. Excel will present the solution, -2.302775, as before, in cell D1.

FIGURE C-36 Entering the constraint $t \geq -2.5$

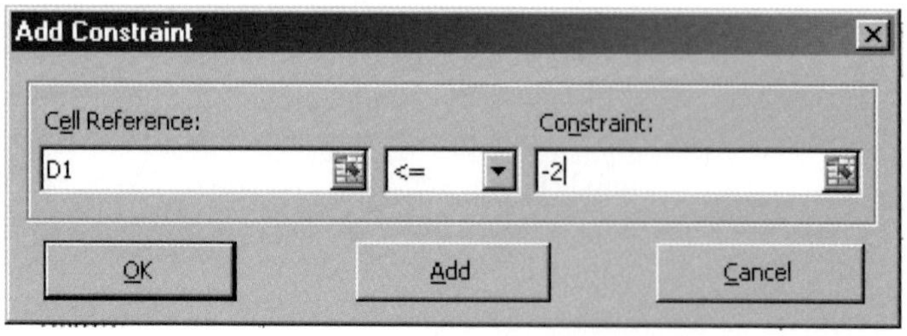

FIGURE C-37 Entering the constraint $t \leq -2$

Let's see how to solve the same equation $t^2 + t - 3 = 0$ using the Goal Seek feature of Excel. Let's find the solution near $t = 1.5$ first. As in the case of Solver we enter the formula we want in cell E1. We can either type =D1^2+D1-3, or, if we have given D1 the name t, we can use =t^2+t-3. As with Solver, it is good practice to give Goal Seek a good starting point. Thus we type 1.5 in cell D1. Next click on E1, and then click on the Data tab. Select Goal Seek from the What-If Analysis menu. Excel will present you with the Goal Seek entry screen. Be sure E1 is in the Set cell box. (If you clicked on E1 before calling Goal Seek, this will be done automatically.) Type 0 in the To value box. Finally in the By changing cell box type t if you have made the appropriate definition; otherwise, type D1. The proper entries are shown in Figure C-38. If we now click on OK twice, we get the solution in cell D1 as shown in Figure C-39. To find the other zero, we type –2 in cell D1 before calling up Goal Seek. We emphasize that if we use either Goal Seek or Solver it is important first to graph the function so that we can give Excel a good place to start its work.

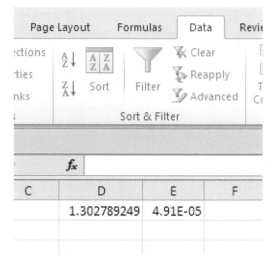

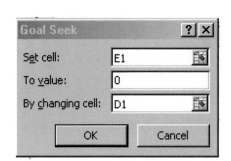

FIGURE C-38 The Goal Seek entry screen   **FIGURE C-39** One solution using Goal Seek

◼ Chapter 2: Optimization

Maxima and *minima*, or *peaks* and *valleys*, of a graph can be located using Excel in much the same way that we solve equations with Solver. To illustrate the method, we will look at the graph of $f = 2^t - t^2$. We have made the graph using a horizontal span of -1 to 4 in Figure C-40. We see one maximum and one minimum. Let's locate the maximum first. Just as we did when solving an equation, either we first give the cell D1 the name t and then type =2^t-t^2 in E1, or, if we prefer not to use name definitions, we type =2^D1-D1^2 in E1. The graph shows that the maximum occurs near $t = 0.5$, so let's enter that number into D1 to give Solver a good starting point. Now select E1 and click on Solver. As before, the Solver constraint page is displayed. Select Max (short for Maximum) and enter the appropriate one of t or D1 in the By Changing Cells box. The properly completed page is in Figure C-41. Click on Solve, and then OK, and in cells D1 and E1 Excel shows us that a maximum value of 1.164361 is reached at $t = 0.48509$.

To find the minimum value, we first note that it occurs near $t = 3$. Thus we type 3 in cell D1 to give Solver a good starting point. Now select E1 and click on Solver. The entries are the same as for the maximum except that we now select Min (short for Minimum). When we finish, Excel shows that a minimum value of -1.050602 is reached at $t = 3.212432$.

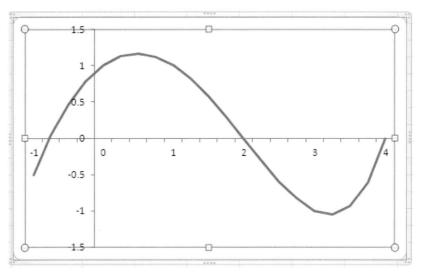

FIGURE C-40 The graph of $2^t - t^2$

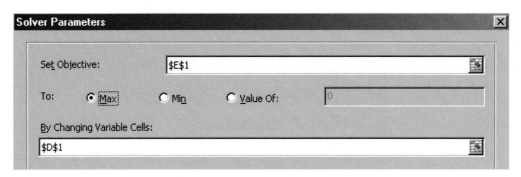

FIGURE C-41 The properly completed Solver parameter page

There may be cases when just giving Excel a good starting point is not enough to get to the maximum or minimum value we are looking for. On such occasions, we must use the Subject to the Constraints box just as we did in solving equations.

Chapter 5: Piecewise-Defined Functions

Sometimes functions are defined by different formulas over different intervals. Such functions are known as *piecewise-defined* functions. For example, we define

$$f(x) = \begin{cases} x + 1, \text{if } x < 1 \\ x - 1, \text{if } x \geq 1. \end{cases}$$

This means that $f(x)$ is $x + 1$ if x is less than 1, but it is $x - 1$ when x is greater than or equal to 1. Thus $f(0) = 0 + 1 = 1$ since $0 < 1$. But $f(3) = 3 - 1 = 2$ since $3 \geq 1$. Excel will graph piecewise-defined functions correctly if we arrange the data right. We start by putting a sequence in column A as usual. Let's go from -1 to 3 in steps of 0.25. Since $-1 < 1$, the function formula there is $x + 1$. Thus we type =A1+1 in B1. This definition only applies for values less than 1. Thus we click on B1 and drag downward only to B8 (corresponding to $x = 0.75$). Leave the remainder of column B blank. Now click on cell C9, corresponding to $x = 1$. From here on, the function is $x - 1$, and so we type =A9−1. We expand this down to A17, leaving the top half of column C blank. The correctly entered data are partially displayed in Figure C-42. Now we click

3	-0.5	0.5	
4	-0.25	0.75	
5	0	1	
6	0.25	1.25	
7	0.5	1.5	
8	0.75	1.75	
9	1		0
10	1.25		0.25
11	1.5		0.5
12	1.75		0.75
13	2		1
14	2.25		1.25

FIGURE C-42 Entering data for a
piecewise-defined function

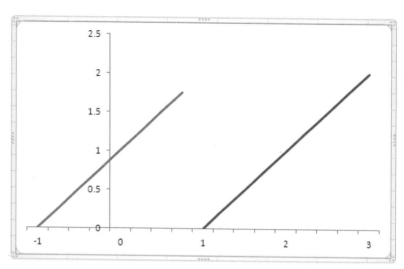

FIGURE C-43 The completed graph

on the Insert tab, look at Charts, select Line as the type of graph, and choose the first
subtype. Next go to Select Data, where we encounter the Select Data Source screen.
On the Data Range tab of that screen, we delete whatever is in the Data range box,
click on the worksheet data, and select the data from both columns B and C. That is,
we include the complete rectangle with corners at B1, B17, C17, and C1, including the
blank cells in the rectangle. For the Horizontal (Category) axis labels we choose A1
through A17. Now complete the graph using appropriate options. Our graph is in Figure
C-43. Note that as might be expected, the graph shows a jump at $x = 1$.

DISCRETE DATA

Many times information about physical or social phenomena is obtained by gathering
individual bits of data and then performing various kinds of analysis on the data. We
will show several of Excel's abilities to handle data.

Chapter 3: Plotting Data

We will show how to make data plots using the following table. It is taken from the *1996
Information Please Almanac*, and it shows median American family income *I* by year
in terms of 1996 dollars. That means the dollar amounts shown have been adjusted to
account for inflation. The variable *d* in the table is years since 1980. Thus, for example,
the $d = 2$ column corresponds to 1982.

d = years since 1980	0	1	2	3	4	5
I = median income	21,023	22,388	23,433	24,580	26,433	27,735

We enter the *d* values in column A and the *I* values in column B. To plot the data,
call up the Insert tab, then Charts. Select Scatter as the chart type, and select the
first chart subtype (the one without lines) as shown in Figure C-44. Then go on to the
Next step. Click in the Data range box and select the data in column B. (If you have
given column B the name *y*, you may simply enter y in the Data range box.) Select the
Series tab, click in the X values box, and select A1 through A6. Complete the graph as
before. Our graph is in Figure C-45. This method works to plot any data set.

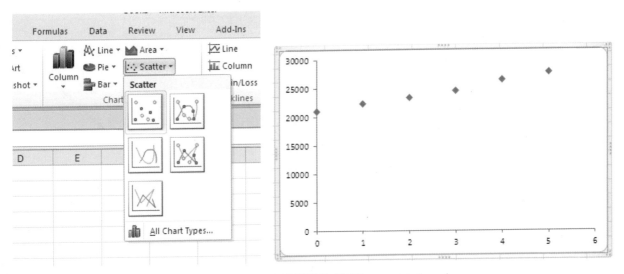

FIGURE C-44 Preparing for a data plot

FIGURE C-45 The completed graph

Chapter 3: Linear Regression

Linear regression is a method of getting a linear function which approximates data that are almost linear. Excel will perform this and other types of regression automatically. We will use the plot from the previous section to show the method. With the chart selected, go to the Chart Tools menu at the top and click on Layout, then pull down the Trendline menu. (Alternatively, you can left-click on the data points, then right-click, and you will get a menu containing Add Trendline.) On the Type tab, choose Linear as seen in Figure C-46. Right-click on the line, select Format Trendline and check the box Display Equation on chart and then Close; the regression line and its equation $y = 1338.3x + 20919$ will be added to the plot as shown in Figure C-47.

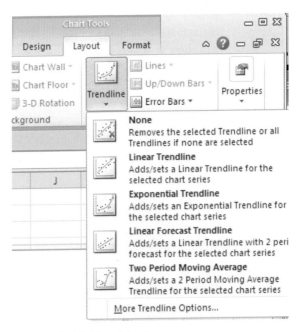

FIGURE C-46 Plot of the data

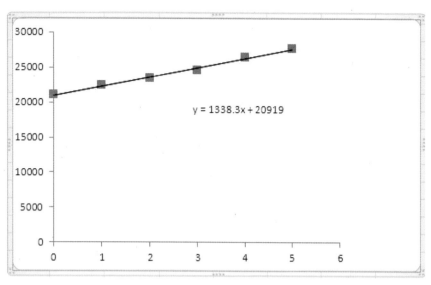

FIGURE C-47 Regression line and equation added

It is often convenient to copy and paste the regression formula into the spreadsheet so that you can do calculations with it. Highlight =1338.3x+20919 (leaving off the y) and Copy. Now click in B7 and Paste. Note that you will need to edit the formula before it will work. You will need to insert the symbol * between 1338.3 and x to indicate multiplication, and, if you did not give column A the name x, you will need to replace x by A7. The properly edited formula is in Figure C-48. In Figure C-49 we have entered 2.7, and the regression equation evaluated there is automatically calculated in B7.

	A	B
1	0	21023
2	1	22388
3	2	23433
4	3	24580
5	4	26433
6	5	27735
7		=1338.3*A7+20919
8		
9		

FIGURE C-48 Pasting the regression equation into the spreadsheet

	A	B
1	0	21023
2	1	22388
3	2	23433
4	3	24580
5	4	26433
6	5	27735
7	2.7	24532.41
8		

FIGURE C-49 Regression equation evaluated at 2.7

 Chapter 4: Exponential Regression

We can use Excel to get an exponential model just as we did for linear models. We will illustrate with the following data table.

x	1	2	3	4	5
y	0.55	0.83	1.5	2.7	4.4

We enter the data in columns A and B. The plot is in Figure C-50. Highlight the plot. Now from the Layout menu, click on the Trendline menu. This time, rather than selecting Linear, we select Exponential. As before, right-click on the curve, select Format Trendline, and check the box Display Equation on chart, then Close. We see from Figure C-51 that Excel displays the regression equation as

$$0.3066e^{0.5338x}.$$

To convert this to the form for exponential functions used in the text, we use the fact that

$$e^{0.5338x} = (e^{0.5338})^x = 1.71^x.$$

Thus the exponential model is 0.31×1.71^x.

As with linear regression, you may wish to copy and paste the regression equation into the spread sheet. Remember to edit the equation using the symbol * for multiplication and if necessary correcting the cell reference.

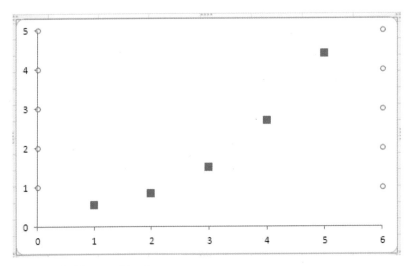

FIGURE C-50 Plot of the data

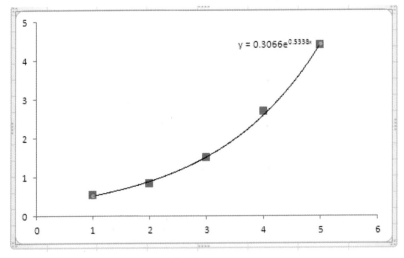

FIGURE C-51 Exponential model added

Chapter 5: Power and Polynomial Regression

Power models can be constructed in much the same way as linear and exponential models. We illustrate using the following data table.

x	1	2	3	4	5
y	0.22	0.54	0.88	1.25	1.91

As before, we put the x data in column A and the y data in B. The plot is in Figure C-52. Highlight the plot. Next we select Layout, then pull down the Trendline menu, but this time we select More Trendline Options, then select Power and check Display Equation on chart, then Close. The result is in Figure C-53.

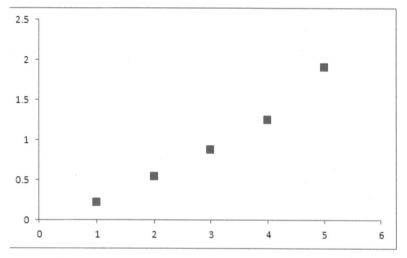

FIGURE C-52 Plot of the data

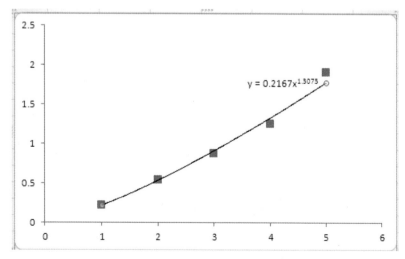

FIGURE C-53 Power model added

Quadratic (or higher-degree polynomial) regression is just as easy as other types of regression with Excel. Suppose that we want to fit the following data with a quadratic model.

x	−2	0	1	3	6
y	3.3	1.7	2.5	10.4	33.2

As usual, we enter the x data in column A and the y data in column B. The plot of the data is in Figure C-54. Next choose Layout, then pull down the Trendline menu. This time we choose More Trendline Options, then select Polynomial and set the order to 2. Check Display Equation on chart, then Close. Figure C-55 shows that the quadratic model (rounded to two decimal places) for this data is $0.78x^2 + 0.61x + 1.43$.

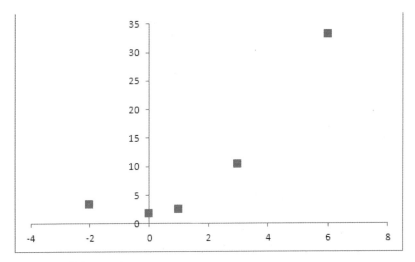

FIGURE C-54 Data plot

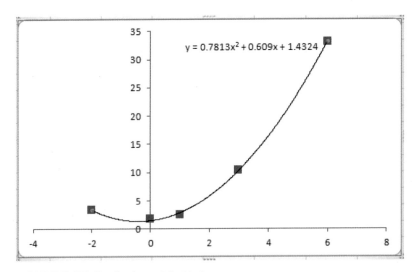

$$y = 0.7813x^2 + 0.609x + 1.4324$$

FIGURE C-55 Quadratic model added

NOTE: The answers presented here are intended to provide help when students encounter difficulties. Complete answers will include appropriate arguments and written explanations that do not appear here. Also, many of the exercises are subject to interpretation, and properly supported answers that are different from those presented here may be considered correct.

PROLOGUE

1. 55.62%   **3.** 67.87 miles   **5.** $74.25   **7.** $669.60
9. 2.70%   **11. a.** $5.99   **b.** 43.08%   **13.** $4464.29
15. a. 5.54 years   **b.** 1.97; no   **c.** $9850; no
17. a. 25,132.74 miles, or about 25,000 miles
   b. 268,082,573,100 cubic miles, or about 2.68×10^{11} cubic miles
   c. 201,061,929.8 square miles, or about 201,000,000 square miles
19. a. 584,336,233.6 miles, or about 584 million miles
   b. 584,336,233.6 miles per year, or about 584 million miles per year
   c. 8760 hours
   d. 66,705.05 miles per hour, or about 67,000 miles per hour
21. 735 newtons; 165.38 pounds
23. 277.19 cycles per second; 293.67 cycles per second
25. Lean body weight is 100.39 pounds.
   Body fat is 31.61 pounds.
   23.95% body fat
27. 28.21 meters   **29. a.** $0.95   **b.** $1.07375 or $1.07
   c. $331.58   **d.** $279.39   **e.** 18.68%; Using the Advantage Cash card allows you to buy 18.68% more food than using cash.
S-1. 2.43   **S-3.** 1.53   **S-5.** 0.12   **S-7.** 1.32   **S-9.** 0.6
S-11. −2.17   **S-13. a.** 0.62   **b.** 1.03   **S-15.** 0.71
S-17. 1.43   **S-19.** 50.39   **S-21.** 0.34   **S-23.** 3.62
S-25. −2.37   **S-27.** 6.96   **S-29.** 0.0029   **S-31.** $750.00
S-33. 89.60 degrees   **S-35.** $1338.23   **S-37.** 2.73 grams

REVIEW EXERCISES

1. −3.33   **2.** 2.40   **3.** 6.97   **4.** 1.8 gallons; 16.67 gallons
5. 3681 million miles; 93 million miles   **6.** 4.92 seconds

CHAPTER 1

SECTION 1.1

1. a. $S(60)$; 39.12 miles per hour
   b. It is the speed, in miles per hour, at which an emergency stop will leave a skid mark of 100 feet.
3. a. $W(14,6)$ **b.** 22.29 pounds
5. a. $T(13,000) = \$930$ **b.** $110   **c.** $110
7. a. $V(1)$; 8 feet per second. The ball is rising.
   b. −24 feet per second. The ball is falling.

   c. The velocity is 0, and the ball is at the peak of its flight.
   d. The velocity changes by −32 feet per second for each second that passes.
9. a. 12 deer
   b. $N(10) = 380$ deer (rounded to the nearest whole number)
   c. $N(15) = 410$ deer
   d. 30 deer
11. a. $C(800) = 4.54$ grams   **b.** 5730 years
13. a. Higher   **b.** Continuous compounding
   c. $190.67 per month. It is 10 cents higher than if interest is compounded monthly as in Example 1.2.
15. a. $P(350, 0.0075, 48)$; $14,064.67
   b. $15,812.54   **c.** $19,478.33
17. Sirius appears about 24.89 times as bright as Polaris.
19. About 4.34 light-years
21. a. $1 - 0.5^1 = 0.5$, giving a yield of 50% of maximum
   b. $Y(3)$; 0.88, or 88% of maximum
   c. 79% of maximum
23. a. 2307.14 watts per square meter   **b.** 5767.85 watts
25. 7.25 kilometers
27. a. $r = 0.005$
   b. $E(240)$; $183,985.66
   c. $E(y) = 400,000 \times \dfrac{(1 + .005)^{12y} - 1}{(1 + .005)^{360} - 1}$
29. a. $10,726.84   **b.** $M = 2135.00$ dollars.
S-1. 0.35   **S-3.** 3.67   **S-5.** 1.06   **S-7.** 1.16   **S-9.** 1.73
S-11. 0.06   **S-13.** −2.21   **S-15.** $C(0) = 2.88$; $C(10) = 0.17$
S-17. 22.06   **S-19.** $R(0) = 45$; $R(30) = 84$ **S-21.** 87.06
S-23. 107.95   **S-25.** $p(2.5)$
S-27. It is the top speed of a fish 13 inches long.
S-29. It is the time required for an investment of $5000 at an APR of 6% to double in value.
S-31. It is the bird population 7 years after observation began.

SECTION 1.2

1. a. $M(2005)$ is *Star Wars Episode III*; $B(2005)$ is 380.27 million dollars   **b.** $B(2003)$
3. a. $B(55)$ is the recommended bat length for a man weighing between 161 and 170 pounds who is 55 inches tall; 31 inches.
   b. $B(63)$

5. a. $P(1989) = 30\%$
   b. Using averages, $P(1999) = 32\%$
   c. 0.2 percentage point per year
   d. 33%
   e. If growth rate from 2009 to 2014 continues at the same rate as from 1989 to 2009, then $P(2014)$ will be about 35%.
7. a. 113 million
   b. 5 million per year
   c. 186 million
9. The limiting value of W is the total amount of water frozen in the snowball. This value is reached when the snowball is completely melted.
11. a. -0.000454 gram per year
   b. About 4.44 grams   **c.** 0
13. a. yearly, semiannually, monthly, daily, hourly, each minute
   b. 12.683%
   c. $1019.76
   d. Answers will vary. About 12.75%
15. a. $H(13)$, about 62.2 inches
   b. **i.** Units for growth rate: inches per year
       ii. From age 0 to age 5
       iii. Answers will vary

Period	Growth Rate
0 to 5	4.2
5 to 10	2.5
10 to 15	2.4
15 to 20	1.3
20 to 25	0.1

   c. Answers will vary.
17. a. Units for rate of change: dollars per dollar

Interval	Rate of Change
16,000 to 16,200	0.09
16,200 to 16,400	0.09
16,400 to 16,600	0.09

   b. Answers will vary.   **c.** Answers will vary.
19. a. 1.43 pounds per centimeter
   b. 3.85 pounds per centimeter
   c. Large
   d. 205.95 pounds
   e. 0.26 centimeter per pound
   f. 171.96 centimeters
21. a. Units for rate of change: thousands of widgets per worker

Interval	Rate of Change
10 to 20	1.25
20 to 30	0.63
30 to 40	0.31
40 to 50	0.15

   b. Answers will vary.
   c. About 49.2 thousand widgets
   d. Too high
23. a. -4.2 points per year   **b.** 201.6
   c. 35 to 45 years old
25. a. Equity after 10 years; $27,734
   b.

Interval	Rate of Change
0 to 5	2361.60
5 to 10	3185.20
10 to 15	4296.60
15 to 20	5795.40
20 to 25	7817.00
25 to 30	10,544.20

   c. Late in the life of a mortgage   **d.** $60,807.80   **e.** No
27. Answers will vary.
S-1. 17.6   **S-3.** 44.6   **S-5.** 53.2   **S-7.** 53.9   **S-9.** 34.2
S-11. 52.3   **S-13.** 53.8
S-15. Average rate of change is 2.08; 38.4
S-17. Average rate of change is 0.19; 51.7
S-19. Average rate of change is 0.02; 53.8
S-21. 5.7   **S-23.** 1.1   **S-25.** -7.9
S-27. Average rate of change is -0.64; 3.0
S-29. Average rate of change is -0.86; -7.0
S-31. Answers will vary.

SECTION 1.3

1. Answers will vary.
3. a. A larger ratio indicates a longer skirt.
   b. About 1969   **c.** No
5. a. $U(1990)$ is the unemployment rate in 1990. Its value is 5%.
   b. From about 1930 to 1940   **c.** About 1982 or 1983
7. a. $v(1970) = \$10,000$, $v(1980) = \$5000$, $v(1990) = \$35,000$, $v(2010) = \$35,000$
   b. Various acceptable graphs
   c. Answers will vary.
9. a. $F(7)$; about 1500 cubic feet per second
   b. At the end of June   **c.** At the end of May
   d. About 0 cubic feet per second
   e. Answers will vary.

11. a. About $14,000 per acre
   b. About 110 years old
   c. About 30 years old
   d. About 60 years old
   e. Answers will vary.
13. a. In 2010 there were about 102 tornadoes reported.
   b. In 2002 there were about 18 tornadoes reported.
   c. About 43 tornadoes per year
   d. About 20 to 22 tornadoes per year
   e. Close to 0 tornado per year
15. Answers will vary.
17. a. About 700 foot-candles
   b. About 800 foot-candles
   c. 80 degrees **d.** 40 degrees
19. a. About 2 P.M.
   b. From about 6 A.M. to about 9 A.M.
   c. From about 9 A.M. to about 2 P.M.
   d. The net carbon dioxide exchange is zero.
21. a. About 100 pounds per acre
   b. From the difference between the yield and cost graphs
   c. About 85 pounds per acre
23. a. Concave up
   b. Equity grows faster and faster.
   c. About $335,000
25. a. 32.3°C since higher beetle mortality
   b. More likely to infest the maize
27. $\log l_x$

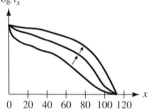

S-1. About 3.3 **S-3.** About 0.5 **S-5.** About $x = 4$
S-7. At $x = 2.4$, f is about 4. **S-9.** From 2.4 to 4.8
S-11. It is concave up. **S-13.** It is concave up.
S-15. At about $x = 0.5$ and $x = 2.5$.
S-17. At about $x = 1.5$ and $x = 3.5$
S-19. At $x = 1$, $x = 2$, and $x = 3$
S-21. It is concave down.
S-23. From $x = 1.5$ to $x = 2$ and from $x = 3.5$ to $x = 4$
S-25. Inflection points
S-27. Answers will vary, but the graph should have the same shape as the figure at the beginning of the Skill Building Exercises, except that the maximum should occur at $x = 3$.
S-29. Answers will vary. One solution is

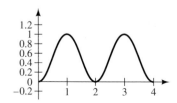

S-31. Answers will vary. One solution is

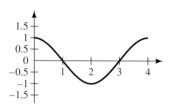

SECTION 1.4

1. a. 195 grams
   b. $F = 24h + 13c + 30f + 25o$
3. a. $N(3) = 186.56$ million
   b. Population is in millions.

Year	Population
1960	180
1961	182.16
1962	184.35
1963	186.56
1964	188.80
1965	191.06

c.

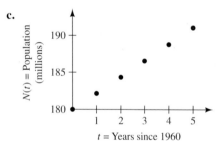

d. Same numbers as in part b
e. 290.06 million
5. a. $A = 200 + 150t$, where t is the time in minutes since takeoff and A is the altitude in feet.
   b. $A(1.5)$; 425 feet
c.

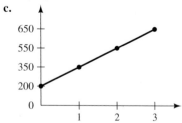

7. a. $123.00
   b. $C(d, m) = 49d + 0.25m$, where C is the rental cost in dollars, d is the number of rental days, and m is the number of miles driven.
   c. $C(7, 500) = $468.00

9. a. 22 cents   **b.** $19.22
   c. $c(p) = 0.16 + 0.03(p - 1)$, where c is the cost in dollars of the stationery and p is the number of pages.
   d. $C(h, p) = 9.25h + 0.03(p - 1) + 0.16 + 0.50$, where C is the cost in dollars of preparing and mailing the letter, h is the hours of labor, and p is the number of pages.
   e. $4.54
11. a. 7   **b.** 7   **c.** 10
   d. $T = \dfrac{S}{50}$, where T is the annual stock turnover rate and S is the number of shirts sold in a year.
13. a. $C = 15N + 9000$, where N is the number of widgets produced in a month and C is the total cost in dollars.
   b. $C(250)$; $12,750
15. a. Answers will vary.
   b. $R = (50 - 0.01N)N$, where R is in dollars
   c. $R(450)$; $20,475
17. a. $81 each for a total amount of $243
   b. Rate $= 87 - 2n$ dollars
   c. $R(n) = n(87 - 2n)$ dollars
   d. $R(9) = $621
19. a.   i. $13
   ii. $C = \dfrac{150}{n} + 10$, where C is the amount charged per ticket, in dollars, and n is the number of people attending.
   iii. $C(65)$, about $12.31.
   b. $P = \dfrac{250}{n} + 10$, where P is the amount charged per ticket, in dollars, and n is the number of people attending.
21. a. Using k for the constant of proportionality, $t = kn$.
   b. k is the number of items each employee can produce.
23. a. $p = 0.434h$
   b. The head is 96 feet. The back pressure is about 41.66 pounds per square inch.
   c. The head is 145 feet. The back pressure is 62.93 pounds per square inch.
25. a. $V = KS$ **b.** 0.00123 meter per day
   b. 1.23 meters per day
27. a. $8940   **b.** $F = 2500 + 0.02M$
29. Answers will vary.
S-1. $312.50   **S-3.** 405   **S-5.** $N = 67s$
S-7. $B = 500 + 37t$ **S-9.** $C = 249 + 0.99s$
S-11. $C = 1.46s + 3.50r + 2.40p$
S-13. $P = 500(t + s) - 300s - 21t$ **S-15.** 16   **S-17.** Yes
S-19. Yes   **S-21.** No   **S-23.** Yes   **S-25.** Yes

REVIEW EXERCISES

1. 447.20   **2. a.** 3.93 million   **b.** $N(20)$ **c.** 7.10 million
3. a. 41.9   **b.** 1.8   **4. a.** 2.47 million graduating in 1989
   b. $N(1988)$; 2.56 million   **c.** −0.09 million per year
   d. 2.02 million
5. a. Increasing   **b.** 2008 to 2012; 2002 to 2008   **c.** 2008
6. a. Answers will vary.   **b.** mid-2009   **c.** 2008
   d. Inflection point

7. $B = 780 - 39t$
8. a. Let t denote the number of text messages and C the charge in dollars.
   b. $C = 39.95 + 0.1(t - 100)$
   c. $C(450)$; $74.95
   d. $C = 39.95$
9. a. Let t denote the number of text messages, m the number of minutes, and C the charge in dollars.
   b. $C = 34.95 + 0.35(m - 4000) + 0.1(t - 100)$
   c. $769.95
   d. $C = 34.95 + 0.35(m - 4000)$
   e. $104.95
10. a. 2.97   **b.** 12   **c.** −1   **d.** 5
11. a. $258.90 is your monthly payment if you borrow $5500 at a monthly rate of 1% for 24 months.
   b. $M(8000, 0.006, 36)$; $247.75
12. a. 1.67   **b.** 2   **c.** 57
13. a. Units for rate of change: Value per cord divided by value per MBF Scribner

Interval	Rate of Change
20 to 24	3.4
24 to 28	3.4
28 to 36	3.4

   b. No   **c.** $25; $71
14. Concave down; concave up
15. a. Answers will vary.   **b.** About 132 feet
   c. Yes, the graph approaches a horizontal line.
   d. Concave down; each year the amount of growth decreases.
16. a. $52   **b.** $156
   c. Let R denote the rental cost in dollars per room; $R = 56 - 2(n - 1)$, or $R = 58 - 2n$.
   d. Let C denote the total cost in dollars; $C = n(56 - 2(n - 1))$, or $C = n(58 - 2n)$.
17. a. $4750
   b. Let C denote the cost in dollars; $C = 3200 + 31(n - 50)$, or $C = 1650 + 31n$.
   c. Up to 124
18. a. No   **b.** Answers will vary.   **c.** No
   d. Answers will vary.

ANSWERS FOR A FURTHER LOOK: Average Rate of Change with Formulas

1. −1/8 or −0.125   **2.** 3   **3.** 1/5 or 0.20   **4.** 2   **5.** h
6. 3   **7.** $2x + h + 1$ **8.** The average rate of change is m.
9. They are the same.   **10.** 1/2 or 0.5 inch per year
11. a. −15/2 or −7.50 grams per minute
   b. The amount of the radioactive substance is decreasing.
12. $2x$ **13.** $3x^2$

ANSWERS FOR A FURTHER LOOK: Areas
Associated with Graphs

1. 27   **2.** 48   **3.** 18   **4.** 16   **5.** 99   **6.** 216   **7.** 32   **8.** 18
9. 48   **10.** 48   **11.** 140   **12.** 92

CHAPTER 2
SECTION 2.1

1. 7 items
3. a. Decreases   **b.** 200 (one word per sentence)
   c. 16 sentences
5. a. Decrease   **b.** $E(200)$; 75   **c.** High average
   d. 166 or lower
7. a. 14.67 million students in 2010
   b. 2007; 14.97 million students
   c. 0.00 million per year; yes
9. The seventh race
11. a. 120 ways   **b.** 7 or more people   **c.** 24 guesses
   d. 8.07×10^{67}
13. a. Larger
   b. Once each year, EAR is 10%. (This is the only circum
   stance under which the EAR and APR will be the same.)
   Monthly, EAR is 10.471%. Daily, EAR is 10.516%.
   c. Monthly, $5523.55. Daily, $5525.80.
   d. Continuously, EAR is 10.517%. The EARs for monthly
   and continuous compounding differ by less than 0.05
   percentage point.
15. a. $159.95
   b. We show the table only for months 18 through 24.

X	Y$_1$
18	934.92
19	782.01
20	627.94
21	472.72
22	316.33
23	158.76
24	0
X=24	

17. a. $E = \left(\dfrac{Q}{2}\right)850 + \left(\dfrac{36}{Q}\right)230$ **b.** $4035
   c. 4 cars at a time   **d.** 9 orders this year
   e. $80 per year per additional car
19. a. $v(2)$; 19.2 feet per second
   b. The average rate of change in velocity during the first
   second is 16 feet per second per second. The average
   rate of change from the fifth second to the sixth second
   is 0.005 foot per second per second.
   c. 20 feet per second
   d. With a parachute, about 3 seconds into the fall. With
   out a parachute, 25 seconds into the fall. A feather will
   reach 99% of terminal velocity before a cannonball.
21. a. $C = 50N + 150$ dollars
   b. $R = 65N$ dollars
   c. $P = 65N - (50N + 150)$ dollars
   d. 10 widgets per month
23. a. $P = 2h + 2w$ inches
   b. $P = 2h + 2\left(\dfrac{64}{h}\right)$ inches

   c. A square 8 blocks by 8 blocks with a perimeter of 32
   inches.
   d. $P = 2h + 2\left(\dfrac{60}{h}\right)$ inches. A rectangle 6 blocks by 10
   blocks giving a perimeter of 32 inches.
25. a. 33.64 centimeters. A 4-year-old haddock is about 33.64
   centimeters long.
   b. 2.00 centimeters per year from age 5 to 10 years
   0.27 centimeter per year from age 15 to 20 years
   Haddock grow more rapidly when young than when
   they are older.
   c. 53 centimeters
27. a. 0.058 or 5.8%   **b.** 0   **c.** 0.379 or 37.9%
   d. 0.621 or 62.1%
29. Answers will vary.

S-1.

X	Y$_1$	
4	15	
6	35	
8	63	
10	99	
12	143	
14	195	
16	255	
X=4		

S-3.

X	Y$_1$	
3	-11	
7	-327	
11	-1315	
15	-3359	
19	-6843	
23	-12151	
27	-19667	
X=3		

S-5.

X	Y$_1$	
.1	1.0818	
.2	1.1887	
.3	1.3211	
.4	1.4795	
.5	1.6642	
.6	1.8757	
.7	2.1145	
X=.1		

S-7.

X	Y$_1$	
5	2.0694	
10	2.8289	
15	3.373	
20	3.8055	
25	4.1667	
30	4.4772	
35	4.7494	
X=5		

S-9.

X	Y$_1$	
1	.75	
2	.5	
3	.32143	
4	.20732	
5	.13525	
6	.08904	
7	.05896	
X=1		

S-11.

X	Y$_1$	
0	-1	
20	.57087	
40	.57129	
60	.57137	
80	.57139	
100	.57141	
120	.57141	
X=0		

   About 0.5714
S-13. About 2.72   **S-15.** 0.69   **S-17.** 0.17   **S-19.** 0.50

S-21.

X	Y₁
0	21
1	14
2	9
3	6
4	5
5	6
6	9
X=0	

f has a minimum value of 5 at $x = 4$.

S-23.

X	Y₁
4	129
5	194
6	261
7	314
8	321
9	218
10	⁻123
X=8	

f reaches a maximum of 321 at $x = 8$.

S-25. f has a maximum value of 1.13 at $x = 3$.

S-27. f has a maximum value of 324 at $x = 10$.

SECTION 2.2

1. a. $W = 52\dfrac{5}{a} = \dfrac{260}{a}$.

b. Horizontal, 0 to 16; vertical, 0 to 100

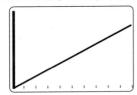

c. Larger

d. The graph is concave up. The expected adult weight decreases at a decreasing rate as the age at which 5 pounds is reached increases.

3. a. $3550

b. Let n be the number of employees and C the weekly cost in dollars. Then $C = 2500 + 350n$.

c. Horizontal, 0 to 10; vertical, 2500 to 7000

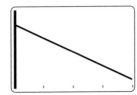

d. 5 employees

5. a. $16,300

b. $V = 18{,}000 - 1700t$

c. Horizontal, 0 to 4; vertical, 10,000 to 20,000

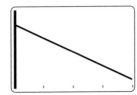

d. $V(3)$; $12,900

7. a. Horizontal, 0 to 120; vertical, 0 to 425

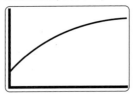

b. 75 degrees

c. Temperature rose more during the first 30 minutes. Average rate of change over first 30 minutes is 4.89 degrees per minute. Over the second 30 minutes it is 2.68 degrees per minute.

d. Concave down

e. After about 46 minutes

f. 400 degrees

9. a. Horizontal, 0 to 25; vertical, −35 to 35

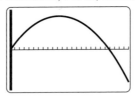

b. $G(4)$; 16 thousand animals

c. −11 thousand animals

d. Increasing from 0 to 10 thousand animals; concave down

11. a. **i.** $Q = \sqrt{\dfrac{800c}{24}}$

Horizontal, 0 to 25; vertical, 0 to 30

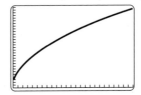

ii. 14 items

iii. The number will increase.

b. **i.** $Q = \sqrt{\dfrac{11200}{h}}$

Horizontal, 0 to 25; vertical, 0 to 125

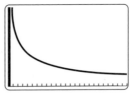

ii. 27 items per order

iii. It should decrease.

iv. About −0.79 item per dollar

v. Concave up

13. a. $P = 200 \times \dfrac{1}{0.01} \times \left(1 - \dfrac{1}{1.01^{t}}\right)$

Horizontal, 0 to 480; vertical, 0 to 25,000

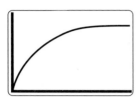

b. $7594.79   **c.** $13,940.10   **d.** $20,000

15. a.   i. $N = \dfrac{30}{\pi} \times \sqrt{\dfrac{9.8}{r}}$

   ii. Horizontal, 10 to 200; vertical, 0 to 10

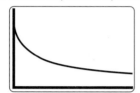

   iii. Decreases
   iv. 2.44 per minute

   b.   i. $N = \dfrac{30}{\pi} \times \sqrt{\dfrac{a}{150}}$

   ii. Horizontal, 2.45 to 9.8; vertical, 1 to 3

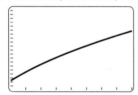

   iii. Increases

17. a.   i. $Y = 339.48 - 0.01535N - 0.00056N^2$
   ii. Horizontal, 0 to 800; vertical, -50 to 400

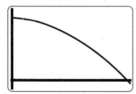

   iii. Decreases
   b.   i. $Y = 260.56 - 0.01535N - 0.00056N^2$
   ii. Horizontal, 0 to 800; vertical, -50 to 400

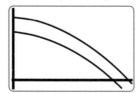

   iii. Decreases
19. a. Horizontal, 0 to 6; vertical, 0 to 55

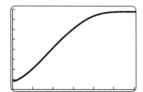

b. $C(1.5)$; 19.67 thousand   **c.** Answers will vary.
d. Answers will vary.   **e.** 52 thousand
21. a. Horizontal, 0 to 30; vertical, 0 to 330

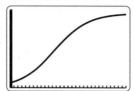

b. 21   **c.** In 2026   **d.** 315
e. Concave up from $t = 0$ to about $t = 11$; concave down
   afterwards
23. Answers will vary.
S-1.

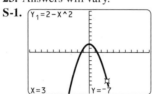

$f(3) = -7$
S-3.

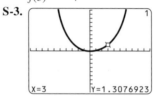

$f(3) = 1.31$
S-5. From a table we chose a vertical span from -0.06 to 0.06.

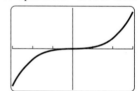

S-7. From a table we chose a vertical span from 0 to 90,000.

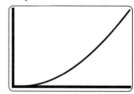

S-9. From a table we chose a vertical span from 0 to 1.

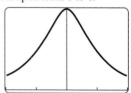

S-11. Vertical span from 0 to 2.5

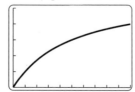

S-13. Vertical span from -300 to 500

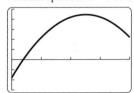

S-15. Vertical span from 0 to 2500

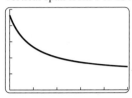

S-17. Vertical span from 0 to 10

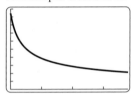

S-19. Vertical span from 0 to 0.6

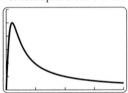

S-21. Vertical span from 0 to 3

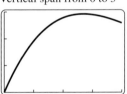

S-23. Vertical span from 0 to 1.2

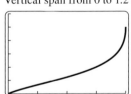

S-25. Vertical span from -1.2 to 0.4

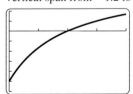

SECTION 2.3

1. a. $s = 2000 - 400w$ **b.** 5 weeks
3. a. $P = C + I + G + E$
   b. **i.** Imports were larger.
     ii. 14,257 billion dollars
   c. $E = P - C - I - G$
   d. 227 billion dollars
5. a. 9.2 thousand dollars
   b. At the start of 2016
   c. $t = \dfrac{12.5 - V}{1.1}$, or $t = 11.36 - 0.91V$
   d. At the start of 2018
7. a. 288 miles
   b. **i.** $m = \dfrac{d}{g}$
     ii. 25.77 miles per gallon
   c. **i.** $g = \dfrac{425}{m}$
     ii. Horizontal, 0 to 30; vertical, 0 to 75

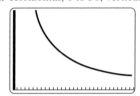

9. $1.96 per bushel
11. a. 153.28 psi   **b.** 85.37 psi
   c. $NP = \dfrac{EP}{1.1 + K \times L}$ **d.** 0.18   **e.** $K = \dfrac{\frac{EP}{NP} - 1.1}{L}$
13. a. $p = (d - c)n - R$ or, equivalently,
     $p = dn - cn - R$.
   b. $2726.10
   c. $d = \dfrac{p + R}{n} + c$ or, equivalently, $d = \dfrac{p + cn + R}{n}$.
   d. $9.43
15. a. 303.15 kelvins
   b. $C = K - 273.15$
   c. $F = 1.8(K - 273.15) + 32$, or $F = 1.8K - 459.67$.
   d. 98.33 degrees Fahrenheit
17. a. $S(30)$; 3.3 centimeters per second
   b. $T = \dfrac{S + 2.7}{0.2}$ **c.** 28.5 degrees Celsius.
19. a. $C = 55N + 200$ dollars
   b. $R = 58N$ dollars
   c. $P = 58N - (55N + 200)$ dollars or, equivalently,
     $P = 3N - 200$ dollars
   d. 66.67 thousand widgets per month
21. a. $R = \dfrac{Y + 55.12 + 0.01535N + 0.00056N^2}{3.946}$
   b. $R = \dfrac{55.12 + 0.01535N + 0.00056N^2}{3.946}$

c. Horizontal, 0 to 800; vertical, 0 to 120

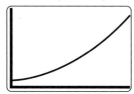

d. Increases **e.** 38.23 millimeters **f.** Die back

23. a. $n = 1 - m$ **b.** $n = \dfrac{1 - 0.7m}{1.2}$

    c. $m = 0.4$, $n = 0.6$ thousand animals

S-1. $x = 7$ **S-3.** $x = -20/9$ **S-5.** $x = -38/43$

S-7. $x = (12 - d)/c$ **S-9.** $k = (m - n)/3$

S-11. $C = (5/9)(F - 32)$ **S-13.** $S = (T - 15)/5$

S-15. $v = (w - t)/u$ **S-17.** $b = (a + e)/(c + d)$

S-19. $b = (a - c)/c$ **S-21.** $b = a/(c + a)$

S-23. $b = a/(c + d)$

SECTION 2.4

1. a. 150.14 cubic inches **b.** 3.51 inches

3. 16.20 knots

5. a. 30 foxes

    b. Horizontal, 0 to 25; vertical, 0 to 160

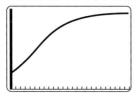

    c. 7.58 years

7. -0.98 and 5.80

9. 7.62 seconds

11. a. Horizontal, 0 to 100; vertical, -5 to 25

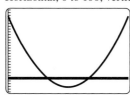

    b. $R(15)$; 9.25 moles per cubic meter per second

    c. 32.68 and 67.32 moles per cubic meter

13. 0.37 liter

15. 9.15 thousand years

17. a. i. Horizontal, 0 to 2000; vertical, -1 to 3

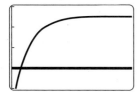

    ii. Concave down

    iii. 163.08 pounds per acre

    iv. 2.5 pounds

b. i. Same scale as above

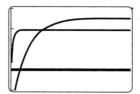

    ii. The red kangaroo

19. a. Horizontal, 0 to 1000; vertical, -1 to 1

    b. 201.18 pounds per acre

21. a. $C = 65N + 700$ dollars

    b. $R = (75 - 0.02N)N$ dollars

    c. $P = (75 - 0.02N)N - (65N + 700)$ dollars

    d. 84.17 and 415.83 thousand widgets per month

23. 7.63 days

25. a. 0.91 second

    b. 0.00096 second

    c. 1.74×10^{15} ergs

27. 276 months, or 23 years

S-1. We used the standard viewing window.

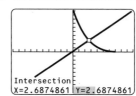

    $x = 2.69$

S-3. We used a horizontal span of -2 to 2 and a vertical span of 0 to 5.

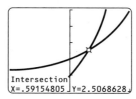

    $x = 0.59$

S-5. For the window we use a horizontal span of -5 to 5 and a vertical span of 0 to 7.

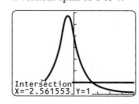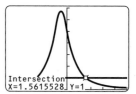

    $x = -2.56$ and $x = 1.56$

S-7. For the window, we use a horizontal span of -2 to 2 and a vertical span of -2 to 2.

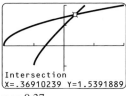

$x = 0.37$

S-9. For the window, we use a horizontal span of 0 to 2 and a vertical span of 0 to 15.

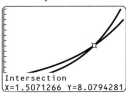

$x = 1.51$

S-11. For the window, we use a horizontal span of 0 to 2 and a vertical span of 0 to 10.

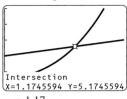

$x = 1.17$

S-13. We graph $\dfrac{20}{1 + 2^x} - x$ using the standard viewing window.

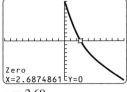

$x = 2.69$

S-15. We graph $3^x - 8$ using a horizontal span of 0 to 3 and a vertical span of -10 to 10.

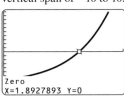

$x = 1.89$

S-17. We graph $x^3 - x - 5$ using a horizontal span of 0 to 3 and a vertical span of -10 to 10.

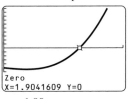

$x = 1.90$

S-19. We graph $\sqrt{x^2 + 1} - 2x$ using a horizontal span of 0 to 1 and a vertical span of -2 to 2.

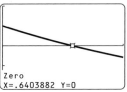

$x = 0.58$

S-21. We graph $\dfrac{x^4}{x^2 + 1} - 0.2$ using a horizontal span of -2 to 2 and a vertical span of -2 to 2.

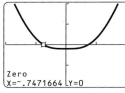 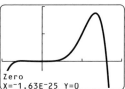

$x = -0.75$ and $x = 0.75$

S-23. We graph $x^4 + x^5 - x^6$ using a horizontal span of -1 to 2 and a vertical span of -1 to 2.

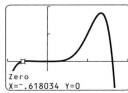 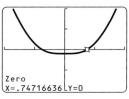

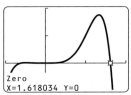

$x = -0.62$, $x = 0$, $x = 1.62$

SECTION 2.5

1. a. Horizontal, 0 to 25; vertical, 0 to 15

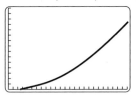

b. Winds from 0 to 14.14 miles per hour

3. a. Horizontal, 2 to 16; vertical, 0 to 300

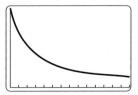

b. Age between 8.32 and 12.48 weeks

5. From 0.56 hour to 2.81 hours after it is administered

7. a. $B = \dfrac{\sqrt{80h}}{60}$ **b.** From 162.45 to 180 centimeters

9. 12:02 to 12:04 A.M.

11. a. Horizontal, 0 to 25; vertical, 0 to 1200. Account balance 2 is the darker line.

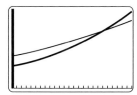

b. After 18.55 years
c. Balance 1 is greater than Balance 2 up to 18.55 years. After that Balance 2 is greater.

13. From 0.81 to 2.32 seconds

15. a. 24.84 thousand to 43.01 thousand
b. From 1.74 thousand to 6.01 thousand and for populations of 50.25 thousand or more

S-1. $x \geq 1.38$ **S-3.** $x \geq 1.61$ **S-5.** $x \geq 2.62$
S-7. $-1.14 \leq x \leq 1.14$
S-9. $-1.30 \leq x \leq 0$ and $x \geq 2.30$
S-11. This inequality is true for all values of x.
S-13. $1.63 \leq x \leq 2.59$ **S-15.** $1.56 \leq x \leq 2.76$

SECTION 2.6

1. a. Horizontal, 0 to 250; vertical, 0 to 8000

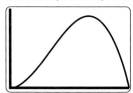

b. 7387.71
c. The beam width is 166.67 feet, so the ship is wider than it is long.
d. 3141.62

3. a. Horizontal, 0 to 4; vertical, 0 to 2

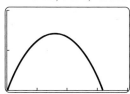

b. 3.22 miles downrange
c. 1.39 miles high at 1.61 miles downrange

5. a. Horizontal, 0 to 1.5; vertical, -0.1 to 0.1

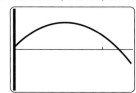

b. $G(0.24)$; 0.04 million tons per year

c. -0.02 million tons per year
d. 0.67 million tons

7. a. Total amount $= 2W + L$ feet.

b. $WL = 100$ **c.** $L = \dfrac{100}{W}$ **d.** $F = 2W + \dfrac{100}{W}$

e. Horizontal, 1 to 15; vertical, 0 to 110

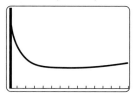

f. 7.07 feet perpendicular, 14.14 feet parallel

9. a. $G = 0.3s(4 - s)$
b. Horizontal, 0 to 4; vertical, 0 to 1.5

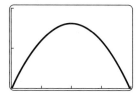

c. 2 thousand dollars   **d.** Answers will vary.
e. Answers will vary.

11. a. Height $= 4.77$ inches. Area $= 36.28$ square inches.
b. Height $= 0.19$ inch. Area $= 163.08$ inches.
c. **i.** Horizontal, 0 to 4; vertical, 0 to 50

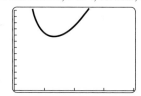

ii. 1.34 inches
iii. 2.66 inches

13. a. $C = 60N + 600$ dollars
b. $R = (70 - 0.03N)N$ dollars
c. $P = (70 - 0.03N)N - (60N + 600)$ dollars
d. 166.67 thousand widgets per month

15. a. Answers will vary.
b. $C = 300L + 500\sqrt{1 + (5 - L)^2}$
c. 4.12 miles under water; $2361.55
d. 2.24 miles under water; $2018.03
e. Horizontal, 0 to 5; vertical, 1000 to 3000

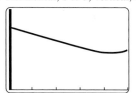

f. $L = 4.25$ miles   **g.** $W = 1.25$ miles
h. $C = 700L + 500\sqrt{1 + (5 - L)^2}$. The entire cable should run under water.

17. **a.** 15.18 years old
  b. $B = (1000e^{-0.1t}) \times (6.32(1 - 0.93e^{-0.095t})^3)$
  Horizontal, 0 to 20; vertical, 0 to 700

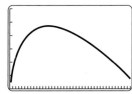

  c. 13.43 years old
  d. **i.** 15.18 years old
     ii. 13.43 years old
19. **a.** Horizontal, 0 to 40; vertical, 0 to 5

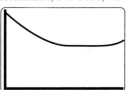

  b. 3.81 pounds per year   **c.** 2.95 pounds per year
21. **a.** Horizontal, 0 to 2400; vertical, 0 to 750

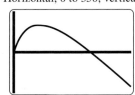

  b. 1675 pupils   **c.** $27.07 per pupil
23. **a.** Horizontal, 0 to 350; vertical, −15 to 15

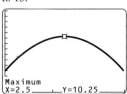

  b. 73.27   **c.** $N = 0$ and $N = 228$ **d.** −7.51 per day
25. Answers will vary.
S-1. A table of values leads us to choose a vertical span of 0 to 15.

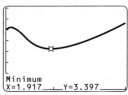

  Maximum value of 10.25 at $x = 2.5$
S-3. A table of values leads us to choose a vertical span of 0 to 7.

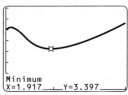

  Minimum value of 3.40 at $x = 1.92$

S-5. We use a vertical span of 0 to 3.

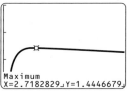

  Maximum value of 1.44 at $x = 2.72$
S-7.

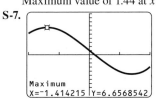

  Maximum $x = -1.41$, $y = 6.66$

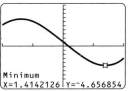

  Minimum: $x = 1.41$, $y = -4.66$
S-9. We use a vertical span of −30 to 300.

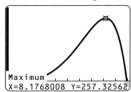

  Maximum: $x = 8.18$, $y = 257.33$
  We use a horizontal span of 0 to 2 and a vertical span of
  −3 to 1.

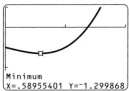

  Minimum: $x = 0.59$, $y = -1.30$
S-11. We use a vertical span of −1 to 3.

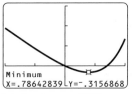

  Minimum: $x = 0.79$, $y = -0.32$
S-13. We use a vertical span of 15 to 40.

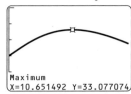

  Maximum: $x = 10.65$, $y = 33.08$

S-15. We use a vertical span of 0.5 to 1.

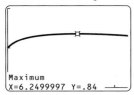

Maximum: $x = 6.25$, $y = 0.84$

S-17. We use a vertical span of −1 to 2.

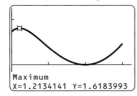

Maximum: $x = 1.21$, $y = 1.62$

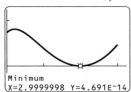

Minimum: $x = 3$, $y = 0$

S-19. We use a vertical span of −1 to 2.

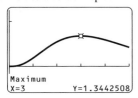

Maximum: $x = 3.00$, $y = 1.34$

S-21. We use a vertical span of −20 to 130.

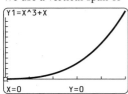

Minimum at endpoint: $x = 0$, $y = 0$

S-23. We use a vertical span of 50 to 250.

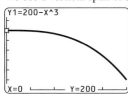

Maximum at endpoint: $x = 0$, $y = 200$

CHAPTER 2 REVIEW EXERCISES

1. $x = 2, f = 3$

2. a. Horizontal, 0 to 20; vertical, 0 to 150

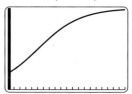

b. 112 foxes after 9 years
c. Concave up for t from 0 to 5; concave down for t from 5 to 20
d. 150 foxes

3. $x = 4$

4. a. Horizontal, 0 to 7.5; vertical, 0 to 16

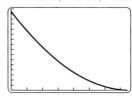

b. $D(3) = 5.72$, so the depth of the water is 5.72 inches above the spigot after the spigot has been open 3 minutes
c. After 7.44 minutes
d. Near the beginning

5. Maximum 101, minimum 40.72

6. a. 6
b. 82 deer after 4 years
c. 177
d. 10.07; 27.78; 30.85; then 12.94 deer per year

7. a. Horizontal, 0 to 30; vertical, 0 to 330

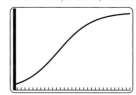

8. a. $F = 2W + 2L$
b. $144 = W \times L$
c. $W = 144/L$
d. $F = 288/L + 2L$
e. Horizontal, 0 to 24; vertical, 0 to 100

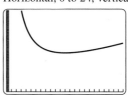

f. $W = L = 12$ feet

9. $t = 23.91$

10. a. Horizontal, 0 to 20; vertical, 0 to 10

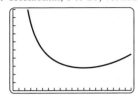

b. 12.01 meters per second
c. Quickly
11. $x = 6$ feet, $f = 11$
12. a. We show the table only from $t = 0$ to $t = 1.5$.

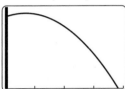

X	Y₁
0	150
.25	154.5
.5	157
.75	157.5
1	156
1.25	152.5
1.5	147

X=0

b. Horizontal, 0 to 4; vertical, 0 to 170

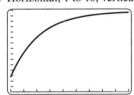

c. $t = 0.69$ second
d. $H = 157.56$ feet high at the peak
e. $t = 0.29$ second and $t = 1.09$ seconds
13. $W = (L - 98.42 + 4.14A)/1.08$
14. a. Horizontal, 1 to 10; vertical, 0 to 15

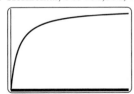

b. $L(4)$, 11.77 inches   **c.** 3 years old
d. 14.8 inches
15. 10.25
16. a. $P(7)$
b. Loss of 6.34 million dollars
c. $N = 0.68$ and $N = 9.32$ million items
d. $N = 5$ million items, profit of 18.66 million dollars
17. Horizontal, 0 to 100,000; vertical, 0 to 0.5

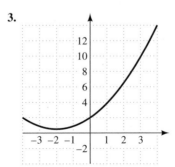

18. a. $C = (F - 32)/1.8$, $C = K - 273.15$
b. $F = 1.8(K - 273.15) + 32$
c. $K = (F - 32)/1.8 + 273.15$
d. 32 degrees Fahrenheit, 273.15 kelvins
e. 22.22 degrees Celsius, 295.37 kelvins

19. a. Horizontal, 0 to 240; vertical, 0 to 30

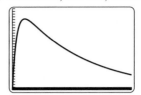

b. At 24.51 minutes   **c.** 196.02 minutes   **d.** No
e. Yes; lethal after 5.54 minutes
20. -1.22, 0, and 1.22
21. $x \geq 1.33$

A FURTHER LOOK: Limits

1. Answers will vary.   **2.** Answers will vary.   **3.** 15
4. $\sqrt{7}$ **5.** 2/3   **6.** 2/3   **7.** 5   **8.** 1   **9.** 3
10. The limit does not exist.   **11.** 0   **12.** 1/2   **13.** k
14. a cubic feet   **15.** 25 pounds

A FURTHER LOOK: Shifting and Stretching

1.

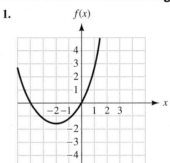

2.

3.

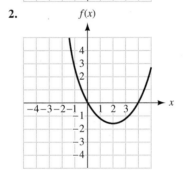

4.

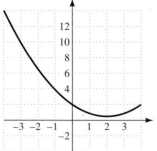

5.

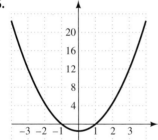

6.

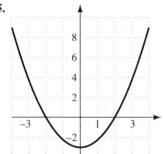

7.

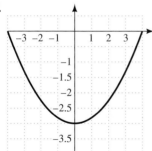

8.

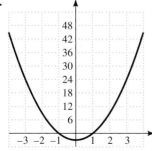

9.

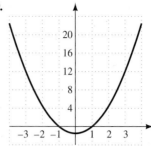

10.

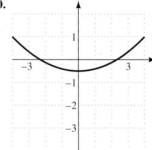

11.

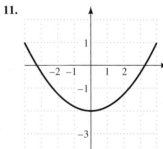

12.

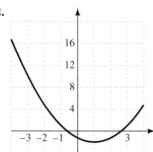

13.

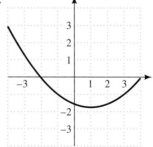

14.

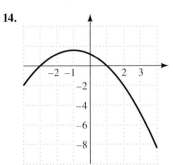

15. The graph is reflected first through the horizontal axis and then through the vertical axis.

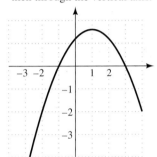

A FURTHER LOOK: Optimizing with Parabolas

1. $(-3, -13)$ is a minimum.   **2.** $(5, -74)$ is a minimum.
3. $(3, 15)$ is a maximum   **4.** $(0, 4)$ is a minimum.
5. a. $100 - x$ yards   **b.** $\dfrac{1}{2}(100 - x)x$ square yards

   c. $x = 50$ yards   **d.** 1250 square yards
6. $x = -1/2$
7. Minimum cost of $2500 with an enrollment of 250 students
8. Minimum temperature of 114 degrees Fahrenheit after 2 minutes
9. At $t = 3$ hours, concentration is 15 milligrams per liter.
10. Maximum height of 45/4 feet after 5/8 second, or 11.25 feet after 0.63 second
11. At $n = 0.5$ ton, the growth rate is 0.125, or about 0.13, ton per year.
12. $500 advertising yields $250,300 in profits.
13. a. Answers will vary.   **b.** Answers will vary.
   c. Answers will vary.
   d. $d = 10/\pi$ or about 3.18 feet; $h = 0$ feet

CHAPTER 3
SECTION 3.1

1.

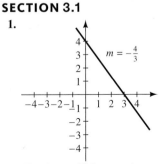

The slope will be negative. $m = -\dfrac{4}{3}$

3. Horizontal intercept is 4.
5.

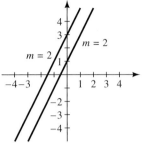

They do not cross. Different lines with the same slope are parallel.
7. 64.2 inches   **9.** 2118 feet
11. a. 0.4 foot high   **b.** 6 feet high
13. a. 0.83 foot per foot   **b.** 22.11 feet   **c.** 4.82 feet
15. h is 4 feet and k is 5 feet 6 inches.
17. At least 38.67 feet
19. a. 276.67 feet per mile   **b.** 5513.35 feet   **c.** 22.30 miles
21. The umbra has a radius of about 2859 miles at this distance. The moon can fit in the umbra, causing a total lunar eclipse.
23. About 430,328 miles
S-1. 4 feet per foot   **S-3.** 7.5 feet   **S-5.** 5.14 feet
S-7. $-\dfrac{2}{3}$ foot per foot   **S-9.** 16.4 feet
S-11. 1.43 feet per foot   **S-13.** $-\dfrac{1}{2}$ or -0.5
S-15. -0.68 **S-17.** $x = 5.08$ **S-19.** $y = 15$
S-21. $y = 5.86$ **S-23.** $(6.6, -1.16)$

SECTION 3.2

1. a. Let C denote the total number of calories burned.
   $$C = 258n + 700$$
   b. 1.55 hours
3. $L = 3t + 12$
5. a. $C = 0.56F - 17.78$ (rounding to two decimal places)
   b. The slope is 0.56.   **c.** They are the same.
7. a. Answers will vary.
   b. Let S be the total amount of storage space (in megabytes) used on the disk drive, and let n be the number of pictures stored. $S = 2n + 6000$.
   c. $S(350)$; 6700 megabytes.
   d. 12,500 pictures on the disk drive. There is room for 384,500 additional pictures.
9. a. Answers will vary. The slope is $20 per widget, and the initial value is $1500; $C = 20N + 1500$.
   b. $1300
   c. Fixed costs: $1100; variable cost: $16 per widget
11. a. $S = -0.746D + 46.26$
   b. $S(10)$; 38.8 miles per hour
13. a. 0.49 pound per dollar   **b.** 63.70 pounds   **c.** $25.14
15 a. The slope is 5 pounds per inch.
   b. Let h be height (in inches), and let w be the weight (in pounds): $w = 5h - 180$.
   c. 66.4 inches   **d.** Light
17. For each pound increase in weight, males will show an increase of 1.08 pounds in lean body weight, but females will show an increase of only 0.73 pound in lean body weight.

19. a. The vertical factor has a constant rate of change.
  b. $V = 40d + 65$ **c.** 82.16 feet **d.** Yes
21. a. 0.5*a* dollars **b.** *g* dollars **c.** $0.5a + g = 5$
  d. $g = -0.5a + 5$
23. 64 days
25. a. Pessimistic, 0.00002; optimistic, 10,000,000
  b. 0.0005 **c.** Answers will vary.
S-1. 4 **S-3.** $f(5) = 12.4$ **S-5.** $x = 2.76$
S-7. $f = 4x - 7$ **S-9.** $f = -1.2x + 12.8$
S-11. An increase of 16.5 units
S-13. A decrease of 16.38 units
S-15. An increase of 0.97 unit
S-17. A decrease of 1.44 units
S-19. -0.33 **S-21.** No
S-23. About a 2.7-unit increase

SECTION 3.3

1. a. The differences are all 73, so the data are linear.
  b. $I = 73t + 200$ **c.** Yes
3. a. Let *t* be the time in months since February 2009, and let *K* be the price in dollars. A table of differences shows a comman difference of -50 dollars every 5 months.
  b. $K = -10t + 349$ **c.** $-\$1$
5. a. Let *d* be the number of years since 2005, and let *T* be the tution in dollars. A table of differences shows a common difference of 416 dollars. $T = 416d + 5965$
  b. 416 dollars per year
  c. 1545 dollars per year
  d. Tuition at public universities increased at a rate of $416 per year, and tution at private universities increased at a rate of $1545 per year.
  e. Tuition at public universities increased by about 7%, which is larger than the increase at private universities (about 6%).
7. a. $p = -0.01N + 45$ **b.** $R = (-0.01N + 45)N$; no
  c. $P = (-0.01N + 45)N - (35N + 900)$; no
9. a. A difference table shows a common difference of 36.
  b. 1.80 degrees Fahrenheit per kelvin
  c. $F = 1.80K - 459.67$ **d.** 310.15 kelvins
  e. An increase of 1 kelvin causes an increase of 1.80 degrees Fahrenheit. An increase of 1 degree Fahrenheit causes an increase of 0.56 kelvin.
  f. -459.67 degrees Fahrenheit
11. a. A difference table shows a constant difference of 1.05; $P = 2.1S - 0.75$
  b. Horizontal, 0 to 3; vertical, 0 to 5

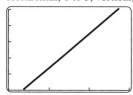

  c. Answers will vary. **d.** 2.21 billion bushels
13. a. Let *V* be the velocity (in miles per hour) and *t* the time (in seconds) since the car was at rest. A table of

differences shows a common difference of 5.9 miles per hour.
  b. 11.8 miles per hour per second **c.** $V = 11.8t + 4.3$
  d. 4.3 mph **e.** 4.72 seconds
15. a. Let *N* be the number graduating (in millions), and let *t* be the number of years since 2001. The slope is 0.07 million graduating per year.
  b. $N = 0.07t + 2.85$ **c.** $N(7)$; 3.34 million
  d. This formula gives 2.36 million, which is much closer than the earlier answer.
17. a. When $S = 36.8$, *c* should be 1517.52.
  b. Answers will vary. **c.** 1520.16 meters per second
19. 250,000 stades; 26,000 miles
21. The average rate of change from 2 to 5 is 1. The average rate of change from 5 to 6 is 3.
S-1. There is a constant change of 2 in *x* and a constant change of 5 in *y*. The data are linear.
S-3. $y = 2.5x + 7$
S-5.

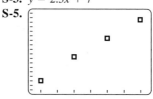

S-7.

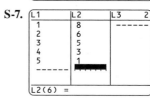

S-9.

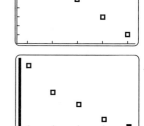

S-11.

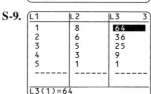

S-13.

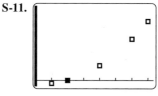

S-15.

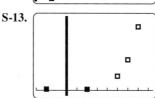

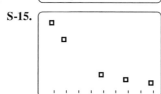

S-17.

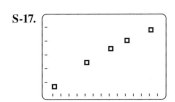

S-19.

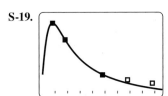

S-21.

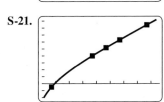

SECTION 3.4

1. $A = -0.15t + 10.86$

3. a. $S = 1.23t + 1.75$ **b.** 14.05 million
   c. DirecTV gained 1.23 million subscribers each year.
   d.

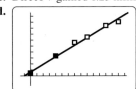

5. a. Let t be the time in hours since the experiment began, and let N be the number of bacteria in thousands.

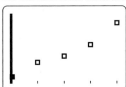

The data do not appear to fall on a straight line, so a linear model is not appropriate.
   b. Let t be the number of years since 2004, and let E be the enrollment in millions.

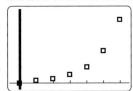

The data are not exactly linear, but they do nearly fall on a straight line. It looks reasonable to approximate the data with a linear model.

7. a. Let t be the number of years since 2005, and let T be the number of tourists in millions.

b. $T = 3.11t + 48.88$

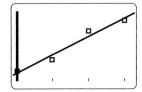

c. Answers will vary.   **d.** $T(4)$; 61.32 million tourists
9. a. Let t be the number of years since 1900, and let L be the length in meters. $L = 0.034t + 7.197$
   b. Answers will vary.
   c.

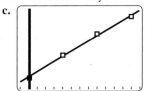

d. Answers will vary.
   e. The regression line model gives 7.88 meters, which is far too long.
11. a.

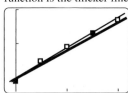

b. $D = 0.88t + 51.98$
   c.

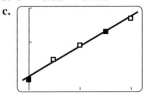

d. The regression line is the thin line, and the true depth function is the thicker line.

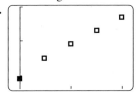

The two lines are very close, but the regression model shows a water level that is a bit too high.
   e. 54.4 feet   **f.** 54.62 feet, or about 54.6 feet

13. a. Let t be the number of years since 2005, and let D be the total sales in millions. $D = -0.32t + 5.68$

b.

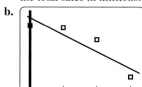

c. The regression model gives a prediction of 4.40 million cars sold in 2009. This estimate is larger than the actual value.

15. The data points show that a larger number of churches yields a larger percentage of antimasonic voting. Furthermore, since the points are nearly in a straight line, it is reasonable to model the data with a linear function.

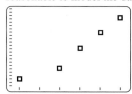

The formula for the regression model is $M = 5.89C + 45.55$. Its graph is added to the data below.

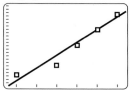

This picture reinforces the idea that the data can be approximated by a linear function.

17. a. $V = 12.57T + 7501.00$
  b. The steam will occupy an additional 1257 cubic feet.
  c. About 12,780 cubic feet   **d.** 532.94 degrees
19. a. $B = 0.29W + 0.01$
  b. The third sample: $W = 13.7$ and $B = 3.7$
  c. 0.29 additional ton per hectare
21. a. $E = 0.34v + 0.37$ **b.** 0.34
  c. Higher cost; less efficient   **d.** 0.37

S-1.

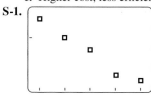

$y = -0.26x + 2.54$

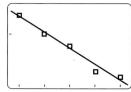

S-3.

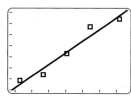

$y = 1.05x + 2.06$

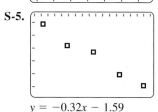

S-5.

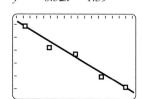

$y = -0.32x - 1.59$

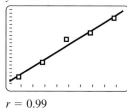

S-7. $y = -x + 4$ **S-9.** $y = 0.5x + 2.33$
S-11. $y = -1.5x + 4$
S-13. $y = 2x + 1.4$

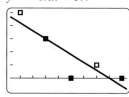

$r = 0.99$
S-15. $y = -0.6x + 5.4$

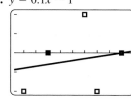

$r = -0.88$
S-17. $y = 0.1x - 1$

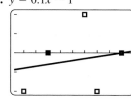

$r = 0.19$

SECTION 3.5

1. If C = number of bags of chips and D = number of drinks, then $2C + 0.5D = 36$ and $D = 5C$. You buy 40 drinks and 8 bags of chips.

3. If c = number of crocus bulbs and d = number of daffodil bulbs, then $0.35c + 0.75d = 25.65$ and $c + d = 55$. You buy 39 crocus bulbs and 16 daffodil bulbs.

5. In 7.55 years there will be about 504 foxes and 504 rabbits.

7. $m = 0.25$, $n = 0.75$ thousand animals

9. a. 0.08 part per million   **b.** Hartsells fine sandy loam

11. At 160 degrees Celsius, the Fahrenheit temperature is 320 degrees.

13. a. $\left(0, \frac{1}{4}\right)$ **b.** Answers will vary.

15. a. $0.1T$ **b.** $0.1T + B = 1.47$ **c.** $0.2B$
   d. $T + 0.2B = 1.47$
   e. $T = 1.2$ million dollars and $B = 1.35$ million dollars
   f. The bonus is 0.27 million dollars, and tax paid is 0.12 million dollars. The remaining profit is 1.08 million dollars.

17. The graphs lie on top of one another. Every point on the common line is a solution to the system of equations.

19. N = number of nickels, D = number of dimes, and Q = number of quarters.

$$0.05N + 0.1D + 0.25Q = 3.35;$$
$$N + D + Q = 21; D = N + 1;$$

There are 5 nickels, 6 dimes, and 10 quarters.

S-1. Each graph consists of points that make the corresponding equation true. The solution of the system is the point that makes both true, and that is the common intersection point.

S-3. We graphed $y = 5 - x$ and $y = x - 1$ using a horizontal span of 2 to 5 and a vertical span of 0 to 3.

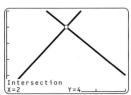

$x = 3, y = 2$

S-5. We graphed $y = 2x$ and $y = \dfrac{14 - 3x}{2}$ using a horizontal span of 0 to 4 and a vertical span of 0 to 5.

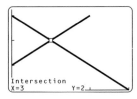

$x = 2, y = 4$

S-7. We graphed $y = \dfrac{6 - 3x}{2}$ and $y = \dfrac{8 - 4x}{-3}$ using a horizontal span of 0 to 5 and a vertical span of -5 to 5.

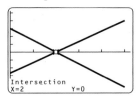

$x = 2, y = 0$

S-9. We graphed $y = -3x$ and $y = x$ using a horizontal span of -5 to 5 and a vertical span of -5 to 5.

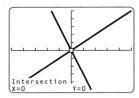

$x = 0, y = 0$

S-11. We graphed $y = x - 1$ and $y = \dfrac{8 - 3x}{2}$ using a horizontal span of 0 to 4 and a vertical span of 0 to 2.

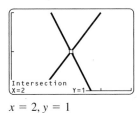

$x = 2, y = 1$

S-13. We graphed $y = 1 - x$ and $y = \dfrac{-2x}{3}$ using a horizontal span of -5 to 5 and a vertical span of -5 to 5.

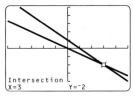

$x = 3, y = -2$

S-15. We graphed $y = \dfrac{6 - 3x}{4}$ and $y = \dfrac{5 - 2x}{-6}$ using a horizontal span of 0 to 5 and a vertical span of -3 to 3.

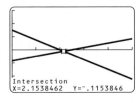

$x = 2.15, y = -0.12$

S-17. We graphed $y = \dfrac{79 + 7x}{21}$ and $y = \dfrac{6 - 13x}{17}$ using a horizontal span of -5 to 0 and a vertical span of 0 to 5.

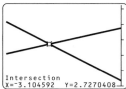

Intersection
X=‾3.104592   Y=2.7270408

$x = -3.10,\ y = 2.73$

S-19. We graphed $y = \dfrac{6.6 - 0.7x}{5.3}$ and $y = \dfrac{1.7 - 5.2x}{2.2}$ using a horizontal span of -2 to 2 and a vertical span of 0 to 5.

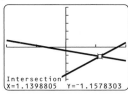

Intersection
X=‾.2117602   Y=1.2732513

$x = -0.21,\ y = 1.27$

S-21. We graphed $y = \dfrac{-1.43 - 2.1x}{3.3}$ and $y = \dfrac{3.78 - 2.2x}{-1.1}$ using a horizontal span of -2 to 2 and a vertical span of -5 to 5.

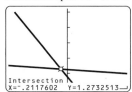

Intersection
X=1.1398805   Y=‾1.1578303

$x = 1.14,\ y = -1.16$

S-23. We graphed $y = \dfrac{29.74 + 4.1x}{5.2}$ and $y = \dfrac{-36.23 - 6.8x}{-3.9}$ using a horizontal span of -5 to 0 and a vertical span of 0 to 5.

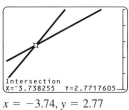

Intersection
X=‾3.738255   Y=2.7717605

$x = -3.74,\ y = 2.77$

CHAPTER 3 REVIEW EXERCISES

1. $-0.5/8 = -0.0625$ foot per foot, or $-6/8 = -0.75$ inch per foot

2. 16.28 feet

3. a. $18/4 = 4.5$ inches per foot, or $1.5/4 = 0.375$ foot per foot
   b. At least 15 feet

4. a. 0.5 foot per foot   **b.** 12.5 feet   **c.** 20 feet

5. -1.72

6. a. $100/\pi$ or 31.83 meters   **b.** 1.22 meters
   c. Lane 4 or higher

7. $g = -1.26x + 4.68$ (may vary due to rounding)

8. a. Answers will vary.   **b.** \$1080
   c. $I = 0.05S + 1000$ **d.** \$7000

9. a. There is a constant change of 0.3 in x and a constant change of -0.6 in f. The data are linear.
   b. $f = -2x + 14$

10. a. There is a constant change of 5 in d and a constant change of 6 in T.
   b. Answers will vary.   **c.** 1.2 patients per day
   d. $T = 1.2d + 35$ **e.** 55

11. a. There is a constant change of 2 in x and a constant change of 0.24 in f. The data are linear.
   b. $f = 0.12x - 1$
   c.

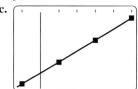

12. a. There is a constant change of \$50 in taxable income and a constant change of \$14 in tax due.
   b. \$0.28   **c.** \$21,913; \$22,193
   d. The tax due is $0.28A + 21{,}913$ dollars.

13. a.

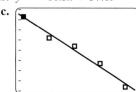

   b. $y = -16.3x + 37.19$
   c.

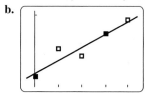

14. Answers will vary.

15. a. If E is life expectancy, in years, and t is time, in years since 2003, then $E = 0.18t + 77.16$.
   b.

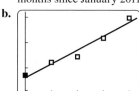

   c. Answers will vary.   **d.** 79.0 years
   e. 1.0 years; 130.6 years

16. a. If P is the stock price, in dollars, and t is time, in months since January 2011, then $P = 0.547t + 43.608$.
   b.

   c. Answers will vary.   **d.** \$50.17; \$56.74

17. We graphed $y = \dfrac{4x - 9}{2}$ and $y = -x$ using a horizontal span of 0 to 2 and a vertical span of -5 to 0.

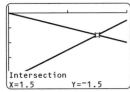

Intersection
X=1.5   Y=-1.5

$x = 1.5, y = -1.5$

18. $x = 1.5, y = -1.5$

19. If T is the number of $20 bills and F is the number of $50 bills, then $20T + 50F = 400$ and $T + F = 11$. There are 5 $20 bills and 6 $50 bills.

20. If B is the number of gallons of blue paint and Y is the number of gallons of yellow paint, then $B + Y = 10$ and $Y = 3B$. Use 2.5 gallons of blue paint and 7.5 gallons of yellow paint.

A FURTHER LOOK: Parallel and Perpendicular Lines

1. Parallel   **2.** Perpendicular   **3.** Neither   **4.** Neither
5. $8/5 = 1.6$ **6.** $y = 3x + 2$ **7.** $y = 5x + 25$
8. $y = -\dfrac{1}{3}x + 4$ **9.** $y = \dfrac{1}{4}x + 3$

A FURTHER LOOK: Secant Lines

1. $y = 13x - 12$ **2.** $y = \dfrac{1}{5}x + \dfrac{6}{5}$ **3.** $y = 5x - 5$
4. $y = -x + 8$ **5.** $y = 11x - 24$
6. a. $y = 4x - 6$ **b.** 6
7. a. $y = -28x + 49$ **b.** -35
8. Answers will vary.
   a. The secant line is below the graph on the interval between the two points that determine the secant line and above the graph outside that interval.
   b. The estimate will be too small on the interval between the two points that determine the secant line and too large outside that interval.

9. Secant line number 3.   **10.** $\dfrac{x^2 - a^2}{x - a} = x + a$ **11.** $2a$
12. $\dfrac{(a + h)^2 + 1 - (a^2 + 1)}{h} = 2a + h$ **13.** $2a$

CHAPTER 4

SECTION 4.1

1. 23×1.4^t. Horizontal, $t = 0$ to 5; vertical, 0 to 130

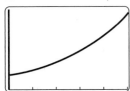

3. a. $w = 104 \times 1.03^t$ **b.** 8.83 weeks

5. a. The weekly decay factor is 0.96. The formula is $P = 300 \times 0.96^w$.
   b. The yearly decay factor is 0.16. The formula is $P = 300 \times 0.16^y$
7. 4 half-lives, or 49 years
9. 6.13 years, with both immigration and extinction rates at 2.69 species per year
11. a. 1.26
   b. To find next year's gold price, we multiply this year's price by 1.26.
13. 9.22×10^{18} grains; about 35.6 trillion dollars
15. a. This is an exponential function because the amount is growing by constant multiples.
   b. $D = D_0 \times 2^t$ **c.** 6.64 years.
17. a. 0.12   **b.** 0.87
S-1. $A = 10 \times 3^t$ **S-3.** $A = 7 \times 0.6^t$
S-5. $A = 8 \times (1/7)^t$
S-7. $f(2) = 17.28$
   $f(x) = 3 \times 2.4^x$
S-9. 1.25   **S-11.** Concave up
S-13. It is proportional to the function value.
S-15. 1.013 or about 1.01   **S-17.** 1.13   **S-19.** 0.98

SECTION 4.2

1. a. $P = 56.61 \times 1.10^t$ **b.** 177.67 thousand
3. a. Let t be time in years and N population in millions. Then $N = 3 \times 1.023^t$.
   b. $N(4)$; 3.29 million
5. a. 231.2 grams   **b.** 32%
   c. Monthly decay factor 0.968; monthly percentage decay rate 3.2%
   d. Percentage decay rate per second 0.00000122%
7. a. $W = 870 \times 0.81^t$ **b.** 17.90 years
9. 23.45 years
11. a. $N = 67.38 \times 1.026^t$, where N is population in millions and t is years since 1980.
   b. $N(3)$; 72.77 million   **c.** $t = 11.28$; sometime in 1991
13. a. Answers will vary.   **b.** 0.98   **c.** 1   **d.** $P = 0.98^Y$
   e. For 10 years: 0.82, or 82%; for 100 years: 0.13, or 13%
   f. $Q = 1 - 0.98^Y$
15. 25.0%
17. a. $D = 60 \times 0.9^t$ **b.** 17.01 months
19. a. 1.1052   **b.** 1.1052   **c.** Answers will vary.
S-1. 6%   **S-3.** 30%   **S-5.** $f(x) = 8 \times 1.05^x$
S-7. $f(x) = 6 \times 0.95^x$ **S-9.** 30×1.04^t, with t in years
S-11. 10×0.96^t, with t in years
S-13. 5×0.97^t, with t in days
S-15. 62.9%   **S-17.** 40.1%   **S-19.** 14.1%   **S-21.** 1.4%
S-23. 21.5%   **S-25.** 64.5%

SECTION 4.3

1. The data show a common ratio of 1.04, rounded to two decimal places, and thus they are exponential.
   $f = 3.80 \times 1.04^t$
3. The successive ratios are not the same.

5. a. Successive ratios are all 1.20, so the data are exponential.
 b. $V = 130 \times 1.20^t$ **c.** 7.39 years after the start
7. From 2005 to 2009, sales grew at a rate of $1.06 thousand per year. From 2009 to 2012, sales grew by 9% per year.
9. a. $1750.00
 b. Let t be the time in months and B the account balance in dollars. The data show a common ratio of 1.012, rounded to three decimal places. $B = 1750.00 \times 1.012^t$
 c. 1.2% **d.** 15.4% **e.** $23,015.94
 f. 58.11 months to double the first time; 58.11 months to double again
11. a. The data show a common ratio of 0.42, so they are exponential. $D = 176.00 \times 0.84^t$
 b. 16% **c.** $V = 176.00 - 176.00 \times 0.84^t$
 d. About 26.41 seconds into the fall
13. a. Let t be the time in minutes and U the number of grams remaining. In the display, the data appear to be linear.

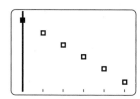

 b. Rounding the calculated parameters to three decimal places, $U = -0.027t + 0.999$.

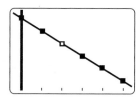

 c. 18.48 minutes
 d. That there would be -0.621 gram remaining
15. The yearly interest rate was 5.0% for the first 3 years, then 4.5% after that.
17. a. Answers will vary. **b.** Answers will vary.
S-1. $N = 7 \times 8^t$ **S-3.** $N = 12 \times 1.803^t$
S-5. $N = 2 \times 2^t$ **S-7.** $N = 4 \times 0.5^t$
S-9. Ratios give a constant value of 2. The data are exponential. $y = 5 \times 1.41^x$
S-11. Ratios give a constant value of 3. The data are exponential. $y = 6 \times 1.73^x$
S-13. The data are not exponential.
S-15. The data are exponential. $y = 2.5 \times 1.26^x$
S-17. $\sqrt{a}$ **S-19.** It multiplies N by a^{20}. **S-21.** 96

SECTION 4.4

1. Linear
3. a. $C = 29.490 \times 1.076^t$ **b.** 7.6%
5. Let t be years since 2008 and N the population, in thousands. $N = 2.300 \times 1.090^t$.

7. a. Let t be years since 2005 and C the number of cell phone subscribers, in millions.

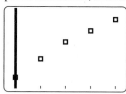

 b. $C = 212.788 \times 1.082^t$
 c.

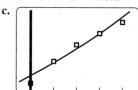

 d. 8.2% **e.** Yes, the model predicts 341.44 million.
9. a. Let t be years since 1970 and H the costs in billions of dollars.

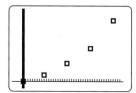

 b. $H = 94.414 \times 1.091^t$ **c.** 9.1%
 d. $H(41)$; 3356 billion dollars
11. a.

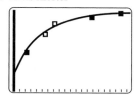

 b. $D = 0.211 \times 0.988^V$
 c. $A = 0.18 - 0.211 \times 0.988^V$
 d. 203.89 pounds per acre
13. Table A is approximately linear, with the model $f = 19.842t - 16.264$. Table B is approximately exponential, with the model $g = 2.330 \times 1.556^t$.
15. a.

 b. 0.0003×1.116^D **c.** It is increased by 11.6%.
17. a. $D = 40.909 \times 0.822^t$ **b.** $L = 53 - 40.909 \times 0.822^t$
 c. A bit shorter

 d. 6.26, or about 6.3 years old

19. a. $N = 16.726 \times 1.021^s$
   b. 71.65 persons injured per 100 accident-involved vehicles
   c. It is increased by 2.1%.
21. a. $P = 84.726 \times 0.999^D$ **b.** 69.36% of pedestrians
   c. The initial value indicates that 84.726% of pedestrians walk at least 0 feet. The correct percentage is 100%.
23. a. $N = 24{,}670{,}000 \times 0.125^M$
   b. About 266   **c.** About 0.5
   d. 0. Earthquakes of large magnitude are very rare.
   e. There are 87.5% fewer earthquakes of magnitude $M + 1$ or greater than of magnitude M or greater.
25. $G = 12.492 \times 1.029^t$ if t is years since 2005 and G is GDP in trillions of dollars.
27. Answers will vary.
S-1. Exponential   **S-3.** Linear
S-5. The exponential model is $y = 51.01 \times 1.04^x$.
S-7. The exponential model is $y = 2.22 \times 1.96^x$.

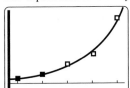

S-9. The exponential model is $y = 6.35 \times 1.03^x$.

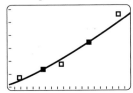

S-11. The exponential model is $y = 2.39 \times 1.40^x$.

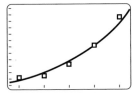

S-13. The exponential model is $y = 3.17 \times 1.15^x$.

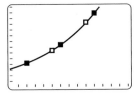

S-15. $y(3.3) = 58.06$; $x = 6.18$
S-17. $y(7) = 246.69$; $x = 3.34$
S-19. $y(10) = 8.53$; $x = 11.80$
S-21. $y(2.2) = 5.01$; $x = 3.59$
S-23. $y(8.8) = 10.84$; $x = 8.22$

SECTION 4.5

1. a. The Alaska earthquake was 2.51 times as powerful as the New Madrid earthquake.
   b. 9.5

3. a. The Veracruz quake was 5.01 times as powerful as the East Coast quake.
   b. 6.8
5. a. 11 times as many   **b.** In the year 2030
   c. In the year 2277, or about 2300
7. a. 63.10 times as acidic   **b.** 3.98 times as acidic
9. 40 decibels
11. a. Horizontal, 0 to 0.4; vertical, 0 to 0.05

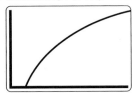

   b. 0.04 unit   **c.** 0.19 unit   **d.** Answers will vary.
13. Horizontal, 25 to 50; vertical, 0 to 15

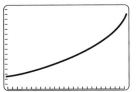

   b. $T(35)$; 4.55 years old   **c.** 46.95 centimeters
15. a. 282   **b.** SDI increases by a factor of 10.   **c.** $SDI = N$
17. a. 2.90   **b.** 17.95 parsecs
   c. Spectroscopic parallax is increased by 5 units.
   d. The spectroscopic parallax for Shaula is 2.89 units larger than that for Atria.
19. a. 9.06 kilometers per second   **b.** Yes
21. a. 0.0289
   b. Horizontal, 0 to 1; vertical, 0 to 1

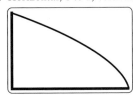

   c. 0.9945   **d.** 37.68
   e. The formula gives 7.33, so 7 would be a reasonable answer.
   f. Horizontal, 0 to 20; vertical, 0 to 40

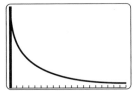

23. The points nearly fall on a straight line, so an exponential model may be appropriate.

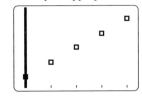

S-1. The second is 1000 times as powerful as the first.
S-3. 8.5
S-5. t units are added to the Richter reading.
S-7. One is 30 decibels more than the other.
S-9. One is $10^{2/10}$ or 1.58 times that of the other.
S-11. 3 **S-13.** −1 **S-15.** 1 **S-17.** −1 **S-19.** e
S-21. t is added to the logarithm. **S-23.** $x = 10$
S-25. $x = e^3$ **S-27.** $x = 1$ **S-29.** $y = 100x$
S-31. $t = \ln 5$ **S-33.** $t = \ln a$

CHAPTER 4 REVIEW EXERCISES

1. 0.37% **2.** 4.89%
3. a. 0.969 **b.** 3.1% **c.** $321.85
4. a. 0.9744 **b.** 2.56% **c.** 1.00485; no
5. Ratios give a constant value of 1.2. The data are exponential.
6. $y = 25 \times 1.2^x$
7. a. $B = 500 \times 0.97^n$ **b.** $500.00 **c.** $240.71
8. a. Ratios give a constant value of 1.03.
  b. Let t be the time in years since the start of 2006 and P the price in dollars; $P = 265.5 \times 1.03^t$.
  c. The year 2013 ($t = 7$)
9. The exponential model is $y = 20.97 \times 1.34^x$.
10. The exponential model is $y = 10.96 \times 0.80^x$.

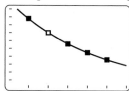

11. a. $A = 8.90 \times 1.02^t$; $B = 2.30 \times 1.27^t$
  b. 2%; 27%
  c. The year 2012 ($t = 7$)
12. a. $B = 520 \times 0.9595^n$
  b. $520.00
  c. 1% ($r = 0.01$)
13. $x = 0.1$
14. $y = 0.1x$
15. The Double Spring Flat quake was approximately 3.16 times as powerful as the Little Skull Mountain quake.
16. a. 7.53 years
  b. 100

A FURTHER LOOK: Solving Exponential Equations

1. 2 **2.** −6 **3.** 5/2 **4.** 18 **5.** −3 **6.** −3 **7.** $t = 1.40$
8. $t = 2.23$ **9.** $t = 5.95$ **10.** $t = 1.18$ **11.** $t = 0.24$
12. $t = 3.21$ **13.** $t = -1.29$
14. $t = \ln 3/(2 \ln a - 4 \ln a)$, or $\ln 3/(-2 \ln a)$
15. $t = (\ln c - \ln a)/\ln b$
16. $t = (-\ln a - \ln b)/(\ln a - \ln b)$, or $(\ln a + \ln b)/(\ln b - \ln a)$
17. $t = (e^4 + 1)/3$, or 18.53
18. $t = (10^{-2} - 4)/6$, or about −0.665
19. $t = (e^c - b)/a$ **20.** $t = (10^c + b)/a$
21. $t = \ln (e^3 - 1)/\ln 2$, or about 4.25

22. $t = \ln (10^2 + 2)/\ln 3$, or about 4.21
23. a. Answers will vary. **b.** $R = 10^{D/10}$
24. a. S increases by 5 log 2, or about 1.51.
  b. $D = 10^{(S+5)/5}$
25. a. $t = \ln 2/\ln 1.005$, or about 138.98 months, or 11.58 years
  b. $t = \ln 2/\ln b$
26. a. $k = \log(2^{1/2})$, or about 0.15.
  b. Answers will vary. **c.** Answers will vary.

CHAPTER 5
SECTION 5.1

1. Answers will vary.
3. $N = \dfrac{1200}{1 + 3e^{-0.182t}}$
5. $N = \dfrac{3600}{1 + 71e^{-0.262t}}$; 16.27 years
7. a. $K = 50$
  b. $b = 1$
  c. $r = 0.095$
  d. $P = 50/(1 + e^{-0.095m})$
9. a. $P = 102.55/(1 + 5.13e^{-0.349m})$
  b. Horizontal, 0 to 15; vertical, 0 to 100

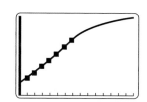

  c. 12 months
11. a. $N = \dfrac{708.42}{1 + 80.12e^{-0.532t}}$
  b. Horizontal 0 to 20; veritcal 0 to 800

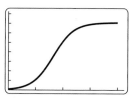

  c. 9.29 days
13. a. $N = \dfrac{12{,}937.03}{1 + 3.14e^{-0.447t}}$ if t is time in years since 1968
  b. At $t = 2.56$ years
15. a. $N = \dfrac{12.14}{1 + 3.84e^{-0.027t}}$ if t is time in years since 1950
  b. $r = 0.027$ per year
  c. $K = 12.14$ billion
  d. When $t = 131$ years after 1950, or the year 2081
17. a. 11.39 years
  b. 17.91 years

19. a. 2.61 per year
  b. 148 thousand tons
  c. 74 thousand tons
  d. Horizontal 0 to 5, vertical 0 to 160

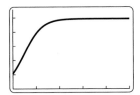

  e. 0.49 year
21. a. Horizontal, 0 to 125; vertical 0 to 0.13

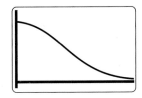

  b. The limiting value is 0. In the long run there will be very few deaths due to tuberculosis.
  c. 1930
S-1. Logistic model
S-3. Theory of maximum sustainable yield
S-5. It is the inflection point on the logistic curve.
S-7. $b = 3$
S-9. 450
S-11. 250
S-13. 35%
S-15. 101%
S-17. 0.104
S-19. 0.495
S-21. $N = \dfrac{2300}{1 + 7e^{-0.77t}}$
S-23. $N = \dfrac{400}{1 + 39e^{-0.44t}}$
S-25. $N = \dfrac{800}{1 + 7e^{-0.20t}}$
S-27. $N = \dfrac{220.42}{1 + 3.95e^{-0.764t}}$
S-29. a. 329.65
  b. $t = 3.23$
S-31. a. 2384.95
  b. $t = 17.84$

SECTION 5.2

1. Larger positive values of c make the power function with a positive power increase faster.
3. 6.47 times as fast
5. Perceived pressure increases by a factor of 2.14.
7. a. 2.12 square meters.
  b. Body surface area will increase by 4.88%.

9. a. 118.63 feet
  b. Yes
  c. Between 42 and 43 miles per hour
11. a. Shorter
  b. 0.007 solar lifetime, or about 70 million years
  c. $E(0.5)$; 5.66 solar lifetimes, or 56.6 billion years
  d. 1.20 solar masses
  e. The more massive star has a shorter lifetime by a factor of 0.18.
13. a. 6.20 feet
  b. The wave is 0.71 times as high.
15. a. The terminal velocity of the man is 6 times that of the mouse.
  b. About 20 miles per hour
  c. About 37.42 miles per hour
17. a. When the distance is halved, the force is 4 times larger. When the distance is one-quarter of its original value, the force is 16 times greater.
  b. $c = 1.8 \times 10^{11}$. When $d = 800$ kilometers, $F = 281,250$ newtons.
  c. Horizontal, 0 to 1000; vertical, 0 to 20,000,000

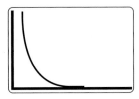

When the asteroids are very close, the gravitational force is extremely large. (In practice, physicists expect that when massive planetary objects get too close, the gravitational forces become so great as to tear the objects apart.) When the asteroids are far apart, the gravitational force is so small that it has little effect.
19. a. By a factor of 125,000,000
  b. By a factor of 250,000   **c.** By a factor of 500
21. About 3.33 seconds
S-1.

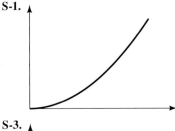

S-3.

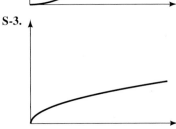

S-5.

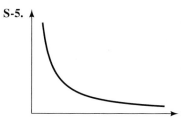

S-7. By a factor of 16
S-9. By a factor of $3^{1/2.53} = 1.54$
S-11. y is 2.03 times as large as z.
S-13. $k = 1.38$
S-15. $c = \dfrac{11}{5^{-1.32}} = 92.05$

SECTION 5.3

1. **a.** $D = 0.89 \times S^{1.42}$
    b. Stopping distance is multiplied by 2.68.
    c.

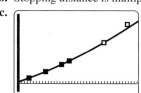

3. **a.** $V = 10.41p^{0.5}$
    b. Answers will vary.
5. **a.** $L = M^{3.5}$ **b.** $L(0.11)$; 0.0004 relative luminosity
    c. About 0.07 solar mass
    d. The larger star is about 47 times as luminous.
7. **a.** Yes
    b. $F = 20.61L^{0.34}$
    c. Horizontal, 0 to 300; vertical, 0 to 150

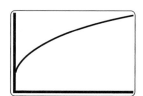

    d. Concave down
    e. 2.19 times as fast
9. **a.** $B = 39.94W^{0.75}$
    b. **i.** Answers will vary.
        ii. 2.14; 37.79
        iii. 0.08; 0.07
11. **a.** $h = 35.91d^{0.66}$
    b. Plains cottonwood
    c. ii. No
13. **a.** Answers will vary.
    b. $w = 4766.06p^{-1.48}$
    c. Weight increases by a factor of 2.79.
    d. Yield increases.

15. **a.** Decreases, generally
    b. $C = 8.58W^{-0.40}$
    c. Answers will vary.
17. **a.** $r = 0.8T^{-1}$. Horizontal, 0 to 10; vertical, 0 to 1

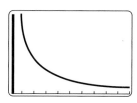

    b. $r = \dfrac{0.002}{L^{0.8}}$

19.

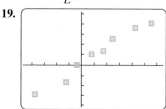

The points nearly fall on a straight line, so a power model may be appropriate.
S-1. Linear
S-3. Answers will vary, but one reasonable choice is a logistic model.
S-5. $y = x^3$
S-7. $f = 3.60x^{1.30}$

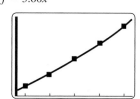

S-9. $f = 2.23x^{4.54}$

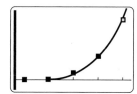

S-11. $f = 2.05x^{-0.90}$

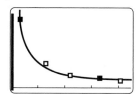

S-13. $f = 2.29x^{-1.59}$

SECTION 5.4

1. a. $F = swd$
  b. 1104 cubic feet per minute
3. a. 176 feet per second
  b. If t is time in seconds since the jump, then
     $D = 176 \times 0.83^t$.
  c. $v = 176 - 176 \times 0.83^t$
  d. $v(4)$; 92.47 feet per second

5. $-\dfrac{v}{r}$

7. a. $D(0) = 17$ inches
  b. $L = 21 - 17 \times 0.82^t$
  c. $W = 0.000293(21 - 17 \times 0.82^t)^3$
  d. $B = 1000e^{-0.2t}0.000293(21 - 17 \times 0.82^t)^3$
  e. 5.89, or about 5.9 years old

9. a. 2.24 seconds **b.** $s = \dfrac{q}{e^{5q} - 1 - 5q}$
  c. 0.42 car per second, or 25.2 cars per minute

11. a. 0.33 or 33% **b.** $Q = \left(\dfrac{d(e^{rt} - 1)}{be^{rt} - d}\right)^k$

  c. $\left(\dfrac{d}{b}\right)^k$ **d.** 0.5^k

  e. 0. If the initial population is large, the probability of extinction is very small.

13. a. t should be replaced by $\dfrac{3}{k}$ to yield the required result.

  b. 1000 years
  c. 47.62 years in the Sierra Nevada mountains compared to 0.75 year in the Congo
15. Answers will vary.
  b. $W = 10.94 \times 1.39^t$, where t is years since 1990 and W is number of wolves
  c. It would be better to use a piecewise-defined function.

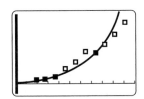

  d. $W = 7.95 \times 1.59^t$ for t years since 1990;
     $W = 112.22 \times 1.24^T$ for T years since 1997
  e. $W = \begin{cases} 7.95 \times 1.59^t & \text{for } 0 \le t \le 6 \\ 112.22 \times 1.24^{t-7} & \text{for } 7 \le t \le 10 \end{cases}$
17. a.

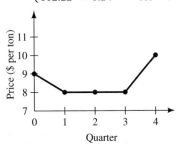

b. $P = -t + 9$ for $t = 0$ to 1
c. $P = 8$ for $t = 1$ to 3
d. $P = 2(t - 3) + 8$ for $t = 3$ to 4
e. $P = \begin{cases} -t + 9 & \text{for } t = 0 \text{ to } 1 \\ 8 & \text{for } t = 1 \text{ to } 3 \\ 2(t - 3) + 8 & \text{for } t = 3 \text{ to } 4 \end{cases}$

S-1. $w = (t - 3)^2 + 1$
S-3. $w = \sqrt{2e^t + 1}$
S-5. $f(g(x)) = \dfrac{6}{x} + 1$; $g(f(x)) = \dfrac{2}{3x + 1}$
S-7. $f(g(x)) = x$; $g(f(x)) = x$
S-9. 7
S-11. $f = t^2 + 3 + \dfrac{t}{t^2 + 1}$
S-13. Answers will vary.
S-15. The initial fee is $30. Books for members cost $17 each.
S-17. $(1 - x)^2 - x^2 + 1$, or $2 - 2x$

SECTION 5.5

1. Horizontal, 0 to 4; vertical, 0 to 1.3

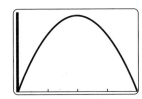

  b. $G(2.26)$; 1.18 thousand dollars per year
  c. 2 thousand dollars
3. a. Horizontal, 0 to 10; vertical, 0 to 400

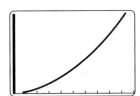

  b. No
  c. 7.18 feet per second
  d. 7.47 inches
5. a. $y = -0.31x^2 + 1.35x + 1037.255$
  b. y is 1037.26, 1038.30, 1038.72, 1038.52, 1037.70, and 1036.26 feet elevation
  c. 217.74 feet along the vertical curve and 1038.72 feet elevation
7. a. $R = 7.79s^2 - 514.36s + 8733.57$
  b. About 2491 per 100,000,000 vehicle-miles
  c. About 33 miles per hour
9. a. Let N be the number, in millions, of women working outside the home and t the number of years since 1942. Then $N = -0.735t^2 + 3.107t + 16.154$.
  b. $N(5)$; 13.31 million
  c. Answers will vary.

11. 4.91 seconds
13. a. 349.07
  b. 26.91 feet
15. a. $C = 0.000422s^3 - 0.0294s^2 + 0.562s - 1.65$
  b. About 1.36 cents per vehicle-mile
  c. About 33 miles per hour
17. a. -0.31 and 0.81; $\lambda_1 = 0.81$
  b. About 65.13%
  c. About 98.52%
19. a. $q = 100$
  b. When immunization levels at birth are near 100%, the average age of contraction is very large.
21. a. Horizontal, 30 to 80; vertical, 4.4 to 5.1

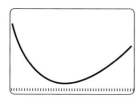

  b. $n(45)$; 4.5 seconds
  c. Answers will vary.
  d. 4.46 seconds
23. a. $S = \dfrac{6.75(300 - F)}{F + 81}$
  b. Horizontal, 0 to 300; vertical, 0 to 25

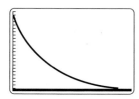

  c. It decreases to zero when $F = 300$.
  d. $S = 0$ when $F = 300$, so the muscle does not contract when the maximum force is applied to the muscle.
  e. S has a horizontal asymptote at $S = -6.75$. This makes no sense in terms of the muscle; as F gets large, it becomes greater than the maximum force.
  f. S has a vertical asymptote at $F = -81$. The equation is not valid for a negative force.

S-1. $x = \frac{1}{2} \pm \frac{\sqrt{11}}{2}$, or $x = -1.16$ and $x = 2.16$
S-3. $0.85x^2 + 0.90x + 0.21$
S-5. $2.47x^2 + 2.93x + 5.77$
S-7. No
S-9. No
S-11. $0.07x^3 - 1.11x^2 + 4.92x - 2.95$
S-13. $-0.01x^4 + 0.20x^3 - 2.00x^2 + 7.26x - 4.58$
S-15. $x = 2$ and $x = 1$
S-17. $y = 2$

CHAPTER 5 REVIEW EXERCISES

1. $N = \dfrac{2500}{1 + 24e^{-0.02t}}$

2. a. $r = 1.4$ per year
  b. $K = 25$ thousand
  c. 12.5 thousand
3. 640
4. f is multiplied by 9.19.
5. 0.30
6. a. 1.19
  b. The flow rate is multiplied by 81.
7. a. 1 second
  b. $c = 1.12$
  c. 2 seconds
8. $f = 3.5x^{-1.2}$
9. a. $D = 22.02g^{-1.0}$
  b. The distance is multiplied by 2/3, or reduced by 33.33%.
  c. About 22 miles per gallon
10. a. $T = 0.20s^{0.5}$
  b. 1.67 seconds
  c. 100 feet
  d. The time is multiplied by 1.41.
11. $y = 3(t - 1)^2 + 5(t - 1)$
12. 120; 2%
13. a. Horizontal, 0 to 200; vertical, 0 to 0.015

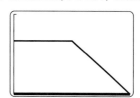

  b. 0.005 per year
14. a. $R = \left(\dfrac{3(1.5 + 0.1t)}{4\pi}\right)^{1/3}$
  b. 0.74 inch
15. $1, -1/2$
16. $-3x^2 + 5x - 2$
17. a. $R = 0.005x^2 - 0.75x + 25$
  b. 9.88 moles per cubic meter per second is the reaction rate at a concentration of 24 moles per cubic meter.
  c. 32.28 moles per cubic meter
18. $0.1x^3 - 0.2x^2 + 3$
19. $x = -1$ and $x = -3$
20. a. $V = 0.03t^3 + 0.50t^2 + 2.51t + 4.19$
  b. $V(5)$; 32.99 cubic inches
21. a. Horizontal, 0 to 70; vertical, 0 to 50

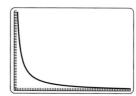

  b. 4 hours is the time required to drive 100 miles if the average speed is 25 miles per hour.
  c. Answers will vary.

A FURTHER LOOK: Fitting Logistic Data Using Rates of Change

1. $G = -0.002N + 0.21$; $r = 0.21$ per year; $K = 105$
2. $G = -0.002N + 0.41$; $r = 0.41$ per year; $K = 205$
3. **a.** –0.05 per day; –0.41 per day
   b. These values are larger than the carrying capacity, so we expect the population to decline.
4. **a.** $r = 0.13$ **b.** $K = 378$
   c. max growth rate: $N = 189$
5. **a.** $G = -0.0037N + 0.034$
   b. 0.034 per year
   c. 9.19 billion
6. **a.** Answers will vary
   b. $N = 0$ and $N = K$
   c. Max value of G is when $N = -r/2(-r/K) = K/2$.
7. **a.** $-0.01764N^2 + 2.61N$
   b. Horizontal 0 to 150, vertical 0 to 160

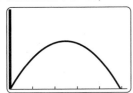

   c. 73.98 thousands tons
   d. They are about the same.
8. **a.** When $N = K/2$, max growth rate
       $= (-r/K)(K/2)^2 + r(K/2) = rK/4$.
   b. Max growth rate for Example 5.1 is 0.20 million tons per year.

A FURTHER LOOK: Factoring Polynomials, Behavior at Infinity

1. The polynomial has degree 5.
2. The smallest degree is 6.
3. $x^3 - 9x^2 + 26x - 24$
4. $3(x - 2)(x + 2)(x - 0)$ or $3x^3 - 12x$.
5. $(x - 1)^5$ or $x^5 - 5x^4 + 10x^3 - 10x^2 + 5x - 1$
6. $5x^{10}$ 7. $5/2$ 8. 2 9. 0 10. a/d 11. $x^2 + x + 3$
12. $8x^2 - 25x + 22$ 13. $3x^2 - 5$ 14. x^2
15. **a.** Answers will vary.
    b. Answers will vary.
    c. Answers will vary.
    d. $(x - 2)(x - 3) - 5(x - 1)(x - 3) + 3(x - 1)(x - 2)$

CHAPTER 6

SECTION 6.1

1. Answers will vary.
3. **a.** Horizontal, 0 to 2; vertical, 0 to 20

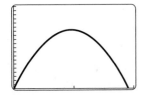

b. 14.06 feet
c. 1.88 seconds after it is tossed
d. Answers will vary.
5. Answers will vary.
7. Answers will vary.
9. Answers will vary.
11. Answers will vary.
S-1. Velocity
S-3. Below the horizontal axis
S-5. It is a horizontal line.
S-7. 60 miles per hour
S-9. **a.** Velocity is positive.
     b. Velocity is zero.
     c. Velocity is negative.
S-11. Velocity switches from positive to negative (passing through 0).

SECTION 6.2

1. Horizontal, -3 to 3; vertical, -10 to 10

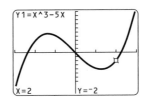

   a. Positive
   b. $x = 0$, or any number between -1.29 and 1.29
3. **a.** Answers will vary.   **b.** Negative
5. Answers will vary.
7. **a.** Horizontal, 0 to 5; vertical, -3 to 5

   b. From 0 to 1.82 years   **c.** 1.82 years
9. Answers will vary.
11. **a.** Answers will vary.
    b. It will be positive.
    c. $\dfrac{dN}{dt}$ would be smaller than before.
    d. $\dfrac{dN}{dt}$ will be near zero.
13. Answers will vary.
15. Answers will vary.
S-1. **a.** Velocity
     b. Acceleration
     c. Marginal tax rate
     d. Marginal profit
S-3. **a.** $\dfrac{df}{dx}$ is positive.   **b.** $\dfrac{df}{dx}$ is zero.

     c. $\dfrac{df}{dx}$ is negative.   **d.** $\dfrac{df}{dx}$ is zero.

S-5. 10
S-7. $13.20
S-9. The graph of *f* is decreasing.
S-11. The rate of change is 0.

SECTION 6.3

1. **a.** 423 reindeer per year
   b. 1524
   c. Too large
3. **a.** -8.98 deaths per 100,000 per year
   b. Answers will vary.
   c. 296.8 deaths per 100,000
   d. -4.53 deaths per 100,000 per year
   e. Answers will vary.
5. **a.** Horizontal, 0 to 10; vertical, 0 to 200

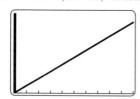

   b. 28 feet   **c.** 19.20 feet per second
7. **a.** Horizontal, 0 to 45; vertical, 0 to 300

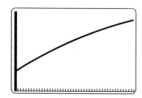

   b. 5.32 degrees per minute
   c. 3.57 degrees per minute
   d. Answers will vary.
9. **a.** Answers will vary.   **b.** Negative
   c. -8000 gallons per minute   **d.** 1,960,000 gallons
S-1. 7
S-3. -5.6
S-5. -0.03
S-7. From the figure below, we get a value of 0.89.

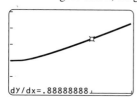

S-9. From the figure below, we get a value of -0.04.

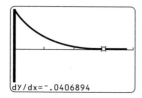

S-11. The rate of change is near 1.

SECTION 6.4

1. **a.** $V = -32t + 40$
   b. 1.25 seconds   **c.** 2.5 seconds
3. **a.** Let *t* be time (in years since the initial investment) and *B* the balance (in dollars). $B = 250e^{0.0575t}$, or $B = 250 \times 1.0592^t$
   b. Alternative, $333.27; standard, $333.30
5. **a.** Answers will vary.   **b.** $B = 10,000e^{0.07t}$
   c. $B = 10,000 \times 1.073^t$ **d.** 9.90 years
7. **a.** $\dfrac{dN}{dt} = 0.04N$
   b. $N = 30e^{0.04t}$ or $N = 30 \times 1.04^t$
   c. About 13 years
9. **a.** Linear   **b.** $\dfrac{dW}{dt} = 5.5$ **c.** $W = 5.5t + 13.5$
11. **a.** -0.05 per day   **b.** $A = 3e^{-0.05t}$ **c.** 13.86 days
S-1. Differential equation
S-3. 5
S-5. $c = 8$
S-7. $\dfrac{dV}{dt} = -\dfrac{1}{3}V$
S-9. $f = 3x + 7$
S-11. $\dfrac{dH}{dt} = 4$

SECTION 6.5

1. **a.** Horizontal, 0 to 225; vertical, -20 to 40

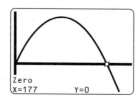

   Equilibrium solutions $N = 0$ and $N = 177$
   b.

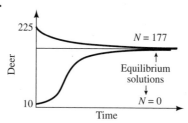

   c. $N = 6$
3. **a.** Answers will vary.   **b.** 8
5. **a.** $F = 0.15$ million tons per year
   b. i. Horizontal, 0 to 3; vertical, -0.15 to 0.15

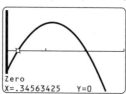

   Two equilibrium solutions, $N = 0.35$ million tons and $N = 2.05$ million tons.

ii. The biomass increases when N is between 0.35 and 2.05 million tons. It decreases when N is less than 0.35 million tons or more than 2.05 million tons.

iii.

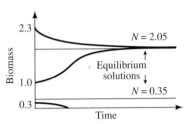

iv. Answers will vary.

7. **a.** $\dfrac{dN}{dt} = -0.338N\left(1 - \dfrac{N}{0.8}\right)\left(1 - \dfrac{N}{2.4}\right)$

b. Horizontal, 0 to 3; vertical, -0.25 to 0.25

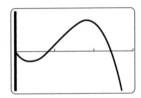

The equilibrium solutions are $N = 0$, the threshold$N = S = 0.8$, and the carrying capacity $N = K = 2.4$.

c. Increasing when N is between 0.8 and 2.4, decreasing otherwise

d. The population will eventually die out.

e. $N = 1.77$ million tons

9. **a.** Yes

b. All bacteria are of type B. There will never be any type A bacteria.

c. Eventually, all type B bacteria will disappear, and all the bacteria will be of type A. Hence P tends to 1.

11. **a.** Answers will vary.

b. Horizontal, 0 to 3; vertical, -0.01 to 0.05

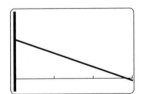

c. 2.8 pounds

S-1. A solution that does not change (remaind constant)

S-3. $f = 3$

S-5. 15 cubic feet

S-7. Increasing

S-9. Increasing

S-11. Decreasing

CHAPTER 6 REVIEW EXERCISES

1. Velocity is 0.

2. Velocity is a constant -3.

3. Answers will vary.

4. **a.** Horizontal, 0 to 3; vertical -12 to 12

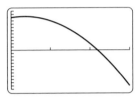

b. Answers will vary.

5. The profit decreases.

6. The rate of change is 0.

7. **a.** Answers will vary.

b. Answers will vary.

c. Negative

8. **a.** Horizontal, 0 to 3; vertical, -5 to 15

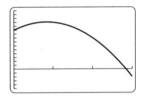

b. From 0 to 2.84 years

c. 2.84 years

9. 6

10. From the figure below, we get a value of 8.31.

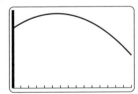

11. **a.** 133.33 cases per day

b. Answers will vary.

c. 240.60 cases per day

12. **a.** Horizontal, 0 to 15; vertical, 0 to 1500

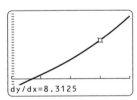

b. -44 dollars per month

c. Answers will vary.

13. $\dfrac{dH}{dt} = -0.5$

14. $f = 3e^{6x}$

15. a. Answers will vary.

   b. $n = 0.02w + 3.3$

16. a. Answers will vary.

   b. Exponential

   c. $P = 1035e^{-0.12h}$ or $P = 1035 \times 0.89^h$

17. $f = 2$

18. Decreasing

19. a. 4 thousand dollars

   b. 1.2 thousand dollars per year

20. a. $\dfrac{dN}{dt} = -0.5N\left(1 - \dfrac{N}{10}\right)$

b. Horizontal, 0 to 12; vertical, -1.5 to 1.5

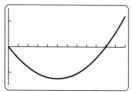

The equilibrium solutions are $N = 0$ and the threshold $N = T = 10$.

c. Increasing when N is greater than 10, decreasing otherwise.

d. Decreases to 0, increases.

Len DeLessio/Getty Images